Photonic and Electronic Properties of Fluoride Materials

Photonic and Electronic Properties of Fluoride Materials

Progress in Fluorine Science Series

Edited by

Alain Tressaud
Emeritus Research Director CNRS
ICMCB-CNRS, University Bordeaux
Pessac, France

and

Kenneth Poeppelmeier
Morrison Professor of Chemistry
Northwestern University
Evanston, IL, United States

ELSEVIER

AMSTERDAM • BOSTON • HEIDELBERG • LONDON • NEW YORK • OXFORD
PARIS • SAN DIEGO • SAN FRANCISCO • SINGAPORE • SYDNEY • TOKYO

Elsevier
Radarweg 29, PO Box 211, 1000 AE Amsterdam, Netherlands
The Boulevard, Langford Lane, Kidlington, Oxford OX5 1GB, UK
50 Hampshire Street, 5th Floor, Cambridge, MA 02139, USA

British Library Cataloguing-in-Publication Data
A catalogue record for this book is available from the British Library

Library of Congress Cataloging-in-Publication Data
A catalog record for this book is available from the Library of Congress

ISBN: 978-0-12-801639-8

Publisher: John Fedor
Acquisition Editor: Katey Birtcher
Editorial Project Manager: Jill Cetel
Production Project Manager: Mohanapriyan Rajendran
Designer: Mark Rogers

Contents

List of Contributors

J. Bendix University of Copenhagen, Copenhagen, Denmark

N.B. Bolotina Shubnikov Institute of Crystallography of Russian Academy of Sciences, Moscow, Russia

G. Calestani University of Parma, Parma, Italy

C. Chen CAS, Beijing, P. R. China

H. Chen Beihang University, Beijing, China; The Hong Kong University of Science and Technology, Kowloon, Hong Kong

L. Clark University of St Andrews, Fife, Scotland

A. Demourgues ICMCB-CNRS, University of Bordeaux, Pessac, France

L. Driscoll University of Birmingham, Birmingham, United Kingdom

P.P. Fedorov Prokhorov General Physics Institute, Russian Academy of Sciences, Moscow, Russia

M. Fichtner Helmholtz Institute Ulm for Electrochemical Energy Storage (HIU), Ulm, Germany

I.N. Flerov Kirensky Institute of Physics (KIP), Krasnoyarsk, Russia

M. Gaudon ICMCB-CNRS, University of Bordeaux, Pessac, France

M.V. Gorev Kirensky Institute of Physics (KIP), Krasnoyarsk, Russia

P. Gredin IRCP, CNRS-UMR 8247, Chimie ParisTech, Paris, France; Université Pierre et Marie Curie, Paris, France

W. Grochala Centre of New Technologies, University of Warsaw, Warsaw, Poland

J. Heo Pohang University of Science and Technology (POSTECH), Gyeongbuk, Republic of Korea

H. Hosono Tokyo Institute of Technology, Yokohama, Japan

M.F. Joubert ILM, UMR5306 CNRS-UCBL, Université de Lyon, Villeurbanne, France

Y. Kamihara Keio University, Yokohama, Japan

P. Kirsch Merck KGaA, Darmstadt, Germany

D. Kurzydłowski Centre of New Technologies, University of Warsaw, Warsaw, Poland; Faculty of Mathematics and Natural Sciences, Cardinal Stefan Wyszynski University in Warsaw, Warsaw, Poland

S.V. Kuznetsov Prokhorov General Physics Institute, Russian Academy of Sciences, Moscow, Russia

N.M. Laptash Institute of Chemistry, Vladivostok, Russia

G. Ledoux ILM, UMR5306 CNRS-UCBL, Université de Lyon, Villeurbanne, France

W. Li Beihang University, Beijing, P. R. China

P. Lightfoot University of St Andrews, Fife, Scotland

C. Liu Wuhan University of Technology, Hubei, P. R. China

L. Liu CAS, Beijing, P. R. China

H. Liu Beihang University, Beijing, P. R. China

Z. Mazej Jožef Stefan Institute, Ljubljana, Slovenia

F. Mezzadri University of Parma, Parma, Italy

S. Mishra IRCELYON, UMR5256 CNRS-UCBL, Université de Lyon, Villeurbanne, France

M.S. Molokeev Kirensky Institute of Physics (KIP), Krasnoyarsk, Russia

M. Mortier IRCP, CNRS-UMR 8247, Chimie ParisTech, Paris, France

E. Murotani AGC Chemicals, Asahi Glass Co., Ltd., Tokyo, Japan

G. Nénert PANalytical B.V., Almelo, Netherland

V.V. Osiko Prokhorov General Physics Institute, Russian Academy of Sciences, Moscow, Russia

S. Pan Xinjiang Technical Institute of Physics & Chemistry, Chinese Academy of Sciences, Urumqi, P. R. China

K.S. Pedersen CRPP-CNRS, Pessac, France; ICMCB-CNRS, Pessac, France

K.R. Poeppelmeier Northwestern University, Evanston, IL, United States

M.A. Reddy Helmholtz Institute Ulm for Electrochemical Energy Storage (HIU), Ulm, Germany

M. Rosseinsky University of Liverpool, Liverpool, United Kingdom

H. Serier-Brault ICMCB-CNRS, University of Bordeaux, Pessac, France

P. Slater University of Birmingham, Birmingham, United Kingdom

B.P. Sobolev Shubnikov Institute of Crystallography of Russian Academy of Sciences, Moscow, Russia

N.I. Sorokin Shubnikov Institute of Crystallography of Russian Academy of Sciences, Moscow, Russia

Y. Takahira Innovative Technology Research Center, Asahi Glass Co., Ltd., Yokohama, Japan

I. Trenque ICMCB-CNRS, University of Bordeaux, Pessac, France

A. Tressaud ICMCB-CNRS, University of Bordeaux, Pessac, France

X. Wang CAS, Beijing, P. R. China

Y. Wang Xinjiang Technical Institute of Physics & Chemistry, Chinese Academy of Sciences, Urumqi, P. R. China

L. Zhu Beihang University, Beijing, P. R. China

Foreword

Fluorine has a special place in chemistry because of its uniquely powerful oxidizing character. This gives fluorine compounds distinctive chemical and physical properties often associated with unusually high oxidation states and strong ionic bonding. As electronic and optical behavior is controlled by fundamental chemical bonding, fluorides are a distinctive class of materials and so an up-to-date overview of their chemistry and physics is both welcome and useful. The editors are highly regarded experts in this area and have ensured broad, high-level coverage spanning synthesis and full scope of properties where fluorides are potentially differentiated from other material classes.

As expected from their high performance in crystalline, liquid crystalline, and glassy forms, there is a strong emphasis on optical properties such as emission near infrared absorption and nonlinear behavior. Here the emergence of mixed anion materials, such as the oxyfluorides and hydroxide fluorides, is noticeable. The predominantly ionic bonding leads to insulating behavior and thus naturally to dielectric and to ferroelectric behavior and to localized electron magnetism in both extended and molecular materials. The emerging science here challenges ancient preconceptions about fluoride properties—the Ag(II) fluoride $KAgF_3$ has a super exchange coupling of over 0.1 eV, which would be strong even in an oxide. The overviews of a range of structure and material types show that the special coordination and bonding characteristics of fluorides offer unique opportunities for multiferroic behavior due to the coupling of the arising structural and magnetic symmetries.

The high energies of fluorine-based bonds make all-fluorine batteries attractive and two contributions deal with the required development of solid fluoride electrolytes.

Fluoride is important not only in stoichiometric compounds, but also as a dopant, particularly in oxides where its similar size but different charge from the host anion leads to an underexplored set of opportunities for chemical control of cooperative properties; here the possibilities are set out by demonstration of the impact of fluoride substitutional chemistry on superconductivity.

This overview of contemporary fluorine-based materials chemistry brings the reader to the forefront of the state-of-the-art in an accessible way. I am sure it will stimulate new and exciting research in this rich area, replete with coupled opportunities in fundamental and applied science.

Matthew Rosseinsky
University of Liverpool, Liverpool, United Kingdom
September 2015

Introduction to the Series "Progress in Fluorine Science"

1

A. Tressaud
ICMCB-CNRS, University of Bordeaux, Pessac, France

The history of science has been currently illustrated by a succession of breakthroughs and regressions. Karl Popper postulated that science goes forward through a process of "conjectures and refutations," meaning that a theoretician should develop a theory, and an empirical scientist should try to test it to refutation. If the theory cannot be properly tested by science, this means that this proposal cannot be called scientific. One of the major points of Popper's postulates was establishing that science is not infallible, and that well-established scientific paradigms may often follow a wrong path, generating incorrect theories. This is true also for the behavior of new materials: the validity of their initially stated "outstanding" properties should be tested in the light of multidisciplinary approaches: purely scientific of course, but also arising from many other aspects: environmental, statistic, economic, sociologic, and even psychologic.

It can be put forward that fluoride products have followed similar schemes. During the last century, they have been in several occasions at the origin of huge discoveries for science and mankind. This is, for instance, the case of the use of fluoride melts to promote the aluminum production, the replacement of ammonia by so-called "freons" as industrial refrigerants, the discovery of Teflon® and fluoropolymers. The last part of the 20th century has been illustrated by the rise of functionalized fluorinated molecules and by their beneficial aspects in agriculture, pharmacy, medicine, and biomedical sciences. One of the best examples is the use of ^{18}F-labeled molecules for positron emission tomography. This imaging technique, which is now proposed in many hospitals and medical centers, allows the detection with a very high accuracy of the proliferation of cancer cells in the body via metastasis process and is thus able to propose therapy solutions at the early stage of the disease.

In these last decades, the development and applications of new materials have been rightly influenced by their effect on our environment. In this scope, although the benefits of fluoride compounds for our society are evident in many fields, the drawbacks caused to our environment by some of these products should not be underestimated. This dual aspect has also peripheral beneficial aspects because it compels teams dealing with fluoride products to launch great challenges for finding solutions able to overcome these problems, as shown below with a couple of examples.

- Nowadays, the ozone-depleting substances such as chlorofluorocarbons (CFCs) and hydrochlorofluorocarbons (HCFCs) are banned, and new replacement compounds with lower *global warming potential* are proposed, such as *hydro*-fluoro-oléfines (HFO). Fluorinated HFO-1234yf (2,3,3,3-tetrafluoropropene) is today a good example suitable for automobile refrigeration.

Photonic and Electronic Properties of Fluoride Materials. http://dx.doi.org/10.1016/B978-0-12-801639-8.00001-5

- Fluorosurfactants are essential in chemistry, biochemistry, and biotechnology. These species may act as detergents, wetting agents, emulsifiers, dispersants, and foaming agents: aqueous film-forming foams are among the most popular fire-fighting foams used for liquid fuel fires. However, since some of these species, such as perfluorooctanesulfonic acid (PFOS) and perfluorooctanoic acid (PFOA), characterized by long carbon chains, exhibit with a widespread occurrence in the blood of populations and wildlife, they have been judged as persistent organic pollutants by the Stockholm Convention. Consequently, the manufacturing and release to the environment of PFOS and PFOA is suspended by 2015. In order to take into account this regulation, companies are now manufacturing fluorosurfactants based upon four-carbon chains, as in perfluorobutanesulfonic acid, instead of longer chains. Multiple studies have indicated that perfluorobutanesulfonic acid (PFBS) as substitute, is neither toxic nor bioaccumulative.

As these examples demonstrate, the global socioeconomic issues associated with a discovery cannot be assessed immediately after the discovery. A true evaluation can only be made after a period of time; once scientists have been able to incorporate accumulated experimental data, including all effects of these new compounds on the environment, nature, human beings, and all life that is present on our planet.

The new series of book *Progress in Fluorine Science* aims to provide useful information based on emerging trends and research in the fields of physics, chemistry, biology of fluoride compounds and to highlight possible applications. Several volumes are expected to bring relevant information in the concerned fields from worldwide contributions proposed from both academia and industry.

Many research fields and applications are indeed impacted by the strategic importance of fluoride materials, or by the use of functionalization processes using fluorination treatments. It is indeed possible to find high-technology fluorine-based products as components in many industrial applications: microelectronics, microphotonics, fluorescent chemical sensors, solid-state lasers, nonlinear optics, optical fibers, special glasses, solar cells, energy storage and conversion, catalysis, textile, cosmetics, plastics, domestic wares, building, automotive, drugs and pharmaceuticals, biomedical materials, prostheses, medicinal and molecular imaging, agrochemistry, superhydrophobic coatings, surfactants, antifire products, etc. Most of these outstanding applications can be correlated to the exceptional electronic properties of the element "fluorine," yielding almost unique types of bonding with the other elements. One of the novelties of this book series is to present, in an individual volume, different aspects of the chemistry and physics of a particular group of fluorine or fluoride compounds. Depending of the subject, the chapters will be specifically focused on inorganic, organic, or polymeric materials and will deal with synthetic pathways, physical—chemical characterizations, and finally applications and issues. The individual volumes of *Progress in Fluorine Science* are thematically independent, comprehensively addressing topical problems in both the science and applications of fluorinated compounds.

The goals of this series are to point out most important features and breakthroughs that have been obtained during this last decade in several key topics of fluorine-related fields. Some outstanding applications of fluorinated materials can be indeed correlated to the unique electronic properties of fluorine and derived materials with respect of other types of materials. In each volume, the interconnections between these different types of compounds will be highlighted with respect to one or several specific aspects and properties: physical–chemical behavior, synthesis routes and methodology, applications. The multiple facets of fluorinated compounds will be illustrated by classes of materials ranging from inorganic materials: eg, ceramics, inorganic nanopowders, nanostructured glasses, coordination complexes, fluorinated nematic liquid crystals, fluoropolymers, organic fluorocompounds, biomaterials, or perfluorinated organic molecules.

The series is composed for the moment of three titles to be appearing soon, each of them having its own editors. It should be noted that these titles and fields have been chosen because of the lack of books currently available on such topics. It can be anticipated that further volumes might deal with *Biologically active fluorinated compounds*, or with *Fluorosurfactants in Industry and Biomedical Sciences*, with recognized experts as editors. The first three titles of the series will be:

- *Photonic & Electronic Properties of Fluoride Materials*, Editors: A. Tressaud, CNRS, Bordeaux, France & K. Poeppelmeier, NW Univ. Evanston, USA;
- *New Forms of Fluorinated Carbon*, Editors: O. Boltalina, Univ. Colorado, USA & T. Nakajima, AIT, Toyota, Japan;
- *Synthesis & Reactivity in Fluorinated Media*, Editors: H. Groult, UPMC, Paris, F. Leroux, Joint Bayer-CNRS Lab., Strasbourg, & A. Tressaud, CNRS, Bordeaux, France.

In the present first volume, the relationships between structural aspects and physical properties will cover optical properties (luminescence, laser, upconversion, UV absorbers, frequency doubling), high-T_c superconductivity, ferro-, ferri-, and molecular magnetisms, multiferroism, ferroelectricity, ionic conduction mechanisms, ab-initio calculations. Important potential applications of the final products will be investigated in photonics: (organic) light-emitting diodes, liquid crystal displays, dye-sensitized solar cell, transparent conducting films, solid-state lasers, nonlinear optics, microelectronics and molecular electronics, imaging devices, electrodes and electrolytes for energy storage and conversion, etc.

It should be anticipated that volume 2 will cover fundamental research and emerging applications of fluorinated compounds of new carbon species such as fullerenes, carbon nanotubes, polycyclic aromatic molecules, carbon nanofibers, and graphene. Applications of these functional materials range from energy storage and energy conversion devices, molecular electronics, photonics, and lubricants.

Volume 3 will focus on the outstanding character of fluorine and fluorinated compounds, through examples taken from all classes of fluorine compounds which illustrate the extreme reactivity of fluorinating media and the peculiarity of their synthetic routes. Among topics to be addressed in the volume are supramolecular chemistry, fluoroorganics, asymmetric syntheses, use of Ruppert–Prakash reagent, CHF_3-based chemistry, fluorous media, photoredox catalysis, fluorosurfactants, synthesis in medicinal chemistry and biology, telomerization processes, photochemical and aHF routes to noble gas fluorocompounds, microwave-enhanced route, sol-gel reactions, superacidic media, ionic liquids, fluorine in the nuclear energy cycle, electrodes and electrolytes for energy storage devices, thin layers, surface modifications, etc.

Therefore the wide scope of entries occurring in this book series—ranging from new synthetic routes, physical and chemical characterizations to applications—will illustrate how fluorides and fluorinated compounds have strong positive impacts on various aspects of modern life.

Finally, I am pleased to thank the members of the "Fluorine pole" of ICMCB-CNRS and also my colleagues of the CNRS "French Fluorine Network" for many exchanges, longstanding friendship, constant support, and well-disposed critical scrutiny of my works during the past decades.

Pessac, October 2015.

Part One

Synthesis

Elaboration of Nanofluorides and Ceramics for Optical and Laser Applications

2

P.P. Fedorov, S.V. Kuznetsov, V.V. Osiko
Prokhorov General Physics Institute, Russian Academy of Sciences, Moscow, Russia

Chapter Outline

1. Introduction

The subject of nanofluorides is a rapidly evolving field (Fig. 2.1). Numerous papers have been published covering this topic with some specific aspects of nanofluoride chemistry and/or material science [33,62,69,89] since the first review had appeared [52].

Nanofluorides are very important photonic materials because of their quite diverse features, for example, Huang et al. [44] and Fedorov et al. [38] discussed their application as luminophores in the solar batteries (panels), light-emitting devices, electronic displays [14,93,101,102], and laser materials, including nanostructured ceramics [24] and glass ceramics [38]. Scientists continue a wide range of studies for biomedical applications of nanofluorides [61,79,90,96]. Nanofluorides as solid-state ionic materials provide enhanced fluoride-ion conductivity and, thus, are very promising objects for electrochemical power sources [4,19,85,86,99]; eg, see studies of nanofluorides as cathode materials for alkaline batteries [59]. Highly developed chemically active surfaces of nanofluorides make them very efficient catalysts [52,77] as well as unique antireflection coatings and other optical film materials [95], etc.

In this paper, we summarize our studies of nanofluorides that have been performed in the Laboratory of Nanomaterial Technology for Photonics at A. M.

Photonic and Electronic Properties of Fluoride Materials. http://dx.doi.org/10.1016/B978-0-12-801639-8.00002-7

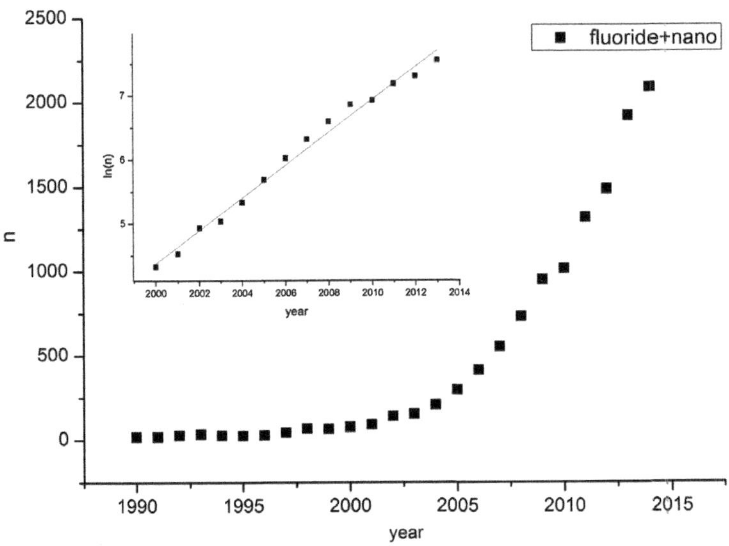

Figure 2.1 Annual numbers of publications on "nano + fluoride" as per Web of Science—all databases records.

Prokhorov General Physics Institute (Russian Academy of Sciences) [8,32,34–36, 50,54–56,63,64,66–68,74,97,98] over the last 10 years, 2005–15. Our work covered three different areas: (1) phase formations in inorganic nanofluoride systems and their comparison with the equilibrium-state phase diagrams; (2) application of various fluoride matrices for the manufacture of upconversion luminophores and the efficiency analyses of such materials; and (3) development of the synthetic precursors for fluoride laser and scintillation ceramics.

2. Phase Formation in Nanofluoride Systems

Phase diagrams are crucial tools in developing technological processes that include the achievement of phase equilibrium at some temperatures followed by removal of the system from that equilibrium in order to provide the target materials with the required functional properties. This method has been used, for example, in metallurgy. With regard to fluoride systems, we would like to mention that the above approach has allowed the selection of conditions for the synthesis of optical-quality single crystals (eg, $M_{1-x}R_xF_{2+x}$ solid solutions of rare earth trifluorides RF_3 in fluorite matrices [26,53], orthorhombic β-YF_3-type solid solutions [83,84] with the use of corresponding MF_2-RF_3 and RF_3-$R'F_3$ phase diagrams. This has also permitted the synthesis of laser materials with compositions corresponding to the extremes and/or saddle points on the liquidus surfaces, eg, $Na_{0.4}Y_{0.6}F_{2.2}$ (NYF) [29], $Ca_{1-y}Sr_yR_xF_{2+x}$ [25], etc., as well as $Ba_2Na_3[B_3O_6]_2F$ [100] birefringence material for the UV part of the spectrum, which was found by the investigation of the ternary reciprocal system Na,Ba//BO_2,F [11,49].

Nanomaterials in general, particularly nanofluorides, belong to the world of the nonequilibrium states and, therefore, to a totally different technology, where nonequilibrium products are obtained by nonequilibrium physical and chemical transformations. Very frequently, the latter leads to the formation of previously unknown phases [57] of variable composition and enhanced reactivity (eg, nanofluorides appear to be prone to facile pyrohydrolysis) [34].

While, such reactivity can create significant problems in interpretation of the obtained results and their reproducibility, thus, challenging the experimenter's art. Conversely, these circumstances open the unmatched opportunities for maneuvering in material design, reactivity of the products, functional features, and architecture of the prepared materials. Very frequently, one can manufacture quite different products using the same starting reagents.

The nanoworld exhibits strikingly unusual discrepancies between composition, structure, and other properties that do not exist in the previously known bulk materials.

Control over nanotechnology processes requires—in addition to a cybernetic component (as per Norbert Wiener [94], cybernetics is the scientific study of control and communication in the animal and the machine)—mandatory accounting for the spontaneously occurring synergetic transformations [70]. The latter includes, in particular, nanoparticle agglomeration that usually demonstrates explicitly clear step-by-step hierarchy [27,34]. Nanoworld is systematically showing nonclassical crystallization by oriented attachment crystal growth [37,45]. However, Ostwald's step rule works unequivocally for the nano-objects, and phase transformation sequences always start with the most nonequilibrium amorphous products that turn into more stable phases in a stepwise manner.

Concluding these introductory notes, we would like to shift to the discussion of the phase formation in MF_2-RF_3 (M = Ca, Sr, Ba, R = rare earth elements) and NaF-RF_3. For the last 10 years, we have carried out a systematic study of the phase compositions of the said binary systems by coprecipitation from aqueous solutions, comparing our results with equilibrium data for the said diagrams, including the ones established at higher temperatures.

Fluoride nanopowders were co-precipitated from the aforementioned aqueous solutions under numerous synthetic conditions (eg, type of the fluorinating agent such as HF, NH_4F, NaF, etc., supplementary organic modifiers, order of reagent addition to the reaction mixture, concentration of starting materials, and so on).

Phase diagrams of the MF_2-YF_3 systems are presented in Figs. 2.2–2.4 (melt areas in the phase diagrams are marked with the character "*L*"). All such systems have broad areas of homogeneity corresponding to $M_{1-x}Y_xF_{2+x}$ solid solutions (*F* phases) with composition limits at $x = 0.38$, 0.41, 0.37 for M = Ca, Sr, Ba, respectively [80]. According to Fedorov [20] and Sobolev [80], different studies carried by various methods indicate that these solid solutions contain Y_6F_{37} clusters as structural units in their crystalline lattices. Coordination polyhedron of yttrium in Y_6F_{37} blocks is Thompson's antiprism, and they can be easily placed in the fluorite lattice (Fig. 2.5). This heterovalent substitution mechanism can be described with the following equation:

$$Ca_6F_{32}^{20-} \leftrightarrow Y_6F_{37}^{19-} + F_{int}^{-} \qquad [i]$$

Lowering temperature under equilibrium conditions results in ordering Y_6F_{37} clusters' positions in $M_{1-x}Y_xF_{2+x}$ solid solutions' crystal lattices and, as a result, series of ordered fluorite-type phases appear in the aforementioned MF_2-YF_3 systems (Figs. 2.2–2.4)

[23,28,41,80]. For concentrations of yttrium fluoride about 70 mol% in calcium and/or strontium systems, the formed high-temperature phases of variable compositions have tysonite (LaF_3) structures (*T* phases) [81]. These YF_3 tysonite-type solid solutions are stabilized by anionic vacancies formed by replacement of Y atoms by Ca or Sr [22,82]:

$$Y^{3+} + F^{-} \leftrightarrow M^{2+} + V_F \quad \text{[ii]}$$

In contrast with Ca and Sr, stoichiometric BaY_2F_8 compound is formed in barium-containing systems [36,88] (Detailed thermal studies of both BaY_2F_8 single

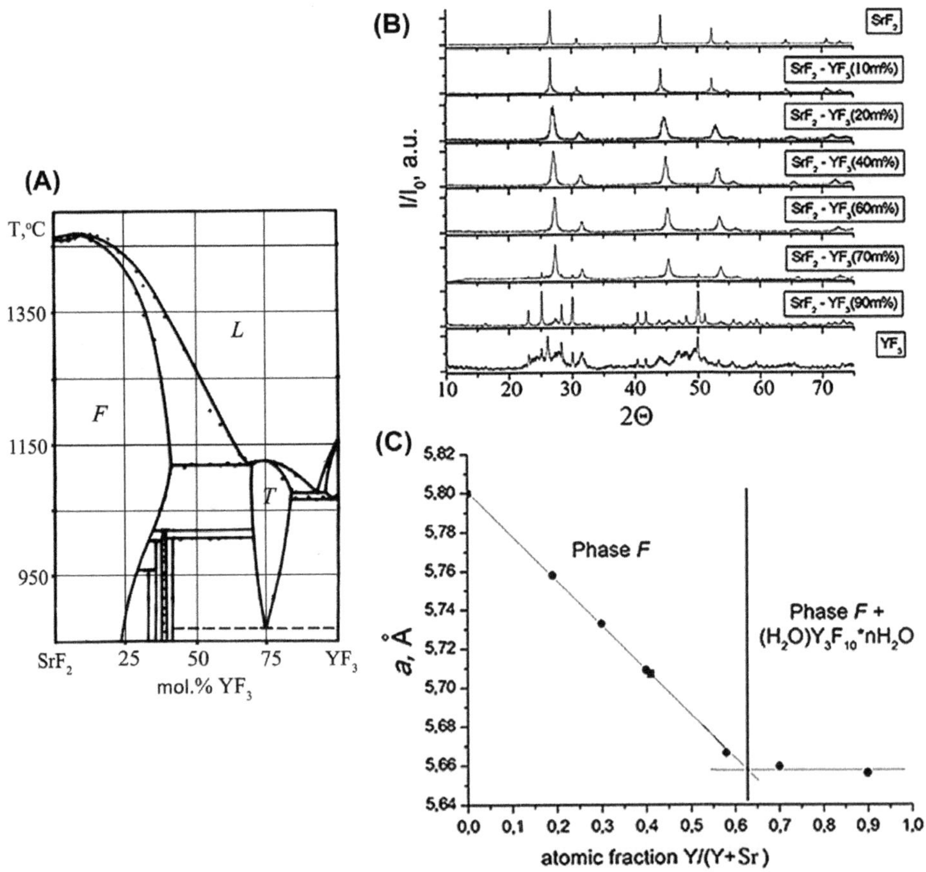

Figure 2.2 (A) The equilibrium phase diagram of the SrF_2-YF_3 system; (B) X-ray diffraction patterns, (C) phase composition, and (D) electron microscopy images of $0.6SrF_2$-$0.4YF_3$; and (E) $0.1SrF_2$-$0.9YF_3$ nanopowders precipitated from aqueous nitrate solutions.
M.N. Mayakova, A.A. Luginina, S.V. Kuznetsov, V.V. Voronov, R.P. Ermakov, A.E. Baranchikov, V.K. Ivanov, O.V. Karban', P.P. Fedorov, Synthesis of SrF_2-YF_3 nanopowders by co-precipitation from aqueous solutions, Mendeleev Commun. 24 (6) (2014) 360−362 and M.N. Mayakova, S.V. Kuznetsov, V.V. Voronov, A.E. Baranchikov, V.K. Ivanov, P.P. Fedorov, Soft chemistry synthesis of powders in the BaF_2-ScF_3 system, Russ. J. Inorg. Chem. 59 (7) (2014) 773−777; with permission of Mendeleev Communications.

Figure 2.2 Continued.

crystals and corresponding melt [36] revealed that there is no BaY_2F_8 phase transition. This is in agreement with [88] phase diagram. Unfortunately, one of the later publications [87] stated otherwise, mentioning that BaY_2F_8 undergoes phase transition, and such error has been reproduced in some other sources [47,80].).

Phase compositions of nanopowders (Figs. 2.2–2.4) were obtained by coprecipitation from aqueous solutions of alkaline earth metal and yttrium nitrates by HF addition (direct precipitation) [36,67].

In all systems, yttrium fluoride precipitated as hydroxonium cation $(H_3O)Y_3F_{10}\cdot nH_2O$ derivative in the form of octahedral 100–200 nm nanoparticles [62].

Results, obtained for strontium systems, were the simplest ones to interpret (Fig. 2.2). First of all, according to the classic chemical analysis data, Sr:Y ratios in the formed precipitates coincided with the same ratios for the starting solutions. Also the said precipitates were fluorite-type $Sr_{1-x}Y_xF_{2+x}$ solid solutions for the relatively wide composition ranges ($x < 0.63$); and composition ranges, where such solid solutions were formed, were much broader than those for the equilibrium samples of the same compositions synthesized at higher temperatures. It is worth noting that correlations between crystal lattice parameters $a(x)$ and specimen compositions were the same for the samples prepared by coprecipitation and products of high-temperature solid phase syntheses. No tysonite—or ordered fluorite—type phases were formed.

Coprecipitation in the BaF_2-YF_3 system results in the formation of solid $Ba_{1-x}Y_xF_{2+x}$ solution with the same dependency between the crystal lattice parameter $a(x)$ and sample composition as established for $Ba_{1-x}Y_xF_{2+x}$ samples synthesized at the higher temperatures. Area of homogeneity of precipitated cubic $Ba_{1-x}Y_xF_{2+x}$ solid solutions also includes compositions corresponding to fluorite-related $Ba_4Y_3F_{17}$ phase with trigonal distortion of initial cubic fluorite lattice, but this area of homogeneity does not reach compositions close to pure barium fluoride; instead of the precipitation of the latter, we have observed formation of $BaF_2 \cdot HF$ hydrofluoride. Barium hydrofluoride decomposed when

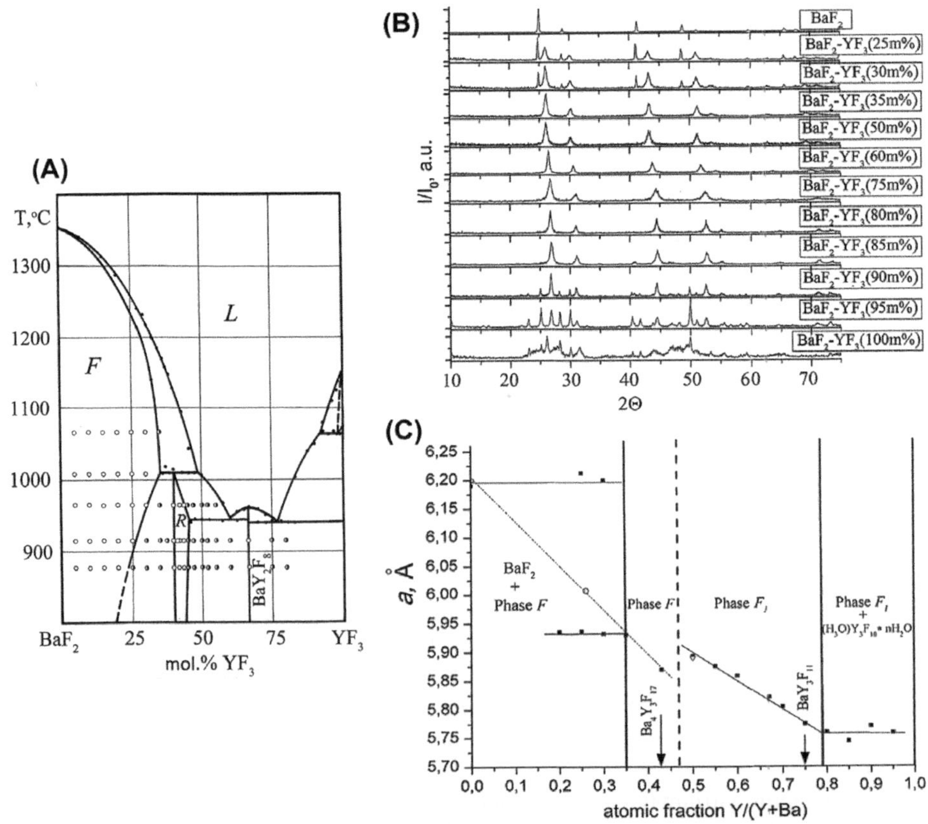

Figure 2.3 (A) The equilibrium phase diagram of the BaF_2-YF_3 system; (B) X-ray diffraction patterns (CuKα radiation), (C) phase composition, and (D) electron microscopy images of BaF_2 and $0.4BaF_2$-$0.6YF_3$; (E) nanopowders precipitated from aqueous nitrate solutions. P.P. Fedorov, M.N. Mayakova, S.V. Kuznetsov, V.V. Voronov, R.P. Ermakov, K.S. Samarina, A.I. Popov, V.V. Osiko, Co-precipitation of yttrium and barium fluorides from aqueous solutions, Mat. Res. Bull. 47 (2012) 1794—1799; with permission of Mater. Res. Bull.

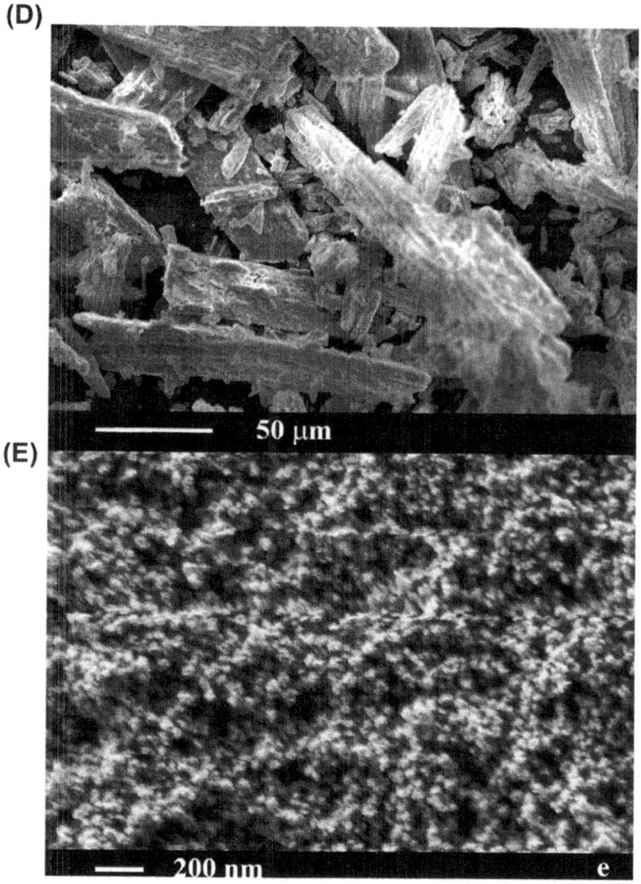

Figure 2.3 Continued.

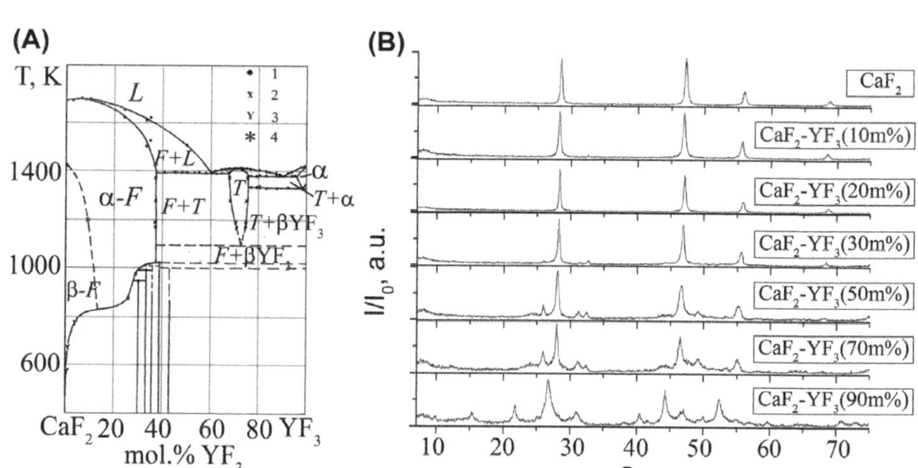

Figure 2.4 (A) The equilibrium phase diagram of the CaF_2-YF_3 system [23] and (B) X-ray diffraction patterns of nanopowders precipitated from aqueous nitrate solutions.

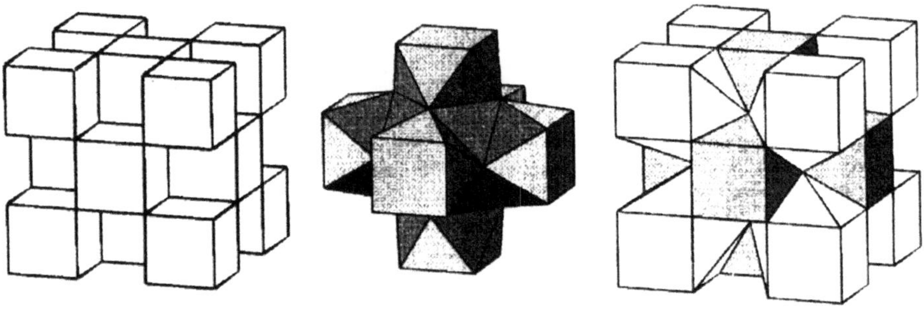

Figure 2.5 R_6F_{37} cluster built in the fluorite matrix.

it was washed with distilled water, and the formed BaF_2 crystals retained the same morphology as starting $BaF_2 \cdot HF$ (crystal size c. 10+ microns, Fig. 2.3D). BaF_2-YF_3 samples retain fluorite-type structures in case of the higher concentrations of yttrium fluoride, too. However, the latter yttrium higher-concentration samples exhibited a different correlation between their compositions and lattice parameters $a(x)$, ie, the formed fluorite-type phase F_1 was different from the yttrium lower-concentration fluorite-type phase F. One of the studied F_1 specimens had low-intensity superstructure lines in its X-ray diffraction pattern, which is quite rare for nanofluorides. This fact allowed concluding that the F_1 phase structure was derived from KY_3F_{10} structural type [57]. All nanophases in the BaF_2-YF_3 system are based on YF_8 Thompson's antiprisms [36].

Coprecipitation in the CaF_2-YF_3 system results in the formation of $Ca_{1-x}Y_xF_{2+x}$ solid solution with the crystal lattice parameters $a(x)$ that coincided with those for the corresponding high-temperature phase of similar composition [78]. However, in contrast with $Sr_{1-x}Y_xF_{2+x}$ and $Ba_{1-x}Y_xF_{2+x}$ in the strontium and barium systems, $Ca_{1-x}Y_xF_{2+x}$ area of homogeneity for precipitated specimens is narrower ($x < 0.2$) than that for the samples synthesized at the higher temperatures. Increase in yttrium content in $Ca_{1-x}Y_xF_{2+x}$ resulted in the appearance of additional lines in its X-ray diffraction patterns in comparison with the lines for cubic fluorite-type solid solution. Calculations of the corresponding parameters for the cubic sublattice became less accurate and reliable, and X-ray diffraction patterns also became more complex with the increase of yttrium content. Whereas our attempts to interpret and index such X-ray diffraction patterns of $Ca_{1-x}Y_xF_{2+x}$ samples co-precipitated from nitrate solutions were unsuccessful; we may not exclude the formation of another fluorite-related nanophase in the CaF_2-YF_3 system with variable Ca:Y ratio.

Thus, despite quite different areas of homogeneity in MF_2-YF_3 systems for M = Ca, Sr, and Ba, all of them had the following common features:

- Said areas of homogeneity were quite broad from composition point of view;
- Formed phases were not ordered as they should be according to the third law of thermodynamics [23];
- All phases were not under equilibrium; and
- Phases formed by coprecipitation were made up of c. 50-nm particles as per electron microscopy and X-ray diffraction (calculations of the sizes of domains of coherent scattering D) data (see Figs. 2.2–2.4).

The results of the nanopowder syntheses by coprecipitation in the BaF_2-BiF_3 system [34] are similar to those observed for the BaF_2-YF_3 system (the difference between these two systems, however, includes the formation of BiOF in the former system due to the strong hydrolysis of the samples with more than 50% BiF_3). At the same time, the BaF_2-ScF_3 and BaF_2-YF_3 systems appear to be much more different: for example, the formation of equilibrium $Ba_3Sc_2F_{10}$ phase has been observed in the BaF_2-ScF_3 system [68].

We studied coprecipitation from aqueous solutions in the NaF-RF_3 systems with the use of sodium fluoride as the fluorinating agent (please see equilibrium phase diagrams for NaF-RF_3 (R = La, Gd, Y, Lu) systems in Fig. 2.6A [21,56]). All systems contained low-temperature hexagonal "$NaRF_4$" *G* phases with small variable compositions and crystal structures derived from UCl_3 or natural mineral gagarinite ($NaCaRF_6$). Instead of commonly used "$NaRF_4$" designation, we recommend a more accurate representation of their composition as $R_{2-x}Na_{3x}F_6$ ($x \sim 0.5$) [51]. Systems with R = Pr-Lu, Y contained fluorite-type (cubic) $Na_{0.5-x}R_{0.5+x}F_{2+2x}$ *F* phases of variable compositions, formed at higher temperature syntheses. For R = Gd-Lu, Y, one may consider these phases as higher temperature modifications of corresponding $NaRF_4$. These systems also contained solid solutions on the base of LaF_3 and α-GdF_3-type (tysonite-type) high-temperature polymorphs, and some fluorite-related ordered phases as well.

Ambient temperature coprecipitations in NaF-RF_3 systems differed from those in MF_2-RF_3 systems by incongruent formation of phases in the NaF-RF_3 systems. Therefore, one needed a significant NaF excess (eg, 5:1) in order to obtain desirable compositions [56]. One can control compositions of cubic NaF-RF_3 precipitates with the use of X-ray diffraction data (eg, please see correlation between sample compositions and fluorite-type crystal lattice parameters in Ref. [30] similarly to the procedure for the MF_2-RF_3 systems).

Formation of gagarinite-type phase in NaF-LaF_3 system was not obtained by coprecipitation from aqueous solutions. Instead, we got a $NaLaF_4$ known to be stable at the higher temperatures (Fig. 2.6A). Our data for $NaLaF_4$ syntheses in nonaqueous media indicated that this phase might be stable only above 330°C.

Whereas coprecipitation in NaF-GdF_3 system produced an equilibrium gagarinite-type *G* phase [21], NaF-YF_3 and NaF-LuF_3 systems produced nonequilibrium fluorite-type *F* phases similar to those obtained at the higher temperatures. Exothermal transformation of such nonequilibrium Na-Y phase to the equilibrium gagarinite phase was observed as a result of heating up to 400°C [32], but corresponding fluorite-type Na-Lu phase could not be converted to the stable gagarinite-type or middle-range temperature $NaScF_4$-type phases (Fig. 2.6).

3. Luminescence

Inorganic fluorides are a unique class of materials for luminophores that can be implemented in various areas of everyday practice. They are more stable than chalcogenides or the other halides, and they have shorter phonon spectrum and larger isomorphic capacity to accommodate active rare earth ion dopants compared to the similar oxide materials. This part of our review is dedicated to luminophores that irradiate in the visible

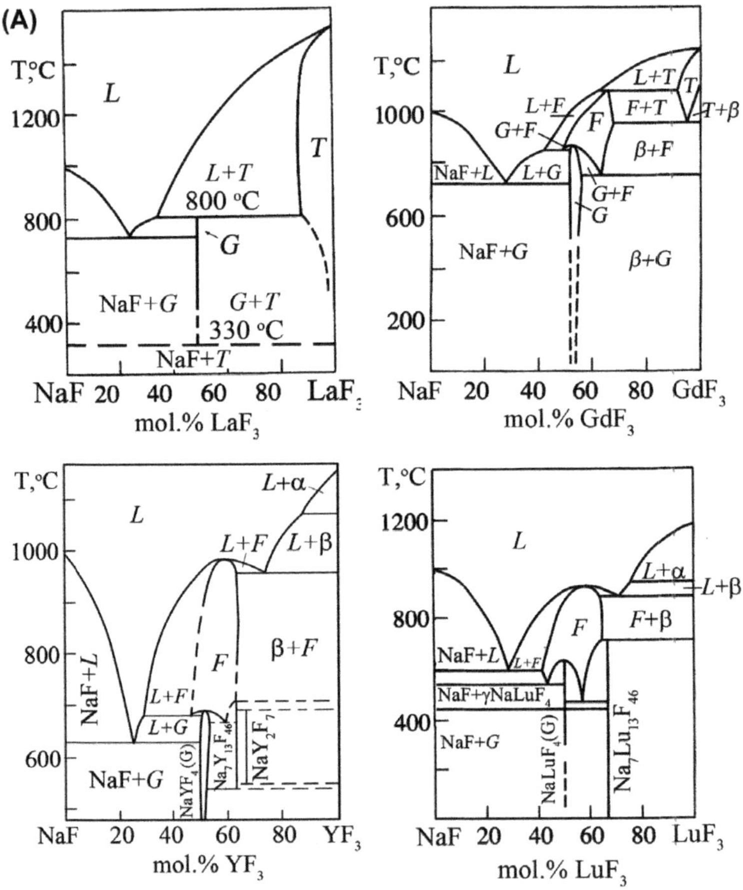

Figure 2.6 (A) The equilibrium phase diagrams for NaF-RF_3 (R = La, Gd, Y, Lu) systems, electron microscopy images for (B) *F*-phase and (C) *G*-phase, and (D) relative stability of different phases under $NaRF_4$ precipitation conditions. *F*, cubic fluorite-type phase $Na_{0.5-x}R_{0.5+x}F_{2+2x}$; *G*, hexagonal gagarinite-type phase $Y_{2-x}Na_{3x}F_6$; *T*, LaF_3-based hexagonal phase; γ, $NaLuF_4$ middle-temperature polymorph with hexagonal type structure. S.V. Kuznetsov, A.A. Ovsyannikova, E.A. Tupitsyna, D.S. Yasyrkina, V.V. Voronov, P.P. Fedorov, N.I. Batyrev, L.D. Iskhakova, V.V. Osiko, Phase formation in LaF_3-$NaGdF_4$, $NaGdF_4$-$NaLuF_4$, $NaYF_4$-$NaLuF_4$ systems: synthesis of powders by co-precipitation from aqueous solutions, J. Fluorine Chem. 161 (2014) 95–101; with permission of J. Fluorine Chem.

part of spectrum via upconversion venue. The latter mechanism was suggested by Ovsyankin and Feofilov [71] and Auzel [6,7]. It includes energy transfer from near-infrared (IR) spectrum region to the visible part of it that occurs when working body matrices are doped with Yb:Er, Yb:Tm, and Yb:Ho pairs of elements. Ytterbium ions play a sensibilizing role and allow increase in the quantum yield (QY) of Er, Tm, and Ho upconversion luminescence. Practical application of such an effect became possible due to the development of compact high-power laser diodes (late 2000s) that

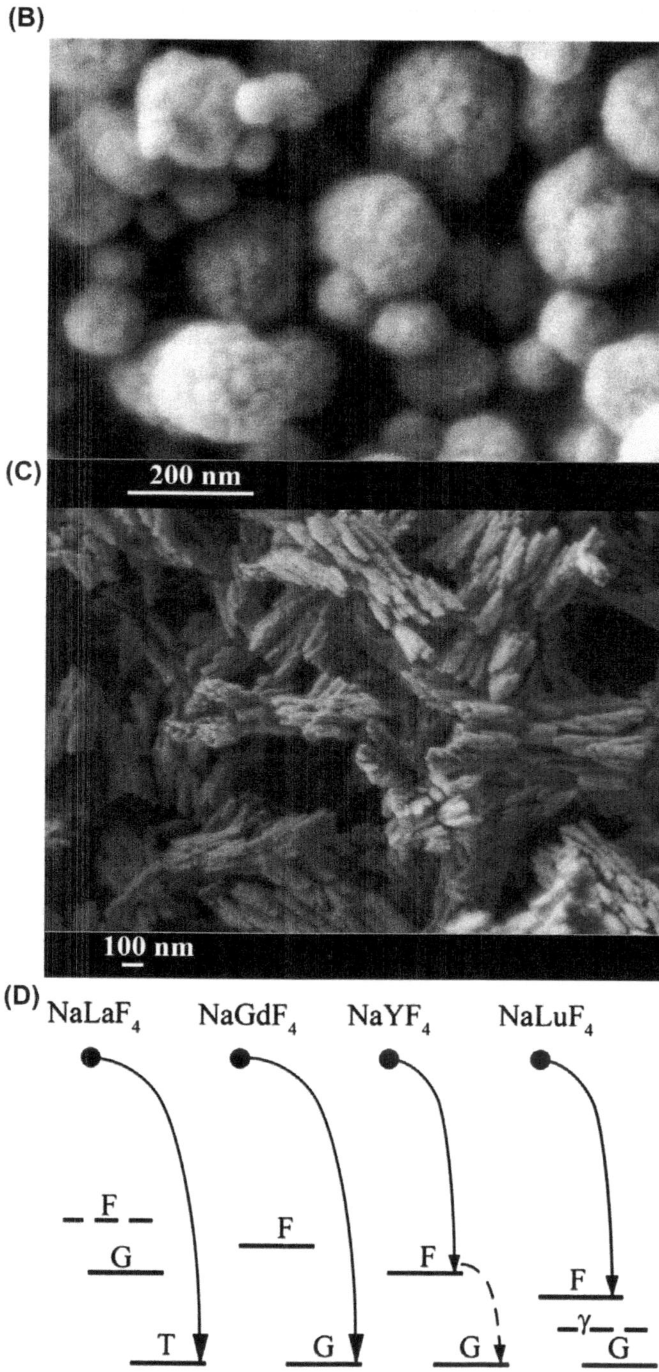

Figure 2.6 Continued.

were capable of efficient pumping ytterbium ions with energy in near-IR parts of spectrum, thus, allowing manufacturing of high-power lasers, eg, CaF_2:Yb-based lasers [24]. Since then, this type of luminescence has been widely implemented in the development of novel technical applications in immunohistochemistry, immunocytochemistry, complex immunological, enzymatic, fluorescent bioanalyses, and flow-type cytometry as well as in vivo, in situ, ex situ biometric visualization [15,43]. Medical application of fluoride materials appears to be much safer than utilizing toxic chalcogenide-based quantum dots or photodegrading photosensibilizers like phthalocyanines [72]. It is quite obvious that transfer of pumping energy from near-IR part of the spectrum to its visible range with higher energy is relatively low, and usually its efficiency does not exceed a few percent. Use of nanoparticles decreases the said upconversion efficiency because of the increase of the particle surface area which is accompanied with the increase of the number of centers quenching the luminescence per nanoparticle volume unit [46]. Modern literature contains thousands of publications on how one can overcome the latter effect. Well-known approaches included creating special particle shells made up of organic or inorganic materials containing, in turn, inactive elements or active rare earth metals [16,60,91,92], attaching the so-called plasmonic nanoparticles (Ag, Au) [13,40], implementing high-power pumping [101], as well as combination of any of these actions. In the 1970s, scientists [48] suggested β-$NaYF_4$ lower temperature modification as one of the most prospective matrices for the efficient upconversion. As a result, numerous studies of β-$NaYF_4$ have followed, but practically all of these publications bear certain similar features:

- Very frequently, researchers use only two compositions with Yb:Er = 20:2 or 17:3 ratios without any reasonable choice justification and without evaluation of the real specimen compositions, ie, scientists do not account for an incongruent crystallization with deviation of the real sample composition from the nominal one for the content of rare earth elements and sodium as well.
- Researchers do not carry out X-ray diffraction analysis in complete and proper manner limiting their studies just by determination of the unit cell parameters based on transmission electron spectroscopy; they do not check for the presence of the second phase.
- While studying luminescence and interpreting luminescence transitions, researchers measure lifetime of excited states very rarely and even less frequently they determine luminescence efficiency via calculating QY or energy yield (B_{en}); also they hardly compare their QY and B_{en} data with the results of other authors.

Of course it is very hard to carry out such comparisons because usually samples are prepared by different techniques from different starting materials of different purity (this includes contaminations from the used matrix solutions, too); particles have different sizes affecting QYs (please see the above remarks about correlation of QY and nanoparticle size); correlations between QY and B_{en} values and pumping power are not linear, so use of different lasers with various pumping powers in different studies makes such result comparisons impossible.

Boyer and van Veggel [12] were among the first ones who attempted QY analysis for $NaYF_4$ nanoparticle luminescence using their own enhanced technique that we have modified and utilized in our own studies (this included registering of upconversion luminescence spectra as well as measuring and calculating upconversion

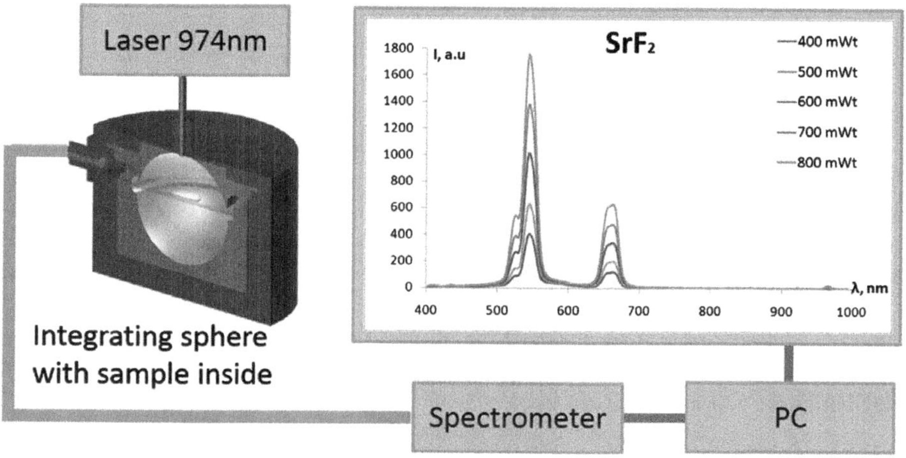

Figure 2.7 Spectroscopy installation for studying energy yield of nanoparticle upconversion luminescence (integrating sphere with the sample holder and sample, spectrometer, 974-nm laser, computer and fiber optics connectors).

luminescence energy yield). Our proprietary equipment for the laser-induced luminescence spectroscopy included LESA-01-BioSpec fiber optics spectrum analyzer and 974 nm/1 W laser (BioSpec, Moscow, Russia). Studied samples were placed in a modified integrating sphere (made by Avantes), IR excitation irradiation and scattered upconverted visible light were transferred via fiber optics, and scattered light was analyzed with Hitachi spectrometer (Fig. 2.7). Studied nanopowders were secured between two flat cover glasses before their placement in the aforementioned integrating sphere. Obtained results were analyzed with Uno Momento software package.

In order to account for the features of the used spectrometer, we have performed preliminary calibrations of our equipment before each experiment.

Energy yield calculations for upconversion spectroscopy were carried out according to the following equation:

$$B_{\mathrm{en}} = \frac{\sum_i^{i=1,2,3} P_{i\_\mathrm{emitted}}^{\mathrm{Sample}}}{P_{974\_\mathrm{absorbed}}^{\mathrm{Sample}}} = \frac{\sum_i^{i=1,2,3} P_{i\_\mathrm{emitted}}^{\mathrm{Sample}}}{P_{974\_\mathrm{scattered}}^{\mathrm{Reference}} - P_{974\_\mathrm{scattered}}^{\mathrm{Sample}}} = \frac{\sum_i^{i=1,2,3} kp_i \cdot S_i^{\mathrm{Sample}}}{kp_{974}\left(S_{974}^{\mathrm{Reference}} - S_{974}^{\mathrm{Sample}}\right)} \qquad [2.1]$$

where: $P_{974\_\mathrm{absorbed}}^{\mathrm{Sample}}$, power of 974-nm laser absorbed by the sample (it is equal to the difference between power of scattered laser radiation for nonabsorbing reference specimen $P_{974\_\mathrm{scattered}}^{\mathrm{Reference}}$ and power of scattered radiation for the studied sample $P_{974\_\mathrm{scattered}}^{\mathrm{Sample}}$); kp_i, calibration coefficient related to spectrometer sensitivity in each i wavelength range; S_i, total (integrated) area below the spectrum curve for each peak. Intrinsic (nondoped by activators) $NaYF_4$ nanoparticles were used as nonabsorbing reference specimens.

In order to compare luminescence properties of the synthesized samples and literature data, we calculated upconversion luminescence QY based on the energy yield data.

We attempted determination of the optimal sample compositions that would provide maximum energy yields for the luminescence of cubic $NaYF_4$ as well as SrF_2, CaF_2, and LaF_3 matrices. All specimens were synthesized using the same high-purity starting materials according to the same coprecipitation protocol. We are leaving low-temperature hexagonal $NaYF_4$ phase outside the scope of the present paper, for this phase luminescence is obstructed due to the relatively large concentration of OH(−) anions when co-precipitated from aqueous solutions. Because the latter OH(−) impurities are one of the most undesirable contamination of luminescence materials, drying/dehydration becomes crucially important in the preparation of luminescent fluorides.

Typical results (X-ray diffraction pattern and electron microscopy image) for $NaYF_4$:Yb:Er nanopowders are presented in Fig. 2.8.

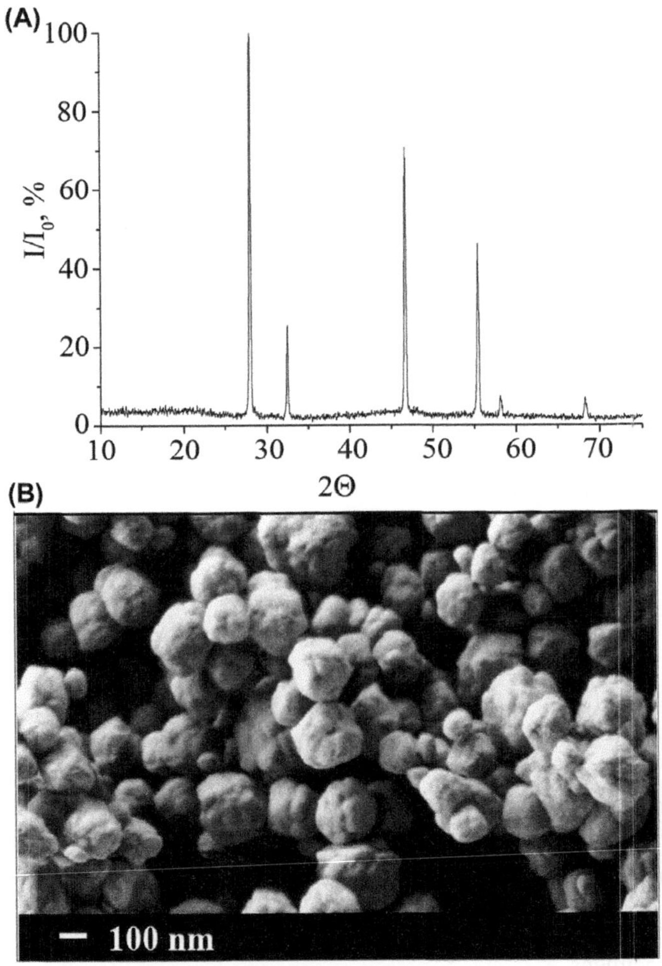

Figure 2.8 (A) X-ray diffraction pattern and (B) scanning electron microscopy image of single-phase fluorite-type $NaY_{0.6}Yb_{0.3}Er_{0.1}F_4$ specimen.

Correlation between upconversion luminescence energy yields and sample compositions for Yb/Er-doped $NaYF_4$ and SrF_2 is presented in Figs. 2.9 and 2.10, respectively. One can see that SrF_2 is an excellent matrix for upconversion Yb,Er luminophore with the energy yield exceeding 5%. However, the same study of LaF_3- and CaF_2-based luminophores revealed that B_{en} values did not exceed 2.5%.

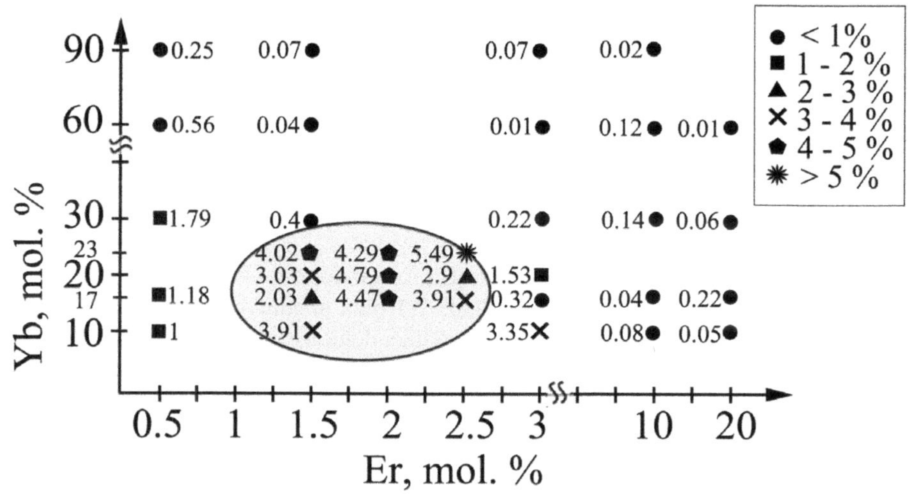

Figure 2.9 Upconversion luminescence energy yield versus Yb/Er content for cubic $NaYF_4$:(Yb,Er) specimens (1 W/cm^2 pumping power).

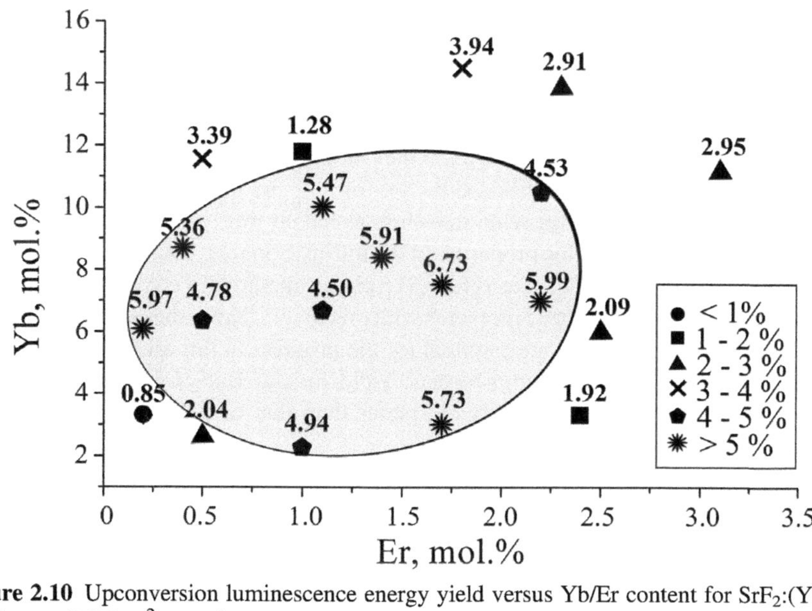

Figure 2.10 Upconversion luminescence energy yield versus Yb/Er content for SrF_2:(Yb,Er) specimens (1 W/cm^2 pumping power).

Table 2.1 Upconversion Luminescence Quantum Yields for $NaYF_4$:Yb (20 mol%):Er (2 mol%) Samples

	Laser Pumping	Quantum Yield, %
Our data	974 nm, 1 W/cm^2	3.06 ± 0.02
	974 nm, 0.5 W/cm^2	0.37 ± 0.02
[12]	977 nm, 22 W/cm^2	7.87 ± 1.2
	977 nm, 3.8 W/cm^2	1.67 ± 0.3
	977 nm, 0.5 W/cm^2	0.21 ± 0.03
[12]	980 nm, 150 W/cm^2	0.005–0.3 (varies with the sample particle size) 3% for the single crystal (bulk sample)

Comparison of our data with literature [12] information for $NaYF_4$(Yb,Er) luminescence QYs is presented in Table 2.1. Table 2.1 data unequivocally indicate that our $NaYF_4$:Yb (20 mol%):Er (2 mol%) nanopowders demonstrate higher QYs for various pumping power values.

4. Laser and Scintillating Ceramics

Laser ceramics is one of the most significant innovations in the area of laser materials at the present time [24,58,76]. The first publication about fluoride laser ceramics [42] described only its lasing effect. Later, our papers [2,3,10,18,24,31,73] also covered such areas as optical transmittance, mechanical properties, thermal conductivity, spectral-luminescence, and lasing properties. This includes fluoride laser ceramics that has spectral and lasing-generation properties similar to the ones for single crystals of the same compositions. Fluoride laser ceramics of these qualities was prepared by hot-forming [18,24]. It belongs to nanomaterials due to its layered nanostructure caused by polysynthetic twinning. Some of the materials synthesized by hot-forming are listed in Table 2.2. Unfortunately, hot-forming does not provide the same excellent mechanical properties for the doped ceramics compared to the properties of the intrinsic one (eg, four- to sixfold increase of stability against cracking K_{1C}, etc.) [1,24,31]: classic methods of ceramics synthesis are not that helpful in overcoming the problems with precursors. Nevertheless, some efforts of French research groups [5,65] are essential for the progress in this area.

It is worth noting that the luminescence yield for our BaF_2:Ce^{3+} scintillation ceramics as well as irradiation stability are better than that of the same parameters for BaF_2:Ce^{3+} single crystals [9,17,39].

5. Self-Fluorinating Agents

Synthesis of ceramic precursors is impaired by water that gets adsorbed on the nanoparticle surfaces. This causes pyrohydrolysis that results in the samples losing their

Table 2.2 **Laser Fluoride Ceramics: Matrixes, Activators, Rare Earth *f–f* Electron Transitions, Lasing Spectral Ranges (μm)**

Matrix	Pr^{3+}	Nd^{3+}	Er^{3+}	Tm^{3+}	Yb^{3+}
CaF_2		$^4F_{3/2} \rightarrow {}^4I_{11/2}$ 1.02–1.08	$^4I_{11/2} \rightarrow {}^4I_{13/2}$ 2.69–2.75	$^3H_4(^3F_4) \rightarrow {}^3H_6$ 1.88–1.92	$^5F_{5/2} \rightarrow {}^2F_{7/2}$ 1.025–1.040
SrF_2	$^3P_0 \rightarrow {}^3F_2$ 0.639	$^4F_{3/2} \rightarrow {}^4I_{11/2}$ 1.036–1.044			$^5F_{5/2} \rightarrow {}^2F_{7/2}$ 1.01–1.09
$(Ca,Sr)F_2$			$^4I_{11/2} \rightarrow {}^4I_{13/2}$ 2.70–2.78	$^3H_4(^3F_4) \rightarrow {}^3H_6$ 1.92–2.04	$^2F_{5/2} \rightarrow {}^2F_{7/2}$ 1.00–1.08
$(Ba,Sr)F_2$				$^3H_4(^3F_4) \rightarrow {}^3H_6$ 1.78–1.97	$^2F_{5/2} \rightarrow {}^2F_{7/2}$ 1.0–1.09

transparency. One of the venues to overcome this problem is developing a self-fluorinating precursor that would generate a fluorinating agent to suppress the said pyrohydrolysis.

We have developed two different types of such precursors.

Precipitation of barium fluoride in concentrated HF results in the formation of $BaF_2 \cdot HF$ micron-sized crystals instead of BaF_2. Heating $BaF_2 \cdot HF$ causes its thermal decomposition with evolution of gaseous HF which is a quite powerful fluorinating agent. Unfortunately, $BaF_2 \cdot HF$ does not dissolve rare earth fluorides, and synthesis of rare earth–doped fluorides with $BaF_2 \cdot HF$ results in the precipitation of the aforementioned rare earth fluorides as separate nanophases [8,64]. BaF_2-based solid solutions form at the later steps when ceramics is prepared by sintering of the formed powders.

Nevertheless, we were able to prepare high-quality $BaF_2{:}Ce^{3+}$ scintillation ceramics with the use of CeF_3-doped $BaF_2 \cdot HF$ as starting material (Fig. 2.11A) via the following chemical transformations:

$$Ba(NO_3)_2 + 3HF \rightarrow BaF_2 \cdot HF\downarrow + 2HNO_3 \qquad [iii]$$

$$R(NO_3)_3 + 3HF \rightarrow RF_3\downarrow + 3HNO_3 \qquad [iv]$$

$$BaF_2 \cdot HF \rightarrow BaF_2 + HF\uparrow \qquad [v]$$

$$(1 - x)BaF_2 + xRF_3 \rightarrow Ba_{1-x}R_xF_{2+x} \qquad [vi]$$

The other type of the aforementioned precursors is represented by $Sr_{1-x-y}R_x(NH_4)_yF_{2+x-y}$ solid solution [74]. The latter was formed when precipitation was carried out with NH_4F: NH_4F dissolved in strontium and rare earth fluorides:

$$(1 - x - y)Sr(NO_3)_2 + xR(NO_3)_3 + (2 + x - y)\,NH_4F \rightarrow$$

$$Sr_{1-x-y}R_x(NH_4)_yF_{2+x-y}\downarrow + (2 + x - 2y)NH_4NO_3 \qquad [vii]$$

Figure 2.11 (A) Optical ceramics, prepared from self-fluorinating precursors by hot-pressing technique BaF:Ce^{3+} *(A.A. Luginina, A.E. Baranchikov, A.I. Popov, P.P. Fedorov, Preparation of barium monohydrofluoride $BaF_2 \cdot HF$ from nitrate aqueous solutions, Mater. Res. Bull. 49 (1) (2014) 199–205; with permission of Mater. Res. Bull.)* and (B) SrF_2-based ceramics.

When heated, $Sr_{1-x-y}R_x(NH_4)_yF_{2+x-y}$ solid solution decomposed producing hydrogen fluoride:

$$Sr_{1-x-y}R_x(NH_4)_yF_{2+x-y} \rightarrow Sr_{1-x-y}R_xF_{2+x-2y} + yNH_3\uparrow + yHF\uparrow \quad \text{[viii]}$$

$Sr_{1-x-y}R_x(NH_4)_yF_{2+x-y}$ has been used for preparation of strontium fluoride-based high efficiency upconversion luminophore and optical ceramics [75]. Fig. 2.11B shows optical ceramics prepared from SrF_2-based precursor.

6. Conclusion

The reviewed literature data about properties of inorganic nanofluorides and the preparation of these compounds by coprecipitation from aqueous solutions unequivocally indicate that this synthesis venue is one of the most prospective ways to obtain various fluoride materials, including upconversion luminophores, optical ceramics, and their precursors.

Acknowledgments

Authors greatly appreciate the invaluable help of A.E. Baranchikov, E.V. Chernova, A.A. Demidenko, M.E. Doroshenko, R.P. Ermakov, E.A. Garibin, V.K. Ivanov, A.A. Luginina, M.N. Mayakova, A.A. Ovsyannikova, A.M. Pak, D.V. Pominova, A.I. Popov, Yu. A. Rozhnova, A.V. Ryabova, R.L. Simoneaux, E.A. Tupitsyna, R.G. Vakhrenev, V.V. Voronov, D.S. Yasyrkina, and many other colleagues and friends without whom this publication would not be possible.

This work was partially supported by the Russian foundation on Basic Research grants RFBR-13-02-12162-ofi-m, RFBR 15-08-02481, RFBR-15-32-21152-mol-a-ved, and grant of the President of the Russian Federation MK-3133-2014-2 to Kuznetsov S.V.

References

[1] M.Sh. Akchurin, R.V. Gainutdinov, P.L. Smolyanskii, P.P. Fedorov, Anomalously high fracture toughness of polycrystalline optical fluorite from the Suran deposit (South Urals), Doklady Phys. 51 (1) (2006) 10–12.

[2] M.Sh. Akchurin, R.V. Gainutdinov, E.A. Garibin, Yu.I. Golovin, A.A. Demidenko, K.V. Dukel'skii, S.V. Kuznetsov, I.A. Mironov, V.V. Osiko, A.N. Smirnov, N.Yu. Tabachkova, A.I. Tyurin, P.P. Fedorov, V.V. Shindyapin, Nanostructure of optical fluoride ceramics, Inorg. Mater. Appl. Res. 2 (2) (2011) 97–103.

[3] M.Sh. Akchurin, T.T. Basiev, A.A. Demidenko, M.E. Doroshenko, P.P. Fedorov, E.A. Garibin, P.E. Gusev, S.V. Kuznetsov, M.A. Krutov, I.A. Mironov, V.V. Osiko, P.A. Popov, CaF_2:Yb laser ceramics, Opt. Mater. 35 (3) (2013) 444–450.

[4] M. Anji Reddy, M. Fichtner, Batteries based on fluoride shuttle, J. Mater. Chem. 21 (2011) 17059–17062.

[5] P. Aubry, A. Bensalah, P. Gredin, G. Patriarche, D. Vivien, M. Mortier, Synthesis and optical characterizations of Yb-doped CaF_2 ceramics, Opt. Mater. 31 (2009) 750–753.

[6] F. Auzel, Compteurquantique par transfertd'énergie entre deux ions de terresraresdansun tungstate mixteetdans un verre, C.R. Acad. Sci. 262 (1966) 1016–1019.

[7] F. Auzel, Upconversion and anti-stokes processes with f and d ions in solids, Chem. Rev. 104 (2004) 139–174.

[8] S.Kh. Batygov, M.N. Mayakova, S.V. Kuznetsov, P.P. Fedorov, X-ray luminescence of $BaF_2:Ce^{3+}$ powders, Nanosystems 5 (6) (2014) 752–756.

[9] Kh.S. Batygov, L.S. Bolyasnikova, A.E. Garibin, V.A. Demidenko, M.E. Doroshenko, K.V. Dykelskii, A.A. Lugingina, V.V. Osiko, P.P. Fedorov, $BaF_2:Ce^{3+}$ scintillation ceramics, Doklady Phys. 53 (9) (2008) 485–488.

[10] T.T. Basiev, M.E. Doroshenko, P.P. Fedorov, V.A. Konyushkin, S.V. Kuznetsov, V.V. Osiko, M.S. Akchurin, Efficient laser based on CaF_2-SrF_2-YbF_3 nanoceramics, Opt. Lett. 33 (5) (2008) 521−523.
[11] T.B. Bekker, P.P. Fedorov, A.E. Kokh, The ternary reciprocal system Na,Ba//BO_2,F, Cryst. Growth Des. 12 (2012) 129−134.
[12] J.-C. Boyer, F.C.J. van Veggel, Absolute quantum yield measurements of colloidal $NaYF_4$:Er^{3+},Yb^{3+}upconverting nanoparticles, Nanoscale 2 (2010) 1417−1419.
[13] Z. Chao, S. Lingdong, Z. Yawen, Y. Chunhun, Rare earth upconversion nanophosphors: synthesis, functionalization and application as biolabels and energy transfer donors, J. Rare Earth 28 (6) (2012) 807−819.
[14] D. Chen, Y. Yu, F. Hang, Y. Wang, Phase transition from hexagonal LnF_3 (Ln = La, Ce, Pr) to cubic $Ln_{0.8}M_{0.2}F_{2.8}$ (M = Ca, Sr, Ba) nanocrystals with enchanced upconversion induced by alkaline-earth doping, Chem. Commun. 47 (2011) 2601−2603.
[15] P.L.A.M. Corstjens, S. Li, M. Zuiderwijk, et al., Infrared up-converting phosphors for bioassays, IEE Proc. Nanobiotechnol. 152 (2) (2005) 64−72.
[16] J. Cybinska, C. Lorbeer, A.-V. Mudrig, Phosphate protected fluoride nano-phosphors, J. Mater. Chem. 22 (2012) 9505−9508.
[17] A.A. Demidenko, E.A. Garibin, S.D. Gain, Yu.I. Gusev, P.P. Fedorov, I.A. Mironov, S.B. Michrin, V.V. Osiko, P.A. Rodnyi, D.M. Seliverstov, A.N. Smirnov, Scintillation parameters of BaF_2 and BaF_2:Ce^{3+} ceramics, Opt. Mater. 32 (10) (2010) 1291−1293.
[18] M.E. Doroshenko, A.A. Demidenko, P.P. Fedorov, E.A. Garibin, P.E. Gusev, H. Jelinkova, V.A. Konyshkin, M.A. Krutov, S.V. Kuznetsov, V.V. Osiko, P.A. Popov, J. Shulc, Progress in fluoride laser ceramics, Phys. Stat. Solidi C 10 (6) (2013) 952−957.
[19] A. Düvel, J. Bednarcik, V. Šepelák, P. Heitjans, Mechanosynthesis of the fast fluoride ion conductor $Ba_{1-x}La_xF_{2+x}$: from the fluorite to the tysonite structure, J. Phys. Chem. C 118 (2014) 7117−7129.
[20] P.P. Fedorov, Association of point defects in non-stoichiometric $M_{1-x}R_xF_{2+x}$ fluorite-type solid solutions, Butll. Soc. Cat. Cien. 12 (2) (1991) 349−381.
[21] P.P. Fedorov, Systems of alkali and rare-earth metal fluorides, Russ. J. Inorg. Chem. 44 (11) (1999) 1703−1727.
[22] P.P. Fedorov, Heterovalent isomorphism and solid solutions with a variable number of ions in the unit cell, Russ. J. Inorg. Chem. 45 (3) (2000) S268−S291.
[23] P.P. Fedorov, Third law of thermodynamics as applied to phase diagrams, Russ. J. Inorg. Chem. 55 (11) (2010) 1722−1739.
[24] P.P. Fedorov, Fluoride laser ceramics, in: B. Denker, E. Shklovsky (Eds.), Handbook on Solid-State Lasers: Materials, Systems and Applications, Woodhead Publishing Limited, UK, 2013, pp. 82−109. Oxford Cambridge Philadelphia New Delhi.
[25] P.P. Fedorov, I.I. Buchinskaya, Spatial inhomogeneity in crystalline materials and saddle-type congruent melting points in ternary system, Russ. Chem. Rev. 81 (1) (2012) 1−20.
[26] P.P. Fedorov, V.V. Osiko, Crystal growth of fluorides, in: P. Capper (Ed.), Bulk Crystal Growth of Electronic, Optical and Optoelectronic Materials. Wiley Series in Materials for Electronic and Optoelectronic Applications, John Wiley & Son, Ltd, 2005, pp. 339−356.
[27] P.P. Fedorov, V.V. Osiko, Formation of nanoparticles aggregates by soft chemistry, Rare Metals (China) 28 (Spec. Issue) (October 2009) 671−673.
[28] P.P. Fedorov, O.E. Izotova, V.B. Alexandrov, B.P. Sobolev, New phases with fluorite-derived structure in CaF_2-(Y, Ln)F_3 systems, J. Solid State Chem. 9 (4) (1974) 368−374.
[29] P.P. Fedorov, B.P. Sobolev, S.F. Belov, Fusibility diagram of the system NaF-YF_3, and the cross-section $Na_{0.4}Y_{0.6}F_{2.2}$-YOF, Russ. J. Inorg. Mater. 15 (5) (1979) 640−643.

[30] P.P. Fedorov, V.B. Alexandrov, O.S. Bondareva, I.I. Buchinskaya, M.D. Val'covskii, B.P. Sobolev, Concentration dependences of the unit-cell parameters of nonstoichiometric fluorite-type phases $Na_{0.5-x}R_{0.5+x}F_{2+2x}$ (R = rare-earth elements), Crystallogr. Rep. 46 (2) (2001) 239–245.
[31] P.P. Fedorov, V.V. Osiko, T.T. Basiev, Yu.V. Orlovskii, K.V. Dykel'skii, I.A. Mironov, V.A. Demidenko, A.N. Smirnov, Optical fluoride and oxysulphide ceramics: preparation and characterization, in: Developments in Ceramic Materials Research, NOVA Science Publishers Inc, 2007, pp. 53–95.
[32] P.P. Fedorov, S.V. Kuznetsov, V.V. Voronov, I.V. Yarotskya, V.V. Arbenina, Soft chemical synthesis of $NaYF_4$ nanopowders, Russ. J. Inorg. Chem. 53 (11) (2008) 1681–1685.
[33] P.P. Fedorov, A.A. Luginina, S.V. Kuznetsov, V.V. Osiko, Nanofluorides, J. Fluorine Chem. 132 (12) (2011) 1012–1039.
[34] P.P. Fedorov, M.N. Mayakova, S.V. Kuznetsov, V.V. Voronov, V.V. Osiko, R.P. Ermakov, I.V. Gontar', A.A. Timofeev, L.D. Iskhakova, Coprecipitation of barium–bismuth fluorides from aqueous solutions: nanochemical effects, Nanotechnol. Russ. 6 (3–4) (2011) 203–210.
[35] P.P. Fedorov, S.V. Kuznetsov, M.N. Mayakova, V.V. Voronov, R.P. Ermakov, A.E. Baranchikov, V.V. Osiko, Coprecipitation from aqueous solutions to prepare binary fluorides, Russ. J. Inorg. Chem. 56 (10) (2011) 1525–1531.
[36] P.P. Fedorov, M.N. Mayakova, S.V. Kuznetsov, V.V. Voronov, R.P. Ermakov, K.S. Samarina, A.I. Popov, V.V. Osiko, Co-precipitation of yttrium and barium fluorides from aqueous solutions, Mat. Res. Bull. 47 (2012) 1794–1799.
[37] P.P. Fedorov, V.V. Osiko, S.V. Kuznetsov, O.V. Uvarov, M.N. Mayakova, D.S. Yasirkina, A.A. Ovsyannikova, V.V. Voronov, V.K. Ivanov, Nucleation and growth of fluoride crystals by agglomeration of the nanoparticles, J. Cryst. Growth 401 (2014) 63–66.
[38] P.P. Fedorov, A.A. Luginina, A.I. Popov, Transparent oxyfluoride glass ceramics, J. Fluorine Chem. 172 (2015) 22–50.
[39] P.P. Fedorov, M.H. Ashyrov, M. Boboyarova, S. Boybobeva, I. Nyritdinov, E.A. Garibin, S.V. Kuznetsov, A.N. Smirnov, Spectral-luminescent properties of BaF_2:CeF_3 crystals and nanoceramics, Inorg. Mater. 52 (2) (2016) 213–217.
[40] S. Fischer, F. Hallermann, T. Eichelkraut, G. Plessen, K.W. Krämer, D. Biner, H. Steinkemper, M. Hermle, J.C. Goldschmidt, Plasmon enhanced upconversion luminescence near gold nanoparticles–simulation and analysis of the interactions, Opt. Express 20 (1) (2012) 271–282.
[41] O. Greis, J.M. Hashke, Handbook on the Physics and Chemistry of Rare Earth, Elsevier, Amsterdam, 1982, 5, 387–460.
[42] S.E. Hatch, W.F. Parsons, R.J. Weagley, Hot-pressed polycrystalline CaF_2:Dy^{2+} laser, Appl. Phys. Lett. 5 (8) (1964) 153.
[43] X. He, K. Wang, Z. Cheng, In vivo near-infrared fluorescence imaging of cancer with nanoparticles-based probes, WIREs Nanomed. Nanobiotechnol. 2 (2010) 349–366.
[44] X. Huang, S. Han, W. Huang, X. Liu, Enhancing solar cell efficiency: the search for luminescent materials as spectral converters, Chem. Soc. Rev. 42 (2013) 173.
[45] V.K. Ivanov, P.P. Fedorov, A.Y. Baranchikov, V.V. Osiko, Oriented aggregation of particles: 100 years of investigations of non-classical crystal growth, Russ. Chem. Rev. 83 (12) (2014) 1204–1222.
[46] T. Jiang, W. Qin, D. Zhao, Size-dependent upconversion luminescence in CaF_2:Yb:Tm nanocrystals, Mater. Lett. 74 (2012) 54–57.

[47] A.A. Kaminskii, B.P. Sobolev, S.E. Sarkisov, G.A. Denisenko, V.A. Fedorov, Physico-chemical aspects of synthesis, spectroscopy, and stimulated emission of $BaLn_2F_8$-Ln^{3+} crystals, Neorg. Mater. 18 (1982) 482–497.
[48] T. Kano, H. Yamamoto, Y. Otomo, $NaLnF_4$:Yb,Er (Ln:Y, Gd, La): efficient green-emitting infrared-excited phosphors, J. Electrochem. Soc. Solid.State Sci. Technol. 119 (11) (1972) 1561–1564.
[49] A.E. Kokh, N.G. Kononova, T.B. Bekker, P.P. Fedorov, E.A. Nigmatylina, A. Ivanova, An investigation of the growth of β-BaB_2O_4 crystals in the BaB_2O_4-NaF system and new fluoroborate $Ba_2Na_3[B_3O_6]_2F$, Crystallogr. Rep. 54 (1) (2009) 146–151.
[50] S.V. Kouznetzov, I.V. Yarotskya, P.P. Fedorov, V.V. Voronov, S.V. Lavristchev, T.T. Basiev, V.V. Osiko, Preparation of nanopowdered $M_{1-x}R_xF_{2+x}$ (M = Ca, Sr, Ba; R = Ce, Nd, Er, Yb) solid solutions, Russ. J. Inorg. Chem. 52 (2007) 315–320.
[51] K.W. Kramer, D. Biner, G. Frei, H.U. Gudel, P. Hehlen, R. Steran, Hexagonal sodium yttrium fluoride based green and blue emitting upconversion phosphors, Chem. Mater. 16 (7) (2004) 1244–1251.
[52] S.V. Kuznetzov, V.V. Osiko, E.A. Tkatchenko, P.P. Fedorov, Inorganic nanofluorides and related nanocomposites, Russ. Chem. Rev. 75 (12) (2006) 1065–1082.
[53] S.V. Kuznetsov, P.P. Fedorov, Morphological stability of solid-liquid interface during melt crystallization of solid solutions $M_{1-x}R_xF_{2+x}$, Inorg. Mater. 44 (13) (2008) 1434–1458 (Supplement).
[54] S.V. Kuznetsov, P.P. Fedorov, V.V. Voronov, K.S. Samarina, R.P. Ermakov, V.V. Osiko, Synthesis of $Ba_4R_3F_{17}$ (R stands for rare earth elements) powders and transparent compacts on their base, Russ. J. Inorg. Chem. 55 (4) (2010) 484–493.
[55] S.V. Kuznetsov, A.V. Ryabova, D.S. Los', P.P. Fedorov, V.V. Voronov, R.P. Ermakov, V.B. Loshchenov, V.V. Volkov, A.E. Baranchikov, V.V. Osiko, Synthesis and luminescent properties of submicron powders based on sodium fluoride, and yttrium-doped rare-earth elements, Nanotechnol. Russ. 7 (11–12) (2012) 615–628.
[56] S.V. Kuznetsov, A.A. Ovsyannikova, E.A. Tupitsyna, D.S. Yasyrkina, V.V. Voronov, P.P. Fedorov, N.I. Batyrev, L.D. Iskhakova, V.V. Osiko, Phase formation in LaF_3-$NaGdF_4$, $NaGdF_4$-$NaLuF_4$, $NaYF_4$-$NaLuF_4$ systems: synthesis of powders by co-precipitation from aqueous solutions, J. Fluorine Chem. 161 (2014) 95–101.
[57] M. Leblanc, V. Maissoneuve, A. Tressaud, Crystal chemistry and selected physical properties of inorganic fluorides and oxide-fluorides, Chem. Rev. 115 (2015) 1191–1254.
[58] J. Li, Y. Pan, Y. Zeng, W. Liu, B. Jiang, J. Guo, The history, development, and future prospects for laser ceramics: a review, Int. J. Refract. Metals Hard Mater. 39 (2013) 44–52.
[59] G. Lieser, C. Dräger, L. De Biasi, S. Indris, H. Geßwein, S. Glatthaar, M.J. Hoffmann, H. Ehrenberg, J.R. Binder, Direct synthesis of trirutile-type $LiMgFeF_6$ and its electrochemical characterization as positive electrode in lithium-ion batteries, J. Power Sources 274 (2015) 1200–1207.
[60] M. Lin, Y. Zhao, S.Q. Wang, M. Liu, Z.F. Duan, Y.M. Chen, F. Li, F. Xu, T.J. Lu, Recent advances in synthesis and surface modification of lanthanide-doped upconversion nanoparticles for biomedical applications, Biotechnol. Adv. 30 (2012) 1551–1561.
[61] Q. Liu, Y. Sun, T. Yang, W. Feng, C. Li, F. Li, Sub-10 nm hexagonal lanthanide-doped $NaLuF_4$ upconversion nanocrystals for sensitive bioimaging in vivo, J. Am. Chem. Soc. 133 (2011) 17122–17125.
[62] B.E.G. Lucier, K.E. Johnston, D.C. Arnold, J.-L. Lemyre, A. Beaupre, M. Blanchette, A.M. Ritcey, R.W. Schurko, Comprehensive solid-state characterization of rare earth fluoride nanoparticles, J. Phys. Chem. C 118 (2014) 1213–1228.

[63] A.A. Luginina, P.P. Fedorov, S.V. Kuznetsov, M.N. Mayakova, V.V. Osiko, V.K. Ivanov, A.E. Baranchikov, Synthesis of ultrafine fluorite $Sr_{1-x}Nd_xF_{2+x}$ powders, Inorg. Mater. 48 (5) (2012) 531–538.
[64] A.A. Luginina, A.E. Baranchikov, A.I. Popov, P.P. Fedorov, Preparation of barium monohydrofluoride $BaF_2 \cdot HF$ from nitrate aqueous solutions, Mater. Res. Bull. 49 (1) (2014) 199–205.
[65] A. Lyberis, A.J. Stevenson, A. Suganuma, S. Ricaud, F. Druon, F. Herbst, D. Vivien, P. Gredin, M. Mortier, Effect of Yb^{3+} concentration on optical properties of Yb:CaF_2 transparent ceramics, Opt. Mater. 34 (2012) 965–968.
[66] M.N. Mayakova, S.V. Kuznetsov, P.P. Fedorov, V.V. Voronov, R.P. Ermakov, K.N. Boldyrev, O.V. Karban', O.V. Uvarov, A.E. Baranchikov, V.V. Osiko, Synthesis and characterization of fluoride xerogels, Inorg. Mater. 49 (11) (2013) 1152–1156.
[67] M.N. Mayakova, A.A. Luginina, S.V. Kuznetsov, V.V. Voronov, R.P. Ermakov, A.E. Baranchikov, V.K. Ivanov, O.V. Karban', P.P. Fedorov, Synthesis of SrF_2-YF_3 nanopowders by co-precipitation from aqueous solutions, Mendeleev Commun. 24 (6) (2014) 360–362.
[68] M.N. Mayakova, S.V. Kuznetsov, V.V. Voronov, A.E. Baranchikov, V.K. Ivanov, P.P. Fedorov, Soft chemistry synthesis of powders in the BaF_2-ScF_3 system, Russ. J. Inorg. Chem. 59 (7) (2014) 773–777.
[69] R. Naccache, Q. Yu, J.A. Capobianco, The fluoride host: nucleation, growth, and upconversion of lanthanide-doped nanoparticles, Adv. Opt. Mater. 3 (4) (2015) 482–509.
[70] G. Nicolis, I. Prigogine, Self-Organization in Nonequilibrium Systems: From Dissipative Structures to Order Through Fluctuations, Wiley, New York, 1977.
[71] V.V. Ovsyankin, P.P. Feofilov, Mechanism of summation of electronic excitations in activated crystals, JETP Lett. 3 (12) (1966) 322–327.
[72] J. Peng, Y. Sun, Q. Liu, Upconversion nanoparticles dramatically promote plant growth without toxicity, Nano Res. 5 (11) (2012) 770–782.
[73] P.A. Popov, K.V. Dykel'skii, I.A. Mironov, V.A. Demidenko, A.N. Smirnov, P.L. Smolyanskii, P.P. Fedorov, V.V. Osiko, T.T. Basiev, Thermal conductivity of CaF_2 optical ceramics, Doklady Phys. 52 (2007) 7–9.
[74] Yu.A. Rozhnova, A.A. Luginina, V.V. Voronov, R.P. Ermakov, S.V. Kuznetsov, A.V. Ryabova, D.V. Pominova, V.V. Arbenina, V.V. Osiko, P.P. Fedorov, White light luminophores based on Yb^{3+}/Er^{3+}/Tm^{3+} – coactivated strontium fluoride powders, Mat. Chem. Phys. 148 (2014) 201–207.
[75] Yu.A. Rozhnova, S.V. Kuznetsov, A.A. Luginina, V.V. Voronov, A.V. Ryabova, D.V. Pominova, R.P. Ermakov, V.A. Usachev, N.E. Kononenko, A.E. Baranchikov, V.K. Ivanov, P.P. Fedorov, New $Sr_{1-x-y}R_x(NH_4)_yF_{2+x-y}$ (R = Yb, Er) solid solution as precursor for high efficiency up-conversion luminophor and optical ceramics on the base of strontium fluoride, Mat. Chem. Phys. (2016), http://dx.doi.org/10.1016/j.matchemphys.2016.01.055.
[76] J. Sanghera, W. Kim, G. Villalobos, B. Shaw, C. Baker, J. Frantz, B. Sadowski, I. Aggarwal, Ceramic laser materials: past and present, Opt. Mater. 35 (2013) 693–699.
[77] L. Schmidt, A. Dimitrov, E. Kemnitz, A new approach to prepare nanoscopic rare earth metal fluorides: the fluorolytic sol–gel synthesis of ytterbium fluoride, Chem. Commun. 50 (2014) 6613–6616.
[78] K.B. Seiranian, P.P. Fedorov, L.S. Garashina, G.V. Molev, V.V. Karelin, B.P. Sobolev, Phase diagram of the system CaF_2-YF_3, J. Cryst. Growth 26 (1) (1974) 61–64.
[79] A.B. Shcherbakov, N.M. Zholobak, A.E. Baranchikov, A.V. Ryabova, V.K. Ivanov, Cerium fluoride nanoparticles protect cells against oxidative stress, Mat. Sci. Eng. C 50 (2015) 151–159.

[80] B.P. Sobolev, The Rare Earth Trifluorides, Parts 1&2, Inst. Estud. Catalans, Barcelona, 2000/2001. www.books.google.ru/books/rare earth trifluorides.

[81] B.P. Sobolev, E.G. Ippolitov, B.M. Zhigarnovskii, L.S. Garashina, Phase composition of the CaF_2-YF_3, SrF_2-YF_3 and BaF_2-YF_3, Inorg. Mater. 1 (1965) 334–340.

[82] B.P. Sobolev, V.B. Aleksandrov, P.P. Fedorov, K.B. Seiranyan, N.L. Tkachenko, Variable-composition phases with the LaF_3 structure in the systems MF_2-$(Y,Ln)F_3$. IV. X-ray characteristics, heterovalent isomorphic substitutions, Sov. Phys. Crystallogr. 21 (1) (1976) 49–54.

[83] D.F. Sobolev, V.S. Sidorov, P.P. Fedorov, D.D. Ikrami, Stabilisation of a structure of the rhombic β-YF_3 type in the systems CdF_3-LnF_3, Sov. Phys. Crystallogr. 22 (5) (1977) 574–577.

[84] B.P. Sobolev, P.P. Fedorov, A.K. Galkin, V.S. Sidorov, D.D. Ikrami, Phase diagrams of binary systems formed by rare earth trifluorides, in: Givargizov (Ed.), Growth of Crystals, vol. 13, Consultants Bureau, New York and London, 1986. ISBN:10: 0306181134.

[85] B.P. Sobolev, I.A. Sviridov, V.I. Fadeeva, S.N. Sul'yanov, N.I. Sorokin, Z.I. Zhmurova, P. Herrero, A. Landa-Canovas, R.M. Rojas, Mechanochemical synthesis of non-stoichiometric fluorite $Ca_{1-x}La_xF_{2+x}$ nanocrystals from CaF_2 and LaF_3 single crystals, Crystallogr. Rep. 50 (2005) 478–485.

[86] B.P. Sobolev, I.A. Sviridov, V.I. Fadeeva, S.N. Sul'yanov, N.I. Sorokin, Z.I. Zhmurova, I.I. Khodos, A.S. Avilov, M.A. Zaporozhets, Mechanochemical synthesis of non-stoichiometric nanocrystals $La_{1-y}Ca_yF_{3-y}$ with a tysonite structure and nanoceramic materials from CaF_2 and LaF_3 crystals, Crystallogr. Rep. 53 (2008) 868–880.

[87] B.P. Sobolev, N.L. Tkachenko, Phase diagrams of the BaF_2-$(Y,Ln)F_3$ systems, J. Less Common Metals 85 (1982) 155–170.

[88] N.L. Tkachenko, M. Svantner, B.P. Sobolev, Phase diagram of the system BaF_2-YF_3, Inorg. Mater. 13 (1977) 693–696.

[89] A. Tressaud (Ed.), Functionalized Inorganic Fluorides, Wiley-Blackwell, Chichester, 2010.

[90] F.C.J.M. Van Veggel, Upconversion of Ln^{3+}-based nanoparticles for optical bio-imaging, in: Luminescence of Lanthanide Ions in Coordination Compounds and Nanomaterials, John Wiley Sons, 2014, pp. 269–302.

[91] F. Vetrone, R. Naccache, V. Mahalingam, C.G. Morgan, J.A. Capobianco, The active-core/active-shell approach: a strategy to enhance the upconversion luminescence in lanthanide-doped nanoparticles, Adv. Funct. Mater. 19 (2009) 2924–2929.

[92] F. Wang, R. Deng, J. Wang, Q. Wang, Yu Han, H. Zhu, X. Chen, X. Liu, Tuning upconversion through energy migration in core-shell nanoparticles, Nat. Mater. 10 (2011) 968–973.

[93] Z. Wang, Z. Lu, D. Ruan, F. Tao, Q. Liu, H. Wang, Y. Sun, Synthesis and enhancement of red UC luminescence properties of ordered hexagonal $NaYF_4$:Yb^{3+}/Er^{3+} nanowire arrays, Appl. Surf. Sci. 320 (2014) 742–745.

[94] N. Wiener, Cybernetics, J. Wiley, New York, 1961.

[95] L. Yan, N. Liu, S. Zhao, H. Yan, H. Lü, X. Yuan, Effect of hydrophobic modification on the durability and environmental properties of porous MgF_2 antireflective films, Acta Metall. Sin. Eng. Lett. 27 (2014) 649–655.

[96] D. Yang, P. Ma, Z. Hou, Z. Cheng, C. Li, J. Lin, Current advances in lanthanide ion (Ln^{3+})-based upconversion nanomaterials for drug delivery, Chem. Soc. Rev. 44 (2015) 1416–1448.

[97] D.S. Yasyrkina, S.V. Kuznetsov, A.V. Ryabova, D.V. Pominova, V.V. Voronov, R.P. Ermakov, P.P. Fedorov, Dependence of quantum yield of up-conversion luminescence on the composition of fluorite-type solid solution $NaY_{1-x-y}Yb_xEr_yF_4$, Nanosyst. Phys. Chem. Math. 4 (5) (2013) 648–656.

[98] D.S. Yasyrkina, S.V. Kuznetsov, P.P. Fedorov, V.V. Voronov, A.V. Ryabova, D.V. Pominova, A.E. Baranchikov, V.K. Ivanov, V.V. Osiko, Effect of the pH on the formation of $NaYF_4$:Yb:Er nanopowders by co-crystallization in presence of polyethyleneimine, J. Fluorine Chem. 158 (2014) 60–64.
[99] L. Zhang, M. Anji Reddy, M. Fichtner, Development of tysonite-type fluoride conducting thin film electrolytes for fluoride ion batteries, Solid State Ionics 272 (2015) 39–44.
[100] H. Zhang, M. Zhang, Sh Pan, Z. Yang, Z. Wang, Q. Bian, X. Hou, H. Yu, F. Zhang, K. Wu, F. Yang, Q. Peng, Z. Xu, K.B. Chang, K.R. Poeppelmeier, $Na_3Ba_2[B_3O_6]_2F$: next generation of deep-ultraviolet birefringent materials, Cryst. Growth Des. 15 (2014) 523–529.
[101] J. Zhao, D. Jin, E.P. Schartner, Y. Lu, Y. Liu, A.V. Zvyagin, L. Zhang, J.M. Dawes, P. Xi, J.A. Piper, E.M. Goldys, T.M. Monro, Single-nanocrystal sensitivity achieved by enchanced upconversion luminescence, Nat. Nanotechnol. 8 (2013) 729–734.
[102] Q. Zhao, B. Shao, W. Lü, Y. Jia, W. Lv, M. Jiao, H. You, Ba_2GdF_7 nanocrystals: solution-based synthesis, growth mechanism, and luminescence properties, Cryst. Growth Des. 14 (2014) 1819–1826.

Part Two

Optical properties

Upconversion Phenomena in Nanofluorides

3

G. Ledoux, M.F. Joubert, S. Mishra
Université Lyon 1, Lyon, France

Chapter Outline

1. Introduction

The chemistry and physical chemistry of fluoride compounds is an active and far reaching area of scientific research as well as application in the world. Fluoride-based materials have special characteristics compared to other dielectric media. Their stability, high ionicity, and low-energy phonons associated with proper combination of the fluoride host and the rare earth dopants make them very interesting for efficient upconversion systems.

Due to high ionicity on one hand and low-energy phonons on the other hand, fluoride-based materials may have quite high-energy bandgap, usually more than 10 eV [1]. For example, the bandgap energy is about 13.6 eV in LiF, 10.8 eV in MgF_2, and 10 eV in $LiYF_4$. This leads to high transparency for these fluoride crystals, from near infrared (NIR), a few micrometers, up to the vacuum ultraviolet (VUV), 150–200 nm. When a rare earth ion is embedded into a fluoride lattice, some of its

Photonic and Electronic Properties of Fluoride Materials. http://dx.doi.org/10.1016/B978-0-12-801639-8.00003-9

well-localized energy states belonging to the $4f^n$ ground electronic configuration and often its lower states of the $4f^{n-1}5d$ first excited configuration are located between the valence band and the conduction band of the host. UV or X-ray photoelectron spectroscopy can estimate the position of the rare earth ion ground state relative to the top of the valence band and it appears that the location of the $4f^n$ ground state in this bandgap is very close to the top of the valence band [2]. Due to the high-energy bandgap, optical excitation of rare earth ions in such fluoride hosts may lead to possibility of luminescence within a large spectrum from NIR to VUV and many papers were devoted to the photoluminescence of fluoride compounds doped with rare earth ions (see Refs [3−8]). Such classical luminescence giving rise to emission of light at energy lower than the exciting photon energy is usually called downconversion.

Upconversion phenomena are due to mechanisms which produce population in an excited state whose energy exceeds that of the excitation photon. In the late 1950s, scientist tried to find a way to produce light by using exciting photons with lower energies than the emitted photons. Originally the goal was to convert infrared (IR) photons that are difficult to detect into visible photons that can be easily counted by photomultipliers. In 1959, Bloembergen [9] first paved in this direction by proposing an energy scheme where it was possible to detect an IR photon by summing it with a visible photon to generate another typical visible photon of higher energy. This was the first published excited state absorption (ESA) energy scheme (see Fig. 3.1). This process was shown experimentally in 1961 by Porter [10] for Pr^{3+} in $LaCl_3$. Finally, ESA giving rise to luminescence at higher energy than the exciting photon energy was shown by Brown and Sand [11,12] in different fluoride hosts doped with praseodymium. Although the principle of upconversion was demonstrated, the efficiency was rather low. A real breakthrough came in 1966 with two publications from Auzel [13,14] where the author demonstrated ETU (initially called APTE (Addition de Photons par Transfert d'Energie)) between ions of different kinds with an increase of upconversion of almost 100 as illustrated in Fig. 3.1. This first results were done on tungstate or germanate and a few years after, in 1972, Menyuk et al. [15] reproduced these results in a $NaYF_4$:Yb,Er phosphor, showing striking upconversion energetic efficiencies of a few percent. Two characteristics of fluoride make them advantageous for upconversion efficiency: they present high stability on one hand and low-energy phonons on the other hand. Indeed, oxide crystals, for example, which are chemically stable and easy to obtain, present phonon energies usually quite high (>500−1000 cm^{-1}), resulting in lower upconversion efficiencies due to the enhancement of nonradiative relaxation pathways. On the contrary, if metal halides like metal chlorides, -bromides, and -iodides generally exhibit very low phonon energies (<300 cm^{-1}), they are hygroscopic and so unstable in air and in water, therefore, not very well suited for use in biological systems. For example, measured phonon energies in stable fluoride crystals are usually between 300 cm^{-1} and 500 cm^{-1} which is quite low: 490 cm^{-1} in $LiYF_4$ and less than 430 cm^{-1} in $LiLuF_4$ [16]; 390 cm^{-1} in BaY_2F_8 [17]; 298, 370, and 418 cm^{-1} in $NaYF_4$ [18]. This permits high metastable intermediate states of rare earth ions embedded in fluorides and then efficient upconversion processes in such compounds. Beside ESA (also called multistep (or sequential) absorption) and ETU the

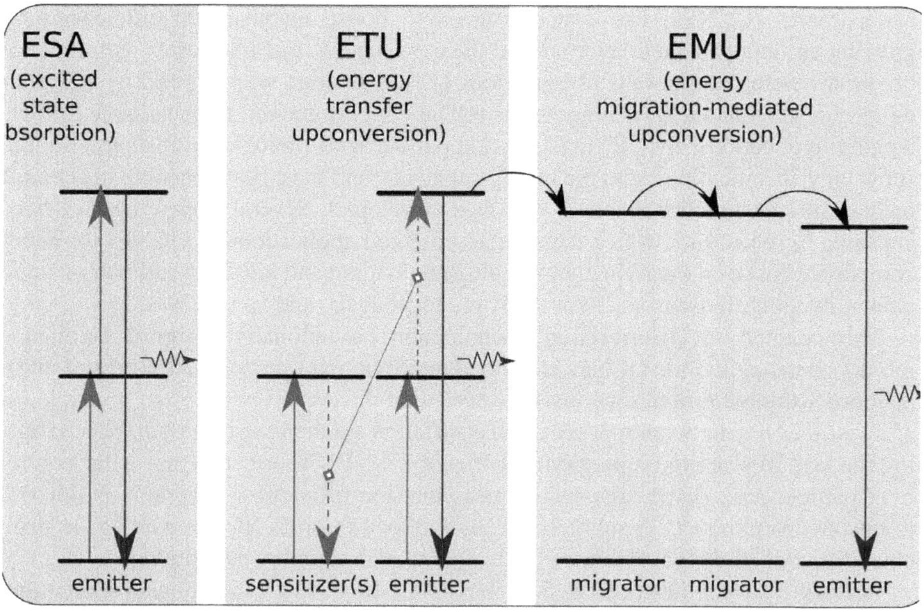

Figure 3.1 Scheme of the different types of upconversion phenomena. In excited state absorption (ESA) a first low-energy photon is absorbed by the emitter to bring it to the first excited state. A second photon can further bring the emitter to a higher excited state from which it will emit an upconverted photon. In energy transfer upconversion (ETU) the access to the higher excited state is mediated by an energy transfer from another ion, the sensititzer. In energy migrated–mediated upconversion (EMU) the first steps are identical from ESA or ETU but once an ion is brought to its higher excited state the energy can be transferred to a series of migrator ions that will transport the excitation to another region where a new kind of emitter is present. This last case allow for emission of light over a broader range of wavelengths.

third main class of upconversion process is photon avalanche, first discovered in Pr^{3+}-based infrared quantum counters [19,20]. An overview of these three main upconversion processes which have been observed in various types of rare earth–doped solids such as crystals and glasses was given in several papers [21–25].

Beside single crystals, glasses, optical fibers, and waveguides, nanocrystals (NCs) appeared at the beginning of the 21st century. Haase and Schäfer [26] presented a review of state of the art in the area of lanthanide-doped NCs. Since 10 years, most of the published papers related to upconversion concern rare earth–doped nanofluorides. Indeed new potential applications appeared for such systems. The first motor for this was biological applications (bioassay, nanolabels, and markers for detection of biomolecules) due to the transparency window of the body. Biological tissue is much more transparent to NIR light compared to UV or visible light and photodegradation is strongly reduced. In addition, since the excitation process is nonlinear there is a natural gain in spatial resolution. As efficiency is lower in simple NC compared to bulk owing to the larger surface-to-volume ratio, core–shell design of upconversion nanoparticles (UCNPs) was conceptualized to boost upconversion efficiencies by growing an undoped shell layer around the core particle that minimizes the influence of surface states on the optical transitions [27]. This idea was inspired by a similar design on semiconductor quantum dots (QDs) [28]. Moreover, to be suitable for bio-tagging applications, the NP must be compatible with biological substrates, should show only specific binding to the biological target, and must be dispersible in aqueous media. Surface functionalization was then developed. Several review articles have appeared in recent time which summarize potential applications of UCNPs (or nano-composites based on them) in controllable drug loading and releasing, cellular or small animal imaging, biosensors, solar cells, photocatalysis, and so on [29–35].

This chapter on upconversion phenomenon on rationally designed lanthanide ion-doped metal fluoride UCNPs is divided into four main sections. After a brief introduction, Section 2 presents a concise overview of the dopant/host design principle for efficient UCNPs. In Section 3, we discuss different synthetic strategies employed during the last 10 years to prepare fluoride-based UCNPs. These strategies offer control over particle size, shape, dispersion, and emission properties. Section 4 is devoted to upconversion results in simple lanthanide-doped fluoride NCs as well as the main characteristics of upconversion in lanthanide-doped complex nanofluorides.

Remark: Among these classical upconversion processes there are also some other different ones occurring in nonlinear optic systems under high-power laser excitation. Under such special conditions they can work with only one kind of dopant and it is not necessary to have real intermediary or upper levels. These are called cooperative luminescence, second harmonic generation, two or multiphotons absorption excitation. Their efficiency is very low and will not be considered further.

2. The Components of the Upconverting Nanoparticles: Dopants and Hosts

The most usual way of preparing efficient upconverting nanomaterials is to select a suitable crystalline host and then dope it with lanthanide/transition metal ion dopants in low concentrations. Following are some general criteria for choosing proper combination of dopants and an efficient host matrix as suited.

2.1 Sensitizer/Activator Combination

Among the various factors that determine upconverting efficiency, the most important is the choice of sensitizer and activator (emitter) ions. Most UCNPs take advantage of ETU in order to generate upconverted luminescence. One of the dopant ion, called sensitizer, absorbs excitation light, then transfers it nonradiatively to a nearby other dopant ion, called activator. Following multiple energy transfers, the activator builds up enough energy to emit energy at a shorter wavelength than the excitation source. In most cases, Yb^{3+} is chosen as the sensitizer due to its single, relatively strong transition in the visible/NIR spectrum. In addition, the $^2F_{7/2} \rightarrow {}^2F_{5/2}$ transition matches very well with several electronic transitions in the usual activators such as Er^{3+} and Tm^{3+} (see Fig. 14 in Ref. [36]). These activators usually have long-lived intermediate excited states and also have multiple transitions with similar energy, which enhances both single-wavelength ESA and ETU. The doping ratio of the sensitizer and activator is also critical. The sensitizer content is usually taken higher (up to 20–25 mol%) than the activator contents (1–2 mol%) to keep the average distance between a sensitizer and an activator close enough to favor the energy transfer but, at the same time, the average distance between two activators is not so close to minimize the cross-relaxation energy loss.

2.2 The Host Matrix

The selection of the host is critical in obtaining intense and efficient upconversion luminescence. Among fluoride-based matrices, there are several considerations for choosing a right host which are described below.

- *Similar ionic size of the dopant lanthanide ions and the matrix cations.* Because trivalent lanthanide ions all possess very similar ionic sizes and chemical properties, they are ideal as inactive host matrix cations. This is particularly true for Y^{3+}, La^{3+}, Lu^{3+}, and Gd^{3+} as they do not have strong electronic transitions in the visible and NIR, though the latter has a few strong transitions in the UV and violet wavelengths. Alkaline earth metals and some transition metals also have similar ionic size to the lanthanide ions. These include Ca^{2+}, Ba^{2+}, Mg^{2+}, Sr^{2+}, Zr^{4+}, and Ti^{4+}. However, due to the charge mismatch, crystals composed of these ions often have numerous defects when doped with lanthanide ions.
- *Crystal structure*: The variation in the crystal structure of the host matrix also influences the optical properties of the NCs. For example, between the cubic and hexagonal phases of one of the most important matrices $NaYF_4$, the hexagonal phase is typically an order of magnitude more efficient for the upconversion process.
  Besides choosing proper combination of dopants and an efficient host matrix, there are several other factors which have direct effect on upconversion efficiency. For example, the capping ligand used to stabilize the UCNPs in solution often diminishes the upconversion intensity as the vibrational modes of the capping ligand can interact with surface ions, resulting in quenching of a part of the luminescence. However, as mentioned in the introduction section, this can be mitigated to an extent by the epitaxial growth of additional metal fluoride shell around the active core.
- *Nature of the host matrix*: One other characteristic that makes fluorides more suitable than oxides is the influence of the host on the rare earth activator lifetime. In oxides, lifetimes are shorter than in fluorides due to a stronger interaction between the luminescent ion and its surroundings. For example, compared to fluorides, in oxide the lifetime of the intermediary $^4F_{11/2}$ level of Er^{3+} is shorter, inducing a lower upconversion process efficiency.

3. Synthetic Strategies

Several synthetic strategies have been employed so far to prepare lanthanide UCNPs with different phases, sizes, shapes, and surface properties, all of which influence the upconversion emission properties and potential applications. These strategies are being described below.

3.1 Thermal Decomposition Method

3.1.1 From Trifluoroacetate Precursors

3.1.1.1 Monometallic Precursors

One important and interesting chemical route to metal fluoride-based UCNPs is the decomposition of the metal trifluoroacetates. The first report that suggested metal trifluoroacetates could potentially be used as precursors for metal fluorides (or metal oxyfluorides) was published by Roberts [37]. Yan's group was the first to use a metal trifluoroacetate precursor for the synthesis of metal fluoride NPs [38]. They decomposed hydrated $La(TFA)_3$ (TFA = trifluoroacetate) in a mixture of oleic acid (OA) and 1-octadecene at 280°C for 1 h to obtain trigonal LaF_3 triangular nanoplates which were not only highly monodisperse, crystalline, and dispersible in organic solvents but also had the tendency to self-align both side-down and face-down. The use of a mixture of solvents was critical for the above results, as the use of OA or 1-octadecene alone afforded a lower degree of uniformity and ill-shaped/aggregated NPs, respectively. Latter, the same group reported the synthesis of Ln^{3+}-doped $NaYF_4$ using NaTFA and hydrated $Ln(TFA)_3$ in a mixture of three solvents: OA, oleylamine (OM) (coordinating solvents), and octadecene (noncoordinating) [39]. They optimized several synthetic conditions to obtain different shapes and sizes of $NaLnF_4$ NPs (Ln = Pr–Lu, Y) and explored the mechanism of the reaction and explained the monodispersibility of the resulting α-$NaYF_4$ NPs on the basis of a small nucleation time (typically less than 1 min) [40]. They further investigated the upconversion properties of doped $NaYF_4$ as a factor of crystal phase, surface ligand, and surface coating and, on the basis of the results obtained, concluded that phase change could result in different upconversion pathways [41]. Using a delayed nucleation pathway, they also prepared $NaYF_4$:Er^{3+} (or Tm^{3+})/Yb^{3+} core–shell NPs in both α- and β-phases, with the particle sizes ranging from 20 to 300 nm. The authors combined the use of upconversion emission spectroscopy, transmission electron microscopy, and X-ray diffraction to propose a four-stage pathway for the nucleation and growth. Formation of the β-phase of $NaYF_4$ via the α-phase through a delayed $\alpha \rightarrow \beta$ transition was also investigated by these authors.

In the same year when Yan's group published the first report on $NaYF_4$ NPs, two other groups described the synthesis of doped $NaYF_4$ using metal trifluoroacetate precursors. While Boyer et al. [42] synthesized the Ln^{3+}-doped α-$NaYF_4$ NPs by using a mixture of two solvents, 1-octadecene and OA, at 300°C, Yi and Chow [43] used OM as a solvent at relatively high temperature (330°C) to obtain β-$NaYF_4$. In the former case, the NPs were polydisperse in size (10–50 nm) but crystalline in nature. Later, Boyer et al. [44] and Park et al. [45] modified the synthesis in terms of mixing the precursors Na(TFA) and $Ln(TFA)_3$ (taken in appropriate solvent) slowly through a cannula at the required reaction temperature to yield highly monodisperse, self-assembled, and crystalline $NaYF_4$ NPs with core–shell structure. The authors explained that the separation of the precursors from the solvents in the reaction flask, and the addition over a prolonged period of time of ~15 min, essentially separated the nucleation and growth processes, leading to monodisperse sizes. Yi and Chow [43], on the other hand, used OM which allowed the phase transition cubic → hexagonal at a higher temperature (above 320°C). The NPs were relatively small (10.5 ± 0.7 nm) which could be made water dispersible by using ligand exchange procedure substituting OM with polyethylene glycol 600 diacid or polyacrylic acid. This also improved the upconversion emission significantly by protecting the shell from the solvent phonons as well as by confining the surface defects [46].

Following these initial reports, there have been several other papers describing synthesis of Ln^{3+}-doped metal fluoride UCNPs using metal trifluoroacetates. The selection of a proper coordinating ligand such as OA, OM, trioctylphosphine (TOP), trioctylphosphine oxide, or ligand combinations such as OA–TOP controls these metal fluoride NPs synthesis and is one of the key factors for achieving monodisperse and size-tunable colloidal NCs. While the majority of works have focused on $NaYF_4$/$NaGdF_4$ or LnF_3 hosts [47–49], there have been few reports on other alternative hosts also, such as LnOF, $LiYF_4$, KYF_4, and $BaYF_5$. Using $Ln(TFA)_3$ and a combination of different solvents (1-octadecene, OA, OM), Sun et al. [50] and Du et al. [51] explored the synthetic conditions for LnOF and LnF_3 NPs (Ln = La to Lu, Y). The authors demonstrated excellent control over morphology of the NPs and their ability to self-assemble over a long range. They concluded that the OM as a fluorophile encouraged the formation of LnOF rather than LnF_3. Quite a few research groups have studied the effect of replacing sodium in $NaYF_4$ with other alkali metals such as Li and K on the structure and upconversion properties. Yi et al. [52] showed the synthesis of monodisperse and crystalline $LiYF_4$ using TFA precursors in the presence of trifluoroacetic acid. The presence of trifluoroacetic acid and a relatively high temperature of 340°C were both necessary for the formation of pure $LiYF_4$, otherwise Yttrium oxyfluoride (YOF) was formed as a side product. Mahalingam et al. [53] prepared the tetragonal phase of $LiYF_4$ NPs which were diamond shaped and of ~80 nm size with a narrow size distribution. Furthermore, these NPs could be coated with silica to make them water dispersible [54]. Du et al. [55] prepared a series of $KLnF_4$ (Ln = La to Gd and Y) or $LiLnF_4$ (Ln = Tb to Lu and Y). For the former, La and Ce formed cubic-phase $KLnF_4$ wormlike nanowires, while the remainder of the ions formed orthorhombic or hexagonal phase nanopolyhedra. In contrast, only tetragonal-phase $LiLnF_4$ NPs were observed predominantly in rhombohedral-like plate morphology. The authors discussed the importance of solvent selection and reaction temperature as key components of the formation of high-quality crystalline and monodisperse NPs. The authors observed that lighter Ln ions preferentially formed $KLnF_4$ nuclei, while heavier ones (Tb and beyond) formed KF and LnF_3 nuclei. The opposite trend was true for the $LiLnF_4$, where heavy Ln ions preferentially formed $LiLnF_4$ nuclei. Mahalingam et al. [56] also synthesized KY_3F_{10}:Er^{3+},Yb^{3+} NPs as a new UCNP by using the trifluoroacetate-based thermal decomposition method at 300°C. They obtained the cubic phase of the 15-nm-sized NPs. In contrast to $NaYF_4$ and $NaGdF_4$, the cubic phase is thermodynamically more stable phase in the case of KY_3F_{10}. Other lanthanide-doped host materials such as MYF_5 (M = Ca, Ba) have also been prepared using [$Ln(TFA)_3(H_2O)_x$] (Ln = Y, Yb, Er, Tm, Eu) and $M(acac)_2$ (M = Ca or Ba) in OA and 1-octadecene [52,57,58].

3.1.1.2 Bimetallic Precursors

Mixed-metal species are a means to provide homogeneity at a molecular level in order to modify the reactivity pattern and solubility. There are very few single source precursors for the synthesis of heterometallic fluoride nanomaterials [59]. As a new development in the trifluoroacetate method, we recently reported novel trifluoroacetate heterometallic precursors [$NaLn(TFA)_4$(diglyme)] and [$Na_2Ln(TFA)_5$(tetraglyme)] (Ln = Y, Gd, Er, Tm, Yb), which contained all the required elements (Na,Y/Ln, F) in a single molecule, for the one-step synthesis of upconverting $NaY(Ln)F_4$ materials [36,60,61]. These heterometallic precursors had multiple advantages over the above-described homometallic $Ln(TFA)_3(H_2O)_3$ for upconverting $NaYF_4:Ln^{3+}$ nanomaterials. These included formation of anhydrous species, lowering of decomposition temperature, no requirement of capping-reagents or surfactants for size control of NCs, and render them monodisperse in organic media, etc. The thermogravimetric data of these heterometallic Na-Ln derivatives under argon atmosphere showed that their final decomposition temperature was considerably lower (265−285°C) than that for the homometallic derivative $Ln(TFA)_3(H_2O)_3$ (310°C). Their decomposition in 1-octadecene (boiling point 315°C) gives $NaY(Gd)F_4$ NCs with controlled size and desirable dispersibility in organic media. Interestingly, the isostructural [$NaY(TFA)_4$(diglyme)] and [$NaGd(TFA)_4$(diglyme)] showed different decomposition patterns. While [$NaGd(TFA)_4$(diglyme)] decomposed at 285°C to give directly the hexagonal $NaGdF_4$ NCs, a phase many folds more efficient than the cubic phase for the upconversion process [60], the yttrium analogue [$NaY(TFA)_4$(diglyme)] afforded only a mixture of the cubic and hexagonal phases of $NaYF_4$, in a relative ratio that depended on the reaction temperature [36,61] (Fig. 3.2). These mixed-phase $NaYF_4$ NCs were required to be calcined at 400°C for getting the hexagonal phase. This difference in the decomposition pattern of the isostructural [$NaLn(TFA)_4$(diglyme)] (Ln = Y and Gd) was attributed to the different ionic radii of the two metal cations Y^{3+} ($r = 0.89$ Å) and Gd^{3+} ($r = 0.94$ Å), which would influence the free energy of the $NaLnF_4$ system [39]. In general, the hexagonal phase $NaLnF_4$ is thermodynamically more stable than the cubic form, and the cubic-to-hexagonal phase transition requires sufficient free energy to overcome the activation barrier. While high temperature is often used to overcome the energy barrier between the two phases of $NaLnF_4$ [39,62,63], there have been few reports on the ionic radius dependence of the energy barriers for such system. For example, Mai et al. [39] found a higher energy barrier for the lanthanides with small ionic radii as compared to those with larger ionic radius. Indeed, Wang et al. [64] have reported the cubic-to-hexagonal phase transition simply by doping $NaYF_4$ with larger Gd^{3+} cation.

3.1.2 From Oleate Precursors

In this method, the metal oleate precursors, often prepared in situ by reacting a metal salt with OA, are decomposed in the presence of a fluoride source [65,66]. The advantage of

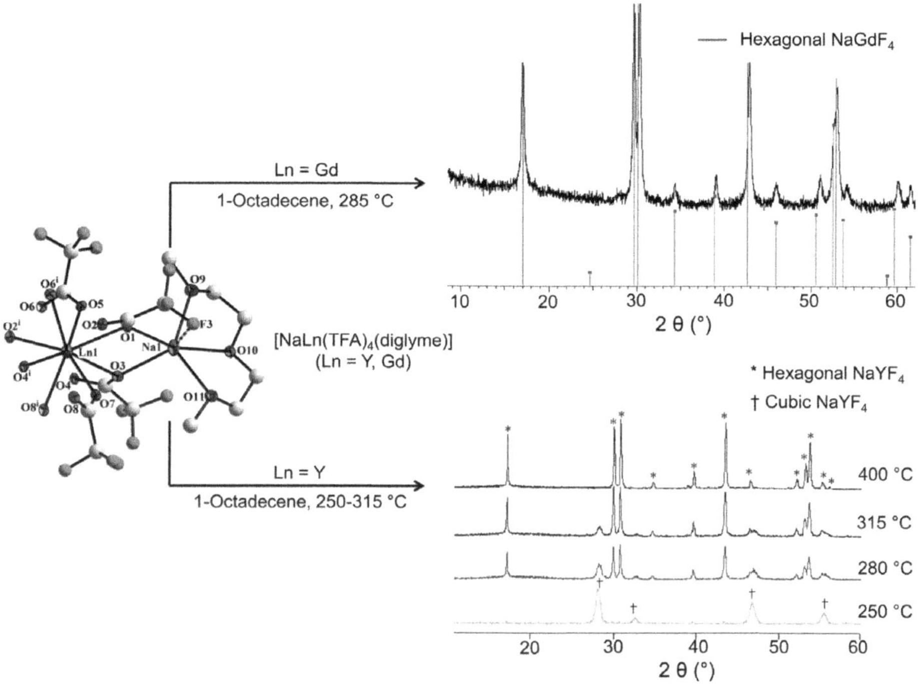

Figure 3.2 Comparison of decomposition of isostructural [NaLn(TFA)$_4$(diglyme)] (Ln = Y and Gd) with NaLnF$_4$ NPs.
Reproduced with permission from Copyright 2010, Royal Chemical Society and 2014, Wiley-VCH.

this process over trifluoroacetate process is the relatively milder conditions used to form the particles, although the reaction time may be longer. The other advantage is the in situ generation of the oleate ligand from the complex which acts as a capping reagent to render the NPs dispersible in common solvents. A mixture of solvents, OA and 1-octadecene, is usually used. The concentration of OA has been reported to influence the shape and size of the particles, with higher concentrations of OA resulting in smaller but elongated particles [67]. Wei et al. [68] synthesized $NaYF_4:Er^{3+}$, Yb^{3+} nanoplates by decomposing NaF and Ln(oleate)$_3$ in 1-octadecene at 260°C over a period of 6 h. The resulting particles were hexagonal shaped and in the hexagonal phase or spherical shaped in the cubic phase depending on the reaction temperature. The cubic shape formed at the lower temperature of 210°C and the hexagonal shape formed at 260°C. The OA bound to the complex also acted as a ligand which imparted solubility to the resultant NPs. Bogdan et al. [69] later used a simple acid treatment process to remove this oleate ligand from the surface of oleate-capped $NaYF_4:Er^{3+}$, Yb^{3+} UCNPs. These ligand-free lanthanide-doped UCNPs remained water dispersible after the acid treatment and showed very high upconversion luminescence intensities in aqueous solutions.

Liu et al. [70] investigated a morphological and phase-controlled synthesis of Ln-doped $NaGdF_4$ NPs using Ln(oleate)$_3$ and NaF as starting materials. They achieved synthesis of the β-phase at 280°C within 15 min of the reaction. In contrast, lower reaction temperatures of 260°C resulted in the formation of the α-phase even after a 5-h reaction time, indicating that there was not enough energy in the system to overcome the energy barrier of the $\alpha \rightarrow \beta$ transition. The morphology of the NPs depended on the reaction time but not on the temperature. For example, reaction times ranging from 25 min to 10 h yielded particles in the following morphologies: nanoplates, dot-like nanorods, nanorods, and sphere-like particles, all of the same hexagonal phase. Using lanthanide oleate precursors, Wang et al. [150] recently synthesized orthorhombic KYb_2F_7:Er^{3+} (2 mol%) at 290°C for 1.5 h under an argon atmosphere as a new class of upconversion NCs containing ytterbium content as high as 98 mol% and in which the lanthanide ions are distributed in arrays of tetrad clusters.

3.2 Hydrothermal and Solvothermal Method

The solvothermal (or hydrothermal) method is one of the most commonly used techniques for the synthesis of highly crystallized lanthanide UCNPs at relatively low temperature. The reaction mixture containing a metal salt and a fluoride source in an appropriate solvent, is subjected to thermal treatment in an autoclave (often Teflon-lined), where solvents can be brought to temperatures well above their boiling points to behave as supercritical fluids. For lanthanide fluoride-based UCNPs, most research groups use Ln nitrates as precursors, often obtaining the nitrate precursor in situ from oxides and nitric acid, though lanthanide chlorides and oxides are also employed occasionally. The choice of fluoride precursor varies from HF, NH_4F, and NH_4HF_2, when synthesizing UCNPs of the type LnF_3, and NaF or KF for fluorides of the type $MLnF_4$. Other, less frequently used, precursors include $NaBF_4$, KBF_4, and 1-butyl, 2-methylimidazolium tetrafluoroborate (an ionic liquid (IL)). Other key factors are pH of the reaction mixture, reaction temperature, and reaction time. Generally speaking, higher temperature increases the rate of reaction and leads to more crystalline particles, whereas a longer time period of the reaction tends to give larger NPs.

Y. D. Li and coworkers have studied extensively the solvothermal synthesis of lanthanide fluoride-based materials LnF_3 and $NaLnF_4$ [71–77]. By using $Ln(NO_3)_3$ and NH_4F as lanthanide and fluoride source, respectively, they obtained hollow structures at pH between 4 and 5 in a temperature range 80–180°C for 12–24 h [73,74]. Wang et al. developed a new synthetic method, mentioned first in 2005 and then in more detail in 2006 based on liquid–solid–solution (LSS) growth, to synthesize a range of lanthanide fluoride materials. The synthesis was based on a phase transfer and separation occurring at the interfaces of the liquid–solid and solid–solution phases (Fig. 3.3). This hydrothermal method used sodium linoleate, linoleic acid, and ethanol added to aqueous solutions of the precursors, lanthanide nitrates, and NaF or NH_4HF_2. Reactions were performed at 100–200°C for about 8–10 h. The obtained $NaYF_4$ NPs were easily dispersible in nonpolar solvents such as cyclohexane and toluene and showed a spherical morphology with a very narrow size distribution, tunable from 6 to 10 nm by increasing the reaction temperature from 100 to 180°C. Although the LSS method brought improvements to the size and morphology of the NPs, the $NaYF_4$ particles obtained were synthesized in the α-phase with relatively

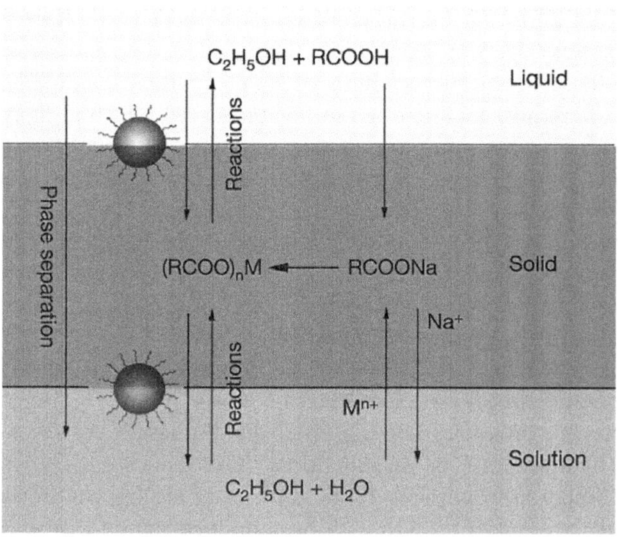

Figure 3.3 Liquid–solid–solution phase transfer synthetic strategy.
Reproduced with permission from Copyright 2005, Nature Publishing Group.

low luminescence intensity. On the other hand, the LnF_3 NPs prepared using similar methods yielded different crystal phases, namely hexagonal and orthorhombic.

The group also used a polar solvent (water, acetic acid, or ethanol) together with the surfactants cetyltrimethylammonuim bromide (CTAB) or ethylenediamine tetraacetic acid (EDTA) [77] as well as OA and alcohol as cosolvents to water along with NaOH [71] for the synthesis of highly crystalline and relatively uniform NPs of $NaYF_4$. Later, they expanded this work to include a range of reaction parameters [72]. The method yielded many different shapes and sizes for $NaYF_4$, including nanorods, hexagonal nanoplates, rectangular particles, and square particles, all of which were crystalline and monodisperse. The authors discussed the role of alcohol chain length, the ratio of F to Y^{3+}, reaction time, and temperature on the shape and size of the NPs. In general, high NaF concentrations, a high Y^{3+}/dopant ion ratio, low reactant concentrations, long synthesis times, and appropriate reaction temperatures favored the formation of nanorods. The authors attributed this to the selective binding of the OA surfactant to the surfaces that favored epitaxial growth along the <001> directions, resulting in a rod shape.

Wang et al. [78] reported the synthesis of polyethylenimine (PEI)-capped $NaYF_4$ NPs. The ligand PEI is a branched hydrophilic polymer with amino groups and possesses an overall net positive charge which helps to stabilize the PEI-coated NPs in polar solutions. The NPs synthesized using high-molecular weight PEI showed an average size of ~50 nm with a spherical morphology, while the ones prepared using low-molecular weight PEI were irregularly shaped with a nonuniform size distribution. Wang and Liu [79] modified slightly the above synthesis and used ethylene glycol instead of ethanol as the solvent. By doing so, they reduced the reaction time significantly from 24 to 2 h for the preparation of $NaYF_4$:Er^{3+}/Tm^{3+}/Yb^{3+} NPs. In addition, the particle size was reduced to 20 nm with a very narrow size distribution. Zhang et al. [80] described the synthesis of $NaYF_4$ particles, and although most particles obtained were not small enough to be considered NPs, the morphologies were very different from any other reported synthesis. The synthesis used NaF and $Y(NO_3)_3$ as precursors, OA, water, and ethanol as solvents and NaOH as a base. Reactant solutions were heated in an autoclave at temperatures 130–230°C for up to 24 h. The authors investigated the effects of metal fluoride precursor concentration, reaction medium basicity, as well as temperature and time of the hydrothermal reaction on the size, morphology, and crystal structure. Fairly uniform disks with diameters of about 800 nm and thicknesses of about 300 nm were obtained, and all the disks featured a flowerlike pattern with a sixfold axis on the top and bottom faces. The effects of NaF concentration were also studied. The previously synthesized nanodisks changed into long nanorods. A mixture of both α-phase and β-phase $NaYF_4$ was observed upon decreasing the hydrothermal temperature to 130°C or reducing the reaction time to 1 h. A phase transformation from the α- to the β-phase of $NaYF_4$ was observed upon prolonging the reaction time from 3 to 24 h or upon increasing the reaction temperature from 130 to 230°C.

Wang et al. [64] recently used codoping of the upconverting $NaYF_4$ NPs with Gd^{3+} ions to control the phase and size of this matrix simultaneously. In general, earlier lanthanide ions with large ionic radii favor the formation of hexagonal phase due to their high tendency toward electron distortion clouds, while the latter ones with small ionic radii favor the formation of a cubic phase. In the case of $NaYF_4$, a dopant that possesses a radius larger than that of Y^{3+} would be expected to affect the arrangement in the crystal structure and hence would induce a phase transfer from the cubic to the hexagonal phase. The authors chose the Gd^{3+} ion as the ideal dopant. At low Gd^{3+} dopant concentration (0–15%), a mixture of cubic and hexagonal phases was observed, whereas doping with a concentration above 30% resulted in the formation of the pure hexagonal phase. They also observed that increasing the Gd^{3+}-doping concentration favored the formation of nanorods over nanocubes with a decrease in size. To achieve nucleation over a very small period of time and thus limiting the initial size distribution, Yi et al. [81] first prepared the lanthanide-chloride complexes with EDTA and injected the solution rapidly into a separate NaF solution. The resulting $NaYF_4$ NPs were uniform in shape and size, most being spherical in shape with diameters in the range 30–45 nm. On increasing the EDTA:Ln^{3+} ratio, the resultant particles decreased in diameter. Sarkar et al. [82] recently reported hydrothermal synthesis of highly water-dispersible BiF_3:Yb^{3+}, Er^{3+} NPs of average size of 6 nm within a polymer matrix poly(vinyl pyrrolidone).

3.3 Coprecipitation Synthesis

Coprecipitation is one of the most convenient synthetic methods that had fast growth rates and does not require elevated reaction temperatures or very costly equipment and reactant materials. Unfortunately, these factors are outweighed by usually low upconversion efficiency for the as-prepared lanthanide-doped LnF_3 and $MLnF_4$ NPs, often requiring additional annealing to achieve good photon upconversion. One of the earliest co-precipitation synthesis of LaF_3 was reported by Zhou et al. [83]. In this synthesis, lanthanide nitrate hexahydrates were mixed with NaF in the presence of the ligand pyridinium di-*n*-hexaoctyldithiophosphate in a water–ethanol solution mixture. The reaction required temperature of only 75°C over a 2 h period. The particles obtained had an average size of 6 nm and could be redispersed in organic solvents such as chloroform, Dichloromethane (DCM), and toluene. However, the particles were not highly crystalline. As an improvement to this method, Stouwdam and van Veggel [84] used a different ligand, namely, ammonium di-*n*-octadecyldithiophosphate, to obtain LaF_3:Eu^{3+} NPs with better dispersibility in nonpolar solvents as well as much higher crystallinity, the size being in the range of sizes from 5 to 10 nm.

Using N-(2-hydroxyethyl)ethylenediamine as the solvent and coordinating ligand that formed complex with the lanthanide precursor, and 1-adamantanecarboxylic acid as a ligand, Heer et al. [85] reported synthesis of highly crystalline upconverting $NaLnF_4$ NPs which were dispersible in Dimethyl sulfoxide (DMSO) but had ill-defined shapes and a large size distribution (5–30 nm). Schafer et al. [86] extended this study by using 1-hydroxyethane 1,1-diphosphonic acid for the synthesis, which not only made the NPs dispersible in water but also increased their upconversion efficiency. Shortly afterward, Ptacek et al. [87] investigated the effects of the Ln^{3+}/F^- ratio on the phase control of $NaGdF_4$:Eu^{3+} NPs, and concluded that the presence of excess of Ln^{3+} favored the formation of the cubic phase, while at a stoichiometric ratio or below, hexagonal phase was obtained. In 2004, Yi et al. used a simplified method to obtain α-$NaYF_4$ NPs which could be transformed to β-phase after annealing at different temperatures. The synthesis involved rapid injection of a lanthanide-EDTA complex into a solution made up of NaF in deionized water, resulting in immediate nucleation. The authors established the relationship between particle size and EDTA:Ln^{3+} ratio, showing that a low ratio favored the formation of large particles and ratios above 1 had no significant additional impact on the resultant particle size. Yi et al. [88] modified the synthesis of Stouwdam and van Veggel [84] and obtained hexagonal-phase particles of LaF_3 with a size of 5.4 ± 0.9 nm and high crystallinity. The NPs synthesized could be dispersed in nonpolar solvents such as chloroform, toluene, and dichloromethane. To render the NPs dispersible in polar solvents, Wang et al. [89] used a naturally occurring biopolymer, chitosan, as a hydrophilic ligand to cap the surface of the LaF_3, $NaYF_4$, and $NaGdF_4$ NPs during the co-precipitation at 75°C for 2 h, yielding particles with a spherical shape and average size of 13–18 nm.

Other organic polymers have also been used to synthesize water-soluble nanoparticles. Polyvinylpyrrolidone (PVP) was utilized by Li and Zhang [90] due to its amphiphilic nature, which makes PVP-coated nanoparticles disperse in both organic and aqueous solutions. The $NaYF_4$ particles were monodisperse in size and crystalline but varied in shape. Another method for making water-soluble $NaYF_4$ NPs with PEI as the capping agent was also reported, utilizing NaCl, NH_4F, and lanthanide chloride as precursors, and water and ethanol as solvents. The reaction time was 24 h and the temperature was 200°C, significantly longer and higher than the previously described synthesis. The same synthesis was employed by Wang et al. [91] for $NaGdF_4$ NPs, though the reaction time was significantly reduced from 24 to 5 h.

The resulting particles were even in shape and size and also crystalline. Li et al. [92] continued working with these NPs and established a multicolor core–shell structure of $NaYF_4$@silica. Using co-precipitation method, core–shell structures of $NaYF_4$@silica and $KGdF_4:Tm^{3+},Yb^{3+}$@$KGdF_4$ were recently reported by Li et al. [92] and Wong et al. [93], respectively, the latter further rendered water-dispersible by capping with a PEG-phospholipid. These core–shell structures showed enhanced upconversion intensity as compared to those of noncoated NPs.

3.4 Microemulsion Synthesis

The microemulsion method utilizes the fact that the small size of reverse micelles, produced by certain combinations of water, oil (usually simple long-chain hydrocarbons), surfactant (normally amphiphilic-containing long-chain organic molecules with a hydrophilic head and lipophilic tail), and an alcohol- or amine-based cosurfactants, subjects them to continuous Brownian motion and thus frequent collisions, even at room temperature. One such collision in every thousand results in the formation of a short-lived dimer, and during the ~100 ns lifetime of the dimer, two reverse micelles exchange the contents of their aqueous cores before decoalescing, resulting in the eventual equilibrium distribution of all the contents. These reverse micelles thus act as nanoreactors, which usually produce NPs of uniform size and shape. Karbowiak et al. [94] reported the synthesis of $NaGdF_4:Eu^{3+}$ NPs, using a water-in-oil microemulsion system containing CTAB, *n*-butanol, *n*-octane, and an aqueous solution of the $GdCl_3$ and NaF. The resultant NPs were spherical in morphology with considerable agglomeration and a certain amorphous phase component. The crystallite size was ~12 nm. Lemyre and Ritcey [95] also attempted to synthesize YF_3 NPs, using the classical microemulsion as well as the single microemulsion methods. They used YCl_3, NH_4HF_2, water, cyclohexane, and polyoxyethylene(5)isooctylphenyl ether. In the classical microemulsion method, the authors found that the NPs were primarily spherical but amorphous. The particle size could be tuned upon variation of the water-to-surfactant ratio. In contrast, using the single microemulsion method, a wide variety of particle morphologies was observed. The resultant particles were crystalline and of the orthorhomic-YF_3 phase. The authors reasoned that the amorphous nature of the particles in the first method was due to the rapid precipitation of YF_3, whereas a slower crystallization process resulted in well-crystallized NPs in the single microemulsion method.

Zhang et al. [96] studied the phase transformation in $NaYF_4$ NPs along with several other parameters such as basicity, fluorination, and lanthanide ion concentration. The reaction of OA and NaOH yielded sodium oleate molecules and induced the formation of reverse micelles. A microemulsion system consisting of Ln^{3+}, NaF, and OA was constructed together with the reverse micelles, where the Ln^{3+} was found in the core of the micelles due to its positive charge. As the reaction temperature increased to 100–130°C, α-$NaYF_4$ NPs formed with the help of the reverse micelles and resulted in a uniform size. Further increase in the hydrothermal temperature to above 160°C led to the formation of β-$NaYF_4$ NPs. The authors proposed a mechanism for the transformation from the α-phase to the β-phase in $NaYF_4$ nanoarrays. Shan et al. [97]

designed an oil–water interface-controlled reaction in normal microemulsions (water/surfactant/*n*-hexane) to produce α-$NaYF_4$:Yb^{3+}, Er^{3+} UCNPs, which were spherical and monodisperse with a uniform size of 20 nm, and showed good dispersibility in nonpolar organic solvents such as *n*-hexane and cyclohexane.

3.5 Microwave-Assisted Synthesis

Because of the rapid heating of the reaction mixtures, particularly those containing water, the microwave processing results in rapid and nearly simultaneous formation of the NPs with small particle sizes and narrow size distributions. Compared to the conventional approaches, the microwave-assisted synthesis also offers the additional benefit of requiring very short reaction times. One of the earliest reports regarding the microwave–assisted synthesis of lanthanide fluorides was provided by Wang and Nann [98] who synthesized rhombohedral GdF_3:Er^{3+}/Yb^{3+}. The rhombic morphology was always observed, regardless of the change in reactant concentration and lithium doping, though the latter had influence on the size of the resultant particles. Ma et al. [99] combined the microwave method with a pH-adjusted hydrothermal method in which the solution containing rare earth nitrates and NaF, with the pH adjusted between 4 and 5, was refluxed in a microwave for 20 min. The resulting particles were around 30 nm in diameter and hollow, approximately spherical and highly crystalline. The authors also analyzed the resulting particles before the microwave heating and found that prior to heating, the particles were small, aggregated, and solid (not hollow).

Starting from the trifluoroacetate precursors, Wang and Nann [100] achieved the synthesis of the oleate-capped NPs of $NaYF_4$, Na(Li)YF_4, and $LiYF_4$ at 290°C for a period of only 5 min. This resulted in highly monodisperse cubic-phase NPs in the case of $NaYF_4$:Er^{3+} (or Tm^{3+})/Yb^{3+} with a particle size of 10 nm. Substitution of the Na metal with Li resulted in the progressive shifting of the crystal phase from cubic to tetragonal. The upconversion emission intensity was also enhanced with the addition of Li. The authors attributed this enhancement to the higher efficiency of the $LiYF_4$ matrix, as well as the enhancement effects observed due to doping with Li ions, even at low concentrations. Wang et al. [101] extended this study further and investigated the effect of reaction parameters such as reaction time and reactant concentrations. In general, a small increase in particle size was observed as a function of time, but the reactant concentration provided a more significant effect on the outcome. At 2, 4, and 20 times the reactant concentrations, the NPs grew significantly, reaching sizes of 20–60 nm depending on the concentrations used. Furthermore, the morphology evolved to become flower-like. Lithium doping also had significant influence on the morphology of these NPs, which was found to be irregular or flower-like, rods and rhombohedral for sub-50, 80 and 100% Li concentration, respectively.

3.6 Ionic Liquid Synthesis

The term "ionic liquid" refers to a salt in the liquid state. These ILs often possess low vapor pressure and high thermal stability, and dissolve most of the organic and inorganic compounds. Jacob et al. [102] investigated the synthesis of metal fluoride NPs, including LaF_3 and YF_3, employing a combined use of ILs and microwave heating. The IL, 1-butyl-3-methylimidazolium tetrafluoroborate ($BMIBF_4$), acted as the solvent and as the F-source. The NPs obtained after 5 min of microwave heating were highly crystalline but polydisperse in shape and size. Nunez and Ocana [103] used the similar IL 1-butyl-2-methylimidazolium tetrafluoroborate ($BMIMBF_4$) as a solvent and F-source to synthesize EuF_3, TbF_3, and YF_3 NPs. Three different forms of Ln precursor were used, nitrate, acetate (OAc), and acetylacetonate (acac). The solvent system was ethylene glycol, the reaction temperature was 120°C, and the reaction time was 15 h. The authors found that $Y(acac)_3$ and $Y(OAc)_3$ yielded NPs which were monodisperse, crystalline, and rhombohedral in shape, whereas $Y(NO_3)_3$ gave particles irregular in shape and size.

Liu et al. [104] investigated the IL synthesis of $NaYF_4:Er^{3+}(Tm^{3+})/Yb^{3+}$. The lanthanide nitrates, sodium chloride, and $BMIBF_4$ were heated in an autoclave for 18 h at 160°C. The resultant NPs were polycrystalline (hexagonal phase) with varying morphologies and sizes ranging from 25 to 70 nm. In this synthesis, $BMIBF_4$ served as the solvent, reagent, and template, which favored the formation of hexagonal-phase particles. The decomposition of the IL provides a large source of F^- anions, decreasing the energy barrier, modifying the Y^{3+} environment, and coordinating to it. The authors explained that the BF_4^- anion directed the hydrophilic property of the IL while the BMI^+ cation exercised a large influence on the chemical properties, including polarity, solvency, and viscosity, directly influencing salvation and diffusion of the lanthanide ions in the IL and, therefore, influencing the nucleation process. Since the IL also acted as a surfactant, it allowed the resulting NPs to be water dispersible. Chen et al. [105] used a series of ILs, 1-butyl-3-methylimidazolium tetrafluoroborate ($BmimBF_4$), 1-butyl-3-methylimidazolium hexafluorophosphate ($BmimPF_6$), and 1-butyl-3-methylimidazolium bromide (BmimBr), for the microwave-accelerated synthesis of spherical $NaYF_4$ nanoclusters. A typical synthesis involved vigorously magnetic stirring of Na and Ln precursors (acetates or trifluoroacetates) in a suitable IL at room temperature for 2 h, followed by microwave-assisted heating to 200°C for 5 min. While the ILs bearing fluorine, ie, $BmimBF_4$ and $BmimPF_6$ produced good quality $NaYF_4$ NPs, the IL with bromide counter ion, BmimBr, failed to do so even with fluorinated precursors (trifluoroacetates of sodium and lanthanides). The IL $BmimBF_4$ proved to be the best one as it yielded spherical $NaYF_4$ nanoclusters with diameters ranging from 200 to 430 nm and formed by the self-assembly of small NPs. The authors concluded that these ILs played crucial key roles, such as the solvents for the reaction, the absorbents of microwave irradiation, and the major fluorine sources during the formation of the $NaYF_4$ NPs.

4. Upconversion in Rare Earth–Doped Nanofluorides

With the advent of nanotechnologies in the 1990s, NPs, and in particular semiconductor QDs, appeared to be an interesting alternative to dyes as markers in biology. However they suffered, in a smaller instance, from similar drawbacks than dyes:

(1) they are more easily excited at wavelength ranges not compatible with biological tissues, (2) their excitation still generates strong autofluorescence from the tissues, (3) they suffer from bleaching and are therefore hard to follow over long times. In 1999 a pioneering work from Zijlmans et al. [106] proposed submicrometer particles of upconverting materials as biomarkers. In 2004, the first fluorides UCNPs were reported [85] and they soon appeared to be interesting alternative since they could overcome the three drawbacks cited above. In the last 10 years, thousands of papers have been devoted to the study of this kind of NPs and they have proven useful as biomarkers (see Refs [34,107–123]), in drug delivery systems (see Refs [124–136]), in photodynamic or photothermal therapy (see Refs [137–145]), or in detection of pollutants [146]. An additional advantage of upconversion is its intrinsic nonlinearity. In the case of a microscopic detection, as for two photon microscopy the excited volume of material is limited to the region where the density of excitation is higher.

If the proof of principles in all this cases has been shown that the system is still far from perfect in particular because the quantum yield of upconversion remains low. Still it has been possible to detect individual particle of 10 nm diameter [147]. In addition the optical properties of upconverting systems are not as straightforward as those of conventional luminescence. All the properties are primarily dependent on the conditions of excitation. For example, it is impossible to define a quantum yield for a given system without specifying the fluence (W/m^2) conditions in which it has been measured. Similarly, the emission spectra can be very different depending on the fluence conditions. Besides, since the process relies on a building up of the excitation by passing through different intermediate levels with different lifetimes, by playing with the time excitation scheme it is possible to also change strongly the emission spectra.

In order to have almost comparable results one has to indicate the quantum yield and the spectra for similar conditions of excitation. The best compromise being to give the value just before saturation of upconversion (but even this is hard to define since the saturation can happen at different degrees of excitation as a function of the transitions involved).

Most results presented in the literature deals with the synthesis and the morphology or structure optimization of UCNPs and those have been presented in the previous part. Here we will focus on works that have dealt with the photophysics of upconversion in fluoride NPs. A rather complete review on this subject was presented by Zheng et al. [148].

The most common scheme used in UCNPs is the ETU by using ytterbium ions as sensitizer and erbium or thulium as emitters. Although many different compositions have been tested for the host material, the highest reported quantum yield are found in tetrafluorides with a general formula $AReF_4$ with A an alkali metal (Na and Li in general) and Re a rare earth ion (Y, Gd, Lu, or Sc). For pure NPs of this composition and after optimizing the concentrations of both Yb and emitters, quantum yield were measured in the range of 0.1–1% near saturation. By adding a passive layer to minimize interactions with the surrounding and passivating the surface of the active core they can be even higher. For instance, Huang et al. [149] reported a quantum yield up to 7.6% for 50 nm particles of $LiLuF_4$:Yb,Tm@$LiLuF_4$.

But upconversion can also be found in many other compositions and if the quantum yield is not the primary concern other hosts can be used. Indeed the local crystal structure of the matrix determines most of the optical properties. The site symmetry and the crystal field strength of the doping ions will determine the strength of the electronic transitions and therefore the efficiency of the transfer between sensitizer and emitter. Generally lower symmetry sites are preferred in the case of upconversion since they increase the 4f-4f transition probabilities. But one has also to take into account the geometry on a slightly larger scale. Indeed it is important to favor the interaction of the emitters with many different sensitizer ions and the more so if one wants to have access to very highly excited states. For instance, Wang et al. [150] showed that in a structure of KYb_2F_7 doped with Er, because it possesses clusters of eight ytterbium ions that are rather distant from one another, the authors observe a significant enhancement of the emission from the higher excited states of Er^{3+} ($^2H_{9/2}$) in comparison to the other excited states ($^4S_{3/2}$ and $^2H_{11/2}$ among others). In this case when exciting the system, the Yb clusters act almost as if they were independent and the excitation is confined in each of them. If one emitter ion is present in the cluster it can be fed by many different ytterbium ions at once, bringing it to higher excited states.

For the smaller NPs some disorder start to appear and the local crystal field is slightly modified leading to small changes in the energy level diagram of the doping ions but also slight changes in the excited state lifetime of this levels. Since upconversion is strongly dependent on both the excited state lifetime and the overlap between emission of the sensitizer and absorption of the emitters, strong differences in behaviors can be observed [151,152]. Once again this can be partly compensated by adding undoped layers at the surface that will allow a better crystallization of the core and avoid energy transfer to the surface of the particles [153].

Going down to single NPs other unusual behavior can also be observed. Gargas et al. [154] realized that it was more convenient for small particles to over-dope with the emitter ion the NP so as to have shorter excited state lifetimes enabling thus a higher number of photons emitted per NPs per second under high energy excitation.

Many efforts were also devoted to try to tune the emission color of the UCNPs. Three approaches were developed, either by changing the phase of the host matrix, by changing the ratio of the different doping ions and by changing the type of ligands at the surface. The three approaches lead to acceptable results but require long systematic research. Besides, the color obtained is valid under certain conditions of excitation only. Another approach relies on the use, for one given composition, of different excitation scheme eventually by using two or more exciting laser. This effect was demonstrated very recently by Deng et al. [155] by using a core—shell structure combining different doping ions and a set of excitation schemes. The authors could cover a wider color gamut than what can be obtained on available high-definition TV sets. However, all this approaches do not allow to have access to all emission wavelengths or to use other excitation wavelength than those efficient for exciting ytterbium and a few other lanthanide ions.

In 2011, a real breakthrough came out with the work of Wang et al. [156]. They showed that by carefully designing a core multishell structure it was possible to transport the excitation of the sensitizers and to bring it to almost any kind of emitter, other lanthanide ions but also organic dyes. They called this kind of upconversion EMU which stands for energy migrated–mediated upconversion and its principle is described in Fig. 3.1. The system they used was $NaGdF_4$ NPs with different shells having different doping ions. In the center Yb and Tm ions were inserted. The core works as a classical ETU upconverter. Then the following shell is undoped and Gd that is a constituent of the matrix serves as a migrator ion through its excited ^6P_n manifolds. Finally the outer layer is doped with any other lanthanide ion (Eu, Tb, Dy, Ho) that can be easily excited through energy transfer from gadolinium. Since gadolinium has no intermediate levels there is no back transfer to the undoped layer and the classical downconversion emission of ions doping the outer layer is observed. Additionally, the author showed that the transfer could also take place toward dye molecules that were placed in close proximity to the NPs.

Shortly after it was also demonstrated that the reverse could also be possible; ie, to transfer the absorption of a dye to the sensitizer and then to the emitters [157]. The authors attached a dye, whose emission matches the absorption of ytterbium, to UCNPs. This way they could generate upconverted emission in the green by using 800-nm excitation thus broadening significantly the possible excitation wavelength. Additionally, since the absorption cross section of dye molecule is much stronger than that of ytterbium ion only a small amount of dyes are needed.

The combination of the two approaches that have just been presented proves that it is possible to engineer almost any kind of excitation and emission scheme by carefully designing a complex nanosized system. The possible geometrical design offered by the use of nanometer-sized material and the combination of molecular design with it is maybe one of the most interesting and less explored contributions of nanoscience.

5. Conclusion

Since the first demonstration of upconversion in the 1960s, this process has attracted a great deal of research activity for bulk and nanocrystalline systems. While the second half of the 20th century research focused on upconversion mechanisms which played a key role in the discovery of new optical devices as IR quantum counter detectors, temperature sensors, solar cells, and compact solid state lasers, past one decade has seen many important steps toward different applications in the fields of biological assay and medical imaging, photovoltaic, photocatalysis, etc., mainly due to extension of this phenomenon in nanomaterials. Nanofluorides emerge as the most suitable host matrices and already reviews have appeared in medical scientific journals such as Nanomedicine (see Refs [140,158–160]). Clearly, the upconversion spectroscopy coupled with nanofluoride technology in scientific research will continue to bring new discoveries not only for biology, diagnostic, and therapy but also for energy harnessing applications. The challenge will be also to use upconversion processes in nanosystems for other societal goals such as the decreasing of pollution of air and water.

References

[1] J.C. Krupa, M. Queffelec, UV and VUV optical excitations in wide band gap materials doped with rare earth ions: 4f–5d transitions, J. Alloys Comp. 250 (1997) 287–292.
[2] D. Bouttet, C. Dujardin, C. Pédrini, W. Brunat, D. Tran Minh Duc, J.Y. Gesland, in: Proceedings of the International Conference on Inorganic Scintillators and Their Applications SCINT95, Delft University Press, The Netherlands, 1996, p. 111.
[3] C. Fouassier, B. Latourrette, J. Portier, P. Haggenmuller, Nature de la fluorescence de l'europium divalent dans les fluorures, Mater. Res. Bull. 11 (1976) 933–938.
[4] A.A. Kaminskii, V.S. Mironov, S.N. Bagaev, N.M. Kaidukov, M.F. Joubert, B. Jacquier, G. Boulon, Spectroscopy and laser action of anisotropic single-centered $LiKYF_5$: Nd^{3+} crystals grown by the hydrothermal method, Phys. Status Solidi A 145 (1994) 177–195.
[5] P. Merchandt, J. Grannec, J.-P. Chaminade, C. Fouassier, Concentration quenching of Nd^{3+} fluorescence in ordered perovskite oxides and fluorides, Mater. Res. Bull. 15 (1980) 1113–1118.
[6] B. Moine, B. Courtois, C. Pedrini, Fluorescence, photoconductivity and polarization properties of divalent rare earth ions in fluoride crystals, J. Lumin. 45 (1990) 248–250.
[7] B. Moine, B. Courtois, C. Pedrini, Photoconductivity and fluorescence properties of divalent ytterbium ions in fluoride crystals, J. Lumin. 48 (1991) 501–504.
[8] J.-J. Videau, J. Fava, C. Fouassier, P. Haggenmuller, Elaboration et proprietes optiques de nouveaux verres fluores actives au neodyme, Mater. Res. Bull. 14 (1979) 499–506.
[9] N. Bloembergen, Solid state infrared quantum counters, Phys. Rev. Lett. 2 (1959) 84–85.
[10] J.F. Porter, Fluorescence excitation by the absorption of two consecutive photons, Phys. Rev. Lett. 7 (1961) 414–415.
[11] M.R. Brown, W.A. Shand, Infrared quantum counter action in Pr-doped fluoride lattices, Phys. Rev. Lett. 11 (1963) 366–368.
[12] M.R. Brown, W.A. Shand, Infrared quantum counter action in Er-doped fluoride lattices, Phys. Rev. Lett. 12 (1964) 367–369.
[13] F. Auzel, Compteur quantique par transfert d'énergie entre deux ions de terres rares dans un tungstate mixte et dans un verre, Compt. Rend. Acad. Sci. Paris 262 (1966) 1016–1019.
[14] F. Auzel, Compteur quantique par transfert d'énergie de Yb^{3+} à Tm^{3+} dans un tungstate mixte et dans un verre germanate, Compt. Rend. Acad. Sci. Paris 263 (1966) 819–821.
[15] N. Menyuk, K. Dwight, J.W. Pierce, $NaYF_4$:Yb,Er – an efficient upconversion phosphor, Appl. Phys. Lett. 21 (1972) 159–161.
[16] E.B. Sveshnikova, A.A. Stroganov, N.T. Timofeev, Role of quasilocalizable vibrations in the deactivation of rare-earth ions in fluoride bases, Opt. Spectrosc. (USSR) 64 (1988) 43.
[17] L.H. Guilbert, J.Y. Gesland, A. Bulou, R. Retoux, Structure and Raman spectroscopy of czochralski-grown barium yttrium and barium ytterbium fluorides crystals, Mater. Res. Bull. 28 (1993) 923–930.
[18] J.F. Suyver, J. Grimm, M.K. van Veen, D. Biner, K.W. Kramer, H.U. Güdel, Upconversion spectroscopy and properties of $NaYF_4$ doped with Er^{3+}, Tm^{3+} and/or Yb^{3+}, J. Lumin. 117 (2006) 1–12.
[19] J.S. Chivian, W.E. Case, D.D. Eden, The photon avalanche – a new phenomenon in Pr/3+/based infrared quantum counters, Appl. Phys. Lett. 35 (1979) 124–125.
[20] A.W. Kueny, W.E. Case, M.E. Koch, Nonlinear-optical absorption through photon avalanche, J. Opt. Soc. Am. B 6 (1989) 639–642.
[21] F. Auzel, in: B. Di Bartolo (Ed.), Radiationless Processes, 1979, p. 213.

[22] F. Auzel, Materials and devices using double-pumped-phosphors with energy transfer, Proc. IEEE 61 (1973) 758–786.
[23] F. Auzel, Upconversion and anti-stokes processes with f and d ions in solids, Chem. Rev. 104 (2004) 139–173.
[24] M.F. Joubert, Photon avalanche upconversion in rare earth laser materials, Opt. Mater. 11 (1999) 181–203.
[25] H. Scheife, G. Huber, E. Heumann, S. Bar, E. Osiac, Advances in up-conversion lasers based on Er^{3+} and Pr^{3+}, Opt. Mater. 26 (2004) 365–374.
[26] M. Haase, H. Schäfer, Upconverting nanoparticles, Angew. Chem. Int. Ed. 50 (2011) 2–24.
[27] X. Chen, D. Peng, Q. Jua, F. Wang, Photon upconversion in core–shell nanoparticles, Chem. Soc. Rev. 44 (2015) 1318–1330.
[28] K. Kompe, H. Borchert, J. Storz, A. Lobo, S. Adam, T. Moller, M. Haase, Green-emitting $CePO_4$:Tb/$LaPO_4$ core–shell nanoparticles with 70% photoluminescence quantum yield, Angew. Chem. Int. Ed. 42 (2003) 5513–5516.
[29] N.M. Idris, M.K.G. Jayakumar, A. Bansal, Y. Zhang, Upconversion nanoparticles as versatile light nanotransducers for photoactivation applications, Chem. Soc. Rev. 44 (2015) 1449–1478.
[30] X. Li, F. Zhang, D. Zhao, Lab on upconversion nanoparticles: optical properties and applications engineering *via* designed nanostructure, Chem. Soc. Rev. 44 (2015) 1346–1378.
[31] F. Wang, X. Liu, Recent advances in the chemistry of lanthanide-doped upconversion nanocrystals, Chem. Soc. Rev. 38 (2009) 976–989.
[32] D. Yang, P. Ma, Z. Hou, Z. Cheng, C. Li, J. Lin, Current advances in lanthanide ion (Ln^{3+})-based upconversion nanomaterials for drug delivery, Chem. Soc. Rev. 44 (2015) 1416–1448.
[33] J. Zhou, Z. Liu, F. Li, Upconversion nanophosphors for small-animal imaging, Chem. Soc. Rev. 41 (2012) 1323–1349.
[34] J. Zhou, X.J. Zhu, M. Chen, Y. Sun, F.Y. Li, Water-stable $NaLuF_4$-based upconversion nanophosphors with long-term validity for multimodal lymphatic imaging, Biomaterials 33 (2012) 6201–6210.
[35] J. Zhou, Q. Liu, W. Feng, Y. Sun, F. Li, Upconversion luminescent materials: advances and applications, Chem. Rev. 115 (2015) 395–465.
[36] S. Mishra, G. Ledoux, E. Jeanneau, S. Daniele, M.F. Joubert, Novel heterometal-organic complexes as first single source precursors for up-converting $NaYF_4$:Yb^{3+}, Er^{3+}/Tm^{3+} nanomaterials, Dalton Trans. 41 (2012) 1490–1502.
[37] J.E. Roberts, Lanthanum and neodymium salts of trifluoroacetic acid, J. Am. Chem. Soc. 83 (1961) 1087–1088.
[38] Y.W. Zhang, X. Sun, R. Si, L.P. You, C.H. Yan, Single-crystalline and monodisperse LaF_3 triangular nanoplates from a single-source precursor, J. Am. Chem. Soc. 127 (2005) 3260–3261.
[39] H.X. Mai, Y.W. Zhang, R. Si, Z.G. Yan, L.D. Sun, L.P. You, C.H. Yan, High-quality sodium rare-earth fluoride nanocrystals: controlled synthesis and optical properties, J. Am. Chem. Soc. 128 (2006) 6426–6436.
[40] H.X. Mai, Y.W. Zhang, L.D. Sun, C.H. Yan, Size- and phase-controlled synthesis of monodisperse $NaYF_4$:Yb,Er nanocrystals from a unique delayed nucleation pathway monitored with upconversion spectroscopy, J. Phys. Chem. C 111 (2007) 13730–13739.

[41] H.X. Mai, Y.W. Zhang, L.D. Sun, C.H. Yan, Highly efficient multicolor up-conversion emissions and their mechanisms of monodisperse $NaYF_4$:Yb,Er core and core/shell-structured nanocrystals, J. Phys. Chem. C 111 (2007) 13721–13729.
[42] J.C. Boyer, F. Vetrone, L.A. Cuccia, J.A. Capobianco, Synthesis of colloidal upconverting $NaYF_4$ nanocrystals doped with Er^{3+}, Yb^{3+} and Tm^{3+}, Yb^{3+} via thermal decomposition of lanthanide trifluoroacetate precursors, J. Am. Chem. Soc. 128 (2006) 7444–7445.
[43] G.S. Yi, G.M. Chow, Synthesis of hexagonal-phase $NaYF_4$:Yb, Er and $NaYF_4$:Yb,Tm nanocrystals with efficient up-conversion fluorescence, Adv. Funct. Mater. 16 (2006) 2324–2329.
[44] J.C. Boyer, L.A. Cuccia, J.A. Capobianco, Synthesis of colloidal upconverting $NaYF_4$: Er^{3+}/Yb^{3+} and Tm^{3+}/Yb^{3+} monodisperse nanocrystals, Nano Lett. 7 (2007) 847–852.
[45] Y.I. Park, S.H. Nam, J.H. Kim, Y.M. Bae, B. Yoo, H.M. Kim, K.S. Jeon, H.S. Park, J.S. Choi, K.T. Lee, Y.D. Suh, T. Hyeon, Comparative study of upconverting nanoparticles with various crystal structures, core/shell structures, and surface characteristics, J. Phys. Chem. C 117 (2013) 2239–2244.
[46] G.S. Yi, G.M. Chow, Water-soluble $NaYF_4$:Yb,Er(Tm)/$NaYF_4$/polymer core/shell/shell nanoparticles with significant enhancement of upconversion fluorescence, Chem. Mater. 19 (2007) 341–343.
[47] W. Niu, S. Wu, S. Zhang, L. Li, Synthesis of colour tunable lanthanide-ion doped $NaYF_4$ upconversion nanoparticles by controlling temperature, Chem. Commun. 46 (2010) 3908–3910.
[48] J. Shan, Y. Ju, Controlled synthesis of lanthanide-doped $NaYF_4$ upconversion nanocrystals via ligand induced crystal phase transition and silica coating, Appl. Phys. Lett. 91 (2007) 123103–123105.
[49] J. Shan, X. Qin, N. Yao, Y. Ju, Synthesis of monodisperse hexagonal $NaYF_4$:Yb, ln (ln = Er, Ho and Tm) upconversion nanocrystals in TOPO, Nanotechnology 18 (2007) 445607–445614.
[50] X. Sun, Y.W. Zhang, Y.P. Du, Z.G. Yan, R. Si, L.P. You, C.H. Yan, From trifluoroacetate complex precursors to monodisperse rare-earth fluoride and oxyfluoride nanocrystals with diverse shapes through controlled fluorination in solution phase, Chem. Eur. J. 13 (2007) 2320–2332.
[51] Y.-P. Du, Y.-W. Zhang, L.-D. Sun, C.-H. Yan, Luminescent monodisperse nanocrystals of lanthanide oxyfluorides synthesized from trifluoroacetate precursors in high-boiling solvents, J. Phys. Chem. C 112 (2008) 405–415.
[52] G.S. Yi, W.B. Lee, G.M. Chow, Synthesis of $LiYF_4$, $BaYF_5$, and $NaLaF_4$ optical nanocrystals, J. Nanosci. Nanotechnol. 7 (2007) 2790–2794.
[53] V. Mahalingam, F. Vetrone, R. Naccache, A. Speghini, J.A. Capobianco, Colloidal Tm^{3+}/Yb^{3+}-doped $LiYF_4$ nanocrystals: multiple luminescence spanning the UV to NIR regions via low-energy excitation, Adv. Mater. 21 (2009) 4025–4028.
[54] B.F. Zhang, M. Frigoli, F. Angiuli, F. Vetrone, J.A. Capobianco, Photoswitching of bis-spiropyran using near-infrared excited upconverting nanoparticles, Chem. Commun. 48 (2012) 7244.
[55] Y.-P. Du, Y.-W. Zhang, L.-D. Sun, C.-H. Yan, Optically active uniform potassium and lithium rare earth fluoride nanocrystals derived from metal trifluroacetate precursors, Dalton Trans. (2009) 8574–8581.
[56] V. Mahalingam, F. Vetrone, R. Naccache, A. Speghini, J.A. Capobianco, Structural and optical investigation of colloidal Ln^{3+}/Yb^{3+} co-doped KY_3F_{10} nanocrystals, J. Mater. Chem. 19 (2009) 3149–3152.

[57] S. Sarkar, C. Hazra, V. Mahalingam, Bright luminescence from colloidal Ln^{3+}-doped $Ca_{0.72}Y_{0.28}F_{2.28}$ (Ln = Eu, Tm/Yb) nanocrystals via both high and low energy radiations, Chem. Eur. J. 18 (2012) 7050–7054.
[58] F. Vetrone, V. Mahalingam, J.A. Capobianco, Near-infrared-to-blue upconversion in colloidal $BaYF_5:Tm^{3+}$, Yb^{3+} nanocrystals, Chem. Mater. 21 (2009) 1847–1851.
[59] S. Mishra, S. Daniele, Metal-organic derivatives with fluorinated ligands as precursors for inorganic nanomaterials, Chem. Rev. 115 (2015) 8379–8448.
[60] Y. Chen, S. Mishra, G. Ledoux, E. Jeanneau, M. Daniel, J. Jhang, S. Daniele, Direct synthesis of hexagonal $NaGdF_4$ nanocrystals from a single-source precursor: upconverting $NaGdF_4:Yb^{3+},Tm^{3+}$ and its composites with TiO_2 for near-IR-driven, Chem. Asian J. 9 (2014) 2415–2421.
[61] S. Mishra, S. Daniele, G. Ledoux, E. Jeanneau, M.F. Joubert, Heterometallic Na-Y(Ln) trifluoroacetate diglyme complexes as novel single source precursors for upconverting $NaYF_4$ nanocrystals co-doped with Yb and Er/Tm ions, Chem. Commun. 46 (2010) 3756–3758.
[62] C.X. Li, J. Yang, Z.W. Quan, P. Yang, D.Y. Kong, J. Lin, Different microstructures of β-$NaYF_4$ fabricated by hydrothermal process: effects of pH values and fluoride sources, Chem. Mater. 19 (2007) 4933–4942.
[63] X. Liang, X. Wang, J. Zhuang, Q. Peng, Y. Li, Synthesis of $NaYF_4$ nanocrystals with predictable phase and shape, Adv. Funct. Mater. 17 (2007) 2757–2765.
[64] F. Wang, Y. Han, C.S. Lim, Y. Lu, J. Wang, J. Xu, H. Chen, C. Zhang, M. Hong, X. Liu, Simultaneous phase and size control of upconversion nanocrystals through lanthanide doping, Nature 463 (2010) 1061–1065.
[65] K.A. Abel, J.C. Boyer, F.C.J.M. van Veggel, Hard proof of the $NaYF_4$/$NaGdF_4$ nanocrystal core/shell structure, J. Am. Chem. Soc. 131 (2009) 14644–14645.
[66] H.S. Qian, Y. Zhang, Synthesis of hexagonal-phase core–shell $NaYF_4$ nanocrystals with tunable upconversion fluorescence, Langmuir 24 (2008) 12123–12125.
[67] Z. Li, Y. Zhang, An efficient and user-friendly method for the synthesis of hexagonal-phase $NaYF_4$:Yb, Er/Tm nanocrystals with controllable shape and upconversion fluorescence, Nanotechnology 19 (2008) 345606–345616.
[68] Y. Wei, F.Q. Lu, X.R. Zhang, D.P. Chen, Synthesis of oil-dispersible hexagonal-phase and hexagonal-shaped $NaYF_4$:Yb, Er nanoplates, Chem. Mater. 18 (2006) 5733–5737.
[69] N. Bogdan, F. Vetrone, G.A. Ozin, J.A. Capobianco, Synthesis of ligand-free colloidally stable water dispersible brightly luminescent lanthanide-doped upconverting nanoparticles, Nano Lett. 11 (2011) 835–840.
[70] C. Liu, H. Wang, X. Zhang, D. Chen, Morphology- and phase-controlled synthesis of monodisperse lanthanide-doped $NaGdF_4$nanocrystals with multicolor photoluminescence, J. Mater. Chem. 19 (2009) 489–496.
[71] L.Y. Wang, Y.D. Li, $Na(Y_{1.5}Na_{0.5})F_6$ single-crystal nanorods as multicolor luminescent materials, Nano Lett. 6 (2006) 1645–1649.
[72] L.Y. Wang, Y.D. Li, Controlled synthesis and luminescence of lanthanide doped $NaYF_4$ nanocrystals, Chem. Mater. 19 (2007) 727–734.
[73] X. Wang, Y.D. Li, Fullerene-like rare-earth nanoparticles, Angew. Chem. Int. Ed. 42 (2003) 3497–3500.
[74] X. Wang, Y.D. Li, Rare-earth-compound nanowires, nanotubes, and fullerene-like nanoparticles: synthesis, characterization, and properties, Chem. Eur. J. 9 (2003) 5627–5635.
[75] X. Wang, J. Zhuang, Q. Peng, Y.D. Li, A general strategy for nanocrystal synthesis, Nature 437 (2005) 121–124.

[76] X. Wang, J. Zhuang, Q. Peng, Y.D. Li, Hydrothermal synthesis of rare-earth fluoride nanocrystals, Inorg. Chem. 45 (2006) 6661–6665.

[77] J.H. Zeng, J. Su, Z.H. Li, R.X. Yan, Y.D. Li, Synthesis and upconversion luminescence of hexagonal-phase $NaYF_4$:Yb, Er^{3+} phosphors of controlled size and morphology, Adv. Mater. 17 (2005) 2119–2123.

[78] F. Wang, D.K. Chatterjee, Z. Li, Y. Zhang, X. Fan, M. Wang, Synthesis of polyethylenimine/$NaYF_4$ nanoparticles with upconversion fluorescence, Nanotechnology 17 (2006) 5786–5791.

[79] F. Wang, X. Liu, Upconversion multicolor fine-tuning: visible to near-infrared emission from lanthanide-doped $NaYF_4$ nanoparticles, J. Am. Chem. Soc. 130 (2008) 5642–5643.

[80] F. Zhang, Y. Wan, T. Yu, F.Q. Zhang, Y.F. Shi, S.H. Xie, Y.G. Li, L. Xu, B. Tu, D. Zhao, Uniform nanostructured arrays of sodium rare-earth fluorides for highly efficient multicolor upconversion luminescence, Angew. Chem. Int. Ed. 46 (2007) 7976–7979.

[81] G.S. Yi, H.C. Lu, S.Y. Zhao, G. Yue, W.J. Yang, D.P. Chen, L.H. Guo, Synthesis, characterization, and biological application of size-controlled nanocrystalline $NaYF_4$: Yb,Er infrared-to-visible up-conversion phosphors, Nano Lett. 4 (2004) 2191–2196.

[82] S. Sarkar, A. Dash, V. Mahalingam, Strong stokes and upconversion luminescence from ultrasmall Ln^{3+}-doped BiF_3 (Ln=Eu^{3+}, Yb^{3+}/Er^{3+}) nanoparticles confined in a polymer matrix, Chem. Asian J. 9 (2014) 447–451.

[83] J. Zhou, Z. Wu, Z. Zhang, W. Liu, H. Dang, Study on an antiwear and extreme pressure additive of surface coated LaF_3 nanoparticles in liquid paraffin, Wear 249 (2001) 333–337.

[84] J.W. Stouwdam, F.C.J.M. van Veggel, Near-infrared emission of redispersible Er^{3+}, Nd^{3+}, and Ho^{3+} doped LaF_3 nanoparticles, Nano Lett. 2 (2002) 733–737.

[85] S. Heer, K. Kompe, H.U. Gudel, M. Haase, Highly efficient multicolour upconversion emission in transparent colloids of lanthanide-doped $NaYF_4$ nanocrystals, Adv. Mater. 16 (2004) 2102–2105.

[86] H. Schafer, P. Ptacek, K. Kompe, M. Haase, Lanthanide-doped $NaYF_4$ nanocrystals in aqueous solution displaying strong up-conversion emission, Chem. Mater. 19 (2007) 1396–1400.

[87] P. Ptacek, H. Schafer, K. Kompe, M. Haase, Crystal phase control of luminescing α-$NaGdF_4$:$Eu^{3+}$and β-$NaGdF_4$:Eu^{3+} nanocrystals, Adv. Funct. Mater. 17 (2007) 3843–3848.

[88] G.-S. Yi, G.-M. Chow, Colloidal LaF_3:Yb,Er, LaF_3:Yb,Ho and LaF_3:Yb,Tm nanocrystals with multicolor upconversion fluorescence, J. Mater. Chem. 15 (2005) 4460–4464.

[89] F. Wang, Y. Zhang, X. Fan, M. Wang, One-pot synthesis of chitosan/LaF_3:Eu^{3+} nanocrystals for bio-applications, Nanotechnology 17 (2006) 1527–1532.

[90] Z. Li, Y. Zhang, Monodisperse silica-coated polyvinylpyrrolidone/$NaYF_4$ nanocrystals with multicolor upconversion fluorescence emission, Angew. Chem. Int. Ed. 45 (2006) 7732–7735.

[91] F. Wang, X.P. Fan, M.Q. Wang, Y. Zhang, Multicolour PEI/$NaGdF_4$:Ce^{3+}, Ln^{3+} nanocrystals by single-wavelength excitation, Nanotechnology 18 (2007) 025701–025706.

[92] Z. Li, Y. Zhang, S. Jiang, Multicolor core/shell-structured upconversion fluorescent nanoparticles, Adv. Mater. 20 (2008) 4765–4769.

[93] H.-T. Wong, F. Vetrone, R. Naccache, H.L.W. Chan, J. Hao, J.A. Capobianco, Water dispersible ultra-small multifunctional $KGdF_4$:Tm^{3+}, Yb^{3+} nanoparticles with near-infrared to near-infrared upconversion, J. Mater. Chem. 21 (2011) 16589–16596.

[94] M. Karbowiak, A. Mech, A. Bednarkiewicz, W. Strek, L. Kepinski, Comparison of different $NaGdF_4:Eu^{3+}$ synthesis routes and their influence on its structural and luminescent properties, J. Phys. Chem. Solids 66 (2005) 1008–1019.

[95] J.L. Lemyre, A.M. Ritcey, Synthesis of lanthanide fluoride nanoparticles of varying shape and size, Chem. Mater. 17 (2005) 3040–3043.

[96] F. Zhang, J. Li, J. Shan, L. Xu, D. Zhao, Shape, size, and phase-controlled rare-earth fluoride nanocrystals with optical up-conversion properties, Chem. Eur. J. 15 (2009) 11010–11019.

[97] S.-N. Shan, X.-Y. Wang, N.-Q. Jia, Synthesis of $NaYF_4:Yb^{3+}$, Er^{3+} upconversion nanoparticles in normal microemulsions, Nanoscale Res. Lett. 6 (2011) 539.

[98] H.-Q. Wang, T. Nann, Monodisperse upconversion GdF_3:Yb, Er rhombi by microwave-assisted synthesis, Nanoscale Res. Lett. 6 (2011) 267.

[99] L. Ma, W.X. Chen, Y.F. Zheng, J. Zhao, Z.D. Xu, Microwave-assisted hydrothermal synthesis and characterizations of PrF_3 hollow nanoparticles, Mater. Lett. 61 (2007) 2765–2768.

[100] H.-Q. Wang, T. Nann, Monodisperse upconverting nanocrystals by microwave-assisted synthesis, ACS Nano 3 (2009) 3804–3808.

[101] H.-Q. Wang, R.D. Tilley, T. Nann, Size and shape evolution of upconverting nanoparticles using microwave assisted synthesis, CrystEngComm. 12 (2010) 1993–1996.

[102] D.S. Jacob, L. Bitton, J. Grinblat, I. Felner, Y. Koltypin, A. Gedanken, Are ionic liquids really a boon for the synthesis of inorganic materials? A general method for the fabrication of nanosized metal fluorides, Chem. Mater. 18 (2006) 3162.

[103] N.O. Nunez, M. Ocana, An ionic liquid based synthesis method for uniform luminescent lanthanide fluoride nanoparticles, Nanotechnology 18 (2007) 455606–455612.

[104] X. Liu, J. Zhao, Y. Sun, K. Song, Y. Yu, C. Du, X. Kong, H. Zhang, Ionothermal synthesis of hexagonal-phase $NaYF_4:Yb^{3+},Er^{3+}/Tm^{3+}$ upconversion nanophosphors, Chem. Commun. (2009) 6628–6630.

[105] C. Chen, L.-D. Sun, Z.-X. Li, L.-L. Li, J. Zhang, Y.-W. Zhang, C.-H. Yan, Ionic liquid-based route to spherical $NaYF_4$ nanoclusters with the assistance of microwave radiation and their multicolor upconversion luminescence, Langmuir 26 (2010) 8797–8803.

[106] H.J.M.A.A. Zijlmans, J. Bonnet, J. Burton, K. Kardos, T. Vail, R.S. Niedbala, H.J. Tanke, Detection of cell and tissue surface antigens using up-converting phosphors: a new reporter technology, Anal. Biochem. 267 (1999) 30–36.

[107] D.E. Achatz, R.J. Meier, L.H. Fischer, O.S. Wolfbeis, Luminescent sensing of oxygen using a quenchable probe and upconverting nanoparticles, Angew. Chem. Int. Ed. 50 (2011) 260–263.

[108] D.K. Chatterjee, A.J. Rufaihah, Y. Zhang, Upconversion fluorescence imaging of cells and small animals using lanthanide doped nanocrystals, Biomaterials 29 (2008) 937–943.

[109] H.Y. Chen, B. Qi, T. Moore, D.C. Colvin, T. Crawford, J.C. Gore, F. Alexis, O.T. Mefford, J.N. Anker, Synthesis of brightly PEGylated luminescent magnetic upconversion nanophosphors for deep tissue and dual MRI imaging, Small 10 (2014) 160–168.

[110] Y.J. Ding, H. Zhu, X.X. Zhang, J.J. Zhu, C. Burda, Rhodamine B derivative-functionalized upconversion nanoparticles for FRET-based Fe^{3+}-sensing, Chem. Commun. 49 (2013) 7797–7799.

[111] N.N. Dong, M. Pedroni, F. Piccinelli, G. Conti, A. Sbarbati, J.E. Ramirez-Hernandez, L.M. Maestro, M.C. Iglesias-de la Cruz, F. Sanz-Rodriguez, A. Juarranz, F. Chen,

F. Vetrone, J.A. Capobianco, J.G. Sole, M. Bettinelli, D. Jaque, A. Speghini, NIR-to-NIR two-photon excited CaF_2:Tm^{3+},Yb^{3+} nanoparticles: multifunctional nanoprobes for highly penetrating fluorescence bio-imaging, ACS Nano 5 (2011) 8665–8671.
[112] Q. Liu, J. Peng, L. Sun, F. Li, High-efficiency upconversion luminescent sensing and bioimaging of Hg(II) by chromophoric ruthenium complex-assembled nanophosphors, ACS Nano 5 (2011) 8040–8048.
[113] H.R. Liu, W. Lu, H.B. Wang, L. Rao, Z.G. Yi, S.J. Zeng, J.H. Hao, Simultaneous synthesis and amine-functionalization of single-phase $BaYF_5$:Yb/Er nanoprobe for dual-modal *in vivo* upconversion fluorescence and long-lasting X-ray computed tomography imaging, Nanoscale 5 (2013) 6023–6029.
[114] Q. Liu, W. Feng, T.S. Yang, T. Yi, F.Y. Li, Preparation of core-shell $NaGdF_4$ nanoparticles doped with luminescent lanthanide ions to be used as upconversion-based probes, Nat. Protoc. 8 (2013) 2033–2044.
[115] Y. Liu, M. Chen, T.Y. Cao, Y. Sun, C.Y. Li, Q. Liu, T.S. Yang, L.M. Yao, W. Feng, F.Y. Li, A cyanine-modified nanosystem for *in vivo* upconversion luminescence bioimaging of methylmercury, J. Am. Chem. Soc. 135 (2013) 9869–9876.
[116] L.C. Ong, L.Y. Ang, S. Alonso, Y. Zhang, Bacterial imaging with photostable upconversion fluorescent nanoparticles, Biomaterials 35 (2014) 2987–2998.
[117] J. Wang, T. Wei, X.Y. Li, B.H. Zhang, J.X. Wang, C. Huang, Q. Yuan, Near-infrared-light-mediated imaging of latent fingerprints based on molecular recognition, Angew. Chem. Int. Ed. 53 (2014) 1616–1620.
[118] H.T. Wong, M.K. Tsang, C.F. Chan, K.L. Wong, B. Fei, J.H. Hao, *In vitro* cell imaging using multifunctional small sized $KGdF_4$:Yb^{3+}, Er^{3+} upconverting nanoparticles synthesized by a one-pot solvothermal process, Nanoscale 5 (2013) 3465–3473.
[119] H.Y. Xing, S.J. Zhang, W.B. Bu, X.P. Zheng, L.J. Wang, Q.F. Xiao, D.L. Ni, J.M. Zhang, L.P. Zhou, W.J. Peng, K.L. Zhao, Y.Q. Hua, J.L. Shi, Ultrasmall $NaGdF_4$ nanodots for efficient MR angiography and atherosclerotic plaque imaging, Adv. Mater. 26 (2014) 3867–3872.
[120] Y.M. Yang, Q. Shao, R.R. Deng, C. Wang, X. Teng, K. Cheng, Z. Cheng, L. Huang, Z. Liu, X.G. Liu, B.G. Xing, *In vitro* and *in vivo* uncaging and bioluminescence imaging by using photocaged upconversion nanoparticles, Angew. Chem. Int. Ed. 51 (2012) 3125–3129.
[121] Y.M. Yang, B. Velmurugan, X.G. Liu, B.G. Xing, NIR photoresponsive crosslinked upconverting nanocarriers toward selective intracellular drug release, Small 9 (2013) 2937–2944.
[122] S.J. Zeng, M.K. Tsang, C.F. Chan, K.L. Wong, J.H. Hao, PEG modified $BaGdF_5$:Yb/Er nanoprobes for multi-modal upconversion fluorescent, *in vivo* X-ray computed tomography and biomagnetic imaging, Biomaterials 33 (2012) 9232–9238.
[123] S.J. Zeng, H.B. Wang, W. Lu, Z.G. Yi, L. Rao, H.R. Liu, J.H. Hao, Dual-modal upconversion fluorescent/X-ray imaging using ligand-free hexagonal phase $NaLuF_4$: Gd/Yb/Er nanorods for blood vessel visualization, Biomaterials 35 (2014) 2934–2941.
[124] J.C. Boyer, C.J. Carling, B.D. Gates, N.R. Branda, Two-way photoswitching using one type of near-infrared light, upconverting nanoparticles, and changing only the light intensity, J. Am. Chem. Soc. 132 (2010) 15766–15772.
[125] Y.L. Dai, H.H. Xiao, J.H. Liu, Q.H. Yuan, P.A. Ma, D.M. Yang, C.X. Li, Z.Y. Cheng, Z.Y. Hou, P.P. Yang, J. Lin, *In vivo* multimodality imaging and cancer therapy by near-infrared light-triggered *trans*-platinum pro-drug-conjugated upconverison nanoparticles, J. Am. Chem. Soc. 135 (2013) 18920–18929.

[126] S.L. Gai, P.P. Yang, C.X. Li, W.X. Wang, Y.L. Dai, N. Niu, J. Lin, Synthesis of magnetic, up-conversion luminescent, and mesoporous core−shell-structured nanocomposites as drug carriers, Adv. Funct. Mater. 20 (2010) 1166−1172.

[127] Z.Y. Hou, C.X. Li, P.A. Ma, G.G. Li, Z.Y. Cheng, C. Peng, D.M. Yang, P.P. Yang, J. Lin, Electrospinning preparation and drug-delivery properties of an up-conversion luminescent porous $NaYF_4$:Yb^{3+}, Er^{3+}@silica fiber nanocomposite, Adv. Funct. Mater. 21 (2011) 2356−2365.

[128] P.A. Ma, H.H. Xiao, X.X. Li, C.X. Li, Y.L. Dai, Z.Y. Cheng, X.B. Jing, J. Lin, Rational design of multifunctional upconversion nanocrystals/polymer nanocomposites for cisplatin (IV) delivery and biomedical imaging, Adv. Mater. 25 (2013) 4898−4905.

[129] M.L. Viger, M. Grossman, N. Fomina, A. Almutairi, Low power upconverted near-IR light for efficient polymeric nanoparticle degradation and cargo release, Adv. Mater. 25 (2013) 3733−3738.

[130] B. Yan, J.C. Boyer, D. Habault, N.R. Branda, Y. Zhao, Near infrared light triggered release of biomacromolecules from hydrogels loaded with upconversion nanoparticles, J. Am. Chem. Soc. 134 (2012) 16558−16561.

[131] D.M. Yang, Y.L. Dai, J.H. Liu, Y. Zhou, Y.Y. Chen, C.X. Li, P.A. Ma, J. Lin, Ultra-small $BaGdF_5$-based upconversion nanoparticles as drug carriers and multimodal imaging probes, Biomaterials 35 (2014) 2011−2023.

[132] D.M. Yang, X.J. Kang, P.A. Ma, Y.L. Dai, Z.Y. Hou, Z.Y. Cheng, C.X. Li, J. Lin, Hollow structured upconversion luminescent $NaYF_4$:Yb^{3+}, Er^{3+} nanospheres for cell imaging and targeted anti-cancer drug delivery, Biomaterials 34 (2013) 1601−1612.

[133] P.P. Yang, S.L. Gai, J. Lin, Functionalized mesoporous silica materials for controlled drug delivery, Chem. Soc. Rev. 41 (2012) 3679−3698.

[134] X. Zhang, P.P. Yang, Y.L. Dai, P.A. Ma, X.J. Li, Z.Y. Cheng, Z.Y. Hou, X.J. Kang, C.X. Li, J. Lin, Multifunctional up-converting nanocomposites with smart polymer brushes gated mesopores for cell imaging and thermo/pH dual-responsive drug controlled release, Adv. Funct. Mater. 23 (2013) 4067−4078.

[135] L.Z. Zhao, J.J. Peng, Q. Huang, C.Y. Li, M. Chen, Y. Sun, Q.L. Lin, L.Y. Zhu, F.Y. Li, Near-infrared photoregulated drug release in living tumor tissue via yolk-shell upconversion nanocages, Adv. Funct. Mater. 24 (2014) 363−371.

[136] L. Zhou, Z.W. Chen, K. Dong, M.L. Yin, J.S. Ren, X.G. Qu, DNA-mediated construction of hollow upconversion nanoparticles for protein harvesting and near-infrared light triggered release, Adv. Mater. 26 (2014) 2424−2430.

[137] Q. Chen, C. Wang, L. Cheng, W.W. He, Z.P. Cheng, Z. Liu, Protein modified upconversion nanoparticles for imaging-guided combined photothermal and photodynamic therapy, Biomaterials 35 (2014) 2915−2923.

[138] L. Cheng, K. Yang, Y.G. Li, J.H. Chen, C. Wang, M.W. Shao, S.T. Lee, Z. Liu, Facile preparation of multifunctional upconversion nanoprobes for multimodal imaging and dual-targeted photothermal therapy, Angew. Chem. Int. Ed. 50 (2011) 7385−7390.

[139] S.S. Cui, D.Y. Yin, Y.Q. Chen, Y.F. Di, H.Y. Chen, Y.X. Ma, S. Achilefu, Y.Q. Gu, *In vivo* targeted deep-tissue photodynamic therapy based on near-infrared light triggered upconversion nanoconstruct, ACS Nano 7 (2012) 676−688.

[140] N.M. Idris, M.K. Gnanasammandhan, J. Zhang, P.C. Ho, R. Mahendran, Y. Zhang, *In vivo* photodynamic therapy using upconversion nanoparticles as remote-controlled nanotransducers, Nat. Med. 18 (2012) 1580−1585.

[141] Y.I. Park, H.M. Kim, J.H. Kim, K.C. Moon, B. Yoo, K.T. Lee, N. Lee, Y. Choi, W. Park, D. Ling, K. Na, W.K. Moon, S.H. Choi, H.S. Park, S.-Y. Yoon, Y.D. Suh, S.H. Lee,

T. Hyeon, Theranostic probe based on lanthanide-doped nanoparticles for simultaneous in vivo dual-modal imaging and photodynamic therapy, Adv. Mater. 24 (2012) 5755–5761.
[142] C. Wang, L. Cheng, Y.M. Liu, X.J. Wang, X.X. Ma, Z.Y. Deng, Y.G. Li, Z. Liu, Imaging-guided pH-sensitive photodynamic therapy using charge reversible upconversion nanoparticles under near-infrared light, Adv. Funct. Mater. 23 (2013) 3077–3086.
[143] M. Wang, Z. Chen, W. Zheng, H.M. Zhu, S. Lu, E. Ma, D.T. Tu, S.Y. Zhou, M.D. Huang, X.Y. Chen, Lanthanide-doped upconversion nanoparticles electrostatically coupled with photosensitizers for near-infrared-triggered photodynamic therapy, Nanoscale 6 (2014) 8274–8282.
[144] Y.H. Wang, H.G. Wang, D.P. Liu, S.Y. Song, X. Wang, H.J. Zhang, Graphene oxide covalently grafted upconversion nanoparticles for combined NIR mediated imaging and photothermal/photodynamic cancer therapy, Biomaterials 34 (2013) 7715–7724.
[145] L. Xia, X.G. Kong, X.M. Liu, L.P. Tu, Y.L. Zhang, Y.L. Chang, K. Liu, D.Z. Shen, H.Y. Zhao, H. Zhang, An upconversion nanoparticle – Zinc phthalocyanine based nanophotosensitizer for photodynamic therapy, Biomaterials 35 (2014) 4146–4156.
[146] Q. Liu, Y. Sun, C.G. Li, J. Zhou, C.Y. Li, T.S. Yang, X.H. Zhang, T. Yi, D.M. Wu, F.Y. Li, ^{18}F-Labeled magnetic-upconversion nanophosphors *via* rare-earth cation-assisted ligand assembly, ACS Nano 5 (2011) 3146–3157.
[147] A.D. Ostrowski, E.M. Chan, D.J. Gargas, E.M. Katz, G. Han, P.J. Schuck, D.J. Milliron, B.E. Cohen, Controlled synthesis and single-particle imaging of bright, sub-10 nm lanthanide-doped upconverting nanocrystals, ACS Nano 6 (2012) 2686–2692.
[148] W. Zheng, P. Huang, D. Tu, E. Ma, H. Zhu, X. Chen, Lanthanide-doped upconversion nano-bioprobes: electronic structures, optical properties, and biodetection, Chem. Soc. Rev. 44 (2014) 1379–1415.
[149] P. Huang, W. Zheng, S.Y. Zhou, D.T. Tu, Z. Chen, H.M. Zhu, R.F. Li, E. Ma, M.D. Huang, X.Y. Chen, Lanthanide-doped $LiLuF_4$ upconversion nanoprobes for the detection of disease biomarkers, Angew. Chem. Int. Ed. 53 (2014) 1252–1257.
[150] J. Wang, R. Deng, M.A. MacDonald, B. Chen, J. Yuan, F. Wang, D. Chi, T.S.A. Hor, P. Zhang, G. Liu, Y. Han, X. Liu, Enhancing multiphoton upconversion through energy clustering at sublattice level, Nat. Mater. 13 (2014) 157–162.
[151] Q. Ju, Y.S. Liu, R.F. Li, L.Q. Liu, W.Q. Luo, X.Y. Chen, Optical spectroscopy of Eu^{3+}-doped BaFCl nanocrystals, J. Phys. Chem. C 113 (2009) 2309–2315.
[152] X.Y. Zeng, J.L. Yuan, Z.Y. Wang, L. Zhang, Nanosheet-based microspheres of Eu^{3+}-doped ZnO with efficient energy transfer from ZnO to Eu^{3+} at room temperature, Adv. Mater. 19 (2007) 4510–4514.
[153] F. Wang, J. Wang, X.G. Liu, Direct evidence of a surface quenching effect on size-dependent luminescence of upconversion nanoparticles, Angew. Chem. Int. Ed. 49 (2010) 7456–7460.
[154] D.J. Gargas, E.M. Chan, A.D. Ostrowski, S. Aloni, M.V.P. Altoe, E.S. Barnard, B. Sanii, J.J. Urban, D.J. Milliron, B.E. Cohen, P.J. Schuck, Engineering bright sub-10-nm upconverting nanocrystals for single-molecule imaging, Nat. Nanotechnol. 9 (2014) 300–305.
[155] R. Deng, F. Qin, R. Chen, W. Huang, M. Hong, X. Liu, Temporal full-colour tuning through non-steady-state upconversion, Nat. Nanotechnol. 10 (2015) 237–242.
[156] F. Wang, R.R. Deng, J. Wang, Q.X. Wang, Y. Han, H.M. Zhu, X.Y. Chen, X.G. Liu, Tuning upconversion through energy migration in core–shell nanoparticles, Nat. Mater. 10 (2011) 968–973.
[157] W.Q. Zou, C. Visser, J.A. Maduro, M.S. Pshenichnikov, J.C. Hummelen, Broadband dye-sensitized upconversion of near-infrared light, Nat. Photonics 6 (2012) 560–564.

[158] D.K. Chatterjee, Yong, Upconverting nanoparticles as nanotransducers for photodynamic therapy in cancer cells, Nanomedicine 3 (2008) 73−82.
[159] L.Y. Ang, M.E. Lim, L.C. Ong, Y. Zhang, Application of upconversion nanoparticles in imaging, detection and therapy, Nanomedicine 6 (2011) 1273−1288.
[160] M. Wang, G. Abbineni, A. Clevenger, C. Mao, S. Xu, Upconversion nanoparticles: synthesis, surface modification and biological applications, Nanomedicine 7 (2011) 710−729.

Optical Properties of Fluoride Transparent Ceramics

4

P. Gredin
IRCP, CNRS-UMR 8247, Chimie ParisTech, Paris, France; Université Pierre et Marie Curie, Paris, France

M. Mortier
IRCP, CNRS-UMR 8247, Chimie ParisTech, Paris, France

Chapter Outline

1. Introduction

A ceramic is a polycrystalline inorganic material excluding metals. The ceramic process was the first "arts of fire" to appear long before the metallurgy and glass making. The word ceramic comes from the ancient Greek and means "clay," "potter's clay." The elaboration and use of ceramic evolved in the 20th century with the industrial development and the research on the chemistry of materials. A new class of ceramic appeared with the technical ceramics. For example, nitride and carbide are essential materials for a lot of industrial applications. Refractory ceramics, which are essential equipment for some sectors of industry as the glass industry or metallurgy, have evolved significantly since the beginning of the 20th century. It is the reason why nowadays the word "ceramic" refers to a polycrystalline material obtained by the

Photonic and Electronic Properties of Fluoride Materials. http://dx.doi.org/10.1016/B978-0-12-801639-8.00004-0

thermal treatment of precursor batches in order to ensure a mechanical cohesion and a continuity of the properties of the manufactured product.

The transparent ceramics were initially developed in the 1960s in order to realize windows for applications in extreme environmental conditions, windows for furnaces, for example. Alumina, spinel, or fluorite was used. For these applications, the mechanical and thermomechanical properties have priority over the optical quality, and logically, oxides represent the major part of the compounds used. However, some works were devoted to elaborate fluoride transparent ceramics for optical elements operating in the infrared (IR) region. In 1964, Hatch et al. elaborate the first transparent ceramic, CaF_2:Dy^{2+}, for a laser application operating at 2.36 μm [17,19,27]. In 1968, Carnall et al. filled a patent about the MgF_2 transparent ceramics for application in the visible and IR regions requiring suitable optical elements for the severe conditions encountered in missiles, projectiles, satellites, and related devices. Nevertheless, only few works are devoted, in the following years, to the transparent ceramics as well as for the oxides or sulfides than for the fluorides [26,57]. The reasons why of this lack of interest for the transparent ceramics were, for application as visible and IR windows, the impossibility to obtain large pieces, the poor optical quality, and the development of chemical vapor deposition which demonstrates its capacity to produce high-performance materials and finally the supremacy of single crystals for which the elaboration was reliable. For the laser applications, the laser characteristics obtained with the transparent ceramics were far away of those obtained with single crystals, and the appearance of very efficient laser systems based on neodymium-doped YAG single crystal pumped with flash lamp did not encourage the development of the research on transparent ceramics as laser materials. In 1995s, A. Ikesue et al. [28,29] developed a new process to obtain Nd^{3+}-doped YAG transparent ceramic with optical quality close to that of the related single crystal. The laser characteristics were identical for single crystals and ceramics at low power but at high power the performances of the ceramics stayed limited by the presence of residual scattering centers. In the same time, T. Yanagitani and H. Yagi from the Konoshima Chemical Company elaborated Nd^{3+}-doped YAG transparent ceramic using a process slightly different of that used by A. Ikesue. Nowadays, the transparent ceramics produced by the Konoshima Chemical Company have the weakest optical losses and have great dimensions (up to 23 cm of length). These works have induced a renewal of the research on transparent ceramics and despite the difficulties inherent in the particular chemistry of the fluorides, the first papers about fluoride transparent ceramics are published in 2007 [2,5,6]. These studies are mainly focused on alkaline earth fluorides CaF_2, SrF_2 except some works devoted to lithium fluoride, cerium fluoride, and the elpasolite-type compound Rb_2NaAlF_6. Nowadays, some fluoride transparent ceramics, essentially rare earth–doped CaF_2, are produced exhibiting excellent optical properties and laser amplification characteristics.

Chemistry of fluoride is tricky. However, the fluorides are excellent materials for a lot of specific optical applications. Because of the highly ionic bond character, fluorides tend to have low phonon energies compared to other compounds as oxides. This low phonon energy causes the IR absorption edge of fluorides to be higher wavelength than other compounds and allows fluorides to be used far in the IR region. The highly ionic bond character of fluorides also leads to large band gaps and therefore ultraviolet

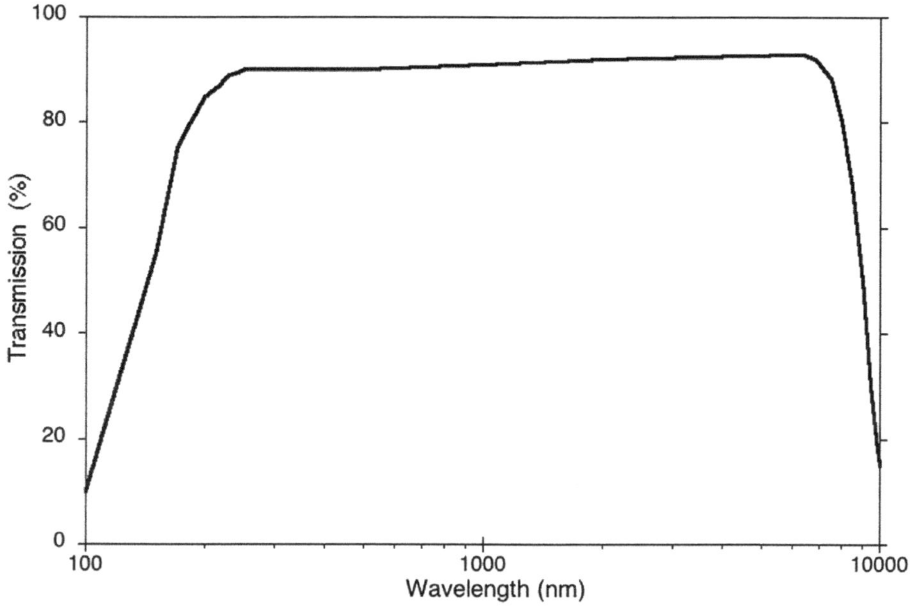

Figure 4.1 Optical transmission of calcium fluoride versus wavelength (nm).

(UV) absorption edges at lower wavelengths than for most other compounds (Fig. 4.1). This property makes fluorides good candidates for laser, window, and lens applications in the UV and vacuum UV wavelength. When doped with rare earth, nonradiative deexcitation probabilities are weaker for fluorides than for most other compounds as oxides allowing particularly efficient emission in the IR region where multiphonon processes increase nonradiative decay rates in many oxide laser hosts, for example. Moreover, weak crystal field in fluorides reduces the probabilities of cross-relaxation between luminescent active ions [48]. Long excited state lifetimes in fluorides facilitate upconversion processes for short wavelength lasers (UV-blue) [54].

Nevertheless, if the chemical nature of the compound used for optical devices is important, the physical nature of the materials (glass, single crystal, glass-ceramic, ceramic) is crucial. Indeed, single crystals exhibit high thermal conductivities but often exhibit macroscopic inhomogeneities of chemical composition when doped. They are fragile and their elaboration requires weeks to grow at high temperature and is thus energy consumer. Glasses can be elaborated within few days and easily formed into fibers or thick films but suffer mainly of low thermal conductivities. The ceramics allow gathering some advantages of the single crystals, and the nature of the materials leads to new interesting features: the ceramic processes need days, and the temperatures required are significantly lower than to produce related single crystals. Ceramics have good mechanical strength and better thermal conductivities than glasses. The chemical composition of doped ceramics is more homogeneous than for the related single crystals, and the ceramic process allows building up complex shapes. Particularly, transparent ceramics are the best choice for applications like high power lasers, which require the best mechanical properties and a good thermal conductivity.

2. Fluoride Transparent Ceramic Preparation Processes

The elaboration of the fluoride ceramics for optical and laser application is discussed in more details in the Chapter 2, and thus this paragraph will be devoted only to a brief summary of the different processes.

To obtain transparent ceramics, it is necessary to form a green body, which must be sintered in order to reach practically a 100% density. If this condition is necessary, it is not sufficient as it will be explained in the following paragraph. Nevertheless, the first key for the transparency of the ceramics is to find a process which allows eliminating completely the porosities inside the ceramic. Two great families of processes are used to achieve the transparency of the ceramics.

- The first process consists in the deformation of heated single crystal under pressure (hot-forming process). Typically, the single crystal is deformed at high temperature and moderate uniaxial pressure, and the as-obtained ceramic is then densified at a higher pressure [25]. This process avoids the difficulties of the synthesis of pure fluoride powders, which are prone to react with water at high temperature (pyrohydrolysis). Nevertheless, the sensitivity of fluorides for water depends on the chemical composition of the fluorides, and for some of them the pyrohydrolysis is not really a brake to work with water. The hot-forming process was successfully used to obtain transparent ceramic of doped CaF_2 [3,44], SrF_2 [10,11], compounds from solid solution between SrF_2 and CaF_2 [8], and BaF_2 [25] or LiF [5,6] with high optical quality for laser application. No secondary oxidized phases due to a contamination by oxygen, which have a negative impact on the optical properties, are observed at the grain boundaries of the as-obtained ceramics on the contrary of some transparent ceramics obtained from nanopowders [45]. Nevertheless, the ceramics produced with this method are heavily textured due to the uniaxial compressive strength applied on the single crystal. Moreover, this method requires long time and a lot of energy to produce single crystal whose size will be finally limited.
- The second great family of processes uses fluoride nanopowders. It can be noticed that the first laser ceramic was a Dy^{2+}:CaF_2 ceramic produced from powder hot-pressed under vacuum in 1960s [17,19,27]. Nowadays, nanopowders are often synthesized by soft chemistry routes in water [4,9,13]. The use of nanopowders is required by the dry shaping process used to produce the green body from the dried powders. Indeed, the nanosized powders promote the sintering and allow reaching a good final density of the ceramic even with the moderate density of the green body. The sintering was often performed under vacuum to avoid reaction with oxygen or moisture. At the end of this step, the ceramics are not transparent in most of the cases, and a posttreatment under pressure at high temperature is necessary [4,45]. The transparency of the ceramic can be also reached applying pressure during the sintering [9]. If this method based on nanopowders has the advantage to be really a low-energy and a fast process to produce transparent ceramics, it suffers the risk of a partial pyrohydrolysis when the powders are dried leading to the presence of oxygenated phases at the grain boundaries inside the final transparent ceramics. Oxygenated phases reduced the optical quality of the transparent ceramic due to the difference of refractive indices of the fluorides and oxides. This kind of process was successfully used particularly to produce transparent ceramics of MgF_2 [15,39], doped CaF_2 and SrF_2 [4,7,9,27,40,42,60]. RF_3 and $R_{0.95}Sr_{0.05}F_{2.95}$ (R = La, Ce, Pr, Nd) transparent ceramics were also prepared using a similar process with a vacuum sintering at high temperature followed by a hot pressing of a green body shaped from dried powder [56]. Nevertheless, RF_3 and $R_{0.95}Sr_{0.05}F_{2.95}$ were synthesized from strontium and rare earth fluorides by mechanical grinding or for $La_{0.95}Sr_{0.05}F_{2.95}$ by fusing LaF_3 and SrF_2 in fluorinating atmosphere.

A new original process to produce Yb-doped CaF_2 transparent ceramic of high optical quality was recently described. This process is counterintuitive because the synthesis and the shaping of the green body are realized in water, and the green body, which is then sintered, is largely wet [50]. Nevertheless, despite the initial presence of water, transparency is reached at the end of the sintering process without any post-treatment under pressure. The quality of the transparent ceramic is excellent allowing the best laser performance obtained with a Yb:CaF_2 ceramic [1].

If the processes, previously described, represent the majority of works about the fluoride transparent ceramics published, some marginal papers open new ways, which can increase the field of applications of the fluoride transparent ceramics. M. Ahrens et al. have obtained transparent ceramic of the elpasolite-type compound Rb_2NaAlF_6 using a sol-gel method to synthesize the fluoride compound and a cold pressing [2]. Recently, W. Li et al. have tried to obtain CeF_3 transparent ceramics using a slip-casting process. Nevertheless, to achieve high transparency, a grain-orientated structure is necessary for CeF_3 ceramic to limit the light scattering due to the birefringence effect related to the hexagonal crystal structure of this fluoride. This kind of grain-orientated structure is obtained by a slip-casting process under a strong magnetic field in order to align the c axis of the grains along the magnetic field direction [37,38].

3. Structural and Optical Properties of Fluoride Transparent Ceramics

3.1 Structural Properties of the Fluoride Transparent Ceramics at the Micro- and Nanoscale and Consequences on Optical Properties

Achieving high transparency of a ceramic is a hard challenge. In order to reach transparency, the ceramic must be theoretically without any defects, with perfect grain boundaries and with a density of 100%. So, the quality of the transparent ceramic depends strongly on the quality of the starting materials, care taken to elaborate the ceramic, and then the microstructure of the ceramics.

Single crystals deformed at high temperature and under pressure exhibit different microstructure versus the degree of deformation, which depends on the process parameters applied. The degree of deformation δ is estimated as followed:

$$\delta = 100 \times \left[\frac{(h_0 - h)}{h}\right]$$

where h_0 is the initial height of the sample and h is the final height of the platelet obtained. For a very low degree of deformation, crack-type structures are observed at the borders of the crystalline domains [23,24]. These structures are not stable and should relax with the sample aging. When the deformation increases and becomes moderate, 300–400 for BaF_2, for example [25], the microstructure is highly nonuniform with grain boundaries irregular in shape and large grain size range. At a high degree of deformation (700), the microstructure of the grains is more uniform and the grain size decreases [25]. At this stage, the microstructure of the ceramic is close to that of a ceramic obtained from hot pressing of a nanopowder. The size of the grains varies as function of the degree of deformation, typically from 1000 μm for low degree of deformation to few tens of micrometers for a high degree of deformation [3,25]. Nevertheless, the uniaxal pressure applied on the sample induces a strong preferred orientation of the grains of the final ceramic. For example, the X-ray powder pattern of a BaF_2 ceramic obtained by hot forming ($\delta = 718$) exhibits a (200) line with the strongest intensity while the (111) lines is almost unobservable [25]. LiF transparent ceramic produced by hot forming shows the same preferred orientation. Basiev et al. [5] have demonstrated that the rate of preferred orientation decreases when the degree of the deformation increases.

The elaboration of transparent ceramics from powder was the historical method. It has been used to produce the first transparent ceramic CaF_2:Dy^{2+} in 1964 and MgF_2 transparent ceramic in 1962 [15,27]. The process consisted of a shaping of a green body by a uniaxial cold pressing followed by a sintering under high pressure. For CaF_2:Dy^{2+}, the initial grain size of the powder is approximately close to 150 μm, and no significant grain growth was observed during the hot pressing [27]. On the contrary, a powder with grain size of about 0.1 μm in size was used to elaborate MgF_2 transparent ceramic and grain have grown during the first step of calcination, up to 1 μm to reach 3 μm after being hot-pressed. This method is largely used since some years to produce transparent ceramics. Basiev et al. [7] have produced CaF_2 ceramics using the hot-pressing method (temperature up to 1300°C and pressure up to 250 MPa) from different sources of commercial powder demonstrating clearly that the purity, size, and morphology of the powder are crucial for the final transparency of the ceramic. In most cases, authors synthesize their own nanosized powders. The advantage of nanosized powder consists of an increasing of the reactivity. Basiev et al. have elaborated Er^{3+}:CaF_2 and Nd^{3+}:BaF_2 transparent ceramics from nanopowders synthesized by reaction between nitrates and hydrofluoric acid [7,9]. The as-synthesized Er^{3+}:CaF_2 powder exhibits grains with size in the range of 20–80 nm. These powders show drastic changes in morphology and size after being dried at 100°C: the shape of the crystal becomes cubic indicating a recrystallization and the size of the crystal increases to reach up to 300 nm. This phenomenon is clearly evidenced by powder X-ray diffraction. The broadened diffraction peaks of the inital powder become narrower after the heating at 100°C. SEM images of the final ceramic show that the material is further recrystallized during the compacting allowing a resulting ceramic with no pores. Recently, some researchers of the *State Key Laboratory of Advanced Technology for Materials Synthesis and Processing* and the *Center of Materials Research and Analysis*, Wuhan, China, have elaborated different transparent ceramics (Nd^3+: CaF_2, Yb^{3+}, Er^{3+}:CaF_2, Er^{3+}:CaF_2) by hot-pressing nanopowders synthesized by reaction between nitrates and potassium fluoride acting as fluorinating agent using a process developed by Aubry et al. [4]. For all these compounds, the process is similar with a powder dried at 80–100°C, slightly crushed in a mortar, filed into a graphite die, and sintered in a hot-pressing furnace at a temperature between 800 and 900°C under vacuum with under a uniaxial pressure of 30 MPa applied on the powder. Er^{3+}:CaF_2

nanopowders used as precursor have grain with size comprised between 15 and 25 nm and are slightly agglomerated due to the drying step with aggregates about 50 nm in size. The as-obtained Er^{3+}:CaF_2 ceramics show grains with wide size distribution between 100 and 1000 nm. SEM micrographs exhibit residual intergranular pores [41]. Identical results are obtained for Yb^{3+}, Er^{3+}:CaF_2 [40], and Nd^{3+}:CaF_2 [42] transparent ceramics. In order to enhance the final quality of the ceramic, a little quantity of lithium fluoride was added to the initial Er^{3+}:CaF_2 nanopowder, and different sintering temperatures were tested [60]. SEM micrographs show that the increase of the grain size is limited at 600°C but increases strongly from 700 to 900°C to reach few tens of micrometers at this final temperature. Many pores are observed at 600°C while the ceramic becomes nearly pore-free up to 800°C. Nevertheless, transparent ceramics obtained by the sintering of nanopowders under high pressure or followed by a posttreatment under high pressure suffer often from residual defects, which reduce the optical quality of the final material and are prejudicial to achieve an efficient laser effect for example. The main defects are the inter- and intragranular porosity, secondary phases, and inhomogeneous chemical composition. The presence of pores is particularly harmful for the transparency of the ceramics. The porosity rate must be the lowest possible. In order to avoid the appearance of porosity, nanopowders used to elaborate transparent ceramic must exhibit a narrow size distribution, without aggregates. A large difference of grain sizes induces a differential sintering favorable to the appearance of porosity. For the same reasons, aggregates, particularly hard aggregates, cannot be reduced by a pressureless sintering process, and cavities remain inside the ceramics. Even if the rate of porosity is very low, there are other possible sources of light scattering. Lyberis et al. [45] have studied the origin of the residual light scattering inside their Yb:CaF_2 transparent ceramics. They have focused their attention particularly on two sources of light scattering localized at grain boundaries: the ytterbium segregation representing nearly 90% of the grain boundaries and secondary grain boundary phase representing the remaining 10%. These Yb:CaF_2 ceramics were elaborated from nanopowders pelletized by uniaxial pressing followed by cold isostatic pressing. The resulting disks of compacted powder were then sintered at a moderate temperature (~600°C) under vacuum. The as-obtained ceramics, which are not transparent at this stage, were hot-pressed at 900°C under vacuum and 60-MPa uniaxial stress. The resulting ceramics exhibit an average grain size of 500 nm without any large cavities. The secondary grain boundary phases are characterized by the presence of a high rate of oxygen identified by energy-dispersive X-ray spectroscopy (EDX) measurements (Fig. 4.2). The presence of oxygen is unexpected in a fluorite structure because calcium fluoride is the least hygroscopic compounds of all fluorides. Thus, investigations on powders have been performed in order to understand the origin of the oxygen among the ceramics. EDX analyses have been performed on single particles of the annealing nanopowders used to produce transparent ceramics. Crystallized shells containing oxygen have been evidenced at the surface of the particles. The chemical compositions of these oxygenated shells are in agreement with the calcium hydroxide $Ca(OH)_2$ which might have been formed during the annealing of the powder, fluoride reacting with the adsorbed water at high temperature. The ytterbium segregation at the grain boundaries is clearly evidenced by a

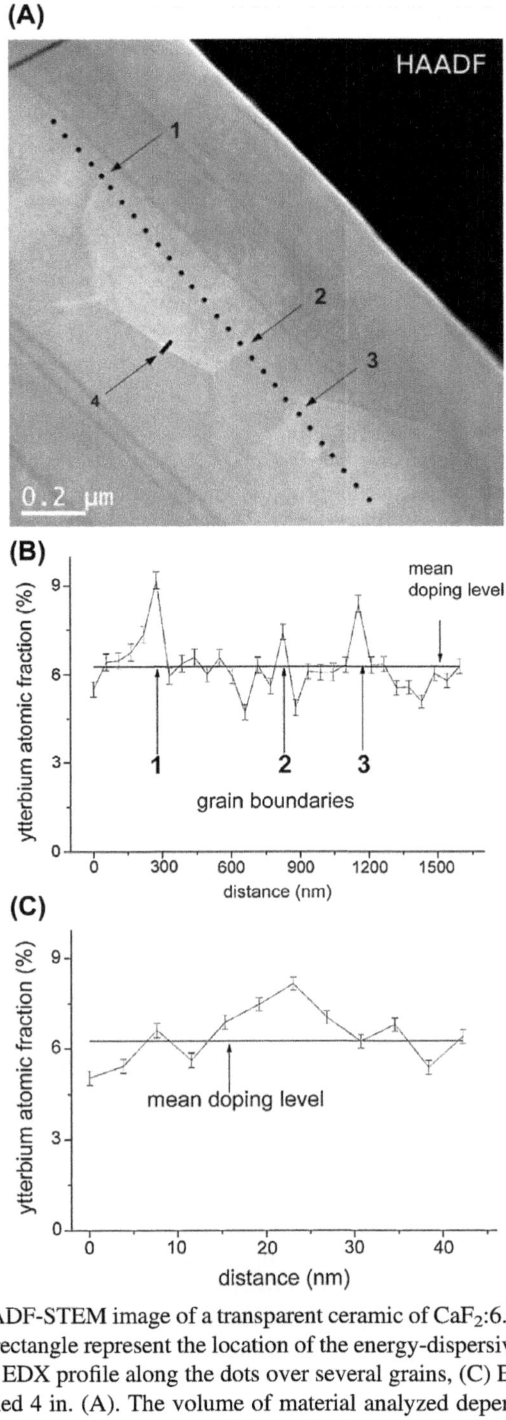

Figure 4.2 (A) HAADF-STEM image of a transparent ceramic of CaF_2:6.4 at% Yb prepared by FIB where dots and rectangle represent the location of the energy-dispersive X-ray spectroscopy (EDX) analyses, (B) EDX profile along the dots over several grains, (C) EDX profile across the grain boundary labeled 4 in. (A). The volume of material analyzed depends on the sample thickness (about 100 nm) and the geometry of the probe (in particular the angle of convergence). For each point, the maximum diameter of the cone of material probed is approximately of 1.5 nm [45].

Figure 4.3 HAADF-STEM image of ytterbium segregation at the grain boundary in CaF_2: 0.5 at% Yb powder along the [211] zone axis of the crystal of the left side from the twin boundary [45].

high-spatial resolution Electron Dispersive X-ray spectroscopy (EDX) and Scanning Transmission Electron Microscope−High-Angle Annular Dark-field (STEM-HAADF) imaging. Analyses highlight the segregation of the rare earth in a thick region (<25 nm) centered on the grain boundary. This region does not correspond to a secondary phase or precipitates fixed at the grain boundary but an increase of the ytterbium incorporation in the calcium fluoride phase (Fig. 4.3). The long range atomic ordering of the fluorite crystal structure is preserved as far as the grain boundary despite a local higher amount of dopant. Both inhomogeneities of grain boundaries, which are source of optical loses in these fluoride ceramics have the same influence. Refractive indices of doped CaF_2 corresponding to the bulk of the grain and to the grain boundary as well as of oxygenated phases have been evaluated. A. Lyberis has studied transparent ceramic Yb:CaF_2 in which the doping rate varies from 6.4 mol% in the bulk of the grain to 9 mol% at the grain boundary. Using the Gladstone and Dale law, he has estimated the refractive indices of two phases: $Ca_{0.936}Yb_{0.064}F_{2.064}$ and $Ca_{0.91}Yb_{0.09}F_{2.09}$. The estimated values at a wavelength of 500 nm are $n_1 = 1.447$ and $n_2 = 1.453$, respectively. The value of the difference between both indices is 0.008. This value must be compared with this obtained when the secondary phases at the grain boundaries are taken into account. Its refractive index is 1.871 at a wavelength of 500 nm. The value of the difference of refractive indices between doped CaF_2 and CaO is in the range of 0.424−0.418, ie, approximately 500 times greater than the difference due to the ytterbium segregation (Table 4.1). Reflectances for both defects at the grain boundary were also calculated at the wavelength of 500 nm, and calculations show clearly that the segregation of the

Table 4.1 Refractive Indexes for Different Compounds [45]

	Compound	*n* Evaluated at 500 nm
a	CaF_2:6.4 at% Yb	1.447
b	CaF_2:9 at% Yb	1.453
c	CaF_2	1.434
d	YbF_3	1.56
e	CaO	1.871

rare earth is negligible (R ~ 0.0004%) while the oxygenized boundary phases have a strong effect (R ~ 1.63%). The presence of oxygenized boundary phase is probably at the origin of the absorption from 220 nm in the UV region. Thus, the oxygenized phase plays a major role in the scattering of the fluoride transparent ceramics.

The fluorides are prone to react with water and/or oxygen at high temperature. Secondary phases observed at the grain boundaries in the transparent fluoride ceramics are mainly oxides due to the use in the process of nanopowders synthesized in aqueous medium.

3.2 The Special Features of Yb:CaF_2 Transparent Ceramics

As amplifying ions, ytterbium (III) ion allows a quasi-three-level laser scheme leading to low quantum defects associated with a low thermal load and exhibits a longer fluorescence lifetime and broader absorption and emission bands than most of rare earth ions operating in the near IR. These features are particularly interesting for femtosecond high-power laser systems. Nevertheless, the nature of matrix is primordial and must be adapted to the specificities of the high-power laser devices in near IR. Among all the possible matrices, crystalline CaF_2 provides a high thermal conductivity (9.7 W/m K^{-1}) similar to the values encountered for usual crystalline optical materials (YAG: 11 W/m K), a good chemical stability, a low cutoff phonon frequency and a weak crystal field. Ytterbium (III)-doped calcium fluoride has a particular position in the field of the fluorinated compounds for laser applications. Indeed, fluorides with rare earth are well known to be interesting for laser application, the fluorite CaF_2 doped with trivalent rare earth has generally poor luminescence properties unfit for its use as laser material except Yb^{3+}:CaF_2. Indeed, the cubic fluorite structure can be described as a packing of fluoride cubes of *a*/2 edge (*a* is the lattice parameter) alternatively empty and occupied by the divalent cations. The substitution of divalent cations by trivalent cations is accompanied of the introduction of interstitial fluorine anions, which take place in the empty cubes of the crystal structure and simultaneously of a migration of a part of the fluoride from normal positions to interstitial positions [21,36]. In the 1980s, some theoretical studies were conducted on the clustering of trivalent cations in the fluorite-type compounds MF_2 (M = Ca, Ba, Sr) doped with different rare earth, calculating their energies of formation from the results of Electronic Paramagnetic Resonance (EPR) experiments [12,20,22,47]. These studies have evidenced that the clusters with different numbers of rare earth cations begin to form at a rate of 0.1 atomic percent. When the doping rate is greater than one atomic percent, the majority of rare earth cations are included inside complex clusters of several ions and particularly hexameric clusters, which induce the greatest stabilization energy. This clustering of rare earth ions is generally harmful for the luminescence properties because the short distances between rare earth ions induce a strong cross-relaxation with consequently short lifetimes of the excited states. It is the reason why Nd^{3+}:CaF_2 is a so poor material for laser application. Nevertheless, the situation is different for Yb:CaF_2 because trivalent ytterbium cation exhibits only a ground state and one excited state level. Therefore, this simple electronic structure forbids cross-relaxation effects encountered in other doped fluorite-type compounds, maintaining a long lifetime of the excited state. The clustering, which is a disadvantage for the fluorite-type structure doped with the most of rare earth, becomes an opportunity with Yb^{3+}:CaF_2. Indeed, clustering of ytterbium anions induces broader absorption and emission bands than most of rare earth ions ensuring efficient energy storage and broader tunability for laser operation. It can be highlighted that isolated ytterbium cations are more stable in CaF_2 than in another fluorite-type compound. Thus, isolated rare earth can be present in CaF_2 at doping rate greater than one atomic percent even if clusters are predominant. One of the first experimental investigations of a single crystal of ytterbium-doped CaF_2 was done as early as in 1969 by Voronko et al. [58]. They have allowed evidencing the formation of the clusters for high doping rates, and a first diagram of energy levels of the Yb^{3+} cations was proposed. These results were confirmed much later by Ito et al. [31]. Petit et al. [52,53] have done a more accurate study of isolated ytterbium as well as clusters in a single crystal of Yb-doped CaF_2, using time-resolved fluorescence spectroscopy at low temperature and Raman spectroscopy. These last studies allowed to compare the values of the measured energy levels to those calculated as function of the crystal field for the three types of the monomers according to the symmetry of the ytterbium site, ie, O_h, C_{4v}, C_{3v}, and for the hexameric clusters. The fluorescence lifetimes measured for each site of isolated ytterbium vary from 10.5 ms for O_h sites to 2.1 ms for C_{3v} sites. Spectra and energy levels of Yb^{3+} ions are not modified in CaF_2 transparent ceramics in comparison with what happens in single crystals [32].

While CaF_2 has raised very early the interest of the laser community [27,55], this matrix was neglected to the 2000s; because CaF_2 exhibits disappointing luminescence performance with most of rare earths except ytterbium, this last rare earth needing powerful monochromatic excitation sources developed from 1980. In the 1990s, the development of powerful diode laser allows having new efficient and compact excitation sources in favor of the development of rare earth and particularly ytterbium-based laser devices. The interest of rare earth—doped CaF_2 for laser application was demonstrated first with single crystals [16,35,43,51]. Very early, Yb:CaF_2 appears to be particularly interesting, and efficiencies greater than 40% were obtained for a continuous-wave laser using a Yb:CaF_2 single crystal. Furthermore, the long lifetimes of the excited states allow not only to realize an inversion of population using rather weak excitation energies but also to generate more powerful pulses, which exhibit shorter duration when the emission band is large (pulse durations are 10 times shorter for Yb:CaF_2 than for Yb:YAG). Thus, a terawatt device with pulses of 197 mJ and 192 fs was developed by Sielbod et al. as early as 2008.

3.3 Ceramic Laser Materials—Laser Effect Performances

Although the compound was particularly interesting, studies about fluoride ceramics doped with rare earth were not restricted to Yb:CaF_2. Several teams in the 2006s have tried to obtain transparent ceramic of alkaline earth fluorides (CaF_2, SrF_2, BaF_2) doped with rare earth, especially Yb, in order to profit of advantages related to the ceramics: thermomechanical strength increased, high-rate doping homogeneity, easiness to realize complex, or/and great size parts. Indeed, single crystals suffer a mechanical and mainly thermomechanical weakness. Moreover, homogeneous high-rate doping concentrations are difficult to obtain due to the segregation of the doping element occurring during the crystal growth, and the access to great parts is costly or even impossible. The transparent ceramics appear to be a solution to get around some drawbacks of the single crystals without losing some of their qualities.

As explained earlier, two main routes to realize fluoride transparent ceramics were used: the hot-forming process applied to a single crystal and the sintering process applied to powders.

3.3.1 Hot-Forming Process Applied to a Single Crystal

LiF with F_2^- color centers was the first transparent ceramic elaborated using a hot-forming protocol in 2007 [6]. F_2^-:LiF is among the most promising laser materials for continuous tunable generation of pico- and femtosecond pulses in near-IR region. LiF single crystals were grown by the Kyropoulos technique in air, and ceramics were obtained by hot forming of single crystals at 600°C. F_2^- color centers were produced irradiating the ceramic samples under a beam of 21 MeV electrons at room temperature. A maximum slope lasing efficiency (lasing output power vs. absorbed pump) of 26% was obtained for a 4-mm thick ceramic, value to compare to this of 18% obtained with a 7-mm single crystal. Moreover, ceramics are characterized by lower lasing thresholds than those encountered with single crystals. In 2008, Basiev et al. [8] reported to have prepared CaF_2—SrF_2—YbF_3 ceramics of high optical quality. Single crystals of a $0.65CaF_2$—$0.30SrF_2$—$0.05YbF_3$ composition were grown by the Bridgman technique in a fluorinating atmosphere and were then hot-pressed to obtain finally transparent ceramics. The fracture toughness of the ceramic samples was estimated to be at least 1.75 times higher than that of the single crystals of the same composition, but the spectroscopic properties are found to be identical between ceramics and single crystals. The lasing properties of the ceramics and single crystals of the same composition were tested under a 972-nm excitation pump from a diode operating in a pulsed mode. Both ceramics and single crystals are characterized by a high slope efficiency (45% for ceramic vs. 50% for single crystal) and a low oscillation threshold (close to 1 W). These studies are followed from 2010 to 2013 by different works on $SrF_2:Nd^{3+}$, $SrF_2:Pr^{3+}$, $CaF_2:Tm^{3+}$, and $CaF_2:Yb^{3+}$. $SrF_2:Nd^{3+}$ [10] and $SrF_2:Pr^{3+}$ [11] are characterized by a low doping level, 0.5% and 0.3–0.5%, respectively, in order to avoid the formation of rare earth clusters. The ceramics exhibit similar optical characteristics with single crystals in laser operation. It can be noted that $SrF_2:Pr^{3+}$ is the first laser ceramic working in the visible spectral range (639 nm). The lasing properties of 4-mol% Tm^{3+}-doped ceramic and 6-mol% Tm^{3+}-doped single crystals have been investigated by Lyapin et al. [44]. A slope efficiency of the 26.7% was demonstrated for the single crystal in continuous mode using diode pumping at 798 nm while the efficiency falls down to 5.5% for the ceramic in pulsed mode using diode pumping at 798 nm (laser oscillation at 1898 nm). This result indicates that the optical quality of the ceramics must be improved; probably residual defects are at the origin of scattering. In 2013, Akchurin et al. [3] have prepared by hot-forming CaF_2:Yb transparent ceramics. (They have also prepared hot-pressing transparent ceramics from powders. See the next section.) They have carefully characterized the properties of the ceramics in comparison with those of the starting single crystals (Fig. 4.4). Some ceramics are hot-formed without any post-treatment while other ceramics are exposed to a fluorinating atmosphere at high temperature in order to reduce the rate of Yb^{2+} cations present in the starting single crystals. The average sizes of the grains of the ceramics are in the range of 40–100 μm. Nevertheless, it is observed that the post-treatment of the ceramics has induced a strong increase of the grain sizes. They have carefully characterized the properties of the ceramics in comparison with those starting single crystals.

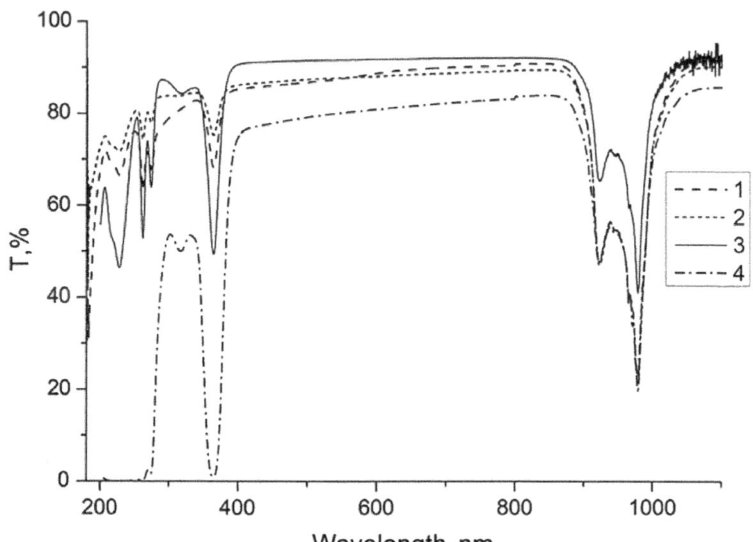

Figure 4.4 Absorption spectra of CaF_2:Yb (3%) single crystal and ceramics: (1) hot-formed ceramic (thickness: 0.40 cm); (2) hot-formed ceramic posttreated at high temperature under CF_4 fluorinating atmosphere (thickness: 0.40 cm); (3) single crystal (thickness: 0.188 cm)—ceramic obtained by hot pressing of a polycrystalline powder (thickness: 0.356 cm) [3].

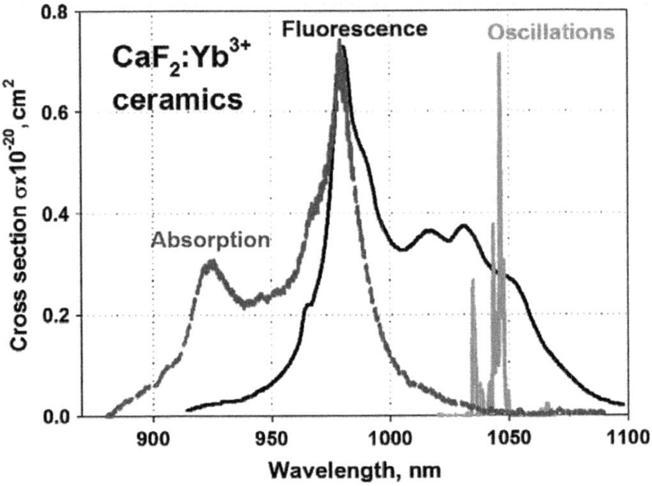

Figure 4.5 Absorption and emission cross section spectra along measured oscillation spectrum of CaF_2:Yb ceramics [3].

Particularly, both ceramics and single crystals possess the same high thermal conductivity, which is strongly dependent on the doping rate: thermal conductivity decreases with the increase of the doping rate. The oscillation spectrum of Yb^{3+} ions in CaF_2 is wide and centered around 1040 nm (maximum at approximately 1045 nm) (Fig. 4.5). The lasing properties of the hot-formed ceramics and single crystals of the same composition were tested in a pulsed mode. Both ceramics and single crystals are characterized by the same high slope efficiency (35% for ceramic vs. 34% for single crystal—mirror reflectivity of 95%) and a low oscillation threshold (close to 2 W) (Figs. 4.6 and 4.7). In continuous-wave mode, the oscillation threshold reaches 1 W with the same slope efficiency of 30%. These results indicate that ceramics and starting single crystals exhibit the same optical quality. Ceramic samples do not exhibit any thermal or optical damage for any pump power used even without any sample cooling.

3.3.2 Sintering Process Applied to Powders

The elaboration of transparent ceramic from powder is in fact the historical process since Hatch et al. [27] have obtained the first fluoride transparent ceramic in 1964 with Dy^{2+}:CaF_2. The starting powder, prepared by grinding a solid obtained by vacuum melting of CaF_2 and DyF_3, was compacted by simultaneous application of heat and pressure in vacuum. Dy^{3+} cations in the as-obtained ceramics are reduced to Dy^{2+} with 0.25-MeV X-ray. Laser oscillations were demonstrated for all samples immersed in liquid nitrogen or helium. Nevertheless, a quantitative comparison of the relative efficiency of the polycrystalline host with a single crystal host could not have been conducted at that time. Examination of the interior of the polycrystalline material using ultramicroscope techniques has revealed very fine isolated scattering centers, most likely CaO, which accounts for the optical loss observed particularly

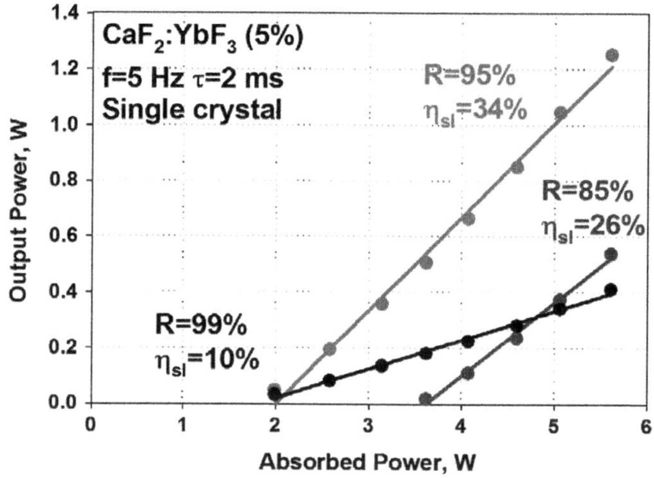

Figure 4.6 Laser efficiency curves of CaF_2:Yb (5%) single crystal versus absorbed pump power for different values of the output mirror reflectivity [3].

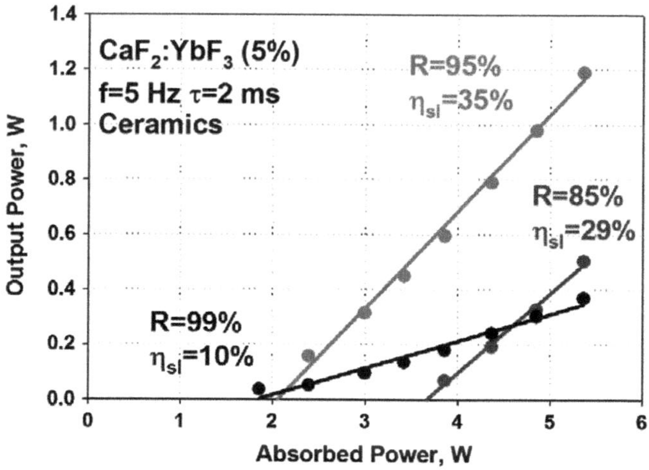

Figure 4.7 Laser efficiency curves of CaF_2:Yb (5%) hot-formed ceramic versus absorbed pump power for different values of the output mirror reflectivity [3].

at short wavelengths. In 2007, Russian and French teams mainly have taken up the idea of Hatch et al. and have tried again to obtain high-optical quality transparent ceramic from powders. Russian and French teams have chosen to synthesize fluoride nanopowders from reaction between nitrates and hydrofluoric acid. The Russian team has realized transparent ceramics at rather high temperature (1200°C) and under high uniaxial pressure (up to 250 MPa) under vacuum [9], particularly LiF-F_2^- and $Ca_{0.65}Sr_{0.30}Yb_{0.05}F_{2.05}$ ceramics. Ceramic samples LiF-F_2^- are characterized by a lower generation threshold compared with analogous single crystal. The maximum slope efficiency was obtained for a 4-mm thick ceramic sample, reaches 26% against 18% for a 7-mm thick single crystal. Analogous laser experiments were carried out with $Ca_{0.65}Sr_{0.30}Yb_{0.05}F_{2.05}$ ceramic. The generation threshold with the ceramic sample was close to this observed for $Ca_{0.95-x}Sr_xYb_{0.05}F_{2.05}$. The slope efficiency for the ceramic reaches 48%, which is somewhat lower than for the single crystal (58%). From 2012, Fedorov et al. have reported result of laser experiments carried out CaF_2 doped with ytterbium [23,3]. The oscillation spectrum of a $Ca_{0.97}Yb_{0.03}F_{2.03}$ ceramic is in the range of 1029.5−1035.5 nm with a maximum of about 1032.5 nm. The slope efficiency of a $Ca_{0.97}Yb_{0.03}F_{2.03}$ ceramic sample tested in a pulsed operation mode (pulse: 2 ms and frequency: 5 Hz) was found to be of 5% with a generation threshold of approximately 35 mW (Fig. 4.8). These results are significantly below those obtained by the same authors for hot-formed ceramics of the same composition. The optical quality of the ceramics obtained from powders is low due to the presence of residual defects (secondary phases at grain boundaries) probably.

French team have obtained transparent ceramics but using a way somewhat different [4,45]. Pelletized annealed powder of Yb:CaF_2 is sintered at a moderate temperature under vacuum and then, to achieve transparency, sample is hot-pressed at temperature and load under vacuum. The authors report to have obtained laser oscillation at 1031 nm with a 1-at% Yb:CaF_2 ceramic under a pump excitation centered at 978 nm [46,49]. The pump wavelength can vary from 973 to 979 nm without affecting the laser output. Nevertheless, slope efficiency was very low. Indeed, despite a low level, optical losses cannot be neglected because they are at the origin of light scattering harmful for an efficient laser effect. The origin of the scattering is attributed mainly to oxygenized secondary phases, which precipitate at grain boundaries [45]. P. Aballea et al. [1] have recently reported to have obtained very good laser performances with a Yb:CaF_2 transparent ceramic obtained with an original process. The

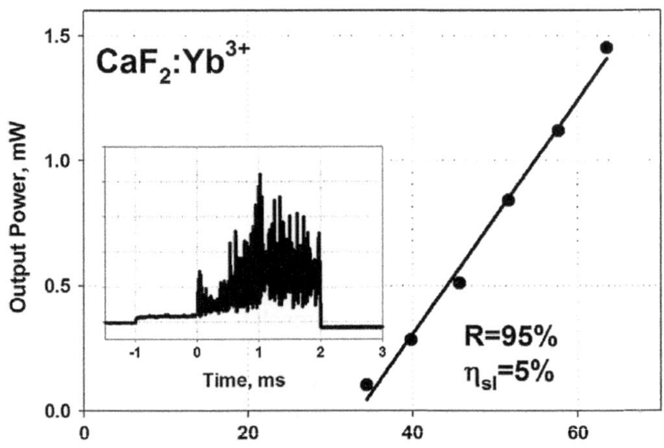

Figure 4.8 Laser efficiency curves of CaF_2:Yb (3%) single ceramic obtained by hot pressing polycrystalline powder versus absorbed pump power for different values of the output mirror reflectivity and temporal behavior of the oscillation pulse [3].

optical and laser properties of a 4-at% Yb:CaF_2 transparent ceramic were investigated. The absorption coefficients of a 2.71-mm thick ceramic sample were found to be rather low with $1.32 \times 10^{-2}\ cm^{-1}$ at 1100 nm, value close to this obtained for a single crystal at this wavelength, testifying the good optical quality of the ceramic. Nevertheless, a significant short wavelength scattering indicates that nanosized residual defects remain in the ceramic. The laser performances of the ceramic were evaluated at room temperature with an excitation pump at 977 nm of a laser diode in a continuous-wave mode. The 4%-doped ceramic absorbs 3.2 W at a full power of 6.9 W. A maximum of 1.6 W was obtained for an output coupler transmission value of 6% with an oscillation threshold of 600 mW. The slope efficiency taking into account the exact absorbed pump power (ie, considering saturation effects) reaches a value of 42.7% for the ceramic, value to compare to this obtained for a 4.5-at% Yb:CaF_2 single crystal of 51.7% (Fig. 4.9). It appears moreover that the ceramic was continuously tunable over a range of 45 nm from 1015 to 1060 nm with a maximum output power at 1051 nm (Fig. 4.10). Nevertheless, the dissipation of the heat inside the 4 at% Yb:CaF_2 ceramic under laser operation is more difficult than in the 4.5 at% Yb:CaF_2 single crystal. This effect is probably due to the microstructure of the ceramic, which exhibits rather small grains. Small grains imply a lot of grain boundaries, which are not in favor of an easy propagation of phonons and thus reduce the thermal conductivity.

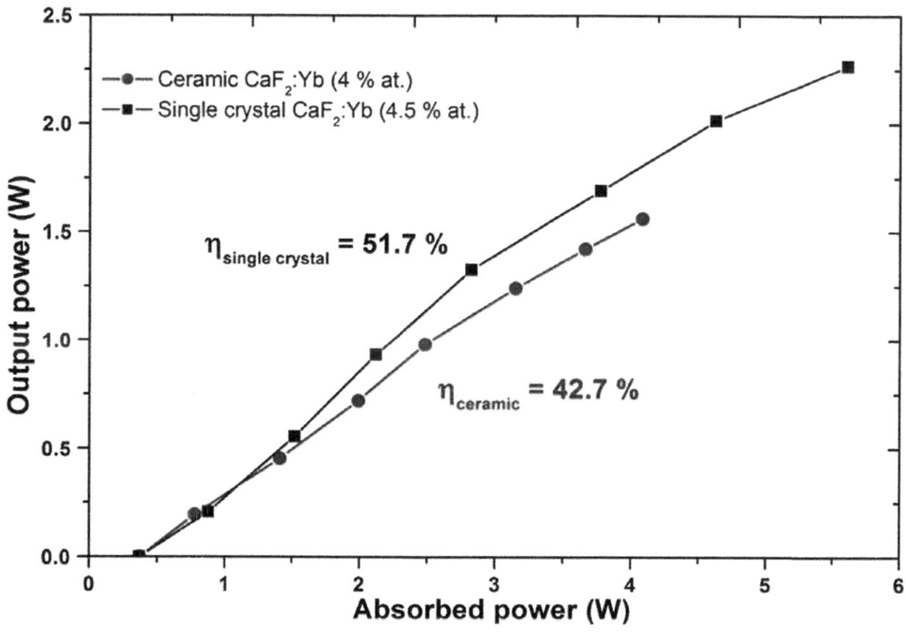

Figure 4.9 Comparison between laser efficiency curves of a 4%-doped CaF_2:Yb ceramic obtained from a green body formed from a polycrystalline nanopowder using a newly strainless packing method, versus real absorbed pump power (ie, considering saturation effects) [1].

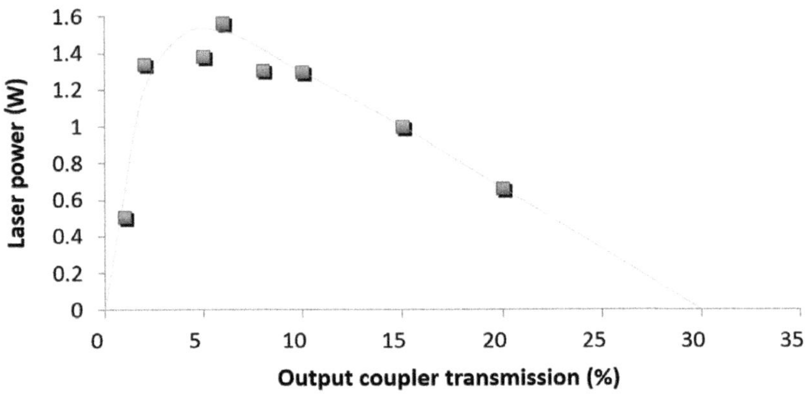

Figure 4.10 Maximum laser power of a 4%-doped CaF_2:Yb ceramic obtained from a green body formed from a polycrystalline nanopowder using a newly strainless packing method, versus output coupler values [1].

4. Conclusion—Future for Fluoride Transparent Ceramics

4.1 Keys for Transparency

Obtaining transparent ceramics with an optical quality close to that of the single crystal of the same composition is a challenge. The ceramic must be the most perfect possible. The doping rate of the grains must be as homogeneous as possible. The grain boundaries must be perfect without any precipitated phases. The size and the number of the residual pores must be as small as possible. The first criterion, ytterbium segregation, is the less harmful than the second criterion, oxygenized phase, as explained earlier. This assertion might suggest that ceramics produced from nanopowders synthesized in aqueous medium are rather unsuitable for applications requiring a high optical quality as laser applications. Indeed, water can react with the fluoride at high temperature to produce oxyfluorides, hydroxides, or oxides particularly at the surface of the particles at the origin of the presence of oxygenized secondary phases at the grain boundaries in the final ceramics. Even if the fluorides are prone to be not very sensitive to the pyrohydrolysis as alkaline earth fluorides [34], the reaction with water could be problematic.

It was shown in a previous paragraph that doped CaF_2 ceramics have demonstrated poor laser effect properties when nanopowders synthesized in aqueous medium are used to produce them. It was particularly shown that oxygenized phases at grain boundaries degrade the optical properties of the final ceramics. Nevertheless, Mortier et al. [1,50] have demonstrated very good laser properties with Yb:CaF_2 produced from nanopowders using a wet process. These results imply that the optical properties of the transparent ceramics are excellent and thus, the rate of the oxygenized phases inside the ceramics is very limited. It is thus possible to eliminate water without degrading the fluoride at least for the less sensitive of them to the pyrohydrolysis.

The third criterion, ie, the residual volume of pores is also very important. The effect of the residual volume of pore was particularly studied for Nd:YAG [14,30]. For example, Boulesteix et al. have shown in 2010 that the volume density of pore must be as low as 50×10^{-6} pores.cm^{-3} with size pores in the range of 0.60–0.76 μm (mean grain size: 30 μm) to have any hope achieving optical and particularly laser properties almost identical with those of the related single crystal. The scattering coefficients increasing exponentially with the porosity and the effect are the more sensitive as the wavelength is short. What is true for oxides as Nd:YAG is also true for the fluorides and the volume density and size of the porosity must be as low as possible.

4.2 Which Applications for Fluoride Transparent Ceramics?

Because of the highly ionic bond character, fluorides are remarkable materials for optical applications from IR to UV domain of wavelengths. As early as 1887, Otto Shott, Ernst Abbe, and Carl and Roderich Zeiss have investigated the possibility to use natural single crystals of calcium fluoride for optical lenses, and the Carl Zeiss company proposed CaF_2 lenses for microscope objectives in 1937. However, the impurities in naturally occurring CaF_2 limited their optical applications, and synthetic single crystal growth methods were developed to produce larger and more pure single crystals. Since 1979, synthetically grown single crystal CaF_2 has been used in some high-performance optical devices (camera objectives, telescopes, microscopes) due to its very low dispersion in the visible and IR wavelength ranges. Thus, lenses made from fluorite exhibit less chromatic aberration than those made of ordinary glass and are commercially available in high-end optics such as those offered by the Takashi Inc. Single crystals often require months to grow and very high temperature. Their production is costly and the substitution of single crystal by ceramic can reduce drastically the cost of the devices. This could only be achieved if the transparent ceramics are produced using a low-temperature process from powder and not single crystal. Recent works seem to indicate that it is possible. Because fluorides have wider optical transparency (190 nm–7 μm for SrF_2 and CaF_2, for example) than most oxide materials, they are materials of choice for windows in the UV or IR regions. Melt growth methods used to produce single crystals may not provide the scalability necessary to achieve large piece to put windows in shape easily on the contrary of the processes used to elaborate ceramics from powders. Versus glasses for which synthesis processes allow the same advantages (scalability, facility to put in shape, cost), ceramics present the advantage of best mechanical and thermomechanical properties. Another field of interest for the fluoride is scintillators. For example, since the 1980s, BaF_2, Ce:BaF_2, Eu:Ce:BaF_2, or Ce:CaF_2 as well as CeF_3 are investigated to be used as fast and efficient scintillators for detection of X-rays, gamma rays, and high-energy particles. Today, Eu:CaF_2 single crystals are commercially available for charged particle and soft gamma ray detection. The possibility to produce fluoride transparent ceramics opens thus the way for the manufacturing of complex devices associating pieces with different doping rates, for example, or/and large pieces with an interesting manufacturing cost. The last great field of application for the fluoride transparent ceramics is the laser application, which merits to be developed.

4.3 Fluoride Transparent Ceramics for Laser Application—Advantages and Limits

Fluoride single crystals are used in some of solid-state lasers, and U^{3+}:CaF_2 was the second material used to demonstrate the laser effect in 1960 [55]. Since then, many fluorinated materials were investigated and used as laser gain media, Use of transparent ceramics instead of single crystals allows solving some difficulties inherent in the single crystals, for example, the heterogeneity of the doping rate, the mechanical fragility, and of course, the energy and time needed to obtain the material. The maximum output power of a solid-state laser is given by [33]:

$$P_{\text{out}} = C\left[\frac{12\sigma(1-\vartheta)K}{\alpha E}\right]\left[\frac{w\ell}{t}\right]$$

where P_{out} is the maximum output power, C is a constant describing the efficiency of the laser system, σ is fracture strength, ϑ is Poisson's ratio, K is thermal shock resistance, α is thermal expansion coefficient, E is Young's modulus, w is the slab width, ℓ is the slab length, and t is its thickness. Clearly, the maximum output power of a gain medium is limited by both the size and thermomechanical properties of the gain medium. Importantly, the thermal shock resistance, K, of ceramics is known to be up to three times higher than single crystals [59], and ceramic processing offers substantially increased scalability relative to single crystal growth methods. Therefore, ceramics may allow higher output powers than single crystals through both increased scalability and increased thermomechanical properties. We can also presume that the thermal properties of a ceramic are close to that of the related single crystal. Nevertheless, it is not really true. Indeed, if the thermal conductivity is the same inside the grain and inside the single crystal, the grain boundaries can act as barriers for the propagation of the heat and this problem must be investigated. The solution can be found in the growth of the grains to limit the number of grain boundaries and an adapted shape of the ceramics to optimize the heat transfer. Thus, the research on the fluoride transparent ceramics is still in fact at the beginning. Some way of elaboration must still be investigated and optimized but the process from powder must be favored.

References

[1] P. Aballea, A. Suganuma, F. Druon, J. Hostalrich, P. Georges, P. Gredin, M. Mortier, Laser performance of diode-pumped Yb:CaF_2 optical ceramics synthesized using an energy-efficient process, Optica 2 (4) (2015) 288–291.

[2] M. Ahrens, K. Schuschke, S. Redmer, E. Kemnitz, Transparent ceramics from sol-gel derived elpasolites by cold pressing, Solid State Sci. 9 (2007) 833–837.

[3] M.S. Akchurin, T.T. Basiev, A.A. Demidenko, M.E. Doroschenko, P.P. Fedorov, E.A. Garibin, P.E. Gusev, S.V. Kuznetsov, M.A. Krutov, I.A. Mironov, V.V. Osiko, P.A. Popov, CaF_2:Yb laser ceramics, Opt. Mater. 35 (2013) 444–450.

[4] P. Aubry, A. Bensalah, P. Gredin, G. Patriarche, D. Vivien, M. Mortier, Synthesis and optical characterizations of Yb-doped CaF_2 ceramics, Opt. Mater. 31 (2009) 750–753.

[5] T.T. Basiev, V.V. Voronov, V.A. Konyushkin, S.V. Kuznetsov, S.V. Lavirishchev, V.V. Osiko, P.P. Fedorov, A.B. Ankudinov, M.I. Alymov, Optical lithium fluoride ceramics, Dokl. Phys. 52 (2007) 677–680.

[6] T.T. Basiev, M.E. Doroshenko, V.A. Konyushkin, V.V. Osiko, L.I. Ivanov, S.V. Simakov, Lasing in diode-pumped fluoride nanostructure F_2^-:LiF colour centre ceramics, Quantum Electron. 37 (2007) 898–990.

[7] T.T. Basiev, V.A. Demindenko, K.V. Dykelskii, P.P. Fedorov, E.I. Gorokhova, I.A. Mironov, Y.V. Orlovskii, V.V. Osiko, A.N. Smirnov, Optical fluoride and oxysulfide ceramics: preparation and characterization, in: D. Rosslere (Ed.), Developments in Ceramic Materials Research, Chapter 3, 2007, pp. 53–95.

[8] T.T. Basiev, M.E. Doroshenko, P.P. Fedorov, V.A. Konyushkin, S.V. Kuznetsov, V.V. Osiko, M.S. Akchurin, Efficient laser based on CaF_2-SrF_2-YbF_3 nanoceramics, Opt. Lett. 33 (2008) 521–523.

[9] T.T. Basiev, M.E. Doroshenko, V.A. Konyushkin, V.V. Osiko, P.P. Fedorov, V.A. Demidenko, K.V. Dukel'skii, I.A. Mironov, A.N. Smirnov, Fluoride optical nano-ceramics, Russ. Chem. Bull. 57 (2008) 877–886.

[10] T.T. Basiev, M.E. Doroshenko, V.A. Konyushkin, V.V. Osiko, SrF_2:Nd^{3+} laser fluoride ceramics, Opt. Lett. 35 (2010) 4009–4011.

[11] T.T. Basiev, V.A. Konyushkin, M.E. Doroshenko, G. Huber, F. Reichert, N.O. Hansen, M. Fechner, First ceramic laser in the visible spectral range, Opt. Mater. Express 1 (2011) 1511–1515.

[12] P.J. Bendall, C.R.A. Catlow, J. Corish, P.W.M. Jacobs, Defect aggregation in anion-excess fluorites II. Clusters containing more than two impurity atoms, J. Solid State Chem. 51 (1984) 159–169.

[13] A. Bensalah, M. Mortier, G. Patriarche, P. Gredin, D. Vivien, Synthesis and optical characterizations of undoped and rare-earth-doped CaF_2 nanoparticles, J. Solid State Chem. 179 (2006) 2621–2629.

[14] R. Boulesteix, A. Maitre, J.F. Baumard, Y. Rabinovitch, F. Reynaud, Light scattering by pores in transparent Nd:YAG ceramics for lasers: correlations between microstructure and optical properties, Opt. Express 14 (2010) 14992–15002.

[15] D.A. Buckner, H.C. Hafner, N.J. Keidl, Hot pressing magnesium fluoride, J. Am. Ceram. Soc. 45 (9) (1962) 435–438.

[16] P. Camy, J.L. Doualan, S. Renard, A. Braud, R. Moncorgé, V. Menard, Tm^{3+}:CaF_2 for 1.9 μm laser operation, Opt. Comm. 236 (2004) 395–402.

[17] J.E. Carnall, S.E. Hatch, W.F. Parsons, Optical studies on hot-pressed polycrystalline CaF_2 with clean grain boundaries, Mater. Sci. Res. 3 (1966) 165.

[18] J.E. Carnall, S.E. Hatch, L.S. Ladd, W.F. Parsons (1968) US patent N°3365271A.

[19] J.E. Carnall, S.E. Hatch, W.F. Parsons, R.J. Weagley (1969) US patent N°3453215A.

[20] C.R.A. Catlow, A.V. Chadwick, J. Corish, The defect structure of anion excess CaF_2, J. Solid State Chem. 48 (1983) 65–76.

[21] A.K. Cheetham, B.E.F. Fender, D. Steele, Defect structure of fluorite compounds containing excess anions, Solid State Comm. 8 (3) (1970) 171–173.

[22] J. Corish, C.R.A. Catlow, P.W.M. Jacobs, S.H. Ong, Defect aggregation in anion-excess fluorites. Dopant monomers and dimers, Phys. Rev. B 25 (1982) 6425–6438.

[23] P.P. Fedorov, V.V. Osiko, S.V. Kuznetsov, E.A. Garibin, Fluoride laser nanoceramics, J. Phys. Conf. Ser. 345 (2012) 012017.

[24] P.P. Fedorov, Fluoride laser ceramics, in: B. Denker, E. Shklovsky (Eds.), Handbook on Solid-State Lasers: Materials, Systems, and Applications, Woodhead, Oxford, 2013, pp. 82−109.
[25] P.P. Fedorov, S.V. Kuznetsov, A.N. Smirnov, E.A. Garibin, P.E. Gusev, M.A. Krutov, K.A. Chernenko, V.M. Khanin, Microstructure and scintillation characteristics of BaF_2 ceramics, Inorg. Mater. 50 (2014) 738−744.
[26] C. Greskovich, J.P. Chernoch, Polycrystalline ceramic lasers, J. Appl. Phys. 44 (1973) 4599−4606.
[27] S.E. Hatch, W.F. Parsons, R.J. Weagley, Hot-pressed polycrystalline $CaF_2:Dy^{2+}$ laser, Appl. Phys. Lett. 5 (1964) 153−154.
[28] A. Ikesue, I. Furusato, K. Kamata, Fabrication of polycrystalline transparent YAG ceramics by a solid-state reaction method, J. Am. Ceram. Soc. 78 (1995) 225−228.
[29] A. Ikesue, K. Yoshida, T. Yamamoto, I. Yamaga, Optical scattering centers in polycrystalline Nd:YAG laser, J. Am. Ceram. Soc. 80 (1997) 1517−1522.
[30] A. Ikesue, K. Yoshida, Influence of pore volume on laser performance of Nd:YAG ceramics, J. Mater. Sci. 34 (1999) 1189−1195.
[31] M. Ito, C. Goutaudier, Y. Guyot, K. Lebbou, T. Fukuda, G. Boulon, Crystal growth, Yb^{3+} spectroscopy, concentration quenching analysis and potentiality of laser emission in $Ca_{1-x}Yb_xF_{2+x}$, J. Phys. Condens. Matter 16 (2004) 1501−1521.
[32] T. Kallel, M.A. Hassairi, M. Dammak, A. Lyberis, P. Gredin, M. Mortier, Spectra and energy levels of Yb^{3+} ions in CaF_2 transparent ceramics, J. Alloys Compd. 584 (2014) 261−268.
[33] W. Koechner, Solid State Laser Engineering, fifth ed., Springer, NY, 1999.
[34] S.V. Kuznetsov, V.V. Osiko, E.A. Tkatchenko, P.P. Fedorov, Inorganic nanofluorides and related nanocomposites, Russ. Chem. Rev. 75 (12) (2006) 1065−1082.
[35] C. Labbe, J.L. Doualan, P. Camy, R. Moncorgé, M. Thuau, The 2.8 μm laser properties of Er^{3+} doped CaF_2 crystals, Opt. Comm. 209 (2002) 193−199.
[36] J.P. Laval, B. Frit, Defect structure of anion-excess fluorite-related $Ca_{1-x}Y_xF_{2+x}$ solid solutions, J. Solid State Chem. 49 (2) (1983) 237−246.
[37] W. Li, H. Kou, M. Chen, Y. Shi, X. Feng, Y. Pan, J. Guo, Fabrication and characterization of grain-orientated cerium fluoride ceramics from a slip-casting process in a magnetic field, J. Mater. Sci. 49 (2014) 5030−5034.
[38] W. Li, Y. Sun, H. Kou, M. Chen, Y. Shi, X. Feng, Y. Pan, J. Guo, Preparation of cerium fluoride suspensions for grain-orientation by a strong magnetic field assisted slip-casting process, Ceram. Int. 40 (2014) 10317−10322.
[39] W.Y. Lin, M.H. Hon, S.J. Yang, Effect of grain growth on hot-pressed optical magnesium fluoride ceramics, J. Am. Ceram. Soc. 73 (3) (1988) 136−137.
[40] Z. Liu, B. Mei, J. Song, W. Li, Fabrication and optical characterizations of Yb, Er codoped CaF_2 transparent ceramic, J. Eur. Ceram. Soc. 34 (2014) 4389−4394.
[41] Z. Liu, B. Mei, J. Song, W. Li, Optical characterizations of hot-pressed calcium fluoride transparent ceramic, J. Am. Ceram. Soc. 97 (8) (2014) 2506−2510.
[42] G. Lu, B. Mei, J. Song, W. Li, R. Xing, Fabrication and properties of highly transparent Nd-doped ceramics, Mater. Lett. 115 (2014) 162−164.
[43] A. Lucca, G. Debourg, M. Jacquemet, F. Druon, F. Balembois, P. Georges, P. Camy, J.L. Doualan, R. Moncorgé, High power diode pumped $Yb^{3+}:CaF_2$ femtosecond laser, Opt. Lett. 29 (2004) 2767−2769.
[44] A.A. Lyapin, P.P. Fedorov, E.A. Garabin, A.V. Malov, V.V. Osiko, P.A. Ryabochkina, S.N. Ushakov, Spectroscopic, luminescent and laser properties of nanostructured CaF_2:Tm materials, Opt. Mater. 35 (2013) 1859−1864.

[45] A. Lyberis, G. Patriarche, P. Gredin, D. Vivien, M. Mortier, Origin of light scattering in ytterbium doped calcium fluoride transparent ceramic for high power lasers, J. Eur. Ceram. Soc. 31 (2011) 1619–1630.
[46] A. Lyberis, A.J. Stevenson, A. Suganuma, S. Ricaud, F. Druon, F. Herbst, D. Vivien, P. Gredin, M. Mortier, Effect of Yb^{3+} concentration on optical properties of Yb:CaF_2 transparent ceramics, Opt. Mater. 34 (2012) 965–968.
[47] S.F. Matar, J.M. Reau, P. Hagenmuller, C.R.A. Catlow, The cubo-octahedral cluster in the fluorite-type lattice: a theoretical approach, 123 (1984) 114–123.
[48] M. Mortier, Y.D. Huang, F. Auzel, Crystal field analysis of Er^{3+}-doped glasses: germanate, silicate and ZBLAN, J. Alloys Compd. 300–301 (2000) 407–413.
[49] M. Mortier, A. Lyberis, A. Suganuma, A.J. Stevenson, D. Vivien, P. Gredin, G. Patriarche, Last advances in Yb^{3+} doped CaF_2 ceramics synthesis, in: Art N°7912R Conference on Solid State Lasers XX, January 23–27, San Jose, CA, Proceeding of SPIE, vol. 7912, 2011, 1R1–1R9.
[50] M. Mortier, P. Gredin, P. Aballea, A. Suganuma, Transparent metal fluoride ceramics (2014) Eur. patent 14.305.495.5
[51] V. Petit, J.L. Doualan, P. Camy, R. Moncorgé, V. Menard, CW and tunable laser operation of Yb^{3+} doped CaF_2, Appl. Phys. B 78 (2004) 681–684.
[52] V. Petit, P. Camy, J.L. Doualan, R. Moncorgé, Refined analysis of the luminescent centers in the Yb^{3+}:CaF_2 laser crystal, J. Lumin. 122–123 (2007) 5–7.
[53] V. Petit, J.L. Doualan, P. Camy, X. Portier, R. Moncorgé, Spectroscopy of Yb^{3+}:CaF_2: from isolated centers to clusters, Phys. Rev. B 78 (2008) 1–12.
[54] H.M. Silva, L.C. Courrol, L. Gomes, R. Bertram, S.L. Baldochi, I.M. Ranieri, Synthesis and characterization of KY_3F_{10}:Yb:Nd:Tm crystals, J. Phys. Conf. Ser. 249 (2010) 012–047.
[55] P.P. Sorokin, M.J. Stevenson, Stimulated infrared emission from trivalent uranium, Phys. Rev. Lett. 12 (5) (1960) 557–559.
[56] N.I. Sorokin, A.N. Smirnov, P.P. Fedorov, B.P. Sobolev, Superionic fluoride ceramics RF_3 and $R_{0.95}Sr_{0.05}F_{2.95}$ (R = La, Ce, Pr, Nd) prepared by hot pressing, Russ. J. Ectrochem. 45 (5) (2009) 606–608.
[57] C.F. Swinehart, H. Packer (1978) US patent N°4089937A.
[58] K. Voronko Yu, V.V. Osiko, I.A. Shcherbakov, Optical centers and interaction of Yb^{3+} ions in cubic fluorite crystals, Sov. Phys. JETP 29 (1969) 86.
[59] K. Ueda, J.F. Bisson, H. Yagi, K. Takaichi, A. Shirakawa, T. Yanagitani, A.A. Kaminski, New Trends Laser Phys. 15 (2005) 927–938.
[60] W.B. Zhou, F.F. Cai, G.L. Zhi, B.C. Mei, Fabrication of highly transparent Er:CaF_2 ceramics by hot-pressing, Mater. Sci. Poland 32 (3) (2014) 358–363.

Synthesis and Characterization of F-Doped Zinc Oxides, Al-Doped Zinc Hydroxyfluoride, and Oxide–Fluoride Composites for Transparent Visible/Infrared Absorbers

5

M. Gaudon, H. Serier-Brault, I. Trenque, A. Demourgues
ICMCB-CNRS, University of Bordeaux, Pessac, France

Chapter Outline

1. Introduction

Transparent conductive oxides (TCOs) are used as transparent electrodes (optoelectronic devices), or as thermal insulator films in smart windows (low-emissive windows) [1–9]. One of the most exciting characteristics of these oxides consists of visible transmission/infrared absorption [10–14], one very useful property leading to improve the heat insulation. Windows glass manufacturers such as Saint-Gobain largely develop these oxides as additional thin films on their commercial glasses. This transparent visible/infrared absorption property is directly linked to one physical parameter: the plasma frequency, ωp, comparable to a "cutoff frequency" with regard to the electronic transport properties and therefore refractive index of the material

Photonic and Electronic Properties of Fluoride Materials. http://dx.doi.org/10.1016/B978-0-12-801639-8.00005-2

[15,16]. In the conduction band, free electron gas can react in two different ways with the electromagnetic wave: (1) if the incident wave is very energetic, ie, with a frequency higher than ωp, electrons "do not have enough time to respond to the perturbation," and thus the wave goes through the material without any interaction; the material is therefore transparent, (2) on the contrary, when the incident wave is less energetic, ie, lower than plasma frequency, the radiation is absorbed. The plasma frequency is related to the carrier concentration but also to their mobility: an adequate TCO should be a compromise between carrier concentration and mobility. Both phenomena (transparency or absorption) involve a peculiar variation of optical indexes with plasma frequency, which is characterized by an abrupt decrease of the index n in combination with an abrupt increase of the absorption coefficient k [16,17].

Especially, Ga-doped ZnO compounds or Al-doped ZnO compounds can be developed as TCOs. These doped zinc oxides are without any doubt the most studied TCO nowadays due especially to its lowest price in comparison with the famous concurrent: doped-In_2O_3 oxides. Indeed, ZnO is a large bandgap semiconductor (3.3 eV at 300 K) [18] with native n-type defects which accepts efficient doping as M^{3+} metallic elements [19–22], located in Zn^{2+} sites and even halogens as F^- in substitution of oxygen in the anion subnetwork as single dopant [23–27].

Hence, doping such oxides with fluorine ions is an adequate manner to tune the charge carrier concentration and so the optical properties, ie, via the plasma frequency. One another effect occurring from the fluorine doping is the decrease of the refractive index in the visible range due to very high electronegativity of the fluorine ion. The more ionic the character of the cation–anion chemical bonding, the lower the refractive index [16,28]. Thus, the visible transparency is easier to achieve for thin films than for pure oxides. Nevertheless, the index refraction variation with fluorine doping remains weak because the low F rate does not impact many Zn^{2+} local environments and consequently the global electronic polarizability. A more ambitious objective could be to transfer the technology developed on TCOs on fluorides, hydroxyl-fluorides, or oxyfluorides materials. It could be estimated by DFT (density functional theory) calculations that Al-doped ZnF_2 should develop same infrared absorption characteristic than Al-doped ZnO but with higher transparency in the visible. None literature report on M^{3+}-doped ZnF_2 was found but numerous studies were performed on one quite analogous oxide: M^{3+}-doped CdF_2 [29–35].

Hence, this double impact of fluorine ions on visible transparent/infrared absorbent material is ambiguous and depends indeed on fluorine rate: (1) the fluorine ions in oxides as dopant contributes to increase the n charge carriers concentration and so the infrared absorption and (2) the use of large content of fluorine ions (as in fluorides or hydroxyl-fluorides) induces a decrease of the n refraction index and so an increase of the visible transparency. Moreover, while fluorine ions are used as dopants, the introduction of fluorine ions gets a weak influence on the compound bandgap. Indeed, for fluorine dopants, as for the aliovalent cationic dopants, the effect on the bandgap is mainly linked to the increase of the carrier charges: Burstein–Moss effect [36,37]. The Burstein–Moss effect is the phenomenon of which the apparent bandgap of a semiconductor is increased as the absorption edge is pushed to higher energies as a result of all states close to the conduction band being populated. This is observed for a degenerate electron distribution such as that found in some degenerate semiconductors and is known as a Burstein–Moss shift (apparent blue-shift). In nominally doped semiconductors, the Fermi level lies between the conduction and valence bands. As the doping concentration is increased, electrons populate states within the conduction band which pushes the Fermi level higher in energy. On the contrary, fluorides or oxifluorides are known to exhibit higher bandgap than the respective oxides. For illustration, DFT calculations and experimental characterizations have already been used to explore correlation between bandgap and anionic subnetwork nature in the TiO-F oxifluoride series [38]. The bandgaps are found to increase with the halide concentration and can be interpreted based on the less important covalence of the cation–fluoride bonds in comparison to the cation–oxygen bonds.

In this chapter, two various stabilizations of F^- ions in zinc oxides and hydroxyfluorides will be investigated in order to tune the optoelectronic properties. Firstly, the literature works about the synthesis and performance of F-doped oxides used as transparent visible/infrared absorbent oxides will be summed up by insisting on F-doped zinc oxides (Section 2). Then, our investigations will be presented focusing on the strategy to obtain infrared absorption properties in zinc hydroxyfluorides and fluorides, using Al^{3+} or Ga^{3+} ions as extrinsic n-type dopants (Section 3). Finally, original investigations have been devoted to zinc oxide–zinc fluoride (or magnesium fluoride) composites, and in particular, core–shell materials with oxide cores and fluoride shells exhibit some synergetic properties and open the possibility to develop a new generation of visible transparent/infrared absorbent material (Section 4). In other words, this chapter deals with synthesis and characterization of inorganic compounds containing mainly Zn^{2+} cations associated to F^- anions and used as transparent visible/infrared absorbent materials.

2. F-Doped Transparent Conductive Oxides: A State of the Art

The TCOs nowadays developed being mainly n-type semiconductors, the substitution of an n-type donor as F^- ions for O^{2-} ions is largely satisfying. Nevertheless, among the three TCOs families (ZnO, In_2O_3, and SnO_2 for which crystalline unit cell is reported in the Fig. 5.1), the tin oxide is the only one for which the fluorine doping has been extensively studied.

Indeed, on the contrary to F-doped zinc oxide or F-doped In_2O_3, fluorine-doped tin oxide (FTO) is a nontoxic, mechanically robust, and temperature-stable TCO composed of earth-abundant elements. Zinc oxide exhibits a low chemical stability and should be protected by additive layers in smart windows application whereas indium is a rare and expensive element [39,40]. The elaboration of FTO films are largely described in the literature, the synthesis varying from chemical processes starting from solution to physical processes with solid targets (magnetron and various sputtering processes). Among these latter techniques, direct current magnetron sputtering (DCMS) of FTO could be qualified nowadays of an industrial relevant method for

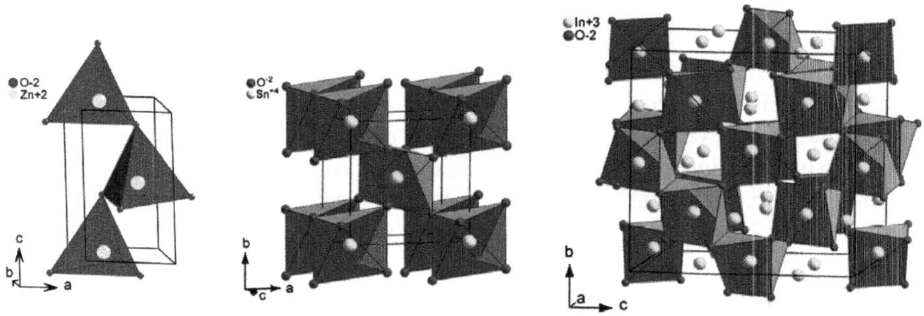

Figure 5.1 Frameworks of the three most studied transparent conductive oxides: ZnO, SnO_2, and In_2O_3 oxides from left to right.

large-area deposition that offers low deposition temperatures and the possibility to combine FTO with other functional layers in the same deposition run. Several approaches for fluorine doping of tin oxide in magnetron sputtering have been investigated: radio-frequency magnetron sputtering (RFMS) from a ceramic SnO_2/SnF_4 target [41,42], reactive RFMS in $Ar/O_2/CF_4$ atmospheres [43], and pulsed DCMS from a metal target [44]. Obviously, besides the physical deposition routes, the chemical solution processes constitute less mature but more promising routes. Indeed, among all the film deposition techniques, chemical solution methods may be the most convenient techniques because of their simplicity, low cost, easy to add other chemical elements, and promising for high rate and mass production capability with uniform large-area coatings in industrial applications [45–47]. Furthermore, our point of view is that only chemical methods are really suitable to explore various film morphologies: crystallite morphology, structure, and surface area of the films. It has been reported that the performances of SnO_2 nanomaterials are greatly affected by their micromorphology [48]. Therefore, advanced nanostructured materials with various dimensionalities obtained by chemical solution method would offer good opportunities to explore new physical and chemical applications including electrodes, gas sensors, solar cells, and photocatalysis. Spray ultrasonic processes [48], ionic solution spray pyrolysis (SP) [49–51], or hydrothermal routes [52,53] are the original techniques developed recently to achieve F-doped SnO_2 films with good optical properties. Finally, it has been added that obviously the thermal stability of F-doped SnO_2 films is decreased in comparison with pure oxides, especially under air atmosphere. Nevertheless, this aspect has been already studied, and it was shown that these compounds can be stable at least at 300°C, allowing hence their introduction in a large number of displays [54].

The F-doped ZnO oxides (FZOs) are less developed, but recent publications are dedicated to this subject. The same competition between physical and chemical processes can be underlined. Fluorine-doped ZnO thin films have been deposited by methods such as SP [55], chemical vapor deposition [56], pulsed laser deposition [57], and magnetron sputtering [58]. Among these techniques, magnetron sputtering has attracted great interest because of the inherent ease with which the deposition conditions can be controlled. Few studies about the effects of ZnF_2 content in sputtering target and hydrogen addition to the sputtering ambient on optoelectronic properties of FZO thin films have been reported [59–64]. Ku et al. have reported that the FZO thin films deposited with 1.3-wt% ZnF_2 in a ZnO target achieved the best optoelectronic properties [59]. Treharne and Durose described that the lowest resistivity was achieved for the FZO thin films sputtered under 5% H_2 and 4% CHF_3 in Ar ambient [61]. Indeed, they attempted to achieve F-doped ZnO by using trifluoromethane (CHF_3), with constant partial pressure of H_2 of 5% and partial pressures of CHF_3 varying in the range of 0–7%. This last study is a school case showing the complexity of the processes used to obtain FZO films and the difficulty to optimize such processes. In a rather easiest way, the FZO thin films could also be deposited on glass slide substrates using direct current reactive magnetron sputter at room temperature, with the substrate holder at the floating potential using an Ar/O_2 (/F_2) gas mixture and a pure metallic Zn as target [64]. Anandhi et al. [65] used $SnCl_2 \cdot 2H_2O$ (0.1 M) and $Zn(CH_3COO)_2 \cdot 2H_2O$ (0.2 M) as host precursors and NH_4F as the dopant precursor for the deposition of the F-doped SnO_2 layer on the FZO layer to get the bilayer thin films by using a simplified SP technique. Besides the sputtering processes, rare are the publications devoted to solution chemistry processes. It can be cited the work of Rozati et al. [66] in which the authors report a study relating to the effect of aging of the starting solution used to fabricate films by SP technique, on the structural, electrical, optical, and morphological properties of ZnO:F thin films. They used 0.4 M solution containing zinc acetate dehydrated as the host solution, which was dissolved in a mixture of double-distilled water:methanol (3:7 volume proportion), and acetic acid. Ammonium fluoride as the doped starting solution has been added, with a fixed [F]/[Zn] molar ratio of 2 at.%. As the aging time increases, an improvement of the optical transmittance was obtained for films using 6 days aged solutions. Beyond this aging period, a monotonous decrease of the transmittance was observed. The best transmittance observed (55%) is unfortunately insufficient and seems to disqualify solution processes in comparison to sputtering ones.

In conclusion to the literature works, one should have to note that the best electrical properties of FZO thin films are still less efficient than that of F-doped SnO_2 or M^{3+}-doped ZnO thin films. Furthermore, the elaboration processes are sometimes very complex and only sputtering processes are of mature technology. Moreover, as it is shown in the recent publication of Wang et al. [67] the best results may be due to desorption of negatively charged oxygen species and defect passivation as a result of hydrogen incorporation. Thus, various interpretations and controversies on the origin of the conductivity in such FZO thin films are still in debate. To conclude, F-doping as n-type extrinsic defect in ZnO large bandgap semiconductor is more difficult to control that M^{3+}-doping. The increase of the refractive index with F-doping is rather limited because of the very low concentration of fluorine ions introduced in the TCOs. Unfortunately, the decrease of the refractive index seems insufficient to get highly transparent films by alternative solution processes (chemical processes starting from aqueous or organic solutions). A most adequate and ambitious study will be to substitute M^{3+} ions for Zn^{2+} ions in low refractive index matrices such as $Zn(OH,F)_2$ or ZnF_2 materials in which the fluorine ions are in high concentration, so allow achieving high transparency.

3. Al^{3+}-Doped $Zn(OH,F)_2$ Compounds/Al^{3+}-Doped ZnF_2 Compounds

3.1 Preliminary Consideration on ZnF_2 Compounds

$Zn(OH,F)_2$ and ZnF_2 are two very ionic-bond compounds in comparison with ZnO. Their optical properties and electronic properties will be impacted by the strong ionic character of the Zn–F bonds. ZnF_2 crystallizes with a rutile structure as the SnO_2. Thus, this compound exhibits suitable structural features to create electrical conductivity. Nevertheless, the question about the concentration and mobility of charge carrier in such inorganic fluoride is opened. Obviously, the decrease of the electronic polarizability, which is required to decrease the refractive index, should drastically also affect the electronic mobility. In order to calculate dispersion curves ($E = f(k)$) and density of states (DOS) versus energy in doped ZnF_2 for transparent conductive properties, DFT calculations (Density Functional Theory) were made by A. Villesuzanne in our laboratory with WIEN2k code. The dispersion curves ($E = f(k)$) and the DOS versus energy of ZnO and ZnF_2 are represented on Figs. 5.2 and 5.3, respectively.

The predicted bandgap for ZnF_2 is about 4 eV. But the comparison with ZnO shows that the DFT calculations lead to underestimated bandgaps, and an equivalent scissor operator can be applied. Hence, knowing the experimental and theoretical bandgap of the ZnO compound, it could be estimated that the real bandgap of ZnF_2 should be about 7–8 eV. This value is in good correlation with literature value [68]. From the $E = f(k)$ dispersion curves, it is possible to calculate and compare the energy dispersion of the conductive band in ZnO and ZnF_2, respectively. The dispersion width is 7 and 6 eV, respectively, for the oxide and the fluoride. Hence, even if the dispersion width of the 4s band is slightly weaker for the fluoride than for the oxide, the effect is not enough significant to drastically decrease the electronic properties and so the capacity of the fluoride to be an infrared absorber.

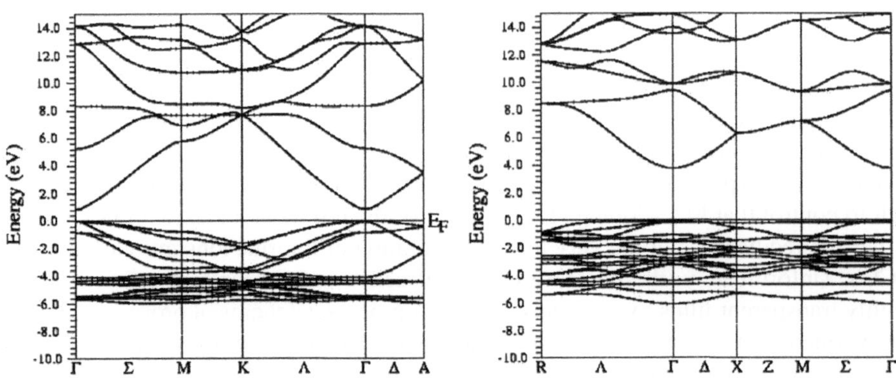

Figure 5.2 Dispersion curves ($E = f(k)$)/band diagrams calculated from density functional theory and for the first Brillouin zone on ZnO (left side) and ZnF_2 (right side) compounds.

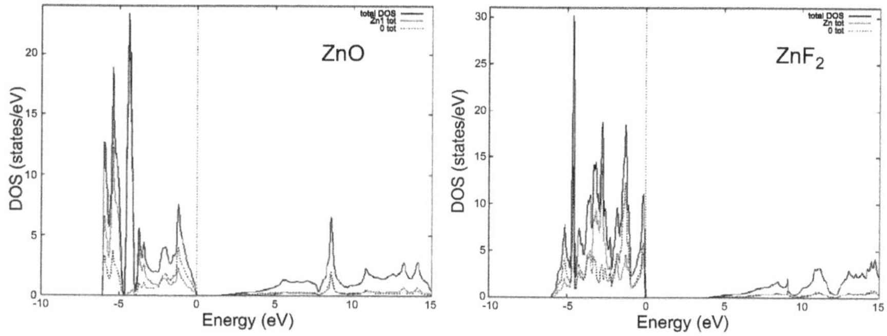

Figure 5.3 Density of states (DOS) versus energy calculated from density functional theory on ZnO and ZnF_2 compounds.

3.2 $Zn(OH,F)_2$ Compounds

Few studies have been performed on basic halides with MOHX chemical formula, where M is a divalent metal (zinc or cadmium) and X is a halogen either chlorine or fluorine atoms [69–73]. These investigations have dealt with the preparation, the determination of the composition, and the crystal structure of MOHX compounds. Some of these compounds were analyzed by ^{19}F NMR or thermal analysis in order to compare their properties with those of homologous $M(OH)_2$ hydroxide and to deduce the impact of the halogen on the chemical reactivity as well as the thermal stability [70,71].

Zinc hydroxyfluoride (ZnOHF) is generally synthesized by hydrolysis and condensation of zinc fluoride tetrahydrate in water followed by evaporation at 100°C to dryness. ZnOHF could also be obtained by thermal degradation under air of zinc fluoride (ZnF_2) [72] or, more recently, could be observed on the surface of ZnO films due to a treatment with a NH_4F aqueous solution [73]. Moreover most of the authors considered that F^- and OH^- groups are in equimolar proportion in ZnOHF, but no accurate chemical analyses were carried out. Our study shows that it is possible to directly obtain a $Zn(OH)_{2-x}F_x$ compound by coprecipitation of a zinc salt into aqueous HF at various pH, and that the amount of fluorine in the compound is tunable by experimental conditions. $Zn(OH)_{2-x}F_x$ hydroxyfluorides were herein prepared from zinc chloride ($ZnCl_2$ 98%, Aldrich), hydrofluoric acid (HF 40%, Panreac), and ammonium hydroxide (NH_4OH 98%, Aldrich). A zinc chloride solution was added drop wise into a continuously stirred fluorinated medium to form a white precipitate. The precipitation was carried out at constant pH by simultaneous addition of a NH_4OH solution (commercialized solution diluted by one-third). After complete addition of the $ZnCl_2$ solution (about 20 mL), the volume is adjusted in order to have a final constant fluorine concentration (about 1.2 mol/L). The mixture was still stirred for another 30 min to favor dissolution/growth mechanisms (Ostwald ripening) and the formation of the "thermodynamically stable" phase. The white powder was then filtered, washed several times with distilled water, and dried at 100°C in an oven during an hour. In the precipitation process, two parameters could be adjusted such as the working pH during the addition of zinc solution and the F/Zn molar precursor ratio. With this process, the formation of a white precipitate occurred for $5 \leq pH \leq 7$ but when the ripening time is increased, a precipitate appears also at pH equal to 4 and 8. For the following, two samples precipitated at pH 5 and 7 are compared.

An electron probe microanalysis (EPMA) using a CAMECA SX 630 was first performed to determine the amount of fluorine and zinc, the hydroxyl content being deduced from the electroneutrality equation. The experiments were carried out with 20-kV accelerating voltage and 10-nA probe current (LaB_6). Each sample was analyzed at five different spots in order to determine an average amount and a standard deviation, the analysis depth and the area of each spot being respectively equal to 1 and 225 μm [2] (15 μm × 15 μm) in order to avoid surface phenomenon. The calibration was performed with a NaF standard for fluorine and with metallic Zn for zinc. The fluorine amount was also determined by a fluoride ion-selective electrode calibrated with different diluted solutions of standard NaF (NaF 1 mg/1 mL, Aldrich). The powders were dissolved in an HCl aqueous solution, and the pH was then adjusted to 5–6 by addition of NH_4OH. The addition of a TISAB III solution enabled to adjust the ionic strength and to favor the ion diffusion during the measurement (support electrolyte). For each sample, the fluorine amount was deduced from nine measurements (with three standard solutions containing different fluorine concentrations and three TISAB III concentrations) in order to accurately evaluate the average value with the standard deviation.

According to the results of the chemical analyses, the fluorine amount is always lower than 1, the F/OH ratio varying between 0.67(3) and 0.83(1) for the EPMA and between 0.80(2) and 0.91(2) for the fluoride ion-selective electrode (for the samples precipitated at pH = 7 and 5, respectively). It should be noted that the pH has actually a relevant impact on the fluorine content: the lower the pH, the higher the final fluorine amount. The F/OH ratio in the final compound is directly linked to the F/OH molar ratio in the solution, which remains almost constant when the pH is fixed whatever the F/Zn molar ratio. Consequently the pH appears the most significant parameter to tune the chemical composition of "ZnOHF" compound. While the pH decreases from 7 to 5, the fluorine content increases relatively in comparison with OH^- groups.

The impact of the fluorine amount on the lattice parameters and crystalline structures was fully investigated by performing a pattern matching process using the Le Bail method and then Rietveld refinements. This work was published in a paper fully devoted to the structural characterization of the as-synthesized $Zn(OH,F)_2$ compounds [74]. It can be remained that from a structural point of view, ZnOHF crystallizes with an orthorhombic symmetry (SG: Pna21), derived from the diaspore α−AlOOH (SG: Pnma), in which all atomic positions correspond to a general 4a position. In this network exhibiting four atomic positions per unit cell, zinc cations are located in a distorted octahedral site, whereas anions are threefold-coordinated to Zn^{2+} ions. The structure provides only one 4a cationic site and two 4a anionic sites, corresponding to sites 1 and 2 (see Fig. 5.4). In the first one, anions are located at the intersection between two edge-shared octahedra linked and a third octahedron corresponding to a corner. In the second site, anions are placed at the boundary of three

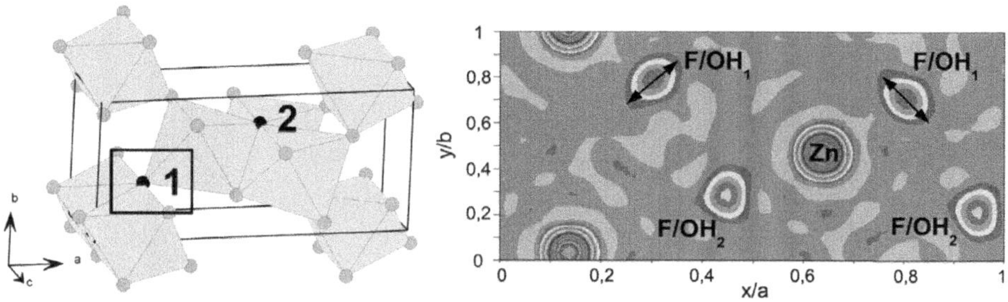

Figure 5.4 Representation of ZnOHF unit cell (SG: Pna21). All anions are represented by spheres with an example of sites 1 and 2 (left side). Fourier map from observed powder X-ray diffraction data (right side).
Figure rebuilt from reference H. Serier, M. Gaudon, A. Demourgues, A. Tressaud, J. Solid State Chem. 180 (12) (2007) 3485. Copyright (2007) Elsevier Ltd.

edge-shared octahedra. The latter exhibits an anion–cation average bond distance longer than for the first site because of stronger Zn–Zn interactions at the origin of a competitive bond.

Le Bail-type refinements were made using anisotropic peak profile function since as shown in SEM the particles exhibit rod or platelet shapes (Fig. 5.5). SEM analyses have been carried out to corroborate the proposed hypotheses concerning the particle shape. In a first step, spherical agglomerates constituted of anisotropic objects getting a particular ordering were observed on SEM pictures, as shown in Fig. 5.5A. The formation of spherical agglomerates can be explained by the easiness for particles to reorganize in a liquid medium, via van der Waals forces, thus minimizing the total surface energy. Once broken by ultrasonics, the agglomerates reveal the presence of three different particle shapes, represented in three insets of Fig. 5.4B. The powders seem to be composed by large isotropic three-dimensional (3D) objects (frame 1), platelet-like particles (frame 2), and needle-like particles (frame 3) even if the former shapes cannot be differentiated by this 2D-technique. This heterogeneity of morphologies allows explaining the difficulty to refine an X-ray diffraction (XRD) diagram which corresponds to the result of an average analysis.

Our main observation from structural investigations has led to consider a splitting of the site 1 into two new positions: one corresponding to fluorine atoms and the other one to hydroxyl atoms. Nevertheless two created positions, indexed F1-1 and F1-2 (Fig. 5.5), were refined from XRD data by considering only the fluorine and zinc atomic form factors. One can notice that the Fourier transform, also provided in Fig. 5.5, corresponding to the observed XRD data shows already that the anionic site 1 exhibits a slight anisotropy represented by a double arrow. Moreover one should have to point out that the Debye Waller value of the site 1 remains high (1.5–2.1) due to an OH/F disorder whereas the one of the site 2 is close to 1 Å^2 (0.8–1.3). Therefore it was consistent to consider a splitting of the former site. Since two new anionic positions were created, two sets of zinc–anion bond distances were deduced for the site 1 with one of the two sets constituted of shorter bonds than the other one, the fluorine atoms corresponding to the smallest distances. The Rietveld refinements were also

Figure 5.5 SEM pictures of sample $Zn(OH)_{2-x}F_x$ powder without preliminary preparation (A), and powder dispersed in ethanol after ultrasounds (B). The frames noted 1, 2 and 3 show, respectively, large particles with isotropic shapes, platelet shapes, and rod shapes.
Figure reproduced from reference H. Serier, M. Gaudon, A. Demourgues, A. Tressaud, J. Solid State Chem. 180 (12) (2007) 3485. Copyright (2007) Elsevier Ltd.

performed considering different amounts of fluorine ions in site 1, ie, various occupancies for each newly created position. Because this occupancy directly impacts the two new atomic positions, this evolution can be discussed considering the whole valence of two anions in their position. The fluorine amount deduced from the XRD refinements is 62% of fluorine and 38% of hydroxyl group for F1-1 and F1-2 atomic positions, respectively (pH = 5), and 52% of fluorine and 48% of hydroxyl groups corresponding to both the previous atomic positions for the other compound prepared at pH = 7. This last titration from X-ray pattern refinement is in good agreement with the previous chemical titrations.

Investigation by ^{19}F MAS-NMR spectroscopy was finally carried out to corroborate the previous results from XRD and chemical analyses. The ^{19}F MAS-NMR spectrum is reported in Fig. 5.6.

The number and the width of the peaks necessary to refine the spectrum clearly show the occurrence of various fluorine ion environments. The two main peaks at −145 and −175 ppm indicate two types of environment associated to two sets of cation−anion bond lengths. Moreover, the spectrum does not show the presence of ZnF_2 with a chemical shift at −205 ppm. The strong variation to high values of the chemical shifts is without any doubt the sign of the presence of hydroxyl groups in the near neighboring of the fluorine ions, as already reported for aluminum or magnesium hydroxyfluorides [75−77]. Thus the two distinctive peaks with various intensities are linked to inequivalent fluorine distribution inside the two crystallographic sites.

Hence, new zinc hydroxyfluoride $Zn(OH)_{2-x}F_x$, synthesized by coprecipitation, can contain varying amounts of fluorine ions depending on experimental conditions because of the high flexibility of the framework. Firstly, combining the chemical analyses performed with an electronic microprobe and a fluorine ion-selective electrode, the fluorine content was determined. The fluorine content is always lower than 1 and can be tuned by varying the pH from 5 to 7. At pH 5, a compound whose chemical formula is $Zn(OH)_{1.17}F_{0.83}$ is obtained, whereas at pH 7, the chemical composition becomes $Zn(OH)_{1.33}F_{0.67}$. The fluorine ions are distributed in two crystallographic sites, as shown by XRD analysis combined with NMR spectroscopy.

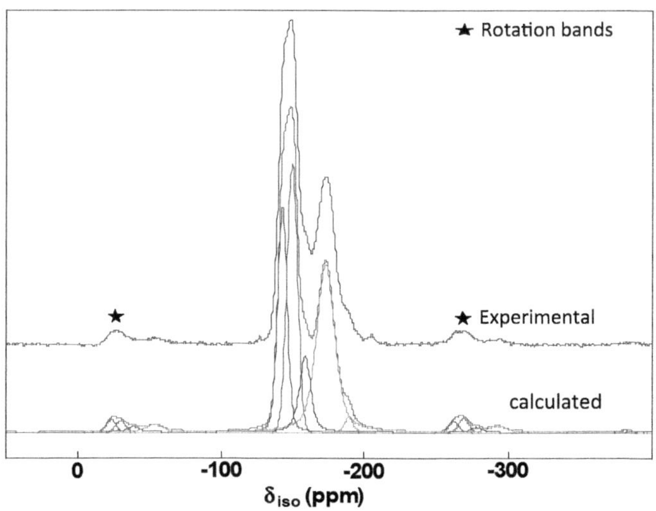

Figure 5.6 ^{19}F MAS-NMR spectra (experimental and simulated) of zinc hydroxyfluoride obtained by precipitation at pH = 7.
Collaboration with Ch. Legein, LOF Le Mans.

3.3 Al^{3+}-Doped $Zn(OH,F)_2$ Compounds

The synthesis protocol was adapted to the introduction of aluminum transition metal in the zinc hydroxyfluorides. In comparison with the experimental process described previously, 6-hydrated aluminum nitrate ($Al(NO_3)_3$,$6H_2O$ 98%, Aldrich) was introduced in stoichiometric proportion in the starting solution in order to achieve a substitution rate varying between 0 and 20 mol%. The introduced concentration was analyzed and compared with the target by inductively coupled plasma (ICP) spectrometer measurements. Whatever the concentration target, the experimental concentration is always about two times smaller than the target.

XRD diagrams are represented in Fig. 5.7. The XRD analyses of the obtained samples have shown that for aluminum content varying between 0 and 5 mol% (experimental content), the ZnOHF structure is preserved. For higher aluminum content, some impurities are observed corresponding to Al-rich phases. Moreover, it can be observed as the aluminum content increases, the crystallite size decreases, ie, an enlargement of the FWHM (full width at half maximum) of the diffraction peaks is so clearly visible. This can be explained by the decrease of the reticular energy of the compound, with consequently a germination critical radius collapse. On the cell parameters evolution versus aluminum concentration, no significant variation could be underlined.

The aim of the study was to try to identify infrared absorption in such low refractive index compound. The diffuse reflectance spectra of the as-synthesized zinc hydroxylfluoride powder with or without Al-doping are reported for comparison in Fig. 5.8.

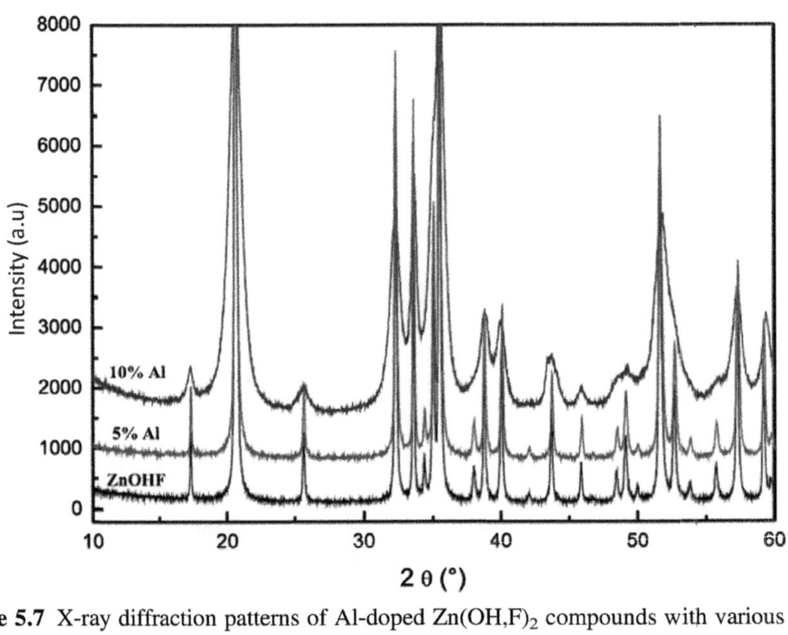

Figure 5.7 X-ray diffraction patterns of Al-doped $Zn(OH,F)_2$ compounds with various Al content.

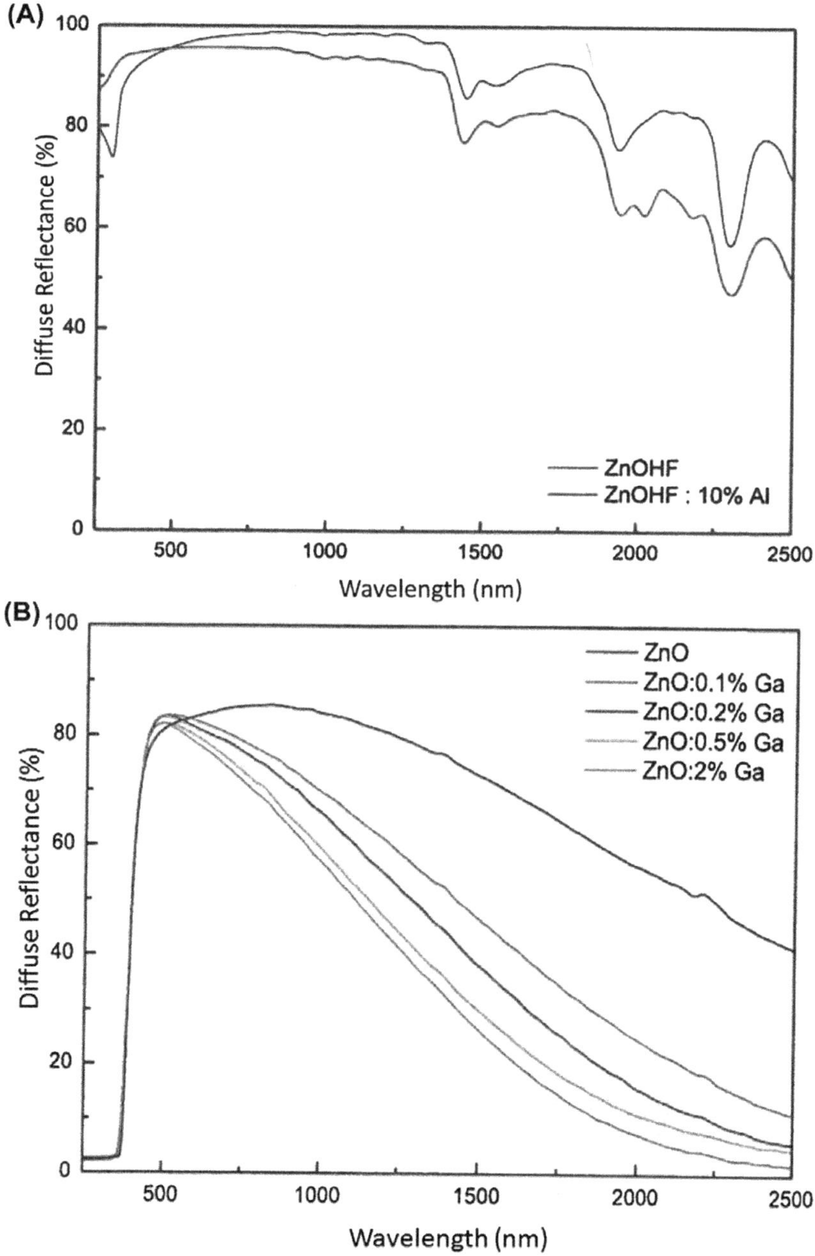

Figure 5.8 Diffuse reflectance spectra of Al-doped $Zn(OH,F)_2$ compounds (A) and comparison with various Ga-doped ZnO compounds (B).

The reflectance spectra show that the Al-doped Zn(OH,F) compound (Fig. 5.8A) exhibits a more important infrared absorption than the undoped compound prepared in the same conditions. It can also be distinguished on the spectra that several Gaussian and narrow absorption bands are associated to OH^- group occurrence. The result is quite positive in the sense the aluminum introduction enhances the charge carrier concentrations. Nevertheless, the near-infrared absorption remains very low in comparison with Al- or Ga-doped ZnO compounds that we have prepared by Pechini routes [78–80], with diffuse reflectance spectra also reported in Fig. 5.8B.

3.4 Al^{3+}-Doped ZnF_2 Compounds

ZnF_2 fluoride crystallizes with the rutile-type network in the P42/mnm space group (136). The unit cell is represented in Fig. 5.9.

Pure ZnF_2 compound and the Al-doped ZnF_2 compounds, with aluminum rate up to 5 mol%, were prepared from the gas fluorination (with anhydrous-HF atmosphere) of the previously described Al-doped zinc hydroxyfluorides. Anhydrous-HF was used instead of F_2 gas in order to avoid oxidative reaction which can lead to the compensation of the n-type charge carriers. The temperature of the fluorination treatment varied between 200 and 800°C for a 5-h dwell time. The aim of this treatment was to achieve the total substitution of F^- ions for OH^- group. Nevertheless, it is reasonable to think that too low treatment temperature or treatment time does not allow the complete substitution of the hydroxyl group leading to a fluorine-rich compound with chemical formulae Al-doped $ZnF_{2-x}OH_x$. Such fluorine-rich compounds would exhibit the same framework as pure ZnF_2 compound, and because of the proximity of the hydroxyl group radius and the F^- ion radius, it would be very difficult to characterize the occurrence of traces of hydroxyl groups from XRD patterns analyses.

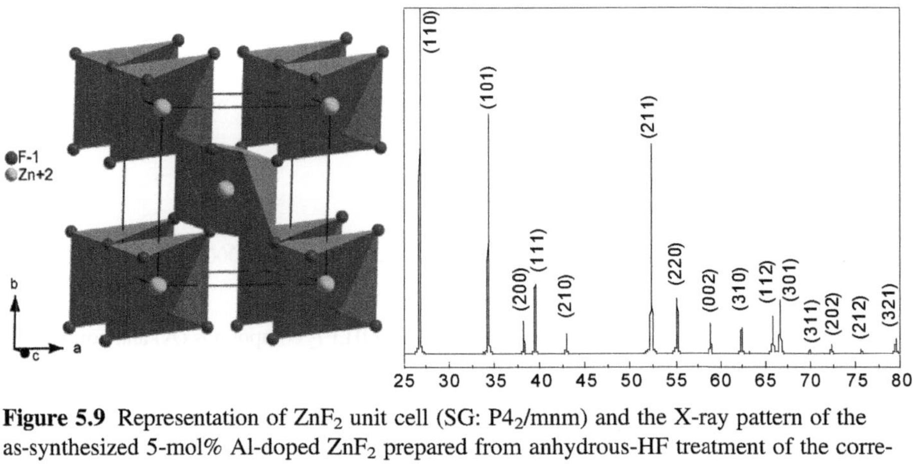

Figure 5.9 Representation of ZnF_2 unit cell (SG: $P4_2/mnm$) and the X-ray pattern of the as-synthesized 5-mol% Al-doped ZnF_2 prepared from anhydrous-HF treatment of the corresponding Al-doped zinc hydroxyfluorides.

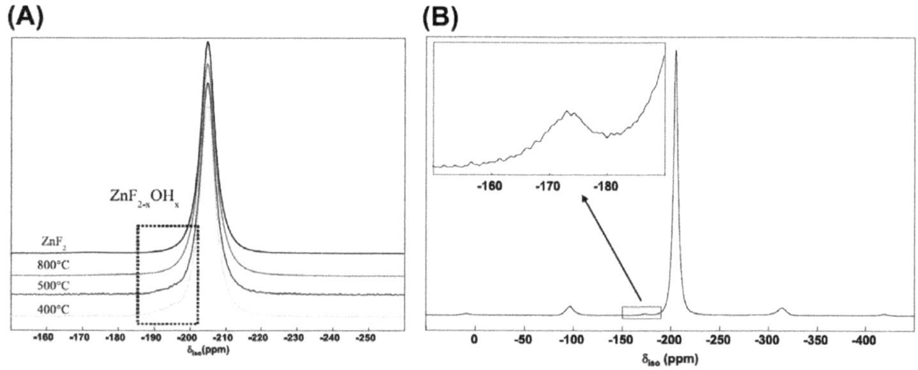

Figure 5.10 Experimental ^{19}F MAS-NMR of the zinc hydroxyfluoride obtained by precipitation at pH = 7 and treated at different temperatures under anhydrous-HF gas (A). Experimental ^{19}F MAS-NMR of the 3-mol% Al-doped zinc hydroxyfluoride obtained by precipitation at pH = 7 and treated at 800°C under anhydrous-HF gas (B).
Collaboration with Ch. Legein, LOF Le Mans.

Hence, to follow the OH^- substitution for F^- ions during the gas treatment versus temperature, ^{19}F NMR analyses were conducted.

Firstly, the study was performed on undoped compounds just to focus on the OH^- group impact. For different HF treatment temperatures, the ^{19}F NMR spectra are reported in the Fig. 5.10A. All the spectra exhibit a main sharp peak at −204.8 ppm, which is the signature of the F^- ions in ZnF_2 compound. Nevertheless, for treatment temperatures up to 500°C, besides the main peak, a shoulder located at about −194 ppm, which is characteristic of OH^- ion traces, appears. The use of high temperature treatment to purify the compounds and to obtain pure fluorides is so necessary.

In a second step, the study was carried out on doped compounds. In Fig. 5.10B the ^{19}F NMR spectrum of a 3-mol% Al-doped compound treated at 800°C under HF gas atmosphere is presented. It can be clearly seen that a new peak (new contribution) is located at −173 ppm in comparison with the undoped compound spectrum. Hence, this new peak which can be attributed, from literature [81,82], to AlF_3 phase demonstrates without any ambiguity that an AlF_3 phase separation involving all the aluminum ions occurs. Unfortunately, no Al^{3+} ions can be incorporated into ZnF_2 matrix by using this synthesis route.

Because of this result, no infrared absorption of the "Al-doped ZnF_2" compounds appears.

4. Zinc Oxyfluoride "Composites"

Our recent investigations have demonstrated that aluminum- or gallium-doped ZnO are promising candidates for getting efficient infrared absorbing powders [74,78–80]. Nevertheless, the refractive index of ZnO ($n = 2.00$) remains too high for ensuring a sufficient transmission of the visible light through the envisaged films/varnishes systems. Our literature review has shown that F^- ions can be used as n-type extrinsic dopant as M^{3+} ions in order to induce infrared absorption. Nevertheless, the concentration is then insufficient to really decrease the refractive index. In a second step, our own investigations have shown that it is possible to obtain promising results by doping with aluminum into zinc hydroxyfluoride compounds. On the contrary, the aluminum substitution into pure ZnF_2 fluoride was unsuccessful, and no infrared absorption was observed on this type of compounds. A new idea was to fabricate inorganic fluoride shells to encapsulate zinc oxide particles. The encapsulation of ZnO particles with an inorganic shell with a lower refractive index should indeed improve the visible light transmission. A part of our investigations was developed in collaboration with E. Kemnitz group from Humboldt university of Berlin and is already reported [83]. For that purpose, metal fluorides should be promising candidates. For instance, magnesium fluoride and zinc fluoride are of particular interest with a refractive index of 1.38 and 1.51 in visible range, respectively [84]. Notice that MgF_2 is commonly used as antireflective thin films on glasses [85–87]. Here the fluoride shells deposited on each individual oxide grains act as "antidiffusive" coatings.

Many models describe light–matter interactions such as Mie scattering by a distribution of particle sizes. Nowadays, free software packages allow the calculation of the total diffuse transmission across multiphasic composite films. Hence simulation of films constituted of dispersed particles of ZnO or ZnO@MF_2 into a standard matrix (with a refractive index about 1.4–1.5 as the majority of the common polymers) can be easily performed. The calculation was performed across three types of films described in Fig. 5.11.

In the first case (Fig. 5.11A), only the particle/matrix interface is considered. For the as-described films with whole thickness of 0.5 mm, in which particles of 100 nm diameter are dispersed and compose 0.1 vol% of the film, a transmission of 69% is calculated. This transmission is a reference for the following. In the case of core–shell particles with a shell layer with an intermediate refractive index, the total transmission increase to 74%. Herein, two types of interfaces have to be considered: the shell–matrix interface and the core–shell interface. Despite the doubling of the interfaces, the antidiffusive layer concept is validated by the Mie scattering calculations. The calculated total transmission for films with dispersed core–shell layers is efficiently raised in comparison with films where only the "cores" are dispersed. In a third model, we have developed this concept by simulating the transmission across a film in which there are dispersed core–shell particles with numerous shells exhibiting a refractive index gradient from the surface to the core. This refractive index gradient was modeled as a multilayer shell composed of five successive shells with refractive index from the surface to the core equal to 1.5, 1.6, 1.7, 1.8, and 1.9, respectively. The total transmission of the as-simulated films is now equal to 90%!

In the second step, we experimentally investigated the preparation and characterization of ZnO–MgF_2 and ZnO–ZnF_2 core–shell particle films. Metal fluoride

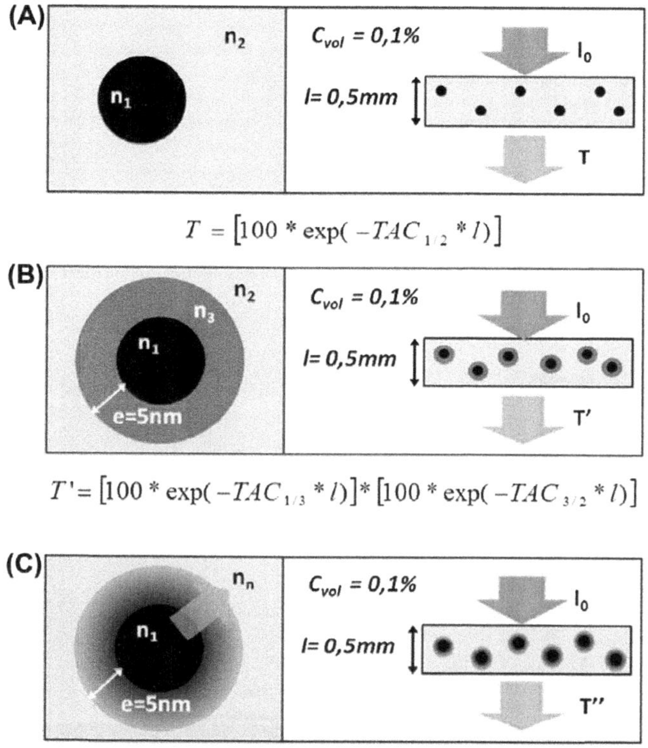

$$T = [100 * \exp(-TAC_{1/2} * l)]$$

$$T' = [100 * \exp(-TAC_{1/3} * l)] * [100 * \exp(-TAC_{3/2} * l)]$$

$$T'' = \prod_{i=1}^{n} [100 * \exp(-TAC_{i/i+1} * l]$$

Figure 5.11 Transmission visible light calculations at a wavelength of 600 nm across various films constituted of a powder dispersed into a matrix. ZnO particles ($n_1 = 2$, Ø = 100 nm) into a matrix with $n_2 = 1.4$ (A); ZnO@MF_2 core–shells ($n_1 = 2$, Ø = 100 nm, $n_3 = 1.7$ and $e = 5$ nm) into a matrix with $n_2 = 1.4$ (B); ZnO@MF_2 core–shells ($n_1 = 2$, Ø = 100 nm, n_3 = index gradient from 1.4 to 2 simulated with five successive layers and $e = 5$ nm) into a matrix with $n_2 = 1.4$ (C). The TAC or total attenuation coefficient (in mm^{-1}) is directly proposed by the software and allows the transmission estimation.
Calculations performed with the Mie calculator free software (© 2012 Scott Prahl).

shells were prepared from sols deposited via a vacuum slip–casting process like [83], ie, from the sol filtration through a ZnO powder bed followed by an annealing stage. Chemical, structural, and morphological characterizations were performed by ICP spectrometer, interferometry, elemental analysis, XRD, and transmission electron microscopy (TEM). Direct transmission measurements were performed through particle films elaborated by a screen-printing process, schematized in Fig. 5.12.

XRD analyses have confirmed the crystallization of the core from the magnesium and the zinc sols into MgF_2 and ZnF_2, respectively. In the case of ZnO–ZnF_2 particles, the comparison of scanning transmission electron microscopy (STEM) images

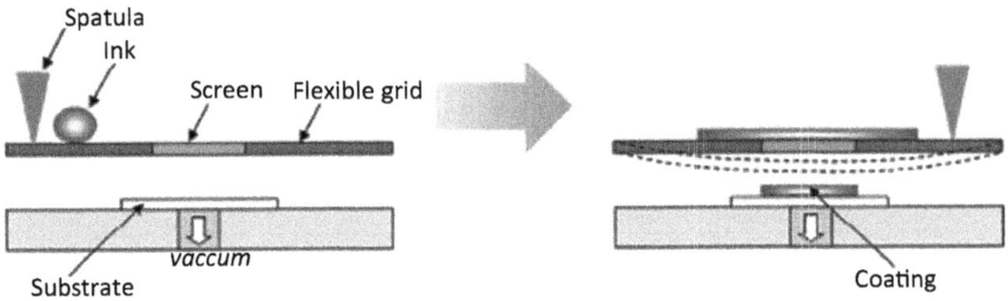

Figure 5.12 Images showing the screen-printing process to get an oxide–fluoride coating.

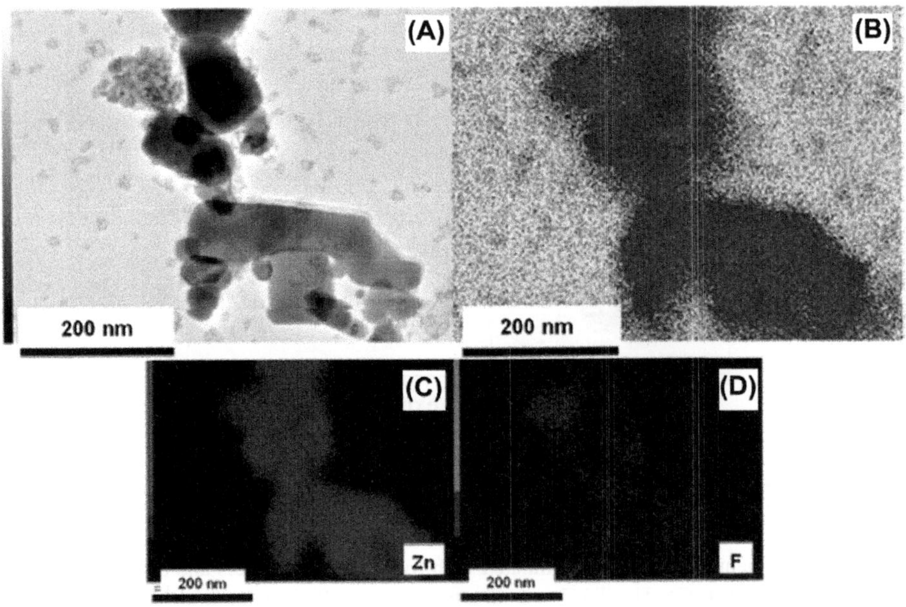

Figure 5.13 TEM image (in scanning mode) of as-prepared $ZnO@ZnF_2$ sample (A). STEM-EDX elemental map with superimposition of the zinc and fluorine signal (respectively in blue and red color) and (C, D) STEM-EDX map of Zn and F signal, respectively. *TEM*, transmission electron microscopy; *STEM*, scanning transmission electron microscopy; *EDX*, energy-dispersive X-ray spectroscopy.

and STEM-EDX maps revealed that the ZnF_2 compound was present but not as a coating onto the ZnO cores, but as core-free particles (Fig. 5.13). A mixture of ZnO and ZnF_2 particles was obtained. On the other hand, TEM images and chemical cartographies obtained with EDX analyses confirmed the presence of MgF_2 as continuous shell on ZnO particles (Fig. 5.14). This various behavior could be due to differences in the wettability behavior of ZnO cores with regards to ZnF_2 and MgF_2 sol dispersions.

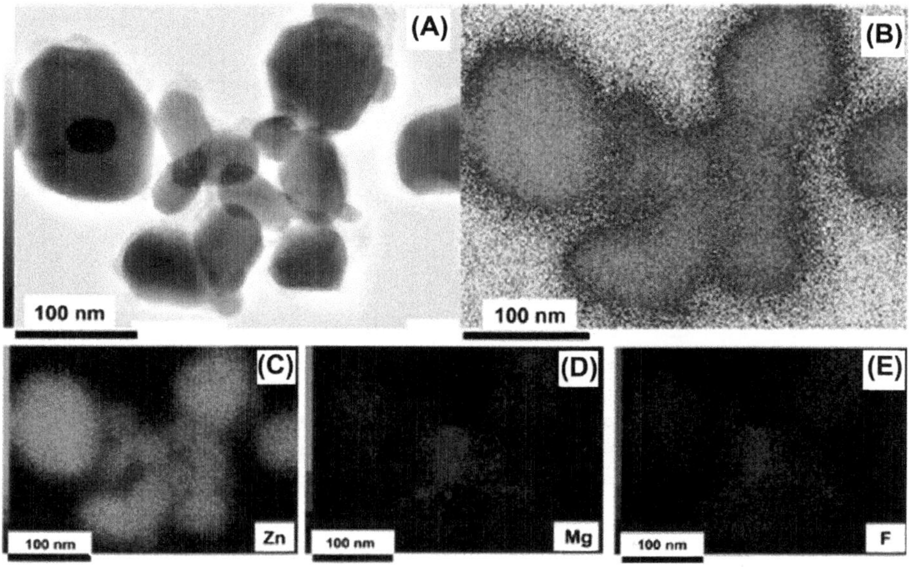

Figure 5.14 TEM image (in scanning mode) of as-prepared $MgO@ZnF_2$ sample (A). STEM-EDX elemental map with superimposition of the zinc, magnesium, and fluorine signals (respectively in green, red, and blue color) and (C, D, E) STEM-EDX map of Zn, Mg, and F signals, respectively. *TEM*, transmission electron microscopy; *STEM*, scanning transmission electron microscopy; *EDX*, energy-dispersive X-ray spectroscopy.
Figure reproduced from reference I. Trenque, S. Mornet, E. Duguet, J. Majimel, A. Brüll, K. Teinz, E. Kemnitz, M. Gaudon, Opt. Mater. 35 (3) (2013) 661. Copyright (2013) Elsevier Ltd.

The transmission properties of ZnO films and $ZnO-MgF_2$ composite films elaborated by screen-printing process were performed. All the films were elaborated in the same experimental conditions. From interferometry measurements (profilometer analyses), the film thickness was estimated to about 3−5 μm with roughness coefficient of about 1 μm (Fig. 5.15). In the case of the $ZnO@MgF_2$ composite coatings, films from additive mixture of ZnO (98.7 wt.%) and MgF_2 (1.3 wt.%) powders and from $ZnO@MgF_2$ core−shell particles (with same weight ratio) were compared. The direct transmissions in the visible range (400−800 nm) of the three compared films are reported in Fig. 5.16. The average transmission was equal to 48% for the ZnO film, 60% for the film made from ZnO and MgF_2 additive mixture, and 80% for the latter made from $ZnO@MgF_2$ core−shell particles. Both film pictures and transmission measurements shown that (1) the ZnO films were largely more opaque than the "mixture" films, (2) the visible transmission was obviously better for the films based on core−shell particles than for the "mixture" films with similar composition. Thus, the encapsulation of particles in shells of low refractive index should be an efficient route to improve the visible light transmission of inorganic particles dispersed in a polymeric matrix. Herein, it was shown for ZnO particles that the use of coating shells of 1.3 wt.% of magnesium fluoride allows a drastic

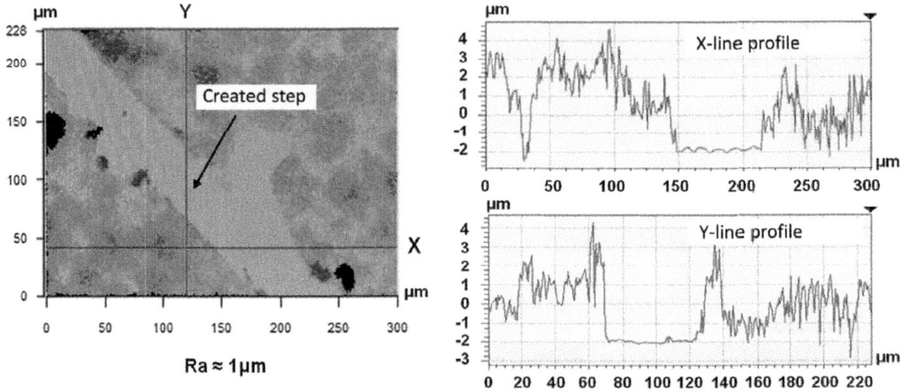

Figure 5.15 Roughness (Ra) and thickness measurements by optical interferometry on ZnO films elaborated by the screen-printing process: aerial map of the film (left side); X-line and Y-line profiles, the two lines crossing a step created by film mechanical abrasion.

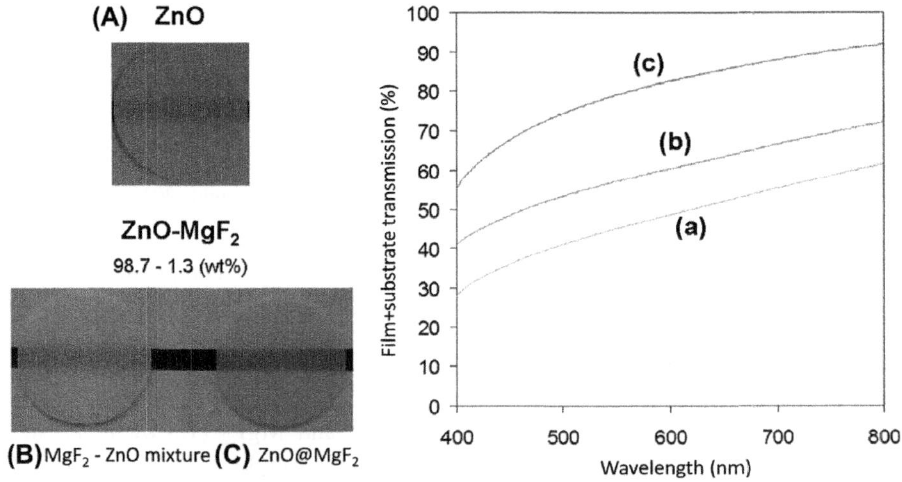

Figure 5.16 Left panel: Pictures of the films elaborated by screen-printing process from ZnO (A), ZnO (98.7 wt.%)—MgF_2 (1.3 wt.%) additive mixture (B), and ZnO@MgF_2 core—shell particles (C). **Right panel**: Diffuse transmission spectra in the visible range of these three films on glass substrate.

Figure rebuilt from reference I. Trenque, S. Mornet, E. Duguet, J. Majimel, A. Brüll, K. Teinz, E. Kemnitz, M. Gaudon, Opt. Mater. 35 (3) (2013) 661. Copyright (2013) Elsevier Ltd.

improvement of the transmission percentage between the composite sheet and the ZnO. This experimental observation validates the simulations performed with the Mie calculator software and so our concept: individual capping of the oxide particles by antidiffusive shells.

5. Conclusion

In this paper related to the next generation of transparent visible/infrared absorption films, three various classes of materials containing mainly fluorine anions and zinc cations are described and the results are evaluated by comparison with standard M^{3+}-doped ZnO films. These transparent conductive oxides or fluorides can find application in numerous displays entering in the orientation of materials research at the interface of the academic and the social-economic worlds. For illustration these materials can be used in glass efficient material besides thermal insulation, photovoltaic panels, transparent optoelectronic display, technical optical lenses (lasers...), etc. These materials are so of a broad interest.

Aluminum- or gallium-doped ZnO are promising candidates for getting efficient infrared absorbing powders. Nevertheless, the refractive index of ZnO ($n = 2.00$) remains too high for ensuring a sufficient transmission of the visible light considering the envisaged films/varnishes systems prepared from low cost routes as chemical processes from oxide particles suspensions dispersed into aqueous/organic solutions. A solution to this key point is to prepare fluorine-doped zinc oxides or M^{3+}-doped zinc fluorides materials instead of pure zinc oxide.

Literature review shows that low concentration of F^- ions used as dopant into zinc oxides does not allow decreasing significantly the refractive index. Finally, the F^- anion seems a nonadequate n-type dopant into ZnO matrices in comparison with M^{3+} cations, since it brings new process difficulties with no strong improvement of the optical properties.

Moreover, our own investigations have shown that the preparation of Al-doped $Zn(OH,F)_2$ compounds is a promising solution. A significant decrease of the refractive index is indeed reached, but their infrared absorption intensity remains very weak in comparison to that obtained for Ga/Al-doped ZnO oxides. Elsewhere, the preparation of Al-doped ZnF_2 fluorides remains still a challenge we did not succeed.

The encapsulation of ZnO particles with an inorganic fluoride shell with a lower refractive index (intermediate between ZnO and the dispersion medium) allows a significant improvement of the visible light transmission across films constituted of dispersed core—shell particles into the medium matrix. The $ZnO@MgF_2$ core—shell particles have been successful prepared. The transmission of visible light of such films containing $ZnO@MgF_2$ core—shell particles is much better than the other particles with pure ZnO particles. This opens new routes for the obtaining of transparent visible/infrared absorbers thick films.

References

[1] R.G. Gordon, MRS Bull. 25 (2000) 52.
[2] K.Y. Cheong, N. Muti, S.R. Ramanan, Thin Solid Films 440 (2002) 142.
[3] D.C. Altamirano-Juarez, G. Torres-Delgado, S. Jimenez-Sandoval, O. Jimenez-Sandoval, R. Castanedo-Perez, Sol. Energy Mater. Sol. Cells 83 (2004) 35.

[4] H. Mizoguchi, P.M. Woodward, Chem. Mater. 16 (2004) 5233.
[5] Z. Ben Ayadi, L. El Mir, K. Djessas, S. Alaya, Nanotechnology 18 (2007) 445702.
[6] H. Gomez, A. Maldonado, R. Castanedo-Perez, G. Torres-Delgado, L. Olivera, Mater. Charact. 58 (2007) 708.
[7] M. Sugiyama, A. Murayama, T. Imao, K. Saiki, H. Nakanishi, S.F. Chichibu, Phys. Status Solidi A 203 (2006) 2882.
[8] S. Major, A. Banerjee, K.L. Chopra, Thin Solid Films 122 (1) (1984) 31.
[9] W.W. Wenas, A. Yamada, K. Takahashi, M. Yoshino, M. Konagai, J. Appl. Phys. 70 (1991) 7119.
[10] V.J. Norman, Aust. J. Chem. 22 (1969) 325.
[11] K. Shirouzu, T. Ohkusa, M. Hotta, N. Enomoto, J. Hojo, J. Ceram. Soc. Jpn. 115 (1340) (2007) 254.
[12] T. Strachowski, E. Grzanka, W. Lojkowski, A. Presz, M. Godlewski, S. Yatsunenko, H. Matysiak, R.R. Piticescu, C.J. Monty, J. Appl. Phys. 102 (2007), 073513/1.
[13] D. Nie, T. Xue, Y. Zhang, X. Li, Sci. China Ser. B Chem. 51 (2008) 823.
[14] T. Tsubota, M. Ohtaki, K. Eguchi, H. Arai, J. Mater. Chem. 7 (1) (1997) 85.
[15] I. Saadeddin, Preparation and Characterization of New Transparent Conducting Oxides Based on SnO_2 and In_2O_3: Ceramics and Thin Films, Ph.D. dissertation, Université Bordeaux, 2007.
[16] K. Nassau, in: The Physics and Chemistry of Color, the Fifteen Causes of Color, John Wiley & Sons, NY, 2001, ISBN 0-471-39106-9.
[17] T. Coutts, D. Young, X. Li, MRS Bull. 25 (8) (2000) 58.
[18] U. Özgür, Y. Alivov, C. Liu, A. Teke, M. Reshchikov, S. Dogan, V. Avrutin, S.-J. Cho, H. Morkoç, J. Appl. Phys. 98 (4) (2005) 1.
[19] L. Schmidt-Mende, J. MacManus-Driscoll, Mater. Today 10 (5) (2007) 40.
[20] G. Luka, T. Krajewski, L. Wachnicki, B. Witkowski, E. Lusakowska, W. Paszkowicz, E. Guziewicz, M. Godlewski, Phys. Status Solidi (A) Appl. Mater. Sci. 207 (7) (2010) 1568.
[21] R. Buonsanti, A. Llordes, S. Aloni, B. Helms, D. Milliron, Nano Lett. 11 (11) (2011) 4706.
[22] R. Mendelsberg, S. Lim, Y. Zhu, J. Wallig, D. Milliron, A. Anders, J. Phys. D Appl. Phys. 44 (23) (2011).
[23] K. Ravichandran, P. Sathish, S. Snega, K. Subha, B. Sakthivel, Powder Technol. 274 (2015) 250.
[24] K. Ravichandran, R. Mohan, N.J. Begum, K. Swaminathan, C. Ravidhas, J. Phys. Chem. Solids 74 (2013) 1794.
[25] Y.-J. Choi, K.-M. Kang, H.-H. Park, Solar Energy Mater. Solar Cells 132 (2014) 403.
[26] Z. Pan, P. Zhang, X. Tian, G. Hu, Z. Wei, J. Alloys Compd. 576 (2013) 31.
[27] R.-X. Ma, M.-K. Wang, B. Kang, Y.-G. Wang, Optoelectron. Lett. 7 (1) (2011) 0045.
[28] R. Tilley, in: Colour and the Optical Properties of Materials: An Exploration of the Relationship Between Light, the Optical Properties of Materials and Colour, Wiley, 2011, ISBN 978-0471851981.
[29] J. Prener, J. Kingsley, J. Chem. Phys. 38 (3) (1963) 667.
[30] M. Voda, S. Nistor, V. Ghiordanescu, J. Crystal Growth 119 (3–4) (1992) 339.
[31] C. Park, D. Chadi, Phys. Rev. Lett. 82 (1) (1999) 113.
[32] P. Fedorov, I. Buchinskaya, S. Ivanov, B. Sobolev, A. Shcheulin, A. Ryskin, Opt. Spectro. (English Translation of Opt. Spektro.) 92 (1) (2002) 125.
[33] S. Kazanskii, A. Shcheulin, A. Ryskin, N. Sobolev, D. Hilger, W. Warren Jr., Phys. B Condens. Matter 401-402 (2007) 282.

[34] S. Kazanskii, A. Shcheulin, A. Ryskin, Opt. Spectro. (English Translation of Opt. Spektro.) 102 (3) (2007) 419.
[35] Y. Uesu, K. Yasukawa, N. Salto, S. Itoh, S. Odoulov, K. Shcherbin, A. Ryskin, J. Opt. Soc. Am. B Opt. Phys. 20 (9) (2003) 1905.
[36] M. Grundmann, in: The Physics of Semiconductors, Springer, 2006, ISBN 978-3-540-25370-9.
[37] E. Burstein, Phys. Rev. 93 (1954) 632. T.S. Moss, Proc. Phys. Soc. Lond. B76 (1954) 775.
[38] X. Rocquefelte, F. Goubin, Y. Montardi, N. Viadere, A. Demourgues, A. Tressaud, M.-H. Whangbo, S. Jobic, Inorg. Chem. 44 (2005) 3589.
[39] G.C. Morris, A.E. McElnea, Appl. Surf. Sci. 92 (9) (1996) 167.
[40] B. Russo, G.Z. Cao, Appl. Phys. A Mater. Sci. Processes 90 (2008) 311.
[41] C. Geoffroy, G. Campet, F. Menil, J. Portier, J. Salardenne, G. Couturier, Act. Passive Electron. Compon. 14 (1991) 111.
[42] T. Maruyama, H. Akagi, J. Electrochem. Soc. 143 (1996) 283.
[43] B. Stjerna, E. Olsson, C.G. Granqvist, J. Appl. Phys. 76 (1994) 3797.
[44] P.-J. Chen, B.-H. Liao, C.-C. Kuo, C.-C. Lee, Proc. SPIE (2010) 77860Q.
[45] K. Murakami, K. Nakajima, S. Kaneko, Thin Solid Films 515 (2007) 8632.
[46] A. Benhaoua, A. Rahal, B. Benhaoua, M. Jlassi, Superlattice. Microst. 70 (2014) 61.
[47] A. Rahal, S. Benramache, B. Benhaoua, J. Semiconduct. 34 (8) (2013) 1.
[48] B. Benhaoua, S. Abbas, A. Rahal, A. Benhaoua, M.S. Aida, Superlattice. Microst. 83 (2015) 78.
[49] G. Rey, C. Ternon, M. Modreanu, V. Consonni, D. Bellet, J. Appl. Phys. 114 (18) (2013) 183713.
[50] A. Muthukumar, G. Giusti, M. Jouvert, V. Consonni, D. Bellet, Thin Solid Films 545 (2013) 302.
[51] A. Muthukumar, G. Rey, G. Giusti, A. Dakshnamoorthy, D. Bellet, AIP Conf. Proc. 1512 (2013) 710.
[52] P. Yao, Desalination 267 (2011) 170.
[53] A. Nadarajah, M.E. Carnes, M.G. Kast, D.W. Johnson, S.W. Boettcher, Chem. Mater. 25 (2013) 4080.
[54] V. Kumar, A. Govind, R. Nagarajan, Inorg. Chem. 20 (2011) 5637.
[55] P.M. Ratheesh Kumar, C. Sudha Kartha, K.P. Vijayakumar, F. Singh, D.K. Avasthi, Mater. Sci. Eng. B 117 (2005) 307.
[56] J. Hu, R. Gordon, Solar Cells 30 (1991) 437.
[57] B.L. Zhu, X.Z. Zhao, F.H. Su, G.H. Li, X.G. Wu, J. Wu, R. Wu, Vacuum 84 (2010) 1280.
[58] B.G. Choi, I.H. Kim, D.H. Kim, K.S. Lee, T.S. Lee, B. Cheong, Y.J. Baik, W.M. Kim, J. Eur. Ceram. Soc. 25 (2005) 2161.
[59] D.Y. Ku, Y.H. Kim, K.S. Lee, T.S. Lee, B. Cheong, T.Y. Seong, Y.H. Kim, J. Electroceram. 23 (2009) 415.
[60] Y.Z. Tsai, N.F. Wang, C.L. Tsai, Mater. Lett. 63 (2009) 1621.
[61] R.E. Treharne, K. Durose, Thin Solid Films 519 (21) (2011) 7579.
[62] Y.Z. Tsai, N.F. Wang, C.L. Tsai, Thin Solid Films 518 (17) (2010) 4955.
[63] H.S. Yoon, K.S. Lee, T.S. Lee, B. Cheong, D.K. Choi, D.H. Kim, W.M. Kim, Solar Energy Mater. Solar Cells 92 (11) (2008) 1366.
[64] X. Noirfalise, T. Godfroid, G. Guisbiers, R. Snyders, Acta Mater. 59 (20) (2011) 7521.
[65] R. Anandhi, K. Ravichandran, R. Mohan, Mater. Sci. Eng. B 178 (1) (2013) 65.
[66] S.M. Rozati, S. Moradi, S. Golshahi, R. Martins, E. Fortunato, Thin Solid Films 518 (4) (2009) 1279.

[67] F.-H. Wang, Y.-H. Lee, T.-K. Kang, H.-W. Liu, Superlattice. Microst. 83 (2015) 289.
[68] Y. Nakanishi, S. Naito, T. Nakamura, Y. Hatanaka, G. Shimoaka, Appl. Surf. Sci. 92 (1996) 400.
[69] W. Feitknecht, Helvetica Chim. Acta 26 (1953) 2196.
[70] O.K. Srivastava, E.A. Secco, Can. J. Chem. 45 (1967) 579.
[71] O.K. Srivastava, E.A. Secco, Can. J. Chem. 45 (1967) 585.
[72] M. Nierlich, P. Charpin, P.M.C.R. Herpin, Acad. Sci. Série C Sci. Chim. 1—3 (1973) 276.
[73] S. Yamabi, H. Imai, Trans. Mater. Res. Soc. Jpn. 28 (2003) 329.
[74] H. Serier, M. Gaudon, A. Demourgues, A. Tressaud, J. Solid State Chem. 180 (12) (2007) 3485.
[75] P.J. Chupas, M.F. Ciraolo, J.C. Hanson, C.P. Grey, J. Am. Chem. Soc. 123 (2001) 1694.
[76] D. Dambournet, A. Demourgues, C. Martineau, E. Durand, J. Majimel, A. Vimont, H. Leclerc, J.-C. Lavalley, M. Daturi, C. Legein, J.-Y. Buzare, F. Fayon, A. Tressaud, J. Mater. Chem. 18 (21) (2008) 2483.
[77] H.A. Prescott, Z.J. Li, E. Kemnitz, J. Deutsch, H. Lieske, J. Mater. Chem. 15 (2005) 4616.
[78] H. Serier, A. Demourgues, M. Gaudon, Inorg. Chem. 49 (15) (2010) 6853.
[79] I. Trenque, S. Mornet, E. Duguet, M. Gaudon, Mater. Res. Bull. 48 (3) (2013) 1155.
[80] I. Trenque, S. Mornet, E. Duguet, M. Gaudon, Inorg. Chem. 52 (21) (2013) 12811.
[81] P.J. Chupas, D.R. Corbin, V.N.M. Rao, J.C. Hanson, C.P. Grey, J. Phys. Chem. B 107 (2003) 8327.
[82] G. Silly, C. Legein, J.Y. Buzare, F. Calvayrac, Solid State Nucl. Mag. Reson. 25 (2004) 241.
[83] I. Trenque, S. Mornet, E. Duguet, J. Majimel, A. Brüll, K. Teinz, E. Kemnitz, M. Gaudon, Opt. Mater. 35 (3) (2013) 661.
[84] D. Popov, Crystallogr. Rep. 47 (2002) 829.
[85] H. Krüger, E. Kemnitz, A. Hertwig, U. Beck, Thin Solid Films 516 (2008) 4175.
[86] T. Murata, H. Ishizawa, I. Motoyama, A. Tanaka, Appl. Opt. 45 (2006) 1465.
[87] F. Perales, J. Herrero, D. Jaque, C. de las Heras, Opt. Mater. 29 (2007) 783.

Fluorine-Containing Beryllium Borates as Nonlinear Optical Crystals for Deep-Ultraviolet Laser Generation

6

C. Chen, L. Liu, X. Wang
CAS, Beijing, P. R. China

Chapter Outline

Photonic and Electronic Properties of Fluoride Materials. http://dx.doi.org/10.1016/B978-0-12-801639-8.00006-4

1. Introduction

Nonlinear optical (NLO) crystals have been rapidly developed, especially for laser frequency generation, ever since the groundbreaking realization of the ruby laser by Maiman in 1960 [26] and the first observation of an NLO effect in quartz crystal by Franken in 1961 [16]. Using NLO crystals is the most efficient way to widen the laser wavelength range from near infrared (IR) to visible, UV, even deep-UV (DUV) region at one end of the spectrum, or to the mid- and far-IR region at the other end. The past 30 years have witnessed tremendous progress both in our fundamental understanding of NLO crystals and in their applications. The field has grown to such a size that no individual can be preeminent in all divisions of the subject.

Most borate crystals have a wide transparency range in the IR to UV region, good chemical stability, and high damage threshold, which has led to hopes of new NLO crystals in borates. For example, β-BaB_2O_4 (BBO) [3] and LiB_3O_5 (LBO) [5] which were developed in Chen's group in the 1980s are the most widely used crystals for frequency conversion in the visible and UV regions. However, they cannot produce shorter wavelengths below 200 nm by a phase-matching second-harmonic generation (SHG) method because of the optical absorption edge for BBO or the relevant small birefringence of LBO in the DUV region.

After the discovery of BBO and LBO, at the beginning of the 1990s, Chen et al. started to explore DUV NLO crystals that can produce laser radiation below 200 nm by direct SHG. They proposed three structure criteria in their search among the borates: (1) the basic anionic group $(BO_3)^{3-}$ in the crystal lattice, with a density per unit volume as large as possible to produce large SHG coefficients [4]; (2) the $(BO_3)^{3-}$ groups must maintain a coplanar configuration to produce a large birefringence [6]; (3) the three terminal oxygen atoms of the $(BO_3)^{3-}$ group ought to be linked with other atoms such as B, Be, and Al which can shift the cutoff wavelength down to 150–160 nm. Thus it was conjectured that beryllium borate would be the best candidate for vacuum-UV NLO or window materials [23]. At that time, most of the crystals used were oxides such as phosphates, niobates, and borates. Chen extended his sight to other compounds such as fluorides or fluoborates since fluoride(s) can replace oxygen without changing the crystal structure significantly due to their similar sizes. This substitution may occur in different types of structures such as rutile, perovskite, pyrochlore, garnet, or fluorite. Moreover, the physical properties are often strongly modified. Based on these considerations, Chen et al. searched for all fluorine compounds ever reported and found that the structure of fluoride-containing beryllium borate $KBe_2BO_3F_2$ (KBBF) matched quite well. Subsequent studies proved that KBBF is an excellent nonlinear crystal in the DUV. Actually, the effect of the F ion in KBBF is debatable, whether the F element is only present in the crystal structure because of its suitable diameter so that the three criteria are met, or whether the electronic structure of the fluoride itself influences the optical properties of the crystal, is still unclear today.

Later, through theoretical calculations it was found that the incorporation of fluorine usually blueshifts the absorption edge of a crystal due to its large electronegativity: a mixture of the 2p orbitals of fluorine and oxygen atoms decreases the energy of the valence band which results in a shorter cutoff wavelength [38]. These studies have helped us to understand the solid-state chemistry of fluorine.

With many fluorine compounds showing NLO effects, fluorine-containing crystals have found their place among the NLO crystals. Of these, KBBF is becoming the most important DUV crystal, and is being used in a growing number of applications.

2. $KBe_2BO_3F_2$ Family Crystals

2.1 $KBe_2BO_3F_2$ Crystal

The $MBe_2BO_3F_2$ family crystals include five compounds: $NaBe_2BO_3F_2$ (NBBF), $KBe_2BO_3F_2$, $RbBe_2BO_3F_2$ (RBBF), $CsBe_2BO_3F_2$ (CBBF), and the newly reported $TlBe_2BO_3F_2$ (TBBF) [15]. In 1968, KBBF was first synthesized by the former Soviet Union scientists [2]; later in 1970, Solov'eva et al. reported that its space structure is C121 [35]. Then, for about 20 years, it attracted no further attention until Chen (at Fujian Institute) et al. picked it out from the database as a potential DUV NLO crystal candidate.

Chen et al. started a systematic study on KBBF at the beginning of the 1990s. They successfully synthesized and grew KBBF crystals with a size of $10 \times 10 \times 0.3$ mm^3 by a spontaneous nucleation method in 1992 [28]. They redetermined the space group as R32 belonging to the uniaxial class, with dimensions of $a = b = 4.427(4)$ Å, $c = 18.744(9)$ Å, $V = 318.3(4)$ Å, and $Z = 3$ [27], and the structure is shown in Fig. 6.1A. It is composed of BO_3 and BeO_3F anionic groups, which form a $(Be_2BO_6F_2)$ network in the a–b plane with all the F atoms above or below the plane (Fig. 6.1B). All the BO_3 groups maintain a coplanar configuration, and their three terminal oxygen atoms are linked with Be atoms to eliminate the three dangling bonds, which is suitable for producing a large birefringence and short cutoff wavelength. The three structure criteria are therefore well satisfied, so KBBF was proposed to be a DUV NLO crystal [46]. The directions of the F ions are upward and downward, which somewhat counteracts their microscopic SHG coefficient. Along the c-axis, it is connected with static forces between the K^+ and F^- ions, which are much weaker than the covalent bonds and the distance between each adjacent layer is as large as 6.25 Å, so the crystal tends to cleave along the c-axis and tends to grow in layers, which makes growing large crystals a very difficult task.

The basic optical properties, especially the NLO properties, were measured after the successful growth of small-sized crystals during 1993–95 [44]. It was then that KBBF was shown to be able to produce SHG in the DUV output range of 200–184.7 nm for the first time [7]. Based on the optical experiments, Chen proposed that KBBF may achieve the sixth-harmonic generation of Nd-based lasers at 177.3 nm in 1996 [7].

Fig. 6.2 shows the differential scanning calorimetry-thermogravimetric (DSC-TG) curve of KBBF crystal from room temperature to 1300°C measured by STA 409CD-SKIMMER. From the curve, we see that the melting point is about 1046°C with a distinct decrease in weight; the remains after melting were checked by powder

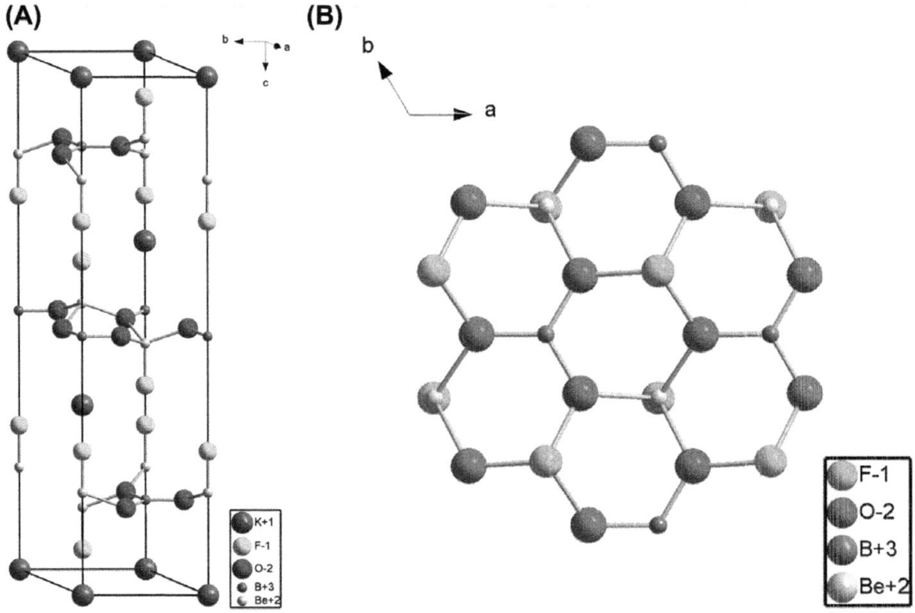

Figure 6.1 Space structure of $KBe_2BO_3F_2$. (A) Space structure in the unit cell. (B) Layer structure in the *a*—*b* plane.

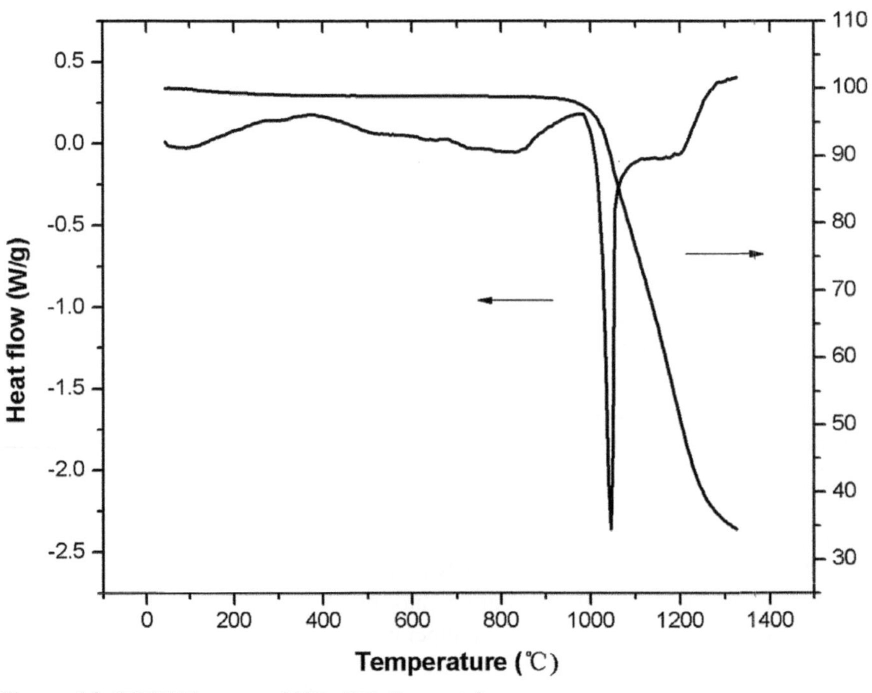

Figure 6.2 DSC-TG curve of $KBe_2BO_3F_2$ crystal.

X-ray diffraction and were found to contain BeO and other glasses. This shows that the crystal underwent incongruent melting, so it should be grown from the flux. As the compound begins to decompose at about 830°C before melting, the growth temperature must be below 830°C. This makes it difficult to find an appropriate flux system.

Polycrystalline KBBF is usually prepared in a high-temperature solid-state reaction according to the following equation:

$$6BeO + 3KBF_4 + B_2O_3 \rightarrow KBe_2BO_3F_2 + BF_3\uparrow$$

Stoichiometric amounts of BeO, KBF_4, and B_2O_3 were weighed and mixed homogeneously, then put into a platinum crucible. The mixture was slowly heated and maintained at 750°C for 48 h. Part of the BF_3 was released to the air as gas. It was then ground and reheated for two or three cycles for complete reaction. As BeO is very toxic, all the operations were performed in a ventilated system.

In the flux method the crystal was grown by spontaneous nucleation from a high-temperature solution. Many attempts were made to grow KBBF by seeding but all failed, as the crystals were always platelike with the prominent face corresponding to the (0001) plane. The reason is probably because the layering always occurs along the *c*-axis, so the growth rate is highly anisotropic and the thickness along the *c*-axis is always less than a millimeter in scale. Before the year 2000, the crystals were all less than 0.5 mm in thickness [36] and by 2003, they were still only up to $20 \times 20 \times 2.0\ mm^3$ in size as reported by Wang et al. [40].

To produce large and thicker crystals, a "localized spontaneous nucleation" method was developed in Chen's group [43]. The key factor is to restrict the nucleation to a narrow region where only one nucleus forms at the beginning, then starts to grow under an appropriate periodic temperature cycle. By this improved localized spontaneous nucleation technique, using KF-B_2O_3 as flux, a large size KBBF crystal was obtained as shown in Fig. 6.3A [43]. Sometimes the thickness along the *c*-axis reached about 3.7 mm, as shown in Fig. 6.3B, which is the maximum ever obtained by the flux method. This thickness is enough for most applications because the optical path for

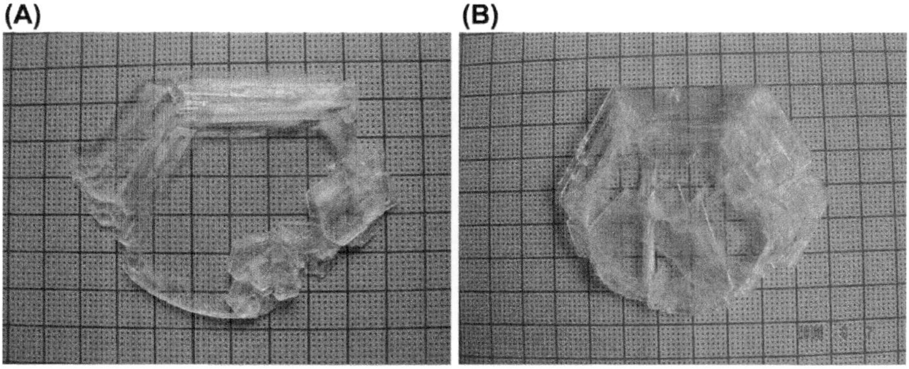

Figure 6.3 Bulk $KBe_2BO_3F_2$ (KBBF) crystal. (A) Bulk KBBF crystal in size of $50 \times 40\ mm^2$. (B) Bulk KBBF crystal with 3.7 mm thickness.

phase-matching will be about twice the thickness. The crystals are usually hexagonal in shape along the $a-b$ plane corresponding to a trigonal symmetry. A few small crystals adhere to the bulk one that spontaneously nucleated in the last growth procedure.

As an alternative to the flux method, a hydrothermal method has also been developed in recent years; the first successful experiment was reported by Ye et al. in 2006 [51]. Their crystal size was up to $12 \times 10 \times 6\ mm^3$ under a growth temperature of 360°C and a pressure of 100 MPa. This method seems more promising than starting with flux to obtain thicker crystals, and Chen et al. also reported successful growth by using (1−2) M KOH as a mineralizer [59]. The crystal can even be grown to a centimeter scale, as shown in Fig. 6.4.

Unfortunately, the optical quality of hydrothermal KBBF is very poor, as indicated by its X-ray rocking curve [52]. Its NLO properties have also been evaluated and compared with flux-grown KBBF, and are much worse. The SHG conversion efficiency is one or two orders of magnitude smaller, for example, the SHG conversion efficiency from 800 to 400 nm for a Ti:sapphire laser (80 MHz, 150 fs) is only 0.045% for a hydrothermal KBBF slice of 1.7 mm thickness, while it is 12% for a flux-grown crystal of the same thickness under the same laser pump conditions. As might be expected, the powder SHG response is also several times lower. To make matters more complicated, for different crystals of the same thickness, the SHG output power varied from sample to sample by an order of magnitude. Chen et al. were the first to point out that this may be related to some kind of structural defect [12] and since then have carried out much work to investigate the reasons. The crystal structure was reanalyzed using powder X-ray diffraction data, and a new $R\overline{3}c$ space group structure with cell parameters of $a = b = 4.422(1)$ Å and $c = 37.524(3)$ Å was obtained [53]. This is a $1 \times 1 \times 2$ superstructure of the standard R32 type, as shown in Fig. 6.5. The BO_3 groups are rotated 60 degrees along each layer which reduces the SHG effect completely. Later, the new structure was further confirmed by single-crystal X-ray diffraction data [32]. As the $R\overline{3}c$ structure is centrosymmetric without any SHG effect, Yu et al. further proposed a stacking fault model to explain the low-SHG efficiency and nonuniformity of the hydrothermal KBBF [53]. Two structures of R32 and $R\overline{3}c$

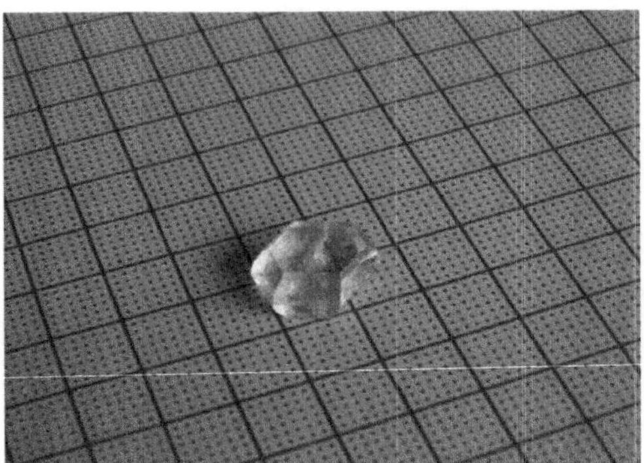

Figure 6.4 A hydrothermally grown $KBe_2BO_3F_2$ crystal of size $18 \times 10 \times 8\ mm^3$.

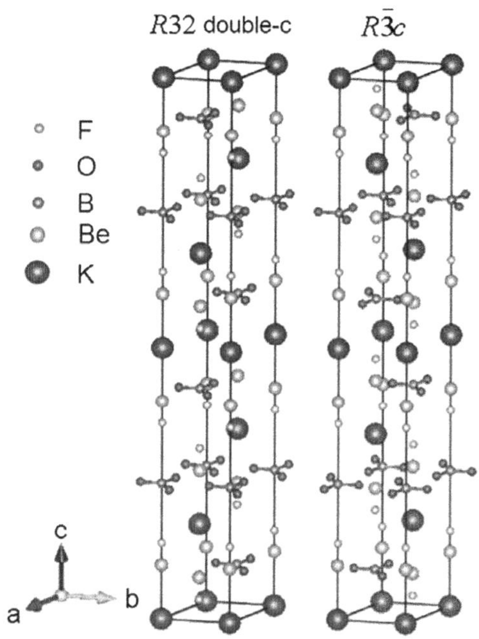

Figure 6.5 Space structures of flux $KBe_2BO_3F_2$ (KBBF) (left) and hydrothermal KBBF (right).

were intergrown along the *z*-axis with a random ratio. As the composition of $R\bar{3}c$ rises, the NLO effect decreases. The nonuniformity can be attributed to the varying density of the stacking faults. Unfortunately, so far it has not been possible to eliminate the stacking faults by improving the hydrothermal growth conditions.

Recently, we employed for the first time, transmission electron microscopy for the direct analysis of hydrothermal KBBF from its microstructure point of view, and observed two kinds of electron diffraction patterns along the [120] direction [48]. The occurrence of the extra spots in the electron diffraction pattern has been related to the new structure of $R\bar{3}c$, which has never been observed before in the flux-grown crystals.

2.2 Linear and Nonlinear Optical Properties

A comprehensive measurement of new materials is the prerequisite to making a correct assessment. For NLO crystals, the most important properties include the transmittance (the cutoff wavelength for applications in the UV region), birefringence, and NLO coefficient.

2.2.1 Transmittance Spectrum

The transmittance spectrum of KBBF in the range of 120–220 nm was recorded by a McPherson VUVaS2000 spectrophotometer, and for the 185–3000 nm range using a

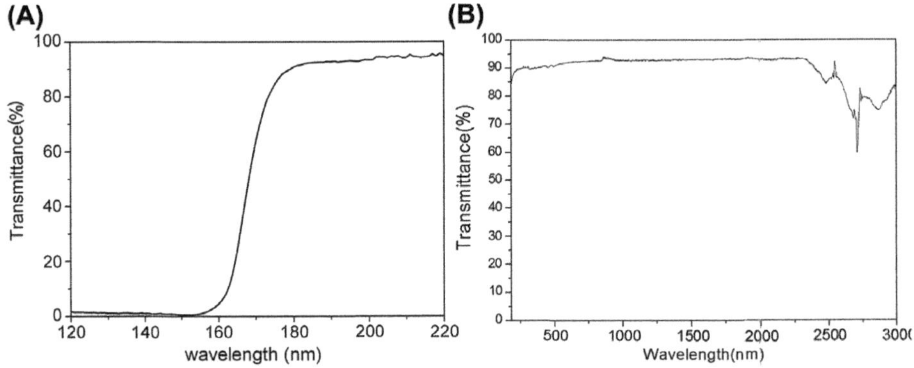

Figure 6.6 The transmittance curves of $KBe_2BO_3F_2$. (A) On the UV side. (b) From UV to infrared side.

PerkinElmer Lambda 900 UV/VIS/NIR spectrophotometer. As shown in Fig. 6.6, the cutoff wavelength on the UV side is located at 150 nm, and on the IR side around 3.5 μm. There is virtually no absorption in this region, while the maximum transmittance can reach up to 90% when surface reflections are included.

2.2.2 *Sellmeier Equations*

By means of a right-angle prism with an apex angle of 30 degrees made from a 2.3-mm-thick KBBF crystal, the refractive index data have been measured from UV to IR at eight wavelengths from 232 to 2325 nm. For the convenience of optical polishing and operation, three glass prisms are attached to the crystal. Table 6.1 lists the measured results, where the accuracy of the data is to four significant figures. These refractive indices are enough to fit the Sellmeier equations. With the new refractive index data, we fit the Sellmeier equations as follows (λ is in units of μm):

$$\begin{aligned} n_o^2 &= 2.1738 + \frac{0.00764}{\lambda^2 - 0.01641} - 0.01174\lambda^2 \\ n_e^2 &= 1.9319 + \frac{0.00623}{\lambda^2 - 0.00400} - 0.00356\lambda^2 \end{aligned} \qquad [6.1]$$

Using these Sellmeier equations, we can calculate the refractive indices to an accuracy of four significant figures, as shown in Table 6.1. They fit quite well with the measured results. The phase-matching angles throughout the whole spectral region calculated according to the Sellmeier equations are shown in Fig. 6.7 and also agree well with the experimental data reported in Ref. [11]. Thus KBBF has the widest phase-matching range from DUV to the visible region, especially in the DUV, and so has many important applications not achievable with any other crystal. The shortest wavelength obtained by SHG can reach 161.1 nm with a phase-matching angle of 88.5 degrees.

Table 6.1 **Measured and Calculated Refractive Indices of $KBe_2BO_3F_2$ With Δ as the Absolute Value of the Difference Between the Measured and Calculated Values**

Wavelength (nm)	n_o			n_e		
	Calculated	Measured	Δ	Calculated	Measured	Δ
254.7	1.52704	1.52662	0.00042	1.42640	1.42634	0.00006
363.0	1.49616	1.49609	0.00007	1.40719	1.40649	0.00070
404.7	1.49121	1.49141	−0.00020	1.40368	1.40345	0.00023
435.8	1.48849	1.48862	−0.00013	1.40169	1.40163	0.00006
546.0	1.48237	1.48224	0.00013	1.39715	1.39750	−0.00035
587.5	1.48087	1.48067	0.00020	1.39604	1.39650	−0.00046
643.8	1.47923	1.47895	0.00028	1.39485	1.39543	−0.00058
706.0	1.47777	1.47737	0.00040	1.39382	1.39445	−0.00063
852.1	1.47514	1.47553	−0.00039	1.39210	1.39386	−0.00176
1014.0	1.47285	1.47316	−0.00031	1.39080	1.38752	0.00328
2325.0	1.45319	1.45314	0.00005	1.38340	1.38367	−0.00027

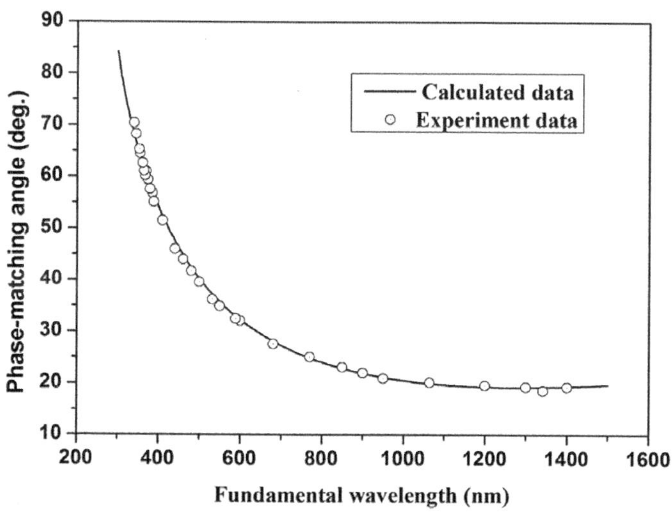

Figure 6.7 Phase-matching angle versus fundamental wavelength for type-I second-harmonic generation with $KBe_2BO_3F_2$ in the whole spectral region.

2.2.3 Determination of the d_{ij} Coefficient

The KBBF crystal belongs to the space group R32, so from the point group symmetry of D3 and the IEEE/ANSI standard definition of the d_{ij} coefficient [31], it has only two nonzero d_{ij} coefficients, namely, d_{11} and d_{14}. The matrix form of the coefficients can be written as follows:

$$\begin{pmatrix} d_{11} & -d_{11} & 0 & d_{14} & 0 & 0 \\ 0 & 0 & 0 & 0 & -d_{14} & -d_{11} \\ 0 & 0 & 0 & 0 & 0 & 0 \end{pmatrix} \qquad [6.2]$$

where d_{14} is very small according to both theoretical calculation and experiments. On the other hand, the effective d_{eff} coefficients of KBBF are:

$$\begin{aligned} d_{eff} &= d_{11} \cos\theta \cos 3\phi \text{ (type I)} \\ d_{eff} &= d_{11} \cos 2\theta \sin 3\phi \text{ (type II)} \end{aligned} \qquad [6.3]$$

Here d_{14} is not included in the expressions, so only d_{11} needs to be determined. A KBBF crystal of size $10 \times 10 \times 2\ mm^3$ was used to determine the Maker fringes precisely. The results are shown in Fig. 6.8 where the dashed line represents the theoretical fringes and envelope, calculated using the refractive indices from the Sellmeier equations. It is clear that the theoretical curves fit the experimental plots very well.

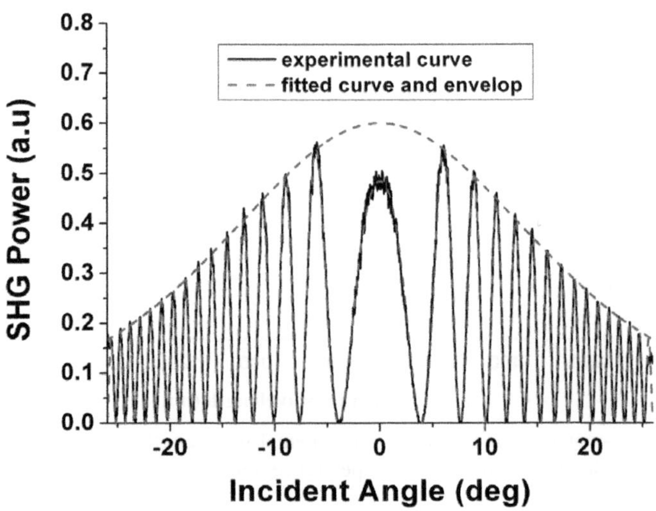

Figure 6.8 Maker fringes of d_{11} coefficient of $KBe_2BO_3F_2$.

By comparing the fringe envelope of the d_{11} coefficient with the d_{36} coefficient in KH_2PO_4 (KDP), the d_{11} of KBBF is deduced to be (0.47 + 0.01) pm/V (d_{36}(KDP) = 0.39 pm/V).

2.2.4 Damage Threshold

The damage threshold of KBBF has been measured and compared with that of BBO under the same laser pump conditions (1064 nm, 1 Hz, 10.5 ns) at the China Academy of Engineering Physics. The value is 35.51 J/cm^2 (3.38 GW/cm^2) for the former, which is more than two times larger than that (13.47 J/cm^2 (1.28 GW/cm^2)) of the latter. With a 200 fs pulse width and 1 kHz repetition rate at 390 nm, the damage threshold of KBBF was found to be as high as 60 GW/cm^2, as measured at the Institute for Solid State Physics, University of Tokyo.

2.3 KBBF Prism-Coupled Device and Deep-UV Laser Generation

2.3.1 KBBF Prism-Coupled Device

Crystal KBBF is capable of producing DUV and even vacuum-UV coherent light by direct SHG. However, a major problem is its platelike growth nature that makes it very difficult to grow samples thicker than 4 mm along the *c*-axis, so it is impossible to cut the crystal along the phase-matching angle. With a directly incident pump, the external incident angle will increase to 90 degrees for a second harmonic wavelength of 235 nm. To solve the problem of producing coherent laser light below 235 nm, Chen and Xu developed a special prism-coupling technique to avoid cutting the crystals along the phase-matching angle [9]. A KBBF crystal is sandwiched between two prisms made of UV-fused silica, which has almost the same refractive indices as KBBF in the visible-to-UV spectral region. The structure of this KBBF prism-coupled device (PCD) is illustrated in Fig. 6.9 for the 177.3 nm harmonic wavelength [8]. When the fundamental beam is incident normal to the prism, the angle of refraction in KBBF is close to that of the prism, so that if the prism's apex angle is equal to the phase-matching angle of KBBF at a given wavelength, phase-matching can be achieved. The surfaces at the sandwich layer are optically contacted. Calcium fluoride

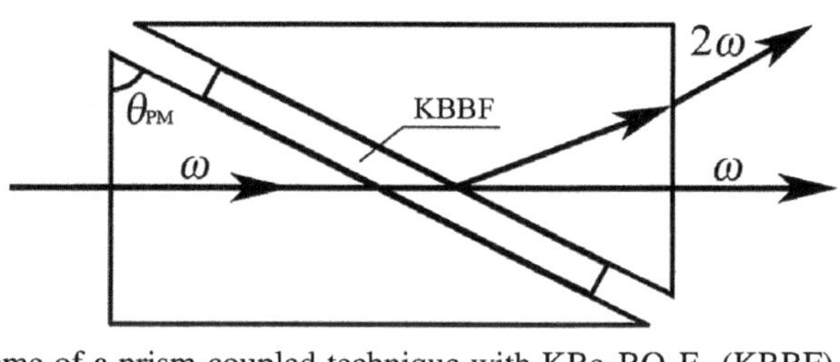

Figure 6.9 Scheme of a prism-coupled technique with $KBe_2BO_3F_2$ (KBBF).

may also be used instead of SiO_2 because their refractive indices are similar in the UV region and it has a higher transmittance than that of SiO_2 in the DUV region. By using a prism angle of 66.5 ± 1 degree in this KBBF-PCD device, DUV laser radiation at 177.3 nm has been produced for the first time [37]. In particular, the device can greatly reduce Fresnel reflection losses in both UV and DUV regions.

In recent years, the need for high-power DUV generation has placed greater demands on the KBBF-PCD that cannot be met. For example, it is unavoidable that there are a few bubbles or impurities inside the interface which can absorb and scatter the incident light, so the device will suffer from interface separation and optical damage at high laser powers due to weak intermolecular bonding between the two materials. To overcome this, an improved "diffusion bonding KBBF-PCD" has been developed in Chen's group [14]. The surfaces of both KBBF and prisms are optically polished and coated with SiO_2 film via electron beam evaporation using a DUV coating machine (Leybold Optics SYRUSpro 1110). Then the device is placed in an oven and maintained at a high temperature of 150°C for two days to allow SiO_2 molecules between the two KBBF prism interfaces to diffuse and form a robust chemical bonding. As the bonding is between the same SiO_2 films, this new technique also eliminated impurities or bubbles between the interfaces. A laser damage threshold, three times higher than that of an ordinary optically contacted KBBF-PCD, might be achieved in this way.

2.3.2 Sixth-Harmonic Generation of Nd-Based Lasers

The first sixth-harmonic generation at 177.3 nm was achieved from an Nd-based laser in 2003 by using an optically contacted KBBF-PCD [37]. The fundamental beam source was a mode-locked frequency-tripled Nd:YVO_4 laser (Vanguard, Spectra-Physics) with a repetition rate of 80 MHz and pulse width of 10 ps at 355 nm. At that time, the conversion efficiency was only about 0.1%, limited by both the fundamental laser power and the crystal thickness, and a maximum output power of 2.5 mW at 177.3 nm was obtained from 3.5 W of the pump power. Later, with the improvement of crystal quality and thickness, this was gradually increased. For example, an average power of 41 mW was obtained with an input power of 7.5 W from the same pump laser by using an optically contacted KBBF-PCD with a 2.1-mm-thick crystal [50]. With a 10-ns, 10-kHz pulsed laser system the output power at 177.3 nm was increased to 120.5 mW when the 355 nm power was increased to 7.17 W using an optically contacted KBBF-PCD with a 1.95-mm-thick crystal (Fig. 6.10) [57].

2.3.3 Even Shorter Wavelength Generation in the VUV Region

The shortest wavelength generated by a phase-matching process in an NLO crystal was obtained at 153.4 nm by using an optical contact KBBF-PCD in 2011 [30]. It was achieved by successive frequency conversion of a fundamental Yb-doped fiber laser centered at 1074 nm using LBO crystals and KBBF-PCDs. The type-I phase-matching SHG process $3\omega\cdot(o) + 3\omega\cdot(o) \rightarrow 6\omega\cdot(e)$ was realized in the

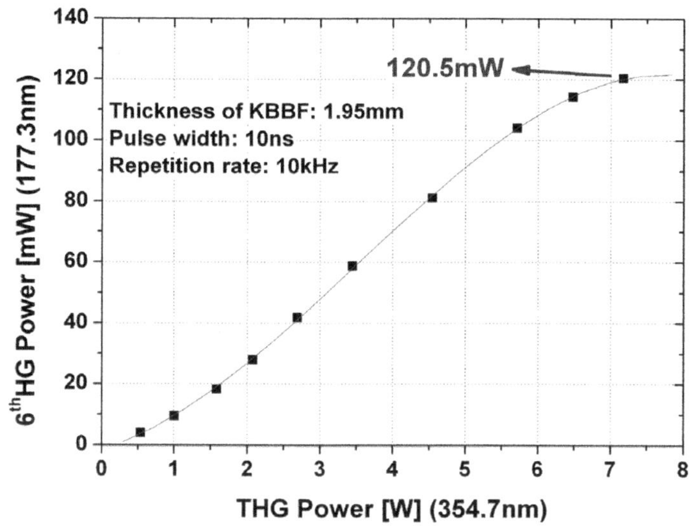

Figure 6.10 Experimental sixth-harmonic generation output power at 177.3 nm versus third-harmonic generation power at 354.7 nm. KBBF, $KBe_2BO_3F_2$.

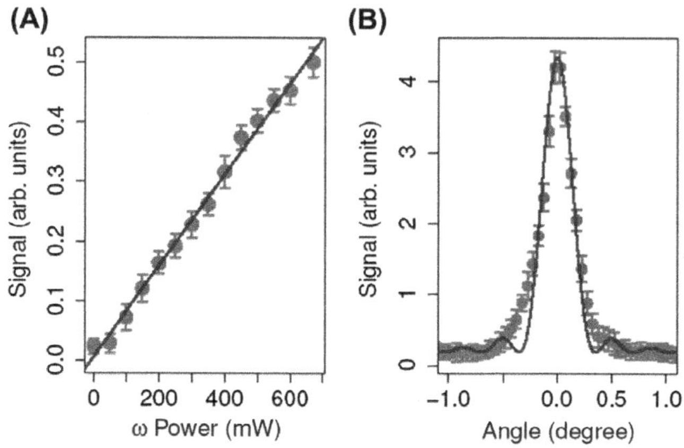

Figure 6.11 (A) The seventh-harmonic signal plotted against the power of the incident infrared beam. *Red circles* show the experimental data, the line is the linear fit. (B) Crystal angle dependence of the seventh-harmonic signal. The *circles* show the seventh-harmonic signal plotted against the angle of $KBe_2BO_3F_2$. The curve is calculated by assuming the phase-matching length of 200 nm.

KBBF-PCD and then $\omega\cdot(o) + 6\omega\cdot(o) \rightarrow 7\omega\cdot(e)$ sum-frequency generation took place in another KBBF-PCD to produce 153 nm. The result is shown in Fig. 6.11. The maximum 7H signal was estimated to be 1 nW. This low output was due to the absorption edge of KBBF itself. Of course higher output power would be desirable, and could be obtained by shifting to longer wavelengths with less absorption, or the absorption could be decreased by cooling the crystal [39]. This shorter wavelength in the vacuum ultraviolet (VUV) region is expected to be used in angle-resolved photoemission spectroscopy (ARPES) measurements.

2.3.4 Continuous Wavelength Generation in the Deep-UV Region

The KBBF crystal was successfully employed to produce continuous-wave (cw) light below 200 nm by a direct SHG method in 2012 [33]. Continuous-wave 191 nm radiation was generated in a two-stage SHG setup, consisting of a bow-tie cavity with a Brewster-cut KBBF-PCD inside. The initial wavelength of 764 nm was first converted to 382 nm in a commercially available TA-SHG system. Then p-polarized 382 nm acted as the fundamental wave for the second SHG process to produce 191 nm output. The dependence of the latter on the pump power was measured to have a maximum output of 1.3 mW. Its bandwidth was estimated to be <100 kHz. This kind of coherent source is very promising for both basic research and industry such as in ARPES, DUV metrology, or seeding of excimer lasers. Soon after, a slightly longer cw wavelength at 193 nm was obtained with more than 15 mW emission by using a KBBF-PCD [34]. This power level is sufficient for replacing excimer lasers in metrological applications.

2.3.5 Tunable Fourth-Harmonic Generation of Ti:Sapphire Lasers

As KBBF crystal has a very wide phase-matching range from visible down to the DUV spectral region, it is conceivable that it could be used to generate the fourth harmonic over the entire tunable wavelength range of the Ti:sapphire laser. This was first achieved from a nanosecond Ti:sapphire laser based on KBBF-PCD in 2008 [55]. The highest output power is only 2.23 mW at 193 nm. Soon after that, by using a Coherent Chameleon Ultra II Ti:Sapphire laser (80 MHz, 150 fs), widely tunable high average power of fourth harmonic from 170 to 232.5 nm was produced [58]. A BBO crystal was used for the first SHG of the tunable fundamental laser from 940 to 680 nm. Then a single KBBF-PCD with a crystal of size $20 \times 6 \times 1.28\ mm^3$ and a prism apex angle of 60 degrees was used to produce tunable fourth-harmonic generation over the entire fundamental wavelength range. Fig. 6.12 shows the tunable fourth- and second-harmonic output power as a function of fundamental wavelength. The output power from 175 to 232.5 nm exceeds 2 mW and the highest power reaches 119.8 mW at 202.5 nm. This result is also very useful for ARPES and other DUV spectrometers.

2.3.6 Fourth-Harmonic Generation of Nd:YAG Lasers

In the DUV region below 200 nm, KBBF is unique and irreplaceable. In the usual UV region it also has several advantages in generating the fourth harmonic of Nd-based lasers at 266 nm compared with currently used BBO and $CsLiB_6O_{10}$ (CLBO) crystals. For example, it has a much higher laser damage threshold than BBO and CLBO, and much smaller two-photon absorption than BBO. It is not hygroscopic compared with

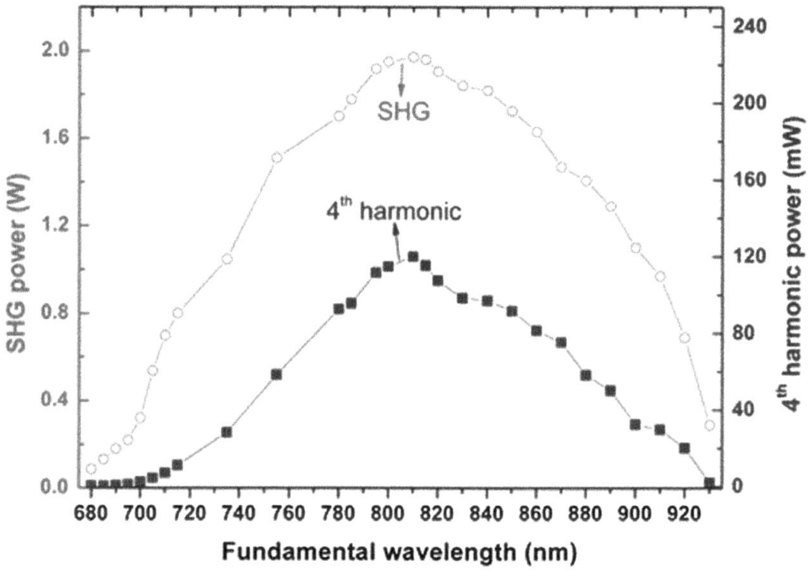

Figure 6.12 The output power curves of tunable fourth- and second-harmonic of Ti:sapphire laser versus the fundamental wavelengths. *Dots* are second-harmonic generation power output and *squares* are fourth-harmonic generation power output.

CLBO. In Ref. [42], a Q-switched Nd:YAG laser (Edgewave IS161-E) at 1064 nm with a pulse width of 10 ns and a repetition rate of 10 kHz was used to pump a 30-mm-thick LBO crystal cut for type-I noncritical phase-matching and produces SHG at 532 nm. An optical contact KBBF-PCD with a crystal of size $10 \times 6 \times 2.92\ mm^3$ was employed to generate the fourth harmonic at 266 nm. A maximum output power of 7.86 W was obtained, corresponding to a conversion efficiency of 13.4%. The next step is to improve the stability at high output powers by applying heat control for commercial applications.

2.4 Applications

All-solid-state DUV coherent light sources based on KBBF-PCD have been successfully used in several advanced scientific instruments in fields such as photoemission spectroscopy, photoemission electron microscopy, and stimulated Raman spectrometry. The following DUV laser sources are all based on KBBF-PCDs.

2.4.1 Photoemission Spectroscopy

Japanese and Chinese scientists have successfully combined the DUV laser ($h\nu = 6.994$ eV) with photoemission spectroscopy, and super-high resolution systems with angle-integrated and angle-resolved modes (in ARPES) have been developed. These systems have several advantages compared with traditional light sources such as synchrotron radiation and discharge lamps (for example, the He Iα line emitted from mercury lamps). Their chief merits include the following: (1) Super-high energy resolution: an energy resolution better than 1 mV has been achieved [22], making it possible to observe explicitly the very small superconducting gap and fine electronic structure produced by electron Cooper pairs; (2) High-momentum resolution: the lower photon energy as well as the super-high energy resolution lead to better momentum resolution; (3) Super-high photon flux: the photon flux is typically $10^{14}-10^{15}$ photons/s, which is nearly two or three orders higher than that of a normal DUV source; (4) Enhanced bulk sensitivity: the escape depth for the photoelectrons is only 5–10 Å in the photon energy range (20–50 eV) that is used in regular ARPES, and can be increased to 30–100 Å for a photon energy of 6.994 eV. By using super-high resolution DUV laser-based ARPES, many new phenomena have been observed in high-temperature superconductors and topological insulators.

For example, the superconducting energy gap of an f-electron $CeRu_2$ superconductor has been directly observed by ARPES for the first time [21]. This study provides direct evidence of an anisotropic superconducting gap in this compound, and demonstrates the potential of DUV laser photoemission spectroscopy in investigating unconventional superconductivity in d- and f-electron superconductors.

Furthermore, scientists have observed the coexistence of Fermi arcs and Fermi pockets in a high-Tc copper oxide superconductor La-Bi2201 for the first time, which was not expected theoretically [29]. This will provide new insight for understanding the anomalous pseudogap state in high-Tc cuprate superconductors. Another notable result is the identification of a new form of electron coupling in Bi2212 [56], which is significant for the understanding of the mechanism of high-Tc superconductivity. By using DUV laser-based spin ARPES, a unique feature of the orbital-dependent spin texture in the topological insulator Bi_2Se_3 has been observed for the first time [47], thus providing definite evidence for the corresponding theory and paving a new way of using light to control the spin structure of topological insulators, which might have important applications in future spin-involved technologies. We can say that the application of DUV lasers promises revolutionary changes in high-resolution photoemission spectroscopy, which should greatly promote our understanding of unusual electron structures in solid materials.

2.4.2 Photoemission Electron Microscopy

The DUV laser source at $\lambda = 177.3$ nm generated by a KBBF-PCD has been successfully used as the excitation source in a photoemission electron microscope (PEEM) developed by Xinhe Bao et al. in the Dalian Institute of Chemical Physics, Chinese Academy of Science [18]. This laser has several advantages compared with other excitation sources like UV light from discharge lamps, synchrotron radiation, or X-rays. The spatial resolution of DUV-PEEM can reach 5 nm, and its energy resolution 0.1 eV. It has been applied in chemistry, physics, and materials science and has produced many new research results.

For example, Jin et al. observed that the growth of graphene can be changed by introducing Ar gas bubbles at the Ru (0001) substrate in situ in a low-energy electron microscope system which comprises a DUV laser ($\lambda = 177.3$ nm) [19]. The results indicate a means to tailor graphene growth on solid substrates.

2.4.3 Deep-UV Raman Spectroscopy

Since most biological substances, such as protein and water, have intense absorption in the DUV region below 200 nm, not only information about the vibrational structure but also the electronic excited states can be obtained by DUV resonance Raman spectroscopy. This technology using excitation at 177.3 nm was first developed by C. Li in the Dalian Institute of Chemical Physics.

For example, experiments on Teflon reveal that the electronic geometry of C_nF_{2n+2} molecules in the $\sigma\sigma^*$ excited state is displaced along directions both parallel and perpendicular to the C—C skeleton, and the excited C—C bond does not dissociate due to delocalization of the excited electron in the σ^* orbital [20]. Indeed, both DUV Raman spectroscopy and DUV resonance Raman spectroscopy are powerful tools for investigating high-energy electronic transition systems.

2.4.4 Future Applications

In the future, all-solid-state DUV lasers may be used in several important fundamental research areas. For example, using a terawatt level 200—195 nm laser based on KBBF-PCD to excite inert gases, it is possible to produce attosecond pulses (10^{-18} s) which can be used to observe the electron orbits in atoms directly. The Al ion-based optical frequency standard may be adjusted to a precision of 10^{-18} s using a vacuum-UV laser at 165—167 nm. The Rydberg constant in H atoms may be remeasured by using a 170—155-nm laser. The development of all-solid-state DUV lasers will undoubtedly have a profound and lasting influence on these research fields.

3. RBBF

3.1 RBBF Crystal Growth

RBBF belongs to the $MBe_2BO_3F_2$ (M = Na, K, Rb, Cs) family which was synthesized by the former Soviet scientists in 1975 [1]. Its structure was at first determined to be C2 (No. 5), belonging to the monoclinic system. In 2005 this was redetermined on the basis of X-ray data after a single crystal became available [45]. Similar to KBBF, its space structure was proved to be R32 (D_3^7), belonging to the uniaxial and hexagonal system, with unit cell $a = b = 4.4341(9)$ Å, $c = 19.758(5)$ Å, and $Z = 3$. The basic building blocks are $(BO_3)^{3-}$ and $(BeO_3F)^{5-}$, which form an infinite lattice network of $(Be_2BO_3F_2)_{n\to\infty}$ along the a—b plane. The distance between each layer is 6.59 Å with a very weak Rb-F electrostatic force. Therefore the structure also exhibits

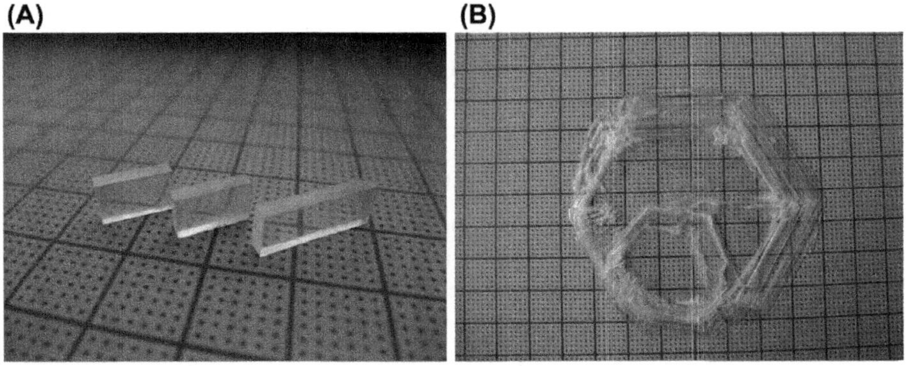

Figure 6.13 (A) RBBF devices cut from the bulk one. (B) Bulk $RbBe_2BO_3F_2$ (RBBF) crystal of size $50 \times 50 \times 1.5$ mm^3.

a strong layering tendency along the *c*-axis which makes it rather difficult to grow thick crystals, similar to the case of KBBF. The two crystals have the same growth habit, since the only difference between them is the different alkali metal. A high-temperature flux method using spontaneous nucleation may also be used to grow RBBF, and details have been reported [25]. Fig. 6.13A shows the as-grown crystal from the RbO-NaF-B_2O_3 flux system with a thickness of 2.5 mm along the *z*-axis. It has been successfully cut into different sizes for device fabrication (Fig. 6.13B).

The crystal has also been grown by the hydrothermal method, and samples thicker than 5 mm have been obtained with 1.0 M RbF as a mineralizer [24]. In recent works, Xu et al. found that some hydrothermally grown crystals have a centrosymmetric structure of $R\overline{3}c$. The unit structure is double in size along the *c*-axis compared with the standard R32 structure. As the crystals show a rather low SHG capability, a stacking fault model has been proposed that suggests the crystal is nonuniform with a mixture of both R32 and $R\overline{3}c$ structures [49]. This is the same as reported for hydrothermal KBBF [53].

3.2 Linear and Nonlinear Optical Properties

The cutoff wavelength of RBBF on the DUV side is located at 150 nm (Fig. 6.14). A small absorption shoulder is observed around 160 nm which may be caused by impurities or defects. The cutoff wavelength of the transmittance spectrum in the IR regions is located at 3550 nm [13].

The Sellmeier equations have been determined by using a right-angle prism with an apex angle of 30 degrees made with a 2.0-mm-thick RBBF crystal [13]:

$$n_o^2 = 1 + \frac{1.18675\lambda^2}{\lambda^2 - 0.00750} - 0.00910\lambda^2$$
$$(\lambda \text{ is in } \mu\text{m}) \qquad [6.4]$$
$$n_e^2 = 1 + \frac{0.97530\lambda^2}{\lambda^2 - 0.00665} - 0.00145\lambda^2$$

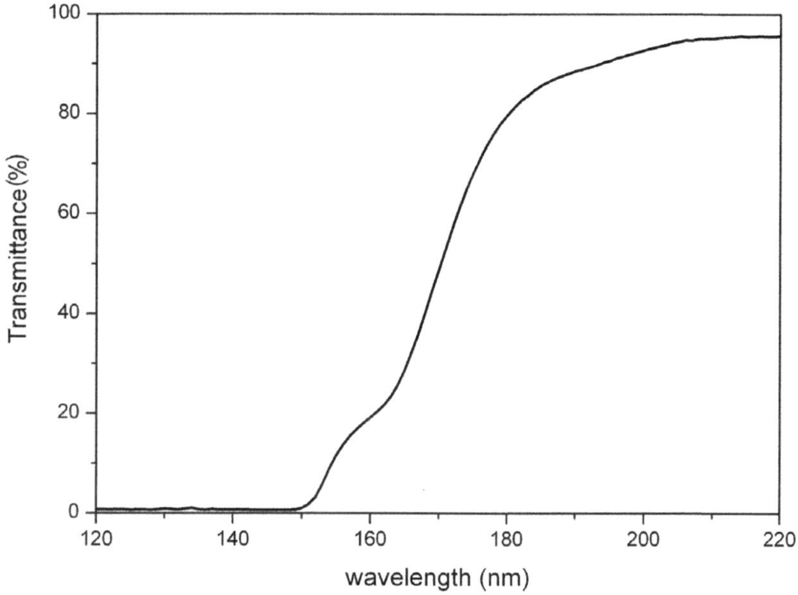

Figure 6.14 The transmittance curve of $RbBe_2BO_3F_2$ in the UV side.

From these equations the refractive indices have been calculated. Combined with the measured and calculated phase-matching angles, we can conclude that the shortest SHG wavelength for RBBF using type-I phase-matching may reach down to 170 nm. Hence, RBBF is the second DUV NLO crystal that can generate DUV laser radiation by a direct SHG method.

By using the same prism, nine refractive indices have been measured from DUV to near IR within a temperature range of 24–160°C [54]. The thermal refractive index coefficients *dn*/*dT* have been determined and their relationship with the wavelength λ may be expressed in the following formula:

$$\frac{dn_o}{dT} = \left(\frac{0.099911}{\lambda^3} - \frac{0.553474}{\lambda^2} + \frac{1.454609}{\lambda} - 13.260115\right) \times 10^{-6}$$

$$\frac{dn_e}{dT} = \left(\frac{0.285633}{\lambda^3} - \frac{2.482927}{\lambda^2} + \frac{6.916728}{\lambda} - 16.153736\right) \times 10^{-6} \quad [6.5]$$

$(0.194\ \mu m \leq \lambda \leq 1.014\ \mu m)$

Based on this formula, the SHG phase-matching angles for RBBF at 266 nm at different temperatures can be derived, and are found to be in good agreement with the experimental results. The angle changes only 0.12 degree when the temperature is varied from 24 to 160°C, which indicates that the crystal has high thermal stability for UV laser generation.

Similar to KBBF, RBBF also has only two nonzero d_{ij} coefficients, ie, d_{11} and d_{14}. Theoretical calculation and experiments both reveal that d_{14} is very small. On the other hand, the effective d_{eff} coefficients are:

$$\begin{aligned} d_{eff} &= d_{11} \cos\theta \cos 3\varphi \quad \text{(type-I)} \\ d_{eff} &= d_{11} \cos 2\theta \sin 3\varphi \quad \text{(type-II)} \end{aligned} \qquad [6.6]$$

Here, d_{14} does not contribute to the d_{eff} coefficients, so it is only d_{11} that needs to be determined. This has been precisely measured by the Maker fringes technique with a $10 \times 10 \times 1.0\ mm^3$ c-cut crystal plate. By comparing the fringe envelope of the d_{11} coefficient of RBBF and that for the d_{36} coefficient of KDP, we can deduce exactly that $d_{11(RBBF)} = (0.45 \pm 0.01)$ pm/V (if d_{36}(KDP) $= 0.39$ pm/V is adopted), which is comparable to that of KBBF [10]. However, due to the intrinsic characteristics of the refractive index, the phase-matching angle of RBBF is always larger than that of KBBF, so its conversion efficiency is always less than the latter (this can be seen from the effective nonlinear coefficient expressions).

3.3 Deep-UV Laser Generation

An RBBF crystal also cannot be cut along the phase-matching angle due to its thin thickness along the *c*-axis. To solve this problem, a special PCD similar to the KBBF-PCD is adopted [8]. The sixth harmonic of an Nd-based laser at 177.3 nm can be realized using an RBBF-PCD device. For example, with a nanosecond, 10-kHz, 355-nm laser source, a maximum average power of 10.8 mW at 177.3 nm has been obtained using an RBBF-PCD with dimensions of $23 \times 6.0 \times 1.0\ mm^3$ [13]. Obviously, the output power can be further increased if a thicker crystal could be used.

Using two RBBF-PCD devices, tunable fourth-harmonic generation of the Ti:sapphire laser has been successfully realized from 175 to 232 nm. The first device with a prism angle of 60 degrees is used for fourth-harmonic generation from 180 to 232 nm, while the second with an angle of 75 degrees is used to produce even shorter wavelengths down to 175 nm. In the experiment, a femtosecond Ti:sapphire laser (Chameleon-ultra II, Physical-spectral, 150 fs, 80 MHz) is used as the pump. A BBO crystal first frequency doubles the tunable fundamental wavelengths of 930–720 nm, then an RBBF-PCD with crystal dimensions of $20 \times 6 \times 0.95\ mm^3$ is used for fourth-harmonic generation. Fig. 6.15A shows the tunable fourth- and second-harmonic power output as a function of fundamental wavelength. Within the tunable DUV range, the maximum output power is 43.3 mW at 202.5 nm with an SHG power of 2.08 W at 405 nm. Fig. 6.15B shows the output power from 175 to 195 nm from the second RBBF-PCD, which may be as high as 1.2 mW at 175 nm.

3.4 UV Laser Generation

Fourth-harmonic generation at 266 nm with an RBBF crystal has also been realized. Using an Nd:YAG laser with a repetition frequency of 10 Hz and pulse width of

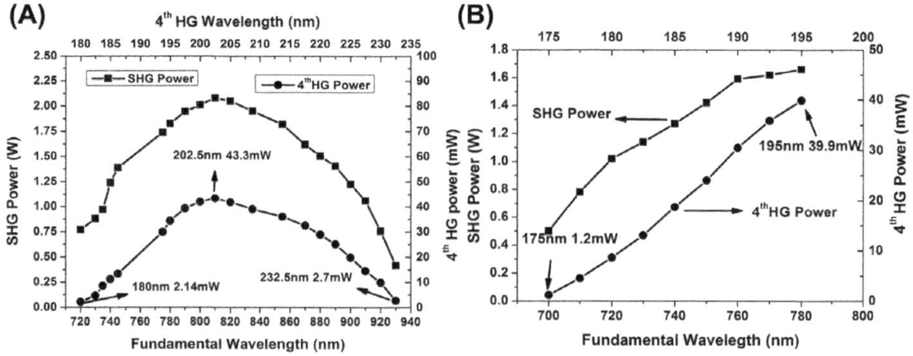

Figure 6.15 The output power curves of tunable fourth- and second-harmonic of Ti:sapphire laser versus the fundamental wavelengths (A) length from 720 to 930 nm and (b) from 700 to 780 nm.

25 ps, a conversion efficiency as high as 25.5% and an output of 3.0 W of 266 nm light has been obtained using a laser source with a high repetition rate of 10 kHz [41]. In summary, RBBF is another DUV crystal suitable for applications, though its NLO performance is inferior to that of KBBF.

4. CBBF

The CBBF crystal, that is $CsBe_2BO_3F_2$, has the same structure of R32 as that of KBBF, and its growth habit is also similar. A size as large as $30 \times 30 \times 2.0\ mm^3$ has been grown using a $CsF\text{-}B_2O_3$ flux system [17], but the crystal can also be grown by the hydrothermal method [15].

Different from KBBF and RBBF, CBBF cannot produce DUV light below 200 nm by a direct SHG method because its birefringence in the DUV region is smaller. Its shortest SHG wavelength is around 201 nm, so it can be used for 266 and 213 nm harmonic generation of Nd-based lasers [17]. In this respect CBBF is a promising UV NLO crystal, but more details will not be discussed here.

5. Conclusions

Two members of the KBBF family of crystals, KBBF and RBBF, are now the only crystals that can produce DUV harmonic generation beyond 200 nm through a simple SHG method. The former shows better performance and has extensive application prospects, while CBBF can only achieve SHG down to 201 nm. Since it is difficult to grow bulk crystals along the *z*-axis and to cut the crystal along the phase-matching angle, a PCD has been developed which has proved to be practical for applications. Stable mW-level 177.3 nm laser radiation based on the KBBF-PCD has been successfully used in many advanced instruments, such as super-high energy resolution laser photoemission spectrographs and DUV laser PEEMs. Now, even shorter wavelength DUV lasers are being designed by a direct SHG method using KBBF-PCDs, which should lead to even wider applications. In a word, the combination of KBBF crystal and PCD technique has promoted the applications of DUV lasers and in a certain extent, opened the door to a new area of DUV laser technology.

References

[1] I.A. Baidina, V.V. Bakakin, L.R. Batsanova, N.A. Palchik, N.V. Podberezskaya, L.P. Soloveva, X-ray structural study of borato-fluoroberyllates with composition $MBe_2(BO_3)F_2$ (M = Na, K, Rb, Cs), J. Struct. Chem. 16 (1975) 963–965.
[2] L.R. Batsanov, V.A. Egorov, A.V. Nikolaev, Fluorideborate of beryllium, Dokl. A.N. SSSR 178 (1968) 1317–1319.
[3] C.T. Chen, B.C. Wu, A.D. Jiang, G.M. You, A new-type ultraviolet SHG crystal β-BaB_2O_4, Sci. Sin. 28 (1985) 235–243.
[4] C.T. Chen, Y.C. Wu, R.K. Li, The relationship between the structural type of anionic group and SHG effect in boron-oxygen compounds, Chin. Phys. Lett. 2 (9) (1985) 389–392.
[5] C.T. Chen, Y.C. Wu, A.D. Jiang, B.C. Wu, G.M. You, R.K. Li, S.J. Lin, New nonlinear-optical crystal-LiB_3O_5, J. Opt. Soc. Am. B Opt. Phys. 6 (1989) 616–621.
[6] C.T. Chen, Y. Wu, R. Li, The development of new crystals in the borate series, J. Cryst. Growth 99 (1990) 790–798.
[7] C.T. Chen, Z.Y. Xu, D.Q. Deng, J. Zhang, G.K.L. Wong, B.C. Wu, N. Ye, D.Y. Tang, The vacuum ultraviolet phase-matching characteristics of nonlinear optical $KBe_2BO_3F_2$ crystal, Appl. Phys. Lett. 68 (1996) 2930–2932.
[8] C.T. Chen, J.H. Lu, G.L. Wang, Z.Y. Xu, J.Y. Wang, C.Q. Zhang, Y.G. Liu, Deep ultraviolet harmonic generation with $KBe_2BO_3F_2$ crystal, Chin. Phys. Lett. 18 (2001) 1081.
[9] C. Chuangtian, X. Zuyan, Prism coupling technique and deep-UV harmonic output of KBBF crystal, J. Synth. Cryst. 31 (2002) 224–227.
[10] C.T. Chen, Recent advances in deep and vacuum-UV harmonic generation with KBBF crystal, Opt. Mater. 26 (2004) 425–429.
[11] C. Chen, G. Wang, X. Wang, Y. Zhu, Z. Xu, T. Kanai, S. Watanabe, Improved sellmeier equations and phase-matching characteristics in deep-ultraviolet region of crystal, IEEE J. Quantum Electron. 44 (2008) 617–621.
[12] C.T. Chen, G.L. Wang, X.Y. Wang, Z.Y. Xu, Deep-UV nonlinear optical crystal $KBe_2BO_3F_2$-discovery, growth, optical properties and applications, Appl. Phys. B Lasers Opt. 97 (2009) 9–25.
[13] C. Chen, S. Luo, X. Wang, G. Wang, X. Wen, H. Wu, X. Zhang, Z. Xu, Deep UV nonlinear optical crystal: $RbBe_2BO_3F_2$, J. Opt. Soc. Am. B Opt. Phys. 26 (2009) 1519–1525.
[14] C. Chen, Y. Zhu, F. Fan, X. Wang, R. Li, Optical Coupling Device Having KBBF Group Crystal Coupled with Prisms and Method for Manufacturing Same, Patent No. US 8773750 B2, 2014.
[15] C.D. McMillen, J.W. Kolis, Hydrothermal crystal growth of $ABe_2BO_3F_2$ (A = K, Rb, Cs, Tl) NLO crystals, J. Cryst. Growth 310 (2008) 2033–2038.
[16] P.A. Franken, G. Weinreich, C.W. Peters, A.E. Hill, Generation of optical harmonics, Phys. Rev. Lett. 7 (1961) 118–234.

[17] H.W. Huang, C.T. Chen, X.Y. Wang, Y. Zhu, G.L. Wang, X. Zhang, L.R. Wang, J.Y. Yao, Ultraviolet nonlinear optical crystal: $CsBe_2BO_3F_2$, J. Opt. Soc. Am. B Opt. Phy. 28 (2011) 2186–2190.
[18] L. Jin, Q. Fu, R. Mu, D. Tan, X. Bao, Pb intercalation underneath a graphene layer on Ru(0001) and its effect on graphene oxidation, Phys. Chem. Chem. Phys. 13 (2011) 16655–16660.
[19] L. Jin, Q. Fu, H. Zhang, R.T. Mu, Y.H. Zhang, D.L. Tan, X.H. Bao, Tailoring the growth of graphene on Ru(0001) via engineering of the substrate surface, J. Phys. Chem. C 116 (2012) 2988–2993.
[20] S. Jin, F. Fan, M. Guo, Y. Zhang, Z. Feng, C. Li, Note: deep ultraviolet Raman spectrograph with the laser excitation line down to 177.3 nm and its application, Rev. Sci. Instrum. 85 (2014).
[21] T. Kiss, F. Kanetaka, T. Yokoya, T. Shimojima, K. Kanai, S. Shin, Y. Onuki, T. Togashi, C. Zhang, C.T. Chen, S. Watanabe, Photoemission spectroscopic evidence of gap anisotropy in an electron superconductor, Phys. Rev. Lett. 94 (2005) 057001.
[22] T. Kiss, T. Shimojima, K. Ishizaka, A. Chainani, T. Togashi, T. Kanai, X.Y. Wang, C.T. Chen, S. Watanabe, S. Shin, A versatile system for ultrahigh resolution, low temperature, and polarization dependent laser-angle-resolved photoemission spectroscopy, Rev. Sci. Instrum. 79 (2008).
[23] R.K. Li, The interpretation of UV absorption of borate glasses and crystals, J. Non Cryst. Solids 111 (1989) 199–204.
[24] L. Liu, H. Zhou, X. He, X. Zhang, X. Wang, F. Lu, C. Zhang, W. Zhou, C. Chen, Hydrothermal growth and optical properties of $RbBe_2BO_3F_2$ crystals, J. Cryst. Growth 348 (2012) 60–64.
[25] L.J. Liu, X.U. Tao, X.Y. Wang, C.T. Chen, Flux growth and thermal properties of $RbBe_2BO_3F_2$ crystals, J. Synth. Cryst. 42 (5) (2013) 779–782.
[26] T.H. Maiman, Stimulated optical radiation in ruby, Nature 187 (1960) 493–494.
[27] L. Mei, X. Huang, Y. Wang, Q. Wu, B. Wu, C. Chen, Crystal-structure of $KBe_2BO_3F_2$, Z. Kristallogr. 210 (1995) 93–95.
[28] L.F. Mei, C.Z. He, C.T. Chen, Flux growth of $KBe_2BO_3F_2$, J. Cryst. Growth 132 (1993) 609–610.
[29] J. Meng, G. Liu, W. Zhang, L. Zhao, H. Liu, X. Jia, D. Mu, S. Liu, X. Dong, J. Zhang, W. Lu, G. Wang, Y. Zhou, Y. Zhu, X. Wang, Z. Xu, C. Chen, X.J. Zhou, Coexistence of Fermi arcs and Fermi pockets in a high-Tc copper oxide superconductor, Nature 462 (2009) 335–338.
[30] Y. Nomura, Y. Ito, A. Ozawa, X. Wang, C. Chen, S. Shin, S. Watanabe, Y. Kobayashi, Coherent quasi-cw 153 nm light source at 33 MHz repetition rate, Opt. Lett. 36 (2011) 1758–1760.
[31] D.A. Roberts, Simplified characterization of uniaxial and biaxial nonlinear optical-crystals – a plea for standardization of nomenclature and conventions, IEEE J. Quantum Electron. 28 (1992) 2057–2074.
[32] Y.H. Sang, D.H. Yu, M. Avdeev, R. Piltz, N. Sharma, N. Ye, H. Liu, J.Y. Wang, X-ray and neutron diffraction studies of flux and hydrothermally grown nonlinear optical material $KBe_2BO_3F_2$, CrystEngComm 14 (2012) 6079–6084.
[33] M. Scholz, D. Opalevs, P. Leisching, W. Kaenders, G. Wang, X. Wang, R. Li, C. Chen, 1.3-mW tunable and narrow-band continuous-wave light source at 191 nm, Opt. Express 20 (2012) 18659–18664.
[34] M. Scholz, D. Opalevs, P. Leisching, W. Kaenders, G. Wang, X. Wang, R. Li, C. Chen, A bright continuous-wave laser source at 193 nm, Appl. Phys. Lett. 103 (2013) 051114.

[35] L.P. Solov'eva, V.V. Bakakin, Crystal structure of potassium boratofluoroberyllate $KBe_2BO_3F_2$, Sov. Phys. Crystallogr. 15 (1970) 802–805.
[36] D.Y. Tang, Y.N. Xia, B.C. Wu, C.T. Chen, Growth of a new UV nonlinear optical crystal: $KBe_2BO_3F_2$, J. Cryst. Growth 222 (2001) 125–129.
[37] T. Togashi, T. Kanai, T. Sekikawa, S. Watanabe, C.T. Chen, C.Q. Zhang, Z.Y. Xu, J.Y. Wang, Generation of vacuum-ultraviolet light by an optically contacted, prism-coupled $KBe_2BO_3F_2$ crystal, Opt. Lett. 28 (2003) 254–256.
[38] Y. Tong, X.Y. Meng, Z.Z. Wang, C. Chen, M.-H. Lee, The mechanism of linear and nonlinear optical effects in fluoride crystals, J. Appl. Phys. 98 (2005) 033504.
[39] F. Urbach, The long-wavelength edge of photographic sensitivity and of the electronic absorption of solids, Phys. Rev. 92 (1953) 1324.
[40] J.Y. Wang, C.Q. Zhang, Y.G. Liu, J.X. Zhang, X.B. Hu, M.H. Jiang, C.T. Chen, Y.C. Wu, Z.Y. Xu, Growth and properties of $KBe_2BO_3F_2$ crystal, J. Mater. Res. 18 (2003) 2478–2485.
[41] L.-R. Wang, G.-L. Wang, X. Zhang, L.-J. Liu, X.-Y. Wang, Y. Zhu, C.-T. Chen, Generation of ultraviolet radiation at 266 nm with $RbBe_2BO_3F_2$ crystal, Chin. Phys. Lett. 29 (6) (2012) 064203.
[42] L. Wang, N. Zhai, L. Liu, X. Wang, G. Wang, Y. Zhu, C. Chen, High-average-power 266 nm generation with a $KBe_2BO_3F_2$ prism-coupled device, Opt. Express 22 (2014) 27086–27093.
[43] X. Wang, X. Yan, S. Luo, C. Chen, Flux growth of large KBBF crystals by localized spontaneous nucleation, J. Cryst. Growth 318 (2011) 610–612.
[44] B. Wu, D. Tang, N. Ye, C. Chen, Linear and nonlinear optical properties of the $KBe_2BO_3F_2$ (KBBF) crystal, Opt. Mater. 5 (1996) 105–109.
[45] X.H. Wen, Nonlinear Optical Crystals MBBF (M = Na, K, Rb, Cs): Growth and Properties, Technical Institute of Physics and Chemistry, Chinese Academy of Science, 2006.
[46] Y.N. Xia, C.T. Chen, D.Y. Tang, B.C. Wu, New nonlinear-optical crystals for UV and VUV harmonic-generation, Adv. Mater. 7 (1995) 79–81.
[47] Z. Xie, S. He, C. Chen, Y. Feng, H. Yi, A. Liang, L. Zhao, D. Mou, J. He, Y. Peng, Orbital-selective spin texture and its manipulation in a topological insulator, Nat. Commun. (2014) 1–9, http://dx.doi.org/10.1038/ncomms4382.
[48] T. Xu, L. Liu, X. Wang, C. Chen, H. Zhou, X. He, C. Zhang, J. Sun, Direct observation of the microstructure and defects in $KBe_2BO_3F_2$ crystals by TEM, Dalton Trans. 43 (2014) 16411–16415.
[49] T. Xu, L. Liu, X. Wang, H. Zhou, X. He, C. Zhang, C. Chen, Superstructure studies in hydrothermal-grown $RbBe_2BO_3F_2$ crystals, J. Alloys Compd. 625 (2015) 118–121.
[50] F. Yang, Z.M. Wang, Y. Zhou, X.K. Cheng, S.Y. Xie, Q.J. Peng, D.F. Cui, J.Y. Zhang, X.Y. Wang, Y. Zhu, C.T. Chen, Z.Y. Xu, 41 mW high average power picosecond 177.3 nm laser by second-harmonic generation in KBBF, Opt. Commun. 283 (2010) 142–145.
[51] N. Ye, D. Tang, Hydrothermal growth of $KBe_2BO_3F_2$ crystals, J. Cryst. Growth 293 (2006) 233–235.
[52] J.Q. Yu, L.J. Liu, X.Y. Wang, H.T. Zhou, X.L. He, C.L. Zhang, W.N. Zhou, C.T. Chen, Study on defects in hydrothermal-grown $KBe_2BO_3F_2$ crystals, J. Cryst. Growth 318 (2011) 621–624.
[53] J.Q. Yu, L.J. Liu, S.F. Jin, H.T. Zhou, X.L. He, C.L. Zhang, W.N. Zhou, X.Y. Wang, X.L. Chen, C.T. Chen, Superstructure and stacking faults in hydrothermal-grown $KBe_2BO_3F_2$ crystals, J. Solid State Chem. 184 (2011) 2790–2793.

[54] N. Zhai, L. Wang, L. Liu, X. Wang, Y. Zhu, C. Chen, Measurement of thermal refractive index coefficients of nonlinear optical crystal $RbBe_2BO_3F_2$, Opt. Mater. 36 (2013) 333–336.

[55] H. Zhang, G. Wang, L. Guo, A. Geng, Y. Bo, D. Cui, Z. Xu, R. Li, Y. Zhu, X. Wang, C. Chen, 175 to 210 nm widely tunable deep-ultraviolet light generation based on KBBF crystal, Appl. Phys. B 93 (2008) 323–326.

[56] W. Zhang, G. Liu, L. Zhao, H. Liu, J. Meng, X. Dong, W. Lu, J.S. Wen, Z.J. Xu, G.D. Gu, T. Sasagawa, G. Wang, Y. Zhu, H. Zhang, Y. Zhou, X. Wang, Z. Zhao, C. Chen, Z. Xu, X.J. Zhou, Identification of a new form of electron coupling in the $Bi_2Sr_2CaCu_2O_8$ superconductor by laser-based angle-resolved photoemission spectroscopy, Phys. Rev. Lett. 100 (2008) 107002.

[57] X. Zhang, L.R. Wang, X.Y. Wang, G.L. Wang, Y. Zhu, C.T. Chen, High-power sixth-harmonic generation of an Nd:YAG laser with $KBe_2BO_3F_2$ prism-coupled devices, Opt. Commun. 285 (2012) 4519–4522.

[58] X. Zhang, Z. Wang, G. Wang, Y. Zhu, Z. Xu, C. Chen, Widely tunable and high-average-power fourth-harmonic generation of a Ti:sapphire laser with a $KBe_2BO_3F_2$ prism-coupled device, Opt. Lett. 34 (2009) 1342–1344.

[59] H.T. Zhou, X.L. He, W.N. Zhou, Z.G. Hu, C.L. Zhang, H.D. Huo, J.L. Wang, S.J. Qin, Y.B. Zuo, F.H. Lu, L.J. Liu, X.Y. Wang, Y.C. Liu, D.P. Li, H.X. Zhang, Y.X. Chen, Hydrothermal growth of KBBF crystals from KOH solution, J. Cryst. Growth 318 (2011) 613–617.

Fluorescence in Nanostructured Oxyfluoride Glasses Doped with Rare Earth Ions

7

J. Heo
Pohang University of Science and Technology (POSTECH), Gyeongbuk, Republic of Korea

C. Liu
Wuhan University of Technology, Hubei, P. R. China

Chapter Outline

1. Introduction

Oxyfluoride glass-ceramics (GC) are composed of fluoride nanocrystals embedded inside the predominantly oxide glass matrix. Wang and Ohwaki first reported the intense visible upconversion emission from oxyfluoride GC containing $Pb_xCd_{1-x}F_2$ cubic fluorite nanocrystals codoped with Er^{3+}/Yb^{3+} [1]. These oxyfluoride GC have many novel features compared to normal glasses or crystals. First of all, oxyfluoride GC combine the advantages of fluoride crystals and oxide glasses. During the thermal treatment, a large portion of rare earth ions is preferentially incorporated into the

Photonic and Electronic Properties of Fluoride Materials. http://dx.doi.org/10.1016/B978-0-12-801639-8.00007-6

fluoride nanocrystals. These fluoride nanocrystals provide low phonon energy environment for rare earth ions, leading to the high emission efficiency and extended lifetimes of the excited metastable states of rare earth ions [1]. The residual glass matrix is predominantly oxide in nature that possesses good mechanical, chemical, and thermal resistance. Second, they can be drawn into fibers that make oxyfluoride GC promising for fiber laser materials [2]. Third, chemical interaction with outside environment can be minimized by keeping these fluoride nanocrystals inside stable oxide glasses. In addition, unusual fluoride nanocrystals such as α-YF_3, normally unstable at room temperature, can be precipitated inside the glasses [3].

Since the pioneering work in 1993 [1], many glasses containing fluoride nanocrystals have been explored. Aluminosilicate glasses were most frequently used to prepare oxyfluoride GC containing wide varieties of rare earth–doped fluoride nanocrystals such as CaF_2, SrF_2, BaF_2, etc. [4–11]. Besides the aluminosilicate glasses, many other host glasses including silicates and germanates were investigated [12–20]. A comprehensive review on the glass systems and types of fluoride nanocrystals precipitated in the oxyfluoride GC can be found from Fedorov et al. [21].

For rare earth–doped oxyfluoride GC, development of efficient upconversion materials has been the main research target and thus, various efficient visible upconversion emissions were realized. For example, upon the formation of $Pb_xCd_{1-x}F_2$ nanocrystals, green emission intensity from Er^{3+} ions increased almost 100 times under the 980-nm laser excitation [1]. Chen et al. reported that the efficient near-infrared to ultraviolet upconversion emission from Tm^{3+}/Yb^{3+} codoped oxyfluoride GC [22,23]. Nd^{3+}:LaF_3 nanocrystals in oxyfluoride GC were developed as laser materials for 1.06-μm laser [24]. GC-containing Er^{3+} were also considered for fiber lasers or fiber-optic amplifier materials operating at 1.54 μm [25–30]. Pr^{3+}-, Ho^{3+}-, and Ho^{3+}-/Tm^{3+}-doped oxyfluoride GC have also been investigated for the laser hosts operating at the wavelengths of 1.3 [30], 1.2–1.7 [31], and 2.0 μm [32], respectively.

Attempts were also reported on the applications of rare earth–doped oxyfluoride GC as solar spectrum convertors to enhance the efficiency of solar cells using both upconversion and downconversion of the frequencies. Upconversion emission at ~525, ~545, ~650, and ~850 nm from Er^{3+}-doped oxyfluoride GC were observed by the 1540-nm laser excitation [33,34]. These upconversion are useful to convert infrared spectrum of solar energy to wavelengths that are readily absorbed by silicon. Infrared quantum cutting from Pr^{3+}/Yb^{3+}- and Tb^{3+}/Yb^{3+}-doped oxyfluoride GC was proposed to convert one high energy photon into two low energy photons with a theoretical efficiency approaching 200% [35–40]. Oxyfluoride GC doped with multiple rare earth ions can also convert the UV and near-infrared excitation light into red, green, and blue with tunable intensity ratios. These materials were investigated to generate white light from blue-emitting light emitting diodes. In fact, M_2ReF_7 (where M = Sr, Ba; Re = La, Gd, etc.) [20,41,42] and YF_3 [43] nanocrystals doped with various rare earth ions in oxyfluoride GC were able to generate white light with color index close to the solar spectrum. Liu et al. achieved white light emission from Ho^{3+}/Tm^{3+}/Yb^{3+} codoped PbF_2 nanocrystals in oxyfluoride GC under a 900-nm laser excitation, and the color index of the emission were tuned by controlling the excitation laser intensities [44].

Emission properties of rare earth ions in oxyfluoride GC are closely related to structure of fluoride nanocrystals, concentration and distribution of rare earth dopants in the fluoride nanocrystals. Although the structure of fluoride nanocrystals and dopant concentration can be controlled, homogeneous distribution of rare earth dopants inside the fluoride nanocrystals remain as a challenge. Conventional characterization methods such as X-ray diffraction (XRD), transmission electron microscopy (TEM), fluorescence line narrowing [45], and energy dispersive X-ray spectroscopy (EDS) can only provide an averaged estimation on the distribution of rare earth ions inside the fluoride nanocrystals. Since emission properties of rare earth ions in the fluoride nanocrystals are strongly dependent on the ion–ion interaction, it is important to analyze and control the distribution of rare earth ions if one were to optimize the emission properties from oxyfluoride GC.

This work reports the use of electron energy loss spectroscopy (EELS) to analyze the distribution of rare earth ions in the GC containing PbF_2 nanocrystals. Changes in absorption and fluorescent emission of Er^{3+} ions upon the preferential incorporation inside PbF_2 crystals are investigated. In addition, small amount of YF_3 was added to improve the emission properties of Er^{3+} ions by decreasing the cross relaxation.

2. Preparation and Characterization of Rare Earth–Doped Oxyfluoride Glass-Ceramics

Host glasses are aluminosilicate glasses containing PbF_2 and their nominal compositions are in Table 7.1. Rare earths were additionally doped up to 1 mol% of host glass composition. Chemical powders with purities of >99.9% were weighed and mixed thoroughly for the melting in alumina crucibles at 1300°C for 30 min under ambient atmosphere. The melts were poured onto the brass mold and pressed with another plate for quenching. The as-quenched glasses were annealed at 380°C for 2 h to relieve the residual thermal stress. All the glasses thus obtained were cut and polished for further treatment and characterization. GC were prepared through thermal treatment at temperatures between the glass transition and crystallization for different durations.

XRD (D/MAX-2500/PC, Rigaku) patterns were recorded to identify the crystalline phases formed in the GC. Cu-Kα radiation with a scanning rate of 2°C/min was used with a resolution of 0.02°. Specimens for TEM analysis were prepared using the

Table 7.1 Nominal Compositions of Glasses (in mol%)

Chemicals	SiO_2	Al_2O_3	PbF_2	TmF_3	NdF_3	ErF_3	YF_3
GN	50	15	35	–	0.5	–	–
GT	50	15	35	0.5	–	–	–
GE	55	10	35	–	–	0.5	–
GE5Y	55	10	35	–	–	0.5	1.0
GE1Y	55	10	35	–	–	0.1	1.0

standard dimple grinding and ion-milling processes. To reduce the damages to the specimens by the electron beam, both sides of the specimens were coated with carbon prior to analysis. A Cs-corrected high-resolution TEM (HR-(S)TEM, JEM-2200FS, JEOL) equipped with a Ω-filter was used.

Absorption spectra of the as-prepared glasses and glass-ceramic specimens were recorded using a UV/Vis/Near-IR spectrophotometer (Lambda 750S, Perkin Elmer). Visible and near-infrared emission spectra and lifetimes of the excited state energy levels of rare earth ions were recorded using a combination of 1/4-m monochromator, photomultiplier tube, InGaAs and InSb detectors, lock-in amplifier, and oscilloscope. An 800 nm laser modulated by a mechanical chopper was used as the excitation source.

3. Nd^{3+} and Tm^{3+} Ions in Oxyfluoride Glass-Ceramics

3.1 Formation of PbF_2 Nanocrystals

The XRD patterns of GN glass, GT glass, and the corresponding GC heat treated at 500°C for 10 h were recorded (Fig. 7.1). For both glasses, only broad halos were observed in the patterns, indicating the noncrystalline nature of the specimens. Upon thermal treatment at 500°C for 10 h, both GN and GT glasses showed several diffraction peaks similar to those of β-PbF_2 crystal (JCPDS No. 72-2304). TEM images in Fig. 7.2 show nanocrystals with sizes of 50–100 nm formed in both GC. For GN GC, both spherical and rectangular nanocrystals were formed in the glass matrix (Fig. 7.2A). For GT GC, most of the nanocrystals were nearly spherical in shape.

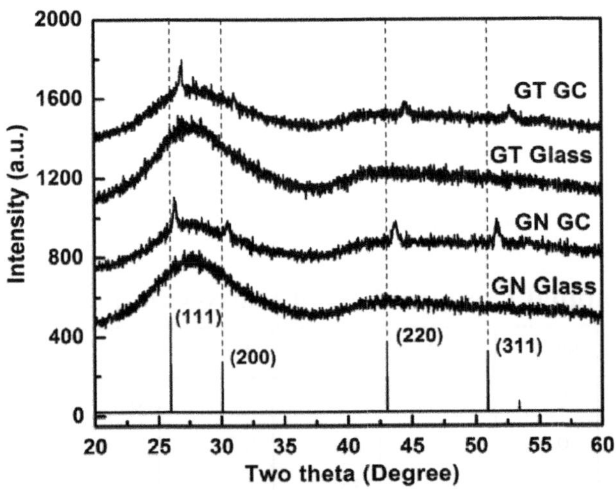

Figure 7.1 X-ray diffraction patterns of GN glass, GT glass, and the glass-ceramics (GN GC and GT GC) obtained upon thermal treatment at 500°C for 10 h. The bottom curve is the diffraction pattern of β-PbF_2 crystal (JCPDS No. 73-2304). The vertical lines extending from the diffraction peaks of β-PbF_2 crystal are drawn as guide for eyes.

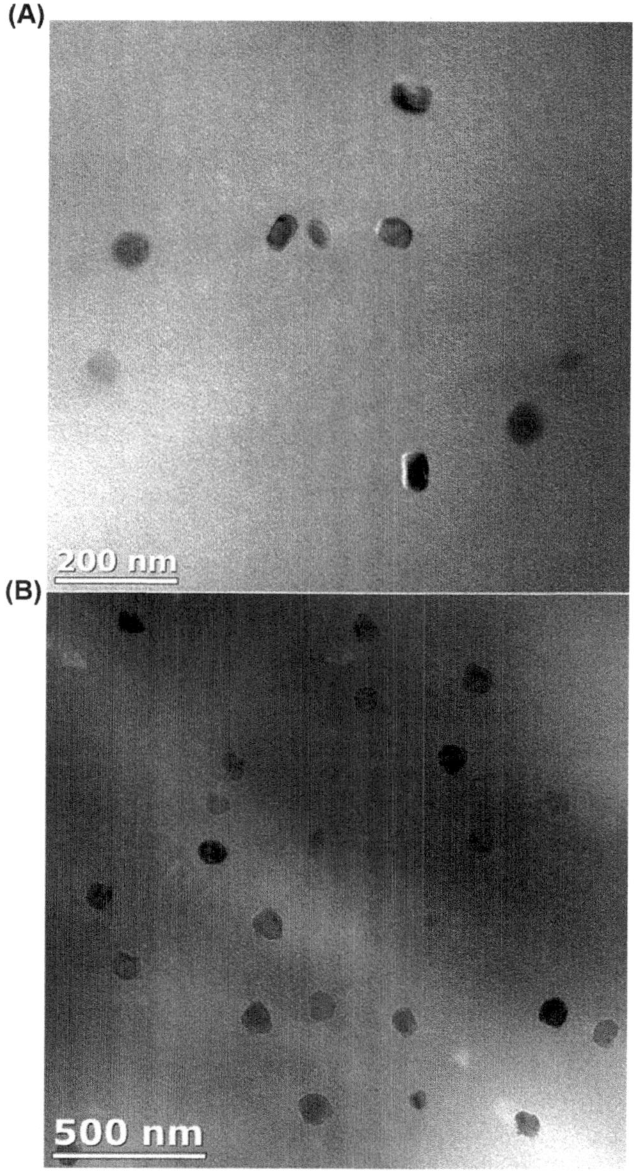

Figure 7.2 Transmission electron microscopy images of (A) GN glass-ceramics (GN GC) and (B) GT glass-ceramics (GT GC). Both specimens were heat treated at 500°C for 10 h.

Appearance of both spherical and rectangular nanocrystals in GN GC in Fig. 7.2A suggested that structures of these two types of nanocrystals are probably different. It is known that structures of $Pb_{1-x}RE_xF_{2+x}$ crystals (RE: rare earth ion) are strongly dependent on the concentration of rare earth ions [46]. For example, a $Pb_{1-x}RE_xF_{2+x}$ crystal has cubic fluorite structure when $x \leq 0.2$, but transforms to tetragonal when $x = 0.5$.

Therefore, difference in the shape of nanocrystals formed in GN GC indicated that concentration of Nd^{3+} ions in the fluoride nanocrystals is probably different. Detailed analysis will be given in the following sections.

3.2 Incorporation of Rare Earth Ions Inside the Crystals

There exists a discrepancy on the distribution of Nd^{3+} ions in the oxyfluoride GC. For example, Pisaska et al. [47,48] and Abril et al. [49] stated that Nd^{3+} ions are difficult to enter the crystalline phase. Pisaska et al. [47,48] showed that lifetime of the Nd^{3+}:$^4F_{3/2}$ level did not change upon thermal treatment since Nd^{3+} ions were not incorporated in the crystalline phase. Abril et al. [49] proved that introduction of rare earth ions in the form of fluoride can facilitate the incorporation of rare earth ions into the fluoride nanocrystals. Yu et al. [50] also suggested that very small number of Nd^{3+} ions is in PbF_2 nanocrystals based on the X-ray diffraction quantitative analysis.

Emission spectra and decay curves of Nd^{3+}- and Tm^{3+}-doped glasses and GC were recorded under 800-nm laser excitation (Fig. 7.3) in order to analyze the incorporation of rare earth ions into the fluoride nanocrystals. Since the emission properties of rare earth ions are sensitive to the local environment [51], emission properties of Nd^{3+} and Tm^{3+} should change if rare earth ions are incorporated into the fluoride nanocrystals by the thermal treatment. There are three emission bands at 890, 1060, and 1330 nm observed from Nd^{3+}-doped glass and GC through the deexcitation from the Nd^{3+}:$^4F_{3/2}$ level. There is no change in intensities of these emissions before and after heat treatment (Fig. 7.3A). Lifetimes of the same level showed only a slight increase from 269 ± 27 μs to 320 ± 32 μs upon thermal treatment (Fig. 7.3C). This observation was consistent with Yu et al. [24] where the lifetime of Nd^{3+}:$^4F_{3/2}$ level increased from 278.8 to 353.4 μs by the formation of fluoride nanocrystals in the GC. These small changes in the lifetimes of Nd^{3+} ions suggested that most of the Nd^{3+} ions still remain in the glass matrix instead of being incorporated into the fluoride nanocrystals [24]. A recent work using the XRD quantitative analysis also supported that most of Nd^{3+} ions remained in the glass after heat treatment [50].

On the other hand, the emission spectra of GT glass and GC in Fig. 7.3 show two bands at 1480 nm (Tm^{3+}:$^3H_4 \rightarrow {}^3F_4$) and 1800 nm ($Tm^{3+}$:$^3F_4 \rightarrow {}^3H_6$). Intensities of these two bands significantly increased with the thermal treatment at 500°C for 10 h. Lifetimes of the Tm^{3+}:$^3F_4$ level also increased from 985 ± 98 to 2536 ± 250 μs (Fig. 7.3D) at the same time similar to those reported previously [52–54]. Change in the lifetimes of Tm^{3+} ions supports that Tm^{3+} ions were incorporated into the fluoride nanocrystals during the heat treatment.

3.3 Local Environment of Rare Earth Ions

Absorption spectra of GN and GT glasses and the corresponding GC after heat treatment at 500°C for 10 h are in Fig. 7.4. Spectra of GN glass and GC in Fig. 7.4A show that absorption bands associated with Nd^{3+} ions showed clear splitting after heat

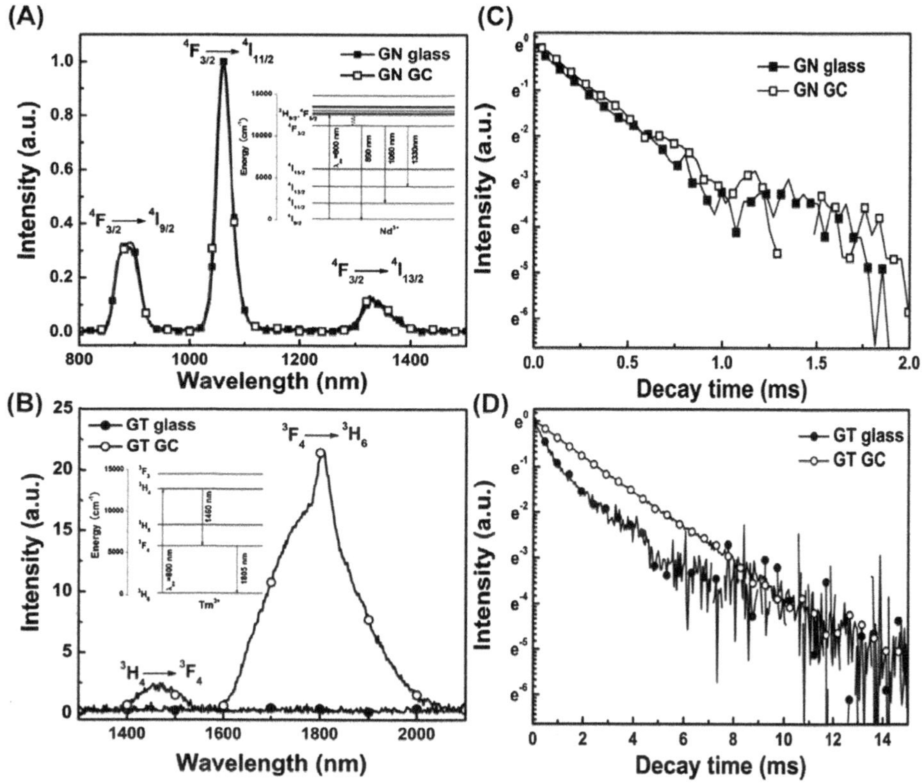

Figure 7.3 Emission spectra of (A) GN glass and glass-ceramics and (B) GT glass and glass-ceramics (GC), and decay curves of the (C) $^4F_{3/2}$ level of Nd^{3+} ions and (D) $^3F_4$ level of Tm^{3+} ions in the glass and GC. The insets in (A) and (B) show the simplified energy level diagrams of Nd^{3+} and Tm^{3+} ions, respectively. All the emission spectra and decay curves were recorded using an 800-nm laser excitation at room temperature.

treatment. Absorption bands from Tm^{3+}-doped glasses and GC in Fig. 7.4B also showed a similar splitting as observed from several previous reports [1,55,56]. Changes in the spectral line-shapes of the absorption bands of Tm^{3+} and Nd^{3+} ions indicate that local environments of rare earth ions were modified after thermal treatment.

XRD peaks observed from GN and GT GC in Fig. 7.1 shifted to the higher diffraction angle compared to those of pure β-PbF_2 crystals. Dantelle et al. also observed the similar shift from the PbF_2 nanocrystals containing Er^{3+} [57]. These shifts in the diffraction bands were attributed to the incorporation of rare earth ions into the PbF_2 crystalline lattice. Since the ionic radii of rare earth ions (99 pm for Tm^{3+} and 112 pm for Nd^{3+}) are smaller than that of Pb^{2+} (129 pm), this incorporation should lead to the decrease of the lattice constants of the doped PbF_2 nanocrystals. The shift in the diffraction peaks supports that rare earth ions are incorporated into the fluoride nanocrystals [57].

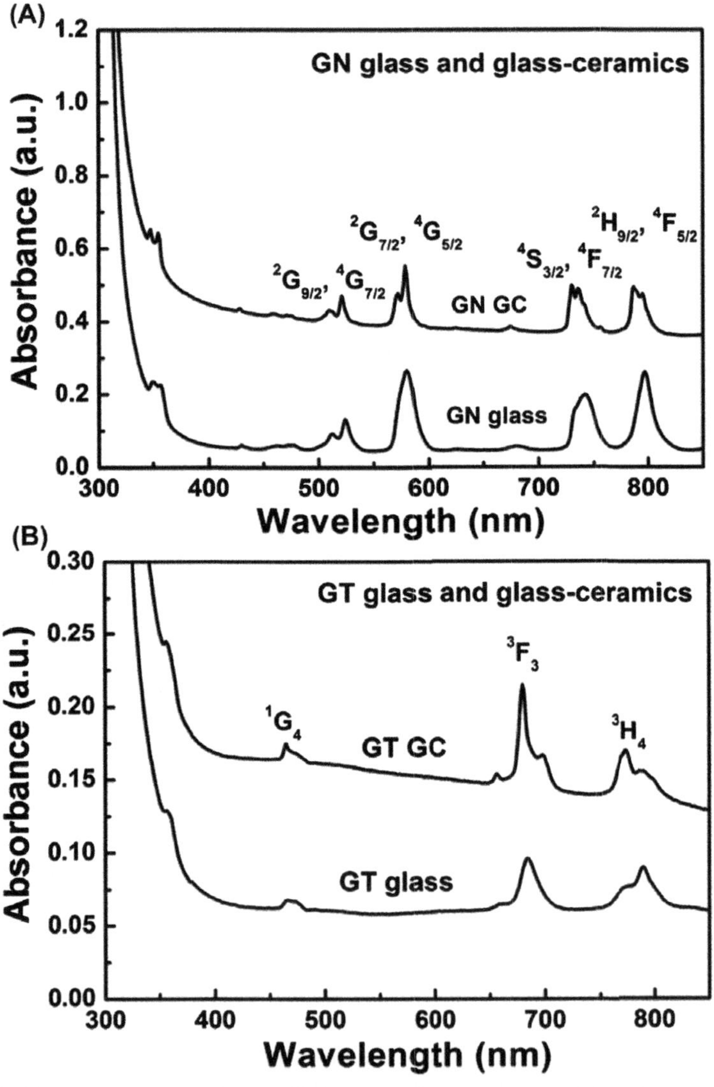

Figure 7.4 Absorption spectra of (A) GN glass and glass-ceramics (GC), and (B) GT glass and GC. The GC were obtained through heat treatment at 500°C for 10 h. Absorption spectra of GN and GT GC were vertically shifted for clear presentation.

Changes in absorption spectra, emission spectra, and XRD patterns all indirectly showed that Nd^{3+} and Tm^{3+} ions are incorporated into the fluoride nanocrystals upon thermal treatment. However, due to the indirect nature of these characterization methods, results still remain controversial [24,50]. It is necessary to obtain the direct evidence on the incorporation of rare earth ions into the fluoride nanocrystals.

4. Electron Energy Loss Spectroscopic Analysis on the Preferential Incorporation of Rare Earth Ions into Nanocrystals

Compositional analyses of GC were attempted using EELS to obtain direct evidences on the preferential orientation of Nd^{3+} and Tm^{3+} ions in nanocrystals. Results of analyses on several nanocrystals in Fig. 7.5A show that they are deficient in O and Si, but rich in F and Pb. These results clearly support that nanocrystals formed in the glass are mainly composed of lead fluoride. In addition, concentrations of Nd^{3+} ions are considerably high inside the fluoride nanocrystals compared to the glass matrix suggesting the preferential incorporation of Nd^{3+} ions into the fluoride nanocrystals (Fig. 7.5D). This incorporation of Nd^{3+} ions into the fluoride nanocrystals led to the shift of XRD peaks to high angles as in Fig. 7.1. On the other hand, only minor changes in the lifetimes of the $Nd^{3+}{:}^4F_{3/2}$ level was observed probably due to the large energy gap ($\sim$5200 cm^{-1}) between this level and the one located immediately below. Lifetimes of rare earth ions in the fluoride nanocrystals are not always good indicators for changes in the local environment change. Elemental analyses of Tm^{3+}-doped GC also showed similar results (Fig. 7.6). O and Si were deficient in the nanocrystals while Pb and F are enriched. Once again, Tm^{3+} ions are highly concentrated inside the fluoride nanocrystals.

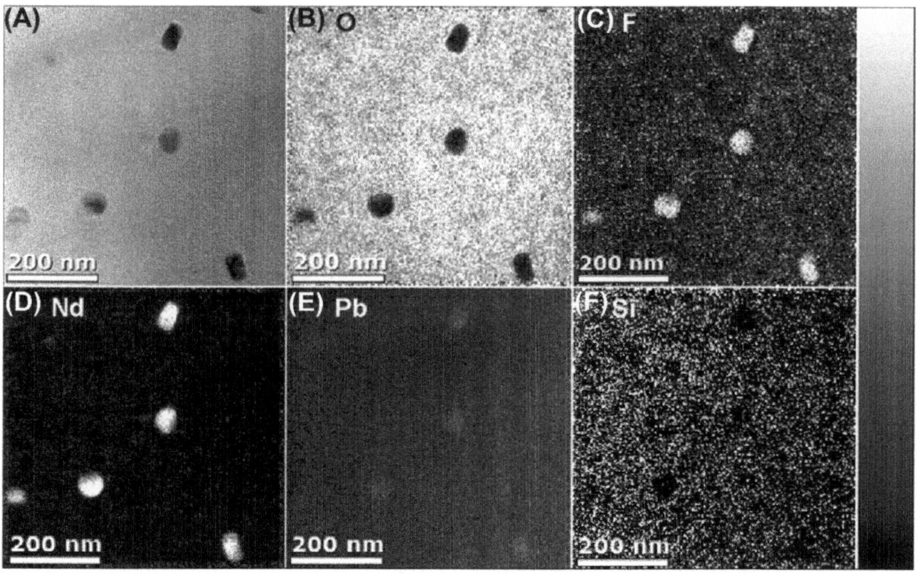

Figure 7.5 Electron energy loss spectroscopic analysis of Nd^{3+}-doped glass-ceramics. (B–F) show the distribution of O, F, Nd, Pb, and Si inside the nanocrystals shown in (A). The gray scale color bar shown on the right side indicates the relative signal intensity of each element.

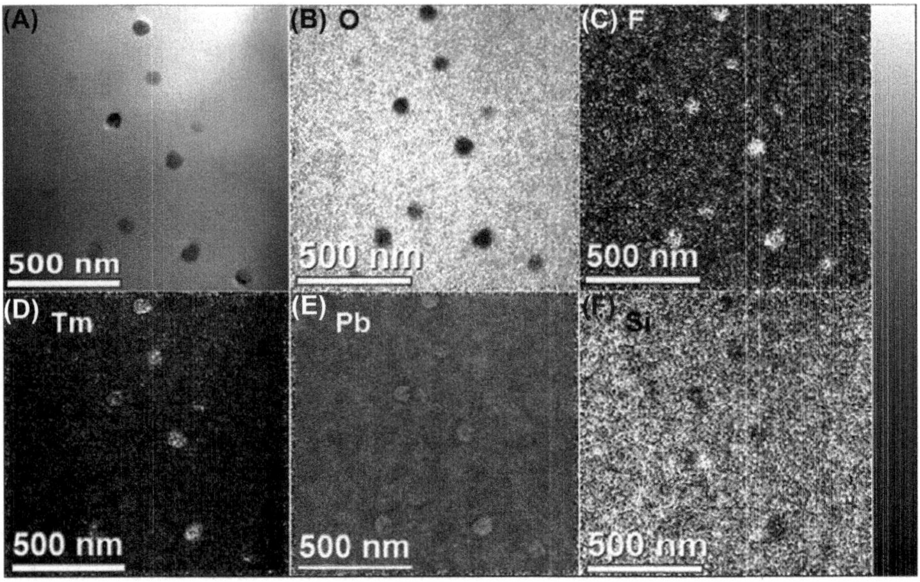

Figure 7.6 Electron energy loss spectroscopic analysis of Tm^{3+}-doped glass-ceramics. (B–F) show the distribution of O, F, Tm, Pb, and Si inside the nanocrystals shown in (A). The gray scale color bar on the right side indicates the relative signal intensity of each element.

5. Distribution of Rare Earth Ions Inside the Fluoride Nanocrystals

Shape of fluoride nanocrystals in GC is either spherical or rectangular as in Fig. 7.2. Structure of cubic fluorite PbF_2 crystal varies with the doping concentration of rare earth ions [46] and therefore, different shape reflects different concentration of rare earth ions in nanocrystals. Out of four fluoride nanocrystals shown in Fig. 7.7A, nanocrystal No. 1 shows the highest concentration of Nd^{3+} ions with nearly homogeneous distribution inside the crystal. Concentrations of Nd^{3+} ions were much lower in nanocrystals No. 2 and No. 4, than No. 1 even though both of them are rectangular in shape. Nanocrystal No. 3 with a spherical shape has the lowest concentration of Nd^{3+} ions. Concentration of Nd^{3+} ions in rectangular-shaped nanocrystal No. 5 in Fig. 7.7B is higher at the top portion of the picture compared to the bottom part. Similar differences were also found from the spherical nanocrystals No. 6 and No. 7. This observation indicates that distribution of Nd^{3+} ions inside the fluoride nanocrystals is inhomogeneous. Similar inhomogeneous distribution of Er^{3+} ions in PbF_2 nanocrystals in the oxyfluoride GC was also reported [7].

Effects of inhomogeneous distribution of Nd^{3+} ions on the lattice constants of the crystals are investigated using the HR-TEM diffraction patterns of one single nanocrystal (Fig. 7.8). Nanocrystal in Fig. 7.8A is the same as nanocrystal No. 7 in Fig. 7.7B. Analyses of the diffraction spots and diffraction rings in Fig. 7.8B supported

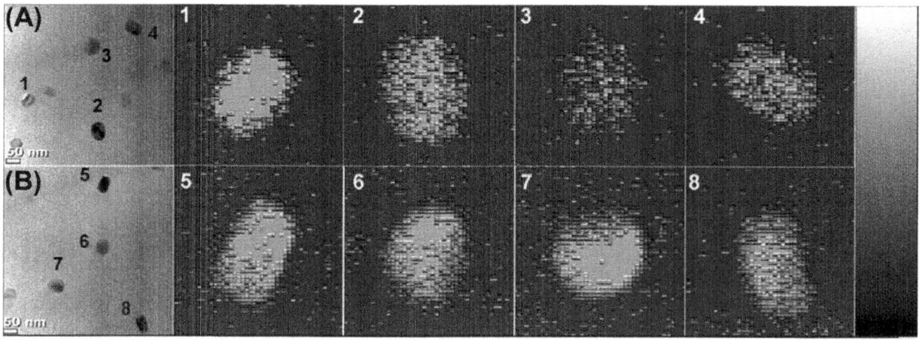

Figure 7.7 Distribution of Nd^{3+} ions in fluoride nanocrystals from two different locations in the bright field images inside GN glass-ceramics. The gray scale color bar on the right side indicates the relative signal intensity of Nd^{3+} ions.

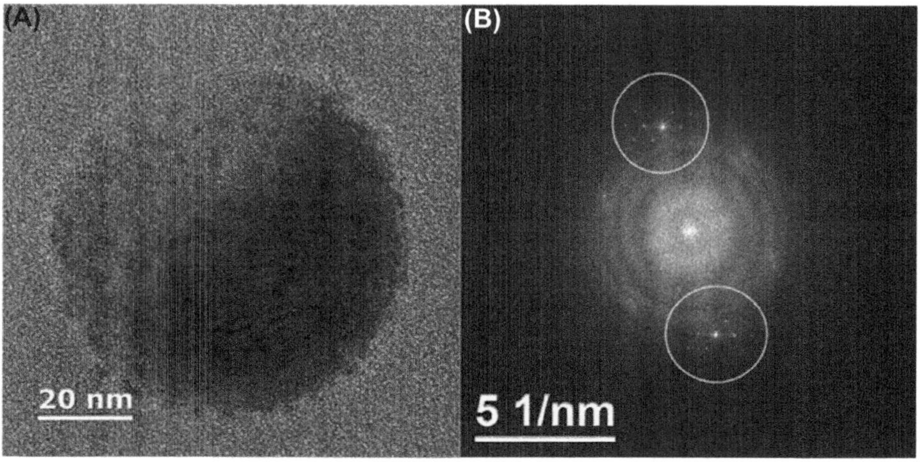

Figure 7.8 (A) High resolution-transmission electron microscopy image of one spherical nanocrystal and (B) its diffraction pattern obtained through Fast Fourier transformation.

the formation of cubic fluorite structure. In addition, several additional diffraction spots were observed outside of the central bright diffraction spot as marked with two white circles. They indicate that there are several other phases with the same crystal structure but with slightly different lattice constants [58]. Similar phenomena were also observed in our previous work [7]. This is most probably originated from the inhomogeneous distribution of Nd^{3+} ions in the fluoride nanocrystals. A certain portion of the nanocrystals with high Nd^{3+} concentration will be severely distorted that lead to the large changes in the lattice constants compared to those with small amount of Nd^{3+} ions.

6. YF_3 Codoping on the Emission of Er^{3+} in Oxyfluoride Glass-Ceramics

6.1 Formation of $Pb_{1-x}Y_xF_{2+x}$ Nanocrystals

Analyses explained in the preceding sections showed that rare earth ions were incorporated into the fluoride nanocrystals, where effective phonon energy was lower than the residual glass matrix. It was also found that distribution of rare earth ions inside the fluoride nanocrystals was inhomogeneous and this inhomogeneity may result in the luminescence quenching. This section reports the effect of YF_3 codoping on the emission properties of Er^{3+} ions in oxyfluoride GC.

Two glasses containing YF_3 and ErF_3 were prepared (GE5Y and GE1Y) and their compositions are given in Table 7.1. All the GC specimens were prepared by heat treatment at 490°C for 10 h. Formation of fluoride nanocrystals in the GC was confirmed from the characteristic peaks in XRD patterns (Fig. 7.9) that match with the cubic fluorite structured crystal of PbF_2 (JCPDS: 72-2304). There also exists a small shift toward the high diffraction angle with Er^{3+} addition indicating the incorporation of Er^{3+} ions into the fluoride nanocrystals. Due to the addition of YF_3, diffraction peaks also appeared even from the as-prepared glasses, indicating that fluoride nanocrystals were formed during quenching. It has to be pointed out that YF_3 can also act as the nucleation agent [30]. Addition of 1.0 mol% YF_3 was large enough to promote the formation of fluoride nanocrystals. These diffraction peaks grew in intensities after thermal treatment and shifted to even high diffraction angle that matches with the peaks of $Pb_{0.52}Y_{0.48}F_{2.48}$ crystal (JCPDS: 37-949). Incorporation of both Er^{3+} ions and Y^{3+} ions into the fluoride nanocrystals were also reported from the EELS measurement [25]. However, results were similar to those of $Pb_{1-x-y}Er_xY_yF_{2+x+y}$ crystal [46,59] and therefore, it was not possible to determine the precise concentration of Er^{3+} ions and Y^{3+} ions in the fluoride nanocrystals.

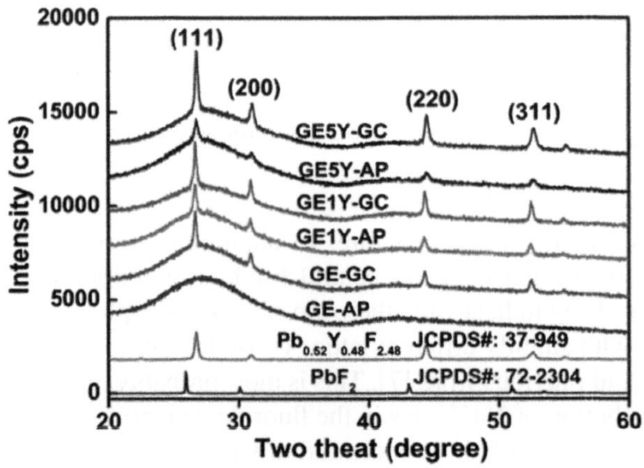

Figure 7.9 X-ray diffraction patterns of the as-prepared glasses and glass-ceramics heat-treated 490°C for 10 h. The bottom two lines show the diffraction pattern of PbF_2 crystal and $Pb_{0.52}T_{0.48}F_{2.48}$ crystal. All these patterns were vertically shifted for clarity.

6.2 Optical Properties

Absorption spectra of all specimens show bands at ~486, ~520, ~540, and ~1525 nm induced by the transition from the ground state ($^4I_{15/2}$) of Er^{3+} ions to the excited state $^4F_{7/2}$, $^2H_{11/2}$, $^4S_{3/2}$, and $^4I_{13/2}$, respectively (Fig. 7.10). Among them, the absorption band at λ ~ 520 nm showed the largest changes upon ceramization. For GE glass, absorption coefficient of this band decreased ~40% after thermal treatment. For GE5Y and GE1Y glasses, absorption coefficient of this band also decreased about ~40.8 and 44.5%, respectively. It is known that this absorption band due to the Er^{3+}:$^4I_{15/2} \rightarrow {}^2H_{11/2}$ transition is sensitive to the local symmetry of ligand field and the covalency of the chemical bonds [60]. Therefore, changes in the absorption coefficient of the ~520-nm bands indicate that Er^{3+} ions are incorporated into the fluoride nanocrystals during thermal treatment processes. Codoping of YF_3 appears to promote the incorporation of Er^{3+} ions into the fluoride nanocrystals but needs further confirmation. In addition, the ~1525-nm absorption bands showed pronounced Stark splitting after thermal treatment, indicating that Er^{3+} ions reside in the crystalline-like environment. No such changes were observed from glasses doped with 0.1 or 0.2 mol% ErF_3 only without YF_3 [25]. Both XRD patterns (Fig. 7.9) and absorption spectra (Fig. 7.10) indicated that YF_3 acted as the nucleation agents for the formation of fluoride nanocrystals and most probably promoted the incorporation of Er^{3+} ions into the fluoride nanocrystals during thermal treatment.

Emission spectra of the as-prepared glasses and GC have three emission bands at ~980 nm (Er^{3+}:$^4I_{11/2} \rightarrow {}^4I_{15/2}$), ~1540 nm ($Er^{3+}$:$^4I_{13/2} \rightarrow {}^4I_{15/2}$), and ~2730 nm ($Er^{3+}$:$^4I_{11/2} \rightarrow {}^4I_{13/2}$) (Fig. 7.11). There are concentration-dependent energy transfer pathways such as ($^4I_{11/2}$, $^4I_{13/2}$) $\rightarrow$ ($^4I_{15/2}$, $^4F_{9/2}$) in Er^{3+} ions that lead to the quenching

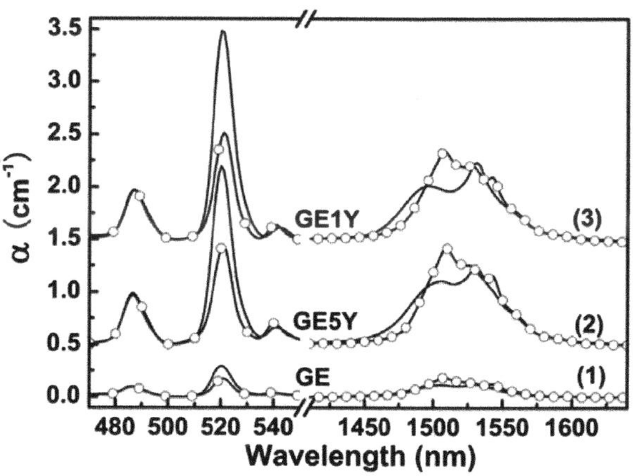

Figure 7.10 Absorption spectra as-prepared glasses and glass-ceramics (GC) heat treated at 490°C for 10 h for compositions of (1) GE, (2) GE5Y, and (3) GE1Y. Solid lines represent absorption of the as-prepared glasses, and the dotted lines represent the absorption of the GC. Absorption spectra of glasses and GC with the same compositions were drawn together for comparison and spectra were vertically shifted for clarity.

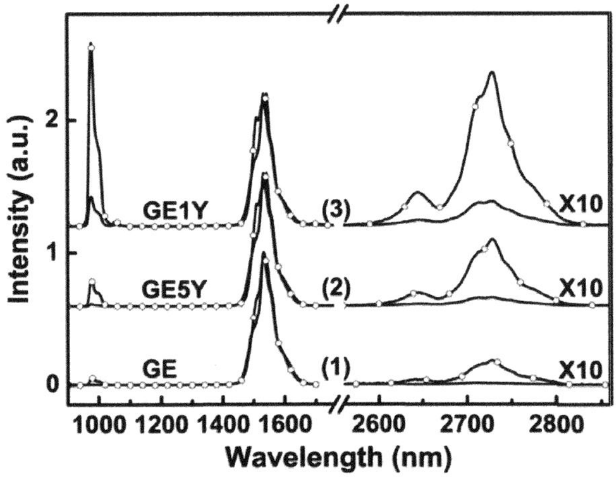

Figure 7.11 Near-infrared emission spectra of the as-prepared glasses and glass-ceramics (GC). The solid lines represent emission of the as-prepared glasses, and the dotted lines represent the emission from the GC. Emission spectra were normalized to the maximum intensities of the ~1540 emission band. Emission spectra of glasses and GC of the same compositions were drawn together for comparison and spectra were vertically shifted for clarity.

of 980- and 2730-nm band emissions [61,62]. Decrease in the clustering of Er^{3+} ions in the fluoride nanocrystals with YF_3 codoping can decrease the concentration quenching of these two bands. For GE glass, intensities of these two bands were low and increased only slightly after heat treatment. This is due to the preferential incorporation of the Er^{3+} ions into the fluoride nanocrystals that led to the strong concentration quenching through the efficient cross relaxation [7]. For the specimens with 0.5 mol% ErF_3 (GE5Y), these two emission bands became clear especially in the GC. As the concentration of Er^{3+} decreased for GE1Y composition, intensities of both 980- and 2730-nm bands increased especially after heat treatment. Ratios of the integrated emission intensity between the 980- and 1540-nm bands and 2730- and 1540-nm bands are in Table 7.2. There are significant increases in intensities of these two bands with the addition of YF_3. These enhancements became larger after the precipitation of nanocrystals. Changes in the emission spectra suggest that Y^{3+} ions in the fluoride nanocrystals can reduce the clustering of the Er^{3+} ions and lead to the decrease of the cross relaxation among Er^{3+} ions.

Effect of YF_3 codoping was further studied using the decay curves of 1540- and 2730-nm band emissions. Decay curve of the 1540-nm emission band recorded from GE GC doped with 0.5 mol% ErF_3 only shows a double-exponential decay. Time constants of the fast and slow decay process are 0.7 and 6.51 ms, respectively. The fast decay process contributed 58.7% to the total emission and the slow one contributed the remainder. A similar decay was observed from GE5Y GC with the fast and slow decay time constants of 2.22 and 5.80 ms, each responsible for 85.9 and 14.1% of the total emission, respectively. When ErF_3 content is smaller as in GE1Y GC, the decay curve was nearly exponential with a time constant of 7.94 ms.

Table 7.2 **Ratios of the Integrated Emission Intensities Between 980/1540-nm and 2730/1540-nm Bands Together With Lifetimes of the 1540-nm ($^4I_{13/2}$) and 2730-nm ($^4I_{11/2}$) Emission**

Specimens	Integrated Emission Intensity Ratios		Lifetimes of $^4I_{13/2}$ Level (ms)	Lifetimes of $^4I_{11/2}$ Level (ms)
	I_{980}/I_{1540}	I_{2730}/I_{1540}		
GE-AP	0.00187	0.00148	6.88 ± 0.02	–
GE-GC	0.02102	0.01654	3.17 ± 0.09	1.74 ± 0.02
GE5Y-AP	0.00755	0.00778	6.65 ± 0.59	1.92 ± 0.07
GE5Y-GC	0.07826	0.04645	3.21 ± 0.03	2.17 ± 0.04
GE1Y-AP	0.10046	0.02152	5.72 ± 0.00	4.10 ± 0.07
GE1Y-GC	0.53188	0.10946	7.94 ± 0.02	3.80 ± 0.08
Polycrystalline PbF_2 [62]				3.9–8.8
Germanate glass-ceramics [62]				2.4–4.75
Fluoroindate glasses [63]			10.0–10.3	8.1–8.5
$Ge_{30}Ga_2As_6S_{62}$ glasses [64]			5.0	1.7

Effective emission lifetimes of the 1540-nm band from all three GC were 3.17 ± 0.09, 3.21 ± 0.03, and 7.94 ± 0.02 ms for GE, GE5Y, and GE1Y GC, respectively. These results indicated that codoping of YF_3 increased the effective lifetimes of the 1540 nm emission.

For the 2730-nm band emission, all the decay curves recorded from different GC were nearly exponential (Fig. 7.12B). Lifetimes of this band increased from 1.74 ± 0.02 ms in GE GC to 2.17 ± 0.04 ms in GE5Y GC. In GE1Y GC, lifetime of 2730-nm band emission further increased to 3.80 ± 0.08 ms. Since the 2730-nm band emission is strongly dependent on the cross relaxation of the ($^4I_{11/2}$, $^4I_{13/2}$) → ($^4I_{15/2}$, $^4F_{9/2}$) process, increase in the lifetimes indicated that codoping of YF_3 decreased the clustering of Er^{3+} ions in fluoride nanocrystals and reduced the deleterious energy transfer processes.

For Er^{3+}-doped PbF_2 polycrystals, when the doping concentration of ErF_3 increased from 1 mol%, 9.1 mol% to 16.6 mol%, lifetimes of Er^{3+}:$^4I_{11/2}$ level decreased from 8.8–7.3 to 3.9 ms [62]. For Er^{3+}-doped fluoroindate glasses, lifetimes of Er^{3+}:$^4I_{11/2}$ level were 8.5 and 8.1 ms when the doping concentrations of ErF_3 were 1 and 4 mol%, respectively. This indicated that effective concentration of Er^{3+} ions in the fluoride nanocrystals formed in GE1Y-GC was comparable to PbF_2 polycrystals doped with 16.6 mol% ErF_3, further confirming that it is necessary to dilute the concentration of Er^{3+} ions in the fluoride nanocrystals through codoping.

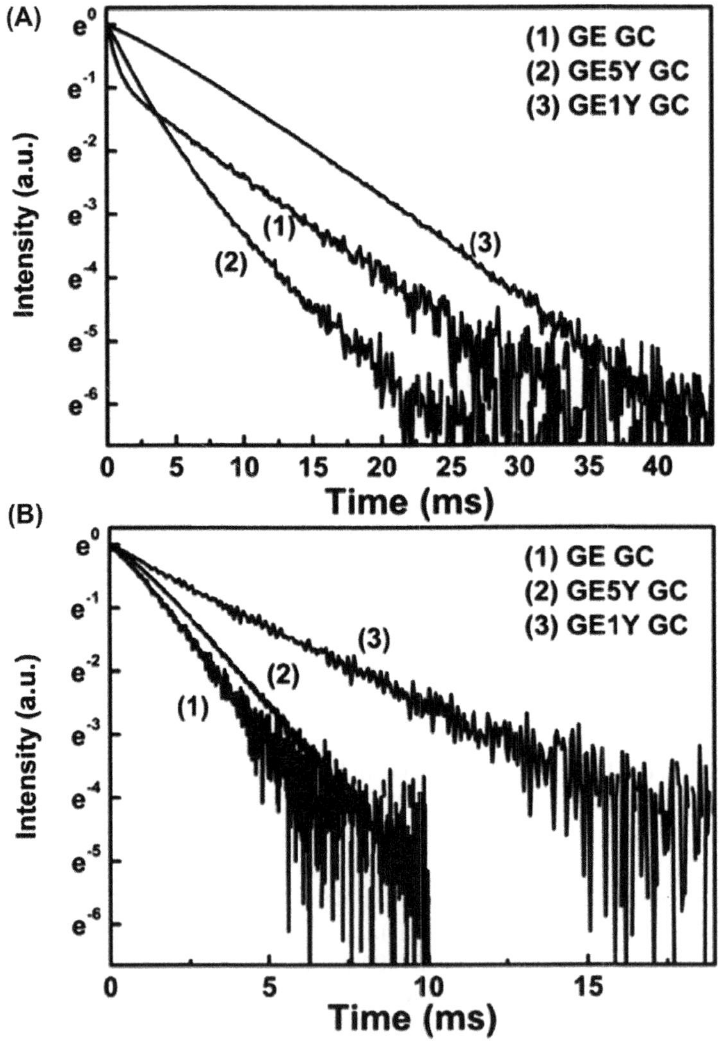

Figure 7.12 Decay curves of (A) 1540-nm band emission and (B) 2730-nm band emission of the glass-ceramics.

7. Summary

Oxyfluoride glasses and GC containing Nd^{3+} and Tm^{3+} ions were prepared. XRD patterns, absorption, and emission spectra indicated that Nd^{3+} and Tm^{3+} ions are incorporated into the fluoride nanocrystals. EELS analysis directly confirmed that Nd^{3+} and Tm^{3+} ions are preferentially incorporated into the fluoride nanocrystals during thermal treatment. This distribution of Nd^{3+} ions in the fluoride nanocrystals led to the continuous variation in the structure of fluoride nanocrystals as evidenced by the electron diffraction patterns. Codoping of YF_3 reduced the cross relaxation process in Er^{3+}, resulting in enhanced 980- and 2730-nm band emission. It suggests that addition of YF_3 reduced the clustering of Er^{3+} ions inside the nanocrystal. It can improve the optical performance of these rare earth–doped oxyfluoride GC.

References

[1] Y. Wang, J. Ohwaki, Appl. Phys. Lett. 63 (1993) 3268–3270.
[2] T. Suzuki, S. Masaki, K. Mizuno, Y. Ohishi, Proc. SPIE 7721 (2010) 77210T.
[3] P.P. Fedorov, Russ. J. Inorg. Chem. 59 (2014) 595–599.
[4] Z. Xue, T.G. Edwards, S. Se, J. Am. Ceram. Soc. 94 (2011) 2092–2098.
[5] M. Reben, J. Non Cryst. Solids 357 (2011) 2653–2657.
[6] S. Bhattacharyya, C. Bocker, T. Heil, J.R. Jinschek, T. Hoche, C. Rüssel, H. Kohl, Nano Lett. 9 (2009) 2493–2496.
[7] C. Liu, X. Zhao, J. Heo, J. Non Cryst. Solids 365 (2013) 1–5.
[8] D.Q. Chen, Y.L. Yu, H. Lin, P. Huang, Z.F. Shan, Y.S. Wang, Opt. Lett. 35 (2010) 220–222.
[9] F. Goutaland, P. Jander, W.S. Brocklesby, G.J. Dai, Opt. Mater. 22 (2003) 383–390.
[10] G. Domoniak-Dzik, R. Lisiecki, W. Ryba-Romanowski, L. Krajczyk, J. Alloys Compd. 511 (2012) 189–194.
[11] D.Q. Chen, Y.L. Yu, P. Huang, F.Y. Weng, H. Lin, Y.S. Wang, Appl. Phys. Lett. 94 (2009) 041909.
[12] K. Biswas, A.D. Sontakke, J. Ghosh, K. Annapurna, J. Am. Ceram. Soc. 93 (2010) 1010–1017.
[13] V.A. Aseev, V.V. Golubkov, A.V. Klementeva, E.V. Kolobkova, N.V. Nikonorov, Opt. Spectrosc. 106 (2009) 691–696.
[14] M. Mortier, A. Bensalah, G. Dantelle, G. Patriarche, D. Vivien, Opt. Mater. 29 (2007) 1263–1270.
[15] L. Dimitrocenko, U. Rogulis, A. Veispals, M. Springis, P. Kulis, A. Fedotovs, A. Mishnev, Phys. Status Solidi C 4 (2007) 753–756.
[16] O.B. Petrova, L.N. Dmitruk, A.V. Popov, V.E. Shukshin, Opt. Spectrosc. 107 (2009) 347–352.
[17] C. Yu, D. He, G. Wang, J. Zhang, L. Hu, Eighth Pacific Rim Conference on Lasers and Electro-Optics, CLEO/PACIFIC RIM, 2009, pp. 1–2.
[18] F.R. Gomes, P. Camila, B. Roberto, F.C. Cassanjes, G. Poirier, J. Non Cryst. Solids 357 (2011) 3345–3350.
[19] Z. Pan, A. Ueda, M. Hays, R. Mu, S.H. Morgan, J. Non Cryst. Solids 352 (2006) 801–806.
[20] F. Xin, S. Zhao, S. Xu, L. Huang, G. Jia, D. Deng, H. Wang, Opt. Mater. 34 (2011) 85–88.
[21] D.P.P. Fedorov, A.A. Luginina, A.I. Popov, J. Fluorine Chem. 172 (2015) 22–50.
[22] Q. Chen, Y.S. Wang, Y.L. Yu, E. Ma, L.H. Zhou, J. Solid State Chem. 179 (2006) 532–537.
[23] H. Lin, D.Q. Chen, Y.L. Yu, A.P. Yang, R. Zhang, Y.S. Wang, Mater. Res. Bull. 47 (2012) 469–472.
[24] Y.L. Yu, D.Q. Chen, E. Ma, Y.S. Wang, Z.J. Hu, Spectrochim. Acta A67 (2007) 709–713.
[25] Z. Zhao, C. Liu, Y. Jiang, J. Zhang, H. Tao, J. Han, X. Zhao, J. Heo, J. Non Cryst. Solids 404 (2014) 37–42.
[26] D.Q. Chen, Y.S. Wang, Y.L. Yu, E. Ma, J. Solid State Chem. 179 (2006) 1445–1452.

[27] F. Liu, E. Ma, D.Q. Chen, Y.S. Wang, Y.L. Yu, P. Huang, J. Alloys Compd. 467 (2009) 317–321.
[28] G. Dantelle, M. Mortier, D. Vivien, G. Patriarche, Opt. Mater. 28 (2006) 638–642.
[29] O. Peron, C. Duverger-Arfuso, Y. Jestin, B. Boulard, M. Ferrari, Opt. Mater. 31 (2009) 1288–1291.
[30] P.A. Tick, N.F. Borrelli, L.K. Cornelius, M.A. Newhouse, J. Appl. Phys. 78 (1995) 6367–6375.
[31] S. Tanabe, H. Hayashi, T. Hanada, J. Am. Ceram. Soc. 85 (2002) 839–843.
[32] W.J. Zhang, Q.Y. Zhang, Q.J. Chen, Q. Qian, Z.M. Yang, J.R. Qiu, P. Huang, Y.S. Wang, Opt. Express 17 (2009) 20952–20958.
[33] A.S. Gouveia-Neto, L.A. Bueno, R.F. do Nascimento, E.A. da Silva, E.B. da Costa, J. Non Cryst. Solids 355 (2009) 488–491.
[34] F.X. Xin, S.L. Zhao, L.H. Huang, D.G. Deng, G.H. Jia, H.P. Wang, S.Q. Xu, Mater. Lett. 78 (2012) 75–77.
[35] L. Zhang, J. Xu, Y. Hu, G. Chen, Z. Wang, J. Wuhan Univ. Technol. Mater. Sci. 28 (2013) 455–459.
[36] Y. Kishi, S. Tanabe, J. Alloys Compd. 408-412 (2006) 842–844.
[37] J. Xu, Y. Hu, Z. Wang, M. Ye, F. Xiang, J. East China Univ. Sci. Technol. 37 (2011) 65–70.
[38] D.Q. Chen, Y.L. Yu, Y.S. Wang, P. Huang, F.Y. Weng, J. Phys. Chem. C 113 (2009) 6406–6410.
[39] Z. Pan, G. Sekar, R. Akrobetu, R. Mu, S.H. Morgan, J. Non Cryst. Solids 358 (2012) 1814–1817.
[40] X.B. Chen, G.J. Yang, S.L. Song, N. Sawanobori, Y.Z. Xu, X.D. Chen, G. Zhou, Acta Phys. Sin. 61 (2012) 227803.
[41] Y. Xu, S. Cui, H.Y. Fu, J.Y. Qian, Q. Luo, X.S. Qiao, X.P. Fan, X.H. Zhang, Chin. Opt. Lett. 10 (2012) 031602.
[42] X.F. Wang, X.H. Yan, Y. Xuan, J. Zheng, W.Y. He, Appl. Phys. A 113 (2013) 41–46.
[43] F.G. Yang, G.T. Chen, Z.Y. You, C.Y. Tu, Mater. Lett. 64 (2010) 824–826.
[44] C. Liu, J. Heo, Mater. Lett. 61 (2007) 3751–3754.
[45] C. Liu, J.W. Hong, J. Heo, J. Non Cryst. Solids 351 (2005) 2317–2323,
[46] A.K. Tyagi, S.J. Patwe, S.N. Achary, M.B. Mallia, J. Solid State Chem. 177 (2004) 1746–1757.
[47] J. Pisaska, W. Ryba-Romanowski, G. Dominiak-Dzik, T. Goryczka, W.A. Pisaski, J. Alloys Compod. 451 (2008) 223–225.
[48] J. Pisaska, W. Ryba-Romanowski, G. Dominiak-Dzik, T. Goryczka, W.A. Pisaski, Opt. Appl. 38 (2008) 211–216.
[49] M. Abril, J. Mendez-Ramos, I.R. Martin, U.R. Rodriguez-Mendoza, V. Lavin, A. Delgado-Torres, V.D. Rodriguez, P. Nunez, A.D. Lozano-Gorrin, J. Appl. Phys. 95 (2004) 5271–5279.
[50] H. Yu, H. Guo, M. Zhang, Y. Lin, M. Liu, L.-J. Zhao, Nanoscale Res. Lett. 7 (2012) 275.
[51] J.-L. Adam, Chem. Rev. 102 (2002) 2461–2476.
[52] M. Mattarelli, V.K. Tikhomirov, A.B. Seddon, M. Montagna, E. Moser, A. Chiasera, S. Chaussedent, G. Nunzi Conti, S. Pelli, G.C. Righini, L. Zampedri, M. Ferrari, J. Non Cryst. Solids 345&346 (2004) 354–358.
[53] J. Qiu, A. Mukai, A. Makishima, Y. Kawamoto, J. Phys. Condens. Matter 14 (2002) 13827–13834.
[54] A. de Pablos-Martin, D. Ristic, S. Bhattacharyya, Th Höche, G.C. Mather, M.O. Ramírez, S. Soria, M. Ferrari, G.C. Righini, L.E. Bausá, A. Durán, M.J. Pascual, J. Am. Ceram. Soc. 96 (2013) 447–457.

[55] G. Wu, S. Fan, Y. Zhang, G. Chai, Z. Ma, M. Peng, J. Qiu, G. Dong, Opt. Lett. 38 (2013) 3071–3074.
[56] V.K. Tikhomirov, D. Furniss, A.B. Seddon, M. Ferrari, R. Rolli, J. Mater. Sci. Lett. 21 (2002) 293–295.
[57] G. Dantelle, M. Mortier, G. Patriarche, D. Vivien, J. Solid State Chem. 179 (2006) 1995–2003.
[58] D.B. Williams, C.B. Carter, Transmission Electron Microscopy, Springer Science+Business Media, LLC, New York, USA, 2009, p. 296 and p. 393.
[59] C. Liu, J. Heo, J. Am. Ceram. Soc. 95 (2012) 2100–2102.
[60] S. Tanabe, H. Hayashi, T. Hanada, N. Onodera, Opt. Mater. 19 (2002) 343–349.
[61] L. Zhang, H. Hu, J. Non Cryst. Solids 326-327 (2003) 353–358.
[62] M. Mortier, P. Goldner, C. Chateau, M. Genotolle, J. Alloys Compd. 323-324 (2001) 245–249.
[63] T. Catunda, L.A. Nunes, A. Florez, Y. Messaddeq, M.A. Aegerter, Phys. Rev. B 53 (1996) 6065–6070.
[64] Y.G. Choi, K.H. Kim, B.J. Lee, Y.B. Shim, Y.S. Kim, J. Heo, J. Non Cryst. Solids 278 (2000) 137–144.

Fluorinated Nematic Liquid Crystals: Design, Synthesis, and Properties

8

P. Kirsch
Merck KGaA, Darmstadt, Germany

Chapter Outline

1. Introduction: Liquid Crystals, Fluorine Chemistry, and Liquid Crystal Display Technology

About 125 years ago, two apparently unrelated discoveries initiated one of the most impressive success stories in applied science. In 1886, H. Moissan in Paris prepared for the first time elemental fluorine, an extremely reactive gas [1]. Two years later, in 1888, F. Reinitzer in Prague discovered serendipitously that cholesteryl benzoate forms a new, until then never described state of matter [2]: a chiral nematic mesophase. Hundred years later, the two stories converged into fluorinated liquid crystals which are the active medium of most current flat-panel displays [3]. In 2014, around 160 km^2 of liquid crystal displays (LCDs) were produced worldwide. The liquid crystal mixtures used in high-resolution, full color LCDs always contain highly fluorinated components. Fluorine shapes, on the one hand, the physical property profile of liquid crystals. On the other hand, it renders the material more robust, thus increasing LCD production yield and reducing cost. With continuously dropping LCD prices the efficiency of the LCD production process—depending on the

Photonic and Electronic Properties of Fluoride Materials. http://dx.doi.org/10.1016/B978-0-12-801639-8.00008-8

robustness of the liquid crystal—moves more and more into the focus of material development. Such extreme requirements can be met only by extensive use of fluorinated materials (such as **4** and **5**) (Scheme 8.1).

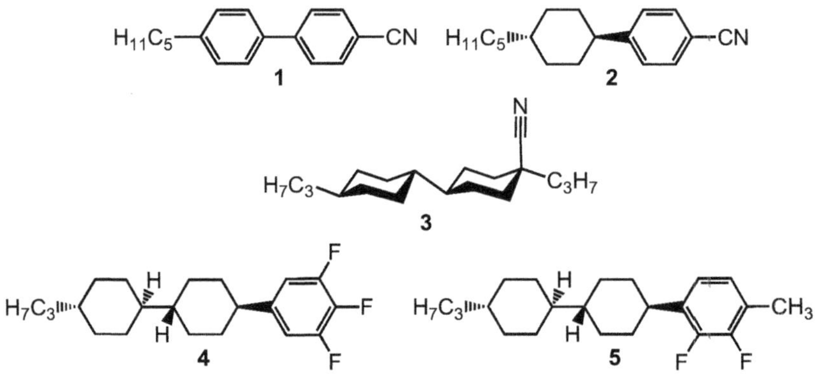

Scheme 8.1 Examples of typical liquid crystals as they are used in liquid crystal displays. Compounds **1**, **2**, and **4** have their dipole moment aligned with the long molecular axis and have positive dielectric anisotropy ($\Delta\varepsilon > 0$). In materials **3** and **5** the dipole moment is oriented perpendicularly to the long molecular axis, resulting in a negative $\Delta\varepsilon$.

The fluorine in liquid crystals can have various functions [4]. As a terminal substituent, it provides the molecule with a dipole moment parallel to the long molecular axis, and accordingly with positive dielectric anisotropy ($\Delta\varepsilon > 0$). As a lateral aromatic substituent it can either augment the dipole moment of the terminal group, or—in combination with a second lateral *ortho*-fluorine substituent—it results in a dipole moment perpendicular to the long axis, resulting in negative dielectric anisostropy ($\Delta\varepsilon < 0$). In a central position of the mesogenic core structure, the lateral fluorine often widens the nematic phase range, therefore dubbed "nematic fluorine."

2. "Nematic Fluorine" and *ortho*-Metalation Technology

The "nematic fluorine" effect was discovered in the 1950s by G.W. Gray [5]. It is typically used in order to improve polarity and mesophase characteristics of dielectrically positive materials. Compounds with a 2,3-difluorophenylene moiety were originally under investigation as ferroelectric liquid crystals [6]. However, in the late 1990s the potential of the so-called vertical alignment (VA) mode was recognized [7], and they are nowadays the major functional component of liquid crystal mixtures used in LCD TV.

Laterally fluorinated arene building blocks do not only provide the corresponding liquid crystals with a unique physical property profile, but they also offer a special synthetic "leverage": originally independently discovered by H. Gilman and G. Wittig [8], and later developed by V. Snieckus and M. Schlosser [9], site-directed *ortho*-lithiation conveniently allows the derivatization of fluorinated arenes. The *ortho*-fluoro lithio arenes can be converted into a variety of other useful intermediates such as halides, carboxylic acids, or boronic acids. However, these intermediates are stable only at temperatures below −40 to −20°C, depending on the particular substitution pattern. Above this temperature they decompose in a violent, exothermic reaction fueled by the high lattice energy of the generated lithium fluoride (1036 kJ/mol). This poses serious safety challenges, especially in a large scale industrial environment. For this reason, much effort was directed into the development of a flow microreactor technology, where each individual reaction unit has only a relatively small active volume as well as short and well-controllable reaction times. Scale-up is easily achieved by parallelization without any scale-related safety concerns [10]. The choice of the lithium base and the optimal reaction temperature depends on the exact fluorination pattern (Scheme 8.2).

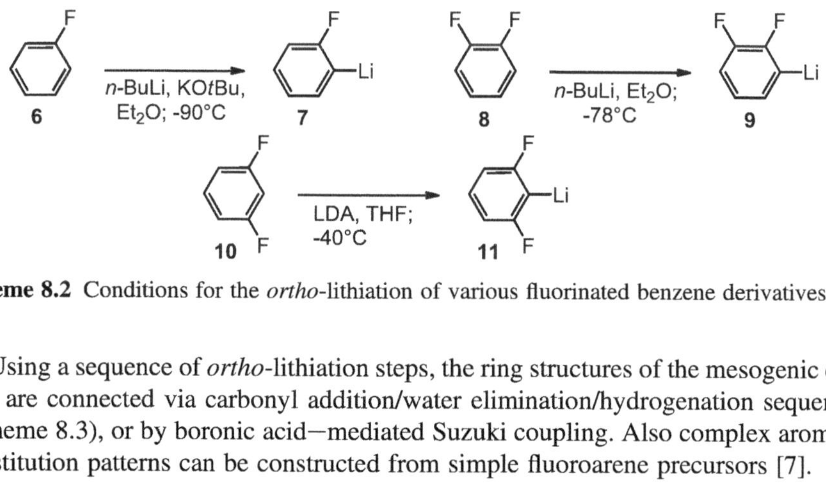

Scheme 8.2 Conditions for the *ortho*-lithiation of various fluorinated benzene derivatives.

Using a sequence of *ortho*-lithiation steps, the ring structures of the mesogenic core unit are connected via carbonyl addition/water elimination/hydrogenation sequences (Scheme 8.3), or by boronic acid–mediated Suzuki coupling. Also complex aromatic substitution patterns can be constructed from simple fluoroarene precursors [7].

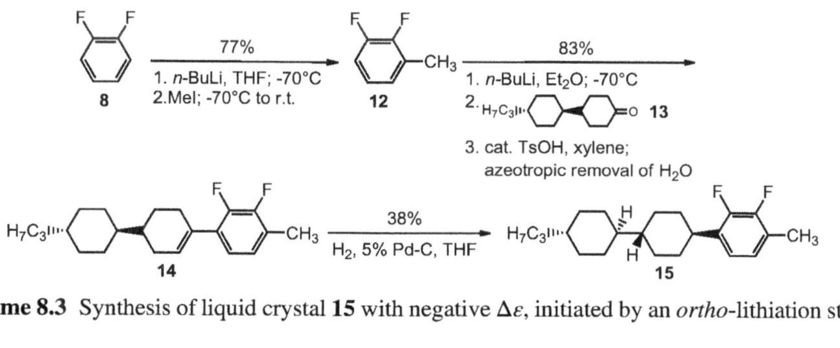

Scheme 8.3 Synthesis of liquid crystal **15** with negative $\Delta\varepsilon$, initiated by an *ortho*-lithiation step.

3. Fluorine in Polar Terminal and Lateral Groups

In the late 1980s it became apparent that the cyano function cannot be used in high-resolution, active matrix (AM) LCD [3a]. The cyano group readily attaches itself by electrostatic interactions to cations. This results in the solvation of ubiquitous traces of ionic impurities in the LCDs, which again allows the solvated ions to drift in the applied electric field [11]. The uncontrolled mobilization of ions leads to a decrease of the so-called voltage holding ratio and becomes visible by flicker, contrast reduction, or image sticking.

The obvious alternative for polar terminal groups is either fluorine itself or more complex fluorine-containing functional groups [12]. Compared to the cyano group, fluorinated polar groups show a drastically reduced tendency to solvate ionic compounds. Fluorine is the most electronegative element, with an electronegativity (EN) of 4.0. Therefore, bonds between fluorine and more electropositive elements, such as carbon (EN = 2.5) or sulfur (EN = 2.6), are strongly polarized and carry a significant dipole moment. Thus, a fluorinated polar terminal group provides the liquid crystal with a dipole moment along the long molecular axis and—correspondingly—with positive dielectric anisotropy $\Delta\varepsilon$. The most common fluorinated polar groups are CF_3, OCF_3, $OCHF_2$, and $OCF{=}CF_2$. These groups are synthesized using classical Swarts fluorination chemistry or difluorocarbene-mediated *O*-difluoromethylation (Scheme 8.4).

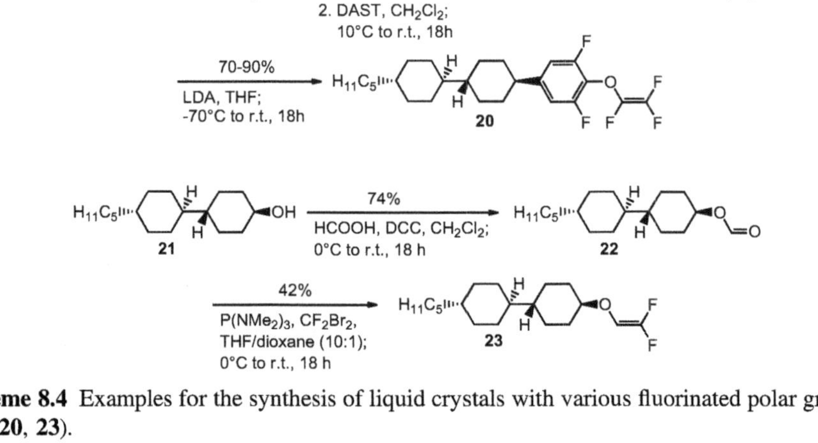

Scheme 8.4 Examples for the synthesis of liquid crystals with various fluorinated polar groups (**17**, **20**, **23**).

The polarity of these groups lags far behind the cyano function, but—in particular when augmented by the additional dipole moments of *ortho* fluorine substituents—suffices for application in LCDs. The dipole moment can be even further enhanced by introduction of lateral fluorine in adjoining rings or by combination with polar dioxane rings. However, Table 8.1 shows that the "price" for increasing lateral fluorination is a general erosion of the mesogenic properties: the clearing point drops by up to 50K for each additional fluorine, the nematic mesophase disappears gradually, and the solubility of the liquid crystals becomes very poor (Table 8.1).

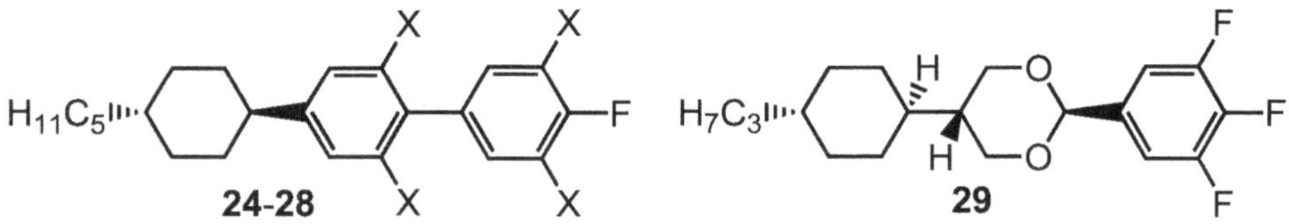

Table 8.1 Examples of Polar Liquid Crystals With Positive Dielectric Anisotropy ($\Delta\varepsilon > 0$), as They Are Used in PC Monitors or Mobile Devices [13]

No.	X^1	X^2	X^3	X^4	Mesophases	$T_{NI,virt}$	$\Delta\varepsilon_{virt}$
24	H	H	H	H	C 102 N 153.9 I	150.7	4.2
25	H	H	F	H	C 55 N 105.4 I	107.0	6.3
26	H	H	F	F	C 25 N 54.8 I	66.0	11.7
27	H	F	F	F	C 63 N (37.0) I	36.9	14.6
28	F	F	F	F	C 114 I	2.0	19.1
4	–	–	–	–	C 64.7 N 93.7 I	74.0	8.3
29	–	–	–	–	C 74 N (51.2) I	69.7	17.0
17	–	–	–	–	C35 S_G? (33) I	−22.0	5.3
20	–	–	–	–	C 57 N 90.9 I	66.2	9.4
23	–	–	–	–	C 34 N (31.0) I	12.7	5.6

Many important video applications, such as LCD-TV and some PC monitors, require materials with a negative dielectric anisotropy ($\Delta\varepsilon < 0$) [14]. As already mentioned, the most widely used class of materials for the VA mode are the 2,3-difluorophenylene derivatives [15], which are found in basically all commercial liquid crystal mixtures. Alternative, newer material concepts, which are still in an experimental stage, are axially fluorinated cyclohexane derivatives [16] and tri- and tetrafluoroindanes (Table 8.2) [17].

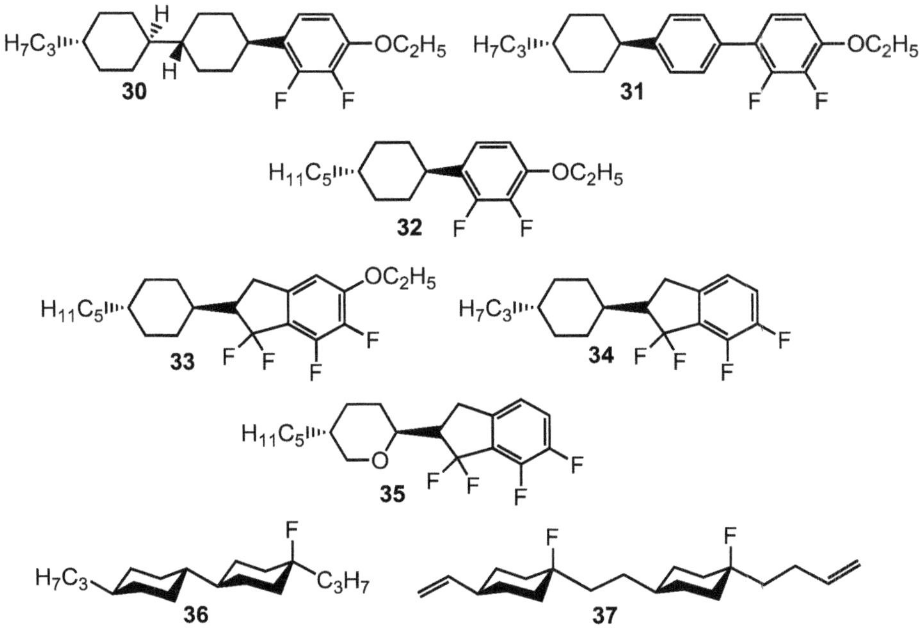

Table 8.2 Examples of Liquid Crystals With Negative Dielectric Anisotropy ($\Delta\varepsilon < 0$). In Particular, Materials 30–32 Are Widely Used in LCD TV [13]

No.	Mesophases	$T_{NI,virt}$	$\Delta\varepsilon_{virt}$	Δn_{virt}	$\gamma_{1,virt}$
30	C 79 S_B (78) N 184.5 I	175.4	−5.9	0.096	413
31	C 80 N 173.3 I	182.0	−5.9	0.156	233
32	C 49 N (12.9) I	16.5	−6.2	0.099	110
33	C 129 I	40.0	−13.5	0.091	514
34	C 85 I	49.4	−8.6	0.085	142
35	C 79 I	20.5	−15.6	0.086	360
36	C 75 S_B 94 I	79.6	−2.2	0.054	–
37	C 74 S_B (70) N 83.0 I	37.3	−3.8	0.061	156

Axially fluorinated cyclohexane derivatives (**36**, **37**) are synthesized from the corresponding olefins by hydrofluorination with 70% HF/pyridine (Olah's reagent) [18]. The selectivity of this reaction is so high that even multiply fluorinated compounds, such as **37**, can be obtained selectively and in reasonable yield (Scheme 8.5). However,

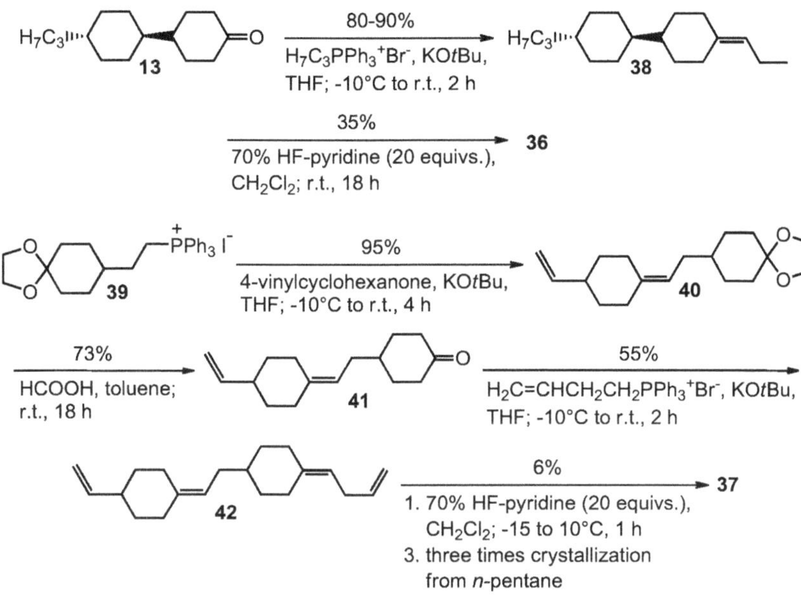

Scheme 8.5 Syntheses of axially fluorinated liquid crystals **36** and **37**.

due to its tendency to autocatalytically eliminate hydrofluoric acid, this material class has not made it into commercial application so far.

In comparison to other classes of dielectrically negative liquid crystals, the synthesis of the indane derivatives (such as **33**–**35**) [17] is much more challenging, and it necessitates several chromatographic purification steps (Scheme 8.6).

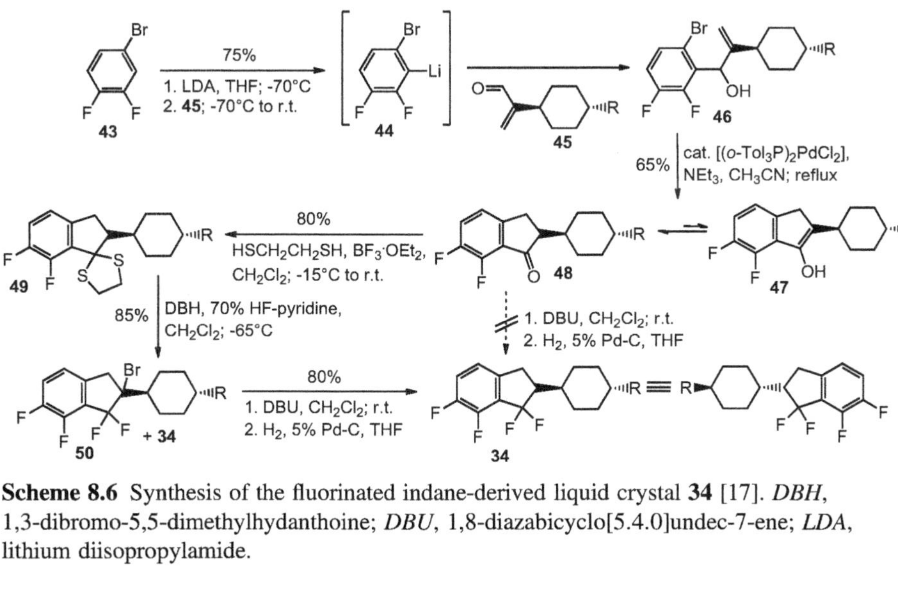

Scheme 8.6 Synthesis of the fluorinated indane-derived liquid crystal **34** [17]. *DBH*, 1,3-dibromo-5,5-dimethylhydanthoine; *DBU*, 1,8-diazabicyclo[5.4.0]undec-7-ene; *LDA*, lithium diisopropylamide.

4. Sulfur-Based Fluorinated Polar Groups

The polarity achievable with terminal fluorine or simple fluorinated groups is sufficient for all LCD modes which require a nematic material and which are related to the original twisted nematic display mode [19]. However, there are also LCD modes under development which are based on other than the nematic mesophase. The most prominent one is the so-called Blue Mode. Blue Mode LCDs require a chiral liquid crystalline material in the blue phase (a cubic phase consisting of a three-dimensional grid of double-twist cylinders) [20], and the liquid crystal has to exhibit a very high dielectric anisotropy. Since the cyano group cannot be used, more "exotic," highly polar fluorinated groups were explored systematically for their usefulness in liquid crystal design: pentafluorosulfanyl (SF_5) [21] and related hypervalent sulfur fluorides (Table 8.3) [22].

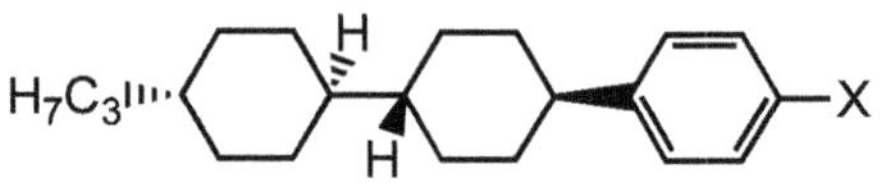

51: X = CF_3
52: X = SF_5

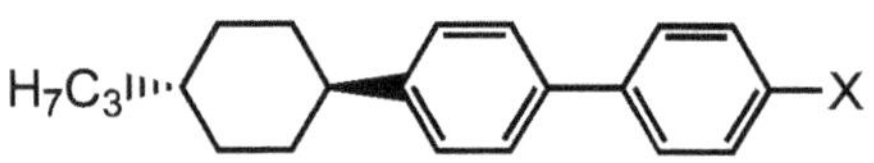

53: X = CF_3 **55**: X = OCF_3
54: X = SF_5 **56**: X = OSF_5

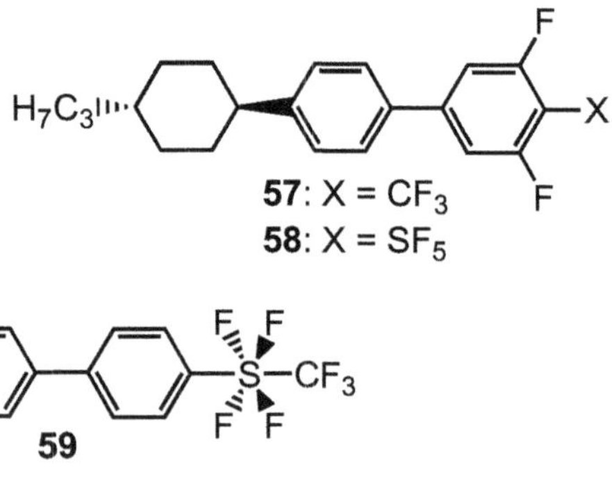

57: X = CF_3
58: X = SF_5

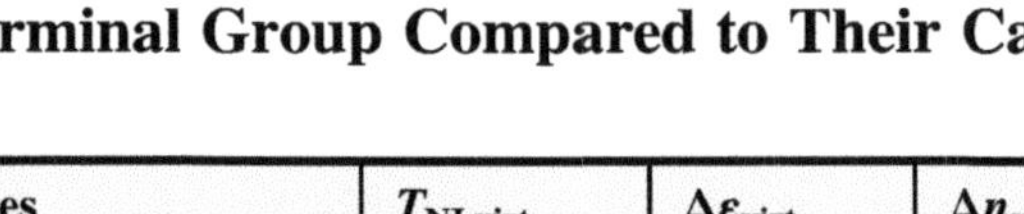

59

Table 8.3 Examples of Liquid Crystals Containing a Hypervalent Sulfur-Based Terminal Group Compared to Their Carbon Analogues [13]

No.	Mesophases	$T_{NI,virt}$	$\Delta\varepsilon_{virt}$	Δn_{virt}	$\gamma_{1,virt}$
51	C 133 I	112.2	9.5	0.091	338
52	C 121 I	95.6	11.6	0.093	612
53	C 134 I	108.6	13.0	0.165	279
54	C 109 N (87.8) I	94.6	14.3	0.154	634
55	C 90 S_B 129 N 151.8 I	134.4	8.7	0.144	137
56	C 147 I	122.9	9.9	0.155	379
57	C 86 I	47.1	21.5	0.149	–
58	C 103 I	49.6	21.4	0.132	–
59	C 197 N 209.7 I	182.6	10.6	0.150	–

The SF_5 group behaves like a "super-trifluoromethyl" function [23]. It has a stronger inductive effect than CF_3, and thus conveys a higher dipole moment to its derivatives: pentafluorosulfanyl benzene has a dipole moment of 3.44 D [24], whereas trifluoromethyl benzene has only 2.86 D. Geometrically, the SF_5 group is sterically far more demanding than CF_3. Its volume is comparable to *tert*-butyl [25]. Similar considerations apply to the OSF_5 [22a] compared to its carba-analogue OCF_3.

The *trans*-SF_4CF_3 group [26] was designed in order to enhance the already strong dipole moment of SF_5 (Scheme 8.7). However, due to the unexpectedly high steric flexibility of the fluorine in hypervalent sulfur fluorides, steric distortion effects actually reduced the dipole moment of *trans*-SF_4CF_3 compared to SF_5 [23] in spite of its enhanced inductive effect [28]. Similar effects are also reducing the effect of lateral *ortho*-fluorination on the polarity of SF_5-based liquid crystals [27] in comparison to the analogous CF_3 compounds. This is illustrated by the comparison of the CF_3-terminated compounds **53**, **55**, and **57** with their SF_5 analogues **54**, **56**, and **58**. One unique property of all hypervalent sulfur fluoride functional groups is their combination of high polarity with extreme hydrophobicity, which is also known as "polar hydrophobicity" [29].

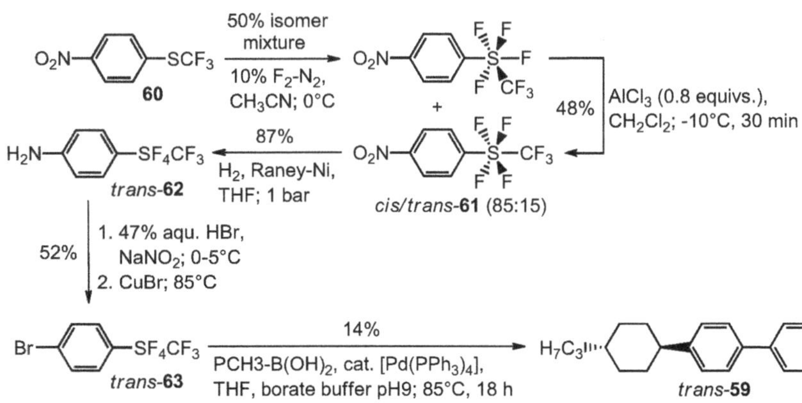

Scheme 8.7 Synthesis of liquid crystal **59** with a *trans*-trifluoromethyltetrafluorosulfanyl terminal group [26]. *PCH-3-B(OH)*$_2$, 4-(propyl-4-*trans*-cyclohexyl)benzeneboronic acid.

5. Fluorinated Bridge Structures

Toward the end of the 1990s, the potential of fluorinated linker groups within the mesogenic core structures was recognized (Table 8.4). First, the α,α-difluoroether function was found to provide the liquid crystal with a wide nematic phase range in combination with exceptionally low rotational viscosity (γ_1) [30]. The difluorooxymethylene group has its most beneficial effect when it is located between a cyclohexane and an arene moiety (Table 8.4).

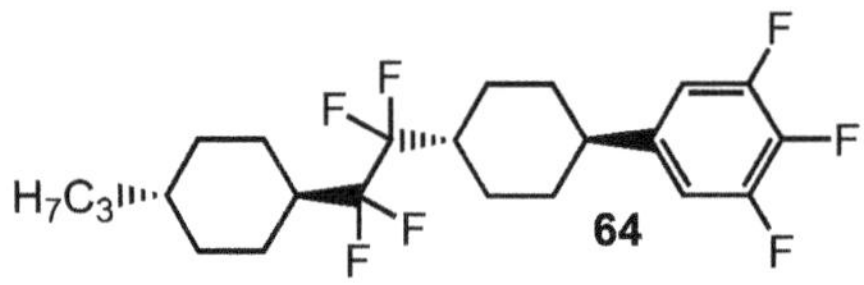

64

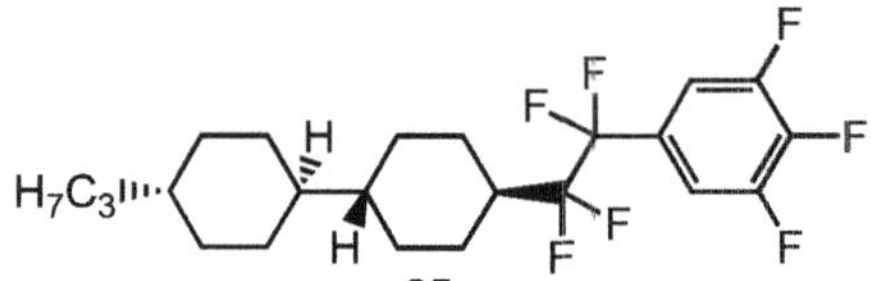

65

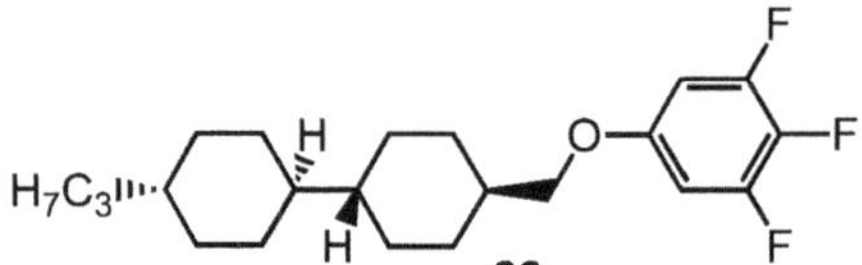

66

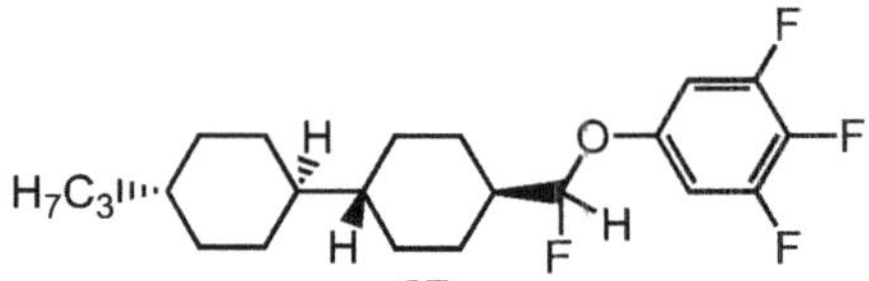

67

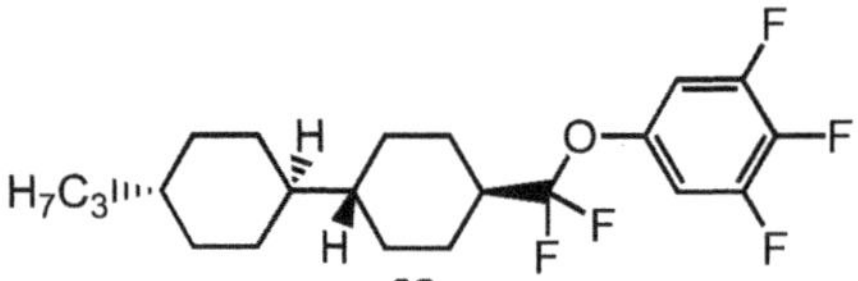

68

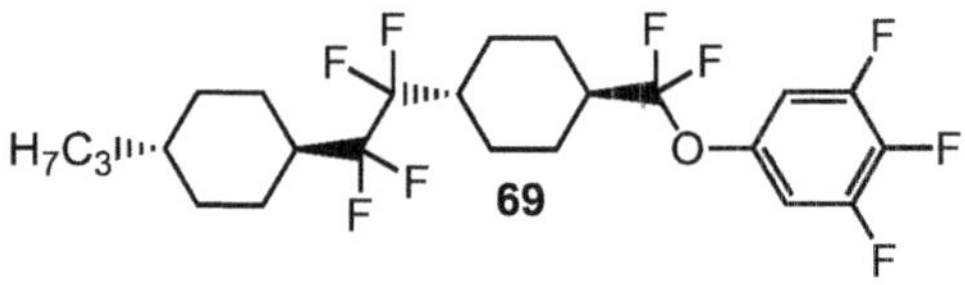

69

Table 8.4 Liquid Crystals With Fluoroaliphatic Spacer Groups in Comparison to Their Analogues Linked by a Single Bond (4) or a Non- (66) or Mono-Fluorinated (67) Ether Group [13]

No.	Mesophases	$T_{NI,virt}$	$\Delta\varepsilon_{virt}$	$\gamma_{1,virt}$
4	C 64.7 N 93.7 I	74.0	8.3	171
64	C 70 S_G 95 S_B 102 N 168.6 I	74.7	9.7	269
65	C 74 S_B 102 N 114.9 I	91.5	8.1	150
66	C 84 N (78.5) I	65.7	8.3	357
67	C 43 N 88.0 I	75.6	8.0	300
68	C 44 N 105.3 I	91.5	10.5	145
69	C 49 S_B 115 N 165.4 I	128.0	9.4	250

The synthesis of the CF_2O group is generally based on the oxidative fluorodesulfuration chemistry, originally developed by T. Hiyama and others in the 1980s [31], but dithianylium salts were introduced as new and readily available intermediates [30d] (Scheme 8.8).

Another fluoroaliphatic linker group is the CF_2CF_2 function. The insertion of this group between two cyclohexane units results in a dramatic increase of the clearing temperature by around 50K [32]. The combination of CF_2CF_2 and CF_2O groups within the same molecule (**69**) results in some of the liquid crystals with the widest ever observed nematic phase range [33].

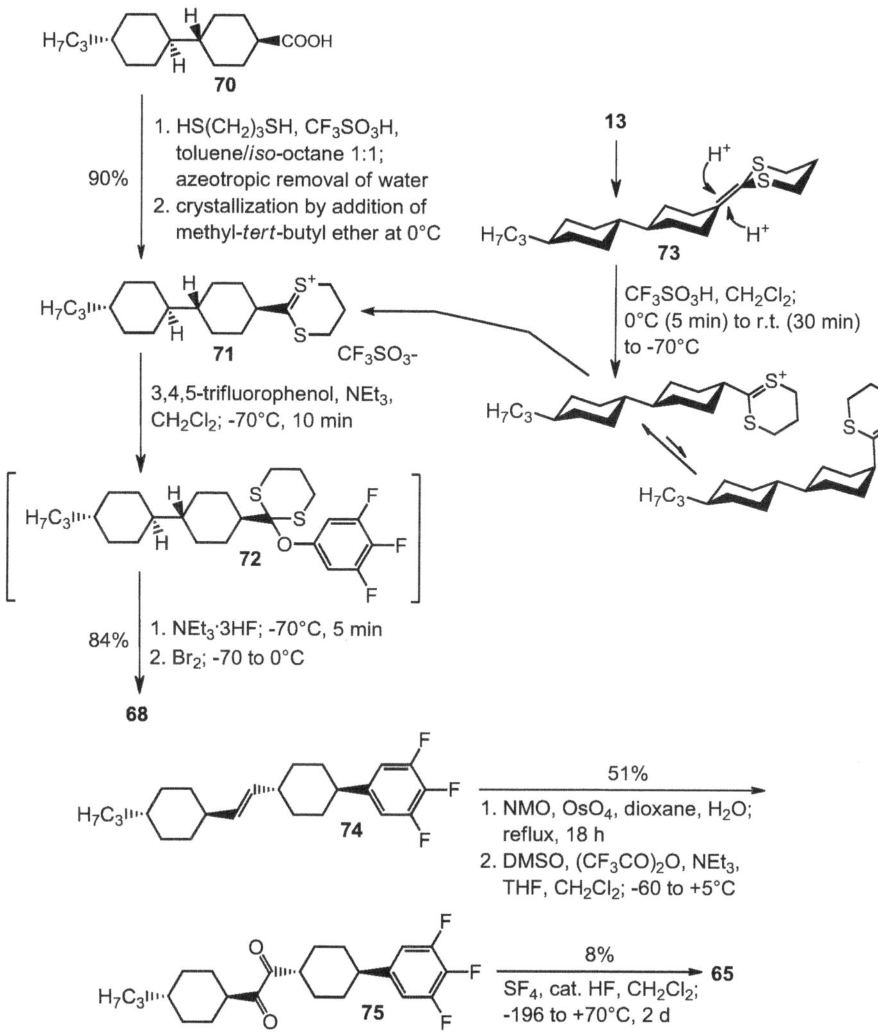

Scheme 8.8 Synthesis of the liquid crystals **68** and **65**, with respective CF_2O and CF_2CF_2 linkers [30d,32].

The beneficial effect of the CF_2O and CF_2CF_2 linkers on the mesophase characteristics of liquid crystals can be explained by stereoelectronic effects: secondary orbital interactions $\sigma_{CH}-\sigma^*_{CF}$ and $n_O-\sigma^*_{CF}$ are rigidifying the linker and thus increasing the effective length of the mesogenic core structure (Fig. 8.1) [34].

Linker groups which are suitable for connecting two arene moieties are the *E*-difluorovinylene (**76**) and the *E,E*-tetrafluorobutadiene (**77**) functions [35]. In particular the difluorovinylene unit leads to materials with a rather unique combination of high clearing point and low rotational viscosity [36]. Due to the conjugation of the adjoining arene moieties through the difluorovinylene linker, this

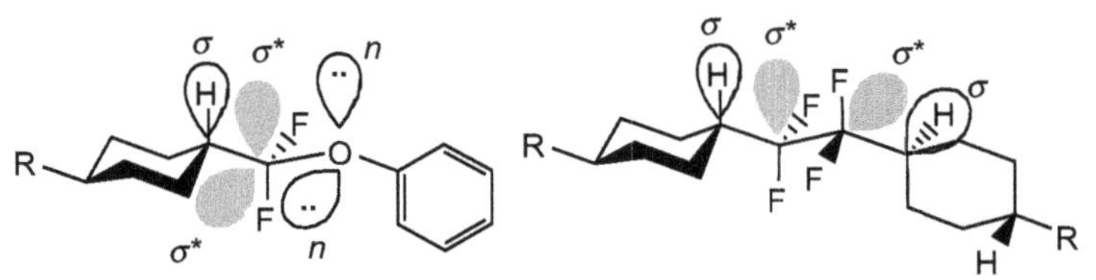

Figure 8.1 Stereoelectronic effects increasing the rigidity of CF_2O- and CF_2CF_2-bridged liquid crystals [34].

bridge element is particularly well suited for the design of polar materials with high birefringence (**78**). The reason, why this material class has so far not been commercialized is its tendency to undergo *E-Z* isomerization under light exposure (Table 8.5).

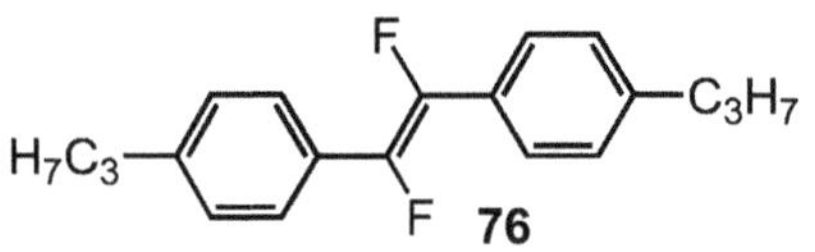

76

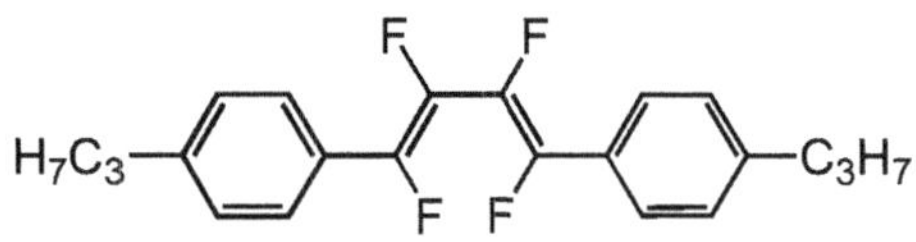

77

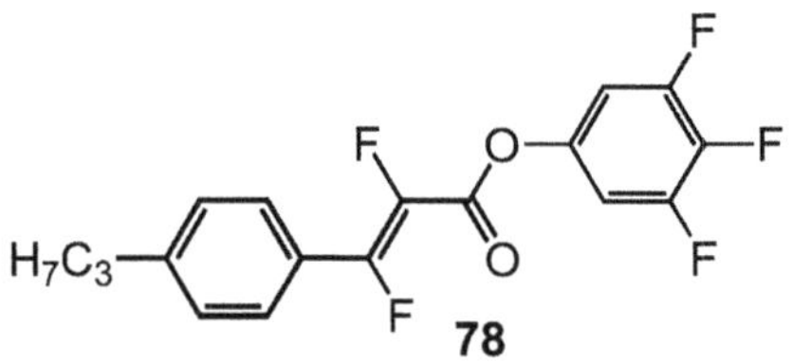

78

Table 8.5 **Liquid Crystals Based on *E*-difluorostilbene (76), 1,4-Diphenyl-*E*,*E*-tetrafluorobutadiene (77), and Difluorocinnamic Acid (78) [13]**

No.	Mesophases	$T_{NI,virt}$	$\Delta\varepsilon_{virt}$	Δn_{virt}	$\gamma_{1,virt}$
76	C 75 N 84.7 I	119.6	3.0	0.265	21
77	C 55 N 115.8 I	126.5	2.8	0.284	42
78	C 49 I	0.4	29.7	0.189	78

Synthetically, difluorostilbenes and related compounds are accessible through a combination of nucleophilic substitution of tetrafluoroethylene, chlorotrifluoroethylene, perfluorobutadiene, and of palladium-catalyzed carbon—carbon coupling reactions [37] (Scheme 8.9).

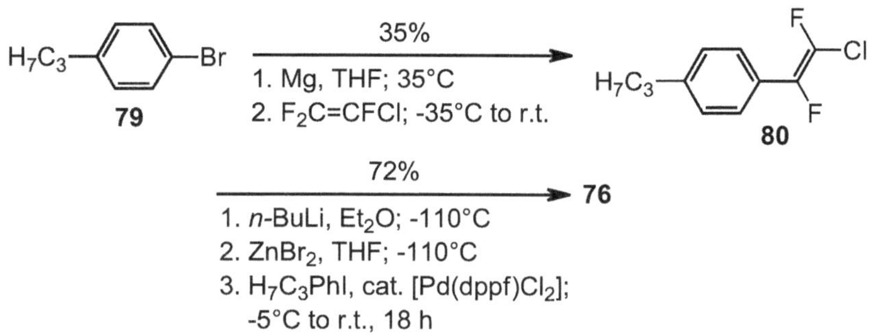

Scheme 8.9 Synthesis of the *E*-difluorovinylene-linked liquid crystal **76**. *dppf*, 1,1′-bis(diphenylphosphino)ferrocene.

6. Computational Methods for Molecular Design

Due to the very short product cycles in electronics and display industry, every year many new liquid crystalline materials are introduced into the market. In order to increase the efficiency of the material development, from the early 1990s on methods for the computational property prediction of liquid crystals were developed [38]. In particular, semiempirical models have been regularly used to do an in silico screening of candidate structures for synthesis. These rather simple and quick methods allow the prediction of the electrooptical parameters: the dielectric anisotropy $\Delta\varepsilon$ and the birefringence Δn. The reason why these two parameters are so well predictable is the fact that they can be conveniently derived from single molecule parameters, such as the dipole moment μ and the polarizability α, using the Maier–Meier theory [39] which links single molecule properties to supramolecular properties in the condensed phase (Eq. [8.1]):

$$\Delta\varepsilon = \frac{NhF}{\varepsilon_0}\left\{\Delta\alpha - F\frac{\mu^2}{2k_BT}\left(1 - 3\cos^2\beta\right)\right\}S \qquad [8.1]$$

For LCD applications not the molecular dipole moment itself but the dielectric anisotropy $\Delta\varepsilon$ is the critical polarity parameter. According to the Maier–Meier theory the supramolecular parameter $\Delta\varepsilon$ can be connected to the molecular dipole moment, taking the angle β between the dipole moment μ and long molecular axis (or—more accurately—the director S of the nematic phase) into account.

The Maier–Meier formula indicates that perfect alignment of dipole moment with the long molecular axis ($\beta = 0$ degree) results in a maximum positive $\Delta\varepsilon$. If the dipole moment is oriented perpendicularly to the molecular axis ($\beta = 90$ degree), maximum negative $\Delta\varepsilon$ is obtained. If $\beta = 54$ degree, $\Delta\varepsilon$ equals always zero, independently from the dipole moment (Fig. 8.2).

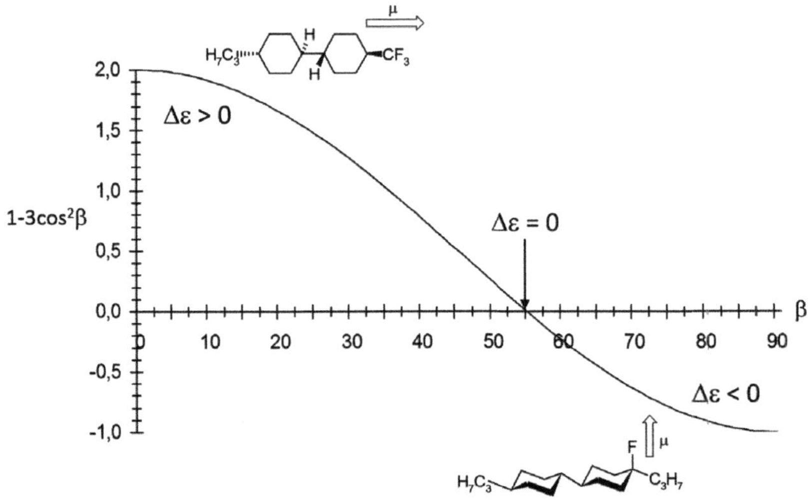

Figure 8.2 The orientation-dependent term of the Maier–Meier formula, describing the relation between $\Delta\varepsilon$ and the molecular dipole moment relative (defined by the angle β) to the nematic phase director S, or—approximately—the long molecular axis.

Fig. 8.2 also illustrates why it is more difficult to obtain materials with a strongly negative $\Delta\varepsilon$: the same dipole moment furnishes twice the absolute $\Delta\varepsilon$, when it is oriented parallel to the long molecular axis, compared to when it has the perpendicular orientation. Commercial liquid crystals fall into three groups: one with β around 0 degree and $\Delta\varepsilon > 0$ mainly for application in mobile devices and PC monitors, one with β around 90 degree and $\Delta\varepsilon < 0$ which is used as polar components for LCD TV application. A third group consists of unpolar compounds ($\Delta\varepsilon \sim 0$). These materials are mainly used as "diluters" in order to reduce the viscosity of liquid crystal mixtures.

Using the Maier–Meier formula, the dielectric anisotropy ($\Delta\varepsilon$) can be predicted easily from the molecular structure, which again can be calculated by semiempirical or density functional theory methods. Using similar methods, the birefringence (Δn) can be predicted with reasonable accuracy as well [40] (Fig. 8.3).

However, such calculations for conformationally flexible liquid crystals only provide a snapshot picture of the electrooptic parameters [38a]. In particular, the dielectric anisotropy ($\Delta\varepsilon$) often depends very much on the conformation, and different, energetically similar conformations can have very different $\Delta\varepsilon$ values. In such cases the whole conformational equilibrium has to be taken into account, and the $\Delta\varepsilon$ values of the various conformers have to be averaged and Boltzmann-weighted according to their relative energies. For example, in the axially difluorinated bicyclohexyl derivative **81**, the two major conformers have a very different dielectric anisotropy [16]. A realistic prediction for the condensed phase has to based on the Boltzmann-weighted average of these two conformers **A** ($\phi = 180$ degree, $\Delta\varepsilon = 0.2$) and **B** ($\phi = 75$ degree, $\Delta\varepsilon = -2.9$) (Fig. 8.4).

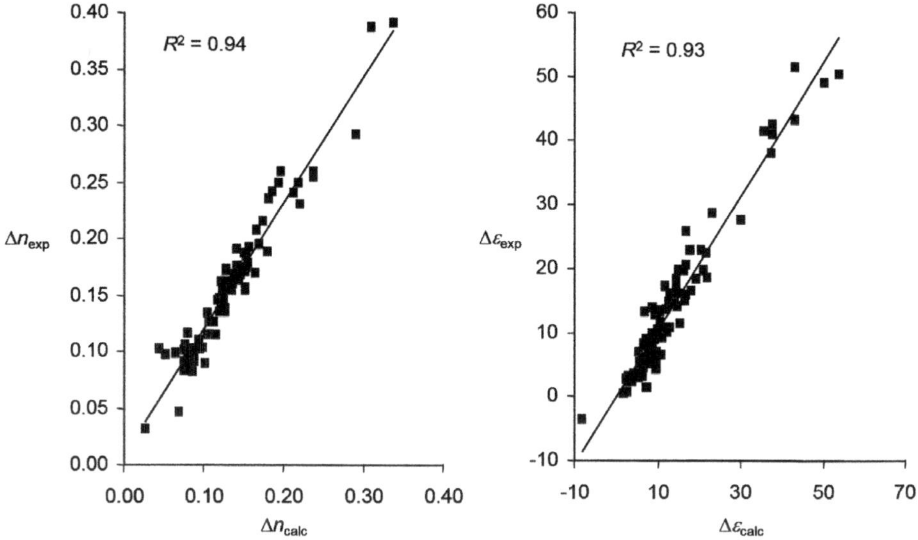

Figure 8.3 Calculated (AM1) and experimental electrooptical data ($\Delta\varepsilon$ and Δn) for a liquid crystal test set in comparison (84 samples).

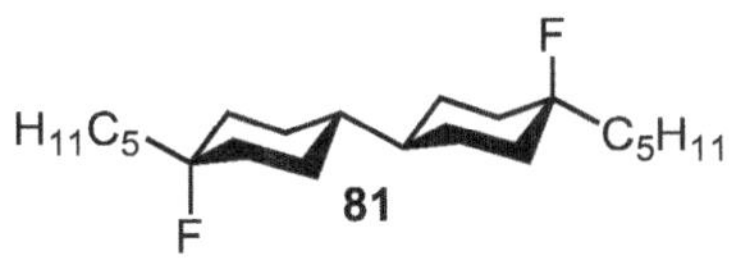

81

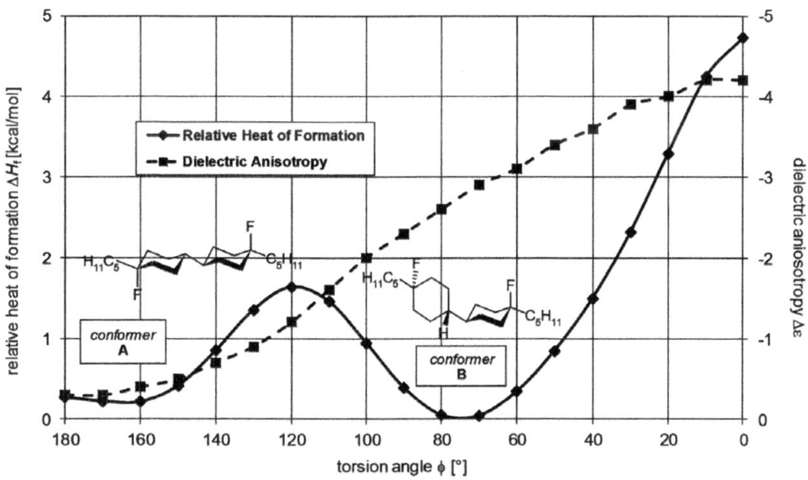

Figure 8.4 Dielectric anisotropy ($\Delta\varepsilon$) versus the central dihedral angle (ϕ) of liquid crystal **81**.

7. The Future of Fluorinated Liquid Crystals

LCDs have been an established technology for more than a decade, and recently most innovations in this field have been directed toward reducing manufacturing cost. However, even for standard LCD application there still is room for technical improvement. Fast switching modes like the Blue Mode are developed in order to facilitate high-frequency driving which is a prerequisite for 3D LCD and ultrahigh-resolution 8–4K TV. Also, true holography (as opposed to current "3D" technology requiring special goggles) has moved into the focus of LCD manufacturers [41]. Other development targets are unbreakable and lightweight plastic LCD and flexible LCD.

Beyond displays, liquid crystals have recently found applications in microwave switching [42] and in "smart windows" [43]. The smart window technology has two major applications: the older one, the so-called "privacy glass," switches between a transparent and a milky-opaque state and is technically based on polymer-dispersed liquid crystal technology [44]. Currently, a new "smart window" technology is introduced into the market, which is based on liquid crystals doped with dichroic dyes [45]. These "liquid crystal windows" automatically shade glass facades and thus reduce the energy required for air-conditioning by up to 40%.

References

[1] [a] H. Moissan, C. R. Acad. Sci. 102 (1886) 1534;
[b] H. Moissan, C. R. Acad. Sci. 103 (1886) 202;
[c] H. Moissan, C. R. Acad. Sci. 103 (1886) 256.

[2] F. Reinitzer, Monatsh. Chem. 9 (1888) 421.

[3] Reviews: [a] P. Kirsch, M. Bremer, Angew. Chem. Int. Ed. 39 (2000) 4216–4235;
[b] M. Bremer, P. Kirsch, M. Klasen-Memmer, K. Tarumi, Angew. Chem. Int. Ed. 52 (2013) 8880–8896.

[4] Review: P. Kirsch J. Fluorine Chem. 177 (2015) 29–36.

[5] [a] G.W. Gray, B. Jones, J. Chem. Soc. (1954) 2556–2562;
[b] G.W. Gray, B.M. Worrall, J. Chem. Soc. (1959) 1545–1550.

[6] N.A. Clark, S.T. Lagerwall, Appl. Phys. Lett. 36 (1980) 899.

[7] Review: P. Kirsch, V. Reiffenrath, M. Bremer Synlett (1999) 389–396.

[8] [a] H. Gilman, R.L. Bebb, J. Am. Chem. Soc. 61 (1939) 109;
[b] G. Wittig, G. Fuhrmann, Chem. Ber. 73 (1940) 1197.

[9] [a] Review: V. Snieckus, Chem. Rev. 90 (1990) 879–933;
[b] M. Schlosser, Pure Appl. Chem. 60 (1988) 1627;
[c] F. Mongin, R. Maggi, M. Schlosser, Chimia 50 (1996) 650–652;
[d] L. Lochmann, J. Pospíšil, D. Lim, Tetrahedron Lett. 7 (1966) 257–262.

[10] C. Stiasny, V. Reiffenrath, Patent EP 1010702, 1999.

[11] S. Naemura, Y. Nakazono, K. Nishikawa, A. Sawada, P. Kirsch, M. Bremer, K. Tarumi, Mater. Res. Soc. Symp. Proc. 508 (1998) 235–240.

[12] P. Kirsch, Modern Fluoroorganic Chemsitry, second ed., Wiley-VCH, Weinheim, 2013.

[13] The phase transition temperatures are given in °C, the γ_1 values in mPa s. C = crystalline, S_x = smectic X, N = nematic, I = isotropic. The "virtual" parameters $T_{NI,virt}$, $\Delta\varepsilon_{virt}$, and

Δn_{virt} were determined by linear extrapolation from a 10% w/w solution in the commercially available Merck mixture ZLI-4792 ($T_{NI} = 92.8°C$, $\Delta\varepsilon = 5.3$, $\Delta n = 0.0964$). The extrapolated values are corrected empirically for differences in the order parameter. For dielectrically negative materials, $\Delta\varepsilon$ was extrapolated from ZLI-2857 ($T_{NI} = 82.3°C$, $\Delta\varepsilon = -1.4$, $\Delta n = 0.0776$). For the pure substances the mesophases were identified by optical microscopy, and the phase transition temperatures by differential scanning calorimetry (DSC); numbers in parentheses denote monotropic phase transitions.
[14] D. Pauluth, K. Tarumi, J. Mater. Chem. 14 (2004) 1219–1227.
[15] Representative examples: [a] S.M. Kelly, Helv. Chim. Acta 67 (1984) 1572;
[b] V. Reiffenrath, J. Krause, H.J. Plach, G. Weber, Liq. Cryst. 5 (1989) 159;
[c] G.W. Gray, M. Hird, D. Lacey, K.J. Toyne, J. Chem. Soc. Perkin Trans. 2 (1989) 2041;
[d] M.E. Glendenning, J.W. Goodby, M. Hird, K.J. Toyne, J. Chem. Soc. Perkin Trans. 2 (1999) 481.
[16] P. Kirsch, K. Tarumi, Angew. Chem. Int. Ed. 37 (1998) 484–489.
[17] M. Bremer, L. Lietzau, New J. Chem. 29 (2005) 72–74.
[18] G.A. Olah, M. Nojima, I. Kerekes, Synthesis (1973) 779–780.
[19] M. Schadt, W. Helfrich, Appl. Phys. Lett. 18 (1971) 127.
[20] H. Kikuchi, M. Ykota, Y. Hisakado, H. Yang, T. Kajiyama, Nat. Mater. 1 (2002) 64–68.
[21] [a] W.A. Sheppard, J. Am. Chem. Soc. 84 (1962) 3064;
[b] P. Kirsch, M. Bremer, M. Heckmeier, K. Tarumi, Angew. Chem. Int. Ed. 38 (1999) 1989;
[c] P. Kirsch, J.T. Binder, E. Lork, G.-V. Röschenthaler, J. Fluorine Chem. 127 (2006) 610.
[22] [a] J.R. Case, R. Price, N.H. Ray, H.L. Roberts, J. Wright, J. Chem. Soc. (1962) 2107;
[b] J.R. Case, H.L. Roberts, Patent GB 928412, 1963.
[23] P. Kirsch, M. Bremer, Chimia 68 (2014) 363–370.
[24] W.A. Sheppard, J. Am. Chem. Soc. 84 (1962) 3072.
[25] M.V. Westphal, B.T. Wolfstädter, J.-M. Plancher, J. Gatfield, E.M. Carreira, ChemMedChem 10 (2015) 461–469.
[26] P. Kirsch, A. Hahn, Eur. J. Org. Chem. (2006) 1125–1131.
[27] P. Kirsch, A. Hahn, Eur. J. Org. Chem. (2005) 3095–3100.
[28] C. Hansch, A. Leo, R.W. Taft, Chem. Rev. 91 (1991) 165–195.
[29] J.C. Biffinger, H.W. Kim, S.G. DiMagno, ChemBioChem 5 (2004) 622.
[30] [a] E. Bartmann, K. Tarumi, Patent DE 19531165, 1995.
[b] T. Ando, K. Shibata, S. Matsui, K. Miyazawa, H. Takeuchi, Y. Hisatsune, F. Takeshita, E. Nakagawa, K. Kobayashi, Y. Tomi, Patent EP 0844229, 1998.
[c] A. Haas, M. Spitzer, M. Lieb, Chem. Ber. 121 (1988) 1329–1340;
[d] E. Bartmann, Adv. Mater. 8 (1996) 570–573;
[e] P. Kirsch, M. Bremer, A. Taugerbeck, T. Wallmichrath, Angew. Chem. Int. Ed. 40 (2001) 1480–1484.
[31] [a] M. Kuroboshi, T. Hiyama, Synlett (1994) 251–252;
[b] J. Kollonitsch, S. Marburg, L.M. Perkins, J. Org. Chem. 41 (1976) 3107–3111;
[c] S.C. Sondej, J.A. Katzenellenbogen, J. Org. Chem. 51 (1986) 3508–3513;
[d] G.K.S. Prakash, D. Hoole, V.P. Reddy, G.A. Olah, Synlett (1993) 691–693;
[e] M. Kuroboshi, T. Hiyama, J. Fluorine Chem. 69 (1994) 127–128;
[f] R.D. Chambers, G. Sandford, M.E. Sparrowhawk, M.J. Atherton, J. Chem. Soc. Perkin Trans. 1 (1996) 1941–1944;
[g] C. York, G.K.S. Prakash, G.A. Olah, Tetrahedron 52 (1996) 9–14;
[h] Review: M. Shimizu, T. Hiyama, Angew. Chem. Int. Ed. 44 (2004) 214–231.

[32] P. Kirsch, M. Bremer, F. Huber, H. Lannert, A. Ruhl, M. Lieb, T. Wallmichrath, J. Am. Chem. Soc. 123 (2001) 5414–5417.
[33] P. Kirsch, F. Huber, M. Lenges, A. Taugerbeck, J. Fluorine Chem. 112 (2001) 69–72.
[34] P. Kirsch, M. Bremer, ChemPhysChem 11 (2010) 357–360.
[35] [a] A.B. Shtarev, M.M. Kremlev, Z. Chvátal, J. Org. Chem. 62 (1997) 3040–3045;
[b] L.M. Yagupolskii, M.M. Kremlev, V.A. Khranovskii, Y.A. Fialkov, J. Org. Chem. USSR 12 (1976) 1365–1366.
[36] [a] P. Kirsch, H. Hirschmann, J. Krause, Patent DE 19906254, 1998.
[b] P. Kirsch, J. Krause, H. Hirschmann, L.M. Yagupolskii, Patent DE 10102630, 2001.
[37] Review: M.M. Kremlev, L.M. Yagupolskii, J. Fluorine, Chem. 91 (1998) 109–123.
[38] [a] M. Bremer, K. Tarumi, Adv. Mater. 5 (1993) 842;
[b] M. Klasen, M. Bremer, A. Götz, A. Manabe, S. Naemura, K. Tarumi, Jpn. J. Appl. Phys. 37 (1998) L945.
[39] The dielectric anisotropy is defined as $\Delta\varepsilon = \varepsilon_{||} - \varepsilon_{\perp}$, the birefringence as $\Delta n = n_{||} - n_{\perp}$, where $_{||}$ stands for parallel and $_{\perp}$ perpendicular to the nematic phase director, which is approximated by the long molecular axis. The correlation between $\Delta\varepsilon$, the dipole moment μ and the angle β between the molecular dipole and the director is the following: $\Delta\varepsilon \sim \Delta\alpha - F(\mu^2/2kT)(1 - 3\cos^2 \beta)S$; $\Delta\alpha$ is the anisotropy of the polarizability, S the order parameter. [a] W. Maier, G. Meier, Z. Naturforsch. 16a (1961) 262–267;
[b] D. Demus, G. Pelzl, Z. Chem. 21 (1981) 1–9.
[40] M.F. Vuks, Opt. Spectrosc. (Engl. Transl.) 20 (1966) 361.
[41] [a] R. Häussler, S. Reichelt, N. Leister, E. Zschau, R. Missbach, A. Schwerdtner, Large real-time holographic displays: from prototypes to a consumer product, in: Stereoscopic Displays and Applications XX, SPIE 7237, 2009, p. 72370S;
[b] S. Reichelt, N. Leister, in: ISDH Proc., 9th Int. Symp. Displ. Hologr, 2012, pp. 1–10.
[42] Review: P. Yaghmaee, O.H. Karabey, B. Bates, C. Fumeaux, R. Jacoby, Int. J. Ant. Prop. (2013) 1–9.
[43] [a] C.M. Lampert, Sol. Energy Mater. 76 (2003) 489–499;
[b] R. Baetens, B.P. Jelle, A. Gustavsen, Sol. Energy Mater. 94 (2010) 87–105.
[44] [a] J.L. Fergason, in: SID Int. Symp. Digest Tech. Papers 16, 1985, p. 68;
[b] Review: D.-K. Yang, Polymer stabilized cholesteric liquid crystal for switchable windows, in: Q. Li (Ed.), Liquid Crystals beyond Displays, John Wiley & Sons, Inc., Hoboken, 2012, p. 505.
[45] http://magazin.merck.de/de/Life_and_Quality/Smart_Windows/Fluessigkristalle1.html (accessed on 03.07.15).

Molecular Engineering of F-Based Iridium(III) Complexes as a Phosphorescent Emitter

9

Y. Takahira
Innovative Technology Research Center, Asahi Glass Co., Ltd., Yokohama, Japan

E. Murotani
AGC Chemicals, Asahi Glass Co., Ltd., Tokyo, Japan

Chapter Outline

1. Introduction

It has been almost two decades since organic light-emitting diodes (OLED) became to be considered as a promising candidate for the application in lighting or display after the remarkable breakthrough by Forrest and Thompson's group [1,2,8,9]. They revealed highly efficient phosphorescent emission from organotransition metal compounds at ambient temperatures. The pioneering work opened the door for a practical use of OLED for commercial applications, followed by various studies on this field (selected reviews, see Refs [18,48,57,69]). The readers can find more in the reviews. Very recently, OLED has been actually introduced into the market of lighting panels and flat display panels.

Photonic and Electronic Properties of Fluoride Materials. http://dx.doi.org/10.1016/B978-0-12-801639-8.00009-X

1.1 Technology Backgrounds

The key material of the phosphorescent emission is needless to say the organometallic small molecules, not like polymeric compounds employed in fluorescent emitting devices (for review, see Ref. [4]). A mechanism of the emission is based on triplet-state excitons of the transition metal complexes, which is totally different from the fluorescent emission in terms of the really high quantum efficiency. As shown in Fig. 9.1, holes and electrons are injected into an emitting layer including the complex as a guest and an organic aromatic compound as a host through hole and electron transporting layers by applying voltage to a device. The recombination of the holes and the electrons in the emitting layer can be classified in two manners: (1) both meet on the host site to give the excitons which then transfer to the guest site or (2) the holes generated directly on the guest site recombine with the electrons passing through the host to provide the excitons. Accordingly, the emission results from the energy releasing when the excitons go back to the ground state.

In OLEDs, the excitons formed by the recombination of the electrically injected charge carriers have singlet and triplet state theoretically in the ratio of 1:3, indicating that the emission from the triplet state excitons is more efficient. Moreover, due to an internal crossing phenomenon of excitons occurred in the metal complex by the heavy atom effect, the singlet state excitons can be converted to those of the triplet state to result in all emission via the phosphorescent process (Fig. 9.2). Hence the maximum of external quantum efficiency of phosphorescent emitters (typically 20−25%) is far higher than that of the corresponding fluorescent ones (typically 5−6%).

OLED devices are consisted of multilayered architecture, mainly including anode, cathode, hole transporting layer, electron transporting layer, emitting material layer, and more in some cases. Each layer is substantially fabricated by thermal evaporation

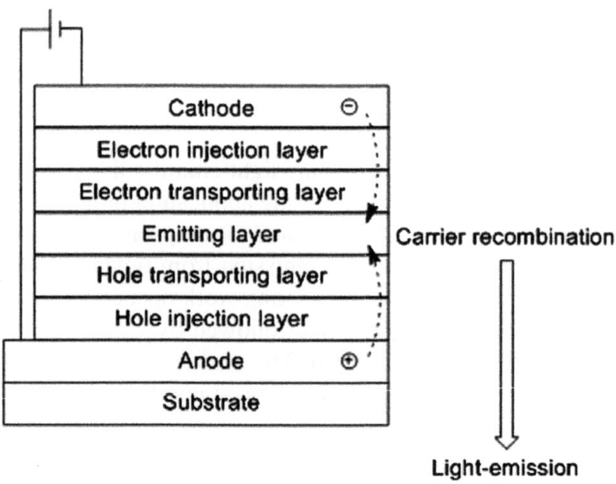

Figure 9.1 Organic light-emitting diode device architecture.

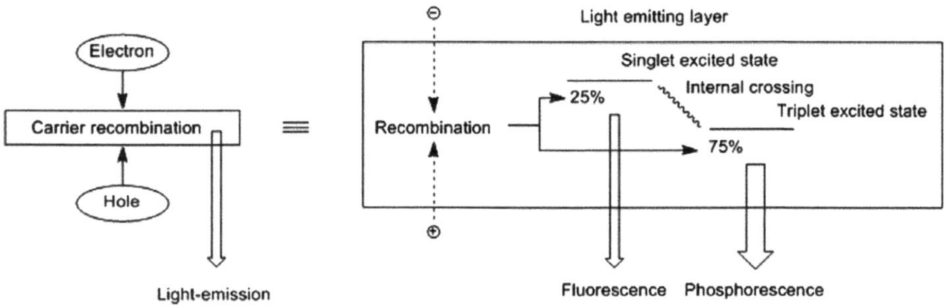

Figure 9.2 Excited state in the light-emitting layer.

(dry process) of small molecules in current commercial products rather than coating or inkjet printing (wet process) of a solution of materials. In the former process, highly pure material thin film can be produced to enable superior quality devices; however, low use efficiency of materials and difficulty of coevaporation of host and guest materials are still considerable. The latter is potentially expected to reduce process cost by high throughput and be suitable for large-area fabrication, whereas there remain several problems to serve enough device quality. Light-emitting electrochemical cell (LEC) has also attracted attention because of the advantage of very simple architecture: only two electrodes and an emitting layer (Fig. 9.3). Transition metal complexes can be applied as an emitter of LEC [27,47], which means that we can share some ideas about materials for the both devices. It could be promising if it overcomes remaining drawbacks, ie, insufficient quantum efficiency or low color purity.

Recently, Adachi's group reported emerging work about thermally activated delayed fluorescence, which is considered as the third emission mode other than described previously [19,22,42,65,71]. It was achieved by molecular design of organic aromatic compounds with a thermally accessible energy gap between singlet and triplet states. Promising aspect of the emission is that it would be cheaper alternatives

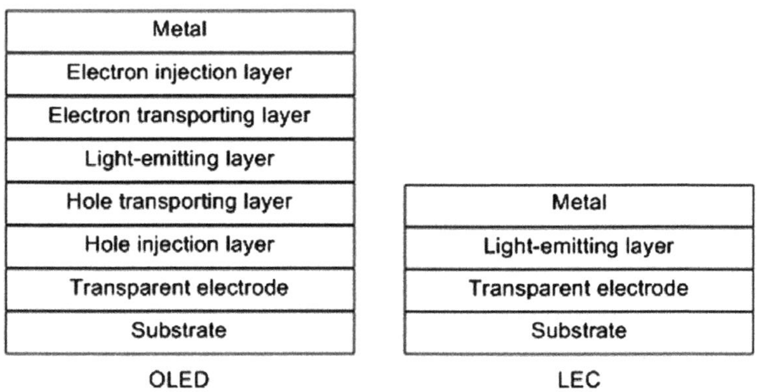

Figure 9.3 Typical device structures of OLED (left) and LEC (right). *OLED*, organic light-emitting diode; *LEC*, light-emitting electrochemical cell.

to phosphorescence and expected to show higher efficiency than that of the conventional fluorescence. Still it is necessary to optimize molecular design or device architecture in order to obtain sufficient device performance.

In this chapter, we then focus on the overview of phosphorescent emission, which is basically much closer to the practical use.

1.2 Phosphorescent Emission Process

Highly efficient phosphorescent emission comes from the triplet state excitons electrically generated in the guest material of the transition metal complexes. The electronic transition generating the excitons in the complex is categorized as the metal-centered (MC) transition, the ligand-centered (LC) transition, the charge transfer (CT) transition, or a combination of them (Fig. 9.4).

Electroluminescence (EL) from the rare earth metal complexes such as Eu [72] or Nd [24] has been assigned to MC transition. On the other hand, the excitons contributing to phosphorescence particularly reviewed in this chapter have proven to be predominantly made by metal-to-ligand charge transfer (MLCT) which is one of the CT transition or MLCT mixed with LC transition. Theoretical studies greatly support the discussion about the electronic transition leading to the exited state [5,12,14]. Molecular orbital diagrams calculated by density functional theory approach indicate that excitons basically exist on the orbital spreading over a metal center to a cyclometalated ligand, which obviously suggests MLCT transition. We believe it will become more important to use the theoretical approach to design emitting materials for further high efficiency or color tuning.

Host materials, typically organic aromatic compounds, play as an important role as guest complexes do. Especially for the blue color luminescence, a choice of a suitable host compound is extremely crucial because it is difficult to prevent triplet excitons

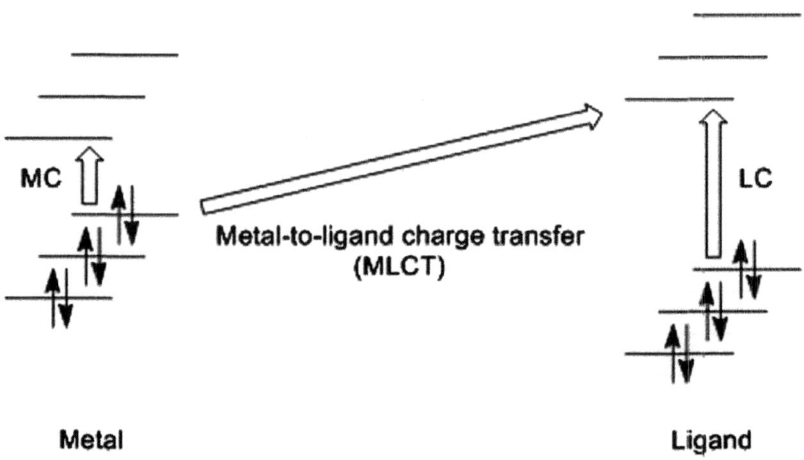

Figure 9.4 Schematic image of electronic transitions. *MLCT*, metal-to-ligand charge transfer; *MC*, metal-centered transition; *LC*, ligand-centered transition.

with high energy level in the guest site from flowing to the host material back. Currently 4,4′-dicarbazolyl-1,1′-biphenyl (CBP) has been most likely employed as a host material because of its good performance. We can also refer to Ref. [70] covering a wide series of host materials for the blue phosphorescent emission to understand more about recent progresses.

1.3 Emitting Materials

The luminescent color which is one of the most important aspects as an emitting device is strongly dependent on metal species, coordinated ligands, and steric complex structure based on them. Hence it is significant to understand the relationship between complex structure and luminescent properties. Besides color tuning, it is of course necessary to consider emission efficiency, chemical stability, processibility, or else. Well-studied transition metal complexes are almost Pt-based or Ir-based compounds and at the point of practical view the latter is substantially standard. The three primary colors (namely red, green, and blue) are actually all realized by designing the structure of Ir-based phosphorescent materials (Fig. 9.5). A tremendous number of scientific publications regarding these light-emitting iridium(III) complexes have already been reported and we will here get an overview of the several work about them in the following part.

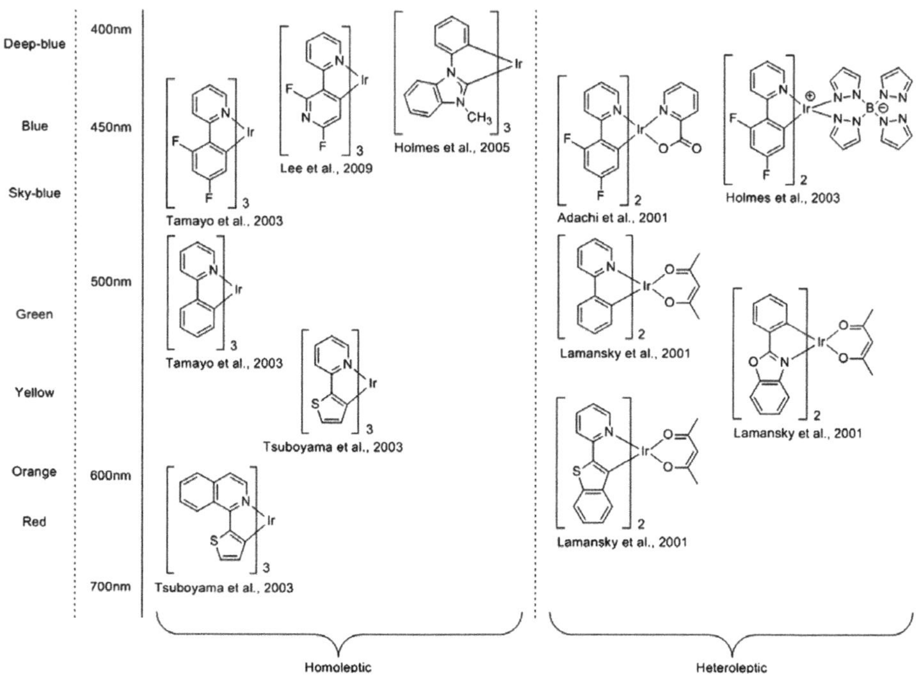

Figure 9.5 Wide variety of colors by Ir-based emitters.

2. Molecular Engineering of F-Based Iridium(III) Complexes as a Phosphorescent Emitter

As described in the introduction, a durable blue emitter is one of the key materials for successful application in future technologies, particularly in lighting and display use. From the standpoint of a synthetic chemist, a reasonable solution to achieve the definitive blue emitter is "minimal-F-based design," that is a synergetic hybrid design of F- and non-F-based emitter which possesses limited numbers of fluorine atoms. We will focus on the "fluorine effect" on light-emitting cyclometalated iridium(III) complexes according to the following Sections 2.1–2.4.

2.1 Molecular Design and Synthesis

The cyclometalated iridium(III) complexes are generally classified into organometallic complexes possessing octahedral molecular geometry wherein six atoms or organic groups (the so-called ligands) coordinate to the central iridium atom (Fig. 9.6). The cyclometalating ligands are generally illustrated using C- and N-terminated arcs and they are commonly abbreviated to as C^N in text. The symbol "C" represents a monoanionic carbon ligation and the symbol "N" represents a neutral nitrogen congener.

A representative cyclometalated iridium(III) complex, tris(2-phenylpyridinato-N,C2′)iridium(III), $Ir(ppy)_3$ was first prepared in 1985 and this has a green photoluminescence based on phosphorescence ($\lambda_{max} = 520$ nm) in solution at room temperature ([33], Fig. 9.7).

The first $Ir(ppy)_3$-based phosphorescent light-emitting device was reported by Baldo et al. [8] and this pioneering work introduced a fundamental strategy for designing organometallic phosphorescent emitters, thereby opening a new field of extensive investigations on its analogous complexes. Note that there are two stereoisomers of tris(cyclometalated) iridium(III) complexes: that is *facial* (*fac*) and *meridional* (*mer*). In these two stereoisomers, the three identical atoms on the ligands are positioned at a face (*fac*-isomer) and at a meridian line (*mer*-isomer) of the octahedron (Fig. 9.8). Both isomers exhibit different properties, for instance, the phosphorescent quantum yields of the latter ones are generally lower than those of the former congeners and this explained why thermal- or light-induced

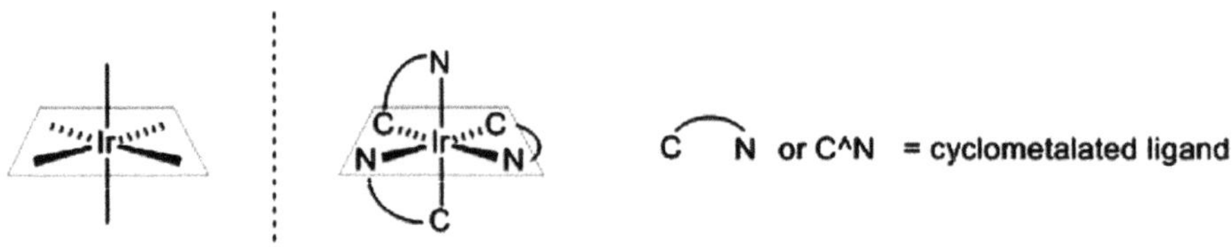

Figure 9.6 Schematic molecular geometry of iridium(III) complex (left) and cyclometalated iridium(III) complex (right).

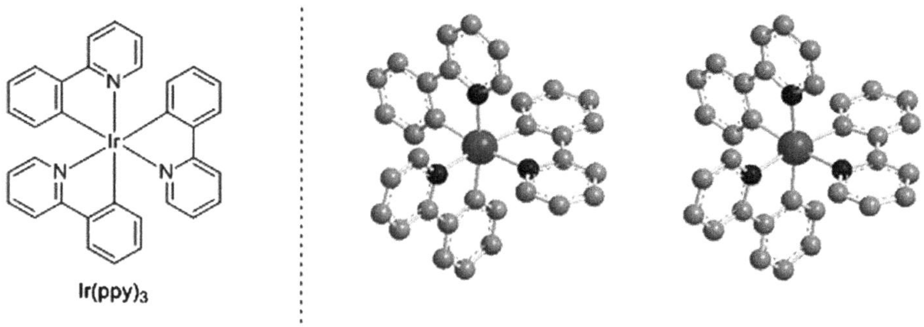

Figure 9.7 Chemical structure of $Ir(ppy)_3$ (left) and its stereo view (right).

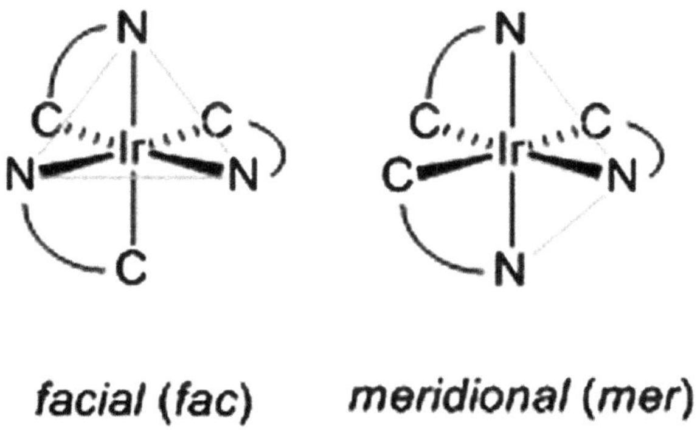

Figure 9.8 Facial and meridional isomers of cyclometalated iridium complexes.

isomerization is a problematic in these emitters. The influences to the properties of this isomerization have been examined by several groups [21,28].

Hundreds of emitting materials based on iridium(III) complexes have been reported to date, and as a result, several classifications are available. Although a detailed discussion is omitted here, the readers can easily understand the overview with some representative structure of the complexes as shown in Fig. 9.9. Dimeric, oligomeric, polymeric materials are known, while the most complexes are monomeric (an exceptional example is dendric complex, see Ref. [44]). These materials are of special interest, with respect to their film-forming properties and processability from solution by inkjet printing. From the viewpoint of the formal charge of the whole complex, noncharged (the so-called neutral) and charged (cationic, anionic, zwitterionic, complex salt [45], etc.) are also prepared. These charged complexes are considered much more promising especially for large-area lighting and display applications due to some of their additional properties: (1) the presence of ionic species should facilitate electric charge injection from electrodes into the emission layers and (2) controllable solubility based on the structure of ligands as well as counterion enables solution process (wet process) fabrication of the devices.

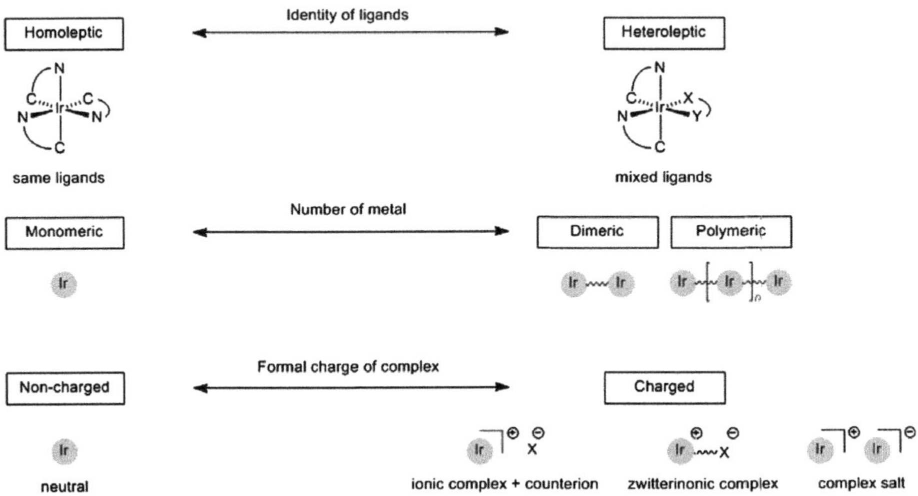

Figure 9.9 Several classification for iridium(III) complexes for phosphorescent emitters.

A brief overview of the most common synthetic strategies toward the various types of cyclometalated iridium(III) complexes is shown in Scheme 9.1. In the first step, a reaction of cyclometalated ligand precursor C^N (eg, 2-phenylpyridine) and iridium(III) source (eg, $IrCl_3 \cdot 3H_2O$) affords a μ-chloro-bridged iridium(III) dimer (Scheme 9.1, step (a)). This step is usually carried out under harsh refluxed conditions in 2-methoxyethanol [49,50] or a mixture (3/1, v/v) of 2-ethoxyethanol and water [56]. In the second step, a treatment with a same cyclometalated ligand C^N in the presence of base splits the dimer complex to give homoleptic $(C^\wedge N)_3Ir$ complexes (Scheme 9.1, step (b)). Although kinetically preferred *mer*-isomer of $(C^\wedge N)_3Ir$ is generally obtained

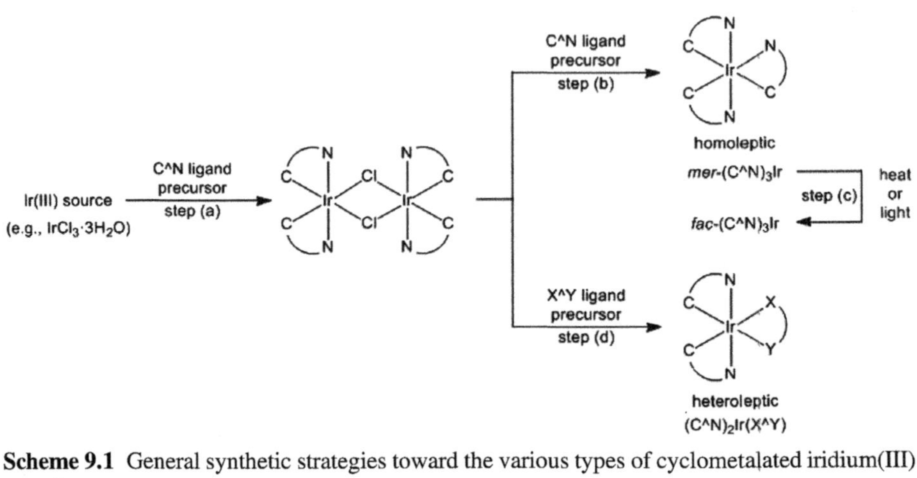

Scheme 9.1 General synthetic strategies toward the various types of cyclometalated iridium(III) complexes.

according to these synthetic procedures, subsequent thermal or photochemical treatments allow to convert into thermodynamically preferred *fac*-isomer (Scheme 9.1, step (c)). Meanwhile, using a different cyclometalated ligand X^Y in the second step, the reaction affords heteroleptic $(C\^{}N)_2Ir(X\^{}Y)$ complex (Scheme 9.1, step (d)). A variety of X^Y ligands, the so-called ancillary ligands, give additional possibility to define structure and tune the physical properties. Various modifications for the general procedures have been investigated in order to evade the classical, harsh reaction conditions. For example, $Ir(acac)_3$ (acac, acetylacetonate [17]), $[Ir(cod)_2(\mu\text{-}Cl)]_2$ (cod, 1,5-cyclooctadiene [11]), and $[Ir(coe)_2(\mu\text{-}Cl)]_2$ (coe, cyclooctene [61]) can be used as starting materials with advantage of shorter reaction times and milder reaction conditions. Alternatively, Konno and Sasaki [34] have reported selective and efficient synthesis of *fac*-$(C\^{}N)_3Ir$ under microwave irradiation conditions, providing a straightforward access from $IrCl_3 \cdot 3H_2O$. Grushin et al. [23] reported a new one-step synthesis of *fac*-$(C\^{}N)_3Ir$ starting from $IrCl_3$ with fluorinated 2-arylpyridines in the presence of CF_3CO_2Ag. A few synthetic attempts of complexes with three different C^N ligands [11,39,52] have been studied, and these researches emphasized the problematic isolation of target complexes (Fig. 9.10). Although most cyclometalated iridium(III) complexes consist of a combination of two or three bidentate ligands as described previously, there is a couple of "exotic" complexes bearing a tridentate ligand. For example, Obara et al. [51] have synthesized a family of iridium(III) complexes with a tridentate ligand (Fig. 9.11). Among these complexes, [Ir(Phbib)(ppy)Cl] shows a bright emission with an excellent photoluminescent quantum yield of 0.95.

Durable emitters corresponding to the primal colors are strongly desired; however, the practical blue emission still remains quite difficult to be accomplished. Several approaches have been suggested to obtain blue emission related to $Ir(ppy)_3$. A well-known approach is the introduction of fluorine atoms to the phenyl ring of the ligand; replacement from hydrogen atom(s) to fluorine atom(s) on phenyl ring increases in the energy gap between the highest occupied molecular orbital (HOMO) and lowest unoccupied molecular orbital (LUMO) levels, leading the blue emission. This is a representative "fluorine effect." Other fluorine effects also improve the various properties of the emitters. For instance, it is generally accepted that (1) replacing C—H bond with C—F bond reduces the rate of nonemissive decay, resulting in enhanced emission; (2) bulky F-based ligands and/or their large electronegativity

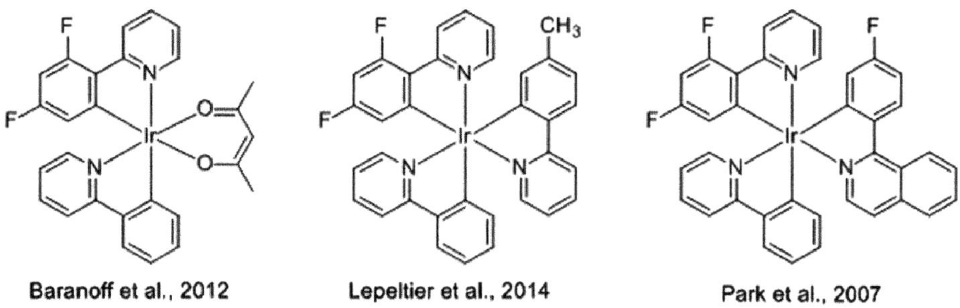

Figure 9.10 Selected example of iridium complexes with three different cyclometalated ligands.

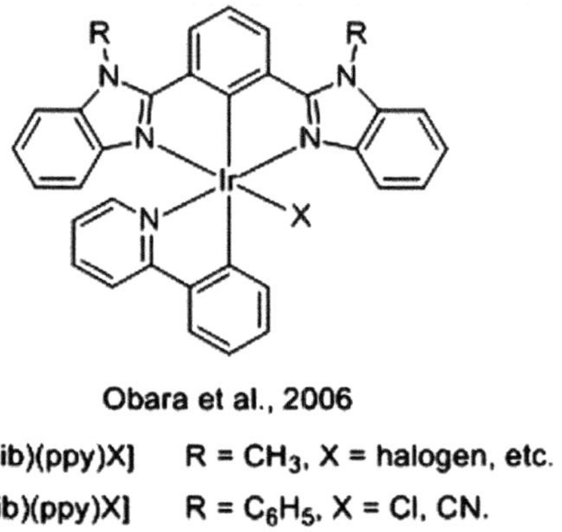

Figure 9.11 "Exotic" cyclometalated iridium complexes with a tridentate ligand.

hinder the molecular packing, minimizing the self-quenching (the so-called concentration quenching); (3) F-based substituents help their deposition ability, enabling easier fabrication of OLED devices. The readers can also find detailed information in some documents shown in the following paper lists (for selected reviews, see Refs [6,13,15,20,57,64,69]).

2.2 F-Based Emitter

After Baldo et al. [8] reported very high-efficient OLED device based on $Ir(ppy)_3$, this class of emitter has been enthusiastically investigated. Synthesis of various *homoleptic* cyclometalated ligands (the so-called main ligand or emissive ligand) eventually enables different color emission, whereas fine color controllability is relatively poor due to the less variety to change structures. On the contrary, properties of the *heteroleptic* complexes can be well-modified by designing both main ligands and an ancillary ligand. For example, Li et al. [42,43] systematically studied variety of the ancillary ligands and concluded that they had successfully elucidated fine color tuning ranging from sky blue to red.

FIrpic is known as a benchmark blue emitter possessing F-based 2-phenylpyridine ligands [1,2]. This successful precedent introduced F-based design of blue emitter, thereby pioneering the intensive investigations on its analogous complexes. Fig. 9.12 summarized selected examples of well-known blue emitters as well as FIrpic.

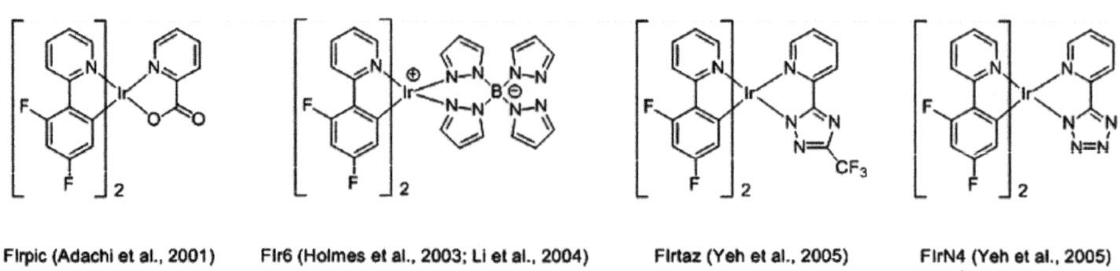

Figure 9.12 Well-known blue-emitters.

It should be mentioned that Baranoff and Curchod [10] have reviewed the synthesis, structural characterizations, and key properties of FIrpic, providing detailed information to the readers. Unfortunately, the standard blue emitters such as FIrpic still have some drawbacks compared to their red and green counterparts. For example, the Commission Internationale de L'Eclairage x,y-coordinates (CIEx,y) of FIrpic is (0.16, 0.29), which correspond to a sky blue (typically CIEy >0.2) and cannot be considered sufficient as deep blue emission (CIEy <0.2). Although tremendous numbers of F-based emitters have been investigated, no definitive material for deep blue emitter is reported so far.

Most F-based ligands are not commercially available and readily prepared via metal-catalyzed aryl—aryl cross-coupling reaction whereas a few ligands are provided by commercial suppliers. Subsequent procedures according to the general synthetic route described above afford target complexes bearing various F-based ligands (Fig. 9.13).

Intensive investigations in regard to F-based substituents (eg, number, position, group) on cyclometalated ligands have revealed that these factors strongly influence the properties of the emitters. Varying the position of F-based substituent is a simple method to color tuning of the emission (Fig. 9.14). In 2001, the synthesis and characterization of a series of CF_3-substituted emitters is reported [23]. These complexes showed good EL efficiency, and the emission wavelength can be tuned from green (c. 500 nm) to yellow-orange (c. 600 nm). The study also pointed out the CF_3-substituted complexes sublime more easily than parent $Ir(ppy)_3$ and they concluded that the F-based substituent would contribute the excellent volatility on fabricating OLED devices via thermal evaporation technique. In analogy to this

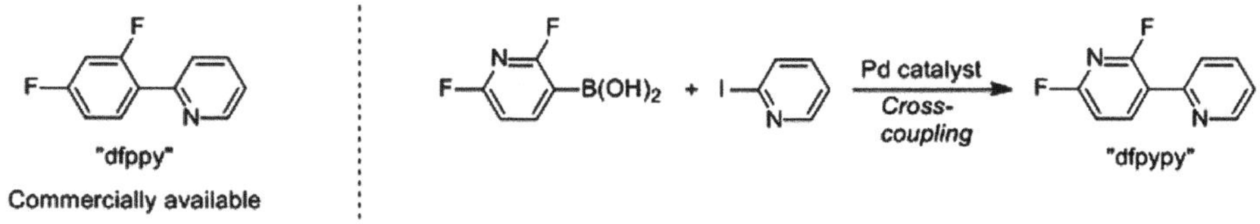

Figure 9.13 Examples of F-based cyclometalated ligands. Commercially available "dfppy" ligand (left) and the general synthetic route for "dfpypy" ligand (right).

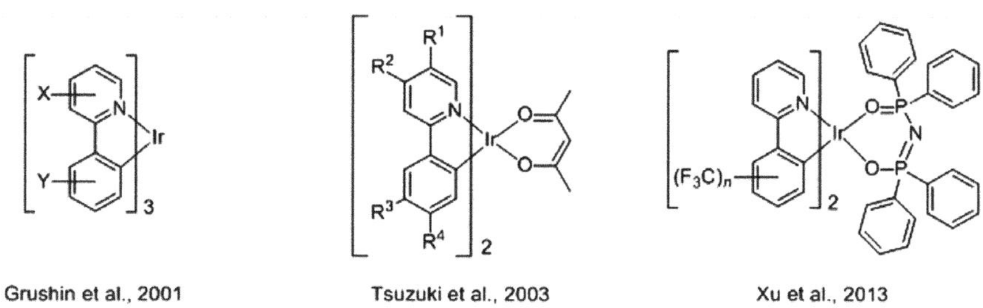

Figure 9.14 Emitters with F-based substituent at different position.

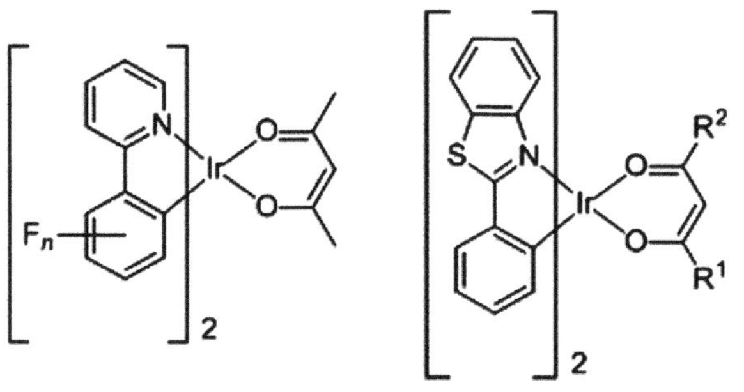

Figure 9.15 Emitters with multiple fluorinated ligands.

precedent, Tsuzuki et al. [63] have reported the color tuning of the emitting material based on C_6F_5-substituted emitter. In their study, the maximum wavelength of EL can be tuned in ranging from green to orange by only changing the position of the substitution. Most recently, Xu et al. [66] prepared five emitters with CF_3 substitution(s) at different positions of phenyl ring. OLEDs based on these emitters were successively fabricated and exhibited various color emission ranging from 480 to 543 nm. A study using relativistic time-dependent density functional theory calculations indicated that the magnitude of radiative rates of the emitters is influenced by the fluorine position in the phenyltriazole skeleton [55].

The number of F-based substituents is also one of the key factors for tuning the emission color of the complexes (Fig. 9.15). Ragni et al. [73] have synthesized a family of [(tri- and tetra-fluoro-2-phenyl)pyridine]-based emitters and evaluated the luminescent properties of these complexes. It should be noted that the introduction of the fourth fluorine atom on the phenyl ring surprisingly gave c. 10-nm bathochromic shift of emission relative to the corresponding trifluorinated counterparts. Moreover, rapid decay within 15 min in the electroluminescence spectrum of *mer*-Ir(F_4ppy)$_3$-based device was observed and these results indicated that large electronegativity of the multiple fluorine substitution may destabilize the complex, leading to potentially short device operational lifetimes. Increasing the number of fluorine atoms on acetylacetonate ancillary ligands resulted in only small changes of emission wavelength probably due to limited contribution of this ligand to emission mechanism [7]. These two studies indicate that it is difficult to obtain deeper blue color just by increasing fluorine atoms on the ligands.

Successful deep blue emitters based on an additional incorporation of other highly fluorinated substituents into the parent sky blue emitter (eg, FIrpic) are one of the notable topics in this research field. As shown in Fig. 9.16, the emission color of the several complexes bearing perfluoroalkyl or pefluoroalkylcarbonyl substituents sufficiently reached deep blue region [32,37,59].

Meanwhile, a recent study has indicated that SF_5 group might be a new candidate for F-based tuning of emission color [74]. The SF_5 group promotes a large blueshift in

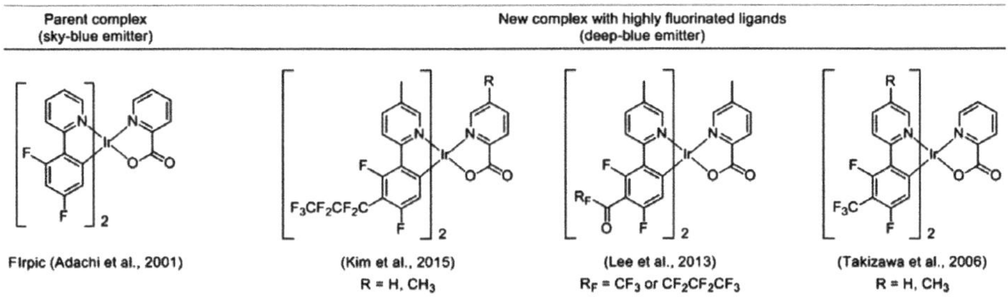

Figure 9.16 Selected examples of deep blue emitter involving highly fluorinated ligands.

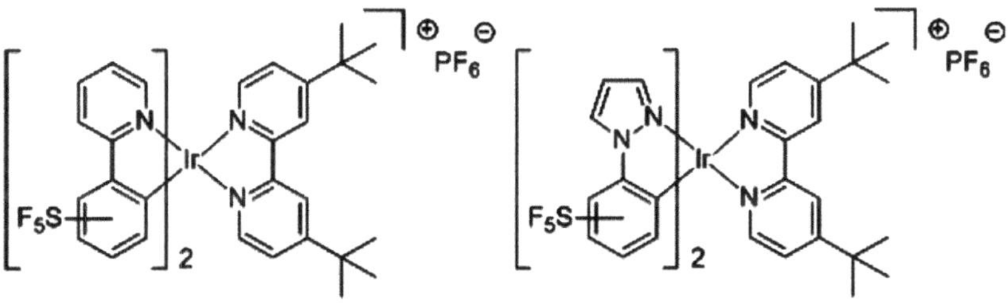

Figure 9.17 SF_5-substituted emitters.

the emission spectrum compared to the nonsubstituted and F- or CF_3-substituted analogues, with the family of complexes emitting brightly in the range of 493–519 nm (Fig. 9.17).

2.3 Non-F-Based Emitter

In the previous section, we mainly discussed the "positive" fluorine effect in the iridium(III) complexes for tuning the emission color, providing straightforward access for achieving blue emission. However, some drawbacks on their instability of F-based emitter have also been underscored for practical application in OLEDs. In this section, we will focus on a couple of "negative" fluorine effects with some recent literatures.

Sivasubramaniam et al. [54] have revealed the unexpected carbon–fluorine bond cleavage of FIrpic molecule analyzed by liquid chromatography coupled with electron spray ionization mass spectrometry. The analysis indicated the loss of one fluorine atom (defluorination reaction) from FIrpic molecule both in device fabrication and operation stages, resulting in insufficient device performance such as short lifetime and emission color change. For instance, the two groups independently reported the similar defluorination reaction during the synthesis of the iridium(III) complex induced by base [42,43] or harsh heating [40] (Scheme 9.2).

Non-F-based (the so-called F-free) emitters that bear alkoxy group(s) instead of fluorine atom(s) are successful examples for evading this defluorination problem (Fig. 9.18). A recent study presented that an alkoxy functionalized cyclometalated iridium(III) complex did not significantly affect emission changes compared to their

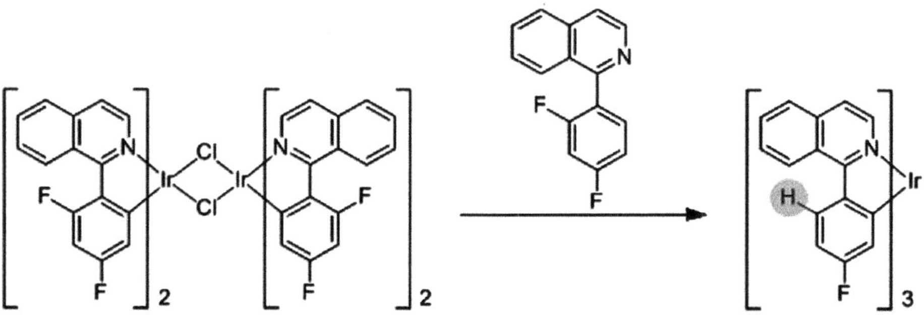

Scheme 9.2 An example of defluorination reaction. The hydrogen atom marked with *gray circle* indicates the event position.

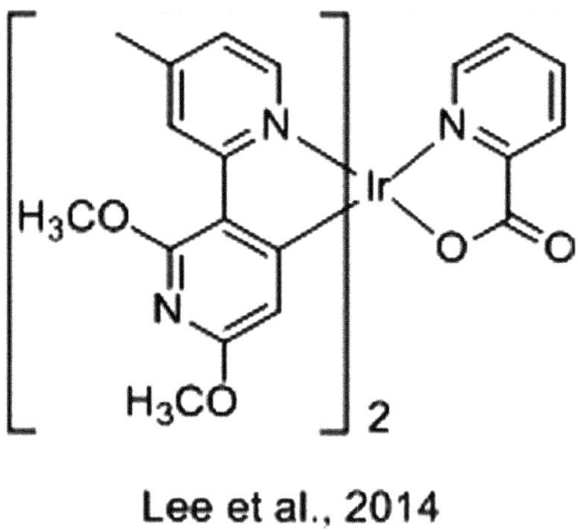

Figure 9.18 An example of non-F-based blue emitter.

F-based congeners [75]. Lee et al. [36] suggested that a strong metal—carbon covalent interaction on the alkoxy-substituted emitter contributes c. 10 nm blueshift of emission wavelength relative to F-based counterparts. The alkoxy substituents would also improve the solubility of the emitter in common organic solvents, enabling the potential application in a solution processable emitter.

Carbene-based complex is another family of non-F-based emitter; tris(phenylmethylimidazolyl)iridium $Ir(pmi)_3$ and tris(phenylmethylbenzoimidazolyl)iridium $Ir(pmb)_3$ are known as pioneering emitters in this field (Fig. 9.19). The OLEDs based on these carbene complexes emitted almost saturated blue (near UV) for both isomers of $Ir(pmb)_3$ [26]. This approach involved the control of the relative energies of emissive and nonemissive exited states; by increasing the energy separation between emissive and nonemissive states, the latter decay is strongly suppressed, leading to their improved emission. While these complexes have no fluorine atoms, note that the synthesis of the emitters is problematic compared to the 2-phenylpyridine-based emitters.

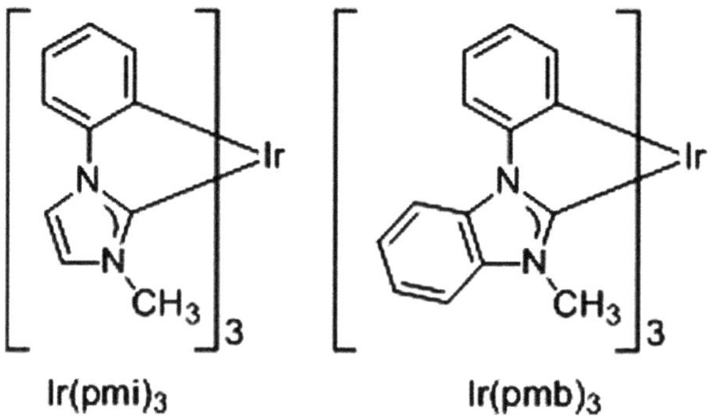

Figure 9.19 Non-F-based deep blue emitters: $Ir(pmi)_3$ and $Ir(pmb)_3$.

The following factors would hinder the efficient and selective synthesis of target emitter: (1) chemical and thermal instability of free carbene ligand and (2) similar physical properties between *fac*- and *mer*-stereoisomers [53].

2.4 Minimal-F-Based Emitter

Previous research studies related to this field have revealed that fluorine atoms not only positively but also negatively influenced the properties of the blue emitters. From the standpoint of a synthetic chemist, a synergetic hybrid design of F- and non-F-based emitter which possesses limited number of fluorine atoms is a reasonable solution to achieve the definitive blue emitter. A persuasive scientific evidence came from a precedent regarding a deep blue emitter $Ir(dfpypy)_3$ developed by Lee et al. [38]. This emitter exhibited remarkable deep blue emission with c. 30 nm hypsochromic shift with respect to its parent sky blue emitter $Ir(dfppy)_3$ [60]. Although these two complexes possess same number of fluorine atoms, drastic blueshift was obtained by just replacing the phenyl ring to pyridyl ring (Fig. 9.20).

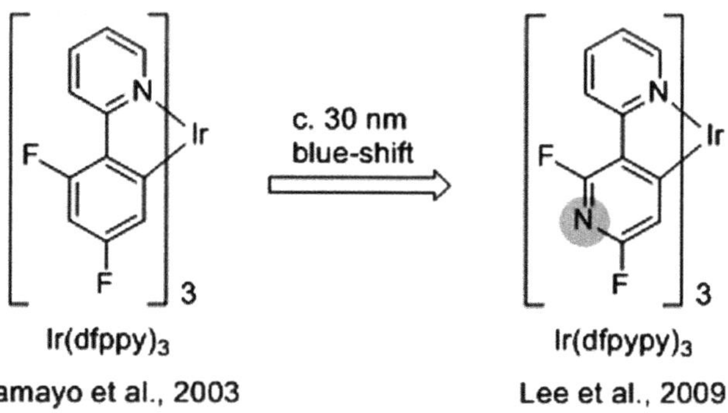

Figure 9.20 $Ir(dfppy)_3$ and $Ir(dfpypy)_3$. The nitrogen atoms marked with *gray circle* play a key role for obtaining blue emission.

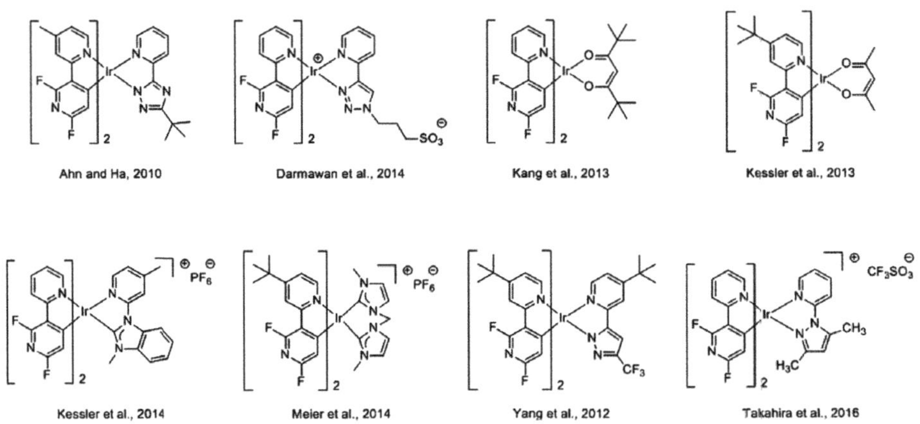

Figure 9.21 Selected examples of $(dfpypy)_2Ir(X$^$Y)$-based emitter.

As shown in Fig. 9.21, $(dfpypy)_2Ir(X$^$Y)$-based emitters were intensively investigated by several groups [3,16,29–31,46,67]. Besides these studies, Takahira's group has reported the synthesis and properties of a new deep blue emitter on the basis of this conceptual blueprint [58].

3. Conclusion

Fluorine plays an extremely important role for adjustment of physical properties of blue light-emitting cyclometalated iridium(III) complexes. Frankly, few decisive materials have been accomplished to date, whereas many studies have been done to reveal a clue of design to improve blue-emitting performances. In this chapter, we have briefly reviewed the current topics in this field and introduced that a plausible solution is a synergetic hybrid design of F- and non-F-based emitter which possesses limited number of fluorine atoms. Further investigations related to this field are expected to be conducted in the future. Needless to say, durable blue emitter is one of the key materials for successive application in future lighting and display technologies. We strongly hope this short contribution will facilitate the material developments toward "brilliant" blue emitter.

References

[1] C. Adachi, M.A. Baldo, M.E. Thompson, S.R. Forrest, Nearly 100% internal phosphorescence efficiency in an organic light emitting device, J. Appl. Phys. 90 (10) (2001) 5048–5051.

[2] C. Adachi, R.C. Kwong, P. Djurovich, V. Adamovich, M.A. Baldo, M.E. Thompson, S.R. Forrest, Endothermic energy transfer: a mechanism for generating very efficient high-energy phosphorescent emission in organic materials, Appl. Phys. Lett. 79 (2001) 2082–2084.

[3] S.Y. Ahn, Y. Ha, Color tuning of blue phosphorescence: new iridium complexes containing a bipyridine derivative and various ancillary ligands, Mol. Cryst. Liq. Cryst. 530 (2010) 116/[272]−122/[278].
[4] M.S. AlSalhi, J. Alam, L.A. Dass, M. Raja, Recent advances in conjugated polymers for light emitting devices, Int. J. Mol. Sci. 12 (2011) 2036−2054.
[5] I. Avilov, P. Minoofar, J. ornil, L.D. Cola, Influence of substituents on the energy and nature of the lowest excited states of heleroleptic phosphorescent Ir(III) complexes: a joint theoretical and experimental study, J. Am. Chem. Soc. 129 (2007) 8247−8258.
[6] F. Babudri, G.M. Farinola, F. Naso, R. Ragni, Fluorinated organic materials for electronic and optoelectronic applications: the role of the fluorine atom, Chem. Commun. (2007) 1003−1022.
[7] W. Bai, On a series of Ir(III) complexes with various numbers of fluorine atoms, J. Lumin. 132 (2012) 2847−2854.
[8] M.A. Baldo, S. Lamansky, P.E. Burrows, M.E. Thompson, S.R. Forrest, Very high-efficiency green organic light-emitting devices based on electrophosphorescence, Appl. Phys. Lett. 75 (1999) 4−6.
[9] M.A. Baldo, D.F. O'Brien, Y.E. You, et al., Highly efficient phosphorescent emission from organic electroluminescent devices, Nature 395 (1998) 151−154.
[10] E. Baranoff, B.F.E. Curchod, FIrpic: archetypal blue phosphorescent emitter for electroluminescence, Dalt. Trans. (2015) 8318−8329.
[11] E. Baranoff, B.F.E. Curchod, J. Frey, R. Scopelliti, F. Kessler, I. Tavernelli, U. Rothlisberger, M. Grätzel, M.K. Nazeeruddin, Acid-induced degradation of phosphorescent dopants for OLEDs and its application to the synthesis of tris-heteroleptic iridium(III) bis-cyclometalated complexes, Inorg. Chem. 51 (2012) 215−224.
[12] H.A. Bronstein, C.E. Finlayson, K.R. Kirov, R.H. Friend, C.K. Williams, Investigation into the phosphorescence of a series of regioisomeric iridium complexes, Organometallics 27 (2008) 2980−2989.
[13] Y. Chi, P.-T. Chou, Transition-metal phosphors with cyclometalating ligands: fundamentals and applications, Chem. Soc. Rev. 39 (2010) 638−655.
[14] Y.C. Chiu, Y. Chi, J.Y. Hung, Y.M. Cheng, Y.C. Yu, M.W. Chung, G.H. Lee, P.T. Chou, C.C. Chen, C.C. Wu, H.Y. Hsieh, Blue to true-blue phosphorescent Ir(III) complexes bearing a nonconjugated ancillary phosphine chelate: strategic synthesis, photophysics, and device integration, ACS Appl. Mater. Interfaces 1 (2) (2009) 433−442.
[15] W.C.H. Choy, W.K. Chan, Y. Yuan, Recent advances in transition metal complexes and light-management engineering in organic optoelectronic devices, Adv. Mater. (2014) 5368−5399.
[16] N. Darmawan, C.-H. Yang, M. Mauro, R. Fröhlich, L. De Cola, C.-H. Chang, Z.-J. Wu, C.-W. Tai, Highly efficient blue and deep-blue emitting zwitterionic iridium(III) complexes: synthesis, photophysics and electroluminescence, J. Mater. Chem. C 2 (2014) 2569−2582.
[17] K. Dedeian, P.I. Djurovich, F.O. Garces, G. Carlson, R.J. Watts, A new synthetic route to the preparation of a series of strong photoreducing agents: fac-tris-ortho-metalated complexes of iridium(III) with substituted 2-phenylpyridines, Inorg. Chem. 30 (1991) 1685−1687.
[18] B. Diouf, W.S. Jeon, R. Pode, J.H. Kwon, Efficiency control in iridium complex-based phosphorescent light-emitting diodes, Adv. Mater. Sci. Eng 2012 (2012) 1−14.
[19] A. Endo, K. Sato, K. Yoshimura, T. Kai, A. Kawada, H. Miyazaki, C. Adachi, Efficient up-conversion of triplet excitons into a singlet state and its application for organic light emitting diodes, Appl. Phys. Lett. 98 (2011) 083302.

[20] G.M. Farinola, F. Babudri, A. Cardone, O. Hassan Omar, C. Martinelli, F. Naso, V. Pinto, R. Ragni, Synthesis of fluorinated organic and organometallic electroluminescent materials: tuning emission in the blue, Adv. Sci. Technol. 75 (2010) 108–117.
[21] J.M. Fernández-Hernández, J.I. Beltrán, V. Lemaur, M.-D. Gálvez-López, C.-H. Chien, F. Polo, E. Orselli, R. Fröhlich, J. Cornil, L. De Cola, Iridium(III) emitters based on 1,4-disubstituted-1H-1,2,3-triazoles as cyclometalating ligand: synthesis, characterization, and electroluminescent devices, Inorg. Chem. 52 (2013) 1812–1824.
[22] K. Goushi, K. Yoshida, K. Sato, C. Adachi, Organic light-emitting diodes employing efficient reverse intersystem crossing for triplet-to-singlet state conversion, Nat. Photonics 6 (2012) 253–258.
[23] V.V. Grushin, N. Herron, D.D. LeCloux, W.J. Marshall, V.A. Petrov, Y. Wang, New, efficient electroluminescent materials based on organometallic Ir complexes, Chem. Commun. (2001) 1494–1495.
[24] Y. Hasegawa, T. Ohkubo, K. Sogabe, Y. Kawamura, Y. Wada, N. Nakashima, S. Yanagida, Luminescence of novel neodymium sulfonylaminate complexes in organic media, Angew. Chem. Int. Ed. 39 (2) (2000) 357–360.
[25] R.J. Holmes, B.W. D'Andrade, S.R. Forrest, X. Ren, J. Li, M.E. Thompson, Efficient, deep-blue organic electrophosphorescence by guest charge trapping, Appl. Phys. Lett. 83 (2003) 3818–3820.
[26] R.J. Holmes, S.R. Forrest, T. Sajoto, A. Tamayo, P.I. Djurovich, M.E. Thompson, J. Brooks, Y.J. Tung, B.W. D'Andrade, M.S. Weaver, R.C. Kwong, J.J. Brown, Saturated deep blue organic electrophosphorescence using a fluorine-free emitter, Appl. Phys. Lett. 87 (2005) 1–3.
[27] T. Hu, L. He, L. Duan, Y. Qui, Solid-state light-emitting electrochemical cells based on ionic iridium(III) complexes, J. Mater. Chem. 22 (2012) 4206–4215.
[28] N. Jung, E. Lee, J. Kim, H.H.-K. Park, K. Park, Y. Kang, Synthesis and crystal structure of blue phosphorescent mer-tris(2′,6′-difluoro-2,3′-bipyridinato-N,$C^{4'}$)iridium(III), Bull. Korean Chem. Soc. 33 (2012) 183–188.
[29] Y. Kang, Y.-L. Chang, J.-S. Lu, S.-B. Ko, Y. Rao, M. Varlan, Z.-H. Lu, S. Wang, Highly efficient blue phosphorescent and electroluminescent Ir(III) compounds, J. Mater. Chem. C 1 (2013) 441–450.
[30] F. Kessler, R.D. Costa, D. Di Censo, R. Scopelliti, E. Ortí, H.J. Bolink, S. Meier, W. Sarfert, M. Grätzel, M.K. Nazeeruddin, E. Baranoff, Near-UV to red-emitting charged bis-cyclometalated iridium(III) complexes for light-emitting electrochemical cells, Dalt. Trans. 41 (2012) 180–191.
[31] F. Kessler, Y. Watanabe, H. Sasabe, H. Katagiri, M.K. Nazeeruddin, M. Grätzel, J. Kido, High-performance pure blue phosphorescent OLED using a novel bis-heteroleptic iridium(III) complex with fluorinated bipyridyl ligands, J. Mater. Chem. C 1 (2013) 1070–1075.
[32] J.-B. Kim, S.-H. Han, K. Yang, S.-K. Kwon, J.-J. Kim, Y.-H. Kim, Highly efficient deep-blue phosphorescence from heptafluoropropyl-substituted iridium complexes, Chem. Commun. 51 (2015) 58–61.
[33] K.A. King, P.J. Spellane, R.J. Watts, Excited-state properties of a triply ortho-metalated iridium(III) complex, J. Am. Chem. Soc. 107 (1985) 1431–1432.
[34] H. Konno, Y. Sasaki, Selective one-pot synthesis of facial tris-ortho-metalated iridium(III) complexes using microwave irradiation, Chem. Lett. 32 (2003) 252–253.
[35] S. Lamansky, P. Djurovich, D. Murphy, F. Abdel-Razzaq, H.-E. Lee, C. Adachi, P.E. Burrows, S.R. Forrest, M.E. Thompson, Highly phosphorescent bis-cyclometalated iridium complexes: synthesis, photophysical characterization, and use in organic light emitting diodes, J. Am. Chem. Soc. 123 (2001) 4304–4312.

[36] J.J.Y. Lee, H. Oh, J. Kim, K.-M. Park, K.S. Yook, Y. Kang, Fluorine-free blue phosphorescent emitters for efficient phosphorescent organic light emitting diodes, J. Mater. Chem. C (2014) 6040–6047.
[37] S. Lee, S. Kim, H. Shin, H. Yun, K. Yang, S. Kwon, J.-J. Kim, Y.-H. Kim, Deep-blue phosphorescence from perfluoro carbonyl-substituted iridium complexes, J. Am. Chem. Soc. 135 (2013) 14321–14328.
[38] S.J. Lee, K.-M. Park, K. Yang, Y. Kang, Blue phosphorescent Ir(III) complex with high color purity: fac-tris(2′,6′-difluoro-2,3′-bipyridinato-N,C$^{4\prime}$)iridium(III), Inorg. Chem. 48 (2009) 1030–1037.
[39] M. Lepeltier, F. Dumur, B. Graff, P. Xiao, D. Gigmes, J. Lalevée, C.R. Mayer, Tris-cyclometalated iridium(III) complexes with three different ligands: a new example with 2-(2,4-difluorophenyl)pyridine-based complex, Helv. Chim. Acta 97 (2014) 939–956.
[40] M. Lepeltier, F. Dumur, J. Marrot, E. Contal, D. Bertin, D. Gigmes, C.R. Mayer, Unprecedented combination of regioselective hydrodefluorination and ligand exchange reaction during the syntheses of tris-cyclometalated iridium(III) complexes, Dalt. Trans. 42 (2013) 4479–4486.
[41] J. Li, P.I. Djurovich, B.D. Alleyne, I. Tsyba, N.N. Ho, R. Bau, M.E. Thompson, Synthesis and characterization of cyclometalated Ir(III) complexes with pyrazolyl ancillary ligands, Polyhedron 23 (2004) 419–428.
[42] J. Li, T. Nakagawa, J. MacDonald, Q. Zhang, H. Nomura, H. Miyazaki, C. Adachi, Highly efficient organic light-emitting diode based on a hidden thermally activated delayed fluorescence channel in a heptazine derivatives, Adv. Mater. 25 (2013) 3319–3323.
[43] L. Li, F. Wu, S. Zhang, D. Wang, Y. Ding, Z. Zhu, A heteroleptic cyclometalated iridium(III) fluorophenylpyridine complex from partial defluorohydrogenation reaction: synthesis, photophysical properties and mechanistic insights, Dalt. Trans. 42 (2013) 4539–4543.
[44] S.C. Lo, R.E. Harding, C.P. Shipley, S.G. Stevenson, P.L. Burn, I.D.W. Samuel, High-triplet-energy dendrons: enhancing the luminescence of deep blue phosphorescent iridium(III) complexes, J. Am. Chem. Soc. 131 (2009) 16681–16688.
[45] M. Mauro, K.C. Schuermann, R. Prétôt, A. Hafner, P. Mercandelli, A. Sironi, L. De Cola, Complex iridium(III) salts: luminescent porous crystalline materials, Angew. Chem. Int. Ed. 49 (2010) 1222–1226.
[46] S.B. Meier, W. Sarfert, J.M. Junquera-Hernández, M. Delgado, D. Tordera, E. Ortí, H.J. Bolink, F. Kessler, R. Scopelliti, M. Grätzel, M.K. Nazeeruddin, E. Baranoff, A deep-blue emitting charged bis-cyclometalated iridium(III) complex for light-emitting electrochemical cells, J. Mater. Chem. C 1 (2013) 58–68.
[47] S.B. Meier, D. Tordera, A. Petegas, C. Roldan-Carmona, E. Orti, H. Bolink, Light-emitting electrochemical cells: recent progress and future prospects, Mater. Today 17 (5) (2014) 217–223.
[48] C. Murawski, K. Leo, M. Gather, Efficiency roll-off in organic light-emitting diodes, Adv. Mater. 25 (24) (2013) 6801–6827.
[49] M. Nonoyama, Benzoquinolin-10-yl-N iridium (III) complexes, Bull. Chem. Soc. Jpn. 47 (1974) 767–768.
[50] M. Nonoyama, Chelating C-metallation of N-phenylpyrazole with rhodium(III) and iridium(III), J. Organomet. Chem. 86 (1975) 263–267.
[51] S. Obara, M. Itabashi, F. Okuda, S. Tamaki, Y. Tanabe, Y. Ishii, K. Nozaki, M.A. Haga, Highly phosphorescent iridium complexes containing both tridentate bis(benzimidazolyl)-benzene or -pyridine and bidentate phenylpyridine: synthesis, photophysical properties, and theoretical study of Ir-bis(benzimidazolyl)benzene complex, Inorg. Chem. 45 (2006) 8907–8921.

[52] G.Y. Park, Y. Kim, Y. Ha, Iridium complexes containing three different ligands as white OLED dopants, Mol. Cryst. Liq. Cryst. 462 (2007) 179–188.
[53] T. Sajoto, P.I. Djurovich, A. Tamayo, M. Yousufuddin, R. Bau, M.E. Thompson, R.J. Holmes, S.R. Forrest, Blue and near-UV phosphorescence from iridium complexes with cyclometalated pyrazolyl or N-heterocyclic carbene ligands, Inorg. Chem. 44 (2005) 7992–8003.
[54] V. Sivasubramaniam, F. Brodkorb, S. Hanning, H.P. Loebl, V. van Elsbergen, H. Boerner, U. Scherf, M. Kreyenschmidt, Fluorine cleavage of the light blue heteroleptic triplet emitter FIrpic, J. Fluorine Chem. 130 (2009) 640–649.
[55] A.R.G. Smith, M.J. Riley, P.L. Burn, I.R. Gentle, S.C. Lo, B.J. Powell, Effects of fluorination on iridium(III) complex phosphorescence: magnetic circular dichroism and relativistic time-dependent density functional theory, Inorg. Chem. 51 (2012) 2821–2831.
[56] S. Sprouse, K.A. King, P.J. Spellane, R.J. Watts, Photophysical effects of metal-carbon bonds in ortho-metalated complexes of Ir(III) and Rh(III), J. Am. Chem. Soc. 106 (1984) 6647–6653.
[57] Y. Suzuri, T. Oshiyama, H. Ito, K. Hiyama, H. Kita, Phosphorescent cyclometalated complexes for efficient blue organic light-emitting diodes, Sci. Technol. Adv. Mater. 15 (2014) 054202.
[58] Y. Takahira, E. Murotani, K. Fukuda, V. Vohra, H. Murata, Design, synthesis, and properties of a series of charged iridium(III) complexes with a neutral bidentate ligand for deep-blue phosphorescent emitter. J. Fluorine Chem. 181 (2016) 56–60.
[59] S. Takizawa, H. Echizen, J. Nishida, T. Tsuzuki, S. Tokito, Y. Yamashita, Finely-tuned blue-phosphorescent iridium complexes based on 2-phenylpyridine derivatives and application to polymer organic light-emitting device, Chem. Lett. 35 (2006) 748–749.
[60] A.B. Tamayo, B.D. Alleyne, P.I. Djurovich, S. Lamansky, I. Tsyba, N.N. Ho, R. Bau, M.E. Thompson, Synthesis and characterization of facial and meridional tris-cyclometalated iridium(III) complexes, J. Am. Chem. Soc. 125 (2003) 7377–7387.
[61] M.W. Thesen, H. Krueger, S. Janietz, A. Wedel, M. Graf, Investigation of spacer influences in phosphorescent-emitting nonconjugated PLED systems, J. Polym. Sci. Part A Polym. Chem. 48 (2010) 389–402.
[62] A. Tsuboyama, H. Iwawaki, M. Furugori, T. Mukaide, J. Kamatani, S. Igawa, T. Moriyama, S. Miura, T. Takiguchi, S. Okada, M. Hoshino, K. Ueno, Homoleptic cyclometalated iridium complexes with highly efficient red phosphorescence and application to organic light-emitting diode, J. Am. Chem. Soc. 125 (42) (2003) 12971–12979.
[63] T. Tsuzuki, N. Shirasawa, T. Suzuki, S. Tokito, Color tunable organic light-emitting diodes using pentafluorophenyl-substituted iridium complexes, Adv. Mater. 15 (2003) 1455–1458.
[64] C. Ulbricht, B. Beyer, C. Friebe, A. Winter, U.S. Schubert, Recent developments in the application of phosphorescent iridium(III) complex systems, Adv. Mater. 21 (2009) 4418–4441.
[65] H. Uoyama, K. Goushi, K. Shizu, H. Nomura, C. Adachi, Highly efficient organic light-emitting diodes from delayed fluorescence, Nature 492 (2012) 234–238.
[66] Q.L. Xu, C.C. Wang, T.Y. Li, M.Y. Teng, S. Zhang, Y.M. Jing, X. Yang, W.N. Li, C. Lin, Y.X. Zheng, J.L. Zuo, X.Z. You, Syntheses, photoluminescence, and electroluminescence of a series of iridium complexes with trifluoromethyl-substituted 2-phenylpyridine as the main ligands and tetraphenylimidodiphosphinate as the ancillary ligand, Inorg. Chem. 52 (2013) 4916–4925.

[67] C.-H. Yang, M. Mauro, F. Polo, S. Watanabe, I. Muenster, R. Fröhlich, L. De Cola, Deep-blue-emitting heteroleptic iridium(III) complexes suited for highly efficient phosphorescent OLEDs, Chem. Mater. 24 (2012) 3684–3695.

[68] S.J. Yeh, M.F. Wu, C.T. Chen, Y.H. Song, Y. Chi, M.H. Ho, S.F. Hsu, C.H. Chen, New dopant and host materials for blue-light-emitting phosphorescent organic electroluminescent devices, Adv. Mater. 17 (2005) 285–289.

[69] K.S. Yook, J.Y. Lee, Organic materials for deep blue phosphorescent organic light-emitting diodes, Adv. Mater. 24 (2012) 3169–3190.

[70] K.S. Yook, J.Y. Lee, Small molecule host materials for solution processed phosphorescent organic light-emitting diodes, Adv. Mater. 26 (2014) 4218–4233.

[71] Q. Zhang, B. Li, S. Huang, H. Nomura, H. Tanaka, C. Adachi, Efficient blue organic light-emitting diodes employing thermally activated delayed fluorescence, Nat. Photonics 8 (2014) 1–7.

[72] Y. Zhang, C. Li, H. Shi, B. Du, W. Yang, Y. Cao, Bright red light-emitting devices based on a novel europium complex doped into polyvinyl carbazole, New J. Chem. 31 (2007) 569–574.

[73] R. Ragni, et al., Blue emitting iridium complexes: synthesis, photophysics and phosphorescent devices. J. Mat. Chem. 16 (12) (2006) 1161.

[74] N.M. Shavaleev, et al., Green phosphorescence and electroluminescence of sulfur pentafluoride-functionalized cationic Iridium(III) Complexes. Inorganic Chemistry, 54 (12) (2015) 5907–5914.

[75] H. Oh, et al., Effective alkoxylation of phosphorescent heteroleptic Iridium(III) compounds bearing fluorinated bipyridine ligands. Organometallics, 32 (21) (2013) 6427–6436.

Zinc Hydroxyfluoride (ZnOHF) Nanostructure as Photoelectrode of Quantum Dot-Sensitized Solar Cells

10

H. Chen
Beihang University, Beijing, China; The Hong Kong University of Science and Technology, Kowloon, Hong Kong

L. Zhu, H. Liu, W. Li
Beihang University, Beijing, P. R. China

Chapter Outline

1. Introduction to Zinc Hydroxyfluoride

Zinc hydroxyfluoride (ZnOHF) is an important zinc-containing material, which has been widely used in the development of new functional nanomaterials. For example, ZnOHF nanostructures have been used as the precursors for the synthesis of ZnO nanostructure materials [10,14,20,29,34,35,39]. Thermal decomposition strategy was commonly used to facilitate the conversion, and ZnO nanostructure would maintain the primary morphology of ZnOHF nanostructure. Therefore the morphology of ZnO nanostructure could be simply controlled by adjusting the morphology of ZnOHF nanostructures. Besides, as the precursor, the semiconducting properties were then observed. And some scientists have employed its semiconducting properties in combination with its nanostructures to act as the photocatalysts for the formation of pyridine from tetrahydrofurfuryl alcohol and ammonia [14], and for the degradation of

Photonic and Electronic Properties of Fluoride Materials. http://dx.doi.org/10.1016/B978-0-12-801639-8.00010-6

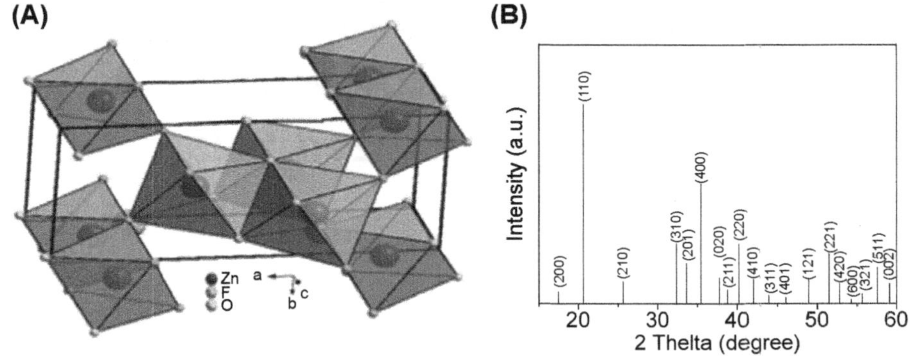

Figure 10.1 Crystal structure of zinc hydroxyfluoride (ZnOHF). (A) ZnOHF unit cell [1] and (B) standard X-ray diffraction pattern.

organic compounds and dyes [1,25,26,31]. Very recently, Chen et al. have applied ZnOHF nanostructures in the photoelectrode of quantum dot-sensitized solar cells (QDSCs) as a wide-band gap semiconductor, which exhibited a promising performance [6].

As reported in the literatures, ZnOHF crystallizes with an orthorhombic symmetry (SG: $Pna2_1$) (Fig. 10.1A) derived from the diaspore, a-AlOOH, (SG: *Pnma*) [28,33]. The standard X-ray diffraction (XRD) pattern of ZnOHF (PDF 32−1469) was drawn and presented in Fig. 10.1B. The characteristic diffraction peaks at 17.5, 20.6, 25.7, 32.4, 33.7, 35.5, 37.8, 38.8, 40.2, 42.0, 43.9, 46.1, 49.0, 51.5, 52.8, 54.3, 55.7, 57.6, and 59.1 degrees correspond to (200), (110), (210), (310), (201), (400), (020), (211), (410), (220), (311), (401), (121), (221), (420), (600), (321), (511), and (002) planes of ZnOHF, respectively. And it was demonstrated by Serier et al. that ZnOHF may exhibit a variable stoichiometry $Zn(OH)_{2-x}F_x$ by controlling the synthesis conditions [28].

2. Introduction to QDSCs

Recently, QDSCs, derived from dye-sensitized solar cells (DSSCs) [12,24], have been widely thought to be one of the most promising third-generation solar cells [13,16,27]. The motivation to develop QDSCs is the obvious advantages of quantum dots (QDs) over dye molecules, such as adjustable band gap, high extinction coefficient, large intrinsic dipole and impact ionization effect, generating multiple electron−hole pairs per photon [2,13,16,23,27]. And as such, the theoretical power conversion efficiency (PCE) of QDSCs is as high as 44% [13,16,19,23,27].

The cell structure and working principle of QDSCs is similar to those of DSSCs, which are illustrated in Fig. 10.2. QDSCs mainly have three parts: photoelectrode, electrolyte, and counter electrode [16,19,27]. And the photoelectrode was usually constructed with the nanostructuring wide-band gap semiconductors (such as TiO_2, ZnO,

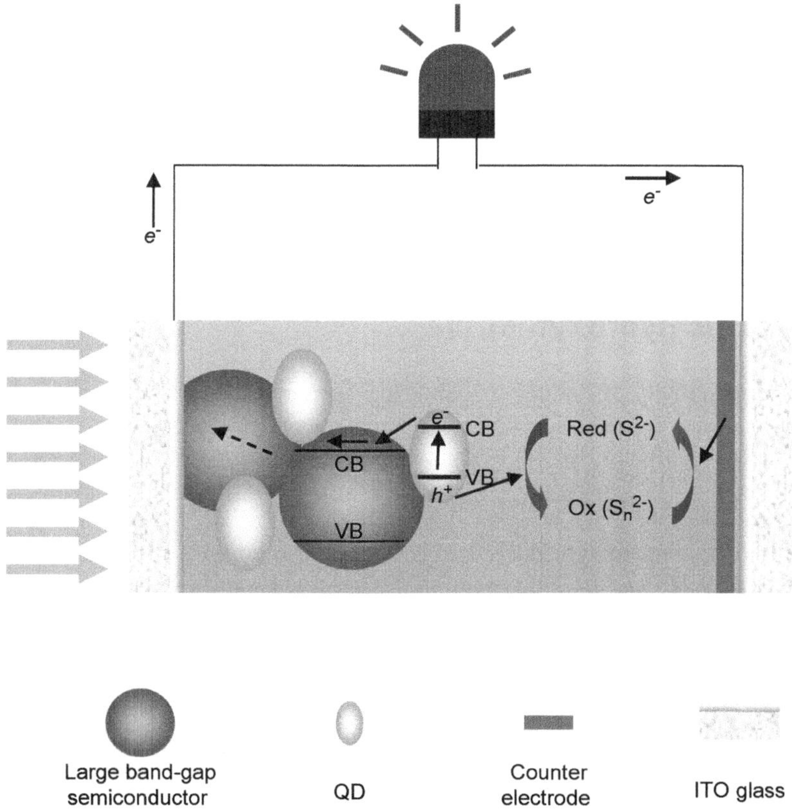

Figure 10.2 Structure and working principle of quantum dot-sensitized solar cells.

SnO_2, WO_3, and Zn_2SnO_4) coated with QDs (such as CdS, CdSe, CdTe, PbS, CuS, and $CuInS_2$) [3,7,16,27]. Incident light will be mainly absorbed by QDs to produce electrons and holes at the conduction band (CB) and valence band (VB) of QDs, respectively. Then, photoelectrons are injected into the CB of the wide-band gap semiconductor, followed by transporting to collection electrode and then to counter electrode via external circuit. The holes in the VB of QDs will transfer into the electrolyte to oxidize the reduced form of the redox couples $S^{2-}/S_n{}^{2-}$. Then the oxidized form of the redox couples will diffuse toward counter electrode to be reduced by the electrons. Since the large-band gap semiconductor serves to accept the photoelectrons from QDs, its CB edge needs to be lower than that of QDs. And in order to enhance the light absorption of the photoelectrode, a nanostructure of large-band gap semiconductor is usually required to increase the deposition amount of QDs.

So far, only about 8% PCE was achieved by the QDSCs [38], which was still considerably lower than those of Si-based solar cells and even DSSCs (13%) [21,36]. Therefore, low PCE is the predominant problem faced by QDSCs, which needed to be tackled in future work. To increase the PCE of QDSCs, improving QDs coverage, exploiting suitable electrolyte and counter electrode are all necessary [16,19,27]. Besides, exploiting more suitable and efficient wide-band gap semiconductor was also very important. However, the researches on exploiting new wide-band gap semiconductors are scarce and hard. And a new semiconductor surpassing TiO_2 or even ZnO has not been reported yet. Our group has been always working on developing new wide-band gap semiconductor for QDSCs [6,8,9]. And ZnOHF was proved to be one of the promising one, which is to be introduced in detail below [6].

3. Physical and Chemical Properties of Zinc Hydroxyfluoride

The thermal stability of ZnOHF materials has been investigated by Wang et al. using Thermogravimetric Analysis (TG)/Differential Scanning Calorimeter (DSC) technique [31]. The DSC curve in Fig. 10.3 exhibited an endothermic peak at about 420°C, corresponding to a 17.2% weight loss at TG curves. This weight loss percentage was very close to the theoretical percentage value of HF in ZnOHF (17%), which well demonstrated the decomposition of ZnOHF to HF and ZnO according to equation Eq. [i] and the removal of HF molecules resulted in the observed weight loss.

$$ZnOHF \rightarrow ZnO + HF\uparrow \quad [i]$$

Therefore, according to TG/DSC results, ZnOHF materials cannot be applied in a device that needs the process temperature or working temperature to be over 350°C. Of course, temperature higher than 450°C is needed to completely convert ZnOHF to ZnO when ZnO is the final target material.

The semiconducting properties were studied by recording the UV—Vis absorption spectra and the room-temperature photoluminescence (PL) spectrum. The absorption onset of ZnOHF was at a wavelength of about 340 nm, which was about 50 nm shorter than that of ZnO (Fig. 10.4A), and the light absorption intensity of ZnOHF is also significantly lower than that of ZnO. This lower light absorption capacity in the UV—Vis range means that less electron—hole pairs would be generated during its

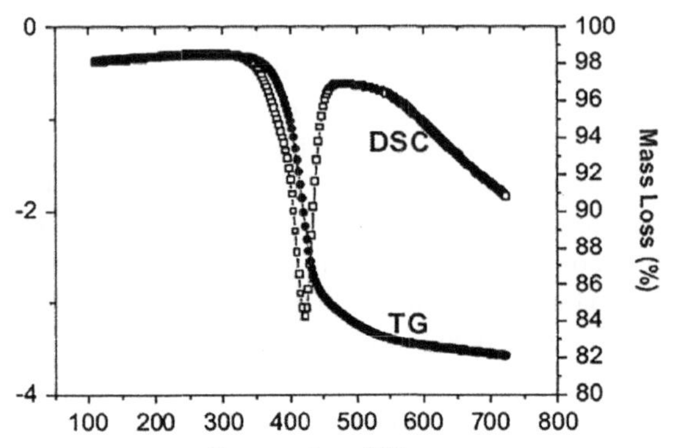

Figure 10.3 TG and DSC characteristic curves of zinc hydroxyfluoride [30].

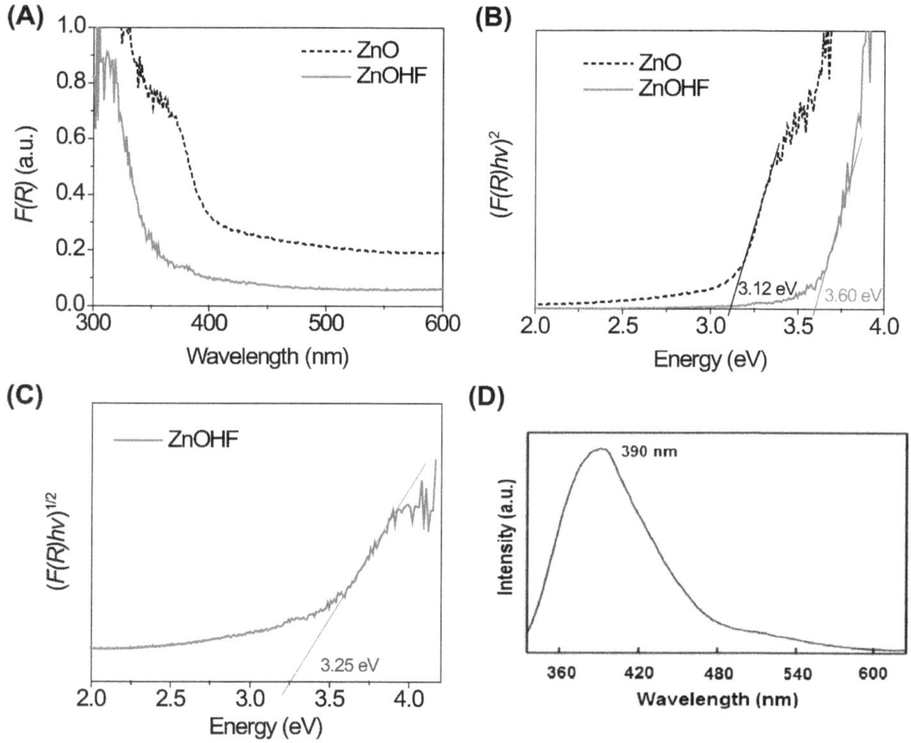

Figure 10.4 (A) UV−Vis absorption spectra and (B) plot of $(F(R)h\nu)^2$ versus $h\nu$ of ZnO and zinc hydroxyfluoride (ZnOHF); (C) Plot of $(F(R)h\nu)^{1/2}$ versus $h\nu$ and (D) the room-temperature photoluminescence spectrum of ZnOHF [1].

use in the photoelectrode of QDSCs and would relieve the photocorrosion by the residual holes in ZnOHF to improve the stability of the photoelectrode. The relationship of the absorption coefficient (α) and incident photon energy ($h\nu$) of a semiconductor is given by Eq. [10.1] [5,15]:

$$\alpha h\nu = C(h\nu - E_g)^n \qquad [10.1]$$

where E_g is the band gap of semiconductor, n is 0.5 and 2 for direct-transition and indirect-transition semiconductors, respectively. Accordingly, α is proportional to $F(R)$; E_g could be calculated based on UV−Vis absorption spectra and Eq. [10.1]. And since no research data report the semiconducting properties of ZnOHF, the E_g for direct-transition and indirect-transition cases are both calculated for ZnOHF. For a direct transition case, the E_g of ZnOHF is about 3.60 eV, obviously larger than that of ZnO (3.12 eV), as shown in Fig. 10.4B. For an indirect transition case, the E_g of ZnOHF is calculated to be about 3.25 eV (Fig. 10.4C). Therefore, for both cases, the E_g of ZnOHF is obviously larger than that of ZnO. The room-temperature

PL spectrum of Zn(OH)F excited with a λ of 320 nm presented a strong-broad emission centered at 392 nm (Fig. 10.4D) [1].

4. Synthesis of Zinc Hydroxyfluoride Nanostructures

ZnOHF has shown many promising physical and chemical properties that are expected to be widely applied in electric- and photoelectric-devices. As a result, various kinds of ZnOHF nanostructures have been prepared by several synthesis methods. These methods included hydrothermal, solution-based method, microwave irradiation, hydrolysis, and electrodeposition.

Hydrothermal method was for the first time applied to synthesize 1D ZnOHF nanostructure by Yu et al., in 2004 [37]. Briefly, ZnO and NaF were mixed in distilled water, which then was placed in an autoclave with a Teflon line for reaction at a certain temperature. Then, Huang et al. have prepared a net-like ZnOHF nanostructure by modifying the hydrothermal processes using a new reaction solution containing $Zn(NO_3)_2 \cdot 6H_2O$, NH_4F, and NaOH [14]. So far, hollow ZnOHF nanostructures [32] and ZnOHF nanorod (NR) arrays [20] have also been successfully synthesized by the hydrothermal method.

In addition to the abovementioned ZnOHF nanostructures prepared by hydrothermal method, ZnOHF microbesoms [25] and hierarchical nanofibers [26] were synthesized by the solution-based method. ZnOHF nanofibers [33], nanobelts [39], and flower-like nanostructures [29] have been prepared by microwave irradiation in the presence of the ionic liquid.

In recent years, electrodeposition method has well been developed to grow ZnOHF nanostructures [6,10,34,35]. Compared with the above methods, electrodeposition is more controllable, efficient, and reusable. Besides, ZnOHF nanostructures could be easily in situ grown on the substrate by electrodeposition, which makes them more suitable for device fabrication. The previously reported aqueous electrolyte used for electrodeposition contained $Zn(NO_3)_2 \cdot 6H_2O$ and KF. During electrodeposition, NO_3^- ions were reduced at the cathode to generate OH^- and NO_2^- ions according to Eq. [10.2]. And the OH^- ions on the cathode surface would react with Zn^{2+} and F^- ions to form ZnOHF precipitate on the substrate according to Eq. [ii] [34].

$$NO_3^- + H_2O + 2e^- \rightarrow NO_2^- + 2OH^- \quad [10.2]$$

$$Zn^{2+} + F^- + OH^- \rightarrow ZnOHF \quad [ii]$$

5. Zinc Hydroxyfluoride Nanostructure-Based QDSCs

Chen et al. have made an attempt to fabricate QDSCs based on ZnOHF nanostructures, which were prepared by the abovementioned electrodeposition method. The aqueous electrolyte contained $Zn(NO_3)_2 \cdot 6H_2O$ and KF and the electrodeposition was conducted in a two-electrode system with Zn sheet and Indium tin oxide (ITO) glass as counter- and working-electrodes, respectively. The deposition potential and temperature were fixed at -1.0 V and 60°C, respectively. Different ZnOHF nanostructures could be obtained by adjusting the electrodeposition durations. Snow-like nanostructures were obtained at short duration ($\leq$20 s), while NR arrays were produced at a long electrodeposition duration ($\geq$1 min). Fig. 10.5A presented the morphology of the ZnOHF NR array at 1 min, indicating the rhombus shape of ZnOHF NR with the diameters of about 30—200 nm, while the inset in Fig. 10.5A demonstrated the length of ZnOHF NR was about 1.0 μm. The transmission electron microscopy (TEM) image in Fig. 10.5B shows the smooth edge of ZnOHF NR, whose electron diffraction pattern could not be obtained due to the instability of ZnOHF NR toward electronic irradiation during recording and high-resolution TEM images.

To prepare the QDSC photoelectrodes based on ZnOHF NR, CdS and CdSe QDs were sequentially deposited on the ZnOHF NR arrays by successive ionic layer adsorption and reaction (SILAR) method. First, ZnOHF NR was successively immersed in the aqueous solutions of $Cd(NO_3)_2$, water, Na_2S, and water for the deposition of CdS

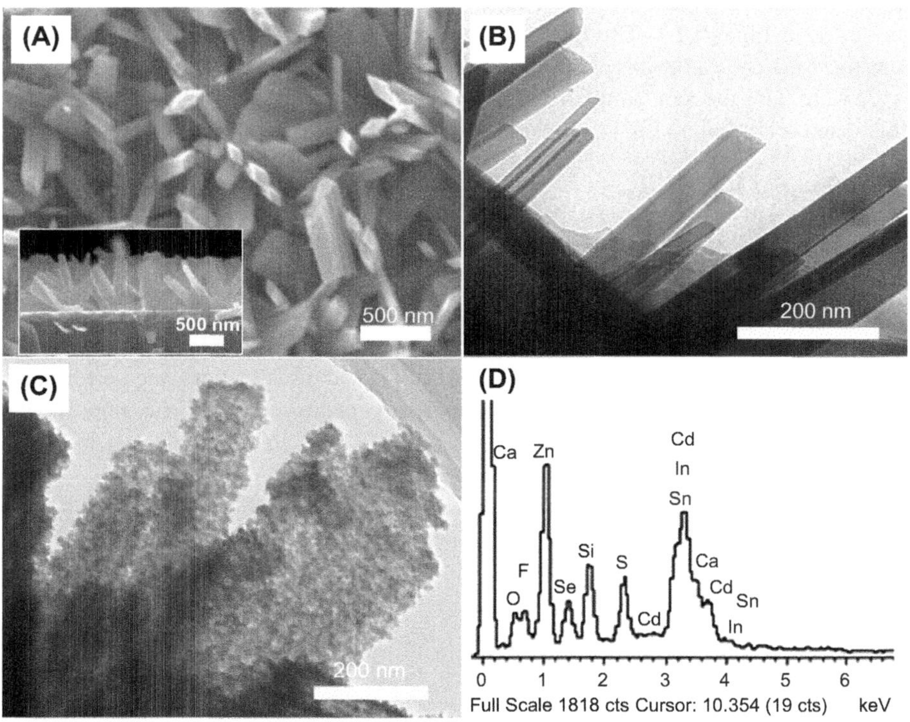

Figure 10.5 Characterizations of zinc hydroxyfluoride (ZnOHF) NR and CdS/CdSe quantum dots (QDs)-sensitized ZnOHF NR. (A) Scanning electron microscopy (SEM) images of the ZnOHF NR on ITO substrate. Inset was the cross-sectional SEM image. Transmission electron microscopy images of (B) ZnOHF NR and (C) CdS/CdSe QDs-sensitized ZnOHF NR. (D) EDS spectrum of CdS/CdSe QDs-sensitized ZnOHF NR [6].

QDs [4]. Then, the ZnOHF NR/CdS QDs sample was successively dipped into the ethanol solution of $Cd(NO_3)_2$ at room temperature and the aqueous solution of Na_2SeSO_3 at 50°C for the deposition of CdSe [18]. The SILAR cycles of CdS and CdSe were 30 and 4 times, respectively. The TEM image in Fig. 10.5C exhibited that after the deposition of CdS/CdSe QDs, the surface of ZnOHF NR became rough and the grain size of CdS and CdSe QDs were about several nanometers. Energy Dispersive Spectrometer (EDS) spectrum in Fig. 10.5D further confirmed the successful deposition of CdS and CdSe QDs with the observation of Cd, S, and Se elements.

After the deposition of CdS and CdSe QDs on ZnOHF NR, the UV−Vis absorption spectra in the wavelength range of 340−670 nm was obviously enhanced (Fig. 10.6). And the two obvious onsets corresponding to the characteristic absorption of CdS and CdSe QDs were easily observed at a wavelength of about 530 nm and 670 nm, respectively.

The CdS/CdSe QDs-sensitized ZnOHF NR was used to construct QDSCs according to the cell structure illustrated in Fig. 10.2 with CdS/CdSe QDs-sensitized ITO glass and ZnO NR as control samples. CuS_2 and polysulfide solution were used as counter electrode and electrolyte, respectively. The photocurrent density−voltage ($J-V$) curves were recorded to evaluate the photovoltaic performance of QDSCs, as presented in Fig. 10.7A. The QDSCs based on ITO glass exhibited very low performance. And the short-circuit photocurrent density (J_{sc}) and open-circuit voltage (V_{oc}) were 2.03 mA/cm^2 and 0.47 V, respectively, while the fill factor (FF) and PCE calculated according to Eq. [10.3] and [10.4] [5,15] were 0.39 and 0.37%, respectively.

$$FF = J_{max}V_{max}/J_{sc}V_{oc} \quad [10.3]$$

$$PCE = J_{max}V_{max}/P_h \quad [10.4]$$

where J_{max} and V_{max} were the photocurrent density and voltage obtained at the maximum power point on the $J-V$ curves, respectively, while P_h is the power density of the incident light. Compared with the QDSCs based on ITO glass, the performance of the QDSCs based on ZnO NR was greatly enhanced. And the photovoltaic parameters of J_{sc} = 4.63 mA/cm^2, V_{oc} = 0.51 V, FF = 0.36, and PCE = 0.86% were

Figure 10.6 UV−Vis absorption spectra of the zinc hydroxyfluoride (ZnOHF) NR and CdS/CdSe quantum dots-sensitized ZnOHF NR [6].

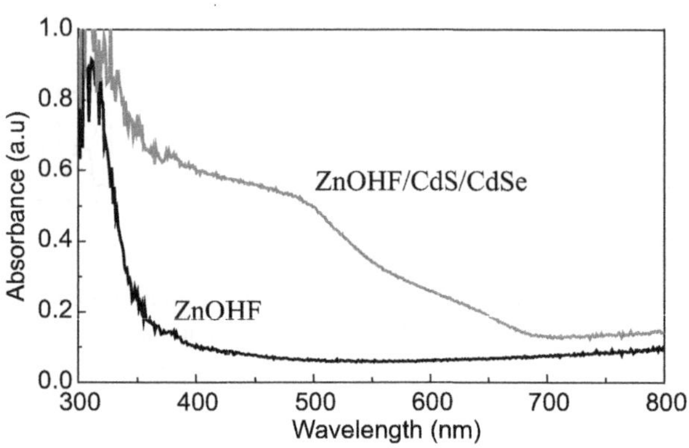

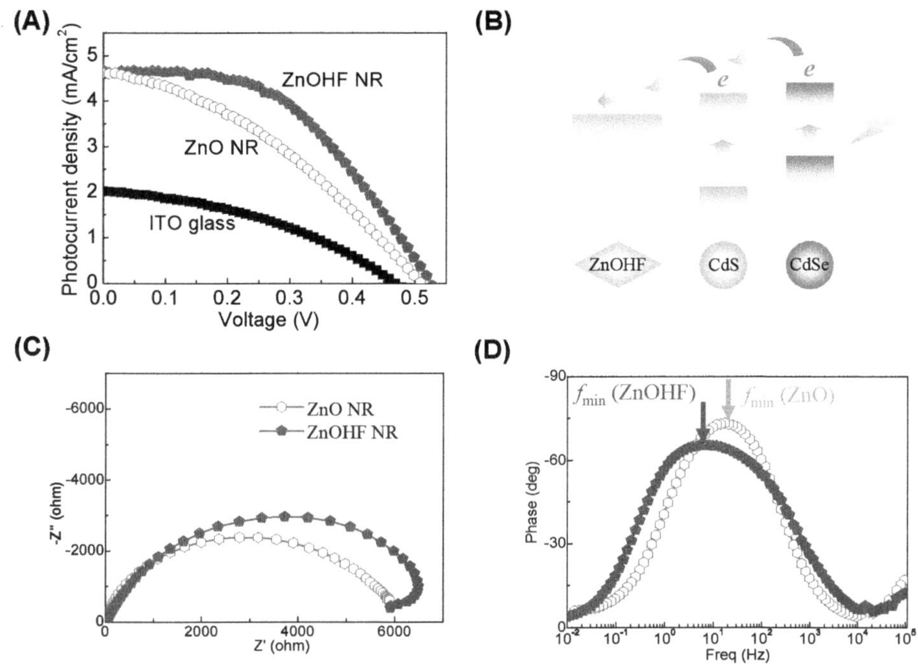

Figure 10.7 Photovoltaic characterizations of the quantum dot-sensitized solar cells (QDSCs). (A) $J-V$ curves of QDSCs for different photoelectrodes. (B) Scheme of the electron generation, transfer, and transport in ZnOHF/CdS/CdSe photoelectrode. IS results of the QDSCs based on ZnO NR and ZnOHF NR obtained at the forward bias potential of 0.3 V in the dark: (C) Nyquist and (D) Bode plots [6].

obtained. The performance enhancement by ZnO NR should be attributed to the increase in surface area for more QDs deposition and the suppression in the charge recombination between ITO and electrolyte.

When the ZnOHF NR with a similar thickness to ZnO NR was used in the photoelectrode, a V_{oc} of 0.53 V was obtained, similar to that of the ZnO NR-based QDSCs, suggesting the CB position of ZnOHF was close to that of ZnO. And a similar J_{sc} of 4.65 mA/cm^2 also achieved by ZnOHF NR well confirmed the suitable CB position of ZnOHF for accepting the injected electrons from QDs and transporting them to the ITO substrate. Therefore, the scheme of the electron generation, transfer, and transport in ZnOHF NR/CdS/CdSe QDs photoelectrode was illustrated in Fig. 10.7B. More importantly, ZnOHF NR/CdS/CdSe QDs photoelectrode exhibited a considerably higher FF (0.48) than ZnO NR-based photoelectrode, which boosted the PCE up to 1.18%. These photovoltaic results well suggested that ZnOHF was a very promising wide-band gap semiconductor for QDSCs.

To gain a deep insight into the performance difference between ZnOHF NR- and ZnO NR-based QDSCs, especially the difference in FF value, impedance spectra (IS) of the ZnO NR- and ZnOHF NR-based QDSCs were measured (Fig. 10.7C and D). Only one semicircle could be observed on the Nyquist plots, which corresponded to the charge transfer at photoelectrode/electrolyte interface [11,22]. This was because the semicircles corresponding to charge transfer at counter electrode/electrolyte interface and diffusion in electrolyte could not be observed in the QDSCs with Cu_2S and polysulfide as counter electrode and electrolyte, respectively [11,22]. Here, the charge transfer resistance for ZnOHF NR-based photoelectrode was higher than that for ZnO NR-based photoelectrode, demonstrating the slower electron recombination in ZnOHF NR-based photoelectrode. Bode plots in Fig. 10.7D show that the frequency peak (f_{min}) of ZnOHF NR-based photoelectrode was lower than that of ZnO NR-based photoelectrode. And as the f_{min} is determined by electron lifetime (τ_n) according to $\tau_n \sim 1/f_{min}$ [17], the τ_n of ZnOHF NR-based photoelectrode was longer than that of ZnO NR-based photoelectrode, suggesting the reduced recombination in ZnOHF-based photoelectrode, which could contribute to the higher FF and PCE.

6. Conclusions and Outlooks

In this chapter, we have systematically introduced the promising ZnOHF material, which has been widely used in several fields. The ZnOHF is crystallized with an orthorhombic symmetry (SG: $Pna2_1$), which tends to be decomposed to ZnO at a temperature higher than 350°C. The band gap (E_g) of ZnOHF was larger than that of ZnO, which was calculated to be about 3.60 and 3.25 eV for direct and indirect transition cases, respectively. Various ZnOHF nanostructures, such as nanonets, nanorod arrays, nanobesoms, nanofibers, nanoflowers, and so on, have been well synthesized by several methods, including hydrothermal, solution-based method, microwave irradiation, and electrodeposition. Finally, we have emphasized the application of ZnOHF in the photoelectrode of QDSCs as a wide-band gap semiconductor, which exhibited a more superior performance than the ZnO counterpart.

Though ZnOHF materials have been mainly used as the precursor for the synthesis of ZnO nanostructure materials, its several successful applications in photocatalystic and QDSCs, demonstrate that its successes in other photoelectrochemical and photovoltaic fields are worth being expected, such as DSSCs, dye-sensitized water splitting cells, perovskite solar cells, organic solar cells, and so on.

Acknowledgment

This work is financially supported by the National Natural Science Foundation of China (No. 51371020).

References

[1] S. Ahmad, P. Rawat, R. Nagarajan, Facile green synthesis of Zn(OH)F from the single source precursor $KZnF_3$, Mater. Lett. 139 (2015) 86–88.

[2] G.H. Carey, A.L. Abdelhady, Z. Ning, S.M. Thon, O.M. Bakr, E.H. Sargent, Colloidal quantum dot solar cells, Chem. Rev. 115 (23) (2015) 12732–12763.
[3] H.-Y. Chen, D.-B. Kuang, C.-Y. Su, Hierarchically micro/nanostructured photoanode materials for dye-sensitized solar cells, J. Mater. Chem. 22 (2012a) 15475–15489.
[4] H.N. Chen, W.P. Li, H.C. Liu, L.Q. Zhu, A suitable deposition method of CdS for high performance CdS-sensitized ZnO electrodes: sequential chemical bath deposition, Solar Energy 84 (2010) 1201–1207.
[5] H.N. Chen, W.P. Li, H.C. Liu, L.Q. Zhu, Performance enhancement of CdS-sensitized TiO_2 mesoporous electrode with two different sizes of CdS nanoparticles, Micropor. Mesopor. Mater. 138 (2011a) 235–238.
[6] H.N. Chen, L.Q. Zhu, Q. Hou, W.T. Liang, H.C. Liu, W.P. Li, ZnOHF nanostructure-based quantum dots-sensitized solar cells, J. Mater. Chem. 22 (2012b) 23344–23347.
[7] H.N. Chen, L.Q. Zhu, H.C. Liu, W.P. Li, ITO porous film-supported metal sulfide counter electrodes for high-performance quantum-dot-sensitized solar cells, J. Phys. Chem. C 117 (2013a) 3739–3746.
[8] H.N. Chen, L.Q. Zhu, H.C. Liu, W.P. Li, Zn-5(OH)(8)Cl-2 center dot H_2O-based quantum dots-sensitized solar cells: a common corrosion product enhances the performance of photoelectrochemical cells, Electrochim. Acta 105 (2013b) 289–298.
[9] H.N. Chen, L.Q. Zhu, M. Wang, H.C. Liu, W.P. Li, Wire-shaped quantum dots-sensitized solar cells based on nanosheets and nanowires, Nanotechnology 22 (2011b).
[10] M. Dai, F. Xu, Y. Lu, Y. Liu, Y. Xie, Synthesis of submicron rhombic ZnO rods via ZnOHF intermediate using electrodeposition route, Appl. Surf. Sci. 257 (2011) 3586–3591.
[11] V. González-Pedro, X. Xu, I. Mora-Seró, J. Bisquert, Modeling high-efficiency quantum dot sensitized solar cells, ACS Nano 4 (2010) 5783–5790.
[12] M. Gratzel, Photoelectrochemical cells, Nature 414 (2001) 338–344.
[13] G. Hodes, Comparison of dye- and semiconductor-sensitized porous nanocrystalline liquid junction solar cells, J. Phys. Chem. C 112 (2008) 17778–17787.
[14] Q.-L. Huang, M. Wang, H.-X. Zhong, X.-T. Chen, Z.-L. Xue, X.-Z. You, Netlike nanostructures of Zn(OH)F and ZnO: synthesis, characterization, and properties, Crystal Growth Design 8 (2008) 1412–1417.
[15] H. Jia, H. Xu, Y. Hu, Y. Tang, L. Zhang, TiO_2@CdS core–shell nanorods films: fabrication and dramatically enhanced photoelectrochemical properties, Electrochem. Comm. 9 (2007) 354–360.
[16] P.V. Kamat, Quantum dot solar cells. Semiconductor nanocrystals as light harvesters, J. Phys. Chem. C 112 (2008) 18737–18753.
[17] R. Kern, R. Sastrawan, J. Ferber, R. Stangl, J. Luther, Modeling and interpretation of electrical impedance spectra of dye solar cells operated under open-circuit conditions, Electrochim. Acta 47 (2002) 4213–4225.
[18] Y.-L. Lee, Y.-S. Lo, Highly efficient quantum-dot-sensitized solar cell based on co-sensitization of CdS/CdSe, Adv. Funct. Mater. 19 (2009) 604–609.
[19] W. Li, X. Zhong, Capping ligand-induced self-assembly for quantum dot sensitized solar cells, J. Phys. Chem. Lett. 6 (2015) 796–806.
[20] L. Luo, W. Tao, X. Hu, T. Xiao, B. Heng, W. Huang, H. Wang, H. Han, Q. Jiang, J. Wang, Y. Tang, Mesoporous F-doped ZnO prism arrays with significantly enhanced photovoltaic performance for dye-sensitized solar cells, J. Power Sources 196 (2011) 10518–10525.
[21] S. Mathew, A. Yella, P. Gao, R. Humphry-Baker, B.F. Curchod, N. Ashari-Astani, I. Tavernelli, U. Rothlisberger, M.K. Nazeeruddin, M. Grätzel, Dye-sensitized solar cells with 13% efficiency achieved through the molecular engineering of porphyrin sensitizers, Nat. Chem. 6 (2014) 242–247.

[22] I. Mora-Seró, S. Giménez, F. Fabregat-Santiago, R. Gómez, Q. Shen, T. Toyoda, J. Bisquert, Recombination in quantum dot sensitized solar cells, Acc. Chem. Res. 42 (2009) 1848−1857.
[23] A.J. Nozik, Quantum dot solar cells, Phys. E 14 (2002) 115−120.
[24] B. O'Regan, M. Gratzel, A low-cost, high-efficiency solar cell based on dye-sensitized colloidal TiO_2 films, Nature 353 (1991) 737−740.
[25] Y. Peng, Controlled-growth of Zn(OH)F microbesoms in the mild aqueous solution and their photocatalytic activity, Chinese J. Chem. 29 (2011) 191−195.
[26] Y. Peng, H.-Y. Zhou, Z.-H. Wang, Synthesis, characterization and photocatalytic activity of Zn(OH)F hierarchical nanofibers prepared by a simple solution-based method, CrystEngComm 14 (2012) 2812−2816.
[27] S. Ruhle, M. Shalom, A. Zaban, Quantum-dot-sensitized solar cells, Chemphyschem 11 (2010) 2290−2304.
[28] H. Serier, M. Gaudon, A. Demourgues, A. Tressaud, Structural features of zinc hydroxyfluoride, J. Solid State Chem. 180 (2007) 3485−3492.
[29] J.-K. Song, M.-B. Zheng, Z.-J. Yang, H.-Q. Chen, H.-Y. Wang, J.-S. Liu, G.-B. Ji, H.-Q. Zhang, J.-M. Cao, Synthesis of novel flower-like Zn(OH)F via a microwave-assisted ionic liquid route and transformation into nanoporous ZnO by heat treatment, Nanoscale Res. Lett. 4 (2009) 1512−1516.
[30] M. Wang, X. Shen, G. Jiang, Y. Shi, Synthesis, characterization and photocatalytic properties of hierarchical fan-shaped ZnOHF and ZnO microcrystals, Mater. Lett. 87 (2012) 54−57.
[31] M. Wang, T. Sun, Y. Shi, G. Jiang, Y. Tang, 3D hierarchical ZnOHF nanostructures: synthesis, characterization and photocatalytic properties, CrystEngComm 16 (2014) 10624−10630.
[32] J. Wu, D. Xue, Hydrothermal crystallization of ZnOHF: an integration of nanoparticles into hollow architectures, Nanosci. Nanotech. Lett. 3 (2011) 400−406.
[33] L. Wu, J. Lian, G. Sun, X. Kong, W. Zheng, Synthesis of zinc hydroxyfluoride nanofibers through an ionic liquid assisted microwave irradiation method, Euro. J. Inorganic Chem. 2009 (2009) 2897−2900.
[34] F. Xu, Y. Lu, L. Sun, L. Zhi, A novel ZnO nanostructure: rhombus-shaped ZnO nanorod array, Chem. Comm. 46 (2010a) 3191−3193.
[35] F. Xu, L. Sun, M. Dai, Y. Lu, Fluorine-ion-mediated electrodeposition of rhombus-like ZnFOH nanorod arrays: an intermediate route to novel ZnO Nanoarchitectures, J. Phys. Chem. C 114 (2010b) 15377−15382.
[36] A. Yella, H.-W. Lee, H.N. Tsao, C. Yi, A.K. Chandiran, M.K. Nazeeruddin, E.W.-G. Diau, C.-Y. Yeh, S.M. Zakeeruddin, M. Grätzel, Porphyrin-sensitized solar cells with cobalt (II/III)−based redox electrolyte exceed 12 percent efficiency, Science 334 (2011) 629−634.
[37] J. Yu, J.C. Yu, W. Ho, L. Wu, X. Wang, A simple and general method for the synthesis of multicomponent $Na_2V_6O_{16}\cdot 3H_2O$ single-crystal nanobelts, J. Am. Chem. Soc. 126 (2004) 3422−3423.
[38] K. Zhao, Z. Pan, I. Mora-Seró, E. Cánovas, H. Wang, Y. Song, X. Gong, J. Wang, M. Bonn, J. Bisquert, X. Zhong, Boosting power conversion efficiencies of quantum-dot-sensitized solar cells beyond 8% by recombination control, J. Am. Chem. Soc. 137 (2015) 5602−5609.
[39] L. Zhu, Y. Zheng, T. Hao, X. Shi, Y. Chen, J. Ou-Yang, Synthesis of hierarchical ZnO nanobelts via Zn(OH)F intermediate using ionic liquid-assistant microwave irradiation method, Mater. Lett. 63 (2009) 2405−2408.

Part Three

Magnetic properties

Molecular Fluoride-Bridged 3d-4f Complexes and Their Magnetic Properties

11

K.S. Pedersen
CRPP-CNRS, Pessac, France; ICMCB-CNRS, Pessac, France

J. Bendix
University of Copenhagen, Copenhagen, Denmark

Chapter Outline

1. Introduction

The interest in simultaneously combining both transition metal and lanthanide ions in molecular assemblies has blossomed due to the observation of interesting magnetic phenomena as single-molecule magnetism and large magnetocaloric effects [1,2]. These properties depend strongly on the nature and magnitude of the magnetic super-exchange interactions between the constituent metal ions. Several problems have therefore arisen for the scientific community: firstly, how to design and engineer specific molecular topologies having the desired physical properties and, subsequently, how to get experimental insight into the underlying microscopic magnetic interactions. Herein, we will focus on our recent endeavors to incorporate fluoride ions as linking motifs between 3d and 4f metal ions in research directed toward single-molecule magnets and, in particular, molecular magnetic coolers. The choice of fluoride is indeed quite obvious taking into account fluoride's preferences for coordination to metal ions being classified as "hard" in the Pearson's classification of hard and soft Lewis bases and acids [3]. However, only extremely few examples of mixed 3d-4f molecular clusters were reported just a few years ago [4]. The reason being the extraordinary high

Photonic and Electronic Properties of Fluoride Materials. http://dx.doi.org/10.1016/B978-0-12-801639-8.00011-8

lattice enthalpy of lanthanide(III) fluorides, which is the main culprit in combining fluoride with lanthanide chemistry. It follows that severe boundary conditions are associated with the preparative chemistry in so far as, synthetic pathways must effectively avoid any sizable concentration of free fluoride ions. One rational approach to come around this issue is the application of kinetically robust transition metal ion complexes as metalloligands toward suitable lanthanide complex precursors. Following this idea, Winpenny et al. and we, have now reported an extensive series of fluoride-bridged 3d-4f complexes of variable nuclearity and with fascinating properties [5,6].

2. Robust Cr^{III} Complexes as Synthons

Kinetic robustness for 3d metal complexes is typically reserved for the d^3 and low-spin d^6 electronic configurations. For the interest of magnetism, only the former is of interest and these systems are largely dominated by Cr(III) complexes. As demonstrated by the Winpenny group, the reaction of $[(R_2NH_2)_3\{Cr_6F_{11}(O_2C^tBu)_{10}\}]_2$ (R = Et, ^{n}Pr) "horseshoes" with lanthanide ions results in aesthetical mixed 3d-4f wheels or condensed wheels featuring Cr–F–Ln bridging moieties supported by carboxylate bridging [5]. Two representative examples of a $\{Cr_6Ce\}$ wheel and a $\{Cr_6Gd_2\}_2$ double wheel are shown in Fig. 11.1.

Although structurally pleasing, the large horseshoes prevent targeting small molecular units and for the analysis and quantification of magnetic interactions a strategy employing simpler, preferentially mononuclear, chromium(III) complexes as building blocks for molecular clusters is desirable. Certainly, the preparative chemistry of fluoride complexes of chromium(III) is quite well developed [7]. In the work by Glerup et al. fluoride is conveniently introduced by reaction of chromium(III) nitrate enneahydrate with aqueous hydrofluoric acid. The remaining coordination sites, not occupied by fluoride ligands, can facilely be substituted by a large variety of organic

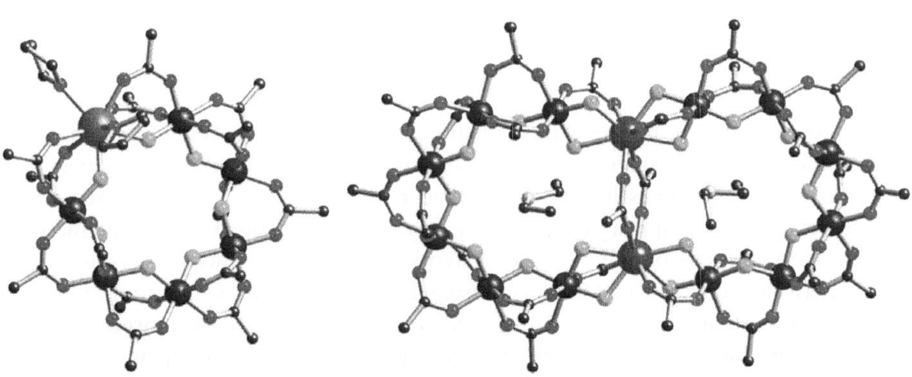

Figure 11.1 Representative examples of mixed 3d-4f systems as described in the main text. Reproduced with permission from A. McRobbie, A.R. Sarwar, S. Yeninas, H. Nowell, M.L. Baker, D. Allan, M. Luban, C.A. Muryn, R.G. Pritchard, R. Prozorov, G. Timco, F. Tuna, G.F.S. Whitehead, R.E.P. Winpenny, Chem. Commun. 47 (2011) 6251.

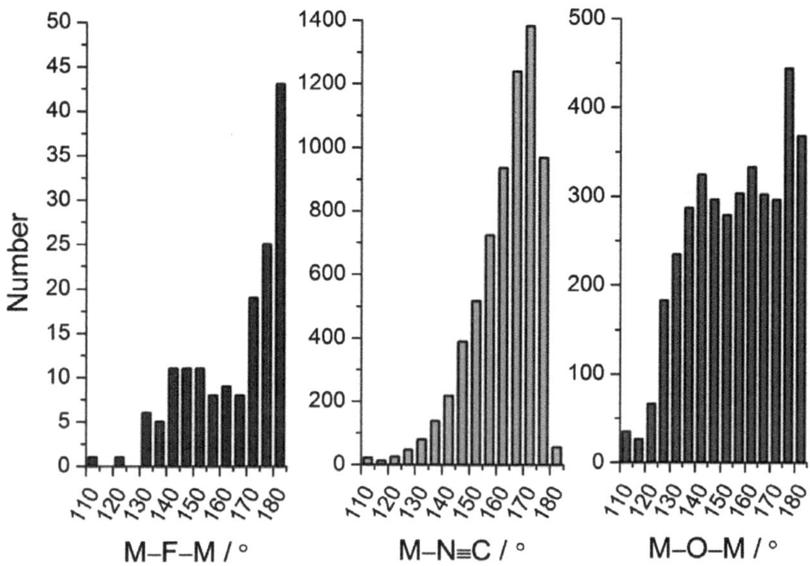

Figure 11.2 Frequency histograms of structurally characterized (Cambridge Structural Database, 2014) unsupported M—F—M′ (left), M—N≡C (middle), and unsupported M—O—M′ (right) bridging angles.
Reproduced with permission from K.S. Pedersen, J. Bendix, R. Clérac, Chem. Commun. 50 (2014) 4396.

mono-, bi-, tri-, and tetradentate ligands [7]. The outcome is an extensive library of paramagnetic, kinetically inert fluoride complexes differing in their local geometric arrangement of the fluoride ligands. Having established suitable metalloligands, it is instructional to consider the properties of fluoride when binding between metal ions. Indeed, μ_2-bridging is vastly dominant but also μ_3, μ_4, and even μ_6 motifs have been reported [8]. Secondly, a common characteristic of intermetallic bridging is the preponderance of nearly linear configuration of the M—F—M′ fragment. This trend is established for many binary main group and transition metal fluorides but is also evident in polynuclear complexes with organic coligands [9]. Fig. 11.2 demonstrates (1) the very limited number of structurally characterized complexes featuring an "unsupported" fluoride bridge in comparison to the more popular cyanide systems and (2) the propensity toward linear bridging in M—F—M′ systems which is, notably, not evident in the isoelectronic M—O—M′ moiety.

The intrinsic preference of linearity in bridging could potentially be employed as a synthetic handle toward specific molecular topologies. For instance, *trans*-$\{CrF_2\}^+$ complexes should dictate linear 3d-4f clusters and *cis*-$\{CrF_2\}^+$ zig-zig clusters or molecular rings. Indeed this conjecture works exceedingly well in predicting and rationalizing synthetic outcomes of reactions between Cr(III) fluoride complexes and lanthanide (Ln) complexes. Some representative examples resulting from the reactions of Cr(III) fluoride complexes with $Ln(NO_3)_3 \cdot xH_2O$ or $Ln(hfac)_3(H_2O)_2$ (hfacH = 1,1,1,5,5,5-hexafluoropentan-2,4-dione) are shown in Fig. 11.3. The

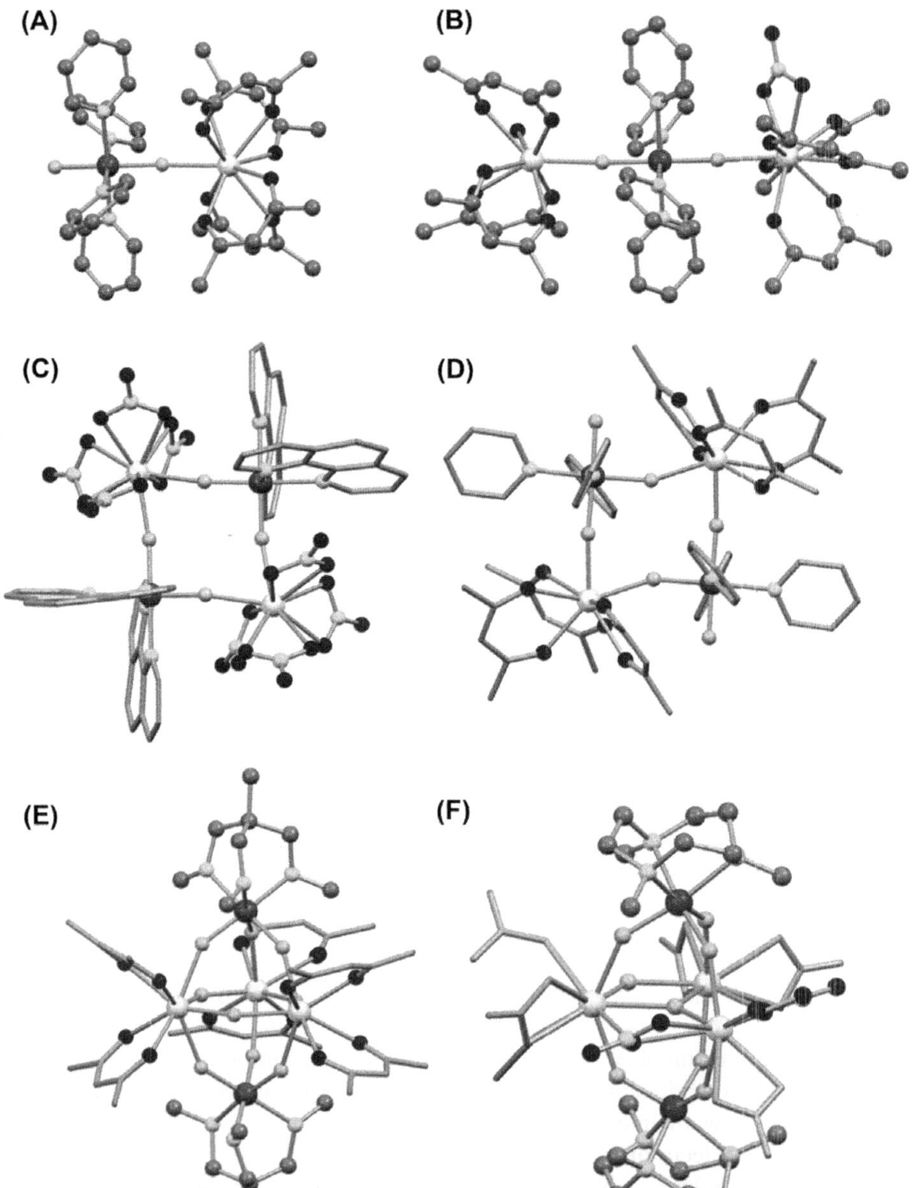

Figure 11.3 Representative examples of fluoride-bridged Cr(III)–Ln(III) of various nuclearities. (A) [{*trans*-$CrF_2(py)_4$}Gd$(hfac)_4$] (py: pyridine; hfacH = 1,1,1,5,5,5-hexafluoropentan-2,4-dione), (B) [Dy$(hfac)_3(H_2O)${*trans*-$CrF_2(py)_4$}Dy$(hfac)_3(NO_3)$], (C) [{*cis*-$CrF_2(phen)_2$}$_2${Gd$(NO_3)_4$}$_2$] (phen = 1,10-phenanthroline), (D) [{*mer*-$CrF_3(py)_3$}$_2${Gd $(hfac)_3$}$_2$], (E) [{*fac*-CrF_3(tmtame)}$_2Gd_3F_3(hfac)_6$], (F) [{*fac*-FeF_3(tmtacn)}$_2Gd_3F_2(NO_3)_7$ $(H_2O)(CH_3CN)$]. Color codes: Ln, pale blue (lightest gray in print versions); Fe, orange

reaction of *cis*-$[CrF_2(phen)_2]^+$ with $Ln(NO_3)_3 \cdot xH_2O$ in methanol results in a hypsochromic shift of the Cr(III) $^4T_2(O) \leftarrow {}^4A_2(O)$ band of 460 cm^{-1} univocally demonstrating ligation of *cis*-$[CrF_2(phen)_2]^+$ to Ln^{3+} in solution [10]. The observation of d−d transitions at Cr(III) shifting toward shorter wavelengths, ie, a strengthening of the ligand field is rather counterintuitive but can be reproduced computationally. The effects can be traced to a breakdown of the conventional transferability assumption for the ligand-field parameters of the auxiliary ligand sphere. Thus, the amine ligands overcompensate in terms of perturbation of the 3d orbitals when the donor strength of the fluoride is reduced by bridging toward metal ions [11]. From relatively concentrated solutions of *cis*-$[CrF_2(phen)_2]^+$ and $Ln(NO_3)_3 \cdot xH_2O$, molecular $\{Cr_2Ln_2\}$ squares (cf. Fig. 11.3C) crystallize in a short amount of time. The Cr−F−Ln angle is approaching linearity with 168.61(9) degrees (Ln = Gd). Interestingly, if the reaction is performed in dilute solution of the precursors, a dinuclear dimethoxide-bridged {CrLn} complex is formed at a much slower rate alongside very insoluble by-products, presumable mainly lanthanide fluoride [12]. The timescale for the completion of this reaction is ~1 year at room temperature and illustrates some limitations of the Cr(III) metalloligand approach when taken to the extremes. However, by optimizing reaction conditions, the fluoride-abstraction reaction between lanthanide ions and robust fluoride complexes can be made preparatively useful as illustrated by the convenient, high-yield synthesis of *cis*-$[CrF(H_2O)(phen)_2]^{2+}$ and *cis*-$[CrF(OH)(phen)_2]^+$ from the difluorido precursor [13].

Whereas the dimethoxide-bridged complexes discussed earlier features two bent methoxide bridges bridging the two metal centers, two fluorides on the same Cr(III) are never found to chelate the same lanthanide ion irrespective of the nuclearity of the system. This empirical fact has interesting implications for the kind of chemistry that can be exploited. For instance, *fac*-$\{CrF_3\}$ complexes would, therefore, be likely to coordinate three lanthanide ions in a triangular (μ_3-κ-F,F′,F″) fashion. The actual experimental outcomes when reacting *fac*-$[CrF_3(tmtame)]$ (tmtame = 1,1,1-tris-((methylamino)methyl)ethane) or *fac*-$[CrF_3(tmtacn)]$ (tmtacn = *N*,*N*′,*N*″-trimethyl-1,4,7-triazacyclononane) with the previously described lanthanide sources are molecular $\{Cr_2Ln_3\}$ trigonal bipyramids formed from bicapped lanthanide triangles (cf. Fig. 11.3E and F) [10]. Notably, for the former case, the lanthanide triangle is further supported by fluoride bridging. As no additional fluoride source than the *fac*-$[CrF_3(tmtame)]$ was added to the reaction, these fluorides result as a decomposition product of *fac*-$[CrF_3(tmtame)]$. The usage on hydrous starting materials is often almost unavoidable due to the complications of dehydrating

(dark gray in print versions); Cr, dark green (darker gray in print versions); F, light green (lighter gray in print versions); O, red (darkest gray in print versions); N, blue (light gray in print versions); C, gray. All F-atoms on $hfac^-$ ligands and H-atoms have been omitted for clarity. For the more complicated structures, the auxiliary ligand sphere is shown as *sticks* for clarity. Cocrystallized solvent molecules have been omitted in the figure and formulas.
Reproduced with permission from K.S. Pedersen, M.A. Sørensen, J. Bendix, Coord. Chem. Rev. 299 (2015) 1.

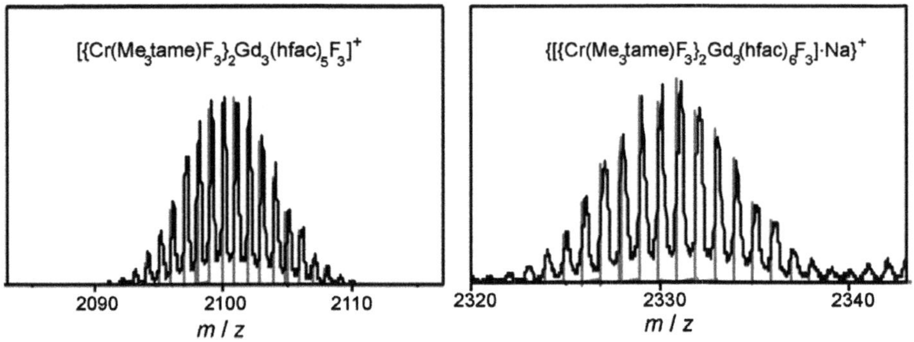

Figure 11.4 Electrospray mass spectra of $[\{CrF_3(tmtame)\}_2Gd_3F_3(hfac)_6]\cdot 7MeCN$ in MeCN (*black trace*) and simulations (*gray*).
Reproduced with permission from T. Birk, K.S. Pedersen, C.Aa. Thuesen, T. Weyhermüller, M. Schau-Magnussen, S. Piligkos, H. Weihe, S. Mossin, M. Evangelisti, J. Bendix, Inorg. Chem. 51 (2012) 5435.

fluorides due to the basicity of fluoride, but this also implies a general risk of hydroxide incorporation in the final products. The crystallographic distinction of fluoride from isolectronic hydroxide (or oxide) is often exceedingly difficult or impossible. It is therefore of central importance to corroborate the nature of all of the bridging ligands by other techniques. For the paramagnetic systems in question ^{19}F-NMR is of little usage, but mass spectrometry has proven very valuable. Benefitting from the significant mass difference (19 u for F^- and 17 u for OH^-), the cluster composition can confidently be unraveled by mass spectrometry even on relatively large molecules. The example of $\{Cr_2Gd_3\}$ (Fig. 11.3E) is shown in Fig. 11.4.

So far the synthetic approach has solely focused on the use of kinetically inert Cr(III) complexes. Not surprisingly, the concept can successfully be applied also to low-spin Co(III) complexes. Being nonmagnetic and significantly more inconvenient to synthesize, the efforts to obtained mixed Co(III)–Ln(III) systems have been modest [14]. Notably, the reaction of *trans*-$[CoF_2(py)_4]^+$ (py = pyridine) with one equivalent of $Dy(hfac)_3(H_2O)_2$ yields the dinuclear $\{CoF_2(py)_4Dy(hfac)_4\}$ featuring a single, unsupported fluoride bridge and being isostructural to the Cr(III)-Gd(III) analog (cf. Fig. 11.3A) [15]. The use of only Cr(III) and Co(III) puts somewhat strong limitations on the diversity of systems and magnetic properties that can be obtained. Therefore, in a second strategy, one may wonder if the use of fluoride complexes with large stability constants may as well serve as metalloligands without formation of LnF_3. For instance, $[FeF_6]^{3-}$ has a large stability constant, thereby fueling the curiosity whether iron(III) fluoride complexes with organic coligands (to facilitate solubility in weakly coordinating organic solvents) could function as synthons. Unfortunately, fluoride complexes of Fe(III), apart from $[FeF_6]^{3-}$, are exceedingly scarce in the literature. In fact, only few mononuclear fluoride complexes with organic coligands have been reported [16]. However, reaction of purported insoluble FeF_3 with tmtacn in boiling dimethylformamide cleanly affords *fac*-$[FeF_3(tmtacn)]\cdot H_2O$. The complex features a distorted octahedral, high-spin Fe(III) center ($S = 5/2$) structurally alike to previously reported Cr(III) and V(III) analogs [7e,17]. Despite the expected lability with respect to fluoride exchange, the reaction of *fac*-$[FeF_3(tmtacn)]\cdot H_2O$ with $Gd(NO_3)_3\cdot 5H_2O$ proceeds cleanly to afford a $\{Fe_2Gd_3\}$ trigonal bipyramid (Fig. 11.3F) without any signs of concomitant gadolinium fluoride formation [18]. In this system, the $\{Gd_3\}$ triangle is further stabilized by a most unusual μ_3-fluoride bridging the three gadolinium centers in a planar fashion. Other metal ions having a strong propensity for fluoride coordination, like Ga(III), were also found possible to isolate as monomeric complexes suited as building blocks. This is exemplified by *fac*-$[GaF_3(tmtacn)]\cdot 3H_2O$, which could be successfully incorporated into a $\{Ga_2Gd_3\}$ trigonal bipyramid. We will return to the discussion of the properties of these latter, unprecedented systems.

3. Magnetic Interactions

Having established a series of polynuclear complexes, the detailed investigation of their properties is mandatory. The fundamental understanding is, however, significantly hampered by the presence of lanthanide ions. In contrast to the 3d transition metal ions, the ligand field in the lanthanide series is much weaker than both the interelectronic repulsion and the spin–orbit coupling energies. Therefore, the lanthanides are reasonably well described by considering the ligand field as a perturbation on the free ion terms ${}^{2S+1}L_J$, where the total angular momentum, J, is given by $J = L - S$ or $L + S$, for less or more than a half-filled 4f shell, respectively. For orbitally nondegenerate systems, magnetic exchange interactions are commonly determined by fitting of an operator of pairwise interactions:

$$\widehat{H} = \widehat{H}_{\text{Zeeman}} + J_{ij}\,\widehat{S}_i \cdot \widehat{S}_j \qquad [11.1]$$

to the temperature and field dependence of the magnetization. The largely unquenched orbital momentum present in many of the lanthanide ions may invoke strong anisotropy in the magnetic exchange interaction thus allowing many more free parameters in the fitting procedure. As powder- and thermally averaged responses provide data of limited detail and information content, the growing number of fitting parameters quickly leads to a vast overparametrization of the experimental data. On top of this problem comes the ligand-field splittings. Although weak, the overall splitting of the ground J multiplet is typically on the order of 10^2–10^3 cm^{-1}, thus a temperature-dependent magnetization between room temperature and liquid helium temperatures will be significantly affected by the population of the various ligand-field states. In conclusion, the fundamental determination of a magnetic exchange coupling constant from magnetism data in most lanthanide-containing systems remains an overwhelming task. We will later return to this problem but initially consider the special case of

Gd(III) ($4f^7$) where the orbital angular momentum is absent. This results in a spin-only $S = 7/2$ ground state that is only very weakly split by the ligand field (second-order effect) and thereby rendering the simple Hamiltonian of Eq. [11.1] applicable. The temperature dependence of the χT ($\chi = M/H$) product of the simple dinuclear $[CrF_2(py)_4Gd(hfac)_4]$ (cf. Fig. 11.3A) is shown in Fig. 11.5. On descending temperature from room temperature, the χT product undergoes a gradual decrease indicative of antiferromagnetic interactions between the Gd(III) and Cr(III) ions through the bridging fluoride ion. The red line of Fig. 11.5 is the best fit obtained with $J = 0.84(4)$ cm^{-1} [19]. The small magnitude of the exchange coupling constant is typical for lanthanide ions due to the shielded nature of the 4f orbitals that, in addition to weak ligand-field effects, proffers only weak magnetic interactions. Further confidence in this value was obtained by inelastic neutron scattering (INS) spectroscopy which is exceedingly sensitive to the magnetic anisotropy of Cr(III) and to the exchange interaction. The best fit of the INS spectra was obtained for $J = 0.87$ cm^{-1} in good agreement with the parameter value extracted from magnetometry. The bridging angle in this dinuclear system is close to linear with $\angle$Gd–F–Cr = 178.3(3) degrees. Considering systems with increasing nuclearity (cf. Fig. 11.3 for selected examples) it is observed that the Cr–F–Gd bridging angle decreases when the nuclearity increases to four and five. Notably, this is paralleled by almost no changes in the Cr–F and Gd–F bond lengths. The magnetic susceptibility data for a series of Gd(III)–Cr(III) complexes of different nuclearity is shown in Fig. 11.5. Irrespective of the nuclearity, all the systems experience a gradual decrease of their χT product when cooling from room temperature to 2K. These members of the Gd(III)–Cr(III) family of clusters are all selected because they feature no other important pathways for magnetic exchange

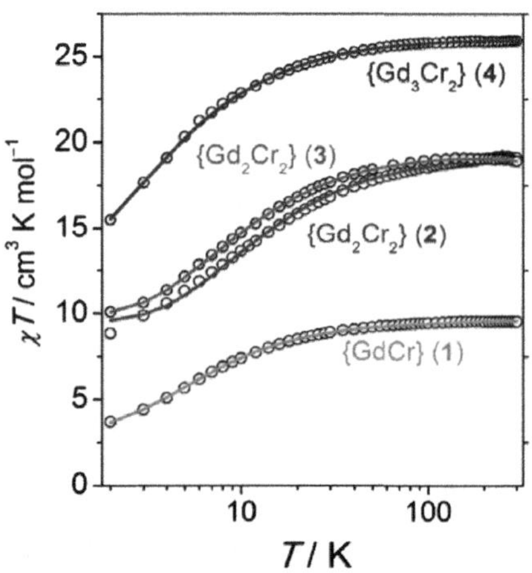

Figure 11.5 Temperature dependence of the χT product for selected Cr(III)–Gd(III) complexes. The *solid lines* are best fits.
Reproduced with permission from S.K. Singh, K.S. Pedersen, M. Sigrist, C.A. Thuesen, M. Schau-Magnussen, H. Mutka, S. Piligkos, H. Weihe, G. Rajaraman, J. Bendix, Chem. Commun. 49 (2013) 5583.

besides via the {Gd—F—Cr} moiety. Thus, it is reasonable to assume that the decreases in the χT products are solely a result of fluoride-mediated Gd(III)-Cr(III) exchange interactions. The solid lines of Fig. 11.5 are the best fits obtained by the same approach as outlined earlier. In Fig. 11.6, the extracted Gd(III)—Cr(III) exchange coupling constants are shown as plotted against the crystallographically determined Gd—F—Cr angles for the different systems.

Obviously, a pronounced dependence of J on ∠Gd—F—Cr is evident and its rationalization was afforded by density functional theory (DFT) calculations. For mixed 3d-4f complexes, several contributions to the magnetic exchange interactions must be taken into account. Firstly, the overlap between the magnetic 3d and 4f orbitals may result in both ferro- and antiferromagnetic contributions depending on the overlap integrals. Herein, the shielded nature of the 4f orbitals renders both of these mechanisms relatively weak. Furthermore, electron transfer excitation from the transition metal 3d orbitals to the 5d orbitals of the 4f ion is a common contribution to ferromagnetic exchange. In the case of Cr(III) having a t_{2g}^3 electronic configuration, all the magnetic orbitals have π symmetry with respect to the intermetallic Gd—F—Cr axis, thereby diminishing the possibilities for a strong (σ) 3d → 5d ferromagnetic contribution [20]. In order to evaluate the magnetic interactions and quantify the importance of the overlap integrals of the 3d and 4f/5d orbitals on Cr(III) and Gd(III), respectively, DFT was employed. For the experimental geometry of [$CrF_2(py)_4Gd(hfac)_4$], the DFT-calculated value is $J = 0.80\ cm^{-1}$, in almost perfect agreement with the experiment (vide supra). Instead of attempting calculations on the complicated polynuclear systems, the effect of the bridging angle at fluoride was examined by bending

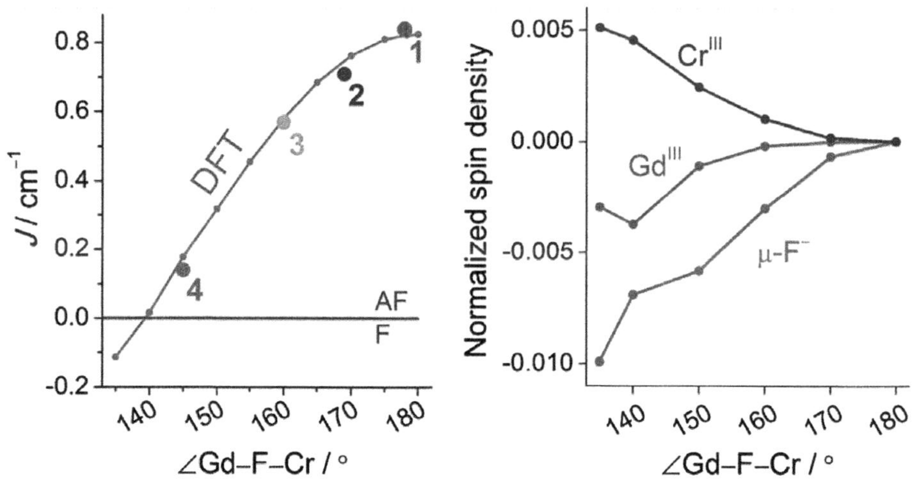

Figure 11.6 Left: Angular dependence of the Gd—Cr exchange coupling constant, J. Right: Normalized spin densities for Gd, Cr, and the bridging fluoride as a function of bridging angle. Reproduced with permission from S.K. Singh, K.S. Pedersen, M. Sigrist, C.A. Thuesen, M. Schau-Magnussen, H. Mutka, S. Piligkos, H. Weihe, G. Rajaraman, J. Bendix, Chem. Commun. 49 (2013) 5583.

$[CrF_2(py)_4Gd(hfac)_4]$ in silico. The DFT-calculated angular dependence of the exchange interaction parameter is shown as the gray trace of Fig. 11.6. The resulting computed exchange couplings at the bending angles corresponding to the experimental structures of the polynuclear systems were found to be in almost perfect concord with the experimentally derived exchange couplings. Another result directly visible in the right-hand panel of Fig. 11.6 is the increased spin delocalization onto the bridging fluoride upon diminishing the bridge angle.

Apart from the good agreement with the experimental results, the vanishing of J at the critical angle of 141 degrees is interesting. If sufficiently bent systems could be engineered, a ferromagnetically coupled ground state should result. Considering that the maximization of the spin ground state has been considered a holy grail for both single-molecule magnets and for the observation of large magnetocaloric effects in molecules, this magnetostructural result has interesting implications. Indeed, in the previously discussed family of $\{M_2Gd_3\}$, $[\{MF_3(tmtacn)\}_2Gd_3F_2(NO_3)_7(H_2O)(CH_3CN)]$ ($M = Cr$, Fe, Ga), the Cr(III) system exhibits an average Gd—F—Cr angle of 141.5 degrees—just at the expected crossing from antiferro- to ferromagnetic Gd—Cr interactions. The magnetic properties of this system, which indeed are reminiscent of weak ferromagnetic Cr(III)—Gd(III) interactions, will be discussed in-depth in the last section of this chapter.

As already emphasized a comparatively more complicated situation arises when magnetic data on systems containing orbitally degenerate lanthanides are considered. For the purpose of molecular magnetism, by far the most fecund and intensively studied lanthanide ion is Dy^{3+} having a $^6H_{15/2}$ ground state. If the ground state is sufficiently well separated energetically from the excited ones, the magnetic moment of Dy^{3+} can be modeled as an effective spin-1/2 ($J_{\mathrm{eff}} = 1/2$) ion with anisotropic g factors. When the population of the excited states becomes sizable this approximation, however, fails. In the aforementioned $\{CoF_2(py)_4Dy(hfac)_4\}$ complex encompassing diamagnetic Co(III), the INS spectra feature a single peak at ca. 13 cm^{-1} corresponding to the excitation from the electronic ground state to the first excited ligand-field state (cf. Fig. 11.7, left panel). Interestingly, in the isostructural $\{CrF_2(py)_4Dy(hfac)_4\}$, a peak at the same energy, but significantly broadened, is present. It is reasonable to assume that the apparent broadening is associated with the introduction of weak Cr—Dy exchange interactions. The magnetic interactions are also evident from comparing the temperature dependence of the χT product (cf. right panel of Fig. 11.7) for $\{CrF_2(py)_4Dy(hfac)_4\}$ with that of $\{CoF_2(py)_4Dy(hfac)_4\}$. Both χT products experience a gradual decrease from room temperature till 2K. In $\{CoF_2(py)_4$-$Dy(hfac)_4\}$, this behavior is attributed to depopulation of excited ligand-field states of the $J = 15/2$ ground multiplet, whereas Cr(III)—Dy(III) exchange interactions likely also influence the χT product for $\{CrF_2(py)_4Dy(hfac)_4\}$. The summed χT product of the isostructural $\{CrF_2(py)_4La(hfac)_4\}$, where La(III) is diamagnetic, and the χT product of $\{CoF_2(py)_4Dy(hfac)_4\}$ is always larger than the χT product of $\{CrF_2(py)_4Dy(h$-$fac)_4\}$. This observation suggests the presence of antiferromagnetic Cr(III)—Dy(III) interactions in $\{CrF_2(py)_4Dy(hfac)_4\}$. As suggested by the strong similarity of the INS spectra of the two compounds and their isostructurality, it is expected that the ligand-field contribution to the χT product is similar for the two compounds. Thus,

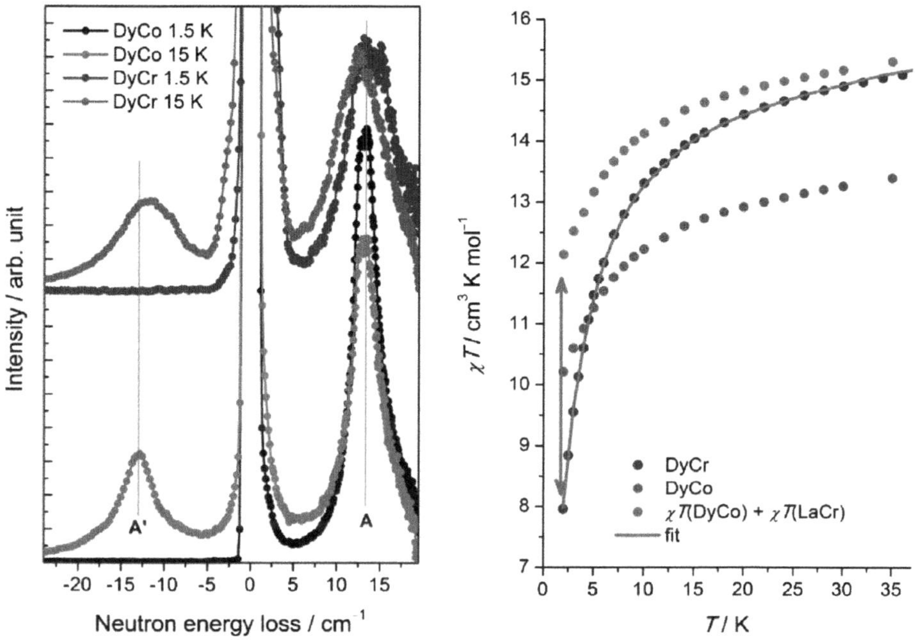

Figure 11.7 Left: Inelastic neutron scattering spectra of $\{CoF_2(py)_4Dy(hfac)_4\}$ ("DyCo") and isostructural $\{CrF_2(py)_4Dy(hfac)_4\}$ ("DyCr") at selected temperatures. Right: temperature dependence of the χT product for "DyCr" and "DyCo" together with $\{CrF_2(py)_4La(hfac)_4\}$ incorporating diamagnetic lanthanum(III).

subtracting the ligand-field contributions allows the fitting of the χT product of $\{CrF_2(py)_4Dy(hfac)_4\}$ to yield $J_{Dy-Cr} \approx 0.1\ cm^{-1}$ (treating the Dy^{3+} ion as a $J = 15/2$).

The temperature dependence of the χT product for a slightly extended version of the dinuclear systems, the trinuclear {DyCrDy} complex, $[Dy(hfac)_3(H_2O)\{CrF_2(py)_4\}Dy(hfac)_3(NO_3)]$ (cf. Fig. 11.3B), is shown in Fig. 11.8. The dominant decrease of the χT product on lowering temperature cannot necessarily, as before, be ascribed to intramolecular antiferromagnetic interactions as it might as well result from depopulation of ligand-field states or merely a combination of the two effects. In this case, isolation of an analog with diamagnetic substitution proved impossible and thereby highlights that the aforementioned approach of estimating magnetic exchange interactions has its inherent limitations. In the following, we will describe an alternative way to acquire information on the sign and magnitude of the magnetic exchange interactions in mixed 3d-4f clusters. Whereas computational approaches are now widely applied to determine exchange coupling constants, with some progression in success, solid experimental evidence for the nature and magnitude or magnetic interactions in such systems is exceedingly scarce in literature. One possible way to obtain direct information on the relative spin orientations of metal ions in heterometallic clusters is by X-ray magnetic circular dichroism (XMCD) spectroscopy. At the expense of requiring synchrotron sources, this technique provides element-specific magnetization data requiring minimal modeling and very small sample sizes. The XMCD is obtained as the difference between the X-ray absorption spectra at an absorption edge obtained with left and right circularly polarized X-ray photons in a magnetic field. Fundamentally, the magnitude of the XMCD is proportional to the magnetization of the probed element. The element-specific XMCD magnetization curves for Dy(III) and Cr(III) are shown in Fig. 11.8B and A, respectively [21]. The large magnetic moment of Dy(III) will align parallel to the polarizing field with the sacrifice of the much smaller magnetic moment of Cr(III). The Cr(III) magnetization has a peculiar "wiggle" behavior where the low-field magnetization is negative and then switches to positive when the polarizing magnetic field becomes sufficiently large. These observations can be rationalized by an antiferromagnetically coupled ground state. At low magnetic fields, the Cr(III) magnetic moment is antiparallel to the Dy(III) magnetic moment resulting in a negative Cr(III) magnetization. As the interactions are relatively weak, the application of a

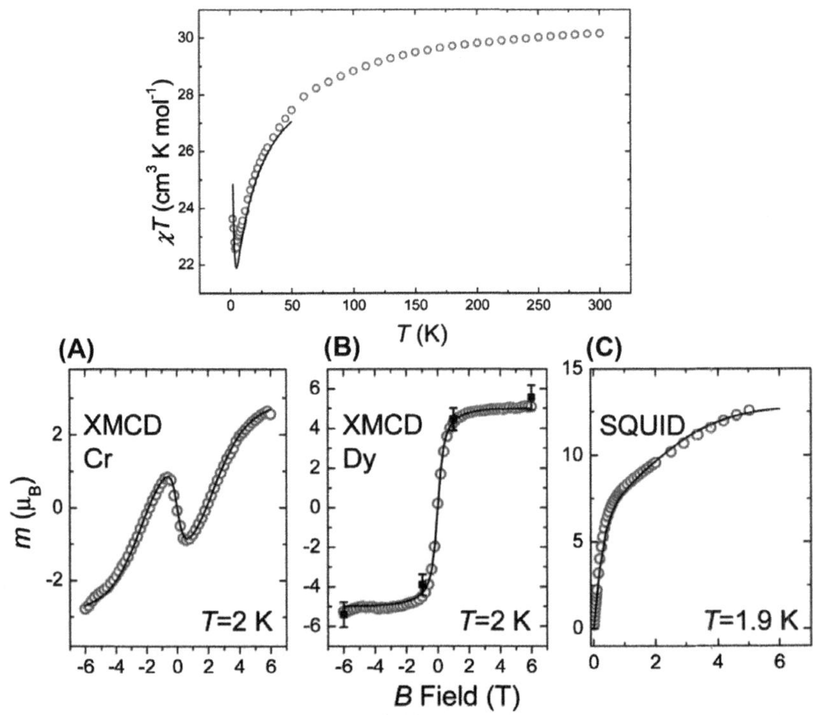

Figure 11.8 Top: Temperature dependence of the χT product for the trinuclear {DyCrDy} complex discussed in the main text. Bottom: Element-specific magnetization data for Cr(III) (A) and Dy(III) (B). In (C) is shown the bulk SQUID magnetization. The *solid lines* are best fits as described in Ref. [21].
Reproduced with permission from J. Dreiser, K.S. Pedersen, C. Piamonteze, S. Rusponi, Z. Salman, M.E. Ali, M. Schau-Magnussen, C.A. Thuesen, S. Piligkos, H. Weihe, H. Mutka, O. Waldmann, P. Oppeneer, J. Bendix, F. Nolting, H. Brune, Chem. Sci. 3 (2012) 1024.

modest magnetic field forces the Cr(III) moment to flip and become parallel to the Dy(III) moments. These rather simple observations directly evidence the presence of antiferromagnetic Dy(III)−Cr(III) interactions and the involved energy in "switching" the magnetization of the Cr(III) ion. These data have also been modeled quantitatively at different levels of sophistication, and the interested reader is referred to Ref. [21].

4. Improving Magnetic Coolers

The adiabatic demagnetization has received steeply growing attention from the molecular magnetism community fueled in part by the practical application for cooling purposes [2]. Recent reports have indicated that molecule-based materials may be superior to the purely inorganic, three-dimensional materials currently employed. The cooling potential of a molecule is intimately linked to the magnetic entropy change following an adiabatic demagnetization. For a general ground-state spin, S_T, the magnetic entropy is given by $S_m = R\ \ln(2S_T + 1)$, where R is the gas constant and $2S_T + 1$ is the degeneracy of the spin ground state. This expression naturally fueled the curiosity of synthetic chemists to synthesize molecular clusters with giant ground state spins. However, as the performance of a magnetic refrigerant is normally quantified by unit of mass, the S_T should be maximized without an explosion of the molar mass. On the other hand, making low-weight molecules favors light-weight bridging ligands like fluoride, but also necessarily means limiting the use of capping ligands separating the molecules. As a result, magnetic long order may result from significant interactions between adjacent molecules causing a precipitous drop in the cooling performance. As a matter of fact, Gd(III) compounds have received the main interest due to its large spin of $S_{Gd(III)} = 7/2$ and its vanishing magnetic anisotropy both of which provide for potent molecular coolers. An extremely large S_T is secured by intramolecular ferromagnetic interactions, thus the aforementioned $\{Cr_2Gd_3\}$ cluster exhibiting Gd(III)−Cr(III) ferromagnetic interactions is a promising molecular cooler. Confident determination of weak interactions in such a relatively complicated system is often difficult. In this regard, the study of the isostructural $\{Ga_2Gd_3\}$ incorporating diamagnetic Ga(III) provides important help. The χT product of $\{Ga_2Gd_3\}$ exhibits a slight increase from room temperature to 2K pointing toward ferromagnetic Gd(III)−Gd(III) interactions within the $\{Gd_3\}$ triangle (Fig. 11.9, $J_{Gd-Gd} = -0.028\ cm^{-1}$) [18]. As all the $\{M_2Gd_3\}$ (M = Cr, Fe, Ga) are isostructural, it is reasonable to believe that this parameter can be considered common for the entire series. The assumption of transferability of the Gd(III)−Gd(III) interaction strength allows for the determination of the magnitude of Gd(III)−Cr(III) interaction yielding $J_{Gd-Cr} = -0.046\ cm^{-1}$. The all-ferromagnetic interactions ensure the stabilization of the maximum possible spin as S_T, being $2S_{Cr} + 3S_{Gd} = 27/2$.

Such a large spin ground state could exhibit a pronounced magnetic cooling potential. Experimentally, the magnetocaloric effect is determined by isothermal magnetic entropy changes $-\Delta S_m$ obtained from the entropy data extracted from heat capacity measurements through $S_m(T) = \int C_m/T\,dT$. It may as well be evaluated from

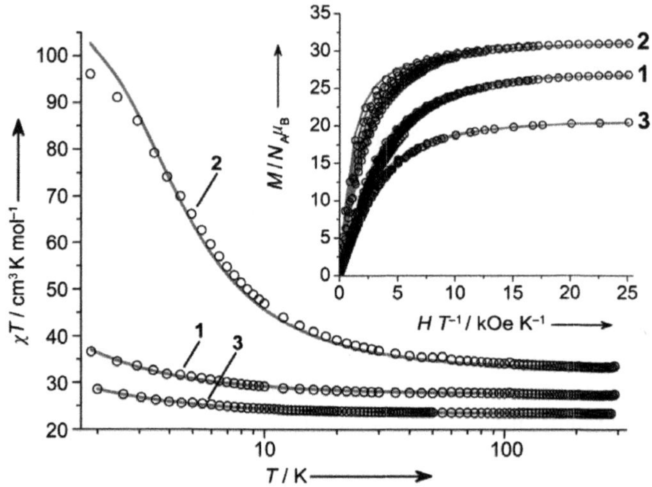

Figure 11.9 Magnetic susceptibility (shown as χT) for the $\{Gd_3M_2\}$ [M = Cr (**1**), Fe (**2**), Ga (**3**)] cluster series. The inset shows the field dependence of the magnetization plotted against the reduced variable HT^{-1}. The *solid lines* are best fits to the spin-Hamiltonian as described in the main text.
Reproduced with permission from K.S. Pedersen, G. Lorusso, J.J. Morales, T. Weyhermüller, S. Piligkos, S.K. Singh, D. Larsen, M. Schau-Magnussen, G. Rajaraman, M. Evangelisti, J. Bendix, Angew. Chem. Int. Ed. 53 (2014) 2394.

magnetization data using the Maxwell relation $\Delta S_m(T) = \int[\partial M(T,B)/\partial T]_B\,dB$. Indeed, the two approaches give almost perfect agreement as evident from Fig. 11.10. The maximum $-\Delta S_m$ for $\{Gd_3Cr_2\}$ of 38.3 J/kg per K is obtained at $T = 2.0$K for the field change 7 T → 0 T. This value is very large, indeed, and still remains one of the highest obtained for a molecular system [22].

Based on these results, the successful preparative extension to the Fe(III) analog $\{Gd_3Fe_2\}$ with its higher spin becomes highly relevant. The χT product of $\{Gd_3Fe_2\}$ is shown in Fig. 11.9 (labeled as **2**). The increase in χT upon descending temperature from room temperature is much more dramatic than for the $\{Gd_3Ga_2\}$ and $\{Gd_3Cr_2\}$ analogs. Qualitatively, the behavior is reminiscent of ferromagnetic Gd(III)–Fe(III) interactions which would lead to a very large spin ground state of $S_T = 31/2$. The expected saturation of χT is therefore $g^2S_T(S_T + 1)/8 = 128\ \text{cm}^3\ \text{K/mol}$ which is slightly higher than the experimental χT of 96 cm^3 K/mol at 2K. Fitting of the χT to the previously described spin-Hamiltonian affords $J_{Gd\text{–}Fe} = -0.25\ \text{cm}^{-1}$. This value is significantly larger than the vanishing $J_{Gd\text{–}Cr}$ in the $\{Gd_3Cr_2\}$ cluster indicating that additional pathways for magnetic exchange coupling are present in $\{Gd_3Fe_2\}$. To illustrate this point we chose an approach similar to the one employed for Gd(III)–Cr(III) systems. However, the simple dinuclear {GdCr} complex has no isolatable Fe(III) analog. Therefore, Cr(III) was exchanged with Fe(III) *in silico*, and a series of DFT calculations as a function of the Fe–F–Gd angle was performed. The results are shown in Fig. 11.11B together with the previously discussed angular

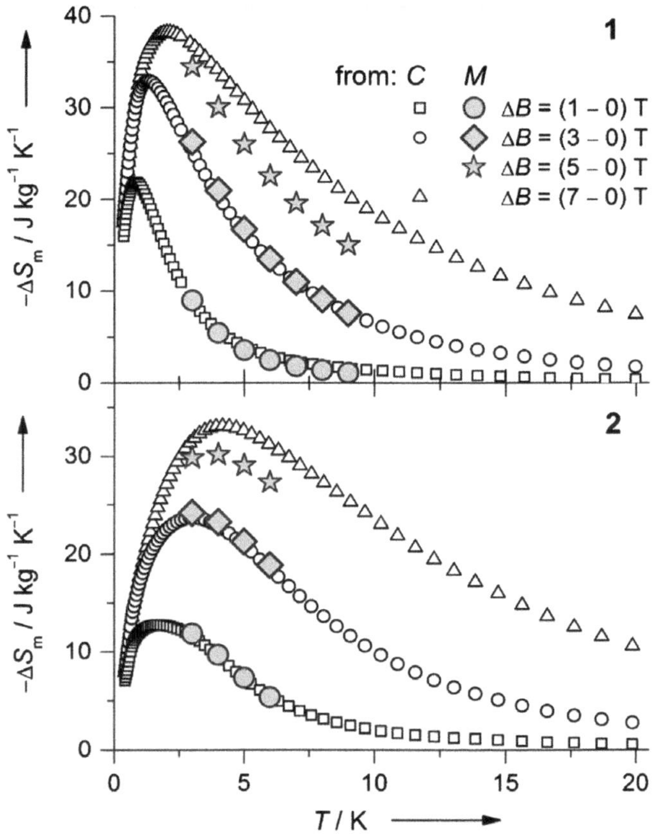

Figure 11.10 Field-inducted magnetic entropy changes for $\{Gd_3Cr_2\}$ (**1**) and $\{Gd_3Fe_2\}$ (**2**) as function of temperature. *Open symbols* are extracted from heat capacity measurements, and the filled symbols are data derived from magnetization measurements.
Reproduced with permission from K. S. Pedersen, G. Lorusso, J.J. Morales, T. Weyhermüller, S. Piligkos, S. K. Singh, D. Larsen, M. Schau-Magnussen, G. Rajaraman, M. Evangelisti, J. Bendix, Angew. Chem. Int. Ed. 53 (2014) 2394.

dependence of the J_{Gd-Cr} coupling constant. The difference in the sign for the exchange coupling between Gd(III) and Fe(III) compared to the Gd(III)–Cr(III) can be traced to the importance of charge transfer from the 3d to the empty 5d orbitals of Gd(III). This mechanism provides an important ferromagnetic contribution to the exchange coupling. It becomes more significant in the Fe(III) case due to the population of the 3d-e_g orbitals in high-spin Fe(III) providing simultaneously better overlap with the Gd(III) 5d orbitals and a smaller energetic separation.

From the larger S_T of 31/2 for $\{Gd_3Fe_2\}$, this system could be expected to exhibit an even larger magnocaloric effect than found in $\{Gd_3Cr_2\}$. Discouragingly, this is not the case. The field dependence of $-\Delta S_m$ is shown in the lower panel of Fig. 11.10. The maximum $-\Delta S_m$ is found to be 33.1 J/kg per K at $T = 4.2$K, significantly lower

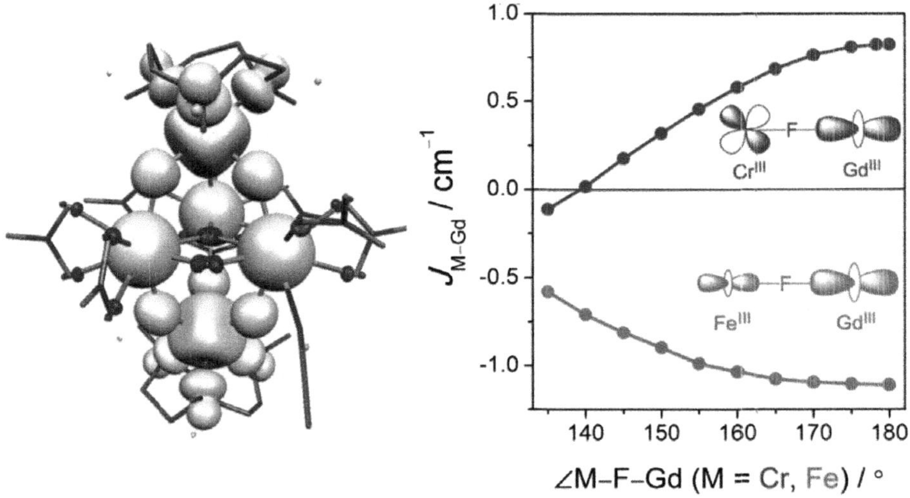

Figure 11.11 Left: Ground-state spin-density plot of $\{Gd_3Fe_2\}$ showing the all-ferromagnetic alignment of the spins. Right: Angular dependence of the Gd(III)–Fe(III) exchange coupling constant. The corresponding correlation for the Gd(III)–Cr(III) exchange interaction is shown for comparison.
Reproduced with permission from K. S. Pedersen, G. Lorusso, J.J. Morales, T. Weyhermüller, S. Piligkos, S. K. Singh, D. Larsen, M. Schau-Magnussen, G. Rajaraman, M. Evangelisti, J. Bendix, Angew. Chem. Int. Ed. 53 (2014) 2394.

than the calculated maximal value of 47.1 J/kg per K. As previously mentioned, the maximization of the S_T has been considered the holy grail for the synthesis of improved molecular coolers, but the story on $\{Gd_3Cr_2\}$ and $\{Gd_3Fe_2\}$ demonstrates an important point. Despite the larger ground state in the latter, its cooling performance is significantly reduced as compared to the former due to the energetic dispersion induced by the nonvanishing Fe(III)–Gd(III) interactions. Conversely, in the former, the vanishing Gd(III)–Gd(III) and Gd(III)–Cr(III) exchange interactions result in a high density of excited states, energetically very close to the ground state, that are easily populated and depopulated by the magnetic field. This demonstrates very palpably how material properties are not solely determined by the electronic ground state but all thermally accessible excited states and that design criteria for magnetic coolants need to be extended beyond maximization of ground-state spin values.

5. Outlook and Conclusions

As emphasized in this chapter, the fluoride ion is a highly useful ligand for rational synthetic assembly of mixed 3d-4f clusters. Monoatomic bridges between transition metal and lanthanide ions in clusters are scarce in literature and essentially limited to chloride and oxide bridges. The high basicity of oxide complicates preparations of such species, and the only examples are limited to transition metal ions in the higher oxidation states. The monoatomic bridges are, however, appealing not only from the viewpoint of the synthetic chemist. Due to the redox-innocence, fluoride and in particular systems with unsupported bridging by fluoride provide ideal possibilities to test theoretical models of magnetic exchange interactions against experimentally determined properties. In systems where the monoatomic bridge is supported by, *eg*, carboxylate bridging, such deconvolution of contributions from different ligands acting in parallel is practically impossible. Starting from the simplest possible, a dinuclear {GdCr} complex, the determining mechanisms leading to magnetic communication between the two paramagnetic centers were determined and shown to have strong implications for the design of molecular magnetic refrigerants. It is highly possible that further chemical tuning of the presented systems could result in recording large cooling effects.

If we for a moment leave the lanthanide series and look upon the role of fluoride in molecular magnetism in general, several interesting results have been reported over the last few years. We recently reviewed the broad literature of fluoride coordination chemistry in the field of molecular magnetism and have no intention to repeat ourselves here [6]. It should, however, be emphasized that the approach of using robust fluoride complexes of *n*d transition element ions toward molecular (0D) or low-dimensional (1D and 2D) magnetic materials is quite generalizable to also encompass other metal ions than the lanthanide series.

Acknowledgments

K.S.P. thanks The Danish Research Council for a DFF Sapere Aude: Research Talent grant (4090-00201).

References

[1] [a] D. Gatteschi, R. Sessoli, J. Villain, Molecular Nanomagnets, Oxford University Press, Oxford, UK, 2006;
[b] R. Winpenny, Single-Molecule Magnets and Related Phenomena, vol. 122, Springer, Heidelberg, Germany, 2006;
[c] J. Bartolomé, F. Luis, J.F. Fernandez, Molecular Magnets, Springer, Heidelberg, Germany, 2014;
[d] C. Benelli, D. Gatteschi, Introduction to Molecular Magnetism. From Transition Metals to Lanthanides, Wiley-VCH, Weinheim, Germany, 2015.
[2] [a] M. Evangelisti, E.K. Brechin, Dalton Trans. 39 (2010) 4672;
[b] J.W. Sharples, D. Collison, Polyhedron 54 (2013) 91;
[c] M. Evangelisti, in: J. Bartolomé, F. Luis, J.F. Fernandez (Eds.), Molecular Magnets, Springer, Heidelberg, Germany, 2014, p. 365.
[3] R.G. Pearson, J. Am. Chem. Soc. 85 (1963) 3533.
[4] [a] F. Perdih, A. Demsar, A. Pevec, S. Petricek, I. Leban, G. Giester, J. Sieler, H.W. Roesky, Polyhedron 20 (2001) 1967;
[b] A. Pevec, M. Mrak, A. Demsar, S. Petricek, H.W. Roesky, Polyhedron 22 (2003) 575.

[5] A. McRobbie, A.R. Sarwar, S. Yeninas, H. Nowell, M.L. Baker, D. Allan, M. Luban, C.A. Muryn, R.G. Pritchard, R. Prozorov, G. Timco, F. Tuna, G.F.S. Whitehead, R.E.P. Winpenny, Chem. Commun. 47 (2011) 6251.
[6] K.S. Pedersen, M.A. Sørensen, J. Bendix, Coord. Chem. Rev. 299 (2015) 1.
[7] [a] J. Glerup, J. Josephsen, E. Michelsen, E. Pedersen, C.E. Schäffer, Acta Chem. Scand. 24 (1970) 247;
[b] P. Andersen, A. Døssing, J. Glerup, M. Rude, Acta Chem. Scand. 44 (1990) 346;
[c] A. Böttcher, H. Elias, J. Glerup, M. Neuburger, C.E. Olsen, H. Paulus, J. Springborg, M. Zehnder, Acta Chem. Scand. 48 (1994) 967;
[d] A. Böttcher, H. Elias, J. Glerup, M. Neuburger, C.E. Olsen, J. Springborg, H. Weihe, M. Zehnder, Acta Chem. Scand. 48 (1994) 981;
[e] K.S. Pedersen, G. Lorusso, J.J. Morales, T. Weyhermueller, S. Piligkos, S.K. Singh, M. Schau-Magnussen, G. Rajaraman, M. Evangelisti, J. Bendix, Angew. Chem. Int. Ed. 53 (2014) 2257.
[8] [a] B.F. Straub, F. Rominger, P. Hofmann, Inorg. Chem. 39 (2000) 2113;
[b] L.F. Jones, C.A. Kilner, M.P. de Miranda, J. Wolowska, M.A. Halcrow, Angew. Chem. Int. Ed. 46 (2007) 4073;
[c] D. Riou, F. Taulelle, G. Férey, Inorg. Chem. 35 (1996) 6392.
[9] K.S. Pedersen, J. Bendix, R. Clérac, Chem. Commun. 50 (2014) 4396.
[10] T. Birk, K.S. Pedersen, C.Aa. Thuesen, T. Weyhermüller, M. Schau-Magnussen, S. Piligkos, H. Weihe, S. Mossin, M. Evangelisti, J. Bendix, Inorg. Chem. 51 (2012) 5435.
[11] T. Birk, M.J. Magnussen, S. Piligkos, H. Weihe, A. Holten, J. Bendix, J. Fluor. Chem. 131 (2010) 898.
[12] J. Dreiser, K.S. Pedersen, T. Birk, M. Schau-Magnussen, C. Piamonteze, S. Rusponi, T. Weyhermüller, H. Brune, F. Nolting, J. Bendix, J. Phys. Chem. A 116 (2012) 7842.
[13] T. Birk, J. Bendix, Acta Cryst. E66 (2010) m121.
[14] [a] J. Glerup, C.E. Schaffer, J. Springborg, Acta Chem. Scand. A 32 (1978) 67;
[b] J. Holzbock, W. Sawodny, U. Thewalt, Z. Anorg. Allg. Chem. 626 (2000) 2563.
[15] K.S. Pedersen, C.A. Thuesen, M. Schau-Magnussen, J. Bendix, unpublished.
[16] [a] K. Anzai, K. Hatano, Y.J. Lee, W.R. Scheidt, Inorg. Chem. 20 (2006) 2337;
[b] S. Christie, S. Subramanian, L. Wang, M.J. Zaworotko, Inorg. Chem. 32 (1993) 5415;
[c] J.F. Berry, E. Bill, R. García-Serres, F. Neese, T. Weyhermüller, K. Wieghardt, Inorg. Chem. 45 (2006) 2027;
[d] M. Tilset, I. Fjeldahl, J.-R. Hamon, P. Hamon, J.-Y. Saillard, K. Costuas, A. Haynes, J. Am. Chem. Soc. 123 (2001) 9984.
[17] P. Knopp, K. Wieghardt, B. Nuber, J. Weiss, Z. Naturforsch. B 46 (1991) 1077.
[18] K.S. Pedersen, G. Lorusso, J.J. Morales, T. Weyhermüller, S. Piligkos, S.K. Singh, D. Larsen, M. Schau-Magnussen, G. Rajaraman, M. Evangelisti, J. Bendix, Angew. Chem. Int. Ed. 53 (2014) 2394.
[19] S.K. Singh, K.S. Pedersen, M. Sigrist, C.A. Thuesen, M. Schau-Magnussen, H. Mutka, S. Piligkos, H. Weihe, G. Rajaraman, J. Bendix, Chem. Commun. 49 (2013) 5583.
[20] [a] S.K. Singh, N.K. Tibrewal, G. Rajaraman, Dalton Trans. 40 (2011) 10897;
[b] S.K. Singh, G. Rajaraman, Dalton Trans. 42 (2013) 3623;
[c] G. Rajaraman, F. Totti, A. Bencini, A. Caneschi, R. Sessoli, D. Gatteschi, Dalton Trans. (2009) 3153.
[21] J. Dreiser, K.S. Pedersen, C. Piamonteze, S. Rusponi, Z. Salman, M.E. Ali, M. Schau-Magnussen, C.A. Thuesen, S. Piligkos, H. Weihe, H. Mutka, O. Waldmann, P. Oppeneer, J. Bendix, F. Nolting, H. Brune, Chem. Sci. 3 (2012) 1024.
[22] [a] J.-L. Liu, Y.-C. Chen, F.-S. Guo, M.-L. Tong, Coord. Chem. Rev. 281 (2014) 26;
[b] J.W. Sharples, D. Collison, Polyhedron 66 (2013) 15.

Unique Silver(II) Fluorides: The Emerging Electronic and Magnetic Materials

12

Z. Mazej
Jožef Stefan Institute, Ljubljana, Slovenia

D. Kurzydłowski
Centre of New Technologies, University of Warsaw, Warsaw, Poland; Faculty of Mathematics and Natural Sciences, Cardinal Stefan Wyszynski University in Warsaw, Warsaw, Poland

W. Grochala
Centre of New Technologies, University of Warsaw, Warsaw, Poland

Chapter Outline

Photonic and Electronic Properties of Fluoride Materials. http://dx.doi.org/10.1016/B978-0-12-801639-8.00012-X

1. Introduction

The properties of chemical elements are largely governed by their position in the periodic table: it is the everyday experience of chemists that elements from the same group of the periodic table show very similar properties. But the coinage metals, copper, silver, and gold, differ markedly in their physical properties, color, and chemical reactivity. Substantial differences persist also for the chemical compounds of these elements in various oxidation states ranging from −1 (eg, CsAu) to +5 (again gold, AuF_6^-). Of particular interest is the divalent (+2) state, which predominates in the chemistry of copper and is incredibly rare for gold. The Ag^{2+} cation is also uncommon and its compounds exhibit a broad range of peculiar physicochemical properties [1]. Silver, in its divalent state, is a very strong oxidizer (the Ag^{2+}/Ag^+ formal oxidation potential, $E^{0\prime}$, reaches +2.90 V in 33% oleum [2]), which can react with the majority of simple anionic ligands; indeed, the value of $E^{0\prime}$ approaches that of the $F_2/2F^-$ redox pair (+2.95 V) and that is one reason why Ag^{2+} is the best F_2-transfer catalyst known [3]. Therefore, not surprisingly, the chemistry of thermodynamically stable compounds of Ag^{2+} is essentially limited to compounds where this cation is found in a fluoride environment.

With the $4d^9$ valence electronic configuration of the Ag^{2+} cation and the significant covalence of the Ag(II)−F bond [4] the fluorides of divalent silver (fluoroargentates(II)) have much in common with oxides of Cu^{2+} [5], precursors of the well-known family of high-temperature superconductors. Both fluoroargentates(II) and oxocuprates(II) are classified as strongly correlated charge-transfer insulators and exhibit diverse magnetic properties owing to the presence of one unpaired electron in the d shell (spin-½ systems). On the basis of these similarities, it has been proposed that metallic conductivity or even superconductivity (SC) could be generated in fluoroargentates(II) [6,7].

In this chapter we present in more detail the structural, magnetic, and electronic properties of various Ag(II) fluorocompounds. The reader is referred to an extensive review about crystal structures and magnetic properties of silver(II) fluorides and theoretical predictions of hypothetical intermediate-valence quaternary fluoride Ag^{II}/Ag^{III} and Ag^{II}/Ag^{I} systems as potential superconductors published in 2001 [6]. The primary intention of this contribution is to summarize results of Ag(II)−F chemistry reported since 2001, with particular emphasis on B_2AgF_4 and $BAgF_3$ (B-alkali metal) systems. These compounds together with the relevant parent (binary) system, AgF_2, are described in Sections 5−7. In the next two sections we give an introduction to the Jahn−Teller effect (JTE) which is of paramount importance for the properties of silver(II) compounds, followed by a short description of magnetic and electronic properties of these systems.

2. The Jahn−Teller Effect

Chemical compounds containing Ag^{2+} or Cu^{2+} cations most often adopt an octahedral (ML_6) coordination. In case of fluorocompounds of Ag(II) there are only a few exceptions such as the pentagonal bipyramid (ML_7) coordination found in $AgFMF_6$ (M = As [8], Au [9], Ir [10]) and $K_3Ag_2Zr_4F_{23}$ [11], or the hexagonal bipyramid (ML_8) coordination exhibited by $AgCd_3M_3F_{20}$ [12] (M = Zr, Hf) as well as $Ag_3Hf_2F_{14}$ [13].

Due to the d^9 electronic configuration of Ag(II) its first coordination sphere is susceptible toward distortions arising from the JTE. In the case of octahedral complexes the JTE usually leads to a 2 + 4 coordination characterized by either elongation ($R_{ax} > R_{eq}$, see Fig. 12.1A) or shortening ($R_{ax} < R_{eq}$) of the ML_6 units. Crystal packing effects often lead to additional small changes in the lengths of equatorial bonds leading to 2 + (2 + 2) coordination, as found in B_2AgF_4 compounds discussed further, as well as departure of the F—Ag—F bond angles from 90 degree (as is the case for AgF_2, see Section 5.

The JTE leads to a considerable plasticity of the first coordination sphere of Ag(II) as shown in Fig. 12.1B. In case of elongated octahedra the difference between the length of the shortest apical and the longest equatorial Ag—F bond spans the range from 10% ($AgTiF_6$ [15]) to 34% for Na_2AgF_4 [16]. This can be compared to values of 8% (α polymorph of K_2AgF_4 [17]) up to 17% ($(AgF)_2AgF_4AsF_6$ [18]) for compressed coordination. The extreme case of elongation is the square-planar coordination

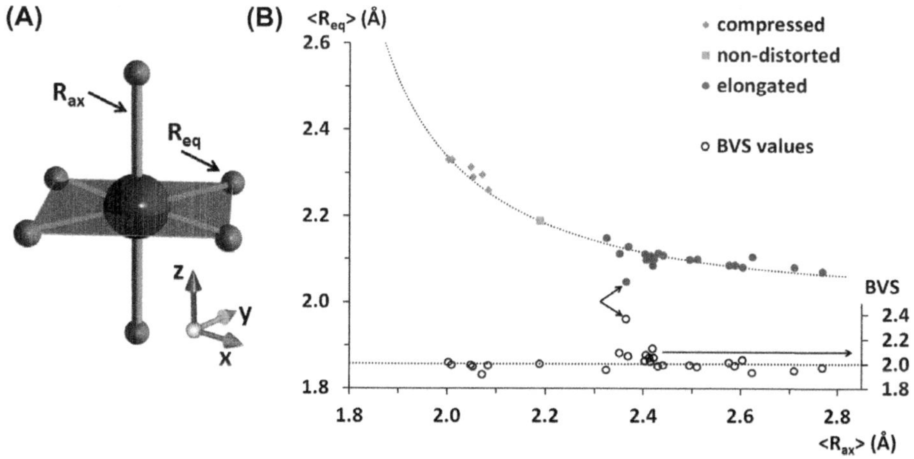

Figure 12.1 Octahedral coordination exhibited by the majority of Ag(II) fluorides (A)—*red* (gray in print versions)/*blue* (dark gray in print versions) *balls* depict F/Ag atoms. (B) Average R_{ax} and R_{eq} values for 23 Ag(II) fluorides (*red* (dark gray in print versions)/*green* (light gray in print versions)/*blue* (gray in print versions) *points* marking systems exhibiting compressed/nondistorted/elongated octahedra) together with the bond valence sum (BVS) values calculated for each Ag^{2+} center assuming R_0 equal to 1.782 Å (for details see text). The *lower dotted line* marks BVS = 2, while the *upper hyperbolic dotted line* corresponds to R_{eq} versus R_{aq} dependence calculated with the assumption of a 2 + 4 coordination of Ag(II) and BVS equal to 2. Two *short arrows* indicate values derived for $Ag(TaF_6)_2$ (Ref. [14]) which depart substantially from the trend lines, calling for the crystal structure of this compound to be revisited.

of Ag(II) cation which can be found in $ZnZr_2Ag_2F_7$ (containing $Ag_2F_7^{3-}$ dimers) [19] or $Ag(MF_4)_2$ (M = Ag, Au) [20]. It is noteworthy to point that the plasticity of the first coordination sphere of Ag(II) (due to *one* unpaired electron) resembles the behavior of principal + secondary interactions (due to *two* paired electrons) typical of hypervalent compounds [21], such as XeF_2 [22]. The JTE for Ag^{2+} thus deserves to be labeled as "strong."

The relationship between axial and equatorial Ag(II)–F bond lengths can be well reproduced with the use of the BVS index [23] (Fig. 12.1B). This index is parameterized by two constants: B and R_0. In most cases B is assumed to be equal to 0.37 while R_0 depends on the types of interacting ions (here Ag^{2+} and F^-). By analyzing the bond lengths of over 30 fluoroargentates(II) we have established the value of R_0 as equal to 1.782 Å, which is somewhat lower than the value previously proposed for the Ag^{2+}/F^- pair (1.8 Å [24]). The BVS indices calculated for various fluoroargentates(II) range from 1.88 to 2.13 which is close to the expected value of 2 for divalent silver. Only in the case of $Ag(TaF_6)_2$ [14] does the BVS value deviate significantly (BVS = 2.37) indicating a need for the redetermination of its crystal structure.

The driving force for the appearance of the JTE is the energy stabilization associated with the distortion of the regular octahedron (E_{JT}). The magnitude of E_{JT} can be derived by spectroscopic measurements and has been estimated at 0.25 eV per Cu^{2+} center in K_2CuF_4 [25] and 0.40 eV/Ag^{2+} for isostructural Cs_2AgF_4 [26]. It has been argued that at high pressures (ie, exceeding 1 GPa) this stabilization may be small as compared to other contributions to the enthalpy, and in consequence the JTE may disappear at such conditions thus leading to metallic behavior of d^9 systems. Recent experiments on $CuWO_4$ show, however, that although for this system the JTE diminishes upon compression, the appearance of a new high-pressure phase leads to a partial restoration of the distortion [27]. This hints at the possible persistence of the JTE even at compression exceeding 40 GPa. Similar robustness of the JTE upon compression has been found for silver(II) sulfate ($AgSO_4$) up to a pressure of ~30 GPa [28].

In this context the apparently regular coordination of Ag^{2+} (ie, JTE quenching) found by X-ray diffraction in $Ag(IF_6)(SbF_6)_3$ [29] is puzzling. As in the case of many Cu^{2+} systems [30] this is most likely the effect of static and/or dynamic disorder. This notion is supported by the fact that X-ray absorption fine structure measurements hint at an elongated coordination of Ag(II) centers with $R_{eq} = 2.11$ Å and $R_{ax} = 2.40$ Å giving BVS(Ag^{2+}) = 2.03.

It is noteworthy to emphasize that the two possible scenarios of the Jahn–Teller distortion in d^9 systems lead to different ground states. Axial elongation leads to a state with B_{1g} symmetry and half-occupation of the $d(x^2-y^2)$ orbital, while compression results in the unpaired electron occupying the $d(z^2)$ orbital (A_{1g} symmetry). As can be seen from Fig. 12.2 the resultant spin density distribution is markedly different. It must be stressed that in the original framework of the Jahn–Teller theory both axial compression and elongation constitute equally probable distortion scenarios. In reality though, the former one is much less frequent in fluoroargentates(II), as evident from Fig. 12.1B. The same applies to the much better-studied compounds of Cu(II) [30]. The reasons for the stabilization of an elongated octahedron with respect to a

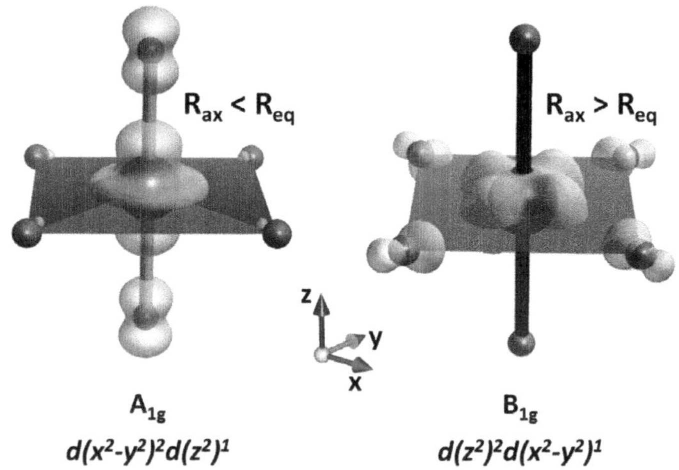

Figure 12.2 Two possible scenarios of the Jahn–Teller distortion: axial compression as found in $AgFBF_4$ [8] (left) and elongation as found in $AgSnF_6$ [15] (right). An isosurface of the spin density (0.1 $Å^{-1}$) is shown in *yellow* (light gray in print versions).

compressed one in d^9 systems can be traced back to the facile mixing between the unoccupied *s* metal orbital and $d(z^2)$ which results in additional stabilization of the $d(z^2)$ orbital with respect to $d(x^2-y^2)$, and thus favors full occupation of the former.

3. The Collective Jahn–Teller Effect

For an isolated AgL_6 complex the Jahn–Teller distortion can take effect with equal probability along each of the fourfold symmetry axes (*x*, *y*, *z* in Fig. 12.1A), leading to an either elongated or compressed octahedron. But when a Jahn–Teller active center constitutes a part of an extended crystal lattice then the JTE is subject to (usually anisotropic) constraints from the crystal lattice. Moreover, in magnetically dense solids which are rich in JT-active centers there are mutual interactions and restrains between the centers, which propagate through the entire lattice. This phenomenon usually leads to stabilization of only one particular type of distortion (elongation/contraction) in one given pattern [30,31]. This is termed as collective (or cooperative) JTE (cJTE).

A special case of the cJTE is found in systems where Jahn–Teller active complexes are interconnected by sharing common ligands. The most classical example is $KCuF_3$ which adopts the perovskite type structure with every $CuF_6{}^{4-}$ octahedron sharing one edge with six other such units thus forming a framework of $[CuF_{6/2}]^-$ stoichiometry. For this compound the operation of the cJTE leads to stabilization of two distinct distortion arrangements [32] ("a" and "d"—see Fig. 12.3). In both of them the direction of the Jahn–Teller elongation is perpendicular for octahedra neighboring in the *ab* plane—an arrangement termed as an *antiferrodistortive* (AD) ordering. This ordering leads effectively to formation of $[CuF_{2/2+2}]^-$ chains running along *c* which are

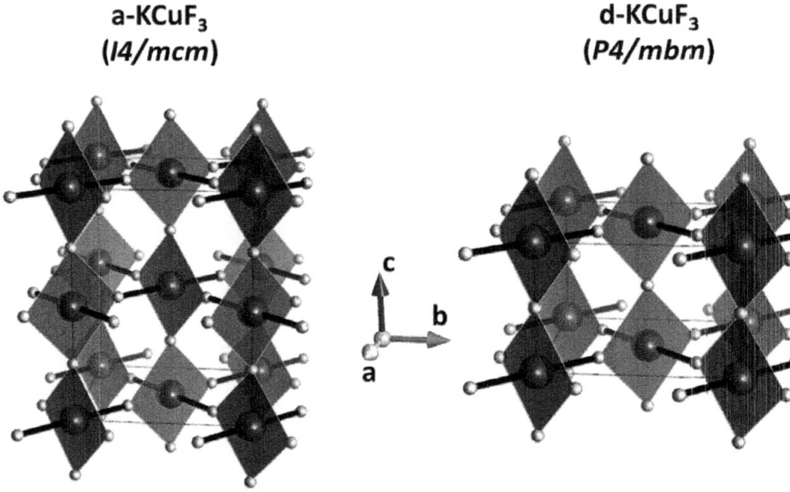

Figure 12.3 Two types of collective (or cooperative) Jahn–Teller effect found in different polymorphs of $KCuF_3$. The axes of the Jahn–Teller elongation are shown in black. Potassium atoms are omitted for clarity.

composed of CuF_4^{2-} plaquettes. The difference between the two phases arises in the stacking along *c*—the elongated axes of neighboring octahedra are parallel for d-$KCuF_3$ (*ferrodistortive* (FD), ordering along *c*) while they are perpendicular in a-$KCuF_3$ (AD ordering along *c*).

Layered perovskites with a general formula B_2MX_4 (B/X—closed shell cation/anion; M—Jahn–Teller active cation) are another class of compounds where cJTE plays an important role. These systems can be described as formed by layers of $MX_{4/2+2}$ stoichiometry separated by B_2 spacers. Among different structural arrangements that are possible for the $MX_{4/2+2}$ sheets three are of considerable interest: (1) FD ordering of elongated octahedra (FD-E); (2) FD arrangement of compressed octahedra (FD-C); and (3) AD ordering of elongated octahedra (AD-E)—see Fig. 12.4.

The FD-E scenario is found in many layered oxocuprates(II), notably in La_2CuO_4 [33], a generic precursor toward copper-based superconductors. On the other hand, layered fluorocuprates(II) (eg, K_2CuF_4 [34]) most often exhibit AD-E ordering. The FD-C is much less common in d^9 layered systems—up to this date it has been proposed to exist only in α-K_2AgF_4 [17], and, as we will show, this assignment is still a matter of debate.

It is noteworthy to point out that the abovementioned idealized arrangements are often modified by buckling of the $M_{4/2+2}$ planes which results from the mismatch between the size of M and B (elegantly described by the Goldshmidt tolerance factor [35]). A varying degree of such tilting is observed both in La_2CuO_4 [33], as well as α-K_2AgF_4 [17].

It is important to note that the structural differences between FD-C and AD-E orderings are subtle and difficult to pinpoint from structural measurements. As both arrangements are characterized by short apical M–X bonds the AD-E scenario can be simply obtained from FD-C by a small displacement (0.15 Å in the case of K_2CuF_4 [34]) of equatorial ligands from the midpoint between two M atoms. However, as

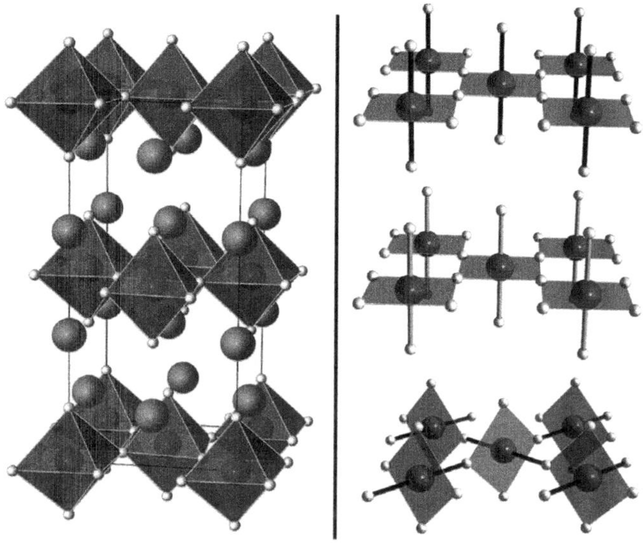

Figure 12.4 The B_2MX_4 layered perovskite (left) together with three collective (or cooperative) Jahn–Teller effect scenarios (right). *Red* (white in print versions)/*green* (gray in print versions)/*blue* (dark gray in print versions) *balls* depict X/B/M atoms. The axis of elongation/compression is marked by a *black/blue* (light gray in print versions) *cylinder.*

will be discussed in the next sections the structural distortions arising from cJTE greatly influence the magnetic properties of d^9 systems. In Section 6 we will also show in greater detail the energetic preference toward different cJTE scenarios in the B_2AgF_4 systems (B = Na–Cs).

4. General Magnetic and Electronic Properties of Fluoroargentates(II)

The magnetic coupling between two spin sites (*i* and *j*) can be described with the use of the Heisenberg Hamiltonian:

$$\boldsymbol{H}_{ij} = -J_{ij}\boldsymbol{S}_i\boldsymbol{S}_j$$

where $\boldsymbol{S}_i$ and $\boldsymbol{S}_j$ are spin operators and J_{ij} is the magnetic coupling constant. The absolute value and sign of this constant (often abbreviated by *J*) is indicative of the type of magnetic interactions between the sites: $J < 0$ indicates antiferromagnetic (AFM) coupling, while $J > 0$ translates to ferromagnetic (FM) interactions.

In the case of insulating compounds of transition metals the interaction between unpaired spins is most often mediated by the superexchange mechanism first proposed by Anderson [37] and later extended by Goodenough and Kanamori. The Goodenough–Kanamori–Anderson (GKA) rules allow to estimate the value of *J* along a given superexchange route based on its geometry. These rules were later reformulated in the language of orbital overlap by Hoffmann and coworkers [38], who showed that good overlap between the orbitals containing the unpaired electron leads to strong AFM interactions while weak overlap most often results in weak ferromagnetism. Based on this principle one should expect both a- and d-$KCuF_3$ (Fig. 12.3) to exhibit 1D AFM properties while 2D AFM coupling should be found for La_2CuO_4 characterized by FD-E cJTE ordering (Fig. 12.4). This is indeed in accordance with experimental findings. The relationship between the structure of fluoroargentates(II) and their magnetic properties will be discussed in length in the following sections.

Turning to electronic properties we recall that fluoroargentates(II) belong to a class of charge-transfer insulators, where the top of the valence band has mainly ligand character, while the bottom of the conduction band is formed by empty metal *d* states [39] (for AgF_2 see Fig. 12.5). This is in analogy to oxocuprates(II), and in contrast to

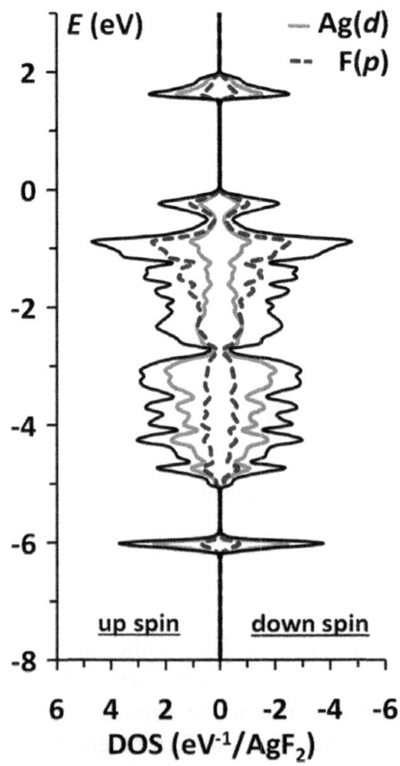

Figure 12.5 Electronic density of states (DOS) of AgF_2 in the vicinity of the Fermi level calculated at the density functional theory + *U* level [36] ($U = 5$ eV, $I = 1$ eV, see text) [39]. The black lines indicate the total DOS, while *red* (gray in print versions)/*blue* (light gray in print versions) ones show the partial contributions from *p* orbitals of F or *d* orbitals of Ag.

fluorocuprates(II) which are best described as Mott–Hubbard insulators [40]. Also, there is a substantial mixing of the F(2*p*) and Ag(4*d*) states in a broad energy window which is at the origin of strong superexchange.

Unfortunately due to the very strong oxidizing power of the Ag^{2+} cation, attempts to measure the electric conductivity of these compounds with *dc* contact methods have so far failed. Even gold wires attached to the samples of the fluoroargentates may be oxidized on their surface. Because of that, the noncontact microwave cavity perturbation method has been used to determine the *ac*-dielectric response of AgF_2 and of $KAgF_3$ [41]. It was found that brown AgF_2 is an insulator at ambient temperature and its electric resistivity slightly increases below the Curie point [41]. On the other hand, $KAgF_3$ has been suggested to be metallic at room temperature but insulating below 68K [41]. A recent impedance spectroscopy study utilizing *ac* conductivity has confirmed that AgF_2 is indeed an insulator with specific conductivity of *ca.* 10^{-10} S/cm at room temperature and an activation energy for conductivity of 56.7 kJ/mol [42]. As we will discuss later, the metallic character of $KAgF_3$ at room temperature is uncertain in view of the most recent findings.

Difficulties in performing resistivity measurements triggered theoretical studies of the band gap in fluoroargentates(II). The electronic structure of these compounds can be calculated with the use of the density functional theory (DFT) method with the inclusion of the on-site Coulomb repulsion parameter (DFT + U). Two main implementations of this method have been proposed by Liechtenstein et al. [36] and Dudarev et al. [43]. In the first both the on-site repulsion (U), as well as the exchange parameter (I, most often marked as J, but here we will use I to distinguish it from the magnetic superexchange constant) parametrize the electronic Hamiltonian, and in the latter approach only $U_{eff} = U - I$ enters into the equations. Calculations utilizing the formulas proposed by Liechtenstein helped to establish that the fair description of the electronic structure of Ag(II) bearing compounds is obtained for $U = 5$ eV and $I = 1$ eV (ie, $U_{eff} = 4$ eV) [39,44,45]. The DFT + U direct bands gaps obtained for fluoroargentates(II) (which largely correspond to the ligand-to-metal charge-transfer optical absorption) range from 1.5/1.6 eV ($KAgF_3$ [44]/AgF_2 [39]) to 2.5 eV (Cs_2AgF_4 [45]). The latter value agrees reasonably well with the optical band gap of 2.2 eV obtained from experiment [46].

Apart from the charge-transfer absorptions the local *d*–*d* electronic transitions are also observed for fluorides of divalent silver. These are most often located at lower energies of the visible spectrum than those of charge-transfer excitations. For d^9 ions in tetragonal (D_{4h}) coordination three *d*–*d* transitions are possible (E_{1-3}—see Fig. 12.6) and their energy can be used for the calculations of both the crystal field splitting parameter (Δ), as well as the energies associated with the Jahn–Teller distortion ($\delta_1 = E_{JT}$, δ_2 [47]). Due to the extreme reactivity of fluoroargentates(II) the UV–VIS spectra of only a handful compounds have been measured [26,46,48]; the obtained δ_1, δ_2, and Δ parameters are summarized in Table 12.1. As can be seen from Table 12.1 the $[AgF_6]$ complexes are characterized by a larger Δ (1.2–1.5 eV) than $[CuF_6]$ ones (*ca.* 0.8–0.9 eV). Also the energy associated with the Jahn–Teller distortion ($\delta_1 = E_{JT}$) is typically larger for Ag(II)/F than for Cu(II)/F bearing systems (0.26–0.58 eV *vs.* 0.25–0.31 eV).

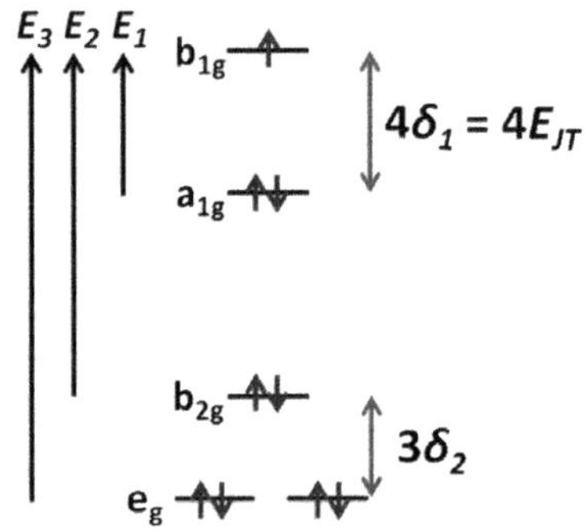

Figure 12.6 Correlation diagram for a d^9 ion in an elongated octahedral environment of D_{4h} symmetry. *Black arrows* indicate possible *d*–*d* electronic transitions.

The experimental reports for an SC in the fluoroargentates(II) are limited to one case only. In 2004, observations were reported of sudden drops in the magnetic susceptibility of a large number of samples in the BeF_2/AgF_2 system, over temperatures ranging from 8.5 to 64K [49]. These magnetic anomalies, strongly reminiscent of the Meissner–Ochsenfeld effect, suggested the presence of SC in a small sample fraction

Table 12.1 Values of δ_1, δ_2, Δ (in eV) for selected fluoroargentates(II), as well as for $KCuF_3$, K_2CuF_4, Ba_2CuF_6, Ba_2CuWO_6, and $Ba_2Cu(OH)_6$

Compound	δ_1	δ_2	Δ	References
$SrAgF_4$	0.58	0.22	1.24	[26]
$BaAgF_4$	0.57	0.23	1.23	[26]
Cs_2AgF_4	0.40	0.18	1.42	[26]
Rb_2AgF_4	0.39	0.15	1.42	[26]
$Ag(SbF_6)_2$	0.31	0.17	1.42	[48]
$AgTiF_6$	0.30	0.17	1.26	[26]
$AgZrF_6$	0.29	0.19	1.49	[26]
$AgSnF_6$	0.26	0.14	1.25	[26]
$KCuF_3$	0.25	0.07	0.81	[47]
K_2CuF_4	0.25	0.16	0.87	[25]
Ba_2CuF_6	0.31	0.11	0.86	[47]
Ba_2CuWO_6	0.27	0.12	0.86	[47]
$Ba_2Cu(OH)_6$	0.46	0.09	0.97	[47]

of the Be/Ag/F system (with a possible small O content). The samples studied did not contain any appreciable copper impurities as determined by the ICP MS (Inductively Coupled Plasma Mass Spectrometry) method and thus, traces of oxocuprate SCs could not be responsible for the observed drops in magnetic susceptibility. The true origin of these magnetic anomalies, the possibly superconducting nature of the samples, and the identification of the minority responsible for the anomalous magnetic response, still await confirmation.

5. Silver(II) Fluoride (AgF_2)—Precursor Toward Fluoroargentates(II)

We begin the description of the properties of fluoroargentates with silver difluoride (AgF_2) which is an important precursor for the preparation of ternary Ag(II) fluorides. Usually it is prepared by high-temperature (~200–250°C) fluorination of various Ag(I) compounds. Very pure AgF_2 can be prepared at ambient temperature by reaction between elemental fluorine and $AgNO_3$ in anhydrous HF (aHF) as a solvent. Freshly prepared AgF_2 is an extremely active fluorinating agent and it slowly releases fluorine even at temperature ~360°C (in inert carrier gas or dynamic vacuum [50]):

$$AgF_2 \rightarrow AgF + \frac{1}{2}F_2$$

This reaction, which corresponds to facile introducing of holes to the s^2p^8 octet of F^- anions again points out to a substantial contribution of F(2p) states at the top of the valence band.

Besides a nondisproportionated dark brown $Ag^{II}F_2$ form (α-AgF_2) there is also a disproportionated metastable red-brown $Ag^{I}[Ag^{III}F_4]$ (β-AgF_2) polymorph which can be obtained by a low-temperature reaction of $AgBF_4$ and $KAgF_4$ in aHF [51]. The diamagnetic β-AgF_2 has been observed to transform exothermally to α-AgF_4 around 0°C, and could not be isolated in crystalline form suitable for structural measurements.

The crystal structure of α-AgF_2 [52] (Fig. 12.7) consists of puckered $AgF_{4/2}$ sheets exhibiting AD-E ordering of the Jahn–Teller distortion. Each Ag(II) is coordinated by four fluorine atoms at 2.068 Å and two apical F atoms at 2.584 Å, originating from neighboring $AgF_{4/2}$ sheets. The atomic arrangement found in α-AgF_2 is quite unique among transition metal difluorides, which adopt either the fluorite or rutile structure [53], but bears many similarities to the rutile-derived structure of CuF_2 [54] which also contains puckered $MF_{4/2}$ sheets with AD-E ordering. In fact, there is one structural report claiming that AgF_2 adopts the CuF_2-type structure [55], but the proposed geometry yields an erroneously high BVS parameter for Ag^{2+} (3.01) which casts the correctness of this assignment into doubt. The polymeric 2D nature of AgF_2 adjoined by marked covalence of the Ag–F bonds in this compound [4] helps to explain why this compound dissolves only in minute amounts in aHF, unlike more ionic CaF_2.

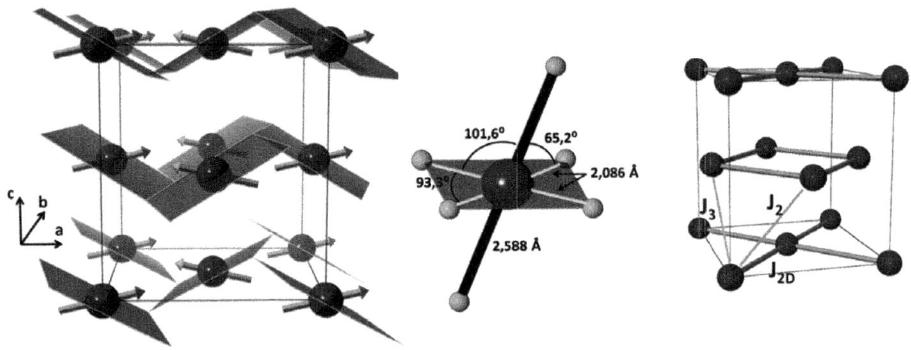

Figure 12.7 The crystal structure of α-AgF_2 (left), the first coordination sphere of Ag^{2+} (middle), and the magnetic coupling pathways between Ag(II) centers (right). *Red arrows* (light gray in print versions) indicate the spin orientation in the magnetic ordered phase.

The AD-E ordering exhibited by α-AgF_2 suggests the presence of intrasheet AFM interactions (due to good overlap in the *ab* plane of the spin-carrying $d(x^2-y^2)$ orbitals) and weak intersheet coupling. Indeed, DFT + *U* calculations yield $J_{2D} = -41$ meV with the absolute values of J_2 and J_3 one order of magnitude smaller [39] thus confirming 2D AFM properties of this compound. The magnetic interactions are somewhat disrupted by the departure of the Ag—F—Ag bond angle from 180°, as exemplified by the fact that in a hypothetical structure of AgF_2 containing flat AgF_2 sheets with linear Ag—F—Ag bridges, the calculated J_{2D} reaches −150 meV—a value comparable to the intrasheet coupling constant of −146 meV measured for 2D oxocuprate, La_2CuO_4 [56]. This feature alone suggests that Ag^{2+}/F compounds hold potential for magnetic superexchange at least as strong as that for the isoelectronic Cu^{2+}/O systems.

The actual magnetic ordering occurring in AgF_2 below 163K is quite complex, as canting of the spins along the ***c*** axis is observed [57] (Fig. 12.7). This leads to a small magnetic moment (*ca.* 0.1 μ_B) along this axis and consequently a drastic rise of the magnetic moment below the Néel temperature, resembling the characteristics of a ferromagnet [41].

As AgF_2 is a fluoroargentate(II) prototype the high-pressure transformations of this compound are of interest. Previous theoretical calculations indicate that at AgF_2 should transform at a modest pressure of 15 GPa to a flat-sheet *Abma* polymorph exhibiting strong 2D AFM superexchange interactions [58,59]. More recent calculations suggest a competitive polymorph adopting a $PbCl_2$-type structure which is characterized by infinite $AgF_{4/2}$ chains built from edge-sharing plaquettes—a structural motif analogous to that found in $CuCl_2$ [60]. The structural transition from the low-pressure *Pbca* to high-pressure *Pnam* orthorhombic form should result in transformation of the magnetic ground state from 2D AFM to a frustrated 1D ordering [60]. This scenario is currently verified by diamond anvil cell experiments—the high chemical reactivity of AgF_2 representing a major technical impediment.

5.1 M(d)/F(2p) Mixing in MF_2 (M = Cu, Ag, Au): X-ray Photoelectron Spectra and Density Functional Theory Picture

The X-ray photoelectron spectra (XPS) of a series of Ag^{1+}, Ag^{2+}, and Ag^{3+} fluorides [4] allow us to understand the strength of the Ag(4*d*)/F(2*p*) mixing taking place for AgF_2 and place it in the broader context of silver compounds at various oxidation states. Due to a much larger cross section for interaction of the Ag(4*d*) rather than F(2*p*) electrons with X-ray radiation, the XPS trace back mostly the partial contribution from the 4*d* orbitals of Ag to the total density-of-states profile. A classical example of a quasi-ionic compound is AgF where the contribution from the 4*d* orbitals to the "metal band" is around nine times larger than that to the "ligand band." For AgF_2 the covalence is substantially increased with many 4*d* states found in the band placed at higher binding energy and the corresponding ratio is now close to 2:1. Remarkably, in Ag^{3+} compound $KAgF_4$, there is an unprecedented inversion of the roles of "metal" and "ligand" with 60% of the 4*d* states sitting in the "ligand band." These experimental results are in good agreement with quantum mechanical DFT calculations for these systems [4].

Substantial Ag(4*d*)/F(2*p*) mixing seen for AgF_2 renders it unique also among difluorides of Group 11 elements. The spin-nonpolarized DFT calculations which offer a simplified yet qualitatively correct view of chemical bonding (Fig. 12.8) reveal substantial differences in M(*nd*)/F(2*p*) mixing ($n = 3$ for Cu, $n = 4$ for Ag, $n = 5$ for Au). This is easily seen if a "weighted average" (in the energy scale, where partial density of states stands for weight) is determined separately for M(*nd*) and F(2*p*) states. It turns out that for CuF_2 the center of gravity of F(2*p*) states is as much as 2.7 eV below the Cu(3*d*) ones. This implies a chemical bonding which is partly ionic in nature. For AgF_2 the metal and nonmetal states come much closer to each other (1.1 eV) thus resulting in a strong Ag(4*d*)/F(2*p*) mixing and enhanced covalence of chemical bonding. However, for hypothetical AuF_2 adopting the α-AgF_2 structure, the metal and nonmetal states are split again (by 1.8 eV), which is mostly due to relativistic destabilization of Au(5*d*) orbitals. Thus, the Ag > Au > Cu trend of the decreasing orbital mixing confirms the well-known peculiarities of coinage group metals and it is in nice agreement with the experimentally established very strong oxidizing properties of Ag^{2+}. This also helps to rationalize the strong superexchange residing in Ag—F—Ag bridges.

Having discussed the AgF_2 binary system, let us now turn to the more complex fluoride derivatives of Ag^{2+}.

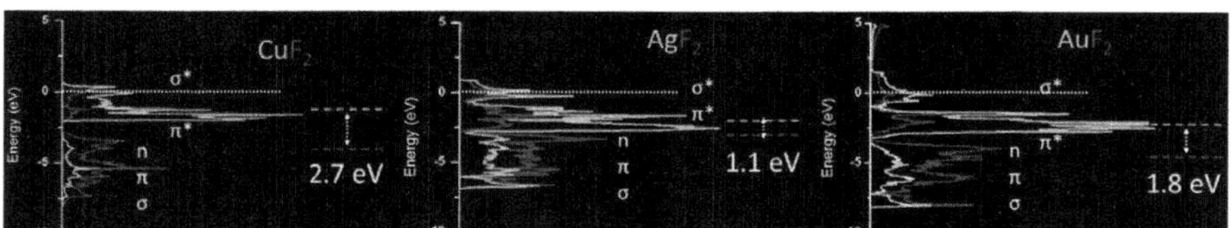

Figure 12.8 The spin nonpolarized electronic structure of CuF_2 (left), α-AgF_2 (middle), and hypothetical AuF_2 adopting the α-AgF_2 structure and fully optimized (right) as calculated with density functional theory (GGA, Generalized Gradient Approximation) [61].

6. Crystal Structures and Magnetic Properties of B_2AgF_4

The first step in the quest for synthesizing novel superconductors based on fluorides of Ag(II) is finding a parent (undoped) compound with suitable properties. Given that, all precursors for oxocuprate superconductors are characterized by strong 2D AFM coupling we have decided to target fluoroargentates(II) with such magnetic structure.

The layered perovskite structure similar to that exhibited by La_2CuO_4 is one structural option for the systems considered. This turned our attention to B_2AgF_4 compounds where B is an alkali metal (Li–Cs). The stability of the perovskite structure for these systems can be assessed with the use of the Goldshmidt tolerance factor [35] (t). As can be seen in Table 12.2 for Cs and Rb the t values close to 1 indicate that the perovskite structure should be stable for both Cs_2AgF_4 and Rb_2AgF_4. On the other hand, Li_2AgF_4 and Na_2AgF_4 with low t values should most probably adopt post-perovskite arrangements, while K_2AgF_4 is at the borderline of the stability of the perovskite structure. As will be shown in subsequent subsections these crude predictions are in strikingly good accordance with experimental observations.

6.1 *Cs_2AgF_4 and Rb_2AgF_4—Two-Dimensional Ferromagnets*

First we will discuss Cs_2AgF_4 and Rb_2AgF_4. Both compounds were first synthesized in 1974 [62], but only in recent years has the cesium derivative been characterized satisfactorily [63,64]. In contrast to AgF_2 none of the members of the B_2AgF_4 can be synthesized by wet chemical methods as these salts undergo solvolysis in aHF with the subsequent formation of BHF_2 and AgF_2. Therefore the typical synthesis procedure for obtaining these compounds includes a solid-state reaction between finely ground AgF_2 and BF powders.

Neutron powder diffraction studies conducted at 6 and 300K indicate that Cs_2AgF_4 adopts the layered perovskite structure with the AD-E arrangement of octahedra [63]. This assignment of the cJTE scenario is in accordance with earlier electron-spin resonance (ESR) studies [26]. Thus the structure of Cs_2AgF_4 differs drastically from that of La_2CuO_4 which exhibits FD ordering of elongated octahedra (FD-E) [33] (Fig. 12.9).

The preference toward the AD-E arrangement of $[AgF_6]^{4-}$ units with respect to the FD-C distortion has been confirmed by DFT + U calculations [45]. These indicated also that the AD-E orbital ordering should result in weak FM intralayer coupling, in

Table 12.2 Goldschmidt tolerance factors calculated for $M^+/Ag^{2+}/F^-$ triads (Calculated assuming $R(F^-) = 1.17$ Å, $R(Ag^{2+}) = 1.08$ Å and cubic Shannon–Prewitt radii of M^+)

M	Li	Na	K	Rb	Cs
t	0.70	0.78	0.89	0.92	0.96

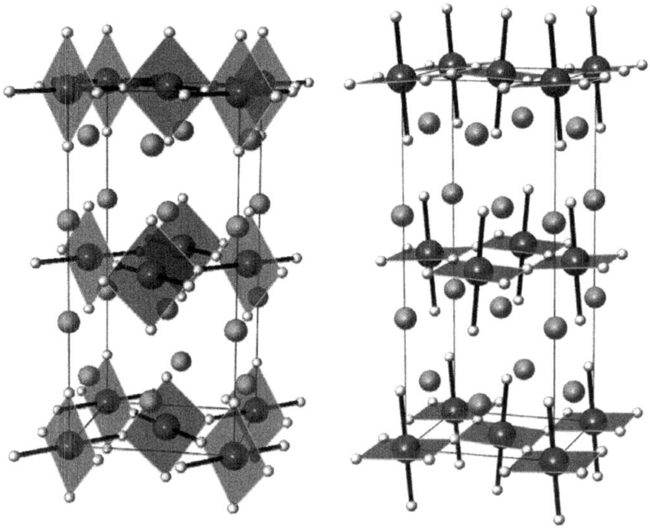

Figure 12.9 Comparison of the crystal structure of Cs_2AgF_4 (left) and La_2CuO_4 (right).

accordance with the GKA rules [65]. Indeed magnetic susceptibility and muon-spin relaxation measurements have proved that Cs_2AgF_4 exhibits 2D FM properties with the Curie temperature (T_C) of 13.95K and the intralayer coupling constant (J_{2D}) of 3.8 meV [63,64].

Similarly to the cesium derivative Rb_2AgF_4 has been assigned the layered perovskite structure [62]. Although no refinement of the crystal structure was made, the AD-E ordering within the $(AgF_{4/2+2})^{2-}$ sheets was inferred from subsequent ESR measurements [26]. This assignment is further supported by the 2D FM properties of this compound ($T_C = 17.5$K, $J_{2D} = 4.8$ meV) [39].

Our own room-temperature X-ray powder diffraction (XPD) measurements for Rb_2AgF_4 could be equally well fitted by an AD-E, as well as FD-C structure (Fig. 12.4), with the difference between the goodness-of-fit parameters of these two models not exceeding 1.5% [39]. This indicates that the XPD technique with laboratory X-ray source is not suitable for distinguishing between different cJTE ordering scenarios in layered fluoroargentates(II), which is the result of the very small X-ray scattering factor of fluorine.

6.2 Polymorphism of K_2AgF_4

As mentioned earlier the rising mismatch between the size of M^+ and Ag^{2+} when moving from cesium to potassium suggests that the layered perovskite polymorph of K_2AgF_4 might be unstable with respect to a post-perovskite phase. Indeed, two polymorphs of this compound were found: layered perovskite α-K_2AgF_4 (first synthesized in 1971 [66]) and the recently obtained post-perovskite β-K_2AgF_4 [67]. Both polytypes can be synthesized directly by annealing AgF_2 with KF, although with different reaction conditions (2 days at 480–500°C for α; 14 days at 270–300°C

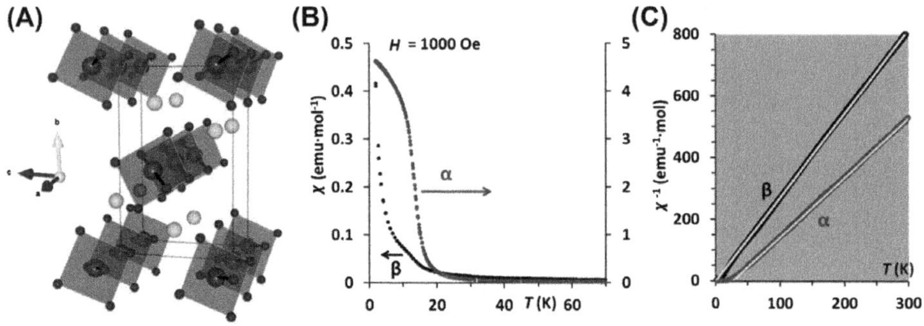

Figure 12.10 The Na_2CuF_4-type crystal structure of β-K_2AgF_4 (A). The temperature dependence of the magnetic susceptibility (χ) for β- (*black points*) and α-K_2AgF_4 (*red* (gray in print versions) *points*) is shown in (B), while the 1/χ versus *T* plots are given in (C). *Yellow* (light gray in print versions) *lines* are fits to the Curie–Weiss law (see text).

for β). These differences point to a slower formation kinetics of β-K_2AgF_4 as compared to α-K_2AgF_4.

Differential scanning calorimetry measurements indicate that α-K_2AgF_4 is metastable at room temperature, as it transforms irreversibly and exothermally to the β polymorph at 190°C. The latter polytype adopts the Na_2CuF_4-type structure in which axially elongated $[AgF_6]^{4-}$ octahedra share two edges thus forming one dimensional $[AgF_{4/2+2}]^{2-}$ chains running along the ***a*** axis (Fig. 12.10A). The elongated coordination of Ag(II) centers has been confirmed by ESR spectroscopy.

This arrangement of $[AgF_6]^{4-}$ units leads to very poor overlap between spin-carrying $d(x^2-y^2)$ orbitals, and consequently the Ag(II) centers interact very weakly. This is confirmed by magnetic susceptibility measurements which can be fitted with the Curie–Weiss law yielding very small positive θ of 1K (indicative of very weak FM coupling) with the Curie constant, C, equal to 0.37 emu K/mol [67], as expected for a spin-½ system (Fig. 12.10B and C).

In comparison, the magnetic susceptibility of α-K_2AgF_4 indicates stronger ferromagnetic coupling ($\theta = 26$K) [17] and a magnetic ordering temperature of 13.4K [39]. As in the case of Cs_2AgF_4 and Rb_2AgF_4 the χ versus *T* dependences can be well fitted with a model of a square lattice of spin-½ FM coupled centers [68] yielding $J_{2D} = 3.4$ meV.

Based on the evident similarities in the magnetic properties, one could expect that α-K_2AgF_4 will be isostructural with B_2AgF_4 (B = Rb, Cs). Single crystal X-ray diffraction indicated that indeed this compound adopts at 200K the layered perovskite structure with buckling of the $(AgF_{4/2+2})^{2-}$ planes [17], but in contrast to heavier analogs these planes exhibit FD ordering of compressed octahedra (Fig. 12.11). This structure of *Cmca* symmetry is characterized by Ag–F bond distances of 2.08 Å (apical F1) and 2.26 Å (equatorial F2) yielding BVS(Ag^{2+}) = 1.98.

This finding is intriguing as the *Cmca* structure of α-K_2AgF_4 would constitute the first example of a layered d^9 system exhibiting compressed coordination of Ag^{2+}. In fact in all other fluoroargentates(II) containing compressed $[AgF_6]^{4-}$ octahedra these

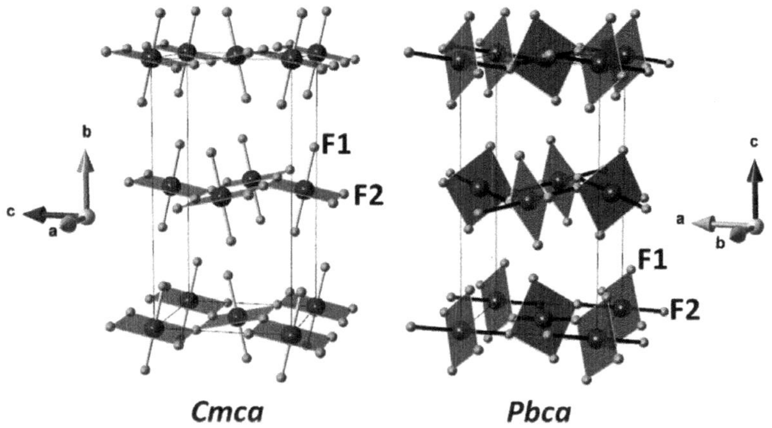

Figure 12.11 Comparison of the crystal structure of *Cmca* and *Pbca* models of α-K_2AgF_4. Potassium atoms have been removed for clarity.

units are connected into 1D chains of $[AgF_{2/2}]^+$ stoichiometry (eg, $AgFBF_4$ [8]), and they do not form extended 2D lattices. The same holds true for analogous Cu^{2+} compounds.

In order to confirm the structure assignment of α-K_2AgF_4 we have refined the single crystal X-ray data with a model that incorporates the same tilt system as *Cmca*, but allows for AD-E ordering within the $[AgF_{4/2+2}]^{2-}$ sheets [69] (*Pbca* structure in Fig. 12.11). The Rietveld refinement of *Pbca* indicated that this model reproduces the experimental data equally well as *Cmca*. The refined Ag−F bond lengths for *Pbca* are 2.08 Å (apical F1), 2.09 Å, and 2.47 Å (equatorial F2) giving $BVS(Ag^{2+}) = 2.06$.

Therefore similar to the case of Rb_2AgF_4, X-ray diffraction measurements could not unequivocally distinguish between AD-E and FD-C orderings in α-K_2AgF_4. We have therefore conducted ESR measurements in order to establish the coordination type (compressed *vs.* elongated) present in α-K_2AgF_4. Unfortunately, due to the similarity between the spectra of the AD-E and FD-C arrangements we were again unable to distinguish between the two cJTE scenarios by employing low resolution X-band measurements. At present high field ESR measurements are under way.

The abovementioned problems led us to perform theoretical DFT + *U* investigation into the stability of various polymorphs of B_2AgF_4. The results of this study are described in the next subsection. Before we move on to them we would like to comment on the driving force of the α → β phase transition in K_2AgF_4. Clearly the buckling of the $[AgF_{4/2+2}]^{2-}$ planes in α indicates the presence of "chemical pressure" which is a result of the too small radius of the K^+ cation. Although this pressure is partly relieved by the puckering, K^+ remains still underbonded as indicated by its BVS parameter of 0.88 [67]. In comparison $BVS(K^+)$ equals to 1.06 in β-K_2AgF_4. Therefore one sees that the improvement in the bonding situation of K^+ is the main thermodynamic incentive for the α → β transition.

6.3 Density Functional Theory + U Calculations on B_2AgF_4 Phases

In order to elucidate the structural preferences found for the B_2AgF_4 system (B = Na–Cs) we have performed DFT + U calculations [39] for these compounds in three different structures: (1) layered perovskite with AD-E ordering (AD-E), (2) layered perovskite with FD-C arrangement, and (3) the Na_2CuF_4-type post-perovskite phase (PP). In case of AD-E and FD-C the possibility of puckering of the $[AgF_{4/2+2}]^-$ was taken into account by adopting suitable symmetry of the structure. Our calculations utilized $U = 5$ eV and $I = 1$ eV.

The calculated energetic differences between various polymorphs (Fig. 12.12A) confirm the experimentally observed preference for the adoption of the AD-E structure in case of Cs_2AgF_4 and Rb_2AgF_4. For both systems FD-C lies high in energy ($\Delta E > 110$ meV $\approx$ 1200K) indicating that this phase is not stable in the experimentally accessible temperature range.

Moving toward smaller cations results in drastic stabilization of the PP structure with respect to layered perovskites. Already for K_2AgF_4 this polymorph is the lowest energy structure followed by AD-E (+47 meV per f.u.) and FD-C (+208 meV per f.u.). The difference in energies between PP and AD-E is in very good accordance with the heat release of 49 meV per f.u. measured for the α-$K_2AgF_4 \rightarrow \beta$-$K_2AgF_4$ transition [67]. This fact, together with the high energy of FD-C, corroborates the notion that α-K_2AgF_4 indeed adopts the *Pbca* structure, but this prediction still awaits unambiguous experimental verification.

Finally for Na_2AgF_4 the layered perovskite phases are severely destabilized with respect to PP by around 120 meV per f.u. ($\approx$ 1400K) and therefore one should expect

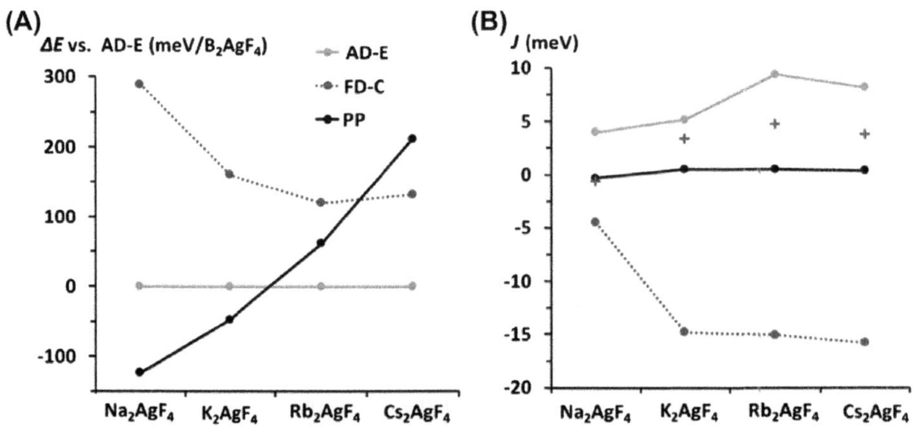

Figure 12.12 Energies (relative to AD-E) of different B_2AgF_4 polymorphs (A); superexchange coupling constants between nearest-neighbor Ag(II) centers calculated for these structures (B). Note that the superexchange route has 1D character for post-perovskite phase and 2D character for AD-E and FD-C. *Blue crosses* (light gray in print versions) denote experimental coupling constants for Na_2AgF_4, α-K_2AgF_4, Rb_2AgF_4, and Cs_2AgF_4.

that this compound will solely adopt the Na_2CuF_4-type PP structure. This is indeed observed, as will be discussed in the subsequent subsection.

Using the "broken-symmetry" method [70] we have calculated at the DFT + U level the intrachain magnetic coupling constants for PP (J_{1D}), as well as the intrasheet coupling for AD-E and FD-C (J_{2D})—see Fig. 12.12B. In all studied systems J_{1D} is close to 0, which is a consequence of the poor orbital interaction along the $[AgF_{4/2+2}]^{2-}$ chains. For the AD-E structure, J_{2D} points to weak ferromagnetism in all studied compounds, in accordance with what was found for Cs_2AgF_4 [45]. On the other hand, calculations indicate that all FD-C polymorphs of B_2AgF_4 should exhibit weak AFM coupling. This results from the fact that the unpaired electron resides not only on the local $d(z^2)$ orbital, as favored by the compressed coordination of Ag(II), but also partly on $d(x^2-y^2)$ as these orbitals exhibit much larger dispersion over the first Brillouin zone [17].

The experimental J_{2D} values for Cs_2AgF_4, Rb_2AgF_4, and α-K_2AgF_4 (Fig. 12.12B) are in good accordance with the theoretical ones under the assumption of AD-E ordering in each of these systems. This again indicates that the α-K_2AgF_4 should adopt the *Pbca* and not *Cmca* structure, as the latter is predicted to exhibit 2D AFM coupling.

6.4 Stabilizing the Perovskite Phase: $K_{1-x}Rb_xAgF_4$ Compounds

In order to more closely investigate the properties of fluoroargentes(II) adopting the layered perovskite structure, we have synthesized and characterized three $K_{1-x}Rb_xAgF_4$ compounds with the fraction of rubidium atoms (x) equal to 0.25, 0.5, and 0.75 [39]. These systems adopt the layered perovskite structure with random distribution of K^+/Rb^+ cations in B sites, as evidence by our room-temperature XPD measurements. Importantly, already for $x = 0.25$ we do not observe a transition to the *PP* structure upon heating up to 300°C. This is in accordance with the predicted stabilization of the B_2AgF_4 perovskite phase with the increasing mean radius of B (see previous section).

All of the synthesized compounds exhibit ferromagnetic properties (Fig. 12.13A) which can be fitted with a 2D FM model yielding J_{2D} values slightly increasing with Rb content (4.2/4.5/4.7 meV for $x = 0.25/0.5/0.75$) and bracketed by that of α-K_2AgF_4 (3.4 meV) and Rb_2AgF_4 (4.8 meV). The value of the saturated magnetic moment in the FM ordered phase corresponds to that of a spin-1/2 d^9 system (Fig. 12.13B).

We have conducted XPD investigation in the 80−350K temperature range and found two types of phase transitions in the $K_{1-x}Rb_xAgF_4$ systems (Fig. 12.14) [39]. Both could be observed for $KRbAgF_4$ (Fig. 12.14), and as can be seen from Fig. 12.13C they are characterized by different behavior of the a and b lattice constants which are parallel to the $[AgF_{4/2+2}]^{4-}$ sheets (compare Fig. 12.11). Therefore most likely both types of phase transition are associated with rearrangement of the octahedral rotation ordering within these sheets (in analogy to what is found for the $La_{2-x}Ba_xCuO_4$ system [71]), although this hypothesis requires confirmation from neutron scattering experiments. We note that temperatures of both structural and magnetic phase transitions could also be confirmed by heat capacity measurements (Fig. 12.13D).

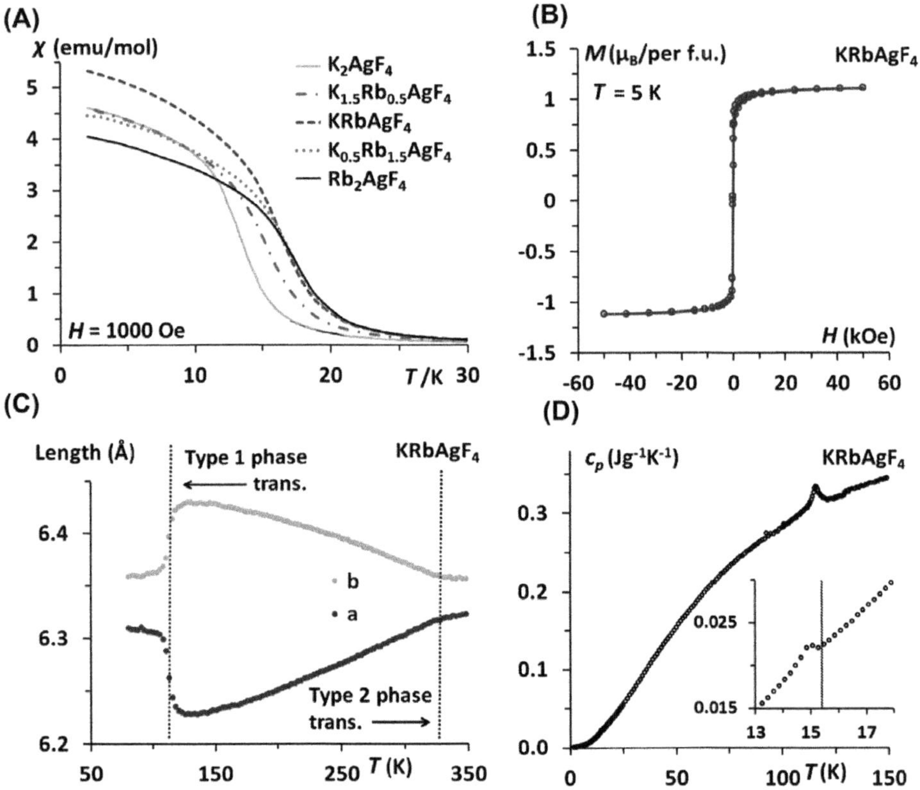

Figure 12.13 Magnetic susceptibility of $K_{1-x}Rb_xAgF_4$ compounds (A) together with the magnetization curve (B), temperature dependence of *a* and *b* cell vectors (C), and heat capacity changes (D) measured for $KRbAgF_4$.

6.5 Na_2AgF_4—1D Antiferromagnetic With Through-Space Magnetic Coupling

In this last subsection devoted to B_2AgF_4 systems we turn to Na_2AgF_4. This compound has been first synthesized in 1988 [72], but its structure has not been determined at that time. We attempted the synthesis of Na_2AgF_4 through the solid-state reaction of NaF and AgF_2, but the final product was contaminated with AgF from partial reduction of AgF_2 and unreacted NaF.

In order to obtain a pure sample of Na_2AgF_4 we followed a novel synthetic pathway proceeding via thermal decomposition of a silver(III) salt, $NaAg^{III}F_4$, in the presence of NaF [16]:

$$NaF + NaAgF_4 \xrightarrow{T=400-550^{\circ}C} Na_2AgF_4 + \frac{1}{2}F_2\uparrow$$

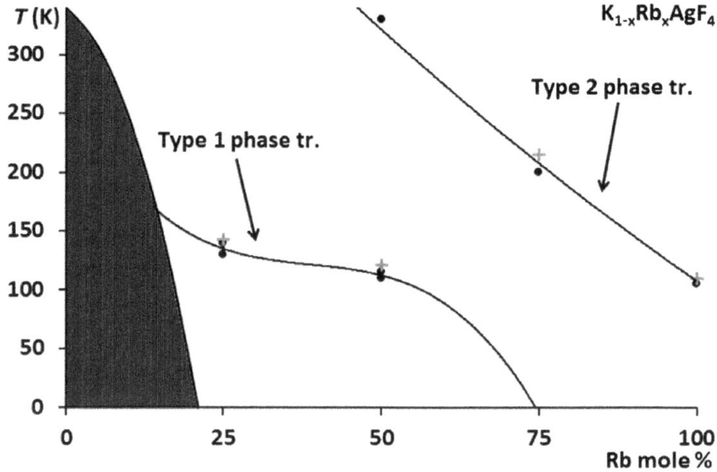

Figure 12.14 Temperature versus Rb mole fraction phase diagram of $K_{1-x}Rb_xAgF_4$. *Lines* indicate proposed phase boundaries, the region for which the post-perovskite phase is thermodynamically stable is indicated in gray. *Black dots/red* (light gray in print versions) *crosses* depict temperature of phase transitions determined by X-ray powder diffraction/heat capacity measurements.

With this synthetic approach we obtained very pure samples of Na_2AgF_4 which could be characterized by both XPD and magnetic measurements (Fig. 12.15).

In accordance with DFT + U calculations Na_2AgF_4 adopts the post-perovskite structure analogous to that of β-K_2AgF_4. There is, however, difference between the arrangements of ions in the two compounds. Most notably the $Ag^{2+}\cdots Ag^{2+}$ contacts

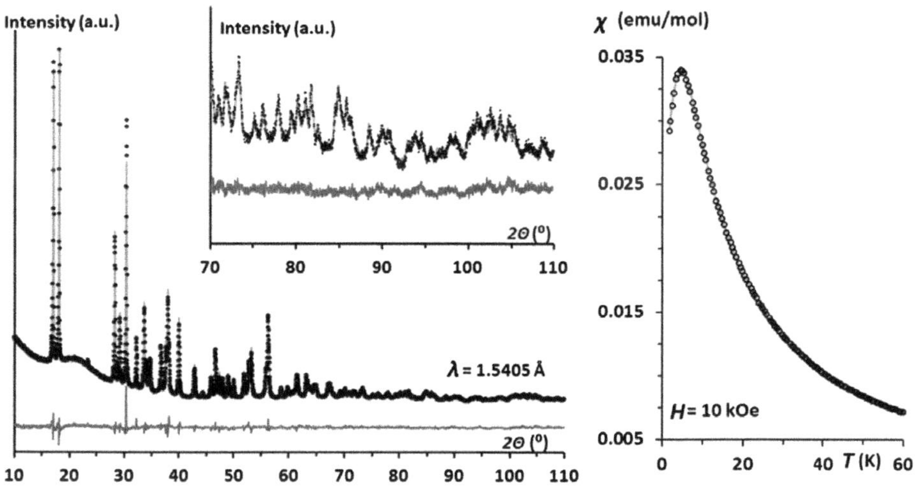

Figure 12.15 X-ray powder diffraction (left) and the χ versus T dependence (right) for Na_2AgF_4. The *red* (gray in print versions)/*orange* (light gray in print versions) *lines* are the fit to the structural Na_2CuF_4-type and magnetic 1D antiferromagnetic models, respectively.

within the $[AgF_{4/2+2}]^{2-}$ chains are over 10% shorter in Na_2AgF_4 (3.342 Å) than in β-K_2AgF_4 (3.717 Å). In fact the separation found in the former system is the smallest found in any Ag(II) compound and is shorter by 3% than the sum of the van der Waals radii of two silver atoms. It must be pointed out, however, that the short contacts between Ag^{2+} cations are certainly not an indication of any chemical bond (such as eg, the genuine Au^{2+}—Au^{2+} bond found in $AuSO_4$ [73]).

The short $Ag^{2+}\cdots Ag^{2+}$ contacts have an influence on the magnetic properties of Na_2AgF_4. By susceptibility measurements it was found that this compound exhibits 1D AFM properties with $J_{1D} = -0.6$ meV [16] (Fig. 12.15), in contrast to FM properties found in β-K_2AgF_4. With the aid of DFT + U calculations it was possible to rationalize the AFM coupling in Na_2AgF_4 as a combination of strong through-space dipolar magnetic coupling coupled with weak superexchange coupling.

7. $BAgF_3$ Systems

Although the perovskite $BAgF_3$ systems are unlikely to exhibit 2D magnetic interactions, investigation of their properties enables direct comparison with analogous fluoride ($KCuF_3$) and oxide (Sr_2CuO_3 [74]) systems. Three members of the $BAgF_3$ family (B = K, Rb, Cs) have been first synthesized by Odenthal and Hoppe [66]. While the structure of $KAgF_3$ was not fully determined both the rubidium and cesium derivatives have been proposed to be isostructural with a-$KCuF_3$ (Fig. 12.16).

7.1 $KAgF_3$—Unusually Strong Antiferromagnetic Coupling

Our studies on $KAgF_3$ [17,75] showed that at low temperature this compound adopts a structure with the same type of the cJTE as found in a-$KCuF_3$ (Fig. 12.3), but with the Ag/F lattice distorted due to tilts of the $[AgF_4]^{2-}$ plaquettes (LT-$KAgF_3$ in Fig. 12.16).

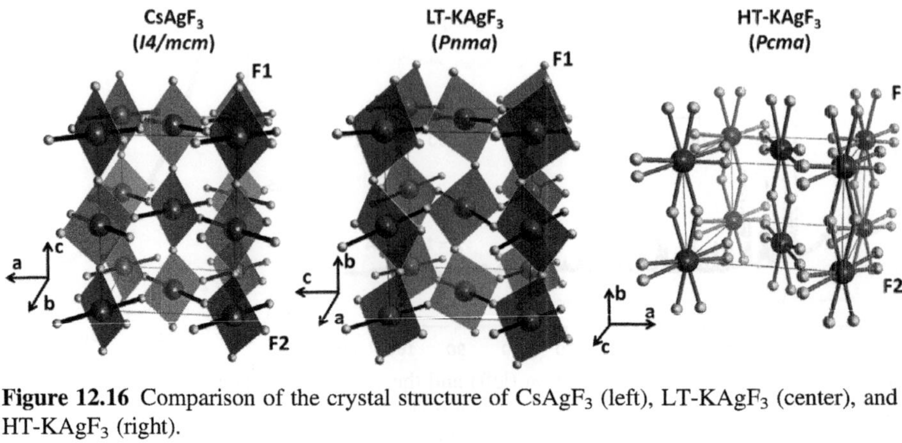

Figure 12.16 Comparison of the crystal structure of $CsAgF_3$ (left), LT-$KAgF_3$ (center), and HT-$KAgF_3$ (right).

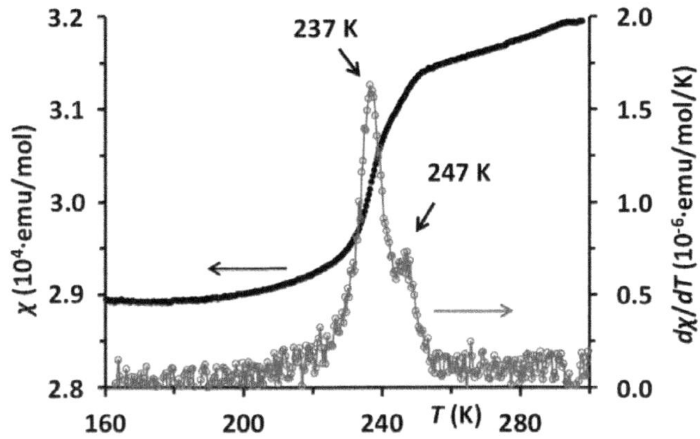

Figure 12.17 Molar susceptibility of $KAgF_3$ and its temperature derivative (*black/red* (gray in print versions) *points*, respectively).

This polymorph adopts the $GdFeO_3$-type structure of *Pnma* symmetry and undergoes a phase transition upon heating to 230K [75] which leads to HT-$KAgF_3$ (*Pcma*). This structure retains the bond connectivity present in LT-$KAgF_3$, but exhibits disorder in the rotation of the $[AgF_4]^{2-}$ plaquettes.

DFT + *U* calculations performed for the *Pnma* structure of $KAgF_3$ point to 1D magnetic properties of this compound with strong AFM interaction within the $[AgF_{2/2+2}]^-$ chains characterized by J_{1D} of -125 meV [75]. This finding has been confirmed by susceptibility measurements which yielded $J_{1D} \approx -100$ meV for both HT and LT-$KAgF_3$—the largest (by absolute value) superexchange coupling constant determined experimentally for fluoroargentates(II). This value also constitutes the largest *J* value among all fluoride anion–bridged magnetic systems [76].

It's noteworthy to point that the LT–HT phase transition results in a drastic rise in the magnetic moment of $KAgF_3$ (Fig. 12.17). This is most probably a result of the partial disruption of the magnetic coupling along the $[AgF_{2/2+2}]^-$ chains upon entering the disordered HT phase. The temperature derivative of the magnetic susceptibility points in fact to two transitions (separated by only 10K), which have been tentatively assigned to a magnetic (237K) and structural (LT-HT, 247K) transition. A similar series of transitions has been observed for LaFeAsO [77].

7.2 $CsAgF_3$ and $RbAgF_3$

DFT + *U* calculations performed on $RbAgF_3$ [39] indicated that this compound should also adopt at low temperature, the *Pnma* structure which is slightly more favorable at 0K than the *I4/mcm* polymorph. In comparison with $KAgF_3$ the *Pnma* polytype of $RbAgF_3$ is characterized by less bent fluorine bridges along the $[AgF_{2/2+2}]^-$ chains (Ag–F–Ag angle of 169 degree compared to 153 degree for $KAgF_3$). This leads to stronger AFM interactions, as shown by the more negative calculated coupling constant

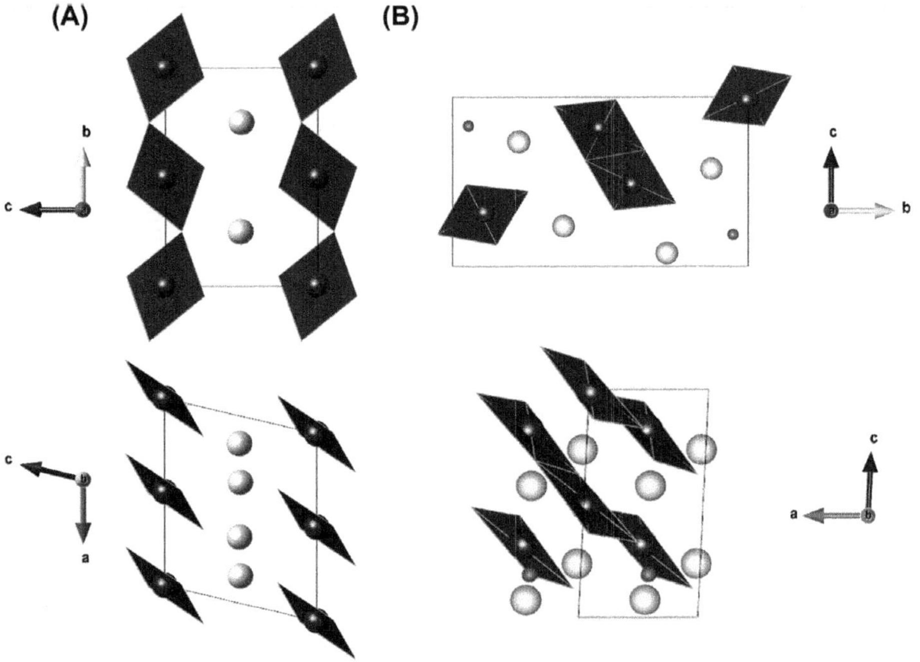

Figure 12.18 Different side views of several possible post-perovskite phases of $RbAgF_3$: distorted $CaIrO_3$-type of $P2_1/m$ symmetry (A), and $KCuCl_3$-type of $P2_1/c$ symmetry (B).

(−167 meV). Finally the value of J_{1D} becomes equal to −189 meV for $CsAgF_3$ which, according to DFT calculations, adopts the *I4/mcm* structure with linear Ag—F—Ag bridges. These theoretical results are now being tested in experiment.

The absolute values of J_{1D} calculated for $BAgF_3$ are much higher than analogous values for $KCuF_3$ (−35 meV [78]), and approach those exhibited by oxocuprates(II). For example, the measured coupling constants for La_2CuO_4 (2D system) and Sr_2CuO_3 (1D system) are −146 meV [56] and −254 meV [74], respectively. The latter absolute value is the largest known value in solid-state physics, but our calculations show that even larger absolute values approaching 300 meV are possible for some Ag(II)—F systems [59].

Possible structural modifications of $RbAgF_3$ have been explored with DFT + *U*. Apart from the perovskite structures (*I4/mcm* and *Pnma*), also two post-perovskite polytypes have been taken into account. The first one is a distorted $CaIrO_3$-type structure of $P2_1/m$ symmetry which contains $[AgF_{2/2+2}]^-$ chains analogous to those found in *Pnma* but these chains are arranged in layers separated by Rb^+ cations (Fig. 12.18A). The $P2_1/m$ structure is characterized by 1D AFM interactions comparable to that of *Pnma* ($J_{1D} \approx -120$ meV). The second postperovskite polymorph is $KCuCl_3$ type with $P2_1/c$ symmetry (Fig. 12.18B). It contains $[Ag_2F_6]^{2-}$ dimers which were not yet found in any other fluoroargentate(II). The coupling between Ag(II)

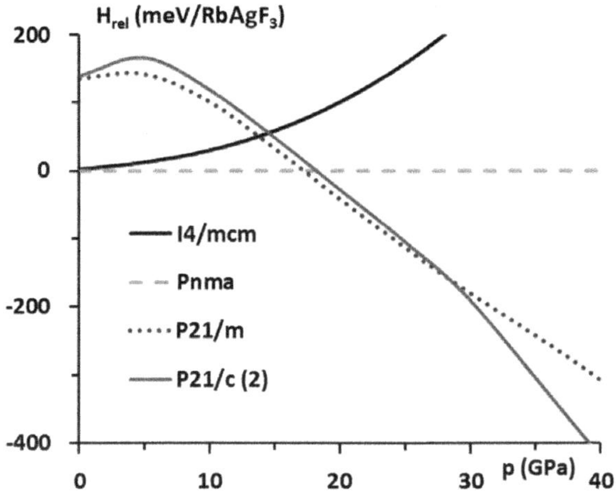

Figure 12.19 Pressure dependence of the relative enthalpy (referenced to that of *Pnma*) of various polymorphs of $RbAgF_3$ (calculated with U = 5 and I = 4 eV).

centers within the dimers is weak ($J_{dim} \approx \pm 5$ meV). Attempting to stabilize this unique structure type constitutes interesting experimental challenge.

The calculated pressure dependence of the enthalpy of $RbAgF_3$ (Fig. 12.19) indicates the perovskite *Pnma* should transform to post-perovskite $P2_1/m$ at about 17 GPa. Interestingly the distorted $CaIrO_3$-type structure is only marginally more stable than the $KCuCl_3$-type structure and becomes less stable above 30 GPa. Thus a drastic change in bonding (chains → dimers) and magnetic properties (1D AFM → isolated dimers) is predicted to occur at such conditions. Moreover the calculations indicated that the JTE persists in $RbAgF_3$ up to at least 50 GPa, similarly to what has been found for $AgSO_4$ [28].

8. Summary and Outlook

We have summarized here the latest advances in the field of physics and chemistry of Ag^{2+} fluoride compounds, with particular emphasis on the electronic and magnetic properties of AgF_2, $BAgF_3$, and B_2AgF_4 systems. The most remarkable feature of these compounds, aside from their strong oxidizing properties and rather narrow band gaps at the Fermi level, is the presence of strong superexchange coupling between Ag^{2+} centers via F^- ligand (Table 12.3). The largest (in terms of absolute value) measured superexchange coupling constant is −106 meV for LT-$KAgF_3$, which exceeds by a factor of 2−3 the value seen for the analogous fluorocuprate(II) ($KCuF_3$), and it approaches *J* values recorded for the record-holding oxocuprate(II) systems. The theoretical calculations suggest that the $BAgF_3$ can host even stronger magnetic superexchange as exemplified by $CsAgF_3$ (the DFT calculations yield *J* value close to −190 meV, Table 12.3).

Table 12.3 Summary of the magnetic properties of AgF_2, B_2AgF_4, and $BAgF_3$ systems. The value of the magnetic superexchange constants (J) are given in meV

		Major coupling constant	
Compound	**Magnetic dimensionality/ordering type**	J_{exp}	J_{theor}
AgF_2	2D/AFM	–	−41
Cs_2AgF_4	2D/FM	3.8	8.2
Rb_2AgF_4	2D/FM	4.8	9.0
$K_{0.5}Rb_{1.5}AgF_4$	2D/FM	4.7	–
$KRbAgF_4$	2D/FM	4.5	–
$K_{1.5}Rb_{0.5}AgF_4$	2D/FM	4.2	–
α-K_2AgF_4	2D/FM	3.4	5.2
β-K_2AgF_4	1D/FM	–	0.6
Na_2AgF_4	1D/AFM	−0.6	−0.3
$CsAgF_3$	1D/AFM	–	−189
$RbAgF_3$	1D/AFM	–	−167
HT-$KAgF_3$	1D/AFM	−97	–
LT-$KAgF_4$	1D/AFM	−106	−125

The unprecedented strong spin polarization of (otherwise quite redox-innocent) fluoride ligands by Ag(II) has prompted one of us (W.G.) to write a short *laudatio*:

I am DIVALENT SILVER

The free radicals' queen!

Take note of my power

Far from routine,

For my coupling's colossal

Even through fluorine… [61]

It remains to be seen whether the unprecedented spin-polarizing power of Ag^{2+} could be used to induce macroscopic magnetic polarization in the Ag^{2+}-doped samples of various host fluorides or oxides—with possible applications in spintronics.

Simultaneously, the quest for superconducting materials based on compounds of divalent silver continues in our collaborating laboratories. Despite many structural relationships uncovered by the studies described above, crystal engineering of fluoroargentates(II) which targets a 2D AFM insulator (a precursor toward a superconducting material) is not straightforward. Therefore it attempts to enlarge the field of possibilities by including other inorganic/organic ligands, L, and targeting various Ag−F−L, M−Ag−F−L (M = metal cation), or Ag−L connections, can be made. But that, as R. Kipling would say, is *an entirely different story*.

Acknowledgments

Research on Ag(II) compounds in W.G.'s laboratory has been supported from the funds of the Polish National Science Center, NCN (UMO-2011/01/B/ST5/06673). Calculations were performed with the support of the "HPC Infrastructure for Grand Challenges of Science and Engineering" project, cofinanced by the European Regional Development Fund under the Innovative Economy Operational Programme. ZM gratefully acknowledges the Ministry of Education, Science and Sport of the Republic of Slovenia for financial support of Research Program P1-0045 (Inorganic Chemistry and Technology).

The authors are grateful to all our collaborators and coauthors of the papers referenced here and other joint contributions to the fascinating chemistry of Ag(II) systems.

References

[1] W. Grochala, Z. Mazej, Phil. Trans. R. Soc. London Ser. A 373 (2015) 20140179-1−20140179−20140212.
[2] P. Połczyński, R. Jurczakowski, W. Grochala, Chem. Commun. 49 (2013) 7480−7482.
[3] G. Cady, A.V. Grosse, E.J. Barber, L.L. Burger, Z.D. Sheldon, Fluorinations carried out in the stream of F_2 gas may be catalytically enhanced upon addition of small amounts of any compound of Ag^{1+} which undergoes an *in situ* fluorination yielding AgF_2, which itself is an active F_2-transfer catalyst, Ind. Eng. Chem. 39 (1947) 290−292.
[4] W. Grochala, R.G. Egdell, P.P. Edwards, Z. Mazej, B. Žemva, ChemPhysChem 4 (2003) 997−1001.
[5] W. Grochala, Scr. Mater. 55 (2006) 811−814.
[6] W. Grochala, R. Hoffmann, Angew. Chem. Int. Ed. Engl. 40 (2001) 2742−2781.
[7] W. Grochala, J. Mater. Chem. 19 (2009) 6949−6968.
[8] W.J. Casteel Jr., G. Lucier, R. Hagiwara, H. Borrmann, N. Bartlett, J. Solid State Chem. 96 (1992) 84−96.
[9] Z. Mazej, E. Goreshnik, G. Tavčar, J. Fluor. Chem. 132 (2011) 686−689.
[10] G. Lucier, J. Muenzenberg, W.J. Casteel, N. Bartlett, Inorg. Chem. 34 (1995) 2692−2698.
[11] D. Koller, B.G. Müller, Z. Anorg. Allg. Chem. 626 (2000) 1429−1433.
[12] O. Graudejus, B.G. Müller, Z. Anorg. Allg. Chem. 622 (1996) 1549−1556.
[13] B.G. Müller, Z. Anorg. Allg. Chem. 553 (1987) 196−204.
[14] B.G. Müller, Angew. Chem. 99 (1987) 685−686.
[15] R. Fischer, B.G. Müller, Z. Anorg. Allg. Chem. 627 (2001) 445−452.
[16] D. Kurzydłowski, Z. Mazej, W. Grochala, Dalton Trans. 42 (2013) 2167−2173.

[17] Z. Mazej, E. Goreshnik, Z. Jagličić, B. Gaweł, W. Łasocha, D. Grzybowska, T. Jaroń, D. Kurzydłowski, P.J. Malinowski, W. Koźmiński, J. Szydłowska, P.J. Leszczyński, W. Grochala, CrystEngComm 11 (2009) 1702–1710.
[18] O. Graudejus, S.H. Elder, G.M. Lucier, C. Shen, N. Bartlett, Inorg. Chem. 38 (1999) 2503–2509.
[19] J. Tong, F. Kraus, J. Köhler, A. Simon, J. Liu, M.-H. Whangbo, Z. Anorg. Allg. Chem. 637 (2011) 1118–1121.
[20] O. Graudejus, A.P. Wilkinson, N. Bartlett, Inorg. Chem. 39 (2000) 1545–1548.
[21] G.A. Landrum, R. Hoffmann, Angew. Chem. Int. Ed. 37 (1998) 1887–1890.
[22] D. Kurzydłowski, P. Zaleski-Ejgierd, W. Grochala, R. Hoffmann, Inorg. Chem. 50 (2011) 3832–3840.
[23] I.D. Brown, Chem. Soc. Rev. 7 (1978) 359–376. N.E. Brese, M. O'Keeffe, Acta Crystallogr. Sect. B Struct. Sci., 47 (1991) 192–197.
[24] http://www.ccp14.ac.uk/ccp/web-mirrors/i_d_brown/bond_valence_param/, (accessed 19.03.15.).
[25] D. Reinen, S. Krause, Inorg. Chem. 20 (1981) 2750–2759.
[26] C. Friebel, D. Reinen, Z. Anorg. Allg. Chem. 413 (1975) 51–60.
[27] J. Ruiz-Fuertes, A. Friedrich, J. Pellicer-Porres, D. Errandonea, A. Segura, W. Morgenroth, E. Haussühl, C.Y. Tu, A. Polian, Chem. Mater. 23 (2011) 4220–4226. J. Ruiz-Fuertes, A. Segura, F. Rodríguez, D. Errandonea, M.N. Sanz-Ortiz, Phys. Rev. Lett., 108 (2012) 166402-1–166402-4.
[28] M. Derzsi, A. Budzianowski, V.V. Struzhkin, P.J. Malinowski, P.J. Leszczyński, Z. Mazej, W. Grochala, CrystEngComm 15 (2013) 192–198.
[29] Z. Mazej, E. Goreshnik, I. Arčon, N. Zabukovec Logar, V. Kaučič, Z. Anorg. Allg. Chem. 636 (2010) 224–229.
[30] M.A. Halcrow, Dalton. Trans. (2003) 4375–4384.
[31] L.R. Falvello, J. Chem. Soc. Dalton Trans. (1997) 4463–4476.
[32] M.W. Lufaso, P.M. Woodward, Acta Crystallogr. Sect. B Struct. Sci. 60 (2004) 10–20 (and references therein).
[33] P. Radaelli, D. Hinks, A. Mitchell, B. Hunter, J. Wagner, B. Dabrowski, K. Vandervoort, H. Viswanathan, J. Jorgensen, Phys. Rev. B 49 (1994) 4163–4175.
[34] M. Hidaka, K. Inoue, I. Yamada, P.J. Walker, Physica B+C 121 (1983) 343–350.
[35] M.W. Lufaso, P.M. Woodward, Acta Crystallogr. B 57 (2001) 725–738 (and references therein).
[36] A.I. Liechtenstein, J. Zaanen, Phys. Rev. B 52 (1995) R5467–R5470.
[37] P.W. Anderson, Solid State Phys. 14 (1963) 99–214.
[38] P.J. Hay, J.C. Thibeault, R. Hoffmann, J. Am. Chem. Soc. 97 (1975) 4884–4899.
[39] D. Kurzydłowski (Ph.D. thesis), University of Warsaw, 2012. Available in Polish at: http://depotuw.ceon.pl/handle/item/171.
[40] P. Olalde-Velasco, J. Jiménez-Mier, J. Denlinger, Z. Hussain, W. Yang, Phys. Rev. B 83 (2011), 241102(R)-1–241102(R)-4.
[41] W. Grochala, P.P. Edwards, Phys. Status Solidi 240 (2003) R11–R14.
[42] W. Adamczyk, P. Połczyński, A. Mika, T. Jaroń, Z. Mazej, K.J. Fijalkowski, R. Jurczakowski, W. Grochala, J. Fluor. Chem. 174 (2015) 22–29 (It is not established whether the measured conductivity is determined by hole/electron or rather ion transport).
[43] S.L. Dudarev, S.Y. Savrasov, C.J. Humphreys, A.P. Sutton, Phys. Rev. B 57 (1998) 1505–1509.
[44] X. Zhang, G. Zhang, T. Jia, Y. Guo, Z. Zeng, H.Q. Lin, Phys. Lett. A 375 (2011) 2456–2459.

[45] D. Kasinathan, K. Koepernik, U. Nitzsche, H. Rosner, Phys. Rev. Lett. 99 (2007) 247210-1–247210-4.

[46] J. Tong, J. Köhler, A. Simon, C. Lee, M.H. Whangbo, Z. Anorg. Allg. Chem. 638 (2012) 1792–1795.

[47] D. Reinen, C. Friebel, Struct. Bond. 37 (1979) 1–60.

[48] Z. Mazej, T. Michałowski, E.A. Goreshnik, Z. Jagličić, I. Arčon, J. Szydłowska, W. Grochala, Dalton Trans. 44 (2015) 10957–10968.

[49] W. Grochala, A. Porch, P.P. Edwards, Solid State Commun. 130 (2004) 137–142.

[50] Z. Mazej, personal communication. Erroneous information is given in various sources, including Wikipedia (https://en.wikipedia.org/wiki/Silver(II)_fluoride), that AgF_2 supposedly melts at 690°C and thermally decomposes above 700°C.

[51] C. Shen, B. Žemva, G.M. Lucier, O. Graudejus, J.A. Allman, N. Bartlett, Inorg. Chem. 38 (1999) 4570–4577.

[52] A. Jesih, K. Lutar, B. Žemva, B. Bachmann, S. Becker, B.G. Müller, R. Hoppe, Z. Anorg. Allg. Chem. 588 (1990) 77–83.

[53] S.M. Dorfman, F. Jiang, Z. Mao, A. Kubo, Y. Meng, V.B. Prakapenka, T.S. Duffy, Phys. Rev. B 81 (2010) 174121-1–174121-13.

[54] P.C. Burns, F.C. Hawthorne, Powder Diffr. 6 (2013) 156–158.

[55] E.A. Baturina, Y.A. Lukyanychev, L.N. Rastorguev, Zh. Strukt. Khim. 7 (1966) 627–628.

[56] R. Coldea, S.M. Hayden, G. Aeppli, T.G. Perring, C.D. Frost, T.E. Mason, S.-W. Cheong, Z. Fisk, Phys. Rev. Lett. 86 (2001) 5377–5380.

[57] P. Fischer, G. Roult, D. Schwarzenbach, J. Phys. Chem. Solids 32 (1971) 1641–1647.

[58] J. Romiszewski, W. Grochala, L.Z. Stolarczyk, J. Phys. Condens. Matter 19 (2007) 116206-1–116206-13.

[59] T. Jaroń, W. Grochala, Phys. Stat. Sol. RRL 2 (2008) 71–73.

[60] M. Schmitt, O. Janson, M. Schmidt, S. Hoffmann, W. Schnelle, S.-L. Drechsler, H. Rosner, Phys. Rev. B 79 (2009) 245119-1–245119-5. M. Banks, R.K. Kremer, C. Hoch, A. Simon, B. Ouladdiaf, J.-M. Broto, H. Rakoto, C. Lee, M.-H. Whangbo, Phys. Rev. B, 80 (2009) 024404-1–024404-15.

[61] W. Grochala, in: 16th European Symposium on Fluorine Chemistry, Ljubljana, 2010 (this and other poems presented at that meeting were refined with help from a native speaker, Dr. Andrew J. Churchard).

[62] R.-H. Odenthal, D. Paus, R. Hoppe, Z. Anorg. Allg. Chem. 407 (1974) 144–150.

[63] S.E. McLain, M.R. Dolgos, D.A. Tennant, J.F.C. Turner, T. Barnes, T. Proffen, B.C. Sales, R.I. Bewley, Nat. Mater. 5 (2006) 561–566.

[64] T. Lancaster, S. Blundell, P. Baker, W. Hayes, S. Giblin, S.E. McLain, F. Pratt, Z. Salman, E. Jacobs, J. Turner, T. Barnes, Phys. Rev. B 75 (2007), 220408-1–220408-4.

[65] W. Grochala, Nat. Mater. 5 (2006) 513.

[66] R.-H. Odenthal, R. Hoppe, Monats. Chem. 102 (1971) 1340–1350.

[67] D. Kurzydłowski, M. Derzsi, A. Budzianowski, Z. Jagličić, W. Koźmiński, Z. Mazej, W. Grochala, Eur. J. Inorg. Chem. 2010 (2010) 2919–2925.

[68] G.A. Baker, H.E. Gilbert, J. Eve, G.S. Rushbrooke, Phys. Lett. A 25 (1967) 207–209.

[69] Z. Mazej, D. Kurzydłowski, E. Goreshnik, W. Grochala (unpublished results).

[70] D. Dai, M.-H. Whangbo, J. Chem. Phys. 114 (2001) 2887–2893. D. Dai, M.-H. Whangbo, J. Chem. Phys., 118 (2003) 29–39.

[71] Y. Zhao, B. Gaulin, J. Castellan, J. Ruff, S. Dunsiger, G. Gu, H. Dabkowska, Phys. Rev. B 76 (2007), 184121-1–184121-7.

[72] A.I. Popov, Y.M. Kiselev, J. Russ, Inorg. Chem. English Transl. 33 (1988) 965.

[73] M.S. Wickleder, Z. Anorg. Allg. Chem. 627 (2001) 2112–2114.

[74] A.C. Walters, T.G. Perring, J.-S. Caux, A.T. Savici, G.D. Gu, C.-C. Lee, W. Ku, I.A. Zaliznyak, Nat. Phys. 5 (2009) 867–872.
[75] D. Kurzydłowski, Z. Mazej, Z. Jagličić, Y. Filinchuk, W. Grochala, Chem. Commun. 49 (2013) 6262–6264.
[76] K.S. Pedersen, M.A. Sørensen, J. Bendix, Coord. Chem. Rev. 299 (2015) 1–21.
[77] R. Klingeler, N. Leps, I. Hellmann, A. Popa, U. Stockert, C. Hess, V. Kataev, H.-J. Grafe, F. Hammerath, G. Lang, S. Wurmehl, G. Behr, L. Harnagea, S. Singh, B. Büchner, Phys. Rev. B 81 (2010), 024506-1–024506-5.
[78] S. Satija, J. Axe, G. Shirane, H. Yoshizawa, K. Hirakawa, Phys. Rev. B 21 (1980) 2001–2007.

Magnetic Properties of Transition Metal Fluoride Perovskites

13

L. Clark, P. Lightfoot
University of St Andrews, Fife, Scotland

Chapter Outline

1. Introduction

This chapter focuses on the magnetic properties of first-row transition metal fluorides with the perovskite-type structure. This structural family is one of the most important in the fields of solid-state chemistry and physics and materials science due to the enormous structural diversity that it is capable of displaying. The aim of this chapter is to elaborate on how this structural diversity leads to the demonstration of a wide variety of interesting magnetic behaviors by transition metal fluorides belonging to this structural class. The following work is organized in terms of different derivatives of perovskite-related structures, beginning with the ideal cubic perovskite, the study of which in terms of transition metal fluorides played a key role in paving out our understanding of magnetic interactions in inorganic solids. In each section, the relevant perovskite-type structure is described in detail in relation to the magnetic properties of several important examples. In doing so, many of the key concepts in solid-state magnetism and the experimental tools required for its study are also covered. For an overview of the magnetic properties of transition metal fluorides in general, the reader is directed toward the works of Dance and Tressaud [15] and Ferey et al. [25] and for a more complete understanding of the structural chemistry of transition metal fluorides there is the recent review by Leblanc et al. [68]. For an excellent introduction to the perovskite family the reader is recommended the monograph by Mitchell [86].

Photonic and Electronic Properties of Fluoride Materials. http://dx.doi.org/10.1016/B978-0-12-801639-8.00013-1

2. Cubic Perovskites

The ideal perovskite has the general formula ABX_3 where A and B are cations, usually with ionic radii, r, such that $r_A > r_B$ and X is an anion with an ionic radius r_X comparable in size to r_A. The charge carried by the X anion dictates which cations can occur within the ideal perovskite structure, such that for a fluoride perovskite ABF_3 with the monovalent F^- anion the A-site cation is monovalent, for example, an alkali group cation or the ammonium NH_4^+ cation, and the B-site cation must be divalent, typically a divalent transition metal cation. The ideal perovskite structure is cubic with space group symmetry $Pm\bar{3}m$ and consists of cubic close-packed layers of anions with an ordered replacement of one quarter of these sites by A cations. The B cations occupy the octahedral interstices adjacent to the anion sites. A common representation of the cubic perovskite unit cell is visualized with the B cations located at the origin, which clearly demonstrates the sixfold octahedral coordination of the B-site cations by anions and the 12-fold cubo-octahedral coordination of the A-site, as shown in Fig. 13.1. An important parameter that plays a role in governing whether a given set of ions will adopt the cubic perovskite structure is known as the tolerance factor t,

$$t = \frac{r_A + r_B}{\sqrt{2}(r_B + r_X)} \qquad [13.1]$$

where once again r_A, r_B, and r_X are the ionic radii of the A, B, and X ions, respectively. In the case of the ideal cubic perovskite structure, $t \sim 1$. The relatively small ionic radius of the fluoride anion (1.33 Å [113]) means that the cubic perovskite structure can tolerate a variety of A- and B-site cations, and a large number of cubic perovskites that are known to be stable under the conditions of ambient temperature and pressure are fluorides.

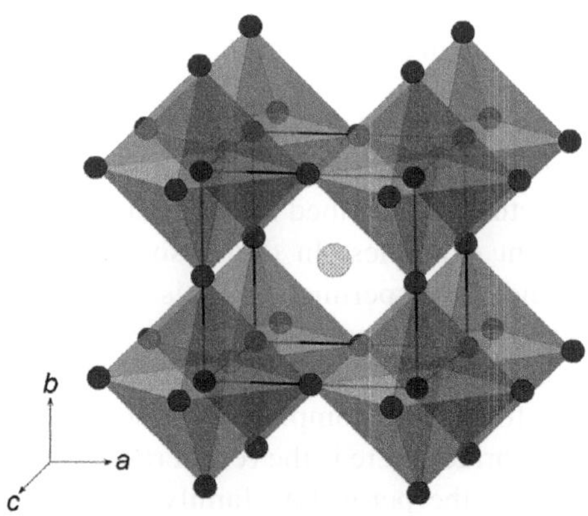

Figure 13.1 The ideal perovskite structure ABX_3 with B-site cations (polyhedra) located at the origin of the cubic unit cell (*solid lines*) to show the corner-sharing network of BX_6 octahedra. The A-site cation (*hatched spheres*) sits at the 12-fold site at the center of the unit cell surrounded by X anions (*solid spheres*).

A well-studied family of magnetic cubic fluoride perovskites are the KMF_3 materials, where M is a divalent 3*d* transition metal cation, M = Mn, Fe, Co, and Ni. The M = Mn−Ni series was first reported in the early 1950s and can be prepared in both polycrystalline and single-crystal forms [92]. The initial magnetic study of these systems gave an important insight into magnetic ordering mechanisms in inorganic fluorides. An early investigation of these materials by magnetic susceptibility measurements was performed by Hirakawa et al. [38], which revealed evidence for antiferromagnetic ordering in polycrystalline KMF_3 M = Mn, Fe, Co, and Ni at high transition temperatures of T_N = 88, 113, 114, and 275K, respectively, implying strong magnetic interactions. A subsequent powder neutron diffraction study by Scatturin et al. [107] allowed for the determination of the antiferromagnetic structures adopted in the spin-ordered states of KMF_3 M = Mn, Fe, Co, and Ni. In the case of the $KFeF_3$, $KCoF_3$, and $KNiF_3$ systems, a comparison of the data that were collected at 4.2K and room temperature showed the appearance of superlattice reflections in the low-temperature data that in each case were indexed on a cubic unit cell with a lattice constant twice that of the corresponding nuclear cell. Furthermore, the intensity ratio of the nuclear Bragg peaks at 298 and 4.2K precluded any structural phase transition upon cooling as well as confirming the absence of any magnetic scattering intensity superimposed on the nuclear Bragg reflections with all even *hkl* indices. The antiferromagnetic spin structure was thus described with nearest-neighbor spins aligned antiparallel with respect to one another in all three directions of a cube with an edge length twice that of the nuclear unit cell, which satisfies the observed *k*-vector and extinction rules of the magnetic lattice. This magnetic structure, with each spin antiferromagnetically coupled to all six of its nearest neighbors, is known as a G-type magnetic structure and is shown in Fig. 13.2. The $KMnF_3$ system showed the appearance of the same magnetic Bragg peaks at low temperature, in addition to two small peaks around the (222) nuclear reflection indicative of a structural distortion from the ideal cubic structure. Subsequent analysis of the structure of $KMnF_3$ has shown that it undergoes a series of crystallographic transitions upon cooling [14].

The G-type magnetic structure was originally observed in the oxide perovskite $CaMnO_3$ with Mn^{4+} $3d^3$ ions located on the *B*-site [122] and can be understood in terms of an indirect superexchange of the magnetic transition metal cations through nonmagnetic anions. As described by the Goodenough−Kanamori−Anderson rules [2,31,53], the sign of a superexchange interaction is closely related to the orbital state of the cation centers as well as the crystallographic structure. In the cubic perovskite structure, chains of anion octahedra run along all three crystallographic directions resulting in strong orbital overlap of neighboring magnetic cations through the nonmagnetic anions that produces strong superexchange interactions and thus the high observed ordering temperatures. For systems based on transition metal cations with the d^3 and d^5−d^8 electronic configurations in this geometry, chains of antiparallel nearest-neighbor spins result and, therefore, the $CaMnO_3$ and the KMF_3 (M = Mn, Fe, Co, and Ni) perovskites all adopt the G-type structure below their respective transition temperatures. For systems based on the d^4 electronic configuration, such as Mn^{3+} in the $LaMnO_3$ perovskite [122] the *p*-orbitals associated with the nonmagnetic anions may overlap with both half-filled and empty e_g orbitals of the transition metal cation,

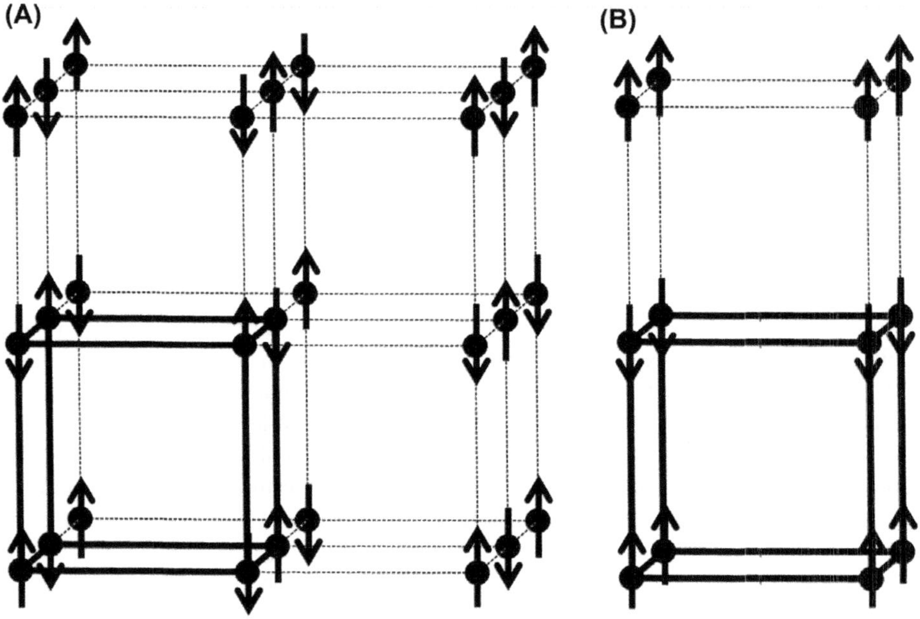

Figure 13.2 (A) The antiferromagnetic G-type structure on a cubic lattice with every magnetic spin arranged antiparallel to each of its six nearest neighbors. This results in a magnetic unit cell (*dotted lines*) with a doubled cell length in all three crystallographic directions in comparison to the nuclear unit cell (*solid lines*). (B) In the A-type structure, layers of ferromagnetically coupled magnetic spins are stacked in an antiferromagnetic fashion, such that the magnetic unit cell is doubled in one direction in comparison to the nuclear unit cell.

which allows for both antiferromagnetic and ferromagnetic exchange couplings and consequently the so-called A-type order can result, which is also shown in Fig. 13.2 for comparison.

3. Distorted Perovskite Derivatives

Despite the fact that the ideal perovskite structure is cubic, many materials with the ABX_3 perovskite-type structure are distorted and hence are of lower symmetry. These distortions most often arise from tilts of the BX_6 octahedra or from distortions of the octahedra themselves. Octahedral tilting occurs as a result of a size mismatch when the A-site cation is too small for the 12-fold coordination site within the BX_6 framework, such that the octahedra must tilt in order to accommodate the cation. In the majority of cases, the BX_6 octahedra can be viewed as rigid units that tilt without distortion or any disruption to their corner-sharing connectivity. The $A-X$ bond length and A-site coordination are affected, however, and as such the symmetry is lowered. Different tilting systems and how they distort the ABX_3 structure have been very well studied [29,30,42,123]. The symmetry of ABX_3 perovskites can also be lowered by distortions inherent to the BX_6 octahedra themselves if, for example, they are based on Jahn–Teller active cations such as $3d^9$ Cu^{2+} or $3d^4$ Cr^{2+}. In the following section we will review some examples of distorted transition metal fluoride perovskites and discuss this in terms of their magnetic behaviors.

As we have outlined in the previous section, the $KMnF_3$, $KFeF_3$, $KCoF_3$, and $KNiF_3$ systems all belong to the ideal perovskite structure type at room temperature, which results in long-ranged three-dimensional antiferromagnetic ordering in these materials upon cooling. $KCuF_3$ is based on Cu^{2+} $3d^9$ ions and is rather unique among its other KMF_3 analogs in terms of its structure, and hence, its magnetic properties. The orbital degeneracy of the Cu^{2+} ion within the octahedral crystal field to which it is subjected in the perovskite structure makes $KCuF_3$ a Jahn–Teller active system. The distortion of the CuF_6 octahedra as a result of the Jahn–Teller effect gives rise to strong Heisenberg coupling of nearest-neighbor spins in one-dimensional chains along the c-axis. Furthermore, the spin $S = ½$ carried by the Cu^{2+} ion means that a description of the system may include strong quantum fluctuations as a result of the small size of the spin moment. Early magnetic studies of $KCuF_3$ showed a broad peak in magnetic susceptibility at 243K [38], and an absence of magnetic Bragg peaks in powder neutron diffraction data collected at 4.2K [107]. Both features were taken as evidence for the low-dimensional quantum nature of $KCuF_3$, such that this distorted perovskite fluoride has become key in experimental studies of the one-dimensional $S = ½$ chain, a complete understanding of which has been a long-standing problem in condensed matter physics [7].

The cooperative Jahn–Teller distortion of $KCuF_3$ at $T_{JT} \sim 800$K results in a significant split of the Cu–F bond lengths within the CuF_6 octahedra and a lowering of the ideal perovskite cubic symmetry [54]. Within its distorted structure, $KCuF_3$ displays a Kugel–Khomskii-type orbital ordering [59,64,65] of the d-orbitals on the Cu^{2+} ions, with short and long Cu–F bonds within each octahedron [18]. The crystal structure of $KCuF_3$ was first determined by Okazaki and Suemune [93] by X-ray diffraction. They described the distorted perovskite superstructure of $KCuF_3$ with tetragonal *I4/mcm* space group symmetry and lattice constants $a = \sqrt{2a_0} = 5.855$ Å and $c = 2c_0 = 7.852$ Å, where a_0 and c_0 are the lattice constants of the fundamental pseudo-perovskite cell. In the tetragonal structure, the Cu^{2+} ion centers are surrounded by a distorted octahedron of F^- ions with Cu–F distances of 2.25(1), 1.960(1), and 1.89(1) Å, described as long, medium, and short bonds, respectively. In $KCuF_3$ antiferrodistortive orbital ordering results, shown in Fig. 13.3, with the filled e_g orbital oriented toward the two F^- p-orbitals situated furthest away, ie, along the longest Cu–F bonds and the Cu^{2+} e_g hole orbital alternating within the ab-plane with each e_g hole orbital orthogonal to that on each of its neighboring octahedra. This so-called antiferrodistortive orbital ordering results in an antiferromagnetic coupling along the c-axis, where the F^- orbital overlaps with two half-filled Cu^{2+} orbitals, and a ferromagnetic coupling in the ab-plane where the orbital overlap is between a half-filled Cu^{2+} orbital on one side and a completely filled orbital on the other. Therefore, in terms of this orbitally ordered structure one expects $KCuF_3$ to adopt an A-type magnetic structure below its Néel temperature, similar to that shown in Fig. 13.2.

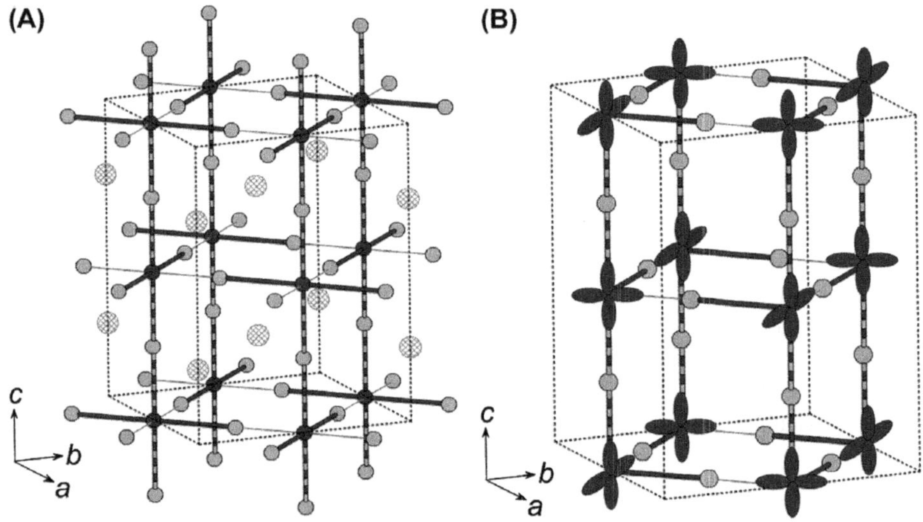

Figure 13.3 The orbital degeneracy of the Cu^{2+} ions in the octahedral crystal field of the $KCuF_3$ perovskite-type structure results in the distortion of the CuF_6 octahedra with (A) three inequivalent Cu–F (*solid spheres*) bond lengths labeled as short (*thin lines*), medium (*dashed solid lines*), and long (*solid lines*). The distorted octahedra are ordered such that the medium length bonds lie along the *c*-axis and the long and short bonds alternate in the *ab*-plane. The K^+ ions are depicted by the *hatched spheres* and the *dashed lines* show the unit cell edges. The (B) antiferrodistortive ordering of the Cu^{2+}*d*-orbitals in $KCuF_3$ results in the hole orbital oriented orthogonally to each of its nearest neighbors giving rise to an antiferromagnetic in-chain interaction along the *c*-axis and a ferromagnetic out-of-chain exchange in the *ab*-plane via the nonmagnetic F^- ions. For clarity the K^+ ions are not shown here.

The tetragonal structure of $KCuF_3$ proposed by Okazaki and Suemune [93] was challenged when two electron spin resonance studies [45,125] supported the existence of an antisymmetric Dzyaloshinsky–Moriya [19,87] (DM) interaction, which requires that the midpoint between two magnetic centers involved in the interaction is not a center of inversion. Despite the F^- ion position between two Cu^{2+} centers in the tetragonal $KCuF_3$ structure obeying this rule, the local symmetry around the Cu^{2+} ions results in the DM term in the spin Hamiltonian being equal to zero, contrary to the electron spin resonance results. Therefore, analysis of a new set of crystals by X-ray diffraction led to the proposal of an orthorhombic structure for $KCuF_3$ with space group symmetry $P2_12_12_1$, which was argued to arise from a cooperative shift of the F^- ions and allows for a DM interaction to exist [37]. However, a more recent study by powder neutron diffraction analyzed the structure of $KCuF_3$ over a wide temperature range by the Rietveld refinement method. Marshall et al. [82] tested both tetragonal *I4/mcm* and orthorhombic $P2_12_12_1$ models against the data and found that all nuclear Bragg peaks could be indexed by the tetragonal structural model. It should also be pointed out that a subsequent analysis of the structure by Okazaki [94] did reveal that there are two polymorphs of the tetragonal structure, shown in Fig. 13.4.

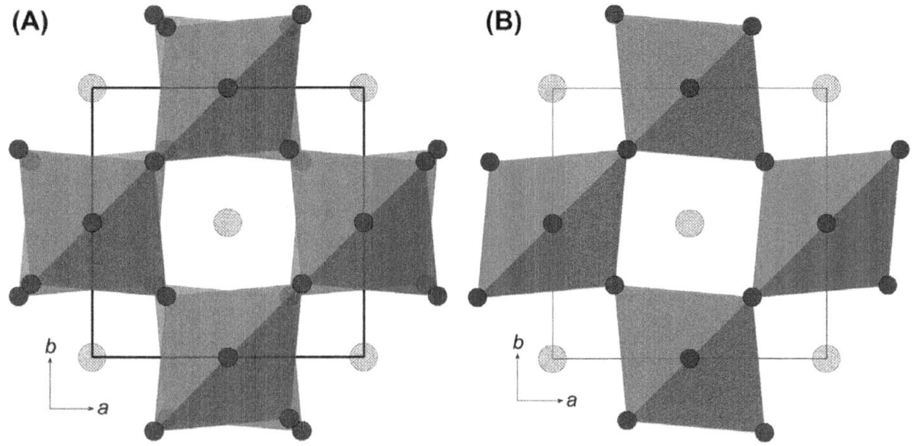

Figure 13.4 The two tetragonal polymorphs of $KCuF_3$ viewed down the *c*-axis to show (A) out-of-phase Jahn-Teller distortions of CuF_6 octahedra, with the sense of the distortion alternating in the *I4/mcm* structure and the (B) in-phase Jahn-Teller distortions occurring in the *P4/mbm* structure.

The displacements of the F^- ions within the distorted CuF_6 octahedra are due to the Jahn-Teller effect. These may either act in the same sense, or in an alternating sense when viewed along the *c*-axis, and the two distinct options mirror those of 'in-phase' and 'out-of-phase' tilts in the Glazer notation, ie, tilt systems $a^0a^0c^+$ and $a^0a^0c^-$.

A second powder neutron diffraction investigation of $KCuF_3$ by Hutchings et al. [43] revealed the onset of long-range A-type order in both of the tetragonal polymorphs. Two single crystals were studied, which both demonstrated the coexistence of the two stable polymorphs. The authors were able to identify a Néel ordering to an A-type ordered state at $T_N = 38(1)$K for the *I4/mcm* phase and at $T_N = 22(4)$K for the second phase, which is thought to have space group symmetry *P4/mbm*. The form of the magnetic scattering in reciprocal space below the Néel temperature is very sharp, which implies three-dimensional long-range order. However, the ordered moment measured at 4.2K was only 0.49(7) μ_B per Cu^{2+} ion, which is considerably smaller than the expected ordered moment of $gS\mu_B = 1.1\ \mu_B$, and this was taken as indirect evidence for the one-dimensional magnetic character of $KCuF_3$.

An alternative view of the magnetic interactions within $KCuF_3$ is to consider the system as a one-dimensional $S = {}^1\!/_2$ chain composed of strongly interacting nearest-neighbor Cu^{2+} cations along the *c*-direction but with very weak interactions in the *ab*-plane. Kadota et al. [49] show that fitting the broad peak observed in the magnetic susceptibility data of $KCuF_3$ at 243K with spin chain theories of Bonner and Fisher [9] and Oguchi [90] allows one to estimate the values of the exchange interaction between nearest-neighbor spins along the chain, $J_c/k_B \sim -190$K, and perpendicular to the chain direction, $J_a/k_B < 0.2$K. This one-dimensional nature is argued to arise from the orbital configuration of the Cu^{2+} ions as a result of the Jahn–Teller distortion, and certainly the antiferrodistortive Kugel–Khomskii-type orbital ordering accounts for the different signs of the in-chain ($J_c < 0$, antiferromagnetic) and out-of-chain ($J_a > 0$, ferromagnetic) exchange interactions as discussed above in terms of the crystal structure. What is puzzling, however, is that this model does not fully account for the very discordant magnitudes of J_c and J_a. Kadota et al. [49] also present heat capacity data for $KCuF_3$ that show the absence of a sharp peak upon cooling through the transition at 243K, which rules out a long-range ordering transition at this temperature. In addition, the authors of this study present torque measurements for single crystals of $KCuF_3$, which reveal that the spin anisotropy in $KCuF_3$ is small and temperature independent, such that it is reasonable to assume Heisenberg spin symmetry for the system. Although Hutchings et al. [43] did not observe any direct evidence for one-dimensionality from their elastic neutron scattering study of $KCuF_3$, a subsequent quasi-elastic neutron scattering experiment by Ikeda and Hirakawa [44] showed that spin correlations along the chain direction persist well above T_N with a correlation length of ~500 Å and an estimate of the ratio $J_c/|J_a| \sim 0.027$, in reasonable agreement with the values of the in-chain and out-of-chain exchange parameters estimated from magnetic susceptibility.

The possibility that a one-dimensional Heisenberg antiferromagnetic $S = ½$ chain is realized in $KCuF_3$ was an exciting one as the complex behavior expected for such a system remains to be completely understood. There is no long-range magnetic order expected for a truly one-dimensional magnetic system, and this disordered ground state is of interest because of the unusual nature of the excitations that are predicted to emerge from it [7]. In a three-dimensional Heisenberg magnet, excitations from the magnetically ordered ground state are bosonic magnons, charge neutral entities carrying a spin $S = 1$. The spin excitations of such a system can be investigated through inelastic neutron scattering measurements, and given that the selection rule for such an experiment is $\Delta S = \pm 1$ a single magnon can interact with a neutron in a neutron spectroscopy experiment to give well-defined spin waves. However, the excitations of a one-dimensional Heisenberg $S = ½$ chain are fermionic species known as spinons, which carry spin $S = ½$ and as such, and are created or annihilated in pairs during an inelastic neutron scattering experiment. The dispersion relation for the spinon excitation is given by [16],

$$\hbar\omega = \pi|J_c \sin(qc)| \tag{13.2}$$

where J_c is the nearest-neighbor exchange along the chain, q is the wavevector, and c is the nearest-neighbor separation. For a pair of spinons, one should observe a continuum of excitations between upper, $\hbar\omega_u$, and lower, $\hbar\omega_l$, limits, respectively.

$$\hbar\omega_u = 2\pi|J_c \sin(qc)| \text{ and } \hbar\omega_l = \pi|J_c \sin(qc)| \tag{13.3}$$

The simplest observable continuum is the two-spinon continuum, shown in Fig. 13.5. The measurement of this characteristic excitation spectrum in $KCuF_3$ by inelastic neutron scattering has been carried out in a number of works

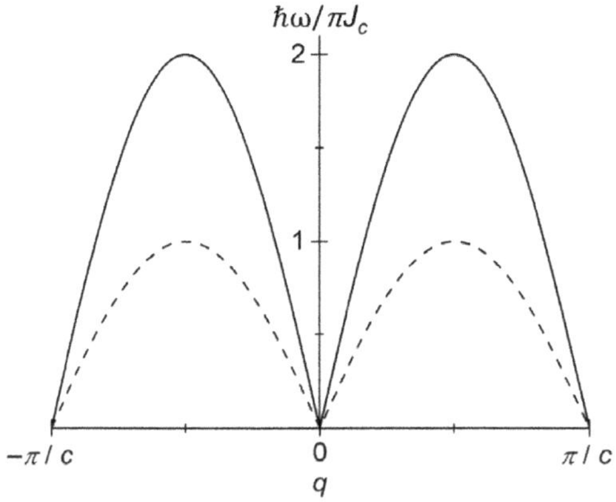

Figure 13.5 The characteristic form of the dispersion relation of the two-spinon excitation that is observed in the inelastic neutron scattering spectra of $KCuF_3$ as a result of its one-dimensional Heisenberg $S = 1/2$ nature. For the two-spinon excitation a continuum of inelastic scattering observed between the upper (*solid line*) and lower (*dashed line*) limits of energy transfer, $\hbar\omega$, as a function of momentum transfer, q.

[66,67,88,106,118–120] and has been key in confirming its one-dimensional magnetic behavior over a wide temperature range. In the work of Lake et al. [66], universal scaling behavior indicative of a one-dimensional quantum system close to a quantum critical point [34,103] was identified in $KCuF_3$. Through careful analysis of inelastic neutron scattering data, Lake et al. [66] were able to map out a magnetic phase diagram of $KCuF_3$ and show that it adopts this universal quantum critical phase over a wide range of temperature. At the lowest temperatures and incident energies of their study, they observe the Néel ordered phase through the strongly dispersive scattering typical of a three-dimensional ordered system. Interestingly, the authors also find a relatively large crossover region between the ordered state and the quantum phase over which the well-defined spin waves of the Néel state begin to fractionalize into $S = 1/2$ spinons.

Certainly, investigations of $KCuF_3$ over the years have proved fruitful with much interest garnered for such an unusual system that displays one-dimensional magnetic behavior despite its three-dimensional structure. Lee et al. [69] have since put forward a proposal based on Raman and X-ray scattering measurements for a two-step orbital ordering process in $KCuF_3$ as an explanation for the distinct regimes in the magnetic phase diagram and the onset of long-range magnetic order at $T_N = 38K$.

$KCrF_3$ is similar to the $KCuF_3$ system in that it adopts a perovskite-type structure at ambient conditions and contains the Jahn–Teller active ion Cr^{2+} with electronic configuration $3d^4$. An early structural investigation revealed that $KCrF_3$ adopts a tetragonally distorted perovskite superstructure at room temperature [21]. Powder neutron diffraction data collected at 4.2K in the study by Scatturin et al. [107] revealed the growth of well-defined magnetic peaks that were indexed on a tetragonal unit cell. The observed magnetic reflections had even h and k indices with $l = 2n + 1$, which implied that the Cr^{2+} spins lie within the (001) plane but with their direction alternating along the [001] direction. A study of the structure of $KCrF_3$ by Margadonna and Karotis [80,81] confirmed that the system is tetragonal at room temperature with space group symmetry *I4/mcm* and lattice constants $a = 6.05230(2)$ Å and $c = 8.02198(4)$ Å. The authors provided high-resolution synchrotron powder X-ray diffraction data that give strong evidence for the same type of cooperative Jahn–Teller distortion of the CrF_6 octahedra and subsequent antiferrodistortive orbital ordering that is also observed in the $KCuF_3$ analog. As such the distorted octahedra in $KCrF_3$ are arranged with medium length Cr–F bonds (2.00549(1) Å) oriented along the *c*-axis and long (2.294(4) Å) and short (1.986(4) Å) bonds alternating in the *ab*-plane. The low-temperature behavior of $KCrF_3$ is considerably different to that observed for $KCuF_3$ and on cooling below 250K $KCrF_3$ undergoes a structural phase transition to a monoclinic *I*112/*m* phase with a pronounced tilting of the CrF_6 octahedra, most probably driven by the reduced tolerance factor of $KCrF_3$ in comparison to its copper-based analog [80]. The orbital ordering observed in the tetragonal phase was shown to remain unchanged, but the low-symmetry, low-temperature structure has two distinct Cr^{2+} sites and has a significantly increased octahedral distortion. On further cooling to below 5K, the system appears to undergo a reentrant phase transition with a reduced monoclinic distortion. The magnetic susceptibility of $KCrF_3$ shows a sharp peak at $T_N = 46$K indicative of antiferromagnetic ordering with a Weiss temperature $\theta = +2.7(1)$K reflecting the ferromagnetic correlations within the A-type ordered state. Interestingly, the field-cooled magnetization of $KCrF_3$ measured in very weak fields shows an increase in the magnetization below T_N with two faint features at 11 and 4K that disappear in stronger applied fields. This behavior is consistent with weak ferromagnetism with the canting of spins argued to arise from single-ion anisotropy effects or the DM interaction [80].

Another important family of magnetic transition metal fluorides with a distorted perovskite-type structure are the $NaMF_3$ (M = Mn, Fe, Co, Ni) materials, which adopt the $GdFeO_3$-type structure at ambient conditions. They are composed of corner-sharing networks of regular MF_6 octahedra with the distortion to the observed orthorhombic superstructure driven solely by the small tolerance factor ($t < 0.9$), which is below the limit required to stabilize the ideal cubic structure. The relatively small ionic radius of the Na^+ ion is such that it causes a tilting of the MF_6 octahedra of the $a^+b^-b^-$ type to give eightfold coordination at the Na^+ site. The $NaMF_3$ systems are of interest due to their propensity to undergo the perovskite to postperovskite structural phase transition under moderate applied pressures [114,127], which has implications in the understanding of the structure and properties of materials in the mantle of the Earth [89]. Furthermore, $NaMnF_3$ is of particular importance due to its potential multiferroicity. A recent first-principles investigation of the ferroelectric instability in fluoride perovskites revealed that although in the majority of cases ABF_3 perovskite-type fluorides are nonferroelectric materials, $NaMnF_3$ is proposed to be an exception with an unusually soft ferroelectric mode, such that multiferroic behavior is predicted in $NaMnF_3$ even in the absence of applied pressure or strain [27]. It has been known for some time that the M = Mn, Co, and Ni systems order antiferromagnetically to adopt a G-type structure with a weak ferromagnetic component below the Néel temperature resulting from a canting of spins from both single-ion anisotropy effects and antisymmetric exchange [26,28,112]. The structural [5] and magnetic

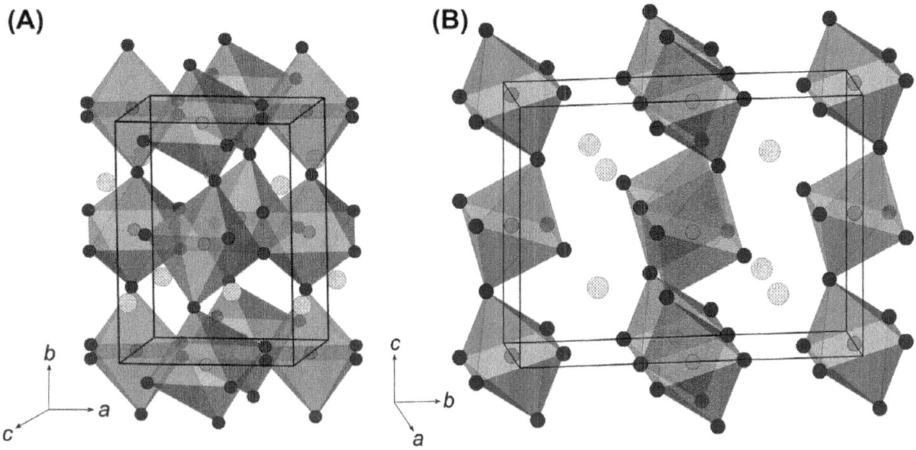

Figure 13.6 The orthorhombic (A) *Pnma* perovskite superstructure of $NaFeF_3$ that results from an $a^+b^-b^-$ tilting of the FeF_6 octahedra due to a reduced tolerance structure and the (B) *Cmcm* postperovskite phase, which contains a network of corner- and edge-sharing octahedra that can be recovered from a high-pressure treatment of the *Pnma* structure.

properties of $NaFeF_3$ were recently investigated through high-resolution synchrotron powder X-ray diffraction and magnetic susceptibility measurements by Bernal et al. [6]. At room temperature and ambient pressure, $NaFeF_3$ adopts the orthorhombic *Pnma* perovskite superstructure shown in Fig. 13.6A. Magnetic susceptibility data show a marked increase below 90K, indicating the onset of long-range magnetic order. A divergence between field-cooled and zero field–cooled susceptibilities was observed below 90K, as well as the opening of a small hysteresis loop in magnetization versus field curves at 4K, which is consistent with the weak ferromagnetic behavior observed in other $NaMF_3$ analogs. In the same study it was also shown that polycrystalline $NaFeF_3$ undergoes a structural transition to the *Cmcm* postperovskite phase, shown in Fig. 13.6B, under application of ~10 GPa pressure at room temperature, with the postperovskite phase fully recovered at ambient pressure. The magnetic susceptibility as a function of temperature for the high-pressure phase shows a sharp peak at $T_N = 46K$, indicating antiferromagnetic ordering but once again, a splitting of the susceptibility measured under field-cooled and zero field–cooled conditions below T_N suggests a canting of spins. This weak ferromagnetic response is abated in stronger applied fields, which results in greater spin alignment and hence a stronger antiferromagnetic response [6].

4. Layered Perovskites

Layered perovskites have received long-standing attention due to their interesting physical properties that stem from their pseudo-two-dimensional crystal structures. The Ruddlesden-Popper family of layered perovskites has a general formula

$A_{n+1}B_nX_{3n+1}$. Their structure can be viewed as being composed of ABX_3 perovskite layers interleaved with AX rocksalt-type units, such that the general formula may also be written in the form $(AX + nABX_3)$ where n gives the number of perovskite layers within a slab. The perovskite layers of corner-sharing BX_6 octahedra are offset and effectively cut at regular intervals along the c-axis. Octahedra that would normally share an anion at their apex in the ABX_3 perovskite structure are now separated and in order to overcome the anion deficiency that is created in the separation of corner-sharing octahedral layers an AX unit is intercalated, thus restoring the BX_6 coordination. The separation of the perovskite layers by nonmagnetic AX units results in two-dimensional magnetic behavior. It also reduces the symmetry of Ruddlesden-Popper phases from the cubic symmetry of the ideal ABX_3 perovskites to tetragonal. The transition metal fluoride K_2NiF_4 is an archetypal $n = 1$ Ruddlesden-Popper phase with *I4/mmm* symmetry and lattice constants $a = 4.006$ Å and $c = 13.076$ Å at room temperature [4]. The K_2NiF_4-type structure, shown in Fig. 13.7, is stable only over a very narrow range of tolerance factors ($0.95 < t < 0.985$), and it is recognized that the structure cannot sustain such large deviations from $t = 1$ as is the case for ABX_3-type perovskites [100]. As a result of its layered structure, it is expected that K_2NiF_4 is a good model for a two-dimensional magnet. From a theoretical point of view, it is known that a two-dimensional Heisenberg magnet does not display spontaneous magnetization at $T > 0$ if the interactions have a finite range [85], but that such a system is very sensitive toward interlayer exchange interactions and spin anisotropy that can result in long-range magnetic ordering at a finite critical temperature [55,74,91]. The existence of materials that may approximate such two-dimensional magnetic behavior was important in order to be able to experimentally investigate their magnetic properties.

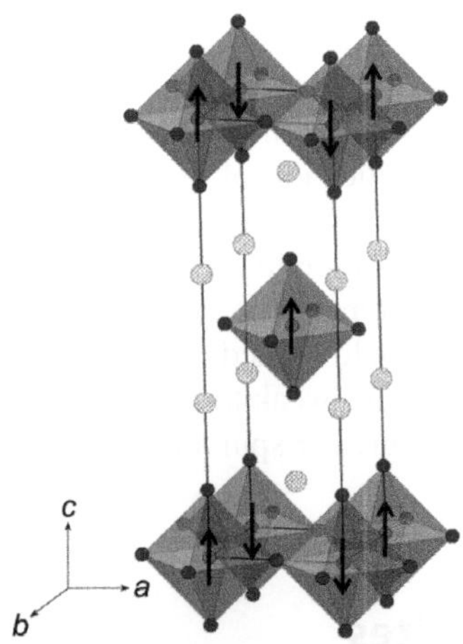

Figure 13.7 The layered $n = 1$ Ruddlesden-Popper structure of K_2NiF_4, which has an antiferromagnetic coupling of Ni^{2+} spins in the perovskite layers reminiscent of $KNiF_3$. The G-type ordering of the spins in each perovskite layer results in an effective decoupling of successive layers which in turn promotes the pronounced two-dimensionality of the magnetic behavior of K_2NiF_4.

The original measurement of the magnetic susceptibility of K_2NiF_4 by Srivastava [117] showed a broad peak with a subsequent discontinuity upon cooling to a temperature approximately half of that at which χ_{max} is observed. An early powder neutron scattering study of K_2NiF_4 found that the perovskite layers order antiferromagnetically at 180K, but no true three-dimensional long-range order was detected to low temperatures [98]. The results of this neutron diffraction study and the characteristic shape of the susceptibility curve were taken as the original evidence for the genuinely two-dimensional magnetic nature of K_2NiF_4. It was determined from neutron diffraction that the spins in K_2NiF_4 are aligned along the c-axis, such that the magnetic structure of K_2NiF_4 can be viewed as shown in Fig. 13.7. The strong antiferromagnetic exchange interaction between neighboring Ni^{2+} spins is such that nearest-neighbor spins are aligned antiparallel with respect to each other, reminiscent of the G-type structure of $KNiF_3$, and the net exchange field at each cation site therefore sums to zero. This effective decoupling of the nearest-neighbor perovskite layers makes the two-dimensionality of K_2NiF_4 far more pronounced that other layered materials. The intralayer nearest-neighbor antiferromagnetic exchange energy in K_2NiF_4 is estimated to be $J/k_B = -100(10)$K with the interlayer next-nearest-neighbor exchange, through which three-dimensional long-range order can be established, at least four orders of magnitude weaker [72]. Unambiguous evidence for the two-dimensional nature of the magnetic correlations in K_2NiF_4 was given by the neutron scattering study of Birgeneau et al. [8]. In this study, the authors concluded that K_2NiF_4 undergoes a three-dimensional magnetic ordering transition at $T_N = 97.1$K, which is evidenced by the growth of well-defined, sharp magnetic Bragg peaks. However, the magnetic diffuse scattering in the critical region above the Néel ordering transition takes the form of rods of scattering, a definitive indication of two-dimensional magnetic correlations. This confirms that K_2NiF_4 behaves as a two-dimensional antiferromagnet with very long-ranged spin correlations with a length scale on the order of 1000 Å within the perovskite layers but no measurable interaction between them. This rod-like diffuse scattering remains well defined up to 200K, indicating that there is a large temperature range over which K_2NiF_4 displays two-dimensional behavior. Furthermore, it was found that the sublattice magnetization in K_2NiF_4 follows a $(T_N - T)^\beta$ temperature dependence with a critical exponent β characteristic of a two-dimensional magnet rather than three-dimensional magnet [33]. K_2MnF_4 [11], Rb_2MnF_4 [33], K_2CoF_4 [12,32], Rb_2CoF_4 [11,12], and Rb_2FeF_4 [71] are isostructural with K_2NiF_4 and as such display two-dimensional antiferromagnetic behavior with subsequent long-range order adopted upon cooling through Néel temperatures of 45K, 38.4K [33], 107K, 101K [12], and 56.3K [71], respectively.

Although K_2CuF_4 essentially adopts the K_2NiF_4 structure type, its magnetic properties are different to those observed for K_2NiF_4. K_2CuF_4 is a two-dimensional ferromagnet. An early structural investigation of K_2CuF_4 demonstrated its adoption of the tetragonal *I4/mmm* structure with lattice constants $a = 4.155$ Å and $c = 12.74$ Å [62], with the first magnetic study indicating the presence of ferromagnetic correlations from the extrapolation of inverse magnetic susceptibility to give a positive Weiss constant [70]. A subsequent magnetic study of K_2CuF_4 was performed by Yamada [124], in which the author presented magnetic susceptibility measurements along the a- and

c-axis that yielded Weiss constants $\theta_a = +18.0$K and $\theta_c = +22.5$K, respectively, from Curie–Weiss analysis of the data. Below 18K it was shown that the magnetic susceptibility is dependent on the external magnetic field, displaying a larger magnetic susceptibility in smaller applied field strengths typical of a ferromagnet. The magnetic heat capacity of K_2CuF_4 shows a sharp λ-peak at $T_C = 6.25$K, which marks the paramagnetic to ferromagnetic phase transition and a much broader feature centered around 9K, which is believed to arise from short-range magnetic correlations. Evidence for the two-dimensional nature of K_2CuF_4 comes from both magnetic and thermodynamic measurements. The magnetic susceptibility of K_2CuF_4 begins to deviate from the Curie–Weiss law at temperatures almost an order of magnitude greater than T_C and the ratio of $|\theta|/T_C$ is unusually large for a three-dimensional magnet. A series expansion fit of the high-temperature magnetic susceptibility gives an estimate for the ferromagnetic intralayer exchange in K_2CuF_4 as $J/k_B = 11.2$K. Heat capacity measurements of K_2CuF_4 also provide evidence to support its two-dimensional nature. The low temperature magnetic heat capacity follows a linear temperature dependence and in addition to this, only 26% of the total magnetic entropy of the system is released at the ordering transition, T_C. Both of these observations are highly unlikely in a system with three-dimensional ferromagnetic order. Fitting the low-temperature magnetic heat capacity gives a value of $J/k_B = 8.0$K, and so often a mean value of the magnetic and thermodynamic estimates of $J/k_B = 10.0$K is quoted. An investigation of K_2CuF_4 by neutron scattering revealed the growth of magnetic Bragg peaks consistent with the ferromagnetic alignment of the Cu^{2+} spins in the ab-plane [39]. Furthermore, the analysis of the critical scattering from the system allowed for an assessment of the interlayer exchange J', with $J'/J = 0.00066$ confirming excellent two-dimensionality of the magnetic interactions within K_2CuF_4.

The ferromagnetic intralayer exchange is argued to result from a Kugel–Khomskii-type orbital ordering of the cooperatively Jahn–Teller-distorted CuF_6, akin to that observed in the Jahn–Teller active $KCuF_3$ and $KCrF_3$ systems discussed in the previous section. The tetragonally distorted CuF_6 octahedra consist of long and short Cu–F bonds, with the elongated bonds oriented within the ab-plane ($Cu{-}F_a = 2.08$ Å and $Cu{-}F_c = 1.95$ Å) with each elongated axis orthogonal to those in adjacent octahedra [35,36] giving rise to a ferromagnetic superexchange through the nonmagnetic F^- ions [58,63]. More recently, the polarization dependence of the soft X-ray absorption spectra of K_2CuF_4 at the Cu L and F K edges has been investigated to show that orbital ordering is indeed the origin the ferromagnetic intralayer exchange. The source of the ferromagnetic interlayer exchange J', however, is different [77]. An analytical and numerical study of the exchange pathways in K_2CuF_4 by Feldkemper et al. [24] revealed that the ferromagnetic interlayer coupling results from interference effects between inequivalent hopping paths. The presence of this nontrivial J' interlayer interaction is thought to bring about the long-range magnetic ordering transition at $T_C = 6.25$K.

The magnetic properties of the related Cu^{2+} systems Rb_2CuF_4 [17] and Cs_2CuF_4 [105] are very similar to those of K_2CuF_4. For instance, the magnetic susceptibility of high-quality single crystals of Cs_2CuF_4 measured with the field aligned perpendicular to the c-axis shows a sharp increase with decreasing temperature until a saturation value is reached at $T_C = 6.05$K. The magnetic susceptibility measured with the field oriented along the *c*-axis shows no such saturated response, but a more gradual increase in susceptibility upon cooling toward T_C, which implies that Cs_2CuF_4 is ferromagnetic below 6.05K with the easy axis in the *ab*-plane [105]. Magnetic susceptibility measurements of Rb_2CuF_4 show an analogous response with a transition to a ferromagnetically ordered state at $T_C = 6.05$K [17]. Due to their K_2NiF_4-type structure, Rb_2CuF_4 and Cs_2CuF_4 also both display two-dimensional magnetic behavior with an estimate of the ferromagnetic intralayer exchange given by $J/k_B = 12.3$K and $J/k_B = 12.6$K, respectively. What is interesting, however, is that despite the similar energy scale and sign of the intralayer exchange in the A_2CuF_4 (A = K, Rb, Cs) systems at ambient pressure, they show remarkably different magnetic behaviors under the application of pressure. Ishizuka et al. [47] performed diamond anvil cell magnetic susceptibility measurements on K_2CuF_4 up to pressures of 13 GPa and found that the signatures of ferromagnetism disappeared at pressures above 9 GPa. From these susceptibility measurements, it was not clear whether the collapse of the ferromagnetic response is due to a change in the intra- or interlayer superexchange pathways, or both. Ensuing high pressure X-ray [46] and Raman [75] structural studies revealed that the high-pressure magnetic transition is concomitant with a change in the orbital ordering of the Cu^{2+} ions such that the elongated axes within the distorted CuF_6 octahedra align along the *a*-axis and thus interrupt the intralayer ferromagnetic exchange. The high-pressure behavior of Rb_2CuF_4 is comparable to that of the K_2CuF_4 system. A high-pressure powder neutron diffraction study of Cs_2CuF_4 revealed surprisingly different behavior. In a study by Manaka et al. [76] the pressure dependence of the ferromagnetic (004) Bragg peak was tracked and was found to disappear above 2.2 GPa. Moreover, at the same critical pressure a magnetic Bragg peak at the $(00\frac{1}{2})$ position was observed in the diffraction data, indicative of an antiferromagnetic spin structure. In contrast to both K_2CuF_4 and Rb_2CuF_4 the Cu^{2+} spins within the layers remain ferromagnetically coupled but the next-nearest-neighbor interlayer interaction is antiferromagnetic as shown in Fig. 13.8.

$K_3Cu_2F_7$ is an $n = 2$ Ruddlesden-Popper phase with the $Sr_3Ti_2O_7$-type structure [101]. It can also be viewed as a double-layer perovskite with $KCuF_3$ double layers with each double-layer offset by $[\frac{1}{2},\frac{1}{2},\frac{1}{2}]$, as shown in Fig. 13.9A [78]. Once again, an antiferrodistortive orbital ordering is expected due to the cooperative Jahn−Teller distortion of the CuF_6 octahedra. Each double layer within $K_3Cu_2F_7$ has been proposed to consist of two simple square lattices of Cu^{2+} ions with two different exchange interactions, shown in Fig. 13.9B, labeled J_{dimer} and J_a along the *c*- and *a*-axis, respectively. The orbital ordering of the distorted CuF_6 octahedra results in a ferromagnetic coupling within the layers such that $J_a > 0$ and the values of $|J_a/J_{dimer}|$ and $|J_{dimer}/J_a|$ are small, in keeping with $KCuF_3$ and K_2CuF_4, such that there is no frustration within the double square lattice. This results in isolated dimers of Cu^{2+} spins along the *c*-axis with a spin dimer gap predicted if the dimer exchange is antiferromagnetic in nature, $J_{dimer} < 0$ [48,50−52,60,104,110]. Analysis of room-temperature powder X-ray diffraction data of $K_3Cu_2F_7$ confirms that it adopts the tetragonal *I4/mmm* structure with $a = 4.17$ Å and $c = 20.63$ Å [78]. The measurement of magnetic susceptibility presented by Manaka et al. [78] shows a broad maximum centered around 250K,

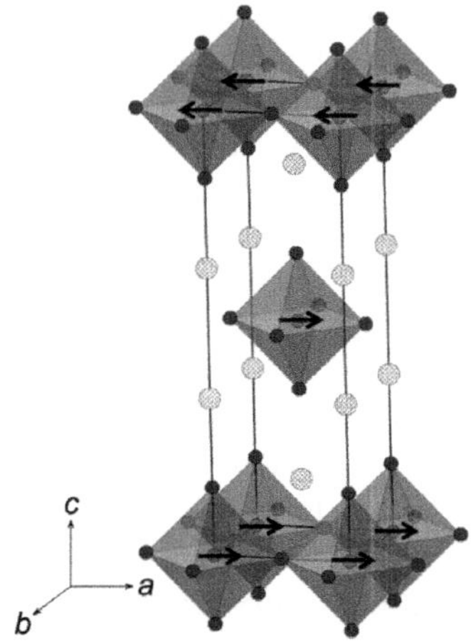

Figure 13.8 The high-pressure magnetic structure of Cs_2CuF_4—at a critical pressure of 2.2 GPa the interlayer exchange interaction becomes antiferromagnetic while the intralayer coupling remains ferromagnetic. This is unlike the magnetic response of the K_2CuF_4 and Rb_2CuF_4 analogs under the external application of pressure, where the ferromagnetic exchange between layers persists up to 9 GPa. The spin direction of the body-centered Cu^{2+} ion is tentative and could not be unambiguously determined from powder neutron diffraction.

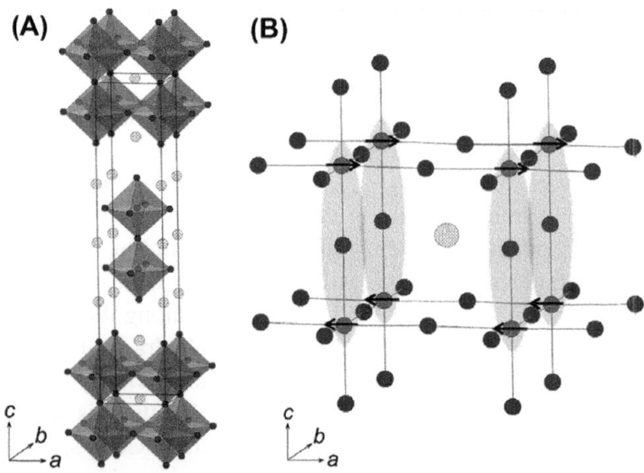

Figure 13.9 The (A) layered structure of the $n = 2$ Ruddlesden-Popper or double-layer perovskite phase of $K_3Cu_2F_7$. Each (B) perovskite double layer can be considered as being composed of two square lattices with a ferromagnetic coupling of the Cu^{2+} spins within one square plane and antiferromagnetic coupling between them, forming nonmagnetic spin $S = 0$ dimers along the c-axis.

with the curvature of this feature characteristic of a large spin gap thus indicating the antiferromagnetic nature of J_{dimer}. The fit of this curve in the magnetic susceptibility of $K_3Cu_2F_7$ by a typical model for isolated spin dimers gives an estimate of the spin dimer gap of $\sim$400K. Unfortunately the nonmagnetic nature of the spin dimer ground state was hidden in this magnetic susceptibility measurement due to a contamination of Cu^{2+} impurity spins giving rise to a large Curie tail below 50K.

The double-layer or $n = 2$ Ruddlesden-Popper perovskites, $K_3Ni_2F_7$ and $K_3Co_2F_7$, have also been reported with the $Sr_3Ti_2O_7$-type structure [3]. Here, however, the spin dimer gap physics observed in the $K_3Cu_2F_7$ analog driven by the cooperative Jahn–Teller distortion of the orbitally degenerated Cu^{2+} ions in an octahedral crystal field is not expected for the Ni^{2+}- and Co^{2+}-based systems. The magnetic susceptibility [79] of $K_3Ni_2F_7$ shows a broad peak around 286K, indicative of low-dimensional magnetic correlations, and a sharper feature at $T_N = 132(4)$K suggestive of an antiferromagnetic phase transition. The magnetic susceptibility above 150K can be well fitted by a model for an $S = 1$ quadratic Heisenberg antiferromagnet [73] with an intraplanar exchange given by $J/k_B = -66(1)$K. The susceptibility data presented for $K_3Co_2F_7$ also reveal a broad peak, in this case centered around 150K, but the curvature of this feature is quite different from the peak observed for the Ni^{2+} analog and that expected for a Heisenberg antiferromagnet. Strong Ising anisotropy is expected for the $K_3Co_2F_7$ system as the orbital angular momentum of Co^{2+} is not fully quenched by the surrounding crystal field. An anomaly in the susceptibility data at $T_N = 126(4)$K reveals the onset of antiferromagnetic order [79].

Another important class of layered perovskite-type transition metal fluorides are the $BaMF_4$ materials with M = Mn, Fe, Co, or Ni, which are of great interest due to their multiferroic behavior. The $BaMF_4$ structure has orthorhombic $A2_1am$ symmetry with buckled sheets of distorted corner-sharing MF_6 octahedra perpendicular to the b-axis, separated by layers of Ba^{2+} cations as shown in Fig. 13.10 [57]. Alternatively, it can be

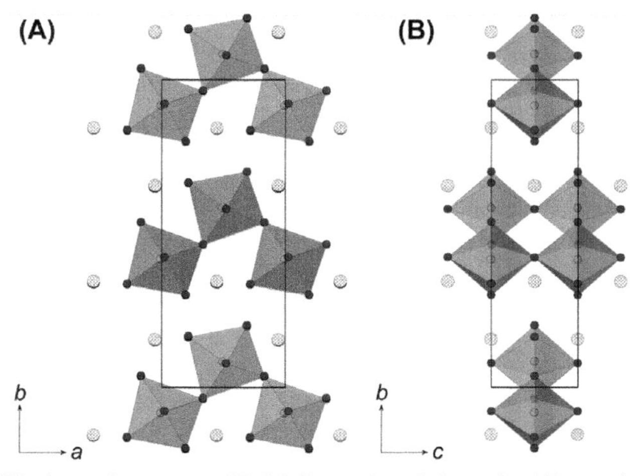

Figure 13.10 The layered structure of $BaMnF_4$ as viewed down the (A) c- and (B) a-axis with puckered sheets of distorted MnF_6 octahedra (polyhedra) separated by layers of Ba^{2+} ions (*hatched spheres*).

viewed as diagonal slabs of perovskite-type octahedral layers [10]. Multiferroic materials are defined by the coexistence of magnetic and dielectric orders, but what is unusual about the $BaMF_4$ systems is that the electric polarization has an entirely geometric origin and is caused by tilting of the MF_6 octahedra [20]. For the M = Co and Ni systems ferroelectric switching has been demonstrated, with the application of an electric field along the *a*-axis causing a rotation of the MF_6 octahedra and the reversal of the electric polarization [22]. In the case of the M = Mn and Fe analogs, the ionic radii are larger such that the rotation of the MF_6 octahedra upon application of an electric field is impeded [57] and, therefore, the $BaMnF_4$ and $BaFeF_4$ systems are pyroelectric rather than ferroelectric [109]. At 250K, $BaMnF_4$ undergoes a second-order transition to an antiferroelectric phase with a structure that is incommensurate along the *a*-axis [116].

The $BaMF_4$ systems display two-dimensional antiferromagnetism characterized by broad maxima in their susceptibility curves upon cooling. In $BaMnF_4$ this broad feature is observed at ~50K with a second transition at 25K indicating the onset of a three-dimensional antiferromagnetically ordered state [41]. Below this ordering temperature, the susceptibility measured with the field aligned along the *b*-axis drops rapidly toward a small but finite value, which was taken to indicate the presence of a weak ferromagnetic component. The magnetic structure of $BaMnF_4$ as determined by neutron diffraction was reported by Cox et al. [13] with the ordered Mn^{2+} moment of 4.8 μ_B aligned at an angle of 9 degrees to the *b*-axis in the *bc*-plane. The weak ferromagnetic behavior observed in early susceptibility studies of $BaMnF_4$ was considered to be an important factor in the origin of an unusual magnetoelectric effect that occurs at the long-range ordering transition marked by a strong first-order dielectric anomaly. Scott [108] proposed that because $BaMnF_4$ is a weak ferromagnet and a pyroelectric there must be terms in the Hamiltonian that are linear in both the applied electric field and in the magnetoelectric tensor and that it is this linear magnetoelectric effect that causes a canting of spins and a ferromagnetic component along the *c*-direction. Scott [108] also argued that this ferromagnetism plays a key role in the appearance of the pyroelectric anomaly at T_N and that without it the anomaly would be a much weaker second-order effect. However, doubt surrounding the origin of the low-temperature pyroelectric anomaly was recently raised as a comprehensive measurement of the susceptibility of $BaMnF_4$ showed no evidence for a ferromagnetic component, suggesting that the pyroelectric anomaly at T_N is not directly associated with the ordering of the Mn^{2+} spins [126]. Further experimental evidence for the absence of a ferromagnetic component in $BaMnF_4$ was given by Poole et al. [96] who used neutron polarization analysis to show that the magnetic structure is antiferromagnetic with the magnetic moment completely in the *bc*-plane with no signature of the ferromagnetic component argued to arise from a linear magnetoelectric effect. Instead, it has been proposed that the unusual dielectric behavior in $BaMnF_4$ stems from a hybridization of Mn^{2+} and F^- *d*- and *p*-orbitals, respectively, which strongly effects the superexchange interactions and causes an electronic modulation of the system [126].

5. Hexagonal Perovskites

The ideal ABX_3 perovskite structure described earlier in this chapter may be viewed as cubic close-packed AX_3 layers with B-site cations located in octahedral interstitial sites to form corner-sharing BX_6 octahedra. The stacking of these cubic close-packed layers follows an ABCABC... sequence, which is commonly referred to as the 3C stacking polytype as the repeat sequence of the cubic close-packed layers of octahedra in the unit cell is *ccc*, where *c* signifies a corner-sharing octahedron [99]. The ABX_3 structure can also be described in terms of a hexagonal unit cell, which consists of three AX_3 layers positioned orthogonally to the [111] direction of the cubic unit cell. In the ideal hexagonal ABX_3 structure ($P6_3/mmc$ space group symmetry) the AX_3 layers are hexagonal close packed (ABABAB...) along the [001] direction of the hexagonal unit cell, with the B-site cations situated within the octahedral interstices that form face-sharing BX_6 octahedra. This hexagonal stacking sequence is known as the 2H polytype due to the *hh* repeat sequence in the unit cell, where *h* denotes a face-sharing octahedron [56]. Hexagonal perovskites tend to form for systems with tolerance factors $t > 1$ when the ionic radius of the A-site cation is too large to stabilize the ideal cubic structure. Between these two ideal stacking polytypes there is a possibility for the formation of a variety of different intermediate structures with varying proportions of cubic and hexagonal stacking. The nomenclature used to describe the different polytypes of hexagonal perovskites is a number signifying the number of AX_3 layers stacked along the c-axis of the unit cell and an upper-case letter designating the symmetry that results from the stacking sequence. A particular stacking polytype that is important in the discussion of the magnetic properties of transition metal fluoride perovskites is the 6*H*-type structure. The 6*H*-type structure is a common polytype constructed from mixtures of cubic (*c*) corner-sharing and hexagonal (*h*) face-sharing octahedra. The structure contains six layers of octahedra in its $P6_3/mmc$ unit cell with a *cchcch* stacking sequence, as shown in Fig. 13.11A. The reason why this hexagonal perovskite polytype is of interest when realized in transition metal fluorides is due to its proclivity to support the ferrimagnetic ordering of spins associated with the magnetic transition metal cations.

Ferrimagnetism was originally observed in the 6*H*-type hexagonal perovskite $RbNiF_3$ [102,111,115]. An early report of the structure of $RbNiF_3$ confirmed that it adopts the hexagonal $P6_3/mmc$ space group symmetry with lattice constants $a = 5.840(2)$ Å and $c = 14.308(4)$ Å [121]. The structure contains two inequivalent Ni^{2+} sites with Wyckoff positions 2*a* and 4*f*, both of which sit at the center of the NiF_6 octahedra. The Ni^{2+} ions on the 4*f* sites sit at the center of pairs of slightly distorted face-sharing octahedra that form Ni_2F_9 dimeric units. The remainder of the Ni^{2+} ions on the 2*a* site form regular NiF_6 octahedra that share corners with the dimeric units. The presence of these two inequivalent sites in the structure of $RbNiF_3$ has important consequences for its magnetic properties and is the source of the ferrimagnetic behavior that it displays. Magnetic measurements of $RbNiF_3$ revealed a considerable net magnetization below the transition temperature $T_N = 139$K [111], which ruled

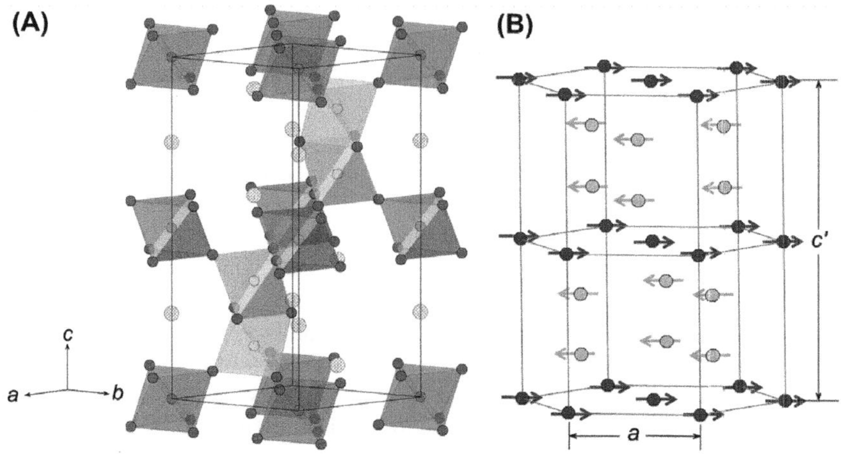

Figure 13.11 The (A) hexagonal 6*H*-type structure of $RbNiF_3$ with two inequivalent Ni^{2+} sites leading to a (B) ferrimagnetic spin structure below the Néel temperature $T_N = 139K$. The ferrimagnetism arises as a result of a 180-degree Ni—F—Ni superexchange pathway between nearest-neighbor Ni^{2+} ions on the 2*a* (light) and 4*f* (dark) sites giving a relatively strong antiferromagnetic coupling and a near-90-degree Ni—F—Ni superexchange pathway between nearest-neighbor Ni^{2+} spins within the 4*f* site polyhedra producing a weaker ferromagnetic coupling.

out previous proposals of an antiferromagnetic or weak ferromagnetic ground state [40]. However, the measured saturated moment below the ordering transition temperature reflects only one-third of the expected value for Ni^{2+} ions. This behavior is indicative of a ferrimagnetic ordering of two magnetic sublattices, which exist in $RbNiF_3$ due to the two inequivalent Ni^{2+} sites in the crystal structure. Curie—Weiss analysis of the high-temperature susceptibility of $RbNiF_3$ gives a Weiss temperature $\theta = -300K$ reflective of strong antiferromagnetic correlations. An early powder neutron diffraction study of $RbNiF_3$ revealed the growth of magnetic Bragg peaks upon cooling that are consistent with a ferrimagnetic model, with an antiparallel spin arrangement of the Ni^{2+} ions on the 2*a* and 4*f* sites [95]. This magnetic structure, shown in Fig. 13.11B, was confirmed by a subsequent neutron scattering study by Als-Nielsen et al. [1] and stems from a 180-degree superexchange pathway between nearest neighbors on the 2*a* and 4*f* sites and a near 90-degree superexchange angle between nearest neighbors on the 4*f* sites resulting in relatively strong and weak antiferro- and ferromagnetic exchange interactions, respectively. Analysis of the spin-wave scattering from $RbNiF_3$ allowed for an estimate of these two exchange parameters yielding a value of −93.2(2)K for the nearest-neighbor antiferromagnetic exchange and 21.1(2)K for the nearest-neighbor ferromagnetic exchange [1]. An interesting study by McGuire et al. [83] showed how the magnetic properties of $RbNiF_3$ vary upon substitution of Ni^{2+} for other magnetic (Co^{2+} and Mn^{2+}) and nonmagnetic (Mg^{2+} and Ca^{2+}) ions. In the case of Mg^{2+} doping, there is little site preference for the Mg^{2+} over the two available cation sites, and the substituted derivative retains the ferrimagnetic nature of $RbNiF_3$ up to $RbNi_{0.33}Mg_{0.66}F_3$. For the $RbNi_{1-x}Ca_xF_3$ system on the other hand, where there is a strong site preference of the Ca^{2+} ions for the $2a$ sites, ferrimagnetic order is destroyed by one-third Ca^{2+} doping as it preferentially disrupts the strong antiferromagnetic exchange between the ferromagnetic layers. It is possible to completely replace Rb^+ for Tl^+ and retain the hexagonal $6H$-type structure for $TlNiF_3$, which has a ferrimagnetic ordering transition at $T_N = 150K$ [61]. $CsFeF_3$ is also isostructural with $RbNiF_3$ and is ferrimagnetic below 60K [23,84,97].

References

[1] J. Als-Nielsen, R.J. Birgeneau, H.J. Guggenheim, Phys. Rev. B 6 (1972) 2030.
[2] P.W. Anderson, Phys. Rev. 115 (1959) 2.
[3] D. Babel, E. Herdtweck, Z. Anorg. Allg. Chem. 487 (1982) 75.
[4] D. Balz, K. Plieth, Z. Elektrochem 59 (1955) 545.
[5] G. Benner, R.J. Hoppe, J. Fluorine Chem. 46 (1990) 283.
[6] F.L. Bernal, K.V. Yusenko, J. Sottmann, C. Drathen, J. Guignard, O. Martin Løvvik, W.A. Crichton, S. Margadonna, Inorg. Chem. 53 (2014) 12205.
[7] H. Bethe, Z. Phys. 71 (1931) 205.
[8] R.J. Birgeneau, H.J. Guggenheim, G. Shirane, Phys. Rev. B 1 (1971) 2211.
[9] J.C. Bonner, M.E. Fisher, Phys. Rev. 135 (1964) A640.
[10] J.K. Brandon, H.D. Megaw, Philos. Mag. 21 (1970) 189.
[11] D.J. Breed, Physica 37 (1967) 35.
[12] D.J. Breed, K. Gilijamse, A.R. Miedema, Physica 45 (1969) 205.
[13] D.E. Cox, S.M. Shapiro, R.A. Cowley, M. Eibschütz, H.J. Guggenheim, Phys. Rev. B 19 (1979) 5754.
[14] M.A. Carpenter, E.K.H. Salje, C.J. Howard, Phys. Rev. B 85 (2012) 224430.
[15] J.-M. Dance, A. Tressaud, Chapter 9, in: P. Hagenmuller (Ed.), Inorganic Solid Fluorides: Chemistry and Physics, Academic Press Inc, Orlando, FL, 1985.
[16] J. Des Cloizeaux, J.J. Pearson, Phys. Rev. 128 (1962) 2131.
[17] C. Dekker, A.F.M. Arts, H.W. de Wijn, Phys. Rev. B 38 (1988) 11512.
[18] J.D. Dunitz, L.E. Orgel, J. Phys. Chem. Solids 3 (1957) 20.
[19] I. Dzyaloshinsky, J. Phys. Chem. Solids 4 (1958) 241.
[20] C. Ederer, N.A. Spaldin, Phys. Rev. B 74 (2006) 024102.
[21] A.J. Edwards, R.D. Peacock, J. Chem. Soc. (1959) 4126.
[22] M. Eibschütz, H.J. Guggenheim, S.H. Wemple, I. Camlibel, M. DiDomenico Jr., Phys. Lett. 29A (1969a) 409.
[23] M. Eibschütz, L. Holmes, H.J. Guggenheim, H.J. Levinstein, J. Appl. Phys. 40 (1969b) 1312.
[24] S. Feldkemper, W. Weber, J. Schulenberg, J. Richter, Phys. Rev. B 52 (1995) 313.
[25] G. Ferey, M. Leblanc, R. de Pape, J. Pannetier, Chapter 10, in: P. Hagenmuller (Ed.), Inorganic Solid Fluorides: Chemistry and Physics, Academic Press Inc, Orlando, FL, 1985.
[26] Z. Friedman, M. Melamud, J. Makovsky, H. Shaked, Phys. Rev. B 2 (1970) 179.
[27] A.C. Garcia-Castro, N.A. Spaldin, A.H. Romero, E. Bousquet, Phys. Rev. B 89 (2014a) 104107.
[28] A.C. Garcia-Castro, A.H. Romero, E. Bousquet, Phys. Rev. B 90 (2014b) 064113.

[29] A.M. Glazer, Acta Cryst. B 28 (1972) 3384.
[30] A.M. Glazer, Acta Cryst. A 31 (1975) 759.
[31] J.B. Goodenough, Phys. Rev. 100 (1955) 564.
[32] T. Goto, K. Nakao, T. Sakakibara, M. Ito, I. Yamada, Phys. B 177 (1992) 373.
[33] P. Heller, Rept. Prog. Phys. 30 (1967) 731.
[34] J.A. Hertz, Phys. Rev. B 14 (1976) 1165.
[35] M. Hidaka, P.J. Walker, Solid State Commun. 31 (1979) 383.
[36] M. Hidaka, K. Inoue, I. Yamada, P.J. Walker, Phys. B 121 (1983) 343.
[37] M. Hidaka, T. Eguchi, I. Yamada, J. Phys. Soc. Jpn. 67 (1998) 2488.
[38] K. Hirakawa, K. Hirakawa, T. Hashimoto, J. Phys. Soc. Jpn. 15 (1960) 2063.
[39] K. Hirakawa, H. Ikeda, J. Phys. Soc. Jpn. 35 (1973) 1328.
[40] W.W. Holloway, M. Kestigian, Phys. Rev. Lett. 15 (1965) 17.
[41] L. Holmes, M. Eibschütz, H.J. Guggenheim, Solid State Commun. 7 (1969) 973.
[42] C.J. Howard, H.T. Stokes, Acta Cryst. B 54 (1998) 782.
[43] M.T. Hutchings, E.J. Samuelsen, G. Shirane, K. Hirakawa, Phys. Rev. 188 (1969) 919.
[44] H. Ikeda, K. Hirakawa, J. Phys. Soc. Jpn. 35 (1973) 722.
[45] T. Ishii, I. Yamada, J. Phys. 2 (1990) 5771.
[46] M. Ishikawa, M. Terai, M. Hidaka, S. Endo, I. Yamada, O. Shimomura, Phys. Rev. B 57 (1998) 64.
[47] M. Ishizuka, I. Yamada, K. Amaya, S. Endo, J. Phys. Soc. Jpn. 65 (1996) 1927.
[48] M. Jaime, V.F. Correa, N. Harrison, C.D. Batista, N. Kawashima, Y. Kazuma, G.A. Jorge, R. Stern, I. Heinmaa, S.A. Zvyagin, Y. Sasago, K. Uchinokura, Phys. Rev. Lett. 93 (2004) 087203.
[49] S. Kadota, I. Yamada, S. Yoneyama, K. Hirakawa, J. Phys. Soc. Jpn. 23 (1967) 756.
[50] H. Kageyama, K. Yoshimura, R. Stern, N.V. Mushnikov, K. Onizuka, M. Kato, K. Kosuge, C.P. Slichter, T. Goto, Y. Ueda, Phys. Rev. Lett. 82 (1999) 3168.
[51] H. Kageyama, T. Kitano, N. Oba, M. Nishi, S. Nagai, K. Hirota, L. Viciu, J.B. Wiley, J. Yasuda, Y. Baba, Y. Ajiro, K. Yoshimura, J. Phys. Soc. Jpn. 74 (2005a) 1702.
[52] H. Kageyama, J. Yasuda, T. Kitano, K. Totsuka, Y. Narumi, M. Hagiwara, K. Kindo, Y. Baba, N. Oba, Y. Ajiro, K. Yoshimura, J. Phys. Soc. Jpn. 74 (2005b) 3155.
[53] J. Kanamori, J. Phys. Chem. Solids 10 (1959) 87.
[54] J. Kanamori, J. Appl. Phys. 31 (1960) 14S.
[55] E.I. Kats, Sov. Phys. JETP 29 (1969) 1098.
[56] L. Katz, R. Ward, Inorg. Chem. 3 (1964) 205.
[57] E.T. Keve, S.C. Abrahams, J.L. Bernstein, J. Chem. Phys. 51 (1969) 4928.
[58] L.D. Khoi, P. Veillet, Phys. Rev. B 11 (1975) 4128.
[59] D.I. Khomskii, K.I. Kugel, Solid State Commun. 13 (1973) 763.
[60] K. Kodama, M. Takigawa, M. Horvati, C. Berthier, H. Kageyama, Y. Ueda, S. Miyahara, F. Becca, F. Mila, Science 298 (2002) 395.
[61] K. Kohn, R. Fukuda, S. Lida, J. Phys. Soc. Jpn. 22 (1967) 333.
[62] K. Knox, J. Chem. Phys. 30 (1959) 991.
[63] H. Kubo, N. Kaneshima, K. Hirakawa, J. Phys. Soc. Jpn. 41 (1976) 1165.
[64] K.I. Kugel, D.I. Khomskii, Sov. Phys. JETP 37 (1973) 725.
[65] K.I. Kugel, D.I. Khomskii, Sov. Phys. Usp. 25 (1982) 231.
[66] B. Lake, D.A. Tennant, C.D. Frost, S.E. Nagler, Nat. Mater. 4 (2005) 329.
[67] B. Lake, D.A. Tennant, J.-S. Caux, T. Barthel, U. Schollwöck, S.E. Nagler, C.D. Frost, Phys. Rev. Lett. 111 (2013) 137205.
[68] M. LeBlanc, V. Maisonneuve, A. Tressaud, Chem. Rev. 115 (2) (2015) 1191.

[69] J.C.T. Lee, S. Yuan, S. Lal, Y.I. Joe, Y. Gan, S. Smadici, K. Finkelstein, Y. Feng, A. Rusydi, P.M. Goldbart, S.L. Cooper, P. Abbamonte, Nat. Phys. 8 (2012) 63.
[70] E. Legrand, A. Van den Bosch, Solid State Commun. 7 (1969) 1191.
[71] H.J. Levinstein, D.N.E. Buchanan, R.C. Sherwood, Phys. Rev. 173 (1968) 614.
[72] M.E. Lines, Phys. Rev. 164 (1967) 736.
[73] M.E. Lines, J. Phys. Chem. Solids 31 (1970) 101.
[74] M.E. Lines, Phys. Rev. B 3 (1971) 1749.
[75] H. Manaka, I. Yamada, T. Kitazawa, M. Kobayashi, M. Ishizuka, S. Endo, J. Phys. Soc. Jpn. 60 (1997) 2989.
[76] M. Manaka, M. Nishi, I. Yamada, High Press. Res. 17 (2000) 171.
[77] H. Manaka, T. Koide, T. Shidara, I. Yamada, Phys. Rev. B 68 (2003) 184412.
[78] H. Manaka, Y. Miyashita, Y. Wantanabe, T. Masuda, J. Phys. Soc. Jpn. 76 (2007a) 044710.
[79] H. Manaka, Y. Miyashita, Y. Wantanabe, T. Masuda, J. Phys. Soc. Jpn. 76 (2007b) 085003.
[80] S. Margadonna, G. Karotsis, J. Am. Chem. Soc. 128 (2006) 16436.
[81] S. Margadonna, G. Karotsis, J. Mater. Chem. 17 (2007) 2013.
[82] L.G. Marshall, J. Zhou, J. Zhang, J. Han, S.C. Vogel, X. Yu, Y. Zhao, M.T. Fernández-Díaz, J. Cheng, J.B. Goodenough, Phys. Rev. B 87 (2013) 014109.
[83] T.R. McGuire, M.W. Shafer, J. Appl. Phys. 39 (1968) 1130.
[84] T.R. McGuire, V.L. Moruzzi, M.W. Shafer, J. Appl. Phys. 41 (1970) 956.
[85] N.D. Mermin, H. Wagner, Phys. Rev. Lett. 17 (1966) 1133.
[86] R.H. Mitchell, Perovskites Modern and Ancient, Almaz Press Inc. Thunder Bay, ON, 2002.
[87] T. Moriya, Phys. Rev. 120 (1960) 91.
[88] S.E. Nagler, D.A. Tennant, R.A. Cowley, T.G. Perring, S.K. Satija, Phys. Rev. B 44 (1991) 12361.
[89] A.R. Oganov, S. Ono, Nature 430 (2004) 445.
[90] T. Oguchi, Phys. Rev. 133 (1964) A1098.
[91] T. Oguchi, J. Phys. Soc. Jpn. 30 (1971) 988.
[92] A. Okazaki, Y. Suemune, T. Fuchikami, J. Phys. Soc. Jpn. 14 (1959) 1823.
[93] A. Okazaki, A. Suemune, J. Phys. Soc. Jpn. 16 (1961) 176.
[94] A. Okazaki, J. Phys. Soc. Jpn. 26 (1969) 870.
[95] S.J. Pickart, H.A. Alperin, J. Appl. Phys. 42 (1971) 1617.
[96] A. Poole, B. Roessli, O. Zaharko, K.W. Krämer, J. Phys. 23 (2011) 266004.
[97] J. Portier, A. Tressaud, R. Pauthenet, P. Haggenmuller, C.R. Hebd. Seances Acad. Sci. 267 (1968) 1329.
[98] R. Plumier, J. Appl. Phys. 35 (1964) 950.
[99] L.S. Ramsdell, Am. Mineral. 32 (1947) 64.
[100] S.N. Ruddlesden, P. Popper, Acta Cryst. 10 (1957) 538.
[101] S.N. Ruddlesden, P. Popper, Acta Cryst. 11 (1958) 54.
[102] W. Rudorff, J. Kandler, D. Babel, Z. Anorg. Allg. Chem. 317 (1962) 261.
[103] S. Sachdev, Science 288 (2000) 475.
[104] Y. Sasago, K. Uchinokura, A. Zheludev, G. Shirane, Phys. Rev. B 55 (1997) 8357.
[105] S. Sasaki, N. Narita, I. Yamada, J. Phys. Soc. Jpn. 64 (1995) 2701.
[106] S.K. Satija, Phys. Rev. B 21 (1980) 2001.
[107] V. Scatturin, L. Corliss, N. Elliott, J. Hastings, Acta Cryst. 14 (1961) 19.
[108] J.F. Scott, Phys. Rev. B 16 (1977) 2329.

[109] J.F. Scott, Rep. Prog. Phys. 42 (1979) 1055.
[110] S.E. Sebastian, N. Harrison, C.D. Batista, L. Balicas, M. Jaime, P.A. Sharma, N. Kawashima, I.R. Fisher, Nature 441 (2006) 617.
[111] M.W. Shafer, T.R. McGuire, B.E. Argyle, G.J. Fan, Appl. Phys. Lett. 10 (1967) 202.
[112] J.R. Shane, D.H. Lyons, M. Kestigian, J. Appl. Phys. 38 (1967) 1280.
[113] R.D. Shannon, Acta Cryst. A 32 (1976) 751.
[114] Y. Shirako, Y.G. Shi, A. Aimi, D. Mori, H. Kojitani, K. Yamaura, Y. Inaguma, M. Akaogi, J. Solid State Chem. 191 (2012) 167.
[115] G.A. Smolensky, V.M. Yudin, P.P. Syroikov, A.B. Sherman, JETP Lett. 8 (1967) 2368.
[116] E.G. Spencer, H.J. Guggenhiem, G.J. Kominiak, Appl. Phys. Lett. 17 (1970) 300.
[117] K.G. Srivastava, Phys. Lett. 4 (1963) 55.
[118] D.A. Tennant, T.G. Perring, R.A. Cowley, S.E. Nagler, Phys. Rev. Lett. 70 (1993) 4003.
[119] D.A. Tennant, R.A. Cowley, S.E. Nagler, A.M. Tsvelik, Phys. Rev. B 52 (1995a) 13368.
[120] D.A. Tennant, S.E. Nagler, D. Welz, G. Shirane, K. Yamada, Phys. Rev. B 52 (1995b) 13381.
[121] J.E. Weidenborner, A.L. Bednowitz, Acta Cryst. B 26 (1970) 1464.
[122] E.O. Wollan, W.C. Koehler, Phys. Rev. 100 (1955) 345.
[123] P.M. Woodward, Acta Cryst. B 53 (1997) 32.
[124] I. Yamada, J. Phys. Soc. Jpn. 33 (1972) 979.
[125] I. Yamada, H. Fujii, M. Hidaka, J. Phys. 1 (1989) 3397.
[126] M. Yoshimura, M. Hidaka, T. Mizushima, J. Sakurai, T. Tsuboi, W. Kleemann, J. Magn. Magn. Mater. 299 (2006) 404.
[127] H. Yusa, Y. Shirako, M. Akaogi, H. Kojitani, N. Hirao, Y. Oishi, T. Kikegawa, Inorg. Chem. 51 (2012) 6559.

Multiferroism in Fluorides

14

G. Calestani, F. Mezzadri
University of Parma, Parma, Italy

Chapter Outline

1. Introduction

Multiferroics are materials in which at least two of the so-called "ferroic" orders (ferromagnetism, ferroelectricity, ferroelasticity, ferrotoroidicity) coexist. Among them magnetoelectrics (MEs), ie, magnetically and electrically polarizable materials [51,60], constitute the most promising subclass in the applicative field, allowing in principle the realization of multifunctional devices. However, this possibility is subordinate to the occurrence of a sizeable coupling between the two ferroic orders, which could in principle open the route to the realization of completely new devices, for example, electric field–controlled magnetic memories. Frequency-dependent devices, for instance, filters, oscillators, phase shifters, and transducers as well as magnetic (and electric) field sensors constitute additional applicative possibilities. However, at present, the most promising field for the application of ME multiferroics is spintronics, where the design of new ferroelectric memories characterized by nondestructive magnetic reading or nonvolatile magnetic random access memories (MRAM) with an electrical writing procedure (MERAM) can be hypothesized in the presence of large enough ME couplings. However, despite the huge efforts provided in the 2000s to engineer matter in order to increase the number of ME compounds, very few ME compounds are known at present and just a handful of them present both magnetic and electric ordering temperatures above room temperature (RT). The primary reason for the rare occurrence of multiferroicity is that magnetism and ferroelectricity seem generally to exclude each other [28]. Indeed, most of the ferroelectric oxides used

Photonic and Electronic Properties of Fluoride Materials. http://dx.doi.org/10.1016/B978-0-12-801639-8.00014-3

for applications show d^0 electronic configuration (eg, Ti^{4+}, Zr^{4+}, and Nb^{5+}), while partially filled orbitals are mandatory for the occurrence of magnetic ordering. In addition, the presence of unpaired electrons in the outer shells of transition metals often induces thermally activated conductive properties, which prevent the ferroelectric properties (if present) to be exploited for applications. Within this framework, fluorides represent an interesting playground for the study and realization of new multiferroic compounds. On one hand indeed fluorine has higher electronegativity with respect to oxygen, giving rise to different hybridization properties; on the other hand the substitution of fluorine for oxygen lowers the negative charges within the structure allowing the replacement of high-valence d^0 transition metals with low-valence d^n transition metals carrying unpaired electrons. As a result a surprisingly large number of fluorides, belonging to different structural classes, are known to be ferroelectric, while some of them are hypothesized to be ME [54]. However, experimental evidence is reported only for two major classes, the $BaMF_4$ family and the transition metal fluorides with tetragonal tungsten bronze structure (TTB fluorides), with the addition of $Ba_2Ni_7F_{18}$. In the latter case, ferroelectricity was ascribed to triclinic *P1* symmetry, while ferrimagnetism is observed below 36K [50]. Based on symmetry analysis, the presence of a large ME coupling in $Ba_2Ni_7F_{18}$ was proposed [42], nonetheless at present this hypothesis has not been confirmed by any direct observation. The other two classes have been much more extensively studied and are described later.

2. The $BaMF_4$ Family

The barium fluorides $BaMF_4$, where M is Mn, Fe, Co, or Ni, attracted considerable interest from the late 1960s until the early 1980s as piezoelectric magnetic materials, but only with the recent revival of multiferroic materials they received a renewed attention. The compounds are usually prepared by reacting in controlled atmosphere stoichiometric amounts of BaF_2 and MF_2, and large crystals can be grown from the melt by Bridgman or Czochralski techniques. The barium fluorides $BaMF_4$, M = Mn, Fe, Co, Ni, Cu, Zn, and Mg, form an isostructural family of compounds with the polar space group $Cmc2_1$ [18,32,52]. The corresponding crystal structure is shown in Fig. 14.1.

The transition metal cations are octahedrally coordinated by the fluorine anions. The MF_6 octahedra share four of the six corners with adjacent octahedra forming puckered sheets perpendicular to the *b* axis that are intercalated by the Ba cations. As pointed out by Brandon and Megaw [8], the structure can also be visualized as being composed of diagonal slabs of a perovskite-type lattice. The structure is polar along the *c* direction, and all the $BaMF_4$ compounds are piezoelectric at RT; owing to their optical transparency, higher than the one observed in most oxides, they were recently used as optical materials [31,59,63]. The dielectric constant shows an increase with temperature that is characteristic of an incoming ferroelectric phase transition, but ferroelectric switching has been demonstrated only for M = Co, Ni, Zn, and Mg, not for M = Mn and Fe [18]. In spite of the high melting temperatures,

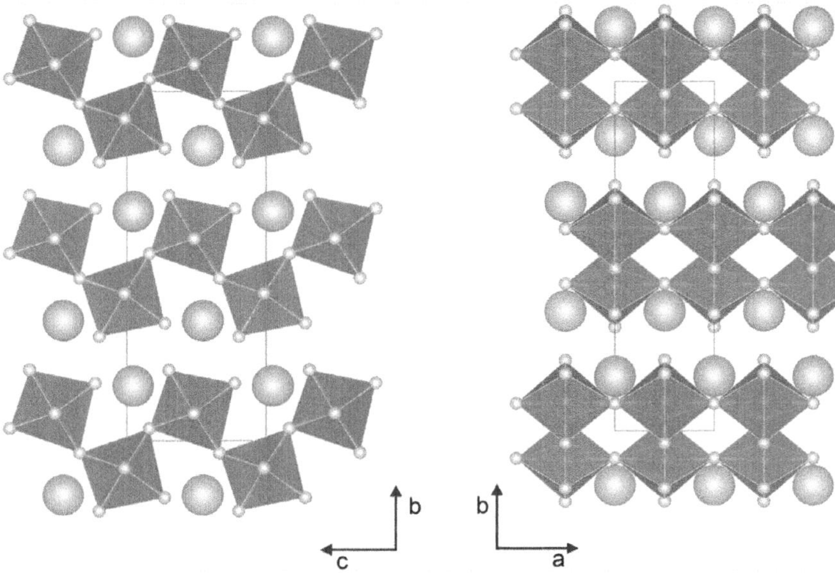

Figure 14.1 Crystal structure of $BaMF_4$ fluorides.

ranging between 720 and 965°C, all crystals melt before the transition to the paraelectric phase. The transition temperatures to a hypothetic paraelectric phase were extrapolated from the temperature dependence of the dielectric constant and range between 810 and 1320°C [13]. While for the other compounds the orthorhombic $Cmc2_1$ structure is stable in the whole temperature range, an additional structural phase transition occurs in $BaMnF_4$ at $\approx$255K [56]. The resulting low-temperature structure is incommensurate along the *c* axis [11,12,47]. The transition is accompanied by a reduction of the symmetry from orthorhombic to monoclinic. The question why $BaMnF_4$ forms a modulated structure while the other two compounds do not, might be answered on the basis of simple geometrical considerations. The structure of $BaZnF_4$ and $BaMgF_4$, the two compounds with smaller octahedral cations, is stable down to 10K and is therefore well adapted to the atomic radius of Ba. In $BaMnF_4$, on the other hand, being Mn^{2+} the largest divalent ion forming this structure, the cavities become too large for the Ba ion at lower temperatures. This induces a structural modulation in which the Ba coordination sphere is adjusted in such a way that the average bonds do not deviate too strongly from their ideal values. The large size of Mn^{2+} can also be considered as the basis of the failure to observe polarization reversal in $BaMnF_4$, as proposed by Keve et al. [32]. In fact, as shown in Fig. 14.2, the polarization reversal would induce a rotation of the MF_6 octahedra that is prevented by the large size of the Mn ion. The same hypothesis can also be reasonably advanced for the iron-based compound.

In addition to the polar distortion, the systems with M = Mn, Fe, Co, and Ni present antiferromagnetic (AFM) ordering below $T_N \approx 50-120$K [17], therefore exhibiting multiferroic behavior. The experimentally observed magnetic structure for M = Ni

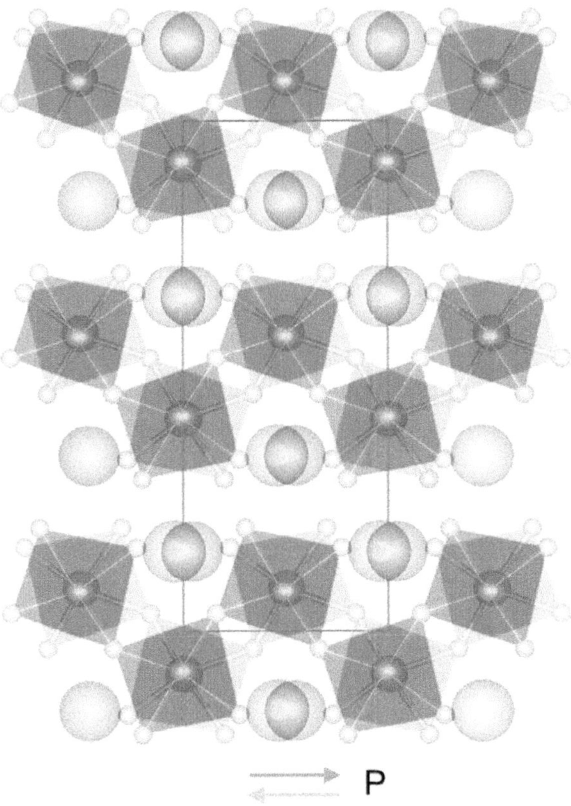

Figure 14.2 Rotation of the MF_6 octahedra connected to the polarization reversal in $BaMF_4$.

and Fe [10,19] is shown in Fig. 14.3. The magnetic cell is doubled along the *a* and *b* directions. The magnetic M cations form quasi two-dimensional puckered layers "parallel" to the *ac* planes. All the moments are oriented along the *b* axis. Within the puckered layers a given moment is coupled antiparallel to the four nearest neighbors: two at a distance of 4.1 Å, and two at 3.8 Å. These intralayer exchange interactions involve Ni–F–Ni configurations which are not far from being linear and are consequently quite strong. However, the coupling between adjacent grids effectively cancels so that the net magnetic resultant given by adjacent layers is zero. The magnetic order along the *b* direction is determined by the weak coupling between the next nearest-neighbor planes that are about 14 Å apart.

Different from these systems, in $BaCoF_4$ the magnetic moments are oriented parallel to the *c* axis [20]. Moreover, two different magnetic phases are observed (Fig. 14.3). In both cases there is AFM ordering within the puckered sheets, and the interactions between nearest-neighbor sheets average out to zero; the structures differ only in the stacking of the next nearest-neighbor sheets, 14 Å apart along the *b* direction, for which the coupling is either parallel (phase B) or antiparallel (phase A). If the different orientation of the moments is not considered, phase A shows the same

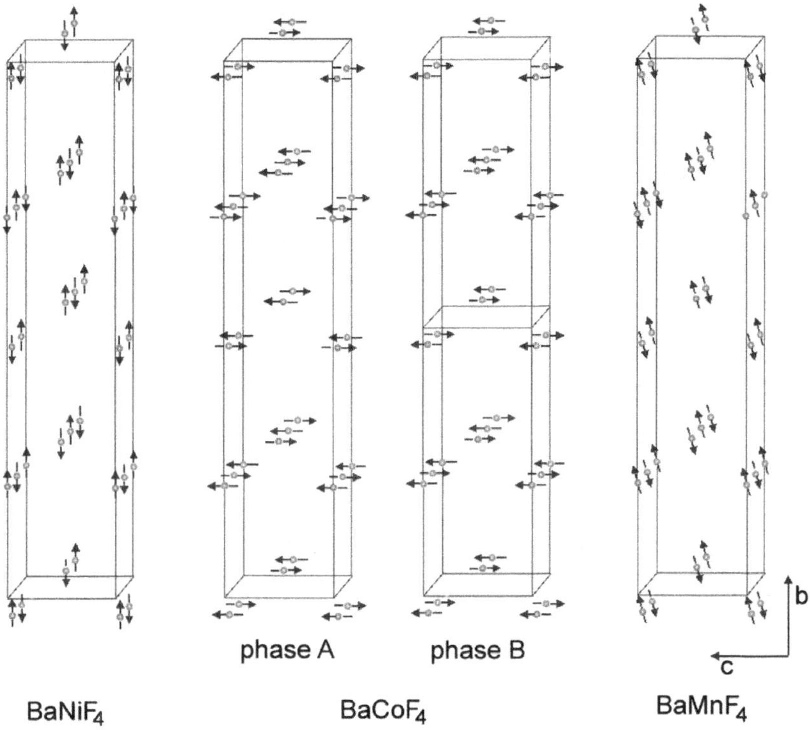

Figure 14.3 Low-temperature antiferromagnetic structures in $BaMF_4$.

three-dimensional order of $BaNiF_4$ and $BaFeF_4$ and the same magnetic cell. On the contrary the lattice parameters of phase B are only doubled along *a*. Since the energy difference between these two phases is very small, it is generally assumed that local microscopic phenomena determine which structure appears in a given region of the samples. The magnetic structure of $BaMnF_4$ (Fig. 14.3) is similar to that of $BaNiF_4$. The ordered magnetic moment was reported to be 4.8 $\mu B/Mn^{2+}$ and to lie in the *ab* plane, 9 degrees away from the *b* axis, with propagation vector $\mathbf{k} = \{0, \frac{1}{2}, \frac{1}{2}\}$ [11]. Therefore, the magnetic and crystallographic unit cells of $BaMnF_4$ can be considered as isomorphic in the *ab* plane of $BaNiF_4$, due to the doubling of *a* and *b*. The coincidence of the magnetic and crystallographic cells in the *bc* plane, joined to the monoclinic crystal symmetry of the low-temperature structure, permits a ferromagnetic (FM) component perpendicular to the AFM axis (ie, roughly along the orthorhombic *c* axis). Although Petrov et al. [45] ruled out the presence of weak ferromagnetism on the basis of their AFM resonance measurements, a subsequent study by Venturini and Morgenthaler [58] revealed a splitting of one mode consistent with the presence of a weak Dzyaloshinsky–Moriya (DM) resultant at 4.2K along the *a* axis [41]. They suggested that this canting might be masked in static magnetization measurements by the presence of domains with oppositely directed weak moments. The weak ferromagnetism would be induced by the spontaneous electric polarization via a linear ME interaction [22,53]. The magnetic structure was reinvestigated by A. Poole et al. [46] with unpolarized and polarized neutron diffraction. The previous AFM structure was confirmed with a slightly increased tilt (12 degrees) of the magnetic moments with respect to the *b* direction, but the FM component, that is indicative of the linear ME effect, was not observed.

The occurrence of weak magnetic order, resulting from the DM interaction, should be expected in magnetic ferroelectrics, since this interaction can occur whenever the midpoint between two magnetic ions is not an inversion center. The linear ME effect, allowed in $BaMnF_4$ by the lowered symmetry, is forbidden in $BaNiF_4$. Nevertheless, the microscopic mechanism governing the ME behavior could be the same in both systems. Ederer and Spalding [16] showed, by using both first-principles calculations and symmetry analysis, that the polar distortion in $BaNiF_4$, chosen as representative of barium fluorides, gives rise to a DM interaction, and that the sign of this interaction is determined by the orientation of the ferroelectric polarization, which can be inverted by an external electric field. This ferroelectricity-induced magnetic order, intimately coupled to the electric polarization, would therefore allow controlled switching of the magnetic order parameter by switching the polarization using an electric field.

3. Fluorides With Tetragonal Tungsten Bronze Structure

3.1 Overview

The term "tungsten bronze" was firstly used by Whöler [61] to identify the compound Na_xWO_3, in which the Na^+ ions are intercalated in a nonstoichiometric way in the vacant A site of the perovskite structure of WO_3, and it is now used to classify a wide number of compounds that share a common structural framework. In particular the term "tetragonal tungsten bronze", originally utilized for K_xWO_3 ($x = 0.4-0.6$), whose structure was elucidated by Magneli [36], is usually utilized to indicate its typical tetragonal structure that can be thought as derived from the perovskite by 45 degrees tilting of some dodecahedral perovskite cages as indicated in Fig. 14.4.

The TTB structure can act as an effective playground for the realization of functional (and multifunctional) materials. In fact the rotation of the perovskite units at the origin of the tetragonal phase gives rise to different channels, having a section ranging from triangular to square and pentagonal, which are able to host cations of different size. The TTB structure, whose general formula can be written as $(A_1)(A_2)_2C_2B_5O_{15}$ by using the nomenclature scheme indicated in Fig. 14.4, shares with perovskites a large tolerance to chemical substitutions. The B site can be occupied by a wide number of transition metals, while alkaline elements, lead, or rare earths can be placed at the A1 and A2 sites. On the contrary the C site (with triangular section) is very often empty and can be occupied only by the smaller Li^+ ion. By substituting tungsten with other high-valence transition metals and potassium with the Li, Na, Ba, Sr, and Pb ions, the TTB family groups today a large number of functional crystals and materials possessing electrooptic, ferroelectric, ferroelastic, pyroelectric and piezoelectric properties. For example, the ferroelectric–ferroelastic niobates

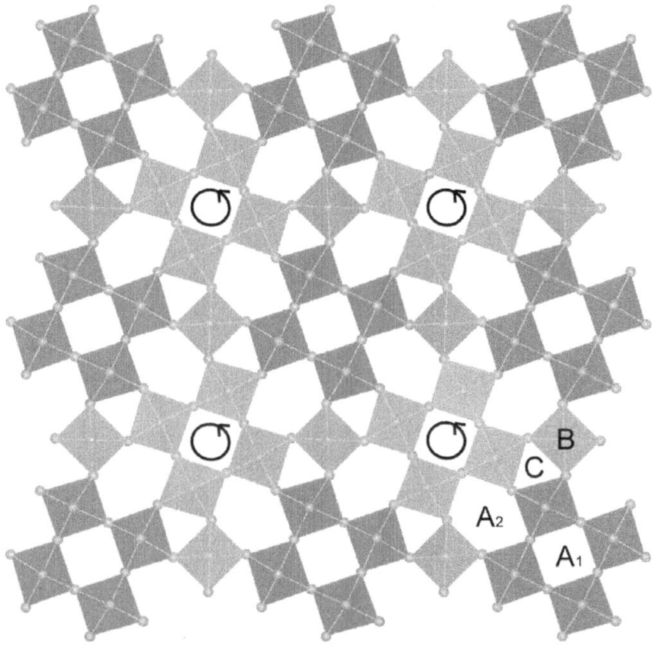

Figure 14.4 Representation of the crystal structure of tetragonal tungsten bronzes viewed along the fourfold stacking axis; the structure can be thought as derived from the perovskite by tilting part of the units by 45 degrees. B indicates the octahedrally coordinated sites forming the framework, while A_1, A_2, and C indicate the extra-framework sites that can be occupied by cations of suitable size to compensate the negative charge of the framework.

crystallizing in variants of the TTB structure [30], whose prototype is represented by $Ba_2NaNb_5O_{15}$ (BNN), show a large nonlinear polarizability leading to useful electro-optical and nonlinear optical properties [26,55]. In addition the insertion of photoluminescent ions allows its application in the lasing field, while photorefractive effects result to be deeply affected by doping [57]. BNN is a good ferroelectric with Curie temperature 853K, high saturation polarization, and coercive field. The polar properties are induced by the shift of the Nb ions from the center of the coordination octahedra in the *c* direction. In addition, ferroelasticity is observed below 573K, being generated by the cooperative tilt of the Nb coordination octahedra. The coexistence of two ferroic orders makes BNN a multiferroic compound, even if obviously no magnetic ordering can be observed owing to the d^0 electronic configuration of Nb^{5+}. The introduction of magnetism in a TTB structure is allowed by the substitution of oxygen with fluorine, characterized by similar ionic radius but higher oxidation state, allowing in this way the replacement of Nb^{5+} with low-valence transition metal ions with partially filled outer electronic shells. The charge neutrality conditions then impose the A_1 and A_2 sites to be occupied by a low-valence ion, for instance, potassium, so that the family of compounds obtained has general formula K_xMF_3 $(0.4 \le x \le 0.6)$, better written as $K_xM_x^{II}M_{5-x}^{III}F_{15}$ $(2.0 \le x \le 3.0)$ in order to point out the mixed valence character of the system. TTB fluorides were first studied by De Pape in 1965. Similar TTB phases have been reported to form for M = Fe, V, and Cu [6,13,21,44,62]; and in the mixed systems Co/Fe, Mn/Fe, Mn/Cr, and Fe/Cr [1,2,7,21,25,44]. These compounds, usually prepared by reacting in controlled atmosphere stoichiometric amounts of KF, MF_2, and MF_3, have been extensively studied because of their frustrated magnetic ordering [3,4,13,23,34,40], which arises at low temperature. In spite of the fact that several members of the series have been synthesized, most of them claimed to be multiferroics, the most extensively studied materials, presenting clear evidences of ME properties, are the members of the $K_xFe_5F_{15}$ series, in particular the composition with $x = 1$.

3.2 $K_xFe_5F_{15}$ Structure

In spite of the fact that the structure of TTB fluorides was investigated starting from the 1970s [1,27], a detailed structural information, highlighting the strong relation with the BNN structure, was obtained only in 2004 on the basis of electron diffraction analysis performed on a series of TTB fluorides and of single crystal X-ray diffraction (XRD) data collected on an untwined tetragonal $K_{2.65}Fe_5F_{15}$ crystal [21]. The synergic use of these diffraction techniques evidenced unique crystallographic features that are typical of all these materials and make them extremely interesting from the structural point of view. In fact, two distinct modulations, charge ordering and cooperative tilting involving the [MF_6] octahedra of the bronze framework, coexist within the structure. While the charge order is peculiar of these mixed valence fluorides, the nature of the tilt modulation involving the [MF_6] octahedra is clearly associable to the structural features of TTB niobates. Each modulation is characterized by its own symmetry and periodicity and, in first approximation, can be considered independent from the other one. Charge ordering involves the perovskite unit of the TTB structure, which is built by [MF_6] octahedra in such a manner that the larger divalent cations alternate with the smaller trivalent cations to reduce the structural stress. This gives rise to the doubling of the c axis leading to a superstructure, indicated as charge order superstructure (COS) which is twice the volume of the fundamental TTB cell. The tilt modulation extends on a larger periodicity, giving rise to an orthorhombic supercell that is eight times in volume of the fundamental TTB cell ($a = 2\sqrt{2}a_{TTB}$, $b = \sqrt{2}a_{TTB}$, $c = 2c_{TTB}$). The superstructure, indicated as ferroelastic superstructure (FES) and showing *Bbm*2 symmetry, is shown in Fig. 14.5. The tilt modulation is strictly similar to that found at RT in ferroelectric BNN [33], where it is related to the appearance of ferroelasticity. In Ref. [14] an FES structure with halved volume ($a = \sqrt{2}a_{TTB}$, $b = \sqrt{2}b_{TTB}$, $c = 2c_{TTB}$) was reported for $K_{2.5}Mn_{2.5}Cr_{2.5}F_{15}$ on the basis of powder synchrotron diffraction. However, these results are in contrast with electron diffraction data, much more powerful in this type of determinations, indicating a conventional FES structure also for this compound [21].

The stability range of the $K_xFe_x^{II}Fe_{5-x}^{III}F_{15}$ system involves a different occupation of the perovskite cage by K^+ ions and the change of the Fe^{2+}/Fe^{3+} ratio within the structure from 0.67 to 1.50. The variation of the chemical composition affects the

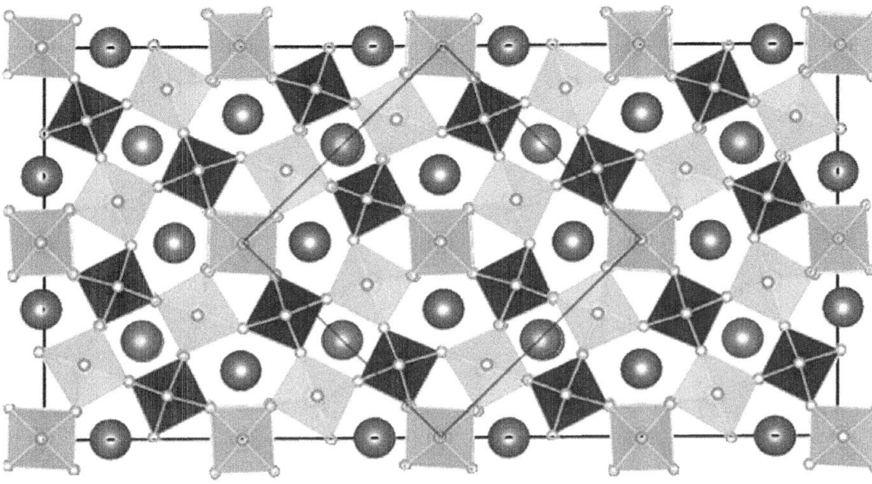

Figure 14.5 Relation between the charge order superstructure (smaller, *blue line* (darkest gray in print versions)) and ferroelastic superstructure (larger, black in print versions) unit cells in the crystal structure of $K_{2.62}Fe_5F_{15}$ determined by single crystal X-ray diffraction data [21]. The *yellow octahedra* (light gray in print versions) represent Fe(III), the *green* (gray in print versions) mixed-valence sites and the *blue* (dark gray in print versions) Fe(II).

structural properties of the compound, giving rise to a slight orthorhombic distortion of the fundamental TTB cell in the range $2.85 \leq x \leq 3.0$. This symmetry lowering involves an orthorhombic–monoclinic transition of the FES superstructure inducing at the same time a change in the charge ordering pattern [38].

For low potassium contents ($2.0 \leq x \leq 2.85$) the COS crystal symmetry is tetragonal $P4_2bc$, and the charge order imposes all the perovskite sites to be occupied alternatively by Fe^{2+} and Fe^{3+} in the *ab* plane and along the *c* axis. This charge order scheme allows an effective reduction of the structural stresses through the alternate stacking of the smaller Fe^{3+} ions over the larger Fe^{2+} ions. For $x = 2$ the extra perovskite site is occupied only by Fe^{3+}, while increasing the potassium content (and consequently the Fe^{2+} concentration), it becomes a mixed-valence site, statistically occupied by the two iron species depending on the composition. For potassium contents higher than 2.85 the orthorhombic distortion removes the diagonal glide planes and reduces the symmetry to *Pba*2. This allows a different charge order pattern, where the doubling of the mixed-valence sites allows the lowering of the structural stresses induced by the presence of adjacent Fe^{2+} coordination octahedra. The ordering schemes in the COS cell for selected compositions, determined by single crystal XRD data, are reported in Fig. 14.6. For what concerns the potassium occupancy of the extra-framework sites, it must be noticed that the larger pentagonal A_2 site is fully occupied in the whole stability range, while the smaller square A_1 site displays partial occupation ranging from 0 to 1. By increasing *x*, in agreement with both the presence of an increased number of potassium ions in the structure and with the substitution of smaller trivalent with larger divalent cations, the cell volume

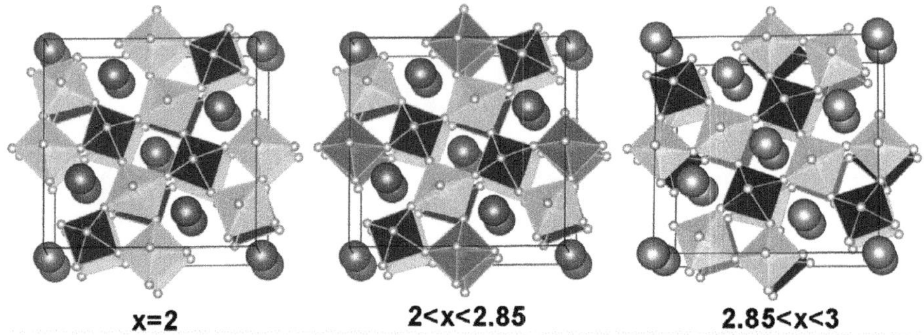

Figure 14.6 Structure of the $K_xFe_x^{II}Fe_{5-x}^{III}F_{15}$ series as a function of composition. *Yellow octahedra* (light gray in print versions): Fe(III); *green* (gray in print versions): mixed valence; *dark blue* (dark gray in print versions): Fe(II).

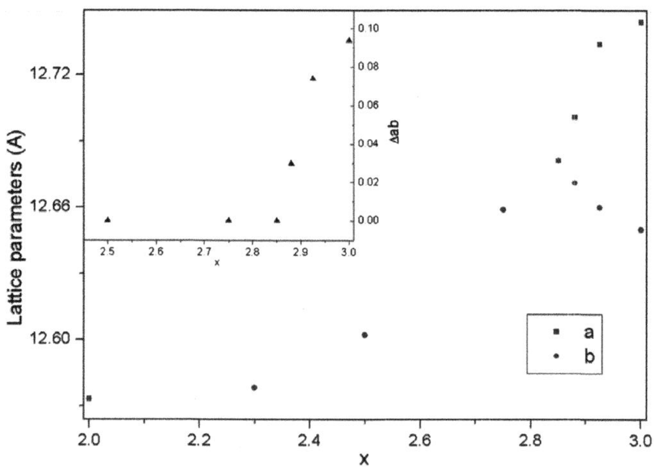

Figure 14.7 Evolution of the a and b lattice parameters as a function of the potassium content. Inset: orthorhombic distortion.

increases, while the orthorhombic distortion, reaching a maximum value for $x = 3$, is highlighted by the cell parameters trend reported in Fig. 14.7.

The polar character of the TTB fluorides is revealed by the structure analysis. Similarly to BNN, all the cations lie, along the *c* axis, above the mean plane of basal fluorine anions, whereas the apical fluorine atoms are displaced in the opposite direction [21]. The two Fe—F apical distances are different for each Fe atom, the shorter pointing toward the positive direction of the polar axis, chosen as the displacement direction of the cations. Differently from BNN, for which the ferroelastic twinning involving FES occurs at the nanometer scale, single FES-domain crystals could be isolated in the TTB fluorides and studied by single crystal XRD without the application of the laborious detwinning procedures that are necessary in the former case. The structural

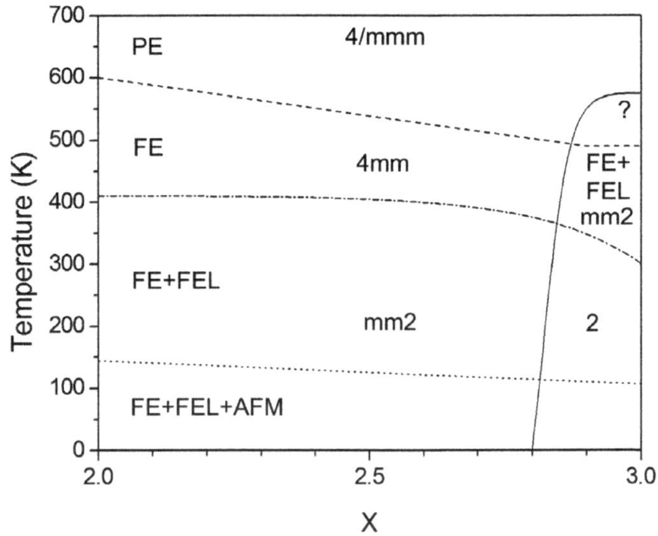

Figure 14.8 Phase diagram of the $K_xFe_x^{II}Fe_{5-x}^{III}F_{15}$ compounds. *PE*, paraelectric phase; *FE*, ferroelectric; *FEL*, ferroelastic; *AFM*, antiferromagnetic.

characterizations indicate that the increase of temperature sequentially removes ferroelasticity, charge order, and ferroelectricity as shown in the phase diagram schematized in Fig. 14.8.

It was reported [38] that the satellite reflections typical of FES, related to the tilt of the octahedral framework, disappear just above RT in $K_3Fe_5F_{15}$, in agreement with a transition observed at 290°C by polarized light microscopy [29]. The transition involves a symmetry change (monoclinic—orthorhombic) and a rotation of the optical indicatrix around the pseudotetragonal *c* axis. Above this temperature, ferroelastic domains persist but their nature changes and becomes simply related to the switching of the orthorhombic *a* and *b* axes. The transition is associated to a large variation of the thermal expansion coefficient. This is not surprising as the thermal compressibility of the structure is larger in the FES, where the tilt of the octahedra allows to effectively reduce the cell parameters (in particular the *c* axis) without substantial variations of the bond lengths. The vanishing temperature of FES increases by decreasing the potassium content. The variation (Fig. 14.8) is significant if related to the orthorhombic—tetragonal transition of the fundamental TTB structure, while it is quite limited in the tetragonal region, indicating the important role of the orthorhombic distortion in relaxing the structural stresses. Although the high-temperature region of the phase diagram for low potassium content has not been studied in great detail in literature, it seems that a connection exists between the disappearance of the charge order and the ferroelectric—paraelectric transition involving the point symmetry change from *4mm* to *4/mmm*. In $K_3Fe_5F_{15}$ the charge order was reported to disappear at 490K [9] on the basis of Mössbauer measurements showing a sharp change in slope of the temperature dependence of the Fe^{3+} quadrupolar splitting and isomer shift. For a long time it has been believed that this temperature would be associated to a dramatic change of the structural properties leading simultaneously to the vanishing of charge order, ferroelectricity, and ferroelasticity. Ravez et al. [48,49] on the basis of a maximum in the relative dielectric permittivity observed at 490K, reported the simultaneous occurrence of the ferroelectric−paraelectric transition, associated to a change in point group symmetry from *mm2* to *4/mmm*, and the disappearance of ferroelasticity at this temperature. However, high-temperature powder XRD data performed unambiguously revealed that the orthorhombic−tetragonal structural transition and the related onset of ferroelasticity are located at 570K [38]. Therefore, no certainty exists on the temperature of the ferroelectric−paraelectric transition, which could be located at 490 or 570K, depending on its association to the order−disorder or to the orthorhombic−tetragonal phase transition. Further work is necessary to clarify this point, but dielectric constant measurements in this temperature region are made difficult by the increase of electrical conductivity related to charge delocalization. For what concerns the permittivity maximum at 490K and its assignment to the ferroelectric−paraelectric transition, it is interesting to note that a temperature dispersion of 10K was observed in the original measurements [48] for the explored frequency range (0.1−10 kHz). Such dispersion, indicative of dielectric relaxation, was interpreted as the signature of a ferroelectric relaxor behavior. However, the simultaneous increase of the dielectric loss at higher temperature, resulting from the semiconducting character of the sample (further enhanced by the order−disorder transition), produces critical experimental conditions that could make very difficult the detection of further anomalies at higher temperatures. No direct measurement of the ferroelectric properties has been provided yet, likely due to the difficulty in synthesizing the fluoride compounds with density levels high enough to avoid electrical breakdowns due to the application of high voltage as well as to the thermally activated character of the compound, making the classical ferroelectric measurements prohibitive.

3.3 Magnetic Properties

The magnetic properties displayed by the TTB fluorides are strongly related to their peculiar crystal structure, where the presence of triangular paths of exchange interactions is similar to the one detected in the so-called Kagomé lattices [37], characterized by the presence of magnetic frustration that gives rise to spin glass, short-range magnetic ordering, ferrimagnetism, or spin canting. TTB fluorides, containing transition metal magnetic ions, are generally characterized by strong AFM interactions that are related to quite high negative values of Θ_W (in most cases <-300K), for which high ordering temperatures are usually expected. However, because of the frustration induced by the triangular interaction paths, the T_N values, graphically reported in Fig. 14.9 for a series of magnetic TTB compounds, are much lower than expected. In spite of similar Θ_W values, the T_N is mostly determined by the nature of the M^{II}−M^{III} pair and does not vary significantly by changing the M^{II}/M^{III} ratio in the whole existence range of the TTB phase. The higher transition temperatures are observed for the Mn^{II}/Fe^{III} and Fe^{II}/Fe^{III} pairs, and the corresponding compounds are most

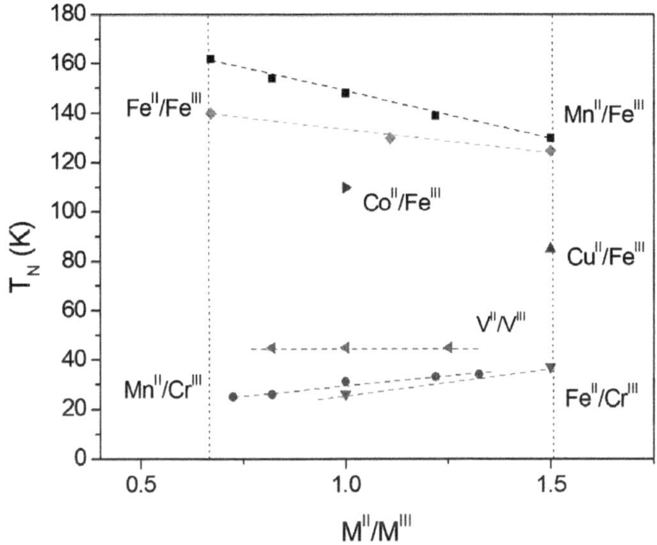

Figure 14.9 Neel temperatures of tetragonal tungsten bronze fluorides for different M^{II}—M^{III} pairs [2,3,5—7,15,23,25,35,38,39,44].

studied from the magnetic point of view. In spite of the fact that Θ_W and T_N are not very sensitive to the composition for a given M^{II}—M^{III} pair, the magnetic susceptibility measurements and the shapes of the Curie—Weiss χ^{-1} versus T plots show in many cases that the nature of the ordering changes significantly, ranging from a classical AFM behavior to a FM behavior. The fact that T_N does not vary substantially is an indication that the overall ordering mechanism is independent from composition. However, as shown in the structural description, the different ordered occupation of the symmetry-independent sites by the smaller trivalent and larger divalent cations depends on the composition and determines the presence or not of a significant FM resultant. In addition, the magnetic susceptibility shows a strong field dependence, zero field cooling (ZFC) measurements usually approaching a classical AFM behavior, while field cooling (FC) measurements emphasize a FM-like response. The understanding of the magnetic behavior is extremely difficult in the absence of a detailed knowledge of the magnetic structure, which has been determined only for $K_{2.5}Mn_{2.5}Fe_{2.5}F_{15}$ (also reported as $KMnFeF_6$) and $K_3Fe_5F_{15}$.

The magnetic structure of $K_{2.5}Mn_{2.5}Fe_{2.5}F_{15}$ was determined by Lacorre et al. [34] by powder neutron diffraction. The description of the spin arrangement requires the lowering of the symmetry from tetragonal to orthorhombic, and the best results were obtained with the magnetic structure described in $Pb'a2'$. All the atomic moments are observed to lie in the *ab* plane (Fig. 14.10), and their arrangement, both in the plane and along the *c* axis, is in agreement with the expected unfrustrated AFM coupling. The spin configuration in the frustrated triangular cycles is not so far from the ideal 120-degree star configuration, the angles ranging from 105 to 139 degrees. The nonzero magnetic resultant detected in magnetization measurements (Fig. 14.11) is

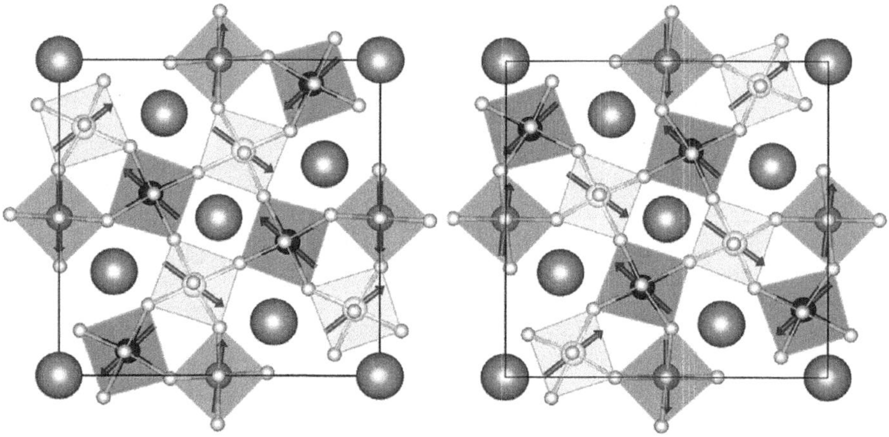

Figure 14.10 Spin arrangements in two adjacent layers of the $K_{2.5}Mn_{2.5}Fe_{2.5}F_{15}$ structure; spins are ordered in an antiferromagnetic way within the single layer as well as between adjacent layers coupled along the stacking axis. *Dark and light coordination octahedra* refer to independent crystallographic sites occupied by Mn^{2+} and Fe^{3+}, respectively, whereas the intermediate sites have a mixed occupancy.

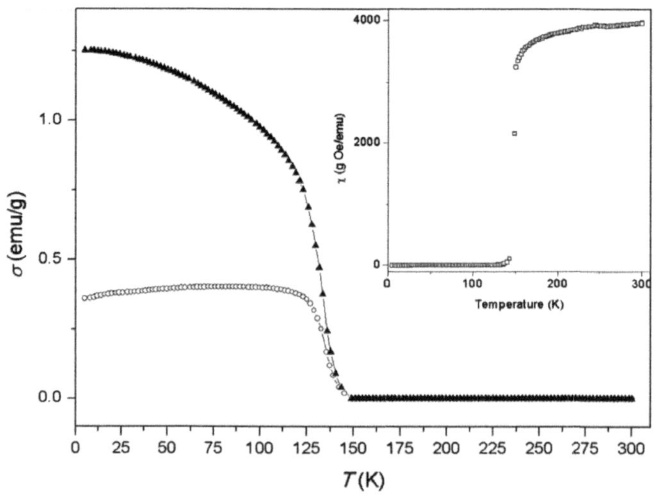

Figure 14.11 Field cooling (FC) and zero field cooling (ZFC) magnetization measurements of $K_{2.5}Mn_{2.5}Fe_{2.5}F_{15}$ with an applied field $H = 10$ Oe (*filled triangles*: FC, *open circles*: ZFC). The inverse susceptibility is shown in the inset.

ascribed to the uncompensated AFM arrangement and, as shown by Mössbauer spectroscopy [4], is along the *a* axis. The presence of long-range magnetic correlations in the whole temperature range from T_N to 12K was confirmed by a further neutron diffraction study [40], whereas the results of a ^{19}F NMR study clearly reveal that

AFM short-range correlations are developed among the magnetic ions below 200K [24]. As shown in Fig. 14.11, an irreversibility between ZFC and FC is observed in magnetization measurements and implies that the ferrimagnetic domains strongly depend upon the cooling process of the sample, since the canting of the spins resulting from the frustration can make the domain wall structure rather complex.

A better insight into the magnetic properties of the system is given by the studies performed on the $K_{3-x}Fe_5F_{15}$ series. Similarly to $K_{2.5}Mn_{2.5}Fe_{2.5}F_{15}$, the magnetization measurements performed on $K_3Fe_5F_{15}$ indicate weak FM properties when the samples are field-cooled, while AFM-like behavior is observed in the ZFC measurements. The data collected on oriented single crystals (Fig. 14.12) show strong magnetic anisotropy of the system, with the orthorhombic *b* axis being the easy axis of magnetization which displays 1.7 emu/g at 5K after a FC in 100 Oe. The *a* direction is similar from both the qualitative and quantitative point of view, although showing lower saturation magnetization (0.5 emu/g). On the other hand, the *c* direction results to be the hard axis of magnetization with saturation magnetization orders of magnitude lower than that of the one detected along *b*, in agreement with the neutron diffraction data of $K_{2.5}Mn_{2.5}Fe_{2.5}F_{15}$, indicating null magnetic component of the spins along this direction.

The inverse susceptibility shows largely negative Curie–Weiss temperatures in all directions and the very high value of $|\Theta_W|/T_N \approx 9.5$ along *a* and *b* reveals the presence of strong frustration. On the other hand, a change in slope of the $1/\chi$ curve is always observed below about 200K, where the linear trend becomes almost temperature independent (Fig. 14.13).

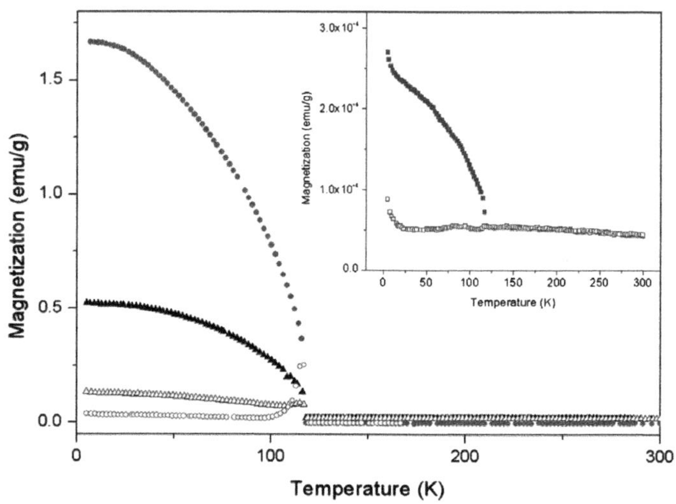

Figure 14.12 Magnetization measurements performed on a $K_3Fe_5F_{15}$ single crystal with the measure field $H = 100$ Oe applied along the *a* (*circles*) and *b* (*triangles*) directions; the measurement with $H = 500$ Oe applied along the *c* direction is shown in the inset; *filled and open symbols* refer to field cooling and zero field cooling measurements, respectively.

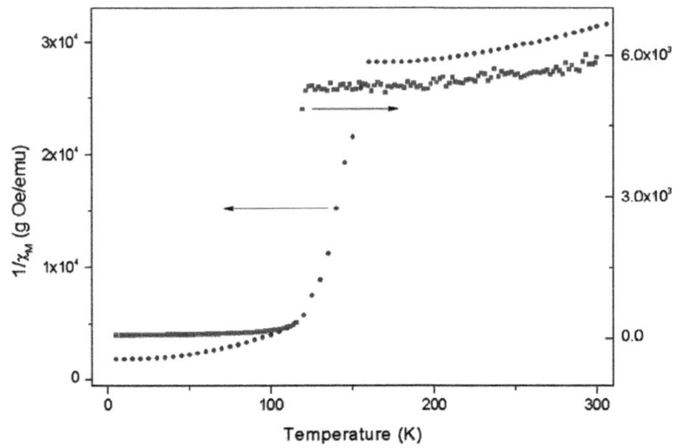

Figure 14.13 Inverse susceptibility measured on a single crystal of $K_3Fe_5F_{15}$ along the *b* direction (*red squares* (dark gray in print versions)) and for a polycrystalline sample of $K_2Fe_5F_{15}$ (*blue circles* (black in print versions)). In both cases the applied field is 100 Oe.

The setting of short-range interactions below this temperature, in agreement with the phenomenology observed in $K_{2.5}Mn_{2.5}Fe_{2.5}F_{15}$, was revealed by the increase of the magnetic background observed in neutron powder diffraction [39]. From the qualitative point of view, the magnetic structure, determined by the complementary use of neutron diffraction and Mössbauer spectroscopy [39], is in agreement with that of $K_{2.5}Mn_{2.5}Fe_{2.5}F_{15}$. Also in this case the magnetic symmetry is orthorhombic *Pb′a2′*, but it is here coincident with the nuclear symmetry, except the time inversion operating on the *b* glide plane and the twofold axis. The refined structure (Fig. 14.14) involves

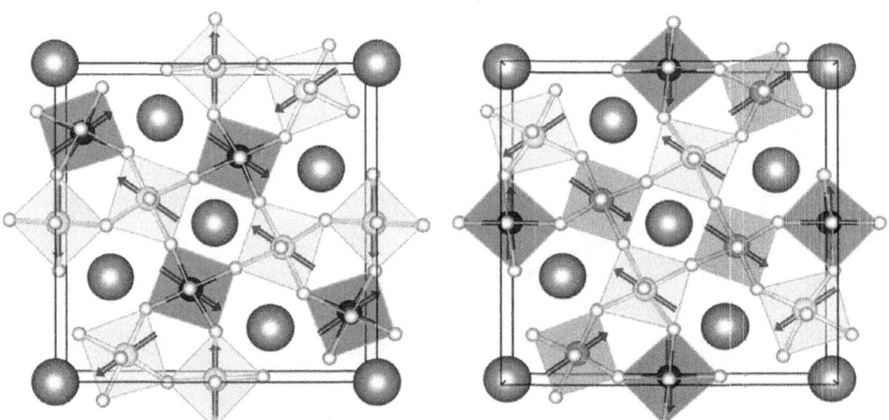

Figure 14.14 Magnetic structure of $K_3Fe_5F_{15}$. *Dark and light coordination octahedra* indicate Fe^{2+} and Fe^{3+}, respectively, intermediate color displays mixed occupancy. The unbalanced charge distribution between the two layers gives the major contribution to the magnetic resultant.

AFM arrangements of the atomic moments both within single layers and between the adjacent layers stacked along *c*. A nonzero resultant along the *b* axis is produced by not only geometrical frustrations but also the nonequivalent distribution, revealed by the structural characterization, of Fe^{2+} and Fe^{3+} in two adjacent layers coupled antiferromagnetically along the *c* axis. The atomic moments of all the iron cations were accurately determined by Mössbauer measurements, evaluating hyperfine fields and quadrupolar parameters together with the measured saturation magnetization value, indicating a strong relation between the ferrimagnetic properties and the different charge distributions at the two adjacent layers produced by the cation ordering.

The knowledge of the magnetic structure allows the full comprehension of the behavior of the magnetization measurements as a function of temperature. A nonzero component along *b* is spontaneously developed on cooling in zero field, but the occurrence of domains with opposite moments gives rise to a zero magnetization. When in the ZFC measurements (performed then on heating) the measuring field is oriented along *b*, the magnetization remains zero until, in the proximity of the ordering temperature, the applied field becomes sufficient to reverse the spontaneous magnetization of the opposite domains. When on the contrary the measuring field is oriented along *a*, a small constant residual magnetization is observed in the whole ordering range. This residual magnetization along *a* is likely produced by spin canting that could be preexistent as well as induced by the measuring field. In the FC measurement, where the field is applied before the cooling process and therefore well above the magnetic transition, the pseudotetragonal character of the crystal structure makes possible to turn the magnetization along *a*, by applying the field in this direction. In case of a TTB structure indeed (for example $K_2Fe_5F_{15}$), two possible solutions with space groups $Pb'a2'$ and $Pb'a2'$, involving the resultant moment along *b* and *a* respectively, would be allowed with equal probability. The structural distortion, leading to the orthorhombic nuclear structure of $K_3Fe_5F_{15}$, removes the degeneracy of the two magnetic solutions, making the former thermodynamically favored, but the latter accessible when an external magnetic field is applied along *a*. M(H) measurements showed that the switch of the magnetization along *a* is permitted also below the transition temperature, but the application of a sufficiently high field is required. In fact M(H) measurements performed at 5K on an oriented crystal reveal an unusual behavior (Fig. 14.15). A "hard" magnetic loop is observed along *b*, with coercive field $H_C = 14$ kOe and saturation magnetization $Ms_b = 2.6$ emu/g. On the contrary, when the data are collected with the field applied along *a*, the presence of a constrained loop is observed. This peculiar behavior corresponds to the need of reaching the coercive field of the *b* component in order to switch the magnetization from its ground state (with resultant along *b*). During the reversal of the magnetic field the magnetization is locked along the easy axis and only the application of a field higher than Hc allows to reverse (if *H* is applied along *b*) or to switch to the 90-degrees solution (if *H* is along *a*).

The magnetic structure of $K_2Fe_5F_{15}$, determined by powder neutron diffraction, is qualitatively identical to that of $K_3Fe_5F_{15}$ and $K_{2.5}Mn_{2.5}Fe_{2.5}F_{15}$: the magnetic space group symmetry is again $Pb'a2'$, leading to the same arrangement of spin orientations. Also in this case an uncompensated moment is allowed along *b*, giving rise to the nonzero magnetization observed in the M(T) measurements. However, relevant

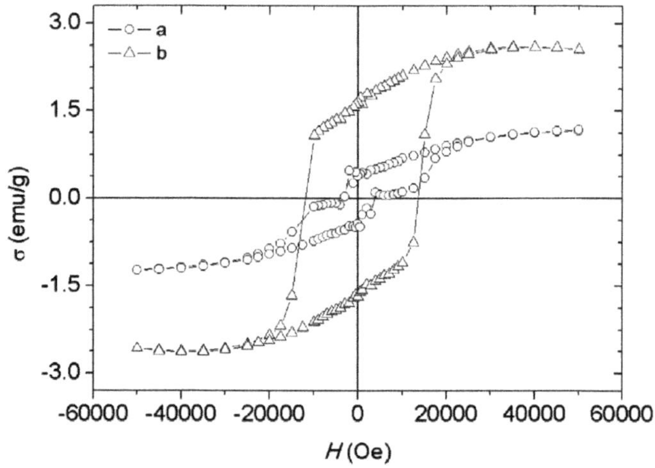

Figure 14.15 Hysteresis loops for of a $K_3Fe_5F_{15}$ single crystal measured with the applied field along *a* and *b*, respectively.

differences are revealed by the comparison with $K_3Fe_5F_{15}$. In fact not only the saturation values in the M(T) curves are much lower (Fig. 14.16), but also the M(H) behavior differs completely, showing at 5K the absence of hysteresis (except a small contribution for low fields) and an almost linear trend for fields higher than 5 kOe never reaching saturation up to 5 T. All these features indicate that the frustrating mechanisms are much less effective in this compound, resulting in an almost AFM behavior of the magnetic phase.

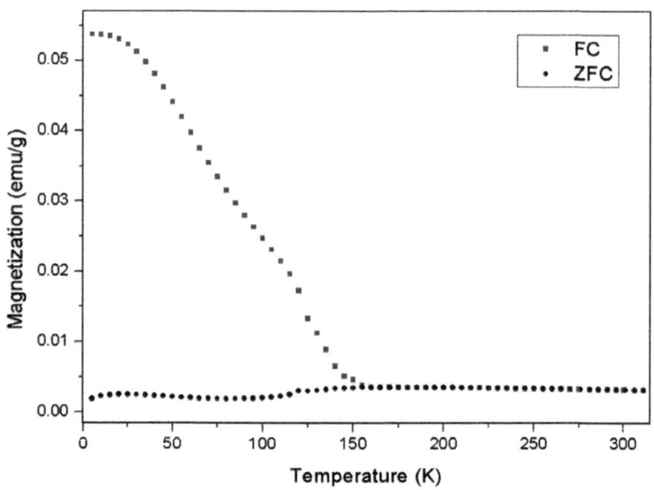

Figure 14.16 Field cooling (FC) and zero field cooling (ZFC) magnetization measurement of $K_2Fe_5F_{15}$ with an applied field $H = 100$ Oe.

The fact that the magnetic behavior of TTB fluorides could differ completely, in spite of the substantial equivalence of the magnetic structure pointed out by the different characterizations, is extremely interesting. In order to understand the occurring of substantial differences, it is necessary to take into account all the mechanisms leading to unbalance the AFM spin order. As discussed before, in ideally nonfrustrated TTB systems the spins are AFM ordered: (1) within a single layer; (2) between two adjacent layers along the stacking *c* axis. The first arrangement is always geometrically frustrated by the nonequivalent spin interactions involving two different cations M^{2+} and M^{3+} distributed in the triangular units; the result is typically a deviation from the ideal starred spin arrangements forming 120-degree angles. The second arrangement is perturbed by the nature of M^{2+} and M^{3+}. By looking for example at Fig. 14.10, it can be easily inferred that M^{2+} cations, alternating with M^{3+} around the perovskite cage in tetragonal compounds, give in each layer the same magnetic resultant oriented in the same direction along the *b* axis. The same is true for the corresponding M^{3+} cations, but the direction is opposite to the previous one. Owing to the different nature of M^{2+} and M^{3+}, the two components are unbalanced and each layer contributes in the same *b* direction with the same magnetic resultant, whose magnitude depends on the nature of the involved cations. This phenomenon is mainly responsible for the FM nature observed in $K_{2.5}Mn_{2.5}Fe_{2.5}F_{15}$, whereas in $K_2Fe_5F_{15}$, where the magnetic contribution of Fe^{2+} with unquenched orbital momentum can be quite similar to that of Fe^{3+}, an almost pure AFM behavior has been observed. When the nuclear structure becomes orthorhombic, passing from *P4₂bc* to *Pba2*, the composition of two adjacent layers is no longer constrained by symmetry. This adds a further degree of freedom to the unbalancing of the AFM coupling between the layers, and again a magnetic resultant is developed along *b*, as in the case of $K_3Fe_5F_{15}$.

4. Conclusions

In this chapter it was shown that fluorides represent a promising playground for the discovery and study of novel multiferroic materials. Indeed, due to different hybridization properties with respect to oxygen, fluorine displays an accentuated tendency to induce polar properties, and this is confirmed by the large number of ferroelectric fluorides reported in literature (see Scott and Blinc [54] for an extensive review). On the other hand, their magnetic properties are usually poorly studied, preventing the effective enlargements to other fluoride families of the potential interest for multiferroic application, which is actually limited to $BaMF_4$ and $K_xM_x^{II}M_{5-x}^{III}F_{15}$.

For what concerns $BaMF_4$, the low magnetic transition temperatures are often considered a strong limitation to their applicability. However, the ferroelectricity–induced magnetic order, intimately coupled to the electric polarization (previously discussed for $BaNiF_4$), would allow controlled switching of the magnetic order parameter by reversing the polarization using an electric field. If confirmed, this mechanism could be considered a very promising route toward useful ME device applications, but unfortunately at present this hypothesis has not been confirmed by any direct observation.

For what concerns the TTB fluorides family, some of its members represent exceptional cases of "complete" multiferroics, displaying the coexistence of ferroelectric, ferrimagnetic, and ferroelastic properties, in addition to charge order related to the presence of TM cations in different oxidation states. Similarly to the TTB oxides, the $K_xM_x^{II}M_{5-x}^{III}F_{15}$ compounds crystallize in nonpolar space groups, related to the inversion symmetry breaking due to the off-centering of the M ions within their coordination octahedra. Despite that so far no direct ferroelectric hysteresis measurement has been provided, the switching of the electrical polarization is expected to be possible in these systems. Ferroelasticity origins from the cooperative tilting of the MF_6 octahedra within two possible variants giving rise to domains, which are suggested to couple not only to polar distortions but also to the magnetic degree of freedom. As reported by Ishihara et al. [29], an anomaly in the spontaneous linear birefringence, independent from the wavelength used in the measurement, is observed in $K_3Fe_5F_{15}$ at the magnetic transition, revealing the occurrence of a related change in the ferroelectric–ferroelastic domains. By the way, it is interesting to point out the anisotropic behavior of the magnetization in the *ab* plane of the same compound, which can be considered quite surprising if the pseudotetragonal nature of the nuclear structure is considered. The fact that the switching of the magnetization from the *b* to the *a* axis requires the application of a field of 14 kOe is a sure indication of a strong microstructural pinning of the magnetic structure, so that the hypothesis of a rearrangement of the ferroelectric–ferroelastic domains induced by the magnetization switch can be considered reasonable in these conditions. On the other hand, based on symmetry considerations [43], the magnetic and electric degrees of freedom are expected to be coupled through the α_{13} α_{32} off-diagonal terms of the ME tensor in $K_{2.5}Mn_{2.5}Fe_{2.5}F_{15}$. A spontaneous (additional) polarization along *c* should be induced by the setting of magnetic order, while a magnetic component along *b* is to be expected in the absence of external stimuli, in agreement with the magnetic and structural characterizations reported in Section 3.3. More interestingly, the application of a magnetic field along *c* and *b* should induce an electrical polarization parallel to *a* and *c*, respectively, while an electric field applied along *c* and *b* would give rise to a magnetic contribution along *a* and *c*, respectively.

The unique experimental clue of ME coupling in these compounds was provided by Blinc et al. [5], through the observation of a double peak in the electrical permittivity at the magnetic ordering temperature in $K_3Fe_5F_{15}$. On the other hand, the actual state of the art, providing the detailed knowledge of the best synthesis conditions as well as the structural and physical properties (including magnetic structure) of the $K_xM_x^{II}M_{5-x}^{III}F_{15}$ series represents the ideal starting point for the performance of new specific ME measurements. The appropriate choice of the M^{II}/M^{III} pairs and their relative composition can allow not only to tune the overall magnetic behavior (spanning from AFM to strongly frustrated FM systems) and to increase the ordering temperatures (up to 150K), but also to assure more effective ME coupling coefficients by wise engineering of the materials.

References

[1] E. Banks, S. Nakajima, G.J.B. Williams, The crystal structure of $K_{0.54}(Mn,Fe)F_3$ at room temperature, Acta Cryst. B 35 (1979) 46–49.

[2] E. Banks, M. Shone, Y.S. Hong, R.F. Williamson, W.O.J. Boo, Ferromagnetic behavior in potassium manganese chromium fluoride ($KMnCrF_6$), Inorg. Chem. 21 (1982) 3894–3897.

[3] E. Banks, M. Shone, R.F. Williamson, W.O.J. Boo, Magnetic properties of the tetragonal phase $K_xMn_xFe_{1-x}F_3$, Inorg. Chem. 22 (1983) 3339–3342.

[4] S.C. Bhargava, S. Singh, A.H. Morrish, Z.W. Li, Mössbauer study of $KMnFeF_6$, Solid State Commun. 116 (2000) 575–580.

[5] R. Blinc, G. Tavčar, B. Žemva, D. Hanžel, P. Cevc, C. Filipic, A. Levstik, Z. Jagličić, Z. Trontelj, N.S. Dalal, V. Ramachandran, S. Nellutla, J.F. Scott, Weak ferromagnetism and ferroelectricity in $K_3Fe_5F_{15}$, J. Appl. Phys. 103 (2008) 074114 (4 pp.).

[6] R. Blinc, G. Tavčar, B. Žemva, E. Goreshnik, D. Hanžel, P. Cevc, A. Potočnik, V. Laguta, Z. Trontelj, Z. Jagličić, J.F. Scott, Electron paramagnetic resonance and Mössbauer study of antiferromagnetic $K_3Cu_3Fe_2F_{15}$, J. Appl. Phys. 106 (2009) 023924 (4 pp.).

[7] R. Blinc, P. Cevc, A. Potočnik, B. Žemva, E. Goreshnik, D. Hanžel, A. Gregorovič, Z. Trontelj, Z. Jagličić, V. Laguta, M. Perović, N.S. Dalal, J.F. Scott, Magnetic properties of multiferroic $K_3Cr_2Fe_3F_{15}$, J. Appl. Phys. 107 (2010) 043511 (5 pp.).

[8] J.K. Brandon, H.D. Megaw, On the crystal structure and properties of $Ca_2Nb_2O_7$, "calcium pyroniobate", Philos. Mag. 21 (1970) 189–194.

[9] Y. Calage, S.C. Abrahams, J. Ravez, R. de Pape, Mössbauer study of Fe^{2+}/Fe^{3+} order–disorder and electron delocalization in $K_3Fe_5F_{15}$ at the 490-K phase transition, J. Appl. Phys. 67 (1990) 430–433.

[10] D.E. Cox, M. Eibschütz, H.J. Guggenheim, L. Holmes, Neutron diffraction study of the magnetic structure of $BaNiF_4$, J. Appl. Phys. 41 (1970) 943–945.

[11] D.E. Cox, S.M. Shapiro, R.A. Cowley, M. Eibschütz, H.J. Guggenheim, Magnetic and structural phase transitions in $BaMnF_4$, Phys. Rev. B 19 (1979) 5754–5772.

[12] D.E. Cox, S.M. Shapiro, R.J. Nelmes, T.W. Ryan, H.J. Bleif, R.A. Cowley, M. Eibschutz, H.J. Guggenheim, X-ray diffraction and neutron diffraction measurements on $BaMnF_4$, Phys. Rev. B 28 (1983) 1640–1643.

[13] R. De Pape, Une série de composés fluorés de type bronze $KxFe_{2+1-x}Fe_{3+x}F_3$, C. R. Acad. Sci. 260 (1965) 4527–4530.

[14] M. Di Domenico Jr., M. Eibschütz, H.J. Guggenheim, I. Camlibel, Dielectric behavior of ferroelectric $BaMF_4$ above room temperature, Solid State Commun. 7 (1969) 1119–1122.

[15] C. Drathen, T. Nakagawa, W.A. Crichton, A.H. Hill, Y. Ohishid, S. Margadonna, Structural transition in $KMnCrF_6$ – a chemically ordered magnetic ferroelectric, J. Mater. Chem. C 3 (2015) 4321–4332.

[16] C. Ederer, N.A. Spaldin, Electric-field-switchable magnets: the case of $BaNiF_4$, Phys. Rev. B 74 (2006) 020401(R) (4 pp.).

[17] M. Eibschütz, H.J. Guggenheim, Antiferromagnetic-piezoelectric crystals: $BaMe_4$ (M = Mn, Fe, Co and Ni), Solid State Commun. 6 (1968) 737–739.

[18] M. Eibschütz, H.J. Guggenheim, S.H. Wemple, I. Camlibel, M. Di Domenico Jr., Ferroelectricity in $BaM^{2+}F_4$, Phys. Lett. A 29 (1969) 409–410.

[19] M. Eibschütz, L. Holmes, H.J. Guggenheim, D.E. Cox, Magnetic behavior of the two-dimensional antiferromagnet $BaFeF_4$, J. Phys. Colloq. 32 (C1) (1971) 759–760.

[20] M. Eibschütz, L. Holmes, H.J. Guggenheim, D.E. Cox, Magnetic structure of the two-dimensional antiferromagnet $BaCoF_4$, Phys. Rev. B 6 (1972) 2677–2683.

[21] S. Fabbrici, E. Montanari, L. Righi, G. Calestani, A. Migliori, Charge order and tilt modulation in multiferroic $K_xM^{II}_xM^{III}_{1-x}F_3$ ($0.4 < x < 0.6$) transition metal fluorides with tetragonal tungsten bronze structure, Chem. Mater. 16 (2004) 3007–3019.

[22] D.L. Fox, J.F. Scott, Ferroelectrically induced ferromagnetism, J. Phys. C: Solid State Phys. 10 (1977) L329–L331.

[23] S. Giri, K. Ghoshray, A. Ghoshray, N. Chatterjee, Magnetic susceptibility of $KMFeF_6$ ($M = Co^{2+}$, Mn^{2+}) systems, J. Magn. Magn. Mater. 157/158 (1996) 411–412.

[24] S. Giri, K. Ghoshray, A. Ghoshray, N. Chatterjee, ^{19}F NMR study of $KMFeF_6$. II. A frustrated magnetic system, Phys. Rev. B 56 (1997) 3347–3352.

[25] S. Giri, K. Ghoshray, Structural and magnetic properties of the mixed fluorides $KMn_{1-x}Co_xFeF_6$, Phys. Rev. B 57 (1998) 5918–5922 (5 pp.).

[26] J.E. Guesic, H.J. Levinstein, S. Singh, R.G. Smith, L.G. Van Uitert, Continuous 0.532-μ solid-state source using $Ba_2NaNb_5O_{15}$, Appl. Phys. Lett. 12 (1968) 306–308.

[27] A.M. Hardy, G. Ferey, Structure cristalline du bronze pseudo-quadratique $K_{0.6}FeF_3$: transition pyrochlore-quadratique pour les composés KMM'X_6, Acta Cryst. B 29 (1973) 1654–1658.

[28] N.A. Hill, Why are there so few magnetic ferroelectrics? J. Phys. Chem. B 104 (2000) 6694–6709.

[29] S. Ishihara, J.P. Rivera, E. Kita, Z.G. Ye, F. Kubel, H. Schmid, Studies of single crystalline tetragonal tungsten bronze type $K_3Fe_5F_{15}$, a potential magnetoelectric material, Ferroelectrics 162 (1994) 51–61.

[30] P.B. Jamieson, S.C. Abrahms, J.L. Bernstein, Ferroelectric tungsten bronze-type crystal structures. I. Barium strontium niobate $Ba_{0.27}Sr_{0.75}Nb_2O_{5.78}$, J. Chem. Phys. 48 (1968) 5048–5057.

[31] C.V. Kannan, K. Shimamura, H.R. Zeng, H. Kimura, E.G. Villora, K. Kitamura, Ferroelectric and anisotropic electrical properties of $BaMgF_4$ single crystal for vacuum UV devices, J. Appl. Phys. 104 (2008) 114113 (7 pp.).

[32] E.T. Keve, S.C. Abrahams, J.L. Bernstein, Crystal structure of pyroelectric paramagnetic barium manganese fluoride, $BaMnF_4$, J. Chem. Phys. 51 (1969) 4928–4936.

[33] P. Labbé, H. Leligny, B. Raveau, J. Schneck, J.C. Toledano, X-ray structural determination of the quasi-commensurate phase of barium sodium niobate, *J. Phys.: Condens. Matter* 2 (1990) 25.

[34] P. Lacorre, J. Pannetier, G. Ferey, Ordered magnetic frustration: XIV. The magnetic structure of the tetragonal bronze $KMnFeF_6$, J. Magn. Magn. Mat. 94 (1991) 331–336.

[35] A. Levstik, C. Filipic, V. Bobnar, E. Goreshnik, B. Žemva, Z. Trontelj, Z. Jagličić, Polarons in magnetoelectric $K_3Fe^{II}_3Cr^{III}_2F_{15}$, J. Appl. Phys. 106 (2009) 073720 (3 pp.).

[36] A. Magneli, The crystal structure of tetragonal potassium tungsten bronze, Ark. Kemi 1 (1949) 213–221.

[37] M. Mekata, Kagome: the story of the basketweave lattice, Phys. Today 56 (2003) 12–13.

[38] F. Mezzadri, S. Fabbrici, E. Montanari, L. Righi, G. Calestani, E. Gilioli, F. Bolzoni, A. Migliori, Structural properties and multiferroic phase diagram of $K_{0.6}Fe^{II}_{0.6}Fe^{III}_{0.4}F_3$ fluoride with TTB structure, Phys. Rev. B 78 (2008) 064111 (6 pp.).

[39] F. Mezzadri, G. Calestani, C. Pernechele, M. Solzi, G. Spina, L. Cianchi, F. Del Giallo, M. Lantieri, M. Buzzi, E. Gilioli, Magnetic and Mössbauer characterization of the multiferroic fluoride $K_3Fe_5F_{15}$, Phys. Rev. B 84 (2011) 104418 (9 pp.).

[40] S.K. Mohammad Yusuf, L. Madhav Rao, R. Mukhopadhyay, S. Giri, K. Ghoshray, A. Ghoshray, Low temperature magnetic phase of $KMnFeF_6$: a neutron depolarisation and diffraction study, Solid State Commun. 101 (1997) 145–148.

[41] T. Moriya, Anisotropic superexchange interaction and weak ferromagnetism, Phys. Rev. 120 (1960) 91–98.

[42] G. Nénert, T.T.M. Palstra, Prediction for new magnetoelectric fluorides, J. Phys. Condens. Matter 19 (2007) 406213 (10 pp.).
[43] G. Nénert, Phenomenological Landau analysis of predicted magnetoelectric fluorides: $KMnFeF_6$ and $Ba_2Ni_7F_{18}$, J. Phys. Condens. Matter 20 (2008) 335229 (6 pp.).
[44] D. Pajić, Z. Jagličić, Z. Trontelj, Slow magnetic dynamics in the $K_3M_3^{II}M_2^{III}F_{15}$ multiferroic system, J. Appl. Phys. 112 (2012) 073908 (9 pp.).
[45] S.V. Petrov, M.A. Popov, L.A. Prozorova, Study of the antiferromagnetic resonance spectrum in $BaMnF_4$, Zh. Eksp. Teor. Fiz. 62 (1972) 1884 (Sov. Phys. JETP 35, 981–983).
[46] A. Poole, B. Roessli, O. Zaharko, K.W. Krämer, The magnetic structure of multiferroic $BaMnF_4$, J. Phys. Condens. Matter 23 (2011) 266004 (5 pp.).
[47] J.M. Posse, A. Grzechnik, K. Friese, Ternary fluorides $BaMF_4$ (M = Zn, Mg and Mn) at low temperatures, Acta Crystallogr. B 65 (2009) 576–586.
[48] J. Ravez, S.C. Abrahams, R. De Pape, Ferroelectric–ferroelastic properties of $K_3Fe_5F_{15}$ and the phase transition at 490 K, J. Appl. Phys. 65 (1989) 3987–3990.
[49] J. Ravez, S.C. Abrahams, A.M. Mercier, L. Rabardel, R. De Pape, Phase transition dependence on composition in ferroelectric–ferroelastic $K_{3-x}Fe_5F_{15}$ for $0 \leq x \leq 0.20$, J. Appl. Phys. 67 (1990) 2681–2683.
[50] J. Renaudin, G. Ferey, A. de Kozak, M. Samouël, P. Lacorre, Crystal and magnetic structures of the ferrimagnet $Ba_2Ni_7F_{18}$, Solid State Commun. 65 (1988) 185–188.
[51] H. Schmid, Multi-ferroic magnetoelectrics, Ferroelectrics 162 (1994) 317–338.
[52] H.G. Von Schnering, P. Bleckmann, Ternäre Fluoride ABF_4 eines neuen Strukturtyps, Naturwissenschaften 55 (1968) 342–343.
[53] J.F. Scott, Phase transitions in $BaMnF_4$, Rep. Prog. Phys. 42 (1979) 1055.
[54] J.F. Scott, R. Blinc, Multiferroic magnetoelectric fluorides: why are there so many magnetic ferroelectrics? J. Phys. Condens. Matter 23 (2011) 113202 (17 pp.).
[55] R.G. Smith, J.E. Guesic, H.J. Levinstein, J.J. Rubin, S. Singh, L.G. Van Uitert, Continuous optical parametric oscillation in $Ba_2NaNb_5O_{15}$, Appl. Phys. Lett. 12 (1968) 308–310.
[56] E.G. Spencer, H.J. Guggenheim, G.J. Kominiak, Barium manganese fluoride, a new crystal for microwave ultrasonic, Appl. Phys. Lett. 17 (1970) 300–301.
[57] Y. Tomita, J. Bergquist, M. Shibata, Photorefractive properties of undoped, Cr-doped, and Cu-doped potassium sodium strontium barium niobate crystals, J. Opt. Soc. Am. B 10 (1993) 94–99.
[58] E.L. Venturini, R.P. Morgenthaler, Proceedings of the 20th conference on magnetism and magnetic materials, San Francisco, California, in: C.D. Graham Jr., G.H. Lander, J.J. Rhyne (Eds.), AIP Conf. Proc. No. 24 (AIP, New York, 1975), 1974, p. 168.
[59] E.G. Villora, K. Shimamura, F. Jing, A. Medvedev, S. Tekekawa, K. Kitamura, Ferroelectric and optical properties of single crystal $BaZnF_4$, Appl. Phys. Lett. 90 (2007) 192909 (3 pp.).
[60] K.F. Wang, J.-M. Liu, Z.F. Ren, Multiferroicity: the coupling between magnetic and polarization orders, Adv. Phys. 58 (2009) 321–448.
[61] F. Whöler, Über das Wolfram, Ann. Phys. 78 (1824) 345–358.
[62] R.F. Williamson, W.O. Boo, Lower valence fluorides of vanadium. 2. Characterization of the tetragonal phase K_xVF_3, J. Inorg. Chem. 16 (1977) 649–651.
[63] H. Zeng, K. Shimamura, E.G. Villora, S. Takekawa, K. Kitamura, Domain growth kinetics and wall strain behavior in $BaMgF_4$ ferroelectric crystal by piezoresponse force microscopy, J. Appl. Phys. 101 (2007) 074109 (5 pp.).

Part Four

Non-linear properties

Frequency-Doubling Oxide Fluorides, Borate Fluorides, and Fluorooxoborates

15

S. Pan, Y. Wang
Xinjiang Technical Institute of Physics & Chemistry, Chinese Academy of Sciences, Urumqi, P. R. China

K.R. Poeppelmeier
Northwestern University, Evanston, IL, United States

Chapter Outline

Photonic and Electronic Properties of Fluoride Materials. http://dx.doi.org/10.1016/B978-0-12-801639-8.00015-5

1. Introduction

The frequency-doubling (FD) materials that are able to expand the frequency ranges of lasers by second harmonic generation (SHG) are the core of the solid-state lasers [1–6]. FD is a nonlinear optical (NLO) process, in which two input laser beams with the same frequency interacting with an FD material "combine" together to generate a new output laser with twice the frequency and half the wavelength of the initial light beams. This phenomenon was first demonstrated by Franken et al. in 1961 [7]. They irradiated a quartz sample using a laser beam with a wavelength of 694 nm, and observed the output light at 347.2 nm. FD is only allowed with coherent light and in mediums without inversion symmetry.

Since the 1980s, oxoborates (henceforth, oxoborates will be simply called borates), owing to their variety of structures and inherent prominent properties, have been generally considered as good candidates for FD materials, especially the ones used in the ultraviolet (UV) and deep-ultraviolet (DUV) wavelength ranges. During the last few decades, many borate-based FD materials, such as β-BaB_2O_4 (BBO) [8], LiB_3O_5 (LBO) [9], CsB_3O_5 (CBO) [10], $CsLiB_6O_{10}$ (CLBO) [11], $YCa_4O(BO_3)_3$ (YCOB) [12], $Sr_2Be_2B_2O_7$ (SBBO) [4], and α-BiB_3O_6 (BiBO) [13], were developed, of which BBO and LBO have been commercially manufactured and used worldwide. However, the availability of FD materials operating in the DUV region ($\lambda < 200$ nm) is still limited, and it remains challenging to efficiently design a suitable DUV FD material. For a DUV FD material to be useful in practice, at least three material requirements have to be satisfied: (1) a moderate to large NLO susceptibility; (2) an appropriate transparent spectral window; and (3) a sufficient birefringence for phase matching (PM).

Among borate materials, a fluoride-containing borate, $KBe_2BO_3F_2$ (KBBF) was found to be the only one that can generate DUV coherent light by direct SHG, which is a significant milestone in nonlinear optics [1,14,15]. The excellent properties of KBBF mainly come from the infinite planar $[Be_2BO_3F_2]_\infty$ layers, which could provide relatively large SHG coefficients, wide bandgap, and sufficient birefringence [1]. Unfortunately, the strong layering tendency of KBBF limits its application. In recent years, it is also reported that the incorporation of an F anion can efficiently widen the transparent window of materials, making them more suitable for UV and DUV applications [1,2,16]. Thus, considerable attention has been attracted to find an optimal composition in fluoride-containing borates. For example, $Ca_5(BO_3)_3F$ (CBF), an apatite derivative, was found to be a potential FD material soon after the discovery of KBBF [17]; $ABe_2BO_3F_2$ (A = Na, Rb, Cs, and Tl) [18,19], $BaAlBO_3F_2$ (BABF) [20], $NaCaBe_2B_2O_6F$ [21], $NaSr_3Be_3B_3O_9F_4$ [22], $M_3B_6O_{11}F_2$ (M = Sr, Ba, Pb) [23–26], $Ba_4B_{11}O_{20}F$ [27] etc. in succession have also been reported as promising FD materials.

Unlike the fixed coordination environment of the Si atom in silicates (SiO_4 tetrahedra), however, crystalline metal borates have a rich structural chemistry originating from the two different connection modes of the B–O bonds (BO_3 triangles and BO_4 tetrahedra). When fluorine atoms are introduced into metal borates, both in anhydrous and hydrated forms, the F atoms can formally replace the O atoms or OH^- groups in the structures, leading to more extensive structural arrangements than other homonuclear polyoxoanions. Generally, on the basis of different connection modes between

B, O, and F atoms, the fluorine-containing borates can be classified as two types: (1) borate fluorides (or fluoride borates) are double salts, in which one (or more) fluorine atom is directly coordinated to the metal atoms, such as KBBF; (2) fluorooxoborates (or simply fluoroborates) are anionic species, where one (or more) fluorine atom is directly coordinated to the B atoms, such as ABF_4 (A = Na, K).

From a survey of the Inorganic Crystal Structure Database (ICSD) and recent publications, over 100 fluorine-containing borates have been reported at present. Among them, 80% of those are borate fluorides while fluorooxoborates are less found, possibly due to their weak thermal stability (releasing BF_3 when heating) [28]. The new borate fluorides and fluorooxoborates exhibit interesting structural types, and some of them have favorable performances as FD materials. Therefore, it is crucial to understand the preparation routing, crystal chemistry, and the structure–property relationship of those materials.

2. Fundamentals of Borate Fluorides and Fluorooxoborates: Synthesis and Structural Chemistry

2.1 The Synthesis of Borate Fluorides and Fluorooxoborates

An effective synthetic way to synthesize materials is needed for the discovery of new FD materials and characterization of NLO properties in succeeding studies. In particular, large single crystals up to centimeter-level are desired. For FD materials, only after the quality large crystals are ready, their linear and nonlinear optical properties, especially, the transparent ranges, SHG coefficients, and refractive indices, can be accurately obtained. Therefore, the synthesis and growth of crystals is one of the key problems to study the application of new FD materials.

In general, similar to traditional borates, the solid-state reaction could provide a cheap and easy way to prepare polycrystalline powders of borate fluorides. For example, polycrystalline KBBF was prepared by a normal solid reaction according to the equation [29]: $3KBF_4 + 6BeO + B_2O_3 \rightarrow KBe_2BO_3F_2 + BF_3\uparrow$. In order to grow large crystals of borate fluorides, although there are many methods for growing crystals, including crystal pulling, recrystallization from the melt/solution, zone refining, chemical vapor transport, hydrothermal/solvothermal, and flux growth, only the flux growth method is commonly used. Because most borate fluorides are not congruently melting compounds and metal fluorides will volatilize at a fairly high temperature, a proper flux is needed for crystal growth at conveniently low temperatures. High-temperature melts of inorganic compounds are used as the solvent for crystallization [30]. The typical flux synthesis route of borate fluoridesis: $MF_n + MO_m + B_2O_3/H_3BO_3 \rightarrow M_xB_yO_zF_n$ (M = metal cations), where metal fluorides MF_n and B_2O_3/H_3BO_3 can be used as both reactant and flux. Some other inorganic compounds, such as $BaCl_2$, Bi_2O_3, KOH, PbO, or Na_2CO_3, could be also introduced as fluxes. The growth rate, crystal habit, and quality are influenced by the choice of different fluxes and conditions.

However, finding a flux or "perfect" solvent is extremely challenging, and numerous attempts need to be carried out.

Different from borate fluorides, fluorooxoborates have weak thermal stability due to the unstable B—F bonds at a relatively high temperature [28]. Therefore, the hydrothermal method was first employed to synthesize fluorooxoborates. In hydrothermal conditions, the fluorine source could be metal fluorides, borofluoride, or hydrofluoric acid. The addition of fluoride ions (normally in the form of HF) has effects on two aspects: low concentrations of fluoride often act as a mineralizer, while F^- ions with larger quantities led to a diverse range of crystal structures where the F^- ions form bridging and terminal bonds involving metal polyhedral units [31]. The fluoride ions can formally replace OH^- groups in the borate structures.

In the last decade, the boric acid flux synthetic technique at ~200°C (also known as low-temperature or molten boric acid flux method) has also been applied. This method, in which excess melt boric acid (H_3BO_3) acts as both the reaction medium and reactant, is very effective for the syntheses of borates and borophosphates [32,33]. It has been demonstrated to be a facile method for preparing novel actinide fluorooxoborates containing the BO_3F unit [34]. In addition, some fluorooxoborates are even sensitive to water or humid air. To avoid decomposition, several lithium fluorooxoborates have been efficiently synthesized by using the standard Schlenk techniques with glove boxes and rigorously dried reaction mixtures in a purified argon atmosphere. The mixture of raw materials is placed in closed silver crucibles, sealed in glass ampoules and followed by heating and slow cooling processes [28,35,36]. In a recent study, however, it is reported that some fluorooxoborates can also be obtained using the flux method in open air under ambient pressure; for example, the crystals of $K_3B_3O_3F_6$ were obtained from a mixture of KBF_4, KF, and H_3BO_3 at 673K in open air [37]. Moreover, solid—gas reaction between Na_3PO_4 and BF_3 has also led to the synthesis of a new fluoridoboratophosphate, $Na_3B_2PO_5F_4$ [38].

2.2 Crystal Chemistry of Borate Fluorides and Fluorooxoborates

The borate chemistry has been discussed exhaustively in many review articles. Owing to the two connection modes of B—O bonds, the crystal structures of borates vary from low-dimensional anionic units to three-dimensional frameworks [39,40]. It is expected that with the introduction of fluorine atoms, borates can embrace more diversity in their structures and would be likely to isolate new phases with interesting stoichiometries, structures, and properties. From a chemical point of view, borate fluorides are a kind of double salt containing metal fluorides and metal borates; whereas fluorooxoborates can be treated as normal metal borates with partial O replacement by F atoms in the anionic units. The concept of a fundamental building unit (FBU), that is, the basic repeat building block of B—O anionic groups in borate chemistry, is also useful for borate fluorides and fluorooxoborates. The FBUs, such as BO_3, BO_4, B_3O_6, B_3O_7, B_6O_{10}, etc. are either isolated or linked with each other by sharing oxygen atoms to generate covalent lattices in borate fluorides, which is the same as typical borates. However, in fluorooxoborates, the partial substitution of O atoms by F atoms in the BO_4 units leads to the formation of new FBUs: BO_2F_2 and BO_3F anionic units

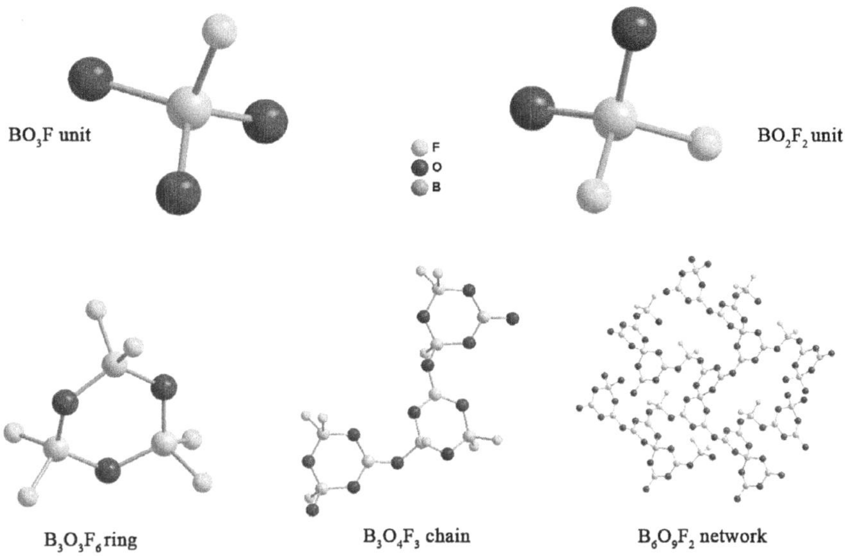

Figure 15.1 The new fundamental building units and extended anionic units found in fluorooxoborates.

(Fig. 15.1). These new units along with other typical FBUs connect to each other generating new structural types, such as $B_3O_3F_6$ boroxine rings [38], $B_3O_4F_3$chains [35], and $B_6O_9F_2$ network [36], and largely expand the crystal chemistry of borates. It is noted that the BOF_3 units and the same F-substitution in triangular BO_3 units, ie, BO_2F or BOF_2, have not yet been found [41].

The structural chemistry of fluorine-containing borates also has some differences from traditional borates. Although F atoms have similar coordination numbers compared with O atoms, the F anions are more likely to be terminal instead of bridging. In some rare cases, the F atoms may also be viewed as located in the hole, channel, or void of the 3D framework [42]. In borate fluorides, the F atoms exclusively connect to metal cations, serving to terminate B—O—B bridges, yielding a less tightly connected framework, which limits the dimensionality of the structure. Therefore low-dimensional structures become more prevalent than in polyhedral oxo- and oxanion compounds [31]. This phenomenon can also be easily explained by the dimensional reduction formalism developed by Tulski and Long for solid structures [43]. A classic example is the dimensional reduction of anionic units from a three-dimensional framework in $Be_2(BO_3)F$ [44] to the two-dimensional layers in $KBe_2(BO_3)F_2$ with KF as the dimensional reduction agent. As for fluorooxoborates, the situations are the same: low-dimensional anionic structures such as chains and layers are more commonly found in fluorooxoborates.

In a broad definition, borate fluorides may also be classified as salt-inclusion solids (SISs), compounds made of a hybrid framework exhibiting integrated covalent (boron oxides) and ionic (metal fluorides) lattices [45—48]. The SISs are commonly found in transition metal oxides and fluorides, in which the salt lattice plays an essential role in

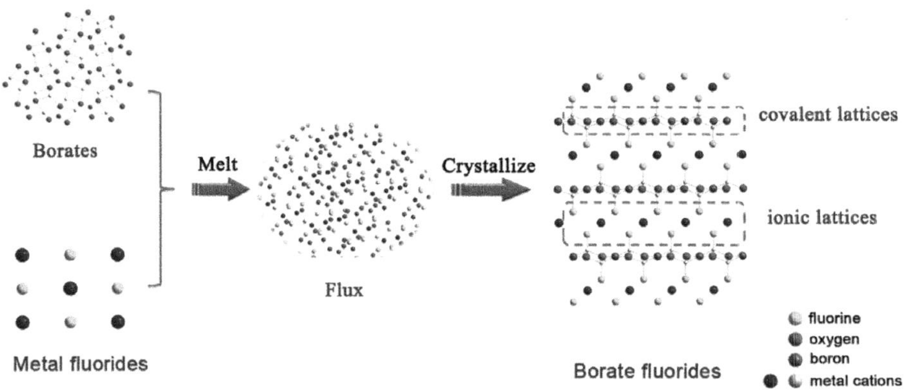

Figure 15.2 The typical preparation routing of borate fluorides.

bulk structural and chemical/physical properties [45,47,49]. Borate fluorides also show two chemically dissimilar lattices: ionic and covalent lattices, between which can reveal a host–guest relationship like organic–inorganic hybrid materials. In particular, the synthesis of borate fluorides shows similar features with typical SISs, which can be thought of as the dissolution of metal borates in molten halide salts followed by cocrystallization process in the flux (Fig. 15.2).

Also interesting is that oxyborates (borates with additional oxygen atoms not connected to boron) can probably be the model or prototype for borate fluorides. It is known that F and O share many similar features in chemistry: ionic radius, electronegativity, and M–O/F bond length (M = metal cation). Herein, the so-called "isolated" oxygen atoms in oxyborates, which only connect to metal cations, could be replaced by fluorine atoms. In order to maintain the charge balance, the cations should also be replaced. For example, from the $RECa_4O(BO_3)_3$ structural model to $Ca_5(BO_3)_3F$, it could be considered that the RE^{3+} is substituted by Ca^{2+}, and O^{2-} by F^-, in respective positions (Fig. 15.3).

The introduction of fluorine also has a positive effect for designing and synthesizing new borates as FD materials. On the one hand, the introduction of F atoms into borates not only helps the diversity of the borate structure but also raises the local distortion increasing the microscopic second-order susceptibility of the anionic group. For borate fluorides, M–F (M = metal cation) bonds are usually longer than M–O bonds, while the B–F bonds are reasonably shorter than other B–O bonds in fluorooxoborates. The difference in bond length would increase the local dipole moment, yielding microscopic polarization. For example, strong SHG intensities originate from F-directed polar displacements in $Ba_4B_{11}O_{20}F$ [27] and the synergistic electron-withdrawing effects of F atoms in BiB_2O_4F [51]. On the other hand, adding fluorine will lead to the blueshift of the transmission spectrum, ie, a larger bandgap. The fluorine atom has a lower electronegativity than the oxygen atom, which means that the M–F (M = cation) bonds may have a wider optical transmission range than that of the M–O bond. Therefore, the bandgap of borate fluorides or fluorooxoborates is usually larger than that of typical borates, and that will be helpful for FD application in DUV optical range.

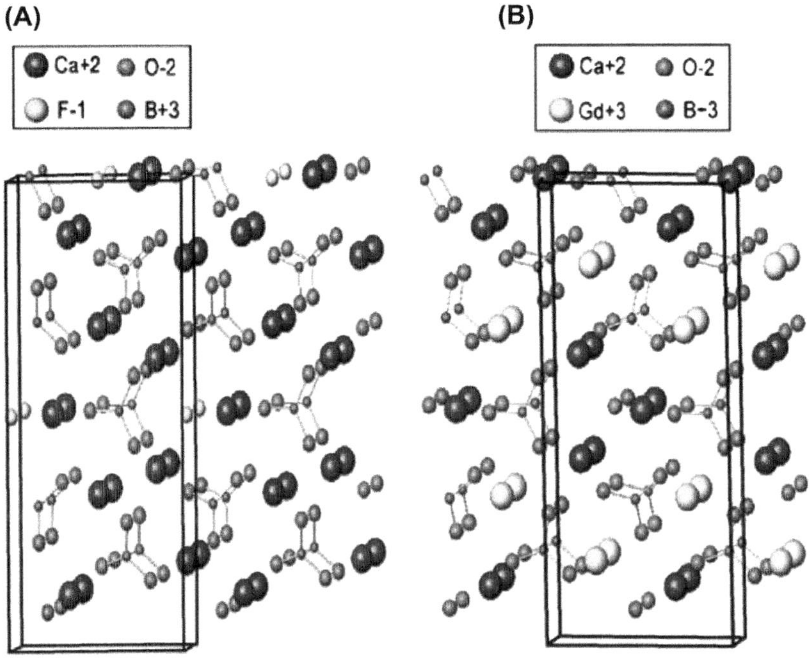

Figure 15.3 View of $Ca_5(BO_3)_3F$ (A) and $GdCa_4O(BO_3)_3$ (B) structures [50].

3. Borate Fluorides as Frequency-Doubling Materials

It is known that a prerequisite for FD materials is the lack of inversion symmetry in crystal structure, ie, noncentrosymmetric (NCS) crystals. In this review, we will concentrate our attention mainly on anhydrous crystals because of their chemical and thermal stabilities, which are beneficial for FD applications. For a better understanding, some typical SHG-active borate fluorides and fluorooxoborates are summarized in Table 15.1. The powder SHG test technique, which was developed by Kurtz and Perry [72], is used to determine the order of SHG effect and whether or not the FD materials can be phase-matchable in powder samples. This technique is widely employed without requiring the growth of single crystals. For FD materials with large single crystals, their linear optical properties can be evaluated in detail by transmission spectrometer and refractometer, while the NLO properties are measured by the Maker fringe method [73] and the PM method.

3.1 $KBe_2BO_3F_2$ and Its Derivatives

3.1.1 $ABe_2(BO_3)F_2$ (A = Na, K, Rb, Cs, and Tl)

In 1990, the discovery of the new NLO crystal KBBF immediately attracted worldwide researchers' attention [15,74]. KBBF is a typical borate fluoride, in which F

Table 15.1 Borate Fluorides and Fluorooxoborates as Frequency-Doubling Materials

Compounds	Space Group	Second Harmonic Generation Intensities[a]	References	Compounds	Space Group	Second Harmonic Generation Intensities	References
Borate Fluorides							
$Be_2(BO_3)F$	*C*2	1/4 × KDP	[44]	α-$Mg_2(BO_3)F$	$Pna2_1$	N/A	[52]
$NaBe_2BO_3F_2$	*C*2	~1.5 × KDP	[18]	$Mg_5(BO_3)_3F$	$Pna2_1$	N/A	[53]
$KBe_2BO_3F_2$ (flux method)	*R*32	~1.5 × KDP	[15]	$Ca_5(BO_3)_3F$	*Cm*	2–3 × KDP	[17,54]
$RbBe_2BO_3F_2$	*R*32	~1.2 × KDP	[19,55]	$Cd_5(BO_3)_3F$	*Cm*	3.8 × KDP	[56]
$CsBe_2BO_3F_2$	*R*32	~1.2 × KDP	[19,57]	$Sr_3B_6O_{11}F_2$	$P2_1$	2.5 × KDP	[24,26]
$TlBe_2BO_3F_2$	*R*32	<1 × KDP	[19]	$Ba_3B_6O_{11}F_2$	$P2_1$	3 × KDP	[23,26]
$NaSr_3Be_3B_3O_9F_4$	*R*3*m*	4–5 × KDP	[22]	$Pb_3B_6O_{11}F_2$	$P2_1$	4 × KDP	[25]
$NaCaBe_2B_2O_6F$	*Cc*	1/3 × KDP	[21]	$Ba_4B_{11}O_{20}F$	$Cmc2_1$	10 × KDP	[27]
$BaAlBO_3F_2$	$P\bar{6}2c$	2 × KDP	[58]	$Rb_3Al_3B_3O_{10}F$	$P3_1c$	1.2 × KDP	[59]
$BaAl_{0.61}Ga_{0.39}BO_3F_2$	$P\bar{6}2c$	2 × KDP	[60]	$Ba_7(BO_3)_3F_5$[b]	$P6_3mc$	1/3 × KDP	[61,62]

α-$BaCaBO_3F$	*R*3	N/A	[63]	$Ba_7(BO_3)_{3.51}F_{3.47}$	$P6_3$	N/A	[64]
β-$BaCaBO_3F$	$P\bar{6}2m$	~0.7 × KDP	[62,65]	$Ba_3Sr_4(BO_3)_3F_5$[b]	$P6_3mc$	0.5 × KDP	[66]
$BaZnBO_3F$	$P\bar{6}$	3 × KDP	[67]	$Ba_{3.12}Sr_{3.88}(BO_3)_{3.65}F_{3.05}$	$P6_3mc$	N/A	[68]
$BaMgBO_3F$	*Cc*	1/10 × KDP	[67]	$Li_3Ca_9(BO_3)_7 \cdot 2[LiF]$	*P*1	1 × KDP	[42]
$Mg_2(BO_3)_{1.14}F_{0.58}$	$Pna2_1$	N/A	[69]				
Fluorooxoborates							
BiB_2O_4F	$P3_2$	10 × KDP	[51,70]	$Li_2B_6O_9F_2$	*Cc*	1 × *a*-SiO_2	[36]
LiB_6O_9F	$Pna2_1$	1 × KDP	[28]	$Li_2B_3O_4F_3$	$P2_12_12_1$	Very weak	[35]
NH_4BPO_4F	$P2_13$	N/A	[33]	$ABAsO_4F$ (A = K, Rb, Cs)	$P2_13$	N/A	[71]

N/A, not available or not been reported.
[a]Recorded using Kurtz and Perry method with 1064 nm laser as fundamental frequency light.
[b]Controversial data.

atoms are only connected to Be and K atoms. In the structure of KBBF, the Be atoms are tetrahedrally coordinated with three oxygen atoms and one fluorine atom, and each BeO_3F tetrahedron is linked with three trigonal BO_3 groups forming the 2D $[Be_2F_2BO_3]_\infty$ layers. The remaining K^+ cations are located between the layers and keeping the charge balance. The direction of the F ion in neighboring BeO_3F tetrahedra is alternately upward and downward (Fig. 15.4).

The cutoff wavelength of KBBF on the UV side is located at 147 nm (Fig. 15.5), and the refractive index data suggest that it has a sufficient birefringence for PM in DUV range. Till now, KBBF possesses the widest SHG PM wavelength range

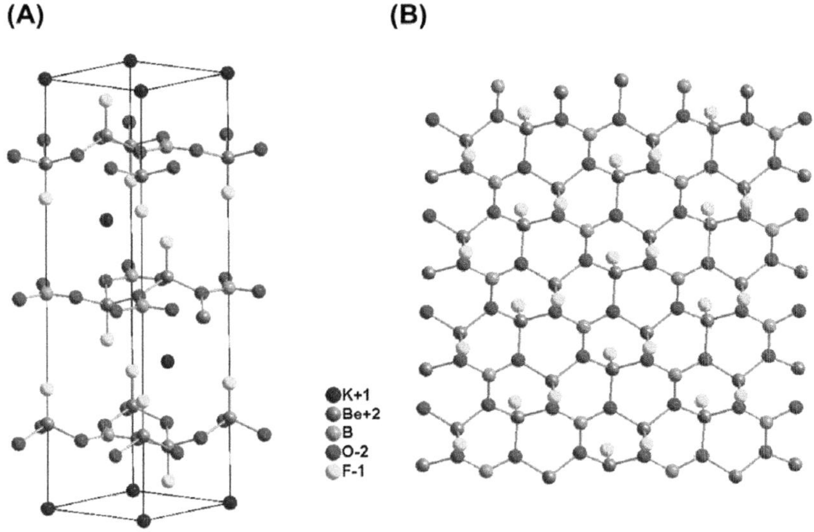

Figure 15.4 (A) Unit cell of $KBe_2BO_3F_2$ and (B) detailed structure of $[Be_2F_2BO_3]^\infty$ layer.

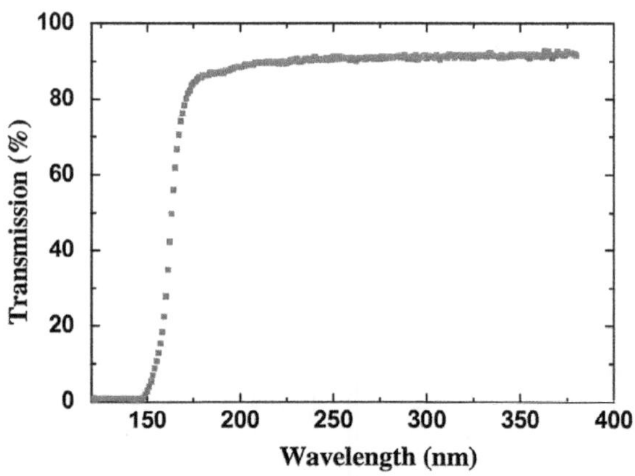

Figure 15.5 Optical transmission spectrum of $KBe_2BO_3F_2$ crystal in UV—DUV region (the crystal, is 0.45 mm thickness) [78].

(1400−340 nm) among all FD materials [3]. It can exactly predict that KBBF crystal can produce SHG output down to wavelength of 161.1 nm [75].

Powder SHG tests show that KBBF has an SHG intensity about 1.5 × KH_2PO_4 (KDP) [19]. The SHG coefficients were precisely measured by the Maker fringe method [73]. KBBF belongs to the space group *R*32, and it has two nonzero d_{ij} coefficients: d_{11} and d_{14}. The d_{11} can be exactly deduced to be $d_{11} = (0.47 \pm 0.01)$ pm/V (d_{36} (KDP) = 0.39 pm/V), which is in agreement with the theoretical value $d_{11} = 0.351$ pm/V calculated by an *ab* initio method [76], while the d_{14} is very small as confirmed by both experiments and theoretical calculations. According to the anionic group theory [77] developed by Chen et al., the FD properties, birefringence, and bandgap of KBBF structure are dominated by the $[Be_2F_2BO_3]_\infty$ planar network structure and the cation has little effect on the above parameters.

Large crystal growth of KBBF was difficult because of its strong layered growth tendency due to the weak interlayer K−F bonds. Chen and coworkers have performed numerous studies on flux growth of KBBF crystals. Detailed descriptions of its growth, optical properties, and important applications were reported [3,79−81]. Bulk KBBF crystals with good optical quality and sizes up to 3.7 mm in thickness have been successfully grown using a KBF_4 and $KF-B_2O_3$ flux system and a new technology-localized spontaneous nucleation (Fig. 15.6) [3]. KBBF has excellent NLO properties; however, the as-grown bulk KBBF crystal is still too thin to be cut along the PM direction. To solve this problem, a specially designed prism-coupling device (PCD) of KBBF is adopted to achieve PM for producing DUV harmonic generation (Fig. 15.7). KBBF is the first practically usable DUV NLO crystal used to generate coherent light below 200 nm [82]. In recent years, it was also reported that high average power 177.3 nm (sixth harmonic of Nd-based lasers) and 200 nm (fourth harmonic of Ti: sapphire laser) lasers can be achieved by using KBBF−PCD [78,83,84] (Fig. 15.8).

In recent years, the hydrothermal growth of KBBF was developed in order to increase the thickness of the crystals along the *c*-axis [19]. However, the SHG conversion efficiency of as-grown crystals was found to be much lower than flux-grown crystals. In 2011, Yu et al. obtained another phase of KBBF with the $R\bar{3}c$ space group by powder X-ray diffraction (PXRD) and Rietveld refinement from hydrothermally grown KBBF crystals, which is a 1 × 1 × 2 superstructure of the previously reported *R*32 structure [85]. It was presumed that the crystal nonuniformity was caused by stacking faults in the $[Be_2F_2BO_3]_\infty$ layer with twisted BO_3 groups.

It is conceivable that the substitution of K^+ by other alkali metals in KBBF will lead to the discovery of other new NLO crystals retaining the same basic framework of KBBF. A series of borate fluorides with the formula of $ABe_2(BO_3)F_2$ (A = Na, K, Rb, Cs and Tl) has been found [18,19,86]. In this series, the Rb, Cs, and Tl analogs are isomorphic to flux-grown KBBF with the space group *R*32, while $NaBe_2(BO_3)F_2$ shows the same basic building unit but with a different space group, *C*2. In $NaBe_2(BO_3)F_2$, two kinds of six-membered $(Be_2BO_3F_2)^-$ rings are connected to form infinite $[Be_2F_2BO_3]_\infty$ layers. Since those compounds share similar structure types, they are all proposed to have favorable NLO properties. At present, only Rb- and Cs-analogs have been systematically characterized; the others are difficult to grow single crystals so that no optical properties have been reported. Large single crystals of

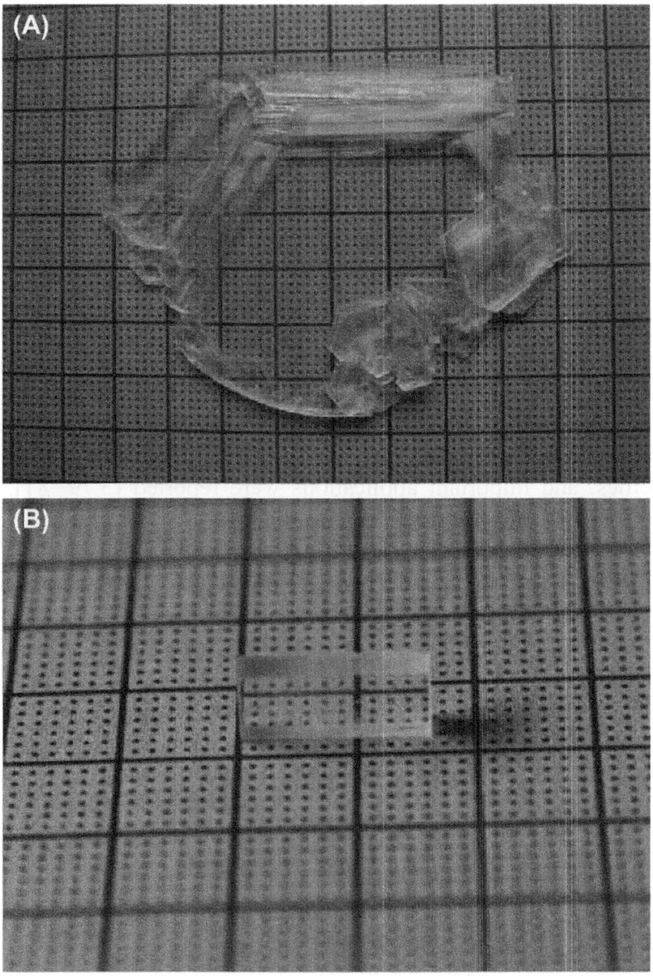

Figure 15.6 Bulk $KBe_2BO_3F_2$ (KBBF) crystal. (A) Bulk KBBF crystal in size of 50 mm × 40 mm × 3.7 mm. (B) Bulk KBBF crystal for device [3].

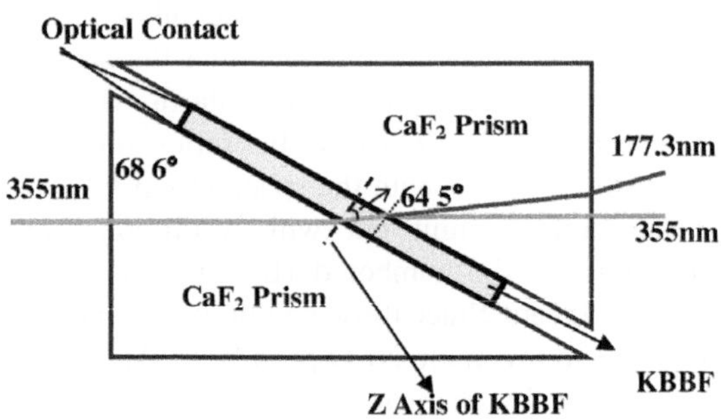

Figure 15.7 Structure of the $KBe_2BO_3F_2$ (KBBF) prism-coupling device [78].

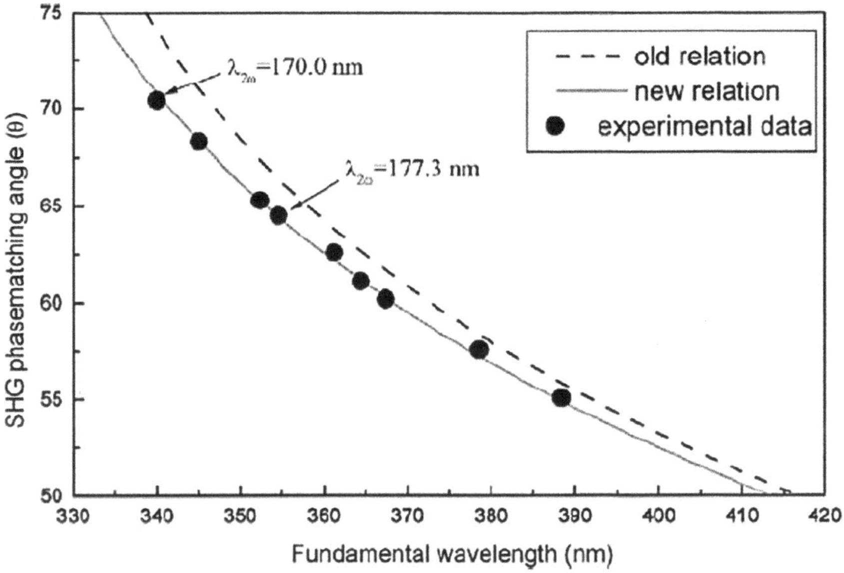

Figure 15.8 Type I second harmonic generation (SHG) phase-matching angles versus wavelength curve [81].

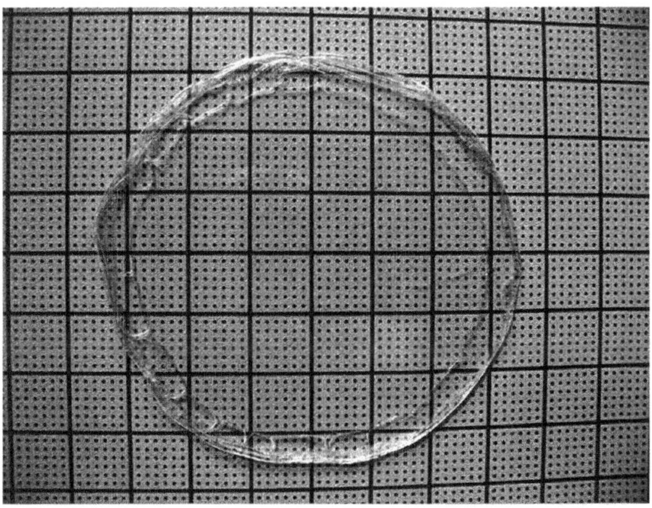

Figure 15.9 Large single crystal of $RbBe_2(BO_3)F_2$ (40 mm × 40 mm) [55].

$RbBe_2(BO_3)F_2$ (RBBF) and $CsBe_2(BO_3)F_2$ (CBBF) have been grown by both the flux method and the hydrothermal method (Figs. 15.9 and 15.10) [19,55,57]. Both of them have powder SHG coefficients comparable to that of KBBF. The differences are that the former has a cutoff edge at 160 nm and can also produce coherent light below 200 nm

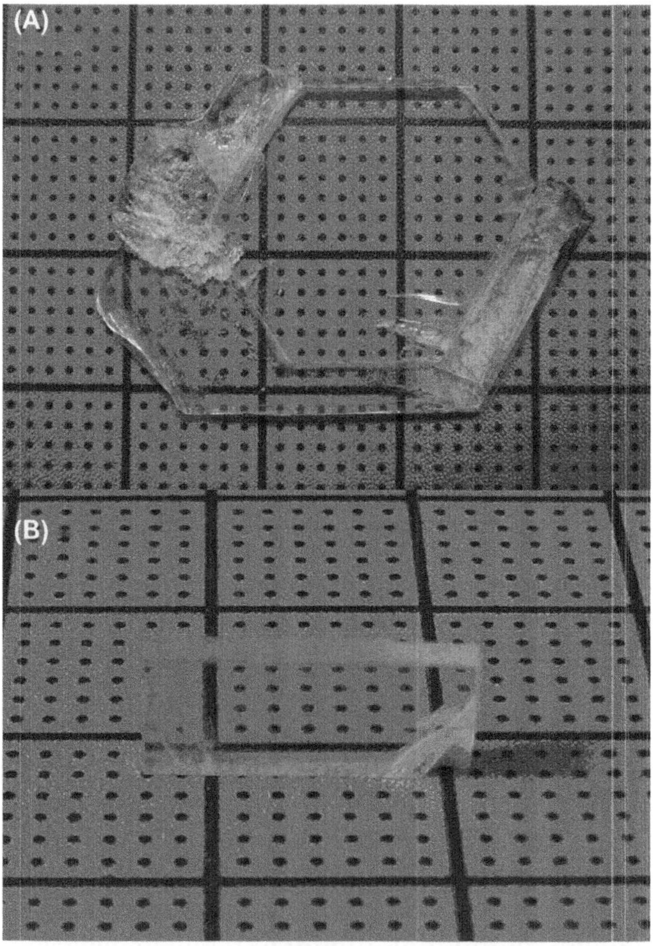

Figure 15.10 (A) Single $CsBe_2(BO_3)F_2$ (CBBF) crystal in relatively regular shape with size 30 mm × 30 mm × 2 mm. (B) Another cut and polished CBBF crystal with thickness of 2.3 mm [57].

through a direct SHG method, while the latter is transparent down to 151 nm but could only achieve SHG PM down to the wavelength of 201 nm [55,57].

3.1.2 $BaAlBO_3F_2$

BABF was first reported with a centrosymmetric space group $P6_3/m$ by Park and Barbier in 2000, which is isotropic to $BaGaBO_3F_2$ [87]. In 2002, Hu et al. revealed that BABF belongs to the NCS space group of $P\overline{6}$ and is a candidate for UV laser generation [20,88]. In 2011, Hu et al. grew large single crystals of BABF and redetermined its structure with a new space group of $P\overline{6}2c$ [58]. The first-principles calculations also verify that the new structure is more energetically favorable [89]. In fact, BABF is supposed to be one of the derivatives of KBBF due to the structural similarity between KBBF and BABF, especially for the anionic groups. From KBBF to BABF, the poisonous Be was substituted by Al, and K was replaced by Ba to maintain charge balance. The microscopic structure of BABF is analogous to the structure of KBBF, which consists of the 2D $(AlF_2BO_3)_n^{2-}$ layers formed by corner-sharing AlO_3F_2 trigonal bipyramids and BO_3 triangles with Ba^{2+} cations located between the layers (Fig. 15.11). In BABF, AlO_3F_2 trigonal bipyramids replace BeO_3F tetrahedral structures in KBBF, and structural analyses show that the distance of interlayers is less than that of KBBF. In addition, the Ba^{2+} cations are coordinated with both O and F atoms; in comparison, the K atoms are sixfold coordinated with F atoms only in KBBF. Therefore, it is expected that BABF has a better growth habit. Indeed, large single crystals of BABF can now be obtained. Hu et al. and many other researchers have obtained large single crystals of BABF using different metal fluorides as the flux, such as LiF, NaF, LiCl, NaCl, and others [20,58,60,90]. In 2010, a BABF single crystal with dimensions of 55 mm × 50 mm × 30 mm was obtained by a modified middle-seeded solution growth method [91] (Fig. 15.12).

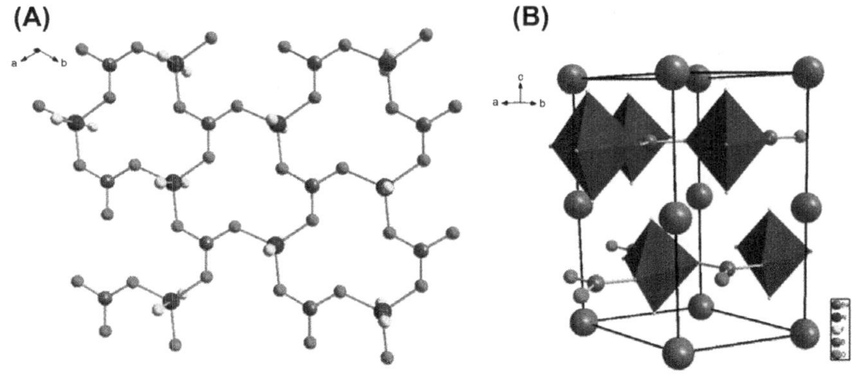

Figure 15.11 The crystal structure of $BaAlBO_3F_2$: (A) the $(AlF_2BO_3)_n^{2-}$ layer and (B) the unit cell, where trigonal bipyramids are AlO_3F_2 groups [58].

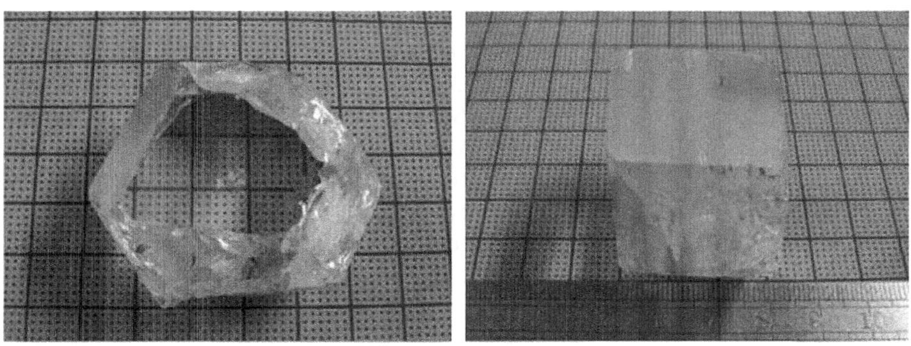

Figure 15.12 $BaAlBO_3F_2$ crystal with size of 55 mm × 50 mm × 30 mm grown by middle-seeded solution growth technique [91].

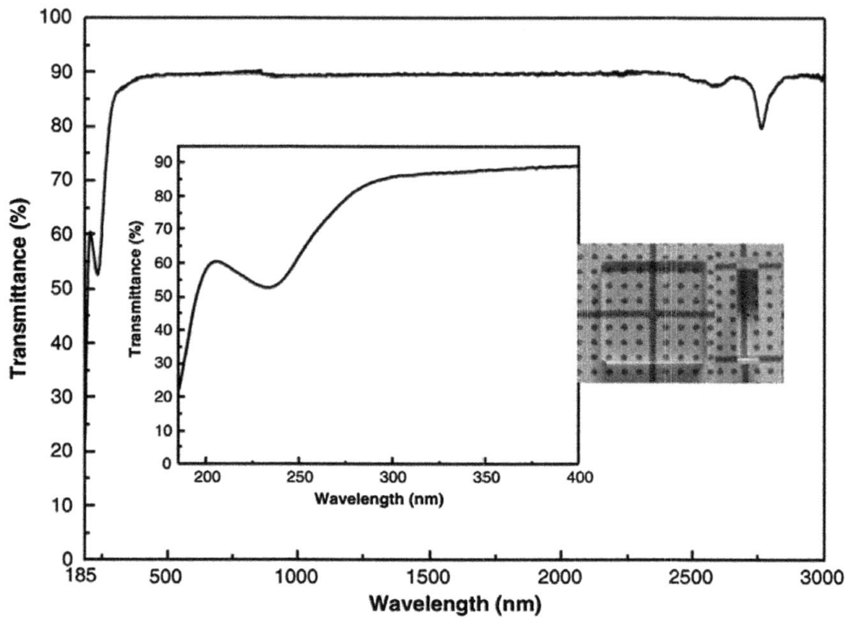

Figure 15.13 Transmission spectrum of $BaAlBO_3F_2$ crystal in the UV−Vis−NIR region [91].

The UV absorption edge of BABF is very short (165 nm), which is only 10 nm longer than that of the KBBF (Fig. 15.13). It was found to have a powder SHG efficiency about twice than that of KDP [20]. The refractive indices of BABF were measured by using the minimum deviation method. The result suggests that BABF also possesses moderate birefringence. Based on the refractive indices and fitted Sellmeier equations, it can be calculated that BABF can achieve SHG with the shortest wavelengths at 273 and 382 nm, for type I and II SHG, respectively [92]. BABF has only one nonzero SHG coefficient d_{22}, and the Maker fringe measured value of d_{22} is $d_{22} = 3.178\ d_{36}$ (KDP) (≈ 1.24 pm/V), which fit the experimental data very well.

It has been demonstrated that BABF is very suitable to produce high-power laser output at the wavelengths of 532 and 355 nm [92,93]. The high-power SHG performance of the high-quality BABF crystals was reported to achieve 8.2 W power output at 532 nm with the energy conversion efficiency of 29.5% [91]. Unfortunately, BABF does not allow realizing 266-nm UV light by direct SHG from the fundamental laser at 1064 nm.

Single crystals of Ga-doped BABF, $BaAl_{1-x}Ga_xBO_3F_2$ ($x < 0.7$) have been obtained by Wang et al. [60] The Ga atoms only occupy the position of Al atoms without changing the BABF structure. The SHG efficiency of $BaAl_{1-x}Ga_xBO_3F_2$ was the same as that of BABF, but its UV cutoff wavelength red-shifted to 265 nm.

3.1.3 $Be_2(BO_3)F$

$Be_2(BO_3)F$ is first reported as the "intermediate" for preparing KBBF [44]. One can immediately realize that there are some correlations between the formula of KBBF

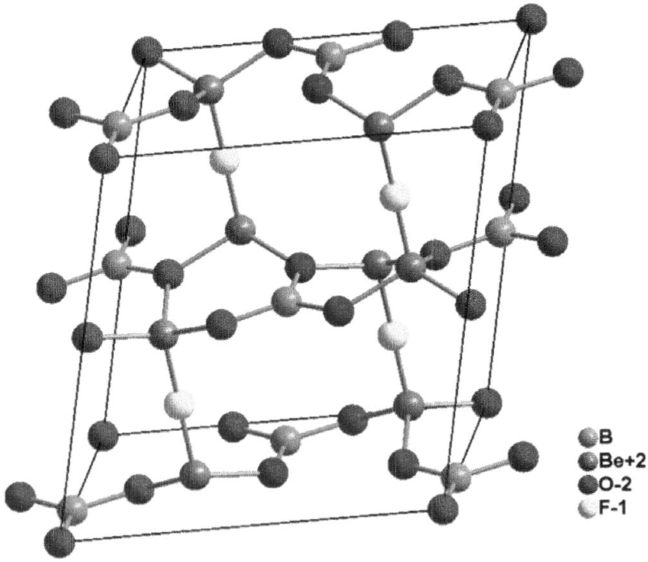

Figure 15.14 The unit cell of $Be_2(BO_3)F$.

and $Be_2(BO_3)F$. The FBU of $Be_2(BO_3)F$ is the six-membered ring $[Be_2BO_3F_2]$, consisting of two BeO_3F tetrahedra and one BO_3 triangle (Fig. 15.14). The rings are connected through the axial F apices of the tetrahedra to generate a 3D network. Interestingly from the structural point of view, it seems that when KF salts are added to $Be_2(BO_3)F$, the network of the latter is broken and then forms the 2D $[Be_2BO_3F_2]_\infty$ layers, with the KF filled in the interlayer void. In this sense, like many SISs, KBBF could be written as "$KF \cdot Be_2(BO_3)F$." Similar to KBBF, each BO_3 group in $Be_2(BO_3)F$ is connected to three BeO_3F tetrahedra, and all BO_3 groups remain a coplanar configuration. $Be_2(BO_3)F$ may have good NLO performance as well and may be superior to KBBF in that it has no layer habit. In 2005, a preliminary evaluation indicated that the absorption edge of $Be_2(BO_3)F$ is around 150 nm, and its powder SHG efficiency is $^1/_4$ of that of KDP [94]. Unfortunately, this material seems to be hard to obtain large single crystals, and detailed NLO properties have not been reported.

3.1.4 $NaCaBe_2B_2O_6F$ and $NaSr_3Be_3B_3O_9F_4$

To overcome the crystal growth problems while retaining the brilliant optical properties of KBBF, Huang et al. have synthesized a novel series of borate fluorides, $NaCaBe_2B_2O_6F$, $KCaBe_2B_2O_6F$, and $KSrBe_2B_2O_6F$ [21]. In this series of compounds, $NaCaBe_2B_2O_6F$ is SHG-active and it consists of the infinite $(Be_6B_6O_{12}F_3)_n^{3-}$ double layers connected by bridging F atoms (Fig. 15.15). The $(Be_6B_6O_{12}F_3)_n^{3-}$ double layer is constructed by two adjacent 2D $(Be_3B_3O_6F_3)_n$ layers reversely through bridging fluorine atoms. The single $(Be_3B_3O_6F_3)_n$ layer is similar to the structure along the *ab* plane in KBBF; however, the connection

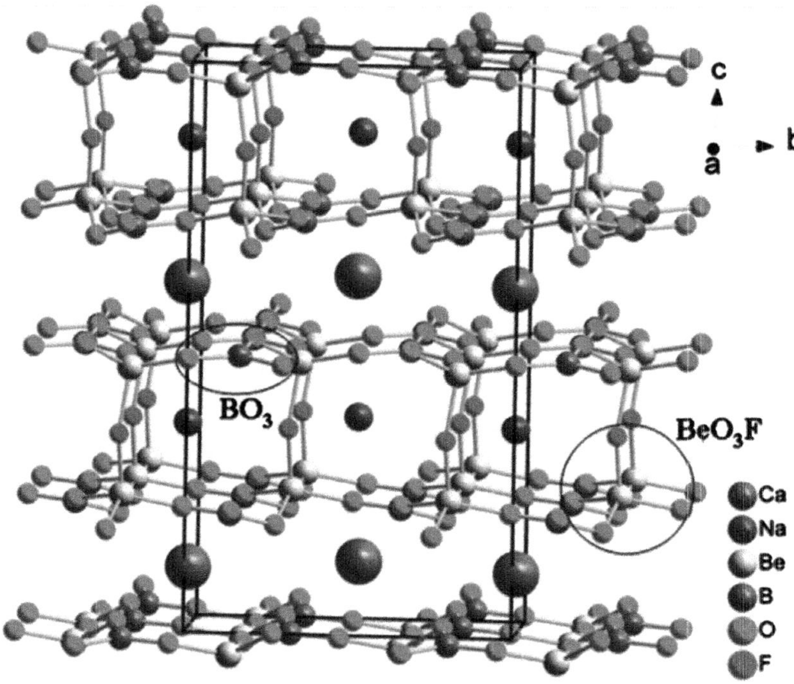

Figure 15.15 Crystal structure of $NaCaBe_2B_2O_6F$ [21].

mode is different. The Na^+ cations reside in the channels and Ca^{2+} cations are located in the space between two double layers. This configuration may greatly improve the serious layering growth habit present in KBBF. The BO_3 groups have dominated contribution to SHG efficiency according to previous research works on borate FD materials. Unfortunately, the BO_3 groups in $(Be_3B_3O_6F_3)_n$ layer are slightly deviated from *ab* plane and aligned in the different orientation so that the resulting SHG effects of $NaCaBe_2B_2O_6F$ are limited (1/3 times that of KDP) (Fig. 15.16). $NaCaBe_2B_2O_6F$ is a PM NLO crystal, and the short-wavelength absorption edge is down to 190 nm. The theoretical calculations on electronic structure suggested that the main optical properties of this material are dominated by the BO_3 groups.

Inspired by the fact that the fluorine beryllium–based borates may have short cutoff wavelengths, moderate birefringence, and relatively large SHG coefficients; continuous research work has been focused on this type of material. In 2011, $NaSr_3Be_3B_3O_9F_4$ was found to be a new promising DUV FD material, which possesses a short-wavelength absorption edge of 170 nm and is phase-matchable with strong SHG signal intensity approximately four to five times as large as that of KDP [22]. The basic building unit of $NaSr_3Be_3B_3O_9F_4$ is a novel anionic group $(Be_3B_3O_{12}F)^{10-}$, which comprises three six-membered rings generated by three BO_3 groups and three BeO_3F tetrahedra with one F atom in the center. Each $(Be_3B_3O_{12}F)^{10-}$ ring is further connected to six neighboring rings through terminal

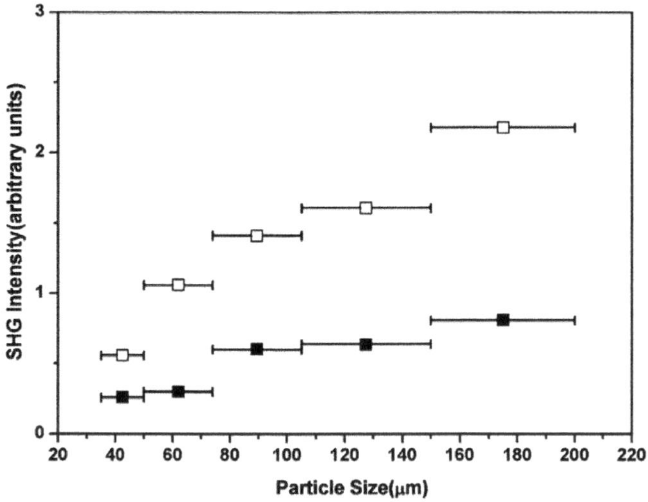

Figure 15.16 Second harmonic generation (SHG) measurements of $NaCaBe_2B_2O_6F$ ground crystals (■) with KDP (□) as a reference [21].

O atoms, and those rings are arranged roughly parallel to each other. That is to say, unlike KBBF, $NaSr_3Be_3B_3O_9F_4$ features a 3D anionic network, with Sr^{2+}, Na^+, and the other isolated F^- ions filled in the cavities. Notably, in $NaSr_3Be_3B_3O_9F_4$, the BO_3 groups are almost aligned in the *ab* plane (see Fig. 15.17). Therefore, the microscopic second-order susceptibilities of BO_3 are positively superimposed, which is favorable for generating a large NLO response and moderate birefringence.

In addition, single crystals with dimensions up to 13 mm × 14 mm × 20 mm have been successfully grown by the top-seeded solution growth method from $LiF-NaF-B_2O_3$ flux (Fig. 15.18) [95]. The new results including refractive indices and SHG PM property suggest that $NaSr_3Be_3B_3O_9F_4$ can achieve type I PM with the shortest wavelength down to 233 nm, and the calculated type I PM angle of 266-nm SHG is $\theta = 61.68$.

3.2 $BaMBO_3F$ (M = Ca, Zn, and Mg)

In this series of compounds, $BaCaBO_3F$ (BCBF) was first reported by Keszler et al. with the NCS space group $P\overline{6}2m$ [62]; but no structural details were given. In 2009, Xu et al. reported the BCBF structure with the same space group from a Rietveld refinement of PXRD pattern [65]. The structure of BCBF consists of $[CaO_2F_2BO_3]$ layers formed by corner-sharing CaO_5F_2 polyhedra and BO_3 triangles with Ba^{2+} cations located between the layers (Fig. 15.19). Different from BABF, the orientations of these BO_3 groups are not perfectly aligned (even 1/3 BO_3 groups are disordered) so that the SHG efficiency of BCBF is less than that of KDP. In 2013, Li et al. revealed that BCBF undergoes a reversible phase transition at 242°C: the α-BCBF (space group *R*3) at room temperature and β-BCBF (space group $P\overline{6}2m$) after the phase transition

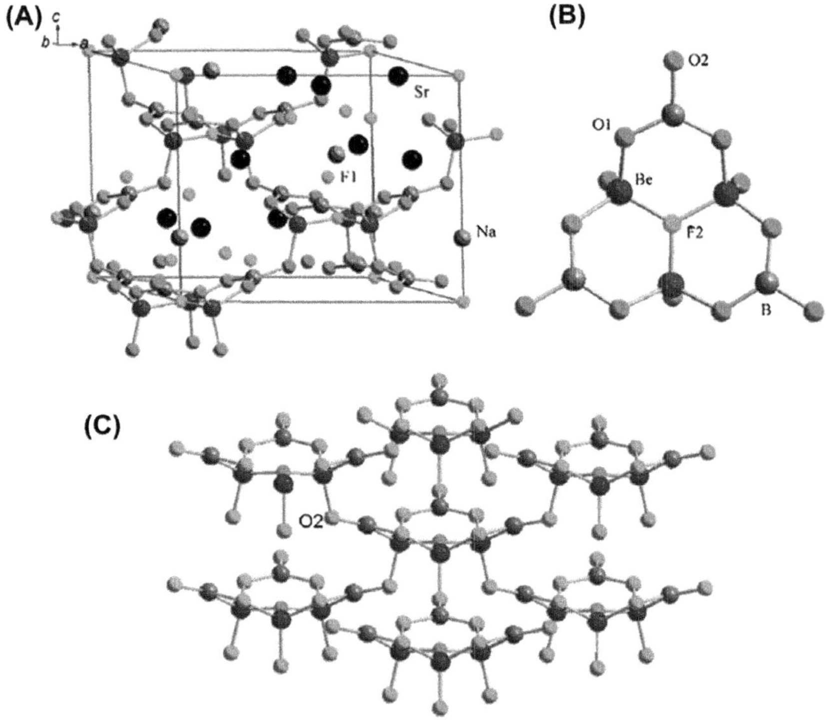

Figure 15.17 (A) Crystal structure of $NaSr_3Be_3B_3O_9F_4$; (B) anionic group $(Be_3B_3O_{12}F)^{10-}$; (C) the connections of the $[Be_3B_3O_{12}F]^{10-}$ rings in the structure [22].

with temperature higher than 242°C [63]. From β- to α-phase, the disordered BO_3 groups at the corner of the unit cell change to long range ordering structure by tripling of the unit cell in all the three directions, which does not involve any bond breaking and reconstruction processes. The previous confusion on the structure of BCBF may due to the β- to α-phase transition when crystals cool down from growth temperature, since the phase transition is both reversible and low energy cost.

In spite of the phase transition, β-BCBF is a congruent-melting compound; therefore, it is feasible to grow large single crystals (Fig. 15.20). For example, single crystals of BCBF with sizes of 39 mm × 23 mm × 12 mm have been obtained by the Kyropoulos method [96]. The linear and nonlinear optical properties of BCBF have been investigated [97]. It has a wide transparent range from 220 to 2500 nm with SHG coefficient $d_{22} = 0.74 \times d_{36}$ KDP. This material can be used for FD of 1064 nm for type I and II PM. It has been used for FD of the 1064 nm Nd:YAG laser with the shortest PM wavelength of 311 nm [97] (Fig. 15.21). BCBF was also reported as a self-FD material when doped with Nd^{3+} or Yb^{3+} [98,99].

$BaZnBO_3F$ and $BaMgBO_3F$ were reported by Li and Chen in 2010 [67]. These two compounds, although crystallized in different NCS space groups ($P\overline{6}$ and Cc), exhibit strong structural relations to BCBF. In $BaZnBO_3F$, the BO_3 groups act as FBUs, and ZnO_3F_2 secondary building units. Each ZnO_3F_2 bipyramid has three basal or

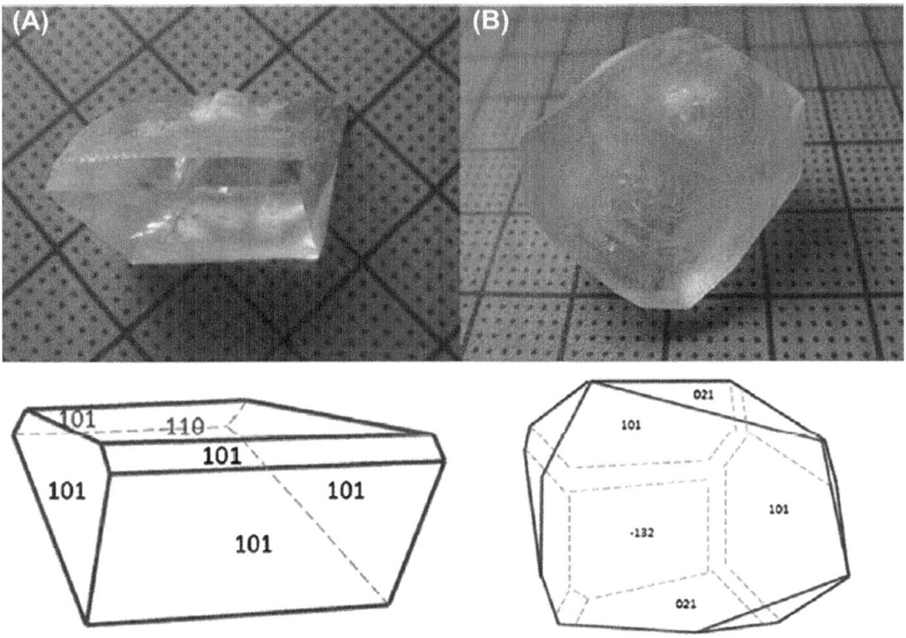

Figure 15.18 Morphology of the as-grown $NaSr_3Be_3B_3O_9F_4$ crystals with different seed directions: (A) [110], (B) [001] [95].

equatorial bonds linking to three BO_3 groups and forcing them to align parallelly in one layer, and two apical F atoms connect with adjacent layers, which is similar to AlO_3F_2 and GaO_3F_2 polyhedra in BABF and $BaGaBO_3F_2$ [87]. It is worth mentioning the coordination environment of Ba atoms. Due to the steric force of the BaO_6F_3 polyhedra, the BO_3 groups are also parallelly aligned in neighboring layers. $BaZnBO_3F$ shows a moderate SHG signal on the order of LBO ($\sim 3 \times$ KDP).

The structure units of $BaMgBO_3F$ are similar to the Zn analog, and the only difference is that the Mg atoms are coordinated with four oxygen and two fluorine atoms. Unlike in the case of $BaZnBO_3F$, each MgO_4F_2 polyherda shares four oxygen atoms with three BO_3 groups, forming a severely distorted equatorial plane. The three BO_3 groups, with two parallel and one antiparallel in one layer, are regarded as main contributors of SHG efficiency. Due to the misaligned fashion of the BO_3 groups, single crystals of $BaMgBO_3F$ with dimensions of 26 mm $\times$ 19 mm $\times$ 4 mm were grown using LiF as flux by the top-seeded solution growth technique [100]. $BaMgBO_3F$ possesses a UV cutoff edge of 190 nm and shows a relatively small SHG response ($\sim 0.6 \times$ KDP).

The three borate fluorides with general formula $BaMBO_3F$ share the strongly correlated MO_nF_2 ($n = 3-5$) FBU, although they have different structures (Fig. 15.22). Following the order Zn, Mg to Ca, the coordination number n of oxygen (in MO_n) increases from 3 to 5. As Li et al. pointed out, the changes in coordination environment have a strong influence on the arrangement of the nearby BO_3 groups as well as the

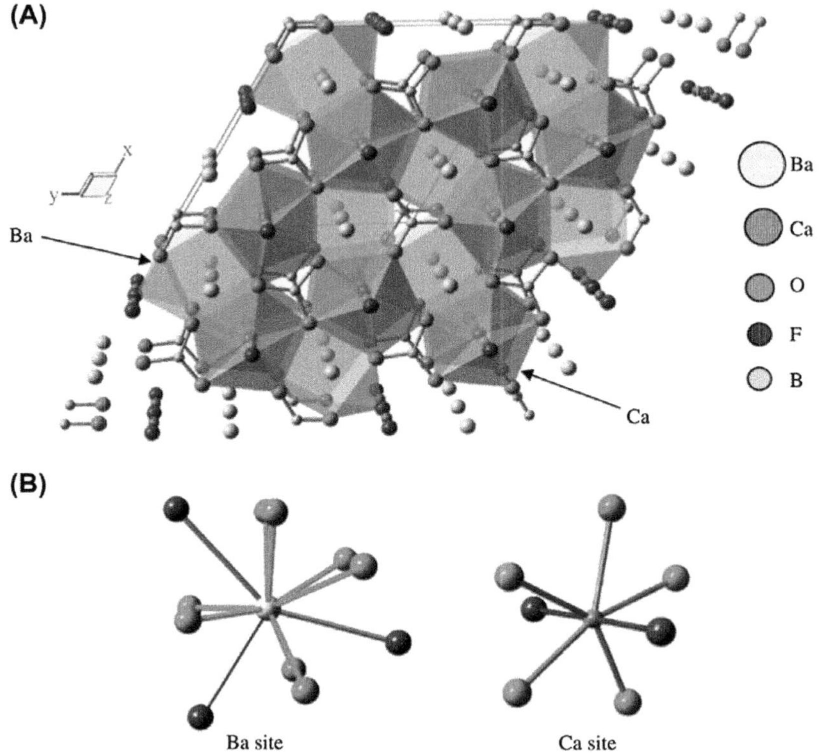

Figure 15.19 (A) Crystal structure of β-$BaCaBO_3F$ (β-BCBF) and (B) polyhedral coordination of Ba and Ca atoms in BCBF [65].

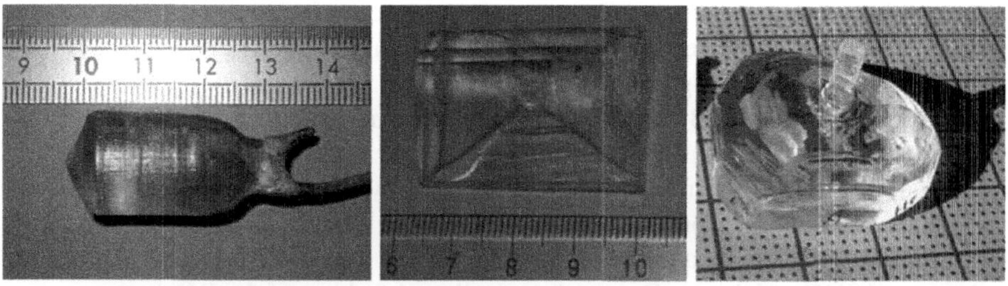

Figure 15.20 Typical grown $BaCaBO_3F$ single crystals grow by (left to right) Czochralski pulling method [65], Kyropoulos method [96], and flux method [63].

NLO effect. When $n = 3$ (ZnO_3F_2), three different BO_3 groups can be forced to align in the same orientation, whereas misaligned BO_3 groups are observed in the case $n = 4$ or 5. That is, the microscopic mechanism of why $BaZnBO_3F$ has the largest SHG efficiency among the three compounds. It is also worth noting that all of these compounds including BABF and $BaGaBO_3F_2$ have one or more M−F layers in their structures.

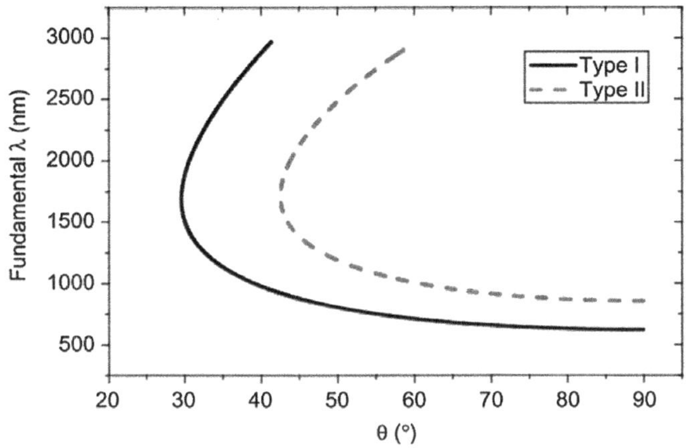

Figure 15.21 Phase-matching angle of $BaCaBO_3F$ for type I and II second harmonic generation as a function of the wavelength of the fundamental wave [65].

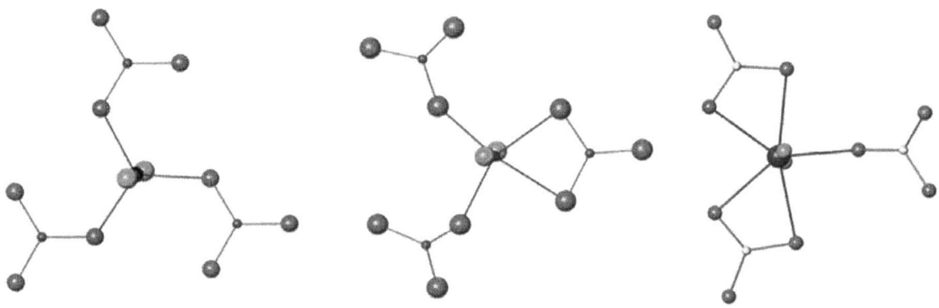

Figure 15.22 Alignments of BO_3 groups by the MO_nF_2 polyhedra in $BaMBO_3F$ structures (*a*, M = Zn; *b*, M = Mg; *c*, M = Ca) [67].

3.3 Apatite-Type Family: $M_5(BO_3)F$ (M = Ca, Mg, and Cd)

Traditionally, apatite refers to a group of phosphate minerals containing high concentrations of OH^-, F^-, or Cl^- ions [101]. The first apatite-type borate fluoride, CBF, was reported in 1989 (space group *Cm*) [17]. It found that CBF has a certain relationship to $Ca_5(PO_4)_3F$: the positional substitution of BO_3 triangles for PO_4 tetrahedra (Fig. 15.23) [54]. The structure of CBF consists of two types of building units: the BO_3 groups and the CaO_4X_2 octahedra (X = O or F). The planar BO_3 groups lie approximately parallel to (001) and are further linked with CaO_4X_2 octahedra forming a 3D network. It is interesting to point out that the structure of CBF is also similar to those of $RECa_4O(BO_3)_3$ (RE = rare earth metal)—a series of promising NLO crystals discovered before CBF [50].

CBF has a wide transmission range with the UV cutoff wavelength near 190 nm. Powder SHG measurement suggested that the SHG efficiency for CBF is about two

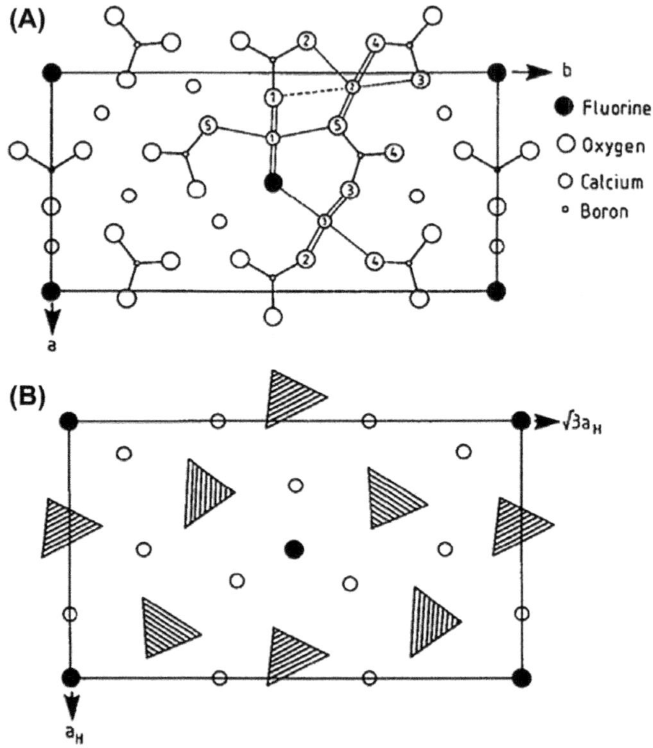

Figure 15.23 (A) $Ca_5(BO_3)_3F$, (001) projection. B—O and Ca—X (X = O, F) contacts are indicated. *Double lines* represent contacts to pairs of atoms related by a *c*-translation. (B) Projection on (001) of the $Ca_5(PO_4)_3F$. PO_4 tetrahedra are shown as shaded triangles [54].

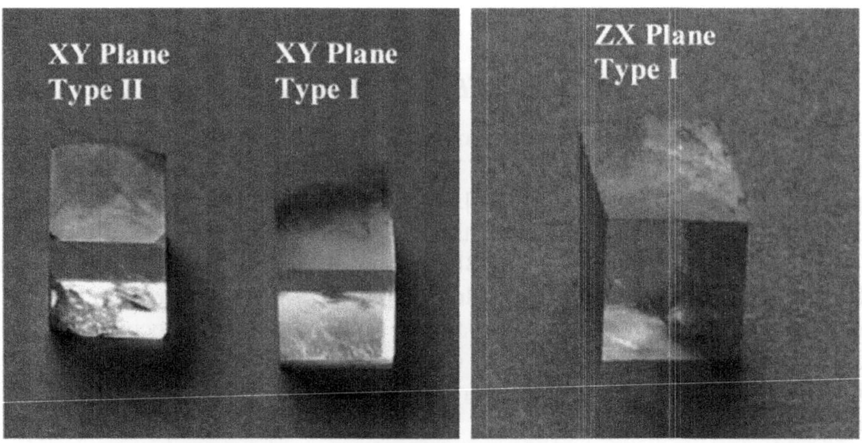

Figure 15.24 Photographs of the uncoated $Ca_5(BO_3)_3F$ crystals oriented at phase-matching conditions for second harmonic generation of 1064 nm [102].

times that of KDP, which is as large as that of $RECa_4O(BO_3)_3$ (RE = rare earth metal), and it is phase-matchable when irradiated by 1064-nm Nd:YAG laser [50]. Xu et al. have grown single crystals of CBF by using 20 wt% of LiF as flux (Fig. 15.24) [102]. CBF is a negative biaxial optical crystal, and its birefringence at 589 nm is 0.053. They also reported that the shortest fundamental wavelengths allowing PM are 577 nm (type I) [103], and the maximum conversion efficiencies for SHG at 1064 nm are 25% and 54% for type I and II SHG, respectively [102]. As compared to YCOB, the UV absorption edge of CBF is about 15 nm shorter than that of YCOB, and the noncritically phase-matching (NCPM) wavelengths of CBF blueshift significantly toward UV than that of YCOB and close to the one of LBO (Fig. 15.25).

Electronic structures and optical properties of CBF have been carried out by Hu et al. The results reveal that the main sources of the SHG response of CBF not only come from the planar BO_3 groups (74–77%) but also to a lesser extent from Ca^{2+} cations (23–26%), which can be attributed to the interband electronic transition from the nonbonding O $2p$ states to the B $2p$ and Ca $4s4p$ states. The calculated SHG coefficients are close to the experimentally measured value.

It is well known that substitutions of both the anion and cation sites are possible for apatite. Before CBF, $Mg_5(BO_3)F$ was reported in 1975 with NCS space group $Pna2_1$, but its crystal data are poor (large R value) [53]; $M_5(BO_3)_3F$ (M = Sr, Ba, and Eu), however, is crystallized in CS $Pnma$ space group [104–106]. Very recently, a new apatite-type borate fluoride, $Cd_5(BO_3)_3F$ was reported [56]. The structure of $Cd_5(BO_3)_3F$ is also isotropic to CBF, with the Ca^{2+} position substituted by Cd^{2+}. However, the SHG responses of $Cd_5(BO_3)_3F$ (powder SHG response ≈ 3.8 ~ KDP) is apparently larger than its isostructural compound CBF (Fig. 15.26). It has been proven that employing polar displacement of the d^{10} cations into borates is a great strategy for

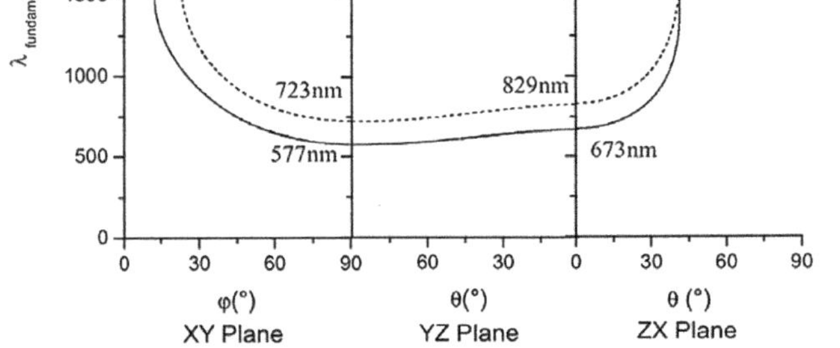

Figure 15.25 Phase-matching angle curves for type I second harmonic generation as a function of the wavelength of the fundamental wave in $Ca_5(BO_3)_3F$ (CBF) by comparison with $YCa_4O(BO_3)_3$ (YCOB) [102].

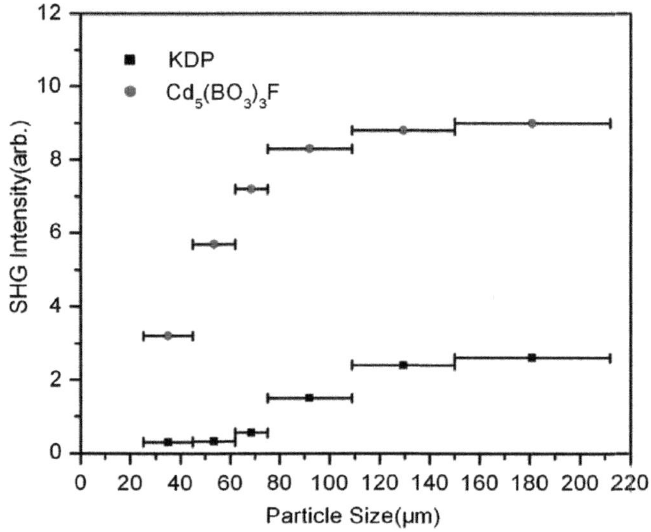

Figure 15.26 Second harmonic generation (SHG) measurements of $Cd_5(BO_3)_3F$ ground crystals (*solid circle*) with KDP (*solid square*) as a reference (1064 nm doubled to 532 nm) [56].

synthesizing new FD materials. In $Cd_5(BO_3)_3F$, according to dipole calculations, the CdO_nF_m polyhedral indeed has considerable contributions to the overall SHG response.

3.4 Fabianite-like Borate Fluorides: $M_3B_6O_{11}F_2$ (M = Ba, Sr, and Pb) [23–26]

The presence of fluorine in the borate framework can help to form SIS compounds with NCS structures. Examples are the isostructural compounds $M_3B_6O_{11}F_2$ (M = Ba, Sr, and Pb). The series of compounds crystallize in NCS chiral and polar space group $P2_1$. $Sr_3B_6O_{11}F_2$ and $Ba_3B_6O_{11}F_2$ have been synthesized by both hydrothermal and the flux method, while the Pb-containing isomorphism was only synthesized by high-temperature flux method. The crystals are stable in air and water and exhibit favorable mechanical and thermal characteristics. The crystal structures of these compounds are complex frameworks formed from two main parts, the $[B_6O_{11}]_\infty$ framework and $[F_2Ba_3]_\infty$ layer (Fig. 15.27). The FBUs of the $[B_6O_{11}]_\infty$ framework share some structural features with those of natural mineral *fabianite* [$CaB_3O_5(OH)$] but different in connection mode [26].

These compounds all possess a moderate SHG response and are phase-matchable. Their SHG responses are about 2.5, 3, and 4 times that of KDP for Sr, Ba, and Pb analogs, respectively. In addition, they also exhibit a wide transparency range; in particular, $Sr_3B_6O_{11}F_2$ and $Ba_3B_6O_{11}F_2$ exhibit a short UV cutoff edge below 190 nm. In order to explore the intrinsic relationship between the electronic structure and optical properties, band structure and densities of states are analyzed by employing first-principles calculations. It is interesting to see that although the B–O groups play

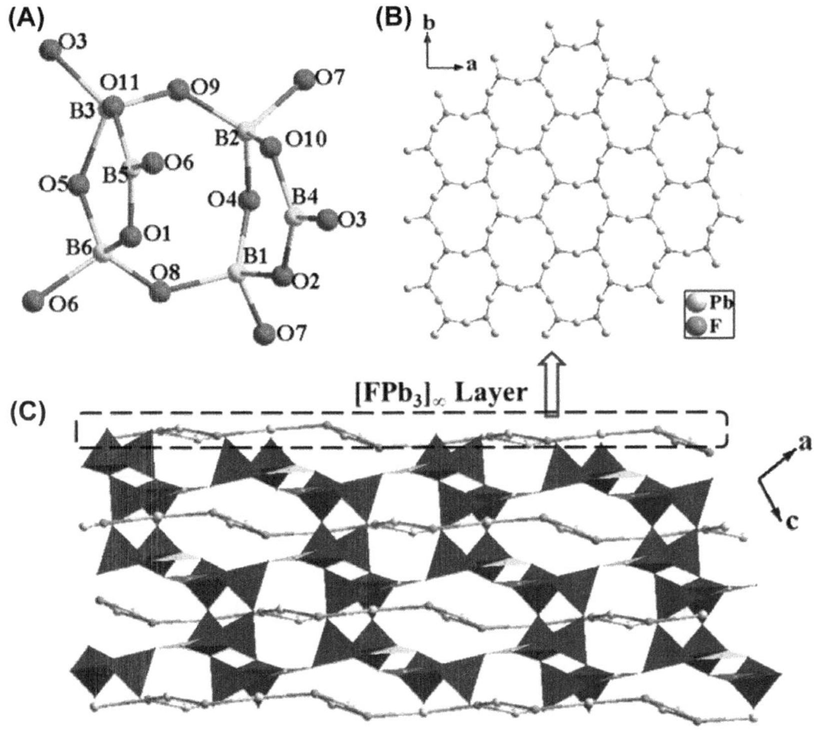

Figure 15.27 (A) The B_6O_{14} group; (B) the novel $[FPb_3]_\infty$ layer; (C) a drawing of the structure of $Pb_3B_6O_{11}F_2$ with Pb–O bonds omitted for clarity viewing along the *b*-axis (BO_3, green (gray in print versions); BO_4, blue (dark gray in print versions)) [25].

a major role in the optical response, the contribution from occupied F-2*p* orbitals just below the Femi level cannot be neglected. Theoretical calculations also exhibit that these compounds have good birefringence that is suitable for PM conditions.

3.5 $Ba_{4-x}Sr_{3+x}(BO_3)_{4-y}F_{2+3y}$ Solid–Solution Series

The BaB_2O_4–BaF_2–BaO ternary system is interesting for optimizing the crystal growth conditions of BBO, where the addition of fluoride BaF_2 is used to reduce the viscosity of the flux. Unexpectedly, some barium borate fluorides are obtained. Two NCS borate fluorides "$Ba_7(BO_3)_3F_5$" and "$Ba_3Sr_4(BO_3)_3F_5$" were first reported by Keszler et al. [62] These two compounds are isotopic, and their crystal structures contain isolated BO_3 groups as their basic structural units, which are predicted to have favorable SHG response. The Ba sites can be partially substituted by Sr atoms while retaining the initial structure. Single crystals of these two barium borate fluorides can be grown by the top-seeded solution growth method (Fig. 15.28) [61,66].

The crystal structure of "$Ba_7(BO_3)_3F_5$" and "$Ba_3Sr_4(BO_3)_3F_5$" was first solved in the space group $P6_3mc$, but some abnormal atomic displacement parameters are observed during the crystal structure refinements [66]. Rashchenko et al. reinvestigated these

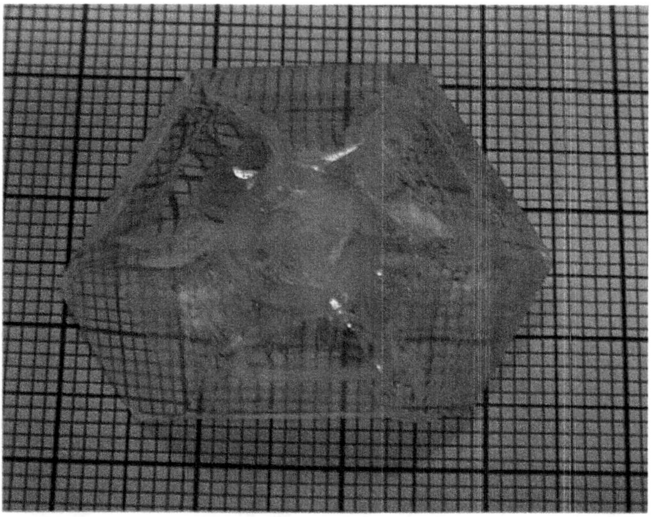

Figure 15.28 Photograph of "$Ba_3Sr_4(BO_3)_3F_5$" crystal [66].

compounds and proposed a new mechanism of anionic $(BO_3)^{3-} \leftrightarrow 3F^-$ substitution [106]. They demonstrated that this substitution, similar to that in $Eu_3(BO_3)_{2+x}F_{3-3x}$ and α-$Mg_2(BO_3)_{1+x}F_{1-3x}$, occurs in the trigonal pyramid anion group "4X" with vertices occupied either by fluorine, or a $(BO_3)^{3-}$ group (Fig. 15.29). By adopting this mechanism, a detailed study of the BaB_2O_4—BaF_2—BaO system resulted in the discovery of the new homogeneity of $Ba_7(BO_3)_{4-y}F_{2+3y}$ solid solution, which can be retained with $0.21 < y < 0.65$ (between $Ba_7(BO_3)_{3.79}F_{2.63}$ and $Ba_7(BO_3)_{3.35}F_{3.95}$ compositions) as a result of $(BO_3)^{3-} \leftrightarrow 3F^-$ anionic substitution [64]. The later synthetic experiments also indicate that the terminal stoichiometry "$Ba_7(BO_3)_3F_5$," which was previously determined by Keszler et al. [62], in fact, is the solid solution compound $Ba_7(BO_3)_{4-y}F_{2+3y}$. The situations are the same for "$Ba_3Sr_4(BO_3)_3F_5$." Rashchenko et al. suggest that $Ba_{4-x}Sr_{3+x}(BO_3)_{4-y}F_{2+3y}$ crystals have been actually grown instead of "$Ba_3Sr_4(BO_3)_3F_5$" reported earlier [68].

Despite structure controversy, $Ba_{4-x}Sr_{3+x}(BO_3)_{4-y}F_{2+3y}$ type compounds do show SHG response. "$Ba_7(BO_3)_3F_5$" and "$Ba_3Sr_4(BO_3)_3F_5$" exhibit SHG intensities of $\sim 1/3$ and $1/2 \times$ KDP, respectively, and their UV cutoff wavelength is about 190 and 210 nm, respectively [61,66]. They also have good thermal stabilities and are insoluble in water.

3.6 Other SHG-Active Borate Fluorides

3.6.1 $Ba_4B_{11}O_{20}F$

The rational synthesis of $Ba_4B_{11}O_{20}F$ (BBOF) was promoted by the introduction of fluorine into BBO [27]. This compound, with a polar $Cmc2_1$ structure, has a complex framework structure consisting of anionic three-dimensional network generated by

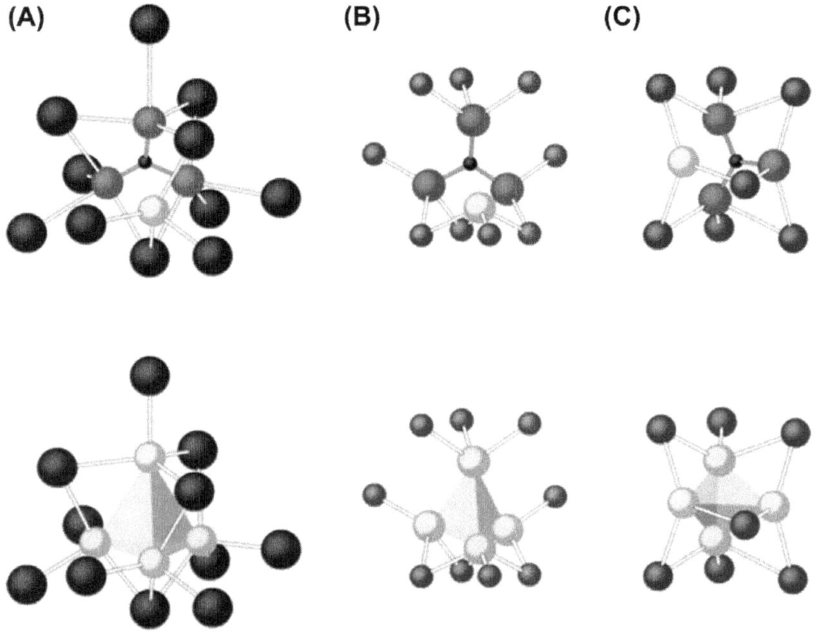

Figure 15.29 Anionic substitution in the $[X_4]^{4-}$ tetrahedral group in fluoride borates. The top row shows $[(BO_3)F]^{4-}$ tetrahedral groups which are replaced by $[X_4]^{4-}$ groups in the bottom row. The fragments correspond to $Ba_{4-x}Sr_{3+x}(BO_3)_{4-y}F_{2+3y}$ (A), α-$Mg_2(BO_3)_{1+x}F_{1-3x}$ (B), and $Eu_3(BO_3)_{2+x}F_{3-3x}$ (C) solid solutions. *Yellow* (light gray in print versions) *spheres* represent F^- anions; $(BO_3)^{3-}$ anions are given as a combination of red (darker gray in print versions) oxygen and black boron atoms; Ba^{2+}, Mg^{2+}, and Eu^{3+} cations are shown in dark blue (darkest gray in print versions), blue (dark gray in print versions), and purple (gray in print versions), respectively [106].

$B_{11}O_{24}$ functional building blocks, and Ba—F—Ba chains. The compound also shows a deep UV absorption edge at ~175 nm that is shifted by the F atoms.

Very interestingly, BBOF exhibits a very large SHG response, ~10 × KDP, even bigger than that of the benchmark BBO (~1.8 times that of BBO) at the same particle size (Fig. 15.30). It is also noted that, for borates with SHG response, BBOF is the largest reported to date containing neither lone-pair-active anions nor Jahn—Teller transition metals. In addition, by applying the conventional anionic group theory [77], it is not possible that the SHG response of BBOF is far larger than that of BBO. Why does it show exceptional SHG response and what is the role of F atoms in this compound? To answer these puzzles, the authors use multiple approaches including dipole moment calculation, mode-polarization vector analysis, specific-acentric-mode displacement analysis, and electronic localization function (ELF) within density functional theory (DFT) method. The results demonstrated that the F-directed polar displacements contribute most to enhanced SHG intensity in BBOF, but the Ba polarizability also has small contribution. The local

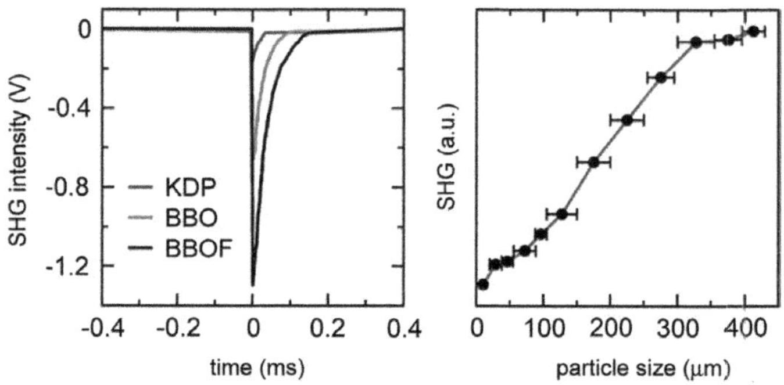

Figure 15.30 (Left) Oscilloscope traces showing second harmonic generation (SHG) intensities for $Ba_4B_{11}O_{20}F$ (BBOF) and for KDP and β-BaB_2O_4 (BBO) as references. (Right) Phase-matching curve for BBOF. The *solid curve* is a guide for the eyes, not a fit to the data [27].

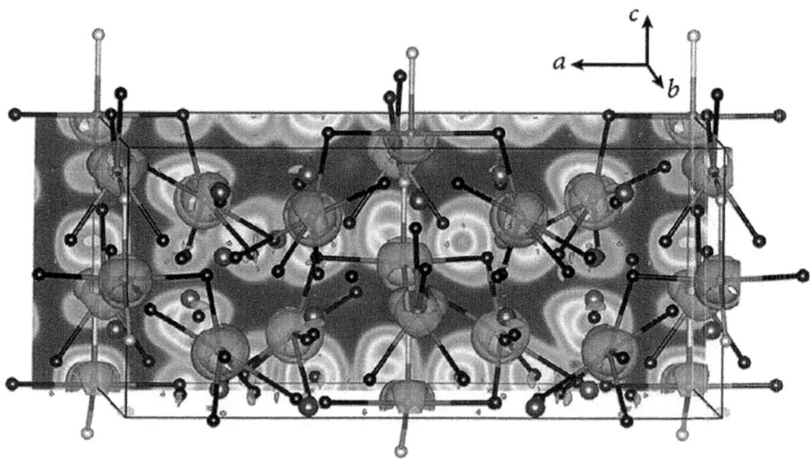

Figure 15.31 Electron localization function (ELF) isosurface plot for $Ba_4B_{11}O_{20}F$ showing asymmetric charge localization (*orange* (dark gray in print versions) *lobes*) mainly about the Ba2 and Ba3 sites. Pink (light gray in print versions), Ba; black, O; gray, F; B has been omitted for clarity [27].

electronic structure has also clearly shown that the asymmetric localization of charge density around the Ba sites contributes to the SHG intensity because of the correlated F atoms (Fig. 15.31), whereas the contribution from the Ba sites in BBO was reported to be very small.

3.6.2 $Rb_3Al_3B_3O_{10}F$

A new beryllium-free borate $Rb_3Al_3B_3O_{10}F$, with a framework structure that preserves the structural merits of KBBF, has been reported [59]. The compound crystallizes in an asymmetric space group of $P3_1c$, and its anionic structure is the $[Al_3(BO_3)_3OF]_\infty$ layers

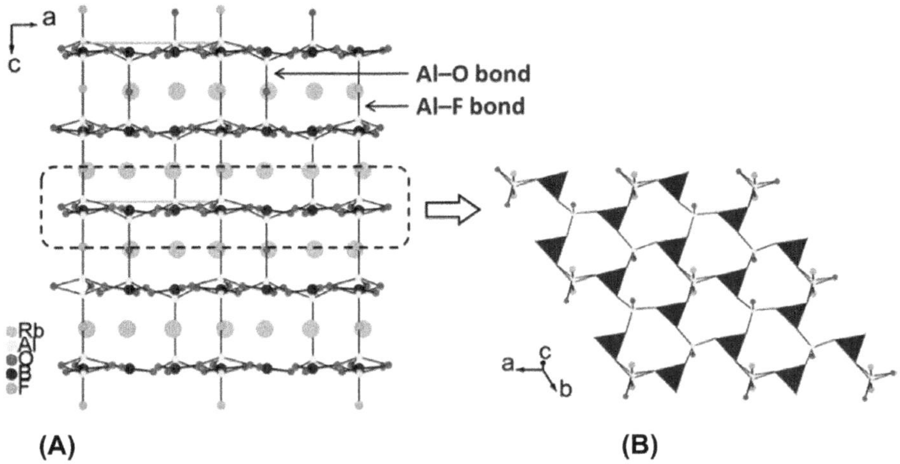

Figure 15.32 Crystal structure of (A) $Rb_3Al_3B_3O_{10}F$ and (B) $[Al_3(BO_3)_3OF]^{\infty}$ layer. The interlayer Al–F and Al–O bonds are indicated in (A). *Blue* (dark gray in print versions) *triangles* in (B) represent BO_3 groups [59].

composed by BO_3 triangles, AlO_4 and AlO_3F tetrahedron, which are further connected by Al–O/F bonds to form a three-dimensional framework (Fig. 15.32). The BO_3 groups in each layer are nearly coplanar and aligned arrangement, although there is a small misalignment between the orientations of the BO_3 groups in adjacent $[Al_3(BO_3)_3OF]_{\infty}$ layers. The structure is very similar to that of KBBF, indicating that it is likely to share the optical advantages of KBBF. Indeed, the SHG efficiency of $Rb_3Al_3B_3O_{10}F$ is about $1.2 \times KH_2PO_4$, which is approximately equal to that of KBBF. It also has a wide transparent range with UV absorption edge down to 200 nm and is chemically stable and thermally stable up to ~1462K.

Remarkably, the as-grown single crystals of $Rb_3Al_3B_3O_{10}F$ are block shaped with no layer growth tendency. By comparing the layered structural features between KBBF and $Rb_3Al_3B_3O_{10}F$, it was found that the weak K–F bonds in KBBF are replaced by strong Al–O or Al–F bonds in $Rb_3Al_3B_3O_{10}F$, providing stable connections between the layered structural units. This was also confirmed by the calculation of electrostatic force of interactions using the Coulomb's law, which suggests that the magnitude of the electrostatic force of interaction of $Rb_3Al_3B_3O_{10}F$ is 9.5 times KBBF. The discovery of $Rb_3Al_3B_3O_{10}F$ not only avoids using the highly toxic beryllium but also points out a facile way to overcome the structural demerits of KBBF.

3.6.3 $Li_3Ca_9(BO_3)_7 \cdot 2[LiF]$

The compound $Li_3Ca_9(BO_3)_7 \cdot 2[LiF]$ was reported as a LiF salt-inclusion borate with the Li–F chains residing in the $Li_3Ca_9(BO_3)_7$ three-dimensional open framework (Fig. 15.33) [42]. It is particularly worth noting that all B–O groups are isolated BO_3 triangles possessing coplanar arrangements in $Li_3Ca_9(BO_3)_7 \cdot 2[LiF]$. According

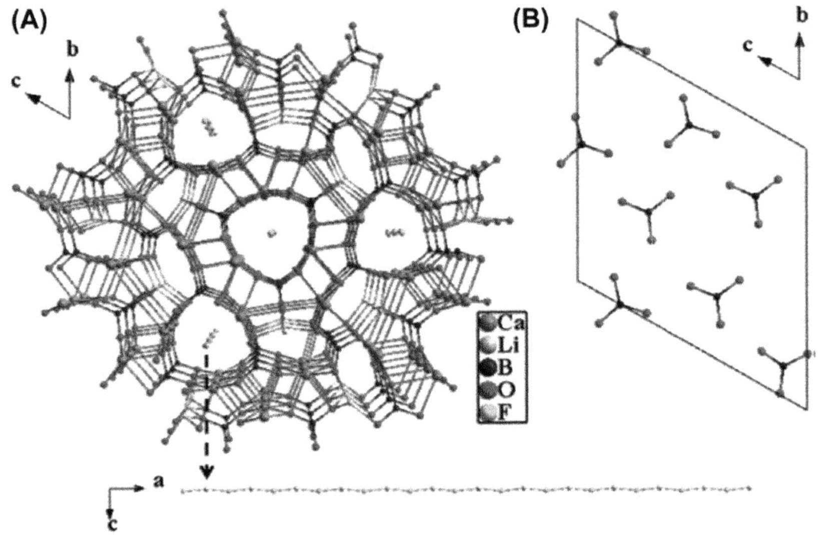

Figure 15.33 (A) The one-dimensional open-framework structure filled with the Li—F chain (Li—O bonds omitted for clarity). (B) The coplanar arrangement of BO_3 triangles in the unit cell [42].

to the anionic group theory [77], the SHG response mainly originates from the B—O units, in particular, from the BO_3 triangles in this compound. However, the coplanar BO_3 triangles do not point to the same direction, and over half of the microscopic SHG response generated by BO_3 triangles is canceled. Nevertheless, the density of the NLO-active BO_3 triangles in $Li_3Ca_9(BO_3)_7 \cdot 2[LiF]$ is higher than that of KBBF so that it also has an SHG response as large as that of the KDP. This compound shows a wide transparent region with a UV cutoff edge at about 230 nm. The DFT calculations suggest that this compound has a relatively large birefringence (0.07; 532 nm), and the largest SHG coefficient $d_{33} = 0.58$ pm/V ($\sim 1.4 \times d_{36}$ KDP).

4. Fluorooxoborates as Frequency-Doubling Materials

Fluorooxoborates are less reported due to their harsh synthesis conditions. However, the partial substitution of O atoms by F atoms will reduce the symmetry of the BO_4 tetrahedra, leading to a NCS structure that could yield a larger NLO response based on the polarization of the B—F bond [34]. In fact, more than half of fluorooxoborates crystallize in NCS space groups. In addition, some fluorooxoborates, such as $K_3B_3O_3F_6$ [37], also have a wide optical transmission range with UV absorption edge down to 200 nm, which is benefit for practical application in UV and DUV area. Although detailed reports on SHG effects of fluorooxoborates are rare, we believe that continuous efforts on discovering new metal fluorooxoborates will open a window for the study of FD materials.

4.1 Lithium Fluorooxoborates: LiB_6O_9F, $Li_2B_6O_9F_2$, and $Li_2B_3O_4F_3$ [28,35,36,107–110]

LiB_6O_9F was reported as the first lithium fluorooxoborate in 2009 [28], and this type of compounds has rapidly expanded to include $Li_2B_6O_9F_2$ [36] and $Li_2B_3O_4F_3$ [35]. In general, the introduction of electronegative fluorine to lithium borates would increase the anisotropy of charge distribution within the borate anions, which might improve the SHG performance.

LiB_6O_9F crystallizes in space group $Pna2_1$, and its FBUs are composed of boroxine rings (B_3O_3) and fluorinated boroxine rings (B_3O_3F) with the ratio of 1:1 (Fig. 15.34A). The fluorooxoborate anion forms a two-dimensional infinite $^2_\infty[B_6O_9F]^-$ layer. The lithium atoms are located inside the cavities. In this structure, the BO_3F tetrahedra draw particular attention, which are oriented approximately along the *c*-direction resulting in a polar structure (Fig. 15.34B).

$Li_2B_6O_9F_2$, with a monoclinic structure (space group *Cc*), displays a three-dimensional $^3_\infty[B_6O_9F_2]^{2-}$ anionic framework. The structure can be described as chains consisting of BO_2F_2 and BO_3 units which are linked by the BO_4 tetrahedron (Fig. 15.35A). It is noted that only a few compounds consisting of a BO_2F_2 tetrahedron are known.

The structure of $Li_2B_3O_4F_3$ is built of $^1_\infty[B_3O_4F_3]^{2-}$ chains interconnected by fourfold-coordinated Li^+ cations (space group $P2_12_12_1$). It is interesting that the $^1_\infty[B_3O_4F_3]^{2-}$ chain contains of BO_3F and BO_2F_2 tetrahedra as well as trigonal planar BO_3 groups (Fig. 15.35B). The geometries of the BO_3F and BO_2F_2 units are similar to building blocks found in LiB_6O_9F and $Li_2B_6O_9F_2$, respectively.

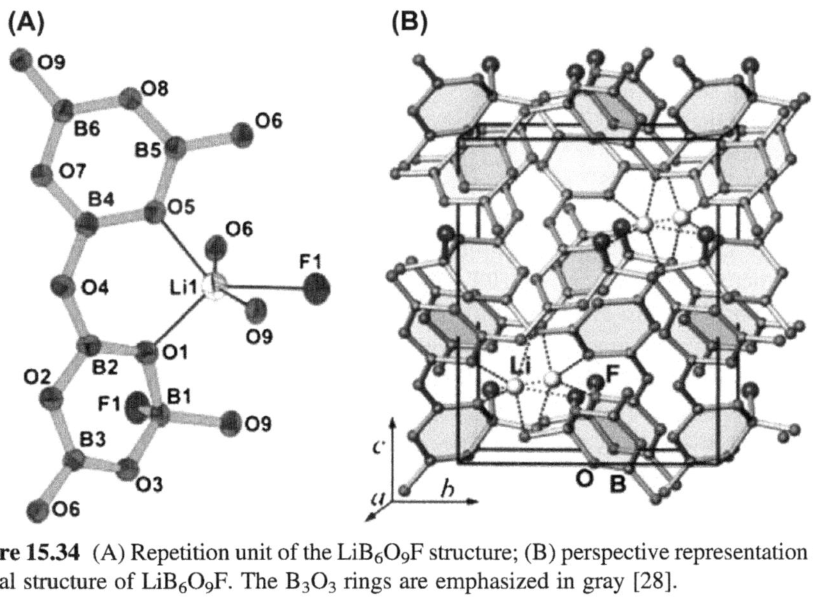

Figure 15.34 (A) Repetition unit of the LiB_6O_9F structure; (B) perspective representation of the crystal structure of LiB_6O_9F. The B_3O_3 rings are emphasized in gray [28].

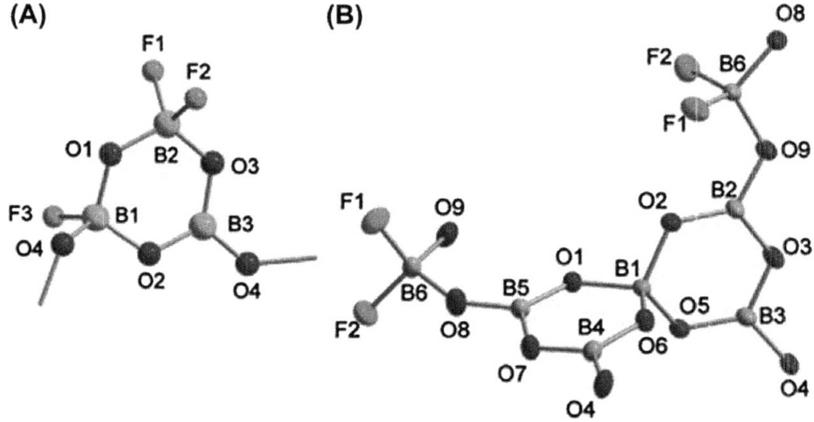

Figure 15.35 Fundamental building blocks of the lithium-containing fluorooxoborates: (A) $Li_2B_3O_4F_3$ and (B) $Li_2B_6O_9F_2$ [109].

These lithium fluorooxoborates are all acentric and therefore are expected to be SHG-active. The powder SHG measurements of LiB_6O_9F suggest that its SHG intensity is slightly larger than that of KDP, and it is phase-matchable at a fundamental wavelength of 1064 nm. The SHG efficiency of $Li_2B_6O_9F_2$ was found to be comparable with that of *a*-SiO_2 [111], while no visible SHG signal was observed for $Li_2B_3O_4F_3$. It should be pointed out that the results might be smaller than the actual values because all lithium fluorooxoborates are sensitive to humid air and decompose slowly at room temperature.

The linear and nonlinear optical properties of these lithium fluorooxoborates have been investigated based on DFT-based *ab* initio studies using two codes: *CRYSTAL* at the B3LYP and LDA levels and *VASP* at the LDA level [107,110]. The calculated bandgaps are 8.83, 8.7, and 10.4 eV for LiB_6O_9F, $Li_2B_6O_9F_2$, and $Li_2B_3O_4F_3$, respectively (on the B3LYP level). In comparison, the calculated bandgap on the same level of a similar compound LBO is 8.6 eV. The results indicated that all these lithium fluorooxoborates have a wide photon energy range of optical transparency and can be used as DUV optical materials. In addition, the maximum values of the calculated second-order susceptibility coefficients follow the order: $Li_2B_6O_9F_2 \approx LBO > LiB_6O_9F > Li_2B_3O_4F_3$. There is a little disagreement with the experimental measurements, where the efficiency of SHG was large for LiB_6O_9F, which might be caused by decomposition of $Li_2B_3O_4F_2$. Furthermore, for all of these lithium fluorooxoborates, the calculated birefringence Δn is relatively large, which is a benefit for PM SHG. The maximal birefringence, $\Delta n_{max} \approx 0.08$ for LiB_6O_9F and 0.07 for $Li_2B_6O_9F_2$ and $Li_2B_3O_4F_3$, in comparison to LBO, $\Delta n_{max} \approx 0.06$. By using the respective relations for biaxial crystals, the calculated PM angles for both the type I (PM1) and type II (PM2) conditions have been obtained. The SHG in $Li_2B_3O_4F_3$ is possible in the corresponding energy range of first photons from 2.0 to 4.4 eV, while the values are ranging from 1.5 to 4.2 eV for $Li_2B_6O_9F_2$ (Fig. 15.36). Therefore, although the theoretical calculations may not be accurate enough, lithium fluorooxoborates appear to be good candidates for FD materials.

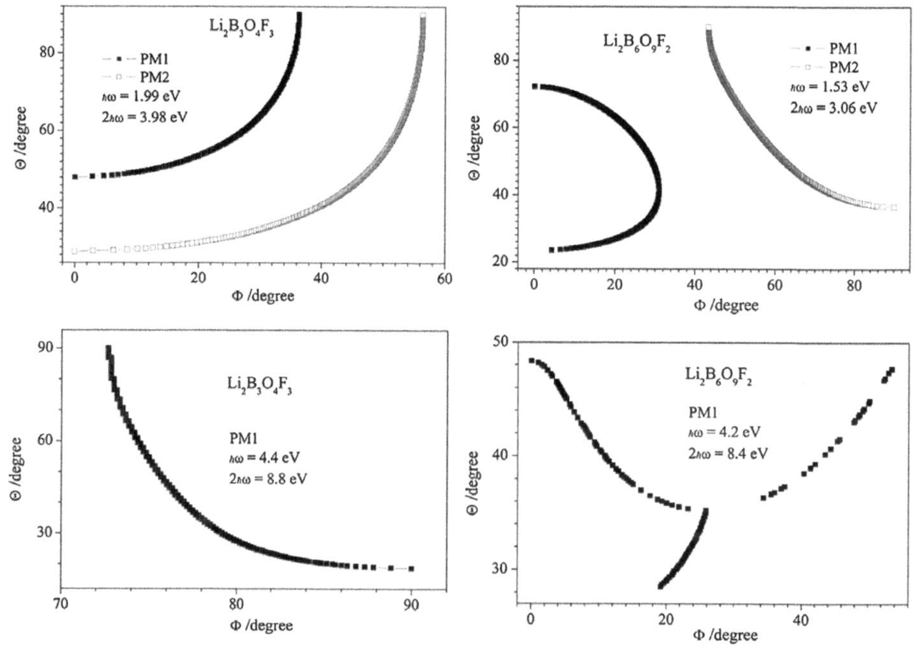

Figure 15.36 Phase-matching angles for second harmonic generation with the first photon energies 1.99 and 4.4 eV in $Li_2B_3O_4F_3$ (left) and first photon energies 1.53 and 4.2 eV in $Li_2B_6O_9F_2$ (right) [107].

4.2 Bismuth Fluorooxoborate: BiB_2O_4F [51,70]

The bismuth fluorooxoborate, BiB_2O_4F, was first obtained using boric acid melts flux method by Li et al. in 2005. It crystallizes in the space group $P3_2$ and was reported as an FD material with SHG efficiency $\sim 1.8 \times$ KDP [70]. Very recently, however, Cong et al. reinvestigated this material by both experiments and theoretical calculations and revealed a very strong SHG effect in BiB_2O_4F ($>10 \times$ KDP), which is even higher than the star compound BIBO (α-BiB_3O_6, $\sim 8 \times$ KDP) [51]. The previous study mainly focused on the structural chemistry of BiB_2O_4F, and the new results seem to be more reliable.

The structure of BiB_2O_4F consists of helical borate chains constructed by corner-shared borate tetrahedra ($2BO_4 + BO_3F$). The bismuth cations occupy the interspaces of the borate helices. The existence of fluorine in the compound is confirmed by EDX analysis, IR spectrum, and bond valence calculations.

In comparison, $BiB_2O_4(OH)$, a highly related structure with BiB_2O_4F, was also synthesized. The main difference between these two compounds is the replacement of F^- by OH^- (Fig. 15.37). As seen from the UV−Vis−NIR absorption spectra (Fig. 15.38), BiB_2O_4F has a lower absorption edge than $BiB_2O_4(OH)$, which proves that introduction of fluorine indeed blueshifts the bandgap. In addition, the SHG measurements of BiB_2O_4F and $BiB_2O_4(OH)$ using BIBO as a reference reveal that

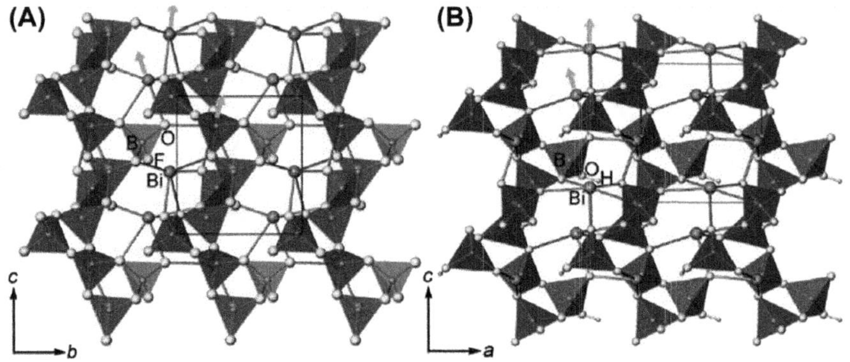

Figure 15.37 View of the structures of (A) BiB_2O_4F and (B) $BiB_2O_4(OH)$. The *pink* (gray in print versions) *arrows* imply the orientation of lone pair electrons on Bi^{3+} cations [51].

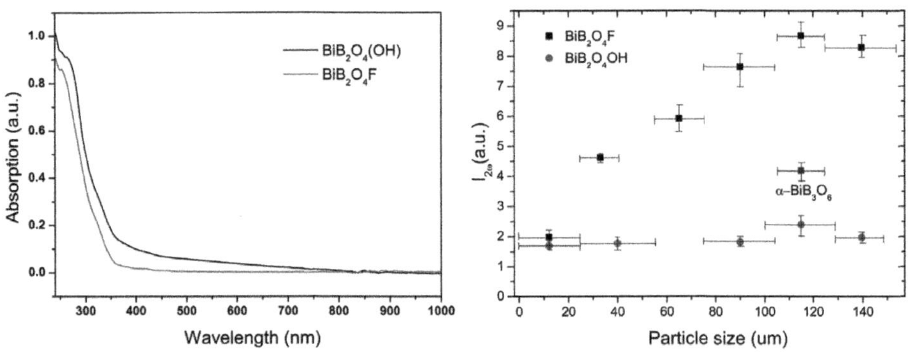

Figure 15.38 UV–Vis–NIR absorption spectra for BiB_2O_4F and $BiB_2O_4(OH)$ (left); dependence of the measured powder SHG signals intensity with the particle size of BiB_2O_4F and $BiB_2O_4(OH)$ (α-BiB_3O_6 as a reference) (right).

BiB_2O_4F has unexceptional large SHG signals while $BiB_2O_4(OH)$ is much smaller (Fig. 15.38). Cong et al. also found that BiB_2O_4F has a low stability under laser irradiation, ie, low laser damage threshold. The SHG signal sharply decreases when the laser power density increased to $\geq$20 mW/cm^2, accompanied with the color change of powder from white to gray. They believe that this is the reason why a significant low SHG effect was reported in previous research.

In order to figure out the significance of the incorporation of fluorine, the structure–property relationships of BiB_2O_4F are investigated in detail. The stereochemical 6*s* lone pair electrons of Bi^{3+} are thought to be the main reason for large SHG coefficient in bismuth FD materials, such as BIBO. At first glance, both BiB_2O_4F and $BiB_2O_4(OH)$ have similar BiO_5 groups with the 6*s* lone pair electrons basically aligning along the *c*-axis. In addition, the real-space atom-cutting analysis based on DFT also confirmed that BiO_5 groups have the dominant contribution of SHG coefficients (giving ~96% of the total value). Why do these two closely related structures differ in SHG responses? A further study suggested that the BO_3F tetrahedron also contributes quite a bit to the overall SHG coefficients in BiB_2O_4F (~26%), which is larger than that of BO_4 groups (~10%). The larger contribution of $[BO_3F]$ tetrahedron is understandable since it exhibits a larger spatial anisotropy than $[BO_4]$ tetrahedron due to the partial substitution of OH^- by F^-. Furthermore, because fluorine is the most electronegative atom, it is deduced that the nonbonding electrons ($6s^2$ electrons) of Bi^{3+} can be withdrawn toward F atoms leading to a higher local polarization. By analyzing the surrounding environment of Bi^{3+} in different bismuth compounds with 6*s* lone pair electrons (Fig. 15.39), it is found that two F atoms are located at the same side of Bi^{3+} in BiB_2O_4F so that the synergistic electron-withdrawing effects will enhance its SHG coefficient. Meanwhile, only moderate SHG response was observed in $BiB_2O_4(OH)$, BIBO, and $Pb_3B_6O_{11}F_2$ without fluorine or with neutralized electron-withdrawing effect.

The analyses on the structure–property relationship of BiB_2O_4F revealed that the F-incorporation not only leads to the more asymmetric unit BO_3F but also withdraws the 6s lone pair electrons of Bi^{3+}, increasing the local polarization and finally enhancing the total SHG efficiency. This finding may offer a new insight on finding FD materials.

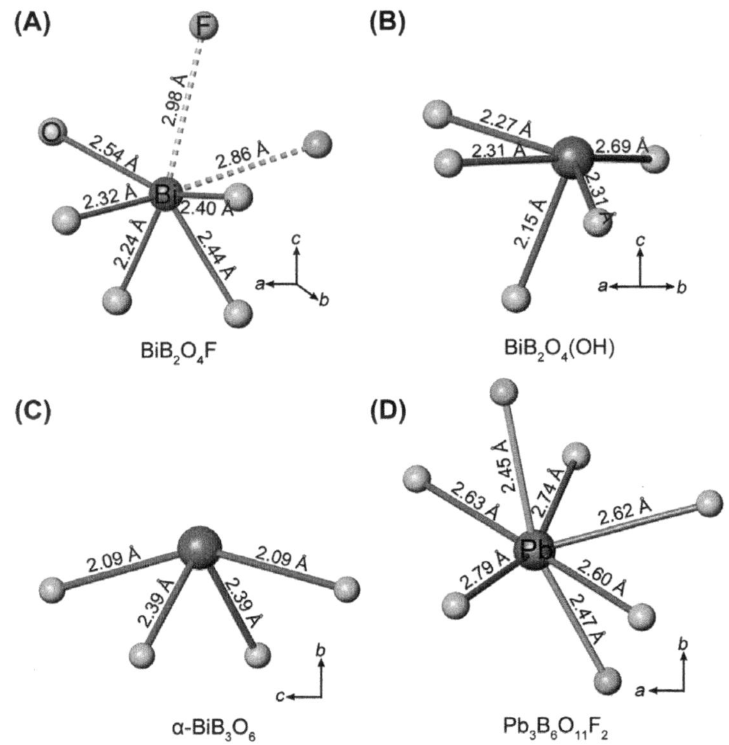

Figure 15.39 Coordination environments of Bi^{3+} cations in (A) BiB_2O_4F, (B) $BiB_2O_4(OH)$, (C) α-BiB_3O_6, and Pb^{2+} cation in (D) $Pb_3B_6O_{11}F_2$.

5. Summary and Outlook

In conclusion, we have summarized FD materials based on borate fluorides and fluorooxoborates. The combinations of borate anionic groups and fluorine atoms in the same anionic framework show rich structural and compositional diversity, and the list of new borate fluorides and fluorooxoborates continues to expand. As for FD applications, we highlight the positive effect of adding fluorine on both structural and SHG properties in borate fluorides and fluorooxoborates. Generally, fluorine contributes to widening the transparency of borates in the UV range, and it is also helpful for structural diversity and increasing local distortion. This observation has also been further confirmed by the theoretical DFT calculations reflecting the structure–property relationship of NLO-active borate fluorides and fluorooxoborates.

Finding an ideal FD material operating in the DUV region is an urgent need. This is not easy to achieve because the demanded FD material should exhibit a large SHG response while also possessing a broad UV transparency window and PM. There has been a breakthrough for recent research focused on borate fluorides and fluorooxoborates. In addition to KBBF, BABF, $NaCaBe_2B_2O_6F$, $NaSr_3Be_3B_3O_9F_4$, $Ba_3B_6O_{11}F_2$ (M = Ba and Sr), $Ba_4B_{11}O_{20}F$, and $Rb_3Al_3B_3O_{10}F$ are the most promising candidates for DUV FD material.

However, the practical application of borate fluorides or fluorooxoborates as FD materials, especially in DUV range, is most limited to the lack of large single crystals. For borate fluorides, the star FD material KBBF suffers from layered crystal growth habit and other borate fluorides are hard to obtain large single crystals with good quality. Because many borate fluorides are not congruent-melting compounds, the Czochralski or Bridgman methods are not suitable. Nowadays, most of the borate fluorides are synthesized by the flux method. Herein, the scope of crystal growth of borate fluorides needs great effort to find suitable fluxes and to avoid volatilization of the metal fluorides at high temperature. As for fluorooxoborates, due to their weak thermal stability, the hydrothermal method might be a better approach at relatively low temperature.

We hope that these investigations may encourage exploratory synthesis of borate fluorides and fluorooxoborates and allow for the rational design of new FD materials, especially for those used in UV and DUV applications.

References

[1] Y.N. Xia, C. Chen, D.Y. Tang, B.C. Wu, New nonlinear optical crystals for UV and VUV harmonic generation, Adv. Mater. 7 (1995) 79–81.

[2] P. Becker, Borate materials in nonlinear optics, Adv. Mater. 10 (1998) 979–992.

[3] C.T. Chen, T. Sasaki, R.K. Li, Y.C. Wu, Z.S. Lin, Y. Mori, Nonlinear Optical Borate Crystals: Principals and Applications, John Wiley & Sons, 2012.

[4] C.T. Chen, Y.B. Wang, B.C. Wu, K.C. Wu, W.L. Zeng, L.H. Yu, Design and synthesis of an ultraviolet-transparent nonlinear optical crystal $Sr_2Be_2B_2O_7$, Nature 373 (1995) 322–324.

[5] D. Cyranoski, China's crystal cache, Nature 457 (2009) 953–955.
[6] S.G. Zhao, P.F. Gong, L. Bai, X. Xu, S.Q. Zhang, Z.H. Sun, Z.S. Lin, M.C. Hong, C.T. Chen, J.H. Luo, Beryllium-free $Li_4Sr(BO_3)_2$ for deep-ultraviolet nonlinear optical applications, Nat. Commun. 5 (2014) 4019.
[7] P.A. Franken, A.E. Hill, C.W. Peters, G. Weinreich, Generation of optical harmonics, Phys. Rev. Lett. 7 (1961) 118–119.
[8] C.T. Chen, B.C. Wu, A.D. Jiang, G.M. You, A new type ultraviolet SHG crystal – β-BaB_2O_4, Sci. Sin. Ser. B 28 (1985) 235–243.
[9] C.T. Chen, G. Liu, Recent advances in nonlinear optical and electro-optical materials, Annu. Rev. Mater. Sci. 16 (1986) 203–243.
[10] Y.C. Wu, T. Sasaki, N. Nakai, A. Yokotani, H.G. Tang, C.T. Chen, CsB_3O_5: a new nonlinear optical crystal, Appl. Phys. Lett. 62 (1993) 2614.
[11] Y. Mori, I. Kuroda, S. Nakajima, T. Sasaki, S. Nakai, New nonlinear optical crystal: cesium lithium borate, Appl. Phys. Lett. 67 (1995) 1818.
[12] I. Makoto, K. Taisuke, F. Hiroyuki, M. Yusuke, S. Takatomo, Crystal growth and optical characterization of rare-earth (Re) calcium oxyborate $ReCa_4O(BO_3)_3$ (Re = Y or Gd) as new nonlinear optical material, Jpn. J. Appl. Phys. 36 (1997) L1562.
[13] H. Hellwig, J. Liebertz, L. Bohaty, Exceptional large nonlinear optical coefficients in the monoclinic bismuth borate BiB_3O_6 (BIBO), Solid State Commun. 109 (1999) 249–251.
[14] C.T. Chen, Y.B. Wang, Y.N. Xia, B.C. Wu, D.Y. Tang, K.C. Wu, W.R. Zeng, L.H. Yu, New development of nonlinear optical crystals for the ultraviolet region with molecular engineering approach, J. Appl. Phys. 77 (1995) 2268.
[15] L. Mei, X. Huang, Y. Wang, Q. Wu, B. Wu, C.T. Chen, Crystal structure of $KBe_2BO_3F_2$, Z. Kristallogr. 210 (1995) 93–95.
[16] C.T. Chen, N. Ye, J. Lin, J. Jiang, W.R. Zeng, B.C. Wu, Computer-assisted search for nonlinear optical crystals, Adv. Mater. 11 (1999) 1071–1078.
[17] L. Shirong, Q.Z. Huang, Y.F. Zheng, A.D. Jiang, C.T. Chen, Structure of calcium fluoroborate, $Ca_5(BO_3)_3F$, Acta Crystallogr. Sect. C 45 (1989) 1861–1863.
[18] L.F. Mei, Y.B. Wang, C.T. Chen, Crystal structure of sodium beryllium borate fluoride, Mater. Res. Bull. 29 (1994) 81–87.
[19] C.D. McMillen, J.W. Kolis, Hydrothermal crystal growth of $ABe_2BO_3F_2$ (A = K, Rb, Cs, Tl) NLO crystals, J. Cryst. Growth 310 (2008) 2033–2038.
[20] Z.G. Hu, M. Yoshimura, K. Muramatsu, Y. Mori, T. Sasaki, A new nonlinear optical crystal-$BaAlBO_3F_2$ (BABF), Jpn. J. Appl. Phys. 41 (2002) l1131.
[21] H.W. Huang, J.Y. Yao, Z.S. Lin, X.Y. Wang, R. He, W.J. Yao, N.X. Zhai, C.T. Chen, Molecular engineering design to resolve the layering habit and polymorphism problems in deep UV NLO crystals: new structures in $MM'Be_2B_2O_6F$ (M = Na, M′ = Ca; M = K, M′ = Ca, Sr), Chem. Mater. 23 (2011) 5457–5463.
[22] H.W. Huang, J.Y. Yao, Z.S. Lin, X.Y. Wang, R. He, W.J. Yao, N.X. Zhai, C.T. Chen, $NaSr_3Be_3B_3O_9F_4$: a promising deep-ultraviolet nonlinear optical material resulting from the cooperative alignment of the $[Be_3B_3O_{12}F]^{10-}$ anionic group, Angew. Chem. Int. Ed. 50 (2011) 9141–9144.
[23] H.W. Yu, H.P. Wu, S.L. Pan, Z.H. Yang, X. Su, F.F. Zhang, A novel deep UV nonlinear optical crystal $Ba_3B_6O_{11}F_2$, with a new fundamental building block, B_6O_{14} group, J. Mater. Chem. 22 (2012) 9665–9670.
[24] Z.J. Huang, X. Su, S.L. Pan, X.Y. Dong, S.J. Han, H.W. Yu, M. Zhang, Y. Yang, S.F. Cui, Z.H. Yang, $Sr_3B_6O_{11}F_2$: a promising polar fluoroborate with short UV absorption edge and moderate second harmonic generation response, Scr. Mater. 69 (2013) 449–452.

[25] H.Y. Li, H.P. Wu, X. Su, H.W. Yu, S.L. Pan, Z.H. Yang, Y. Lu, J. Han, K.R. Poeppelmeier, $Pb_3B_6O_{11}F_2$: the first non-centrosymmetric lead borate fluoride with a large second harmonic generation response, J. Mater. Chem. C 2 (2014) 1704–1710.
[26] C.D. McMillen, J.T. Stritzinger, J.W. Kolis, Two novel acentric borate fluorides: $M_3B_6O_{11}F_2$ (M = Sr, Ba), Inorg. Chem. 51 (2012) 3953–3955.
[27] H.P. Wu, H.W. Yu, Z.H. Yang, X.L. Hou, X. Su, S.L. Pan, K.R. Poeppelmeier, J.M. Rondinelli, Designing a deep ultraviolet nonlinear optical material with large second harmonic generation response, J. Am. Chem. Soc. 135 (2013) 4215–4218.
[28] G. Cakmak, J. Nuss, M. Jansen, LiB_6O_9F, the first lithium fluorooxoborate – crystal structure and ionic conductivity, Z. Anorg. Allg. Chem. 635 (2009) 631–636.
[29] J.Y. Wang, C.Q. Zhang, Y.G. Liu, Z. Hu, M.H. Jiang, C.T. Chen, Y.C. Wu, J. Xu, Growth and properties of $KBe_2BO_3F_2$ crystal, J. Mater. Res. 18 (2003) 2478–2485.
[30] D.E. Bugaris, H.-C. zur Loye, Materials discovery by flux crystal growth: quaternary and higher order oxides, Angew. Chem. Int. Ed. 51 (2012) 3780–3811.
[31] J. Rouse, K.V. Redrup, E. Kotsapa, M.T. Weller, Controlling dimensionality in templated layer, chain and framework structures by combining metal fluorides with oxotetrahedra, Chem. Commun. (2009) 7209–7211.
[32] L.Y. Li, P.C. Lu, Y.Y. Wang, X.L. Jin, G.B. Li, Y.X. Wang, L.P. You, J.H. Lin, Synthesis of rare earth polyborates using molten boric acid as a flux, Chem. Mater. 14 (2002) 4963–4968.
[33] M.R. Li, W. Liu, M.H. Ge, H.H. Chen, X.X. Yang, J.T. Zhao, $NH_4[BPO_4F]$: a novel open-framework ammonium fluorinated borophosphate with a zeolite-like structure related to gismondine topology, Chem. Commun. (2004) 1272–1273.
[34] S.A. Wang, E.V. Alekseev, J. Diwu, H.M. Miller, A.G. Oliver, G. Liu, W. Depmeier, T.E. Albrecht-Schmitt, Functionalization of borate networks by the incorporation of fluoride: syntheses, crystal structures, and nonlinear optical properties of novel actinide fluoroborates, Chem. Mater. 23 (2011) 2931–2939.
[35] T. Pilz, H. Nuss, M. Jansen, $Li_2B_3O_4F_3$, a new lithium-rich fluorooxoborate, J. Solid State Chem. 186 (2012) 104–108.
[36] T. Pilz, M. Jansen, $Li_2B_6O_9F_2$, a new acentric fluorooxoborate, Z. Anorg. Allg. Chem. 637 (2011) 2148–2152.
[37] H.P. Wu, H.W. Yu, Q. Bian, Z.H. Yang, S.J. Han, S.L. Pan, Borate fluoride and fluoroborate in alkali-metal borate prepared by an open high-temperature solution method, Inorg. Chem. 53 (2014) 12686–12688.
[38] M. Jansen, T. Pilz, Novel fluorido-oxoanions via Lewis acid-base reactions – synthesis and crystal structure determination of $Na_3B_2PO_5F_4$, Z. Kristallogr. 228 (2013) 476–482.
[39] S. Filatov, R. Bubnova, Borate crystal chemistry, Phys. Chem. Glasses 41 (2000) 216–224.
[40] M. Touboul, N. Penin, G. Nowogrocki, Borates: a survey of main trends concerning crystal-chemistry, polymorphism and dehydration process of alkaline and pseudo-alkaline borates, Solid State Sci. 5 (2003) 1327–1342.
[41] S. Wang, T.G. Parker, D.J. Grant, J. Diwu, E.V. Alekseev, W. Depmeier, L. Gagliardi, T.E. Albrecht-Schmitt, Elucidation of tetraboric acid with a new borate fundamental building block in a chiral uranyl fluoroborate, Inorg. Chem. 51 (2012) 11211–11213.
[42] H.W. Yu, H.P. Wu, S.L. Pan, Y. Wang, Z.H. Yang, X. Su, New salt-inclusion borate, $Li_3Ca_9(BO_3)_7 \cdot 2[LiF]$: a promising UV NLO material with the coplanar and high density BO_3 triangles, Inorg. Chem. 52 (2013) 5359–5365.
[43] E.G. Tulsky, J.R. Long, Dimensional reduction: a practical formalism for manipulating solid structures, Chem. Mater. 13 (2001) 1149–1166.

[44] I.A. Baidina, V.V. Bakakin, N.V. Podberezskaya, V.I. Alekseev, L.R. Batsanova, V.S. Pavlyuchenko, Crystal structure of beryllium fluoride borate $Be_2(BO_3)F$, J. Struct. Chem. 19 (1978) 105–108.
[45] Q. Huang, M. Ulutagay, P.A. Michener, S.-J. Hwu, Salt-templated open frameworks (CU-2): novel phosphates and arsenates containing $M_3(X_2O_7)_2^{2-}$ (M = Mn, Cu; X = P, As) micropores 5.3 Å and 12.7 Å in diameter, J. Am. Chem. Soc. 121 (1999) 10323–10326.
[46] Q. Huang, S.-J. Hwu, High-temperature synthesis of two new aluminophosphates, $NaRb_2Al_2(PO_4)_3$ and $NaCs_2Al(PO_4)_2$, in molten alkali metal chloride media, Chem. Mater. 13 (2001) 1794–1799.
[47] S.-J. Hwu, M. Ulutagay-Kartin, J.A. Clayhold, R. Mackay, T.A. Wardojo, C.J. O'Connor, M. Krawiec, A new class of hybrid materials via salt inclusion: novel copper(II) arsenates $Na_5ACu_4(AsO_4)_4Cl_2$ (A = Rb, Cs) composed of alternating covalent and ionic lattices, J. Am. Chem. Soc. 124 (2002) 12404–12405.
[48] J. Palmer West, S.-J. Hwu, J.P. West, Noncentrosymmetric salt inclusion oxides: role of salt lattices and counter ions in bulk polarity, J. Solid State Chem. 195 (2012) 101–107.
[49] X.H. Mo, E. Ferguson, S.-J. Hwu, Salt-inclusion synthesis of two new polar solids, $Ba_6Mn_4Si_{12}O_{34}Cl_3$ and $Ba_6Fe_5Si_{11}O_{34}Cl_3$, Inorg. Chem. 44 (2005) 3121–3126.
[50] G.J. Chen, Y.C. Wu, P.Z. Fu, Growth and characterization of a new nonlinear optical crystal $Ca_5(BO_3)_3F$, J. Cryst. Growth 292 (2006) 449–453.
[51] R.H. Cong, Y. Wang, L. Kang, Z.Y. Zhou, Z.S. Lin, T. Yang, An outstanding second-harmonic generation material BiB_2O_4F: exploiting the electron-withdrawing ability of fluorine, Inorg. Chem. Front. 2 (2015) 170–176.
[52] A.A. Brovkin, L.V. Nikishova, The crystal structure of alpha-Mg_2BO_3F and the isomorphic substitution $(3F)^{3-}$ – $(BO_3)^{3-}$, Kristallografiya 20 (1975) 740–745.
[53] A.A. Brovkin, L.V. Nikishova, Structure of $Mg_5(BO_3)_3F$ and its relation to the struliure of leucophoenicite $Mn_7(SiO_4)_2(Si_4(OH)_2)$, Kristallografiya 20 (1975) 415–418.
[54] J.G. Fletcher, F.P. Glasser, R.A. Howie, Pentacalcium triborate fluoride and its relationship to fluorapatite, Acta Crystallogr. Sect. C 47 (1991) 12–14.
[55] C.T. Chen, S.Y. Luo, X.Y. Wang, G.L. Wang, X.H. Wen, H.X. Wu, X. Zhang, Z.Y. Xu, Deep UV nonlinear optical crystal: $RbBe_2(BO_3)F_2$, J. Opt. Soc. Am. B 26 (2009) 1519–1525.
[56] G.H. Zou, L.Y. Zhang, N. Ye, Synthesis, structure, and characterization of a new promising nonlinear optical crystal: $Cd_5(BO_3)_3F$, CrystEngComm 15 (2013) 2422–2427.
[57] H.W. Huang, C.T. Chen, X.Y. Wang, Y. Zhu, G.L. Wang, X. Zhang, L.R. Wang, J.Y. Yao, Ultraviolet nonlinear optical crystal : $CsBe_2BO_3F_2$, J. Opt. Soc. Am. B 28 (2011) 2186–2190.
[58] Z.G. Hu, Y.C. Yue, X.A. Chen, J.Y. Yao, J.N. Wang, Z.S. Lin, Growth and structure redetermination of a nonlinear $BaAlBO_3F_2$ crystal, Solid State Sci. 13 (2011) 875–878.
[59] S.G. Zhao, P.F. Gong, S.Y. Luo, S.J. Liu, L. Li, M. Adnan, T. Khan, M.C. Hong, Z.S. Lin, J.H. Luo, Beryllium-free $Rb_3Al_3B_3O_{10}F$ with reinforced interlayer bonding as a deep-ultraviolet nonlinear optical crystal, J. Am. Chem. Soc. 137 (2015) 2207–2210.
[60] J.N. Wang, Y.C. Yue, J.Y. Yao, Z.G. Hu, Growth and characterization of Ba(Al,Ga) BO_3F_2 crystal, J. Cryst. Growth 318 (2011) 962–965.
[61] S.L. Pan, W.W. Zhao, Compound Barium Boron Oxyfluoride, Barium Boron Oxyfluoride Nonlinear Optical Crystal, and Preparation Methods and Applications Thereof, China, 2013 (Chinese Patent).
[62] D.A. Keszler, A. Akella, K.I. Schaffers, T.A. Alekel, New borate structures for NLO applications, Mater. Res. Soc. Symp. Proc. (1993) 15–22.

[63] R.K. Li, Q.D. Zeng, Crystal growth, structure and phase transition of the nonlinear optical crystal $BaCaBO_3F$, J. Cryst. Growth 382 (2013) 47–51.
[64] T. Bekker, S. Rashchenko, V. Bakakin, Y. Seryotkin, P.P. Fedorov, A.E. Kokh, S.Y. Stonoga, Phase formation in the BaB_2O_4–BaF_2–BaO system and new non-centrosymmetric solid-solution series $Ba_7(BO_3)_{4-x}F_{2+3x}$, CrystEngComm 14 (2012) 6910–6915.
[65] K. Xu, P. Loiseau, G. Aka, $BaCaBO_3F$: a nonlinear optical crystal investigated for UV light generation, J. Cryst. Growth 311 (2009) 2508–2512.
[66] G.C. Zhang, Z.L. Liu, J.X. Zhang, F.D. Fan, Y.C. Liu, P.Z. Fu, Crystal growth, structure, and properties of a non-centrosymmetric fluoride borate, $Ba_3Sr_4(BO_3)_3F_5$, Cryst. Growth Des. 9 (2009) 3137–3141.
[67] R.K. Li, P. Chen, Cation coordination control of anionic group alignment to maximize SHG effects in the $BaMBO_3F$ (M = Zn, Mg) series, Inorg. Chem. 49 (2010) 1561.
[68] S.V. Rashchenko, T.B. Bekker, V.V. Bakakin, Y.V. Seryotkin, V.S. Shevchenko, A.E. Kokh, S.Y. Stonoga, New fluoride borate solid–solution series $Ba_{4-x}Sr_{3+x}(BO_3)_{4-y}F_{2+3y}$, Cryst. Growth Des. 12 (2012) 2955–2960.
[69] A.A. Brovkin, L.V. Nikishova, $Mg_2(BO_3)_{1.14}F_{0.58}$, Kristallografiya 20 (1975) 740.
[70] L. Li, G.B. Li, Y.X. Wang, F.H. Liao, Bismuth borates: one-dimensional borate chains and nonlinear optical properties, Chem. Mater. 47 (2005) 4174–4180.
[71] S.B. Wiggin, M.T. Weller, Boroarsenates: a framework Motif and family templated on cations and anions, J. Am. Chem. Soc. 127 (2005) 17172–17173.
[72] S.K. Kurtz, T.T. Perry, A powder technique for the evaluation of nonlinear optical materials, J. Appl. Phys. 39 (1968) 3798.
[73] D. Maker, E.W. Terhune, M. Nisenoff, C.M. Savage, Mixing of dispersion and focusing on the production of optical harmonics, Phys. Rev. Lett. 8 (1962) 21–22.
[74] L.F. Mei, Y.B. Wang, C.T. Chen, B.C. Wu, Nonlinear optical materials based on $MBe_2BO_3F_2$ (M = Na,K), J. Appl. Phys. 74 (1993) 7014.
[75] C.T. Chen, G.L. Wang, X.Y. Wang, Y. Zhu, Z.Y. Xu, T. Kanai, S. Watanabe, Improved Sellmeier equations and phase-matching characteristics in deep-ultraviolet region of $KBe_2BO_3F_2$ crystal, IEEE J. Quantum Electron. 44 (2008) 617–621.
[76] Z.S. Lin, Z.Z. Wang, C.T. Chen, S.K. Chen, M.H. Lee, Mechanism for linear and nonlinear optical effects in $KBe_2BO_3F_2$ (KBBF) crystal, Chem. Phys. Lett. 367 (2003) 523–527.
[77] C.T. Chen, Y.C. Wu, R.K. Li, The anionic group theory of the non-linear optical effect and its applications in the development of new high-quality NLO crystals in the borate series, Int. Rev. Phys. Chem. 8 (1989) 65–91.
[78] F. Yang, Z.M. Wang, Y. Zhou, X.K. Cheng, S.Y. Xie, Q.J. Peng, D.F. Cui, J.Y. Zhang, X.Y. Wang, Y. Zhu, C.T. Chen, Z.Y. Xu, 41 mW high average power picosecond 177.3 nm laser by second harmonic generation in KBBF, Opt. Commun. 283 (2010) 142–145.
[79] C.T. Chen, Z.Y. Xu, D.Q. Deng, J. Zhang, G.K.L. Wong, B.C. Wu, N. Ye, D.Y. Tang, The vacuum ultraviolet phase-matching characteristics of nonlinear optical $KBe_2BO_3F_2$ crystal, Appl. Phys. Lett. 68 (1996) 2930–2932.
[80] C.T. Chen, G.L. Wang, X.Y. Wang, Z.Y. Xu, Deep-UV nonlinear optical crystal $KBe_2BO_3F_2$—discovery, growth, optical properties and applications, Appl. Phys. B 97 (2009) 9–25.
[81] C.T. Chen, Recent advances in deep and vacuum-UV harmonic generation with KBBF crystal, Opt. Mater. 26 (2004) 425–429.
[82] T. Kanai, X.Y. Wang, S. Adachi, S. Watanabe, C.T. Chen, Watt-level tunable deep ultraviolet light source by a KBBF prism-coupled device, Opt. Express 17 (2009) 8696–8703.

[83] G.L. Wang, X.Y. Wang, Y. Zhou, C.M. Li, Y. Zhu, Z.Y. Xu, C.T. Chen, High-efficiency frequency conversion in deep ultraviolet with a $KBe_2BO_3F_2$ prism-coupled device, Appl. Opt. 47 (2008) 486–488.
[84] C.T. Chen, T. Kanai, X.Y. Wang, Y. Zhu, S. Watanabe, High-average-power light source below 200 nm from a $KBe_2BO_3F_2$ prism-coupled device, Opt. Lett. 33 (2008) 282–284.
[85] J.Q. Yu, L.J. Liu, S.F. Jin, H.T. Zhou, X.L. He, C.L. Zhang, W.N. Zhou, X.Y. Wang, X.L. Chen, C.T. Chen, Superstructure and stacking faults in hydrothermal-grown $KBe_2BO_3F_2$ crystals, J. Solid State Chem. 184 (2011) 2790–2793.
[86] C.D. McMillen, J. Hu, D. VanDerveer, J.W. Kolis, Trigonal structures of $ABe_2BO_3F_2$ (A = Rb, Cs, Tl) crystals, Acta Crystallogr. Sect. B 65 (2009) 445–449.
[87] H. Park, J. Barbier, Crystal structures of the new borate fluorides $BaMBO_3F_2$ (M = Ga, Al), J. Solid State Chem. 155 (2000) 354–358.
[88] Z.G. Hu, K. Maramatsu, N. Kanehisa, M. Yoshimura, Y. Mori, T. Sasaki, Y. Kai, Reinvestigation of the crystal structure of barium aluminum borate difluoride, $BaAlBO_3F_2$, a new nonlinear optical material, Z. Kristallogr. 218 (2003) 1–2.
[89] H. Huang, Z.S. Lin, L. Bai, Z.G. Hu, C.T. Chen, Ab initio calculations on the borate nonlinear optical crystal $BaAlBO_3F_2$, J. Appl. Phys. 106 (2009) 103107.
[90] Y.C. Yue, Z.G. Hu, C.T. Chen, Flux growth of $BaAlBO_3F_2$ crystal, J. Cryst. Growth 310 (2008) 1264–1267.
[91] Y.C. Yue, Z.G. Hu, Y. Zhou, J.N. Wang, X. Zhang, C.T. Chen, Z.Y. Xu, Growth and nonlinear optical properties of $BaAlBO_3F_2$ crystal, J. Opt. Soc. Am. B 28 (2011) 861–866.
[92] Y. Zhou, Y.C. Yue, J.N. Wang, F. Yang, X.K. Cheng, D.F. Cui, Q.J. Peng, Z.G. Hu, Z.Y. Xu, Nonlinear optical properties $BaAlBO_3F_2$ crystal, Opt. Express 17 (2009) 20033–20038.
[93] Y. Zhou, G.L. Wang, Y.C. Yue, C.M. Li, Y.F. Lu, D.F. Cui, Z.G. Hu, Z.Y. Xu, High-efficiency 355 nm generation in barium aluminum borate diflouride $BaAlBO_3F_2$, Opt. Lett. 34 (2009) 746–748.
[94] W.B. Xia, Z.S. Lin, Y.C. Wu, C.T. Chen, A new type of potential nonlinear optical material-beryllium fluoride borate, J. Synth. Cryst. 34 (2005) 597–600.
[95] X.S. Wang, L.J. Liu, X.Y. Wang, C.T. Chen, L. Bai, Growth and optical properties of a novel nonlinear optical crystal $NaSr_3Be_3B_3O_9F_4$, CrystEngComm 17 (2015) 925–929.
[96] G.C. Zhang, H.J. Liu, X. Wang, F.D. Fan, P.Z. Fu, Growth and characterization of nonlinear optical crystal $BaCaBO_3F$, J. Cryst. Growth 289 (2006) 188–191.
[97] X.A. Wang, G.C. Zhang, Y. Zhao, F.D. Fan, H.J. Liu, P.Z. Fu, Optical properties of $BaCaBO_3F$ crystal, Opt. Mater. 29 (2007) 1658–1661.
[98] K.I. Schaffers, L.D. DeLoach, S.A. Payne, Crystal growth, frequency doubling, and infrared laser performance of Yb^{3+}:$BaCaBO_3F$, IEEE J. Quantum Electron. 32 (1996) 741–748.
[99] W. Zhao, W.W. Zhou, M.J. Song, G.F. Wang, J.M. Du, H.J. Yu, J.X. Chen, Polarized spectroscopic properties of a potential self-frequency doubling crystal, Nd^{3+}:$BaCaBO_3F$, Opt. Mater. 33 (2011) 647–654.
[100] J. Zhao, M. Xia, R.K. Li, Flux growth of a potential nonlinear optical crystal $BaMgBO_3F$, J. Cryst. Growth (2011) 971–973.
[101] T.J. White, Z.L. Dong, Structural derivation and crystal chemistry of apatites, Acta Crystallogr. Sect. B 59 (2003) 1–16.
[102] K. Xu, P. Loiseau, G. Aka, R. Maillard, A. Maillard, T. Taira, Nonlinear optical properties of $Ca_5(BO_3)_3F$ crystal, Opt. Express 16 (2008) 17735–17744.
[103] K. Xu, P. Loiseau, G. Aka, J. Lejay, A new promising nonlinear optical crystal for ultraviolet light generation: $Ca_5(BO_3)_3F$, Cryst. Growth Des. 9 (2009) 2235–2239.

[104] K. Kazmierczak, H.A. Höppe, Synthesis, crystal structure and optical spectra of Europium borate fluoride $Eu_5(BO_3)_3F$, Eur. J. Inorg. Chem. 2010 (2010) 2678–2681.
[105] T. Alekel, D.A. Keszler, New strontium borate halides: $Sr_5(BO_3)_3X$ (X = F, Br), Inorg. Chem. 32 (1993) 101–105.
[106] S.V. Rashchenko, T.B. Bekker, V.V. Bakakin, Y.V. Seryotkin, A.E. Kokh, P. Gille, A.I. Popov, P.P. Fedorov, A new mechanism of anionic substitution in fluoride borates, J. Appl. Crystallogr. 46 (2013) 1081–1084.
[107] B. Andriyevsky, T. Pilz, J. Yeon, P.S. Halasyamani, K. Doll, M. Jansen, DFT-based ab initio study of dielectric and optical properties of bulk $Li_2B_3O_4F_3$ and $Li_2B_6O_9F_2$, J. Phys. Chem. Solids 74 (2013) 616–623.
[108] B. Andriyevsky, K. Doll, T. Pilz, M. Jansen, DFT-based ab initio study of electronic band structure and elastic properties of $Li_2B_3O_4F_3$ and $Li_2B_6O_9F_2$ crystals, J. Phys. Chem. Solids 74 (2013) 624–629.
[109] T. Bräuniger, T. Pilz, C.V. Chandran, M. Jansen, Full differentiation and assignment of boron species in the electrolytes $Li_2B_6O_9F_2$ and $Li_2B_3O_4F_3$ by solid-state ^{11}B NMR spectroscopy, J. Solid State Chem. 194 (2012) 245–249.
[110] B. Andriyevsky, K. Doll, G. Cakmak, M. Jansen, A. Niemer, K. Betzler, DFT-based ab initio study of structural and electronic properties of lithium fluorooxoborate LiB_6O_9F and experimentally observed second harmonic generation, Phys. Rev. B 84 (2011) 125112.
[111] Landolt-Börnstein: Numerical Data and Functional Relationship in Science and Technology, New Series, Group III, vol. 16, Springer, Berlin; Heidelberg; New York; Tokyo, 1984.

Ferroelastic and Ferroelectric Phase Transitions in Fluoro- and Oxyfluorometallates

16

I.N. Flerov, M.V. Gorev, M.S. Molokeev
Kirensky Institute of Physics (KIP), Krasnoyarsk, Russia

N.M. Laptash
Institute of Chemistry, Vladivostok, Russia

Chapter Outline

1. Introduction

Inorganic fluorinated compounds attract a great attention of scientists and engineers thanks to a possibility to design a rich variety of materials with crystal structures of different dimensionality and symmetry that enables one to realize a great diversity of useful physical chemical properties. So far a series of books [63,64,81,83] and

Photonic and Electronic Properties of Fluoride Materials. http://dx.doi.org/10.1016/B978-0-12-801639-8.00016-7

reviews [7,29,50] have dealt with functional properties and/or structures showing by fluorinated compounds mainly at room temperature. However, it is known that some rather wide crystallographic families of fluorides and oxide-fluorides can significantly change the symmetry of structure undergoing phase transitions (PTs) under changes of temperature and/or pressure [20,73].

The problem of PTs of different physical nature in solids belongs to one of the most interesting and promising areas of solid-state physics and chemistry as well as materials science. The best prominent peculiarity of these phenomena is that they are accompanied by strong change in the physical properties of materials due to an anomalous behavior of the order parameter (OP) (polarization, magnetization, elastic deformation, etc.) which can occur not only at PT point (strong first order transformations) but very often also in a wide temperature range (second-order and first-order transformations close to the tricritical point). High sensitivity of PTs to variations in external (temperature, pressure, magnetic and electric fields, etc.) and internal (doping, size and shape of ions, cationic–anionic substitution) parameters allows one to elucidate their physical nature as well as the mechanism of changing stability of different crystallographic phases, sequence of several transformations, temperature interval of their existence, and closeness to tricritical point. A comprehensive study of PTs reveals the interrelationships between chemical composition, structure, and physical properties which can be considered as criteria for design of materials with controlled optimal characteristics in the required temperature range.

More than 15 years have passed since the consideration of some aspects of ferroelectricity and ferroelasticity in certain inorganic fluorides and oxide-fluorides [19,20,71–73] and a lot of new detailed, useful and interesting information has appeared in this line.

We suggest a brief overview of ferroelastic and ferroelectric PTs in materials belonging to the family of inorganic fluoro- and oxyfluorometallates containing octahedral or quasi-octahedral anionic groups.

The ordered and statically or dynamically disordered structures as well as displacive and order–disorder PTs are presented here as the limiting states and phenomena associated with the low and high anisotropic/anharmonic vibrations of ions taking active part in the mechanism of structural distortions at the symmetry change under temperature and/or pressure variation. Entropy change associated with the OP variation is discussed as a basic quantitative parameter characterizing the degree of the structural disorder change at PT point. Several series of perovskite-like crystals designed by the cationic and anionic substitution as well as some crystals with isolated octahedral oxide-fluoride anionic species are considered demonstrating a great effect of external and internal pressure on stability of structure and mechanism of phase transformations. The nature of structural distortions associated with the appearance or change of ferroic properties (ferroelectricity, ferroelasticity) at PT has also been paid significant attention. Fluorinated materials showing a great barocaloric effect appear to have a considerable promise as solid-state refrigerants.

2. Mechanism of Structural Phase Transitions in Fluorides and Oxide-Fluorides

One of the most important aims of studies dedicated to structural transformations is to determine what ions and/or ionic groups are potentially responsible for the collapse of the prototype phase stability. The term mechanism of PT implies the character and value of the atom displacements and a change of interatomic distances and thermal parameters in one phase of material compared to other. These characteristics can be revealed by an experimental study of the change of atomic positions and/or ionic polyhedral orientations associated with the symmetry variation. The second stage of the mechanism study is associated with analysis of structural distortions in the framework of the group theory to find critical modes as well as OPs driving structural transformation.

Crystal lattice distortions caused by structural PTs are often considered in the framework of two extreme mechanisms, namely, the displacive and order–disorder types. The most adequate definition of these phenomena was given in [76]: In the case of displacements, the instability arises from long-range cooperative interactions and brings about local distortions, and, in the case of order–disorder, the local distortions stem from local instabilities. In the reality, experiments very often exhibit that the mechanism of the symmetry change can be more complicated including both processes of the small displacements of atoms and the change of the orientational disordering degree of ionic groups. In this respect, the main question is which mechanism dominates in each particular case.

2.1 The Regularity and Rigidity of Structural Units in Perovskite-like Crystals

One of the important structural peculiarities of compounds containing fluoride and/or mixed oxide-fluoride ligands combined into six-, seven-, and eight-coordinated anionic polyhedra is associated with a rather high symmetry of the initial phase. Many fluorides with the general chemical formula $AM^{2+}F_3$ crystallize in the cubic (space group (sp. gr.) *Pm-3m*, $Z = 1$) or distorted perovskite structure [3,5,6]. The striking flexibility of perovskite crystal lattice permits to vary the composition and valency of central atom over a wide range to design the related compounds with the three- and two-dimensional crystal lattices [3,5,6,20,50].

The most common types of perovskite-like crystals are elpasolites $A_2A'MO_xF_{6-x}$ and cryolites $A_2AMO_xF_{6-x}$ containing fluoride and oxide-fluoride six-coordinated polyhedra with different atoms M and A′. Fig. 16.1 shows the relation between perovskite (sp. gr. *Pm-3m*, $Z = 1$) and elpasolite (sp. gr. *Fm-3m*, $Z = 4$) crystal lattices. In the latter case, two kinds of nonequivalent six-coordinated groups, $A'X_6$ and MX_6, alternate along the three fourfold axes.

Fluorides and oxide-fluorides with seven-coordinated anionic polyhedra can also exhibit a cubic crystal lattice of the elpasolite type [27,28,32,85].

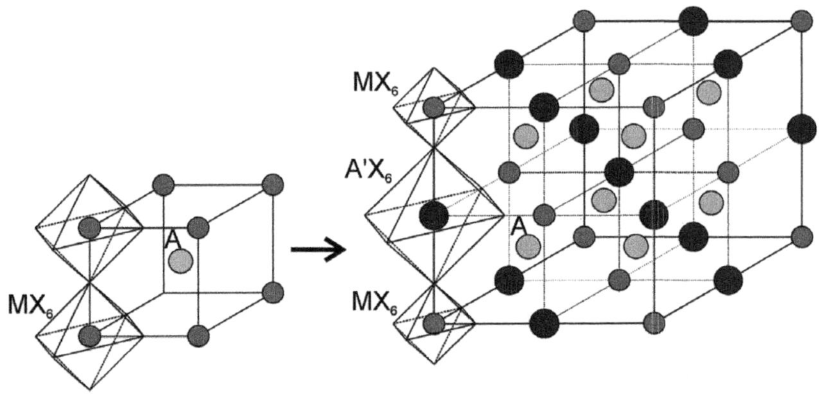

Figure 16.1 The transformation from perovskite crystal lattice (sp. gr. *Pm*-3*m*, Z = 1) to elpasolite crystal lattice (sp. gr. *Fm*-3*m*, Z = 4).

And what about the regularity of six-coordinated polyhedra in initial and distorted phases in perovskite-like crystals? This question was discussed many times on the ground of numerous experimental data on X-ray diffraction, Mössbauer, EPR-, IR-, and Raman spectra for such series of fluoride elpasolites as $A_2A'FeF_6$ (A = K, Rb, Cs; A′ = Na, K) [33,67], $Rb_2KM^{3+}F_6$ (M^{3+} = Sc, In, Y) [15], K_2NaScF_6:Cr^{3+} [77], $Cs_2NaM^{3+}F_6$ (M^{3+} = Al, Ga) [39]. Analysis of the data has allowed one to think that the hypothesis of regular and rigid fluoride octahedra with the O_h local symmetry seems to be fully justified. Moreover, the distortion of octahedra, for example, in $Rb_2KM^{3+}F_6$ and K_2NaScF_6:Cr^{3+} elpasolites, remains very weak in tetragonal phase even far below PT temperature [15,77].

The local symmetry of six-coordinated oxide-fluoride polyhedra is always lower than cubic, and elpasolite-cryolite cubic structure is a consequence of static or dynamic disorder of polyhedra.

Upon cooling, a lot of fluorides and oxide-fluorides can undergo ferromagnetic, ferroelectric, and/or ferroelastic PTs accompanied by the appearance or change of magnetization, electric polarization, or/and elastic deformation. In accordance with a group theory analysis [4,20], a variety of the *Fm*-3*m* structure distortions is possible due to octahedral anionic groups rotations or/and polar displacements of atoms.

In a wide series of halide perovskites including fluorides, structural distortions at ferroelastic PTs are associated with diversity of the ligand displacements, which can be presented as the octahedra tilting as a whole on a small angle around one or simultaneously several main axes of the crystal lattice [3]. This is an important peculiarity of halides compared to oxides with perovskite structure where ferroelectric polarization appears due to alternative displacements of A, M, and O atoms [78]. Although perovskites are studied actively over many years, up to now there is no consensus concerning both the character of atoms' motion in initial cubic phase and the mechanism of PT.

The most fruitful discussion of structural distortions in ferroelectric and ferroelastic perovskites was performed in [41,42,52,76]. The following experimental results proved the interpretation of PTs in these groups of crystals as the displacive-type transformations. A soft mode in the initial and distorted phases and large Curie constant were characteristic for the ferroelectrics. Ferroelastic compounds, apart from the soft modes, exhibited insignificant changes in the entropy.

On the other hand, the experimental data obtained for some ferroic compounds were interpreted as the evidence for the existence of local structural distortions above the PT point which is reflected in the central peak in the light scattering spectra, diffuse X-ray scattering, the birefringence, and the excess heat capacity. These findings gave rise to the development of some theoretical speculations. For example, according to [11,14], the cooperative motion of ions in the structure is associated with the possible "crossover" from the displacive-type mechanism to the order–disorder mechanism at a certain temperature above the PT point T_0. This is caused by a critical increase in the correlation length as the T_0 temperature is approached, resulting in the formation and growth of dynamic clusters, which are small groups of bound atoms, which are displaced from high-symmetry equilibrium positions in the same direction.

2.2 Anisotropy and Anharmonicity of Vibrations of Structural Units

2.2.1 Entropy: Displacive and Order–Disorder Phase Transitions

For a long time the refinement of the cubic structure of fluoride perovskites was performed using isotropic or/and anisotropic models of atomic motion with the assumption that ligands occupy the edge of a cubic cell. Later analysis of the precise structural data in the framework of the cluster and/or anharmonic phonon models performed for a number of ferroelastic perovskites including some fluorides $KMnF_3$, $RbCaF_3$, etc. [41,42], permitted one to give the following conclusions:

1. A model with a multiple-well potential for F atoms inadequately described the initial structure: the R factor (15%) was very large even as compared to the harmonic model ($R = 7\%$).
2. A substantial decrease in the R factor (4%) was observed in the model of the anharmonic atom vibrations.
3. It is difficult to decide between the cluster and anharmonic phonon models, when the predicted cluster distribution function does not exhibit a pronounced behavior inherent in the order–disorder systems.

Unfortunately, the PT entropy ΔS was not usually taken into account in discussions. However, it will be shown later that this parameter is very sensitive to the change of the mechanism of the atom and/or ionic group motion characterized by the anisotropy and anharmonicity.

The role of anisotropy and anharmonicity in the critical atom vibrations in fluorides with the elpasolite structure was considered for several series of crystals obtained by the cationic substitution. An anisotropic model for K_2NaCrF_6 and K_2NaFeF_6 crystals showed a significant decrease in the R factor compared to isotropic [45]. The cationic $K \rightarrow Na$ substitution in the $Rb_2A'FeF_6$ compounds leads to strong increase of the anisotropy parameter B_{33}/B_{11} of ligands and, as a result, to the instability of cubic crystal lattice accompanied by the appearance of PT at 170K in Rb_2KFeF_6 compared to Rb_2NaFeF_6 [33,53], which, according to [20], does not undergo PT down to 0K.

Very interesting and important results were obtained in the refinement of the structure of the Rb_2KScF_6 elpasolite with fluorine atoms occupying one of the crystallographic positions 24*e*, 96*j*, or 192*l* [16]. A substantial decrease in the thermal parameter B_{iso} was found for disordered ligands over four (96*j*) or eight (192*l*) positions compared to the 24*e* site. Very close values of the thermal parameters as well as the reliability *R*-factors made difficult to decide between two models of the ligand disorder, 96*j* or 192*l*. Such a situation could be connected with the considerable anharmonic ligand vibrations or dynamical disordering when the time of the ligand life in one position from the possible four (96*j*) or eight (192*l*) positions is equal to the time of jumping between positions. This supposition was confirmed by observation of soft modes in Raman spectra of distorted phases [49] as well as by small value of entropy change at the successive PTs in Rb_2KScF_6 [17].

As it was shown in [89], anharmonicity of critical ion vibrations, on the one hand, is a necessary condition for realization of PT and, on the other, affects the value of PT entropy as

$$\Delta S \sim (3/2 + b_{12}/b_{11})(1 - u/\mathrm{v}), \qquad [16.1]$$

where u and v are the constants describing the repulsive and attractive effects and the b_{12} and b_{11} are the constants of anharmonic part of potential. The parameter of anharmonicity $\delta = \langle\bar{x}\rangle^2/a^2$ was suggested [89] as linking the amplitude of ion vibrations $\bar{x}$ with the average interatomic distance a.

The role of anisotropy and anharmonicity of thermal vibrations of halogen atoms is considerably more pronounced in elpasolites compared to perovskites in the case of displacive PTs accompanied with almost the same values of octahedra tilts (00ϕ) and/or (00ψ) [9,20,41,74]. Here (00ϕ) and (00ψ) mean the rotation of octahedra in adjacent layers in one and opposite direction, respectively [20]. A lot of experiments have shown that in such a situation the entropy change for halide perovskites AMX_3 and elpasolites $A_2A'MX_6$ (X = F, Cl, Br) with the spherical monovalent cations is also different and universal value equal to $\Delta S \approx 0.1R$ and $\Delta S \approx 0.22R$, respectively [18,20,21].

A good example of the entropy dependence on the anharmonicity of critical ion vibrations is demonstrated by the $Rb_2KM^{3+}F_6$ elpasolites. The increase of the M^{3+} ion size was accompanied by rather strong elevating of the PT temperatures connected with either successive *Fm*-3*m* → *I*4/*m* → *P*12$_1$/*n*1 (M^{3+} = Sc, In, Lu) or direct *Fm*-3*m* → *P*12$_1$/*n*1 (M^{3+} = Er, Y, Ho, Dy, Tb) transformations [19] (Fig. 16.2A). Fig. 16.2B and C shows that the behavior of $\delta = \langle\bar{x}\rangle^2/a^2$ and the total entropy change $\Sigma\Delta S_i$ are correlated with Eq. [16.1], namely, both values increase. Rather large value $\Sigma\Delta S_i \approx 0.7R$ even for crystals with small M^{3+} ion size is the result of additivity of entropies for separate transformations associated with final tilts of octahedra around three main axes of a cubic unit cell: *Fm*-3*m* (000) → *I*4/*m*(00ϕ) → *P*2$_1$/*n*($\psi\phi\phi$), $\Sigma\Delta S_i \approx 3\cdot 0.22R$.

A prominent example of strong disordered *Fm*-3*m* structure was demonstrated by ammonium cryolites $(NH_4)_3M^{3+}F_6$ (M^{3+} = Al, Cr, Ga, V, Fe, Sc) [46,60,61,82].

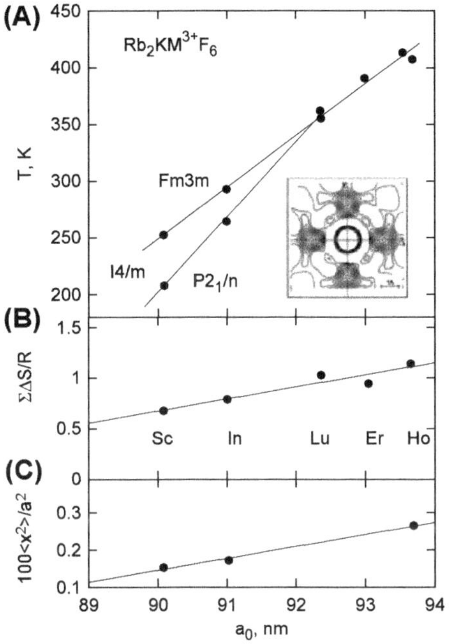

Figure 16.2 Dependences of phase transition temperature (A, D), entropy (B, E), and parameter of anharmonicity (C) on the unit cell parameter for two series of elpasolite-like fluorides. Insets show the electron density distribution of fluorine atoms.

Depending on the size of M^{3+} ion size, these crystals also undergo successive or unique PTs, but the final low-temperature phase in this case is triclinic (Fig. 16.2D). One can see that the total entropy values were found very large compared to $Rb_2KM^{3+}F_6$ elpasolites (Fig. 16.2E) and ranged in rather narrow region $\Delta S = (2.75 \pm 0.24)R \approx R\ln 16$ [31]. In accordance with the entropy change, $(NH_4)_3ScF_6$ is a good example of existence of the order–disorder (*Fm*-3*m* ↔ *P*2$_1$/*n* ↔ *I*2/*m*; $\Delta S = R\ln 16$) and displacive (*I*2/*m* ↔ *I*-1; $\Delta S = 0.1\ R$) transformations in one crystal. The electron density distribution in fluorine atoms presented in the inserts in Fig. 16.2 shows that the ligand vibrations in ammonium compounds are characterized by stronger anisotropy than in crystals with the spherical cations.

More complicated situation with the entropic parameters was observed in oxide-fluorides $A_2A'MO_xF_{6-x}$ with the cryolite–elpasolite structure [24]. Fig. 16.3A and B demonstrates the values of entropy change at successive and unique PTs in compounds with constant relation between O and F atoms and different set of cations. Strong variation of the ΔS value was found as a result of cationic substitution.

On the other hand, compounds with O/F = 1/5 and 2/4 are characterized by the same large value $\Delta S = R\ln 8$, which was found in $(NH_4)_3WO_3F_3$. One more interesting point is connected with large difference between entropic parameters in compounds with spherical cations, cryolite $K_3WO_3F_3$ and elpasolite Rb_2KTiOF_5. It is necessary also to point out that the largest value of $\Delta S = R\ln 8$ determined in more than 10

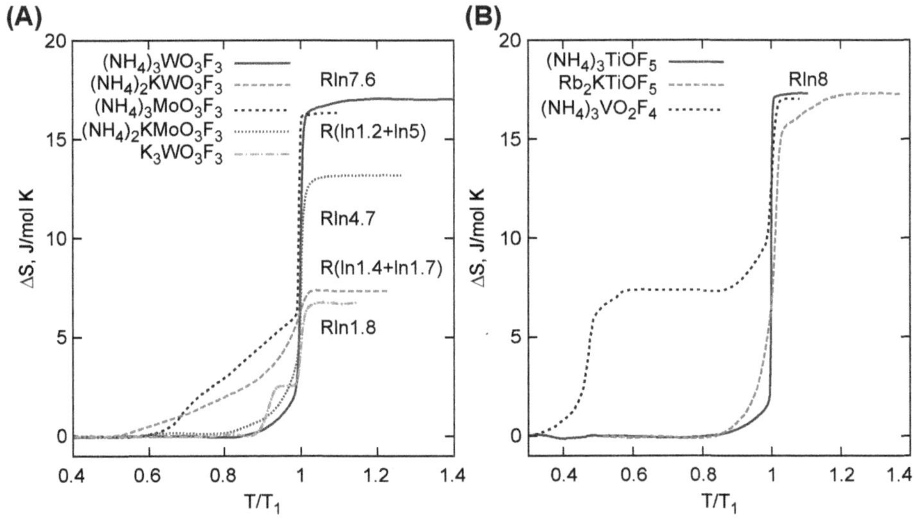

Figure 16.3 Dependences and values of the anomalous entropy on the reduced temperature T/T_1 for some oxide-fluorides. T_1—temperature of phase transition from cubic *Fm*-3*m* phase.

six-coordinated oxide-fluorides including $(NH_4)_3Ta(O_2)_2F_4$ [25] is considerably less than the value of entropy $\Delta S = R\ln 16$ in related fluorides $(NH_4)_3M^{3+}F_6$ (Fig. 16.2E). Therefore it was surprising that the Nb → Ta substitution led to strong increase of entropy $\Delta S = R\ln 48$ associated with succession of three transformations in $(NH_4)_3Nb(O_2)_2F_4$ [25].

2.3 Dynamic Orientational Disorder in Ammonium Fluoro- and Oxyfluoroelpasolites

One of the first attempts to explain rather large values of entropy changes at PTs in ammonium fluorides with the cryolite structure was undertaken in Moria et al. (1977) [60], suggesting an orientational disorder both of MF_6^{3-} octahedra and NH_4^+ tetrahedra in the cubic system *Fm*-3*m*. Fluorine octahedra were assumed to be disordered with eight possible orientations in the general 192*l* position, and one of the ammonium ions at 4*b* site was disordered in two distinct orientations. The total entropy change connected with both octahedra and tetrahedra ordering could be given as $\Delta S = R\ln 2 + R\ln 8$ which is in a good agreement with the experimental values [31,46,60,61,82].

However, the crystal structures refinement of a series of ammonium fluoro-cryolites $(NH_4)_3AlF_6$, $(NH_4)_3TiOF_5$, $(NH_4)_3FeF_6$, and $(NH_4)_3WO_3F_3$ by localizing anions (F^-, O^{2-}) in four acceptable positions of the *Fm*-3*m* system (24*e*, 96*k*, 24*e* + 96*j*, 192*l*) revealed that fluorine (oxygen) atoms should be preferably distributed in mixed 24*e* + 96*j* position. For $(NH_4)_3FeF_6$, the general 192*l* position was also acceptable [84].

Recent study of the $(NH_4)_3GaF_6$ structure by ^{19}F and $^{69,71}Ga$ magic-angle spinning NMR in comparison with X-ray Rietveld refinement supported the model of rigid GaF_6

octahedra. In spite of the very low deviation of the GaF_6 octahedra from exact cubic symmetry, it was concluded that the fluorine 192*l* position is much subjected to errors. Both NH_4^+ ions and GaF_6^{3-} octahedra undergo rapid reorientation rotations [48].

PTs and ionic motions in hexafluoroaluminates $(NH_4)_{3-x}K_xAlF_6$ ($x = 1, 2$) studied by ^1H, ^{19}F, and ^{27}Al magnetic resonance were interpreted in terms of overall reorientations of NH_4^+ (8*c* position) and AlF_6^{3-} ions. Rapid reorientations of both ions freeze into an ordered state below the PT points [37].

Particular attention should be paid to disorder of the NH_4^+ group in the 4*b* position of ammonium cryolites. A neutron diffraction analysis of the disorder in seven-coordinated ammonium heptafluorozirconate $(NH_4)_3ZrF_7$ with elpasolite-like structure confirmed a dynamic nature of disorder in this crystal [40]. It is impossible to fit an ordered tetrahedral ion into a location with point symmetry *m*-3*m*. The observed hydrogen scattering density around the N atom in 4*b* consisted of six peaks along (100) axes. Hydrogen atoms H_{0yz} were placed in 96*j* positions of *Fm*-3*m* with occupancy factors of 16.7%. This picture is consistent with eight equivalent orientations of ammonium ions as it was clearly demonstrated for the $(NH_4)_3MoO_3F_3$ and $(NH_4)_3WO_3F_3$ cryolites [86]. Three H2 atoms of $N2H_4$ group form three hydrogen bonds under tetrahedral angles with the nearest fluorine atoms owing to the displacement of N2 from symmetry center, and the forth H3 atom lies on a threefold axis in the 32*f* position and does not form a hydrogen bond. Each vertex of the observed octahedron is the electron density superposition from four H2 atoms of eight equivalent orientations of the $N2H_4$ group (*Notations of different atoms from Udovenko and Laptash paper*).

It should be noted that substitution of fluorine atoms for oxygens having similar ionic radii brings significant specificity to the problem of orientational disorder. Pauling [66] was puzzled as to how the MO_3F_3 groups can be arranged in a manner completely concordant with the high symmetry of the larger unit. Recent additional refinement of several structures including the displacement of the central atom M into 32*f* position has elucidated this question and helped to determine a separated orientation of MX_6 [86]. A coordination polyhedron of a *fac*-configuration was obtained with three short distances corresponding naturally to M—O bonds and three others corresponding to M—F. It should be noted that one M—O distance is significantly longer than the other two (1.79 and 1.67 Å for $(NH_4)_3MoO_3F_3$, and 1.83 and 1.73—1.76 Å for $(NH_4)_3WO_3F_3$, respectively). The M atoms are shifted on 0.313 Å (Mo) and 0.194 Å (W) from the octahedral center toward the face occupied by O atoms. The true geometry of this polyhedron permits one to explain the observed vibrational spectra of the compounds examined and to eliminate the contradiction in interpretation of the vibrational spectra of $MO_3F_3^{n-}$ species reported in the literature.

The arrangement of the ligand atoms in the mixed 24*e* + 96*j* position leads to six equivalent orientations of MX_6 (two orientations along each axis of the unit cell). Taking into account the identification of O and F atoms, the number of MX_6 equivalent orientations rises to 48 (16 orientations along each axis). Considering eight orientations of the ammonium cation and 48 orientations of the octahedron, the total entropy change must be *R*ln384 under full ordering of structural units. In fact, it is significantly less: ca $R\ln 8$ for $(NH_4)_3WO_3F_3$ [22] and ca $R\ln 5$ for $(NH_4)_3MoO_3F_3$ [23]. Most likely, ammonium cations and octahedra do not move so independently.

The temperature dependence of the second moment of ^{19}F NMR spectra for $(NH_4)_3WO_3F_3$ [90] showed the transition from reorientation motions of octahedral anions at room temperature to a rigid lattice at lower temperature. The absence of the second moment saturation even at the lowest temperature of the experiment (135K, which is lower than the temperature of PTs) meant that reorientations were not "frozen" completely.

Dynamic nature of the disorder in oxyfluoroelpasolites makes it possible to distinguish between O and F atoms on local scale and to determine the real polyhedral geometry. Until now, it is assumed that in disordered structures of oxide-fluoride compounds it is impossible to differentiate oxygen and fluorine atoms by X-ray diffraction. Indeed, many transition metal oxide-fluorides incorporating inherently polar pseudo-octahedral MO_xF_{6-x} ($x = 1-3$) anions give rise to crystal structures, in which no oxide/fluoride ordering appears due to numerous local configurations as in the case of cubic *Fm*-3*m* $A_2{}^+B^+M^{VI}O_3F_3$ (A, B = alkali metal; M = Mo, W) family [91]. During dynamic reorientations a central atom can be shifted toward the edge, face, or apex of octahedron giving its disordering on cuboctahedron, cube, or octahedron, respectively (Fig. 16.4).

The Ti atoms in $(NH_4)_3TiOF_5$ and Rb_2KTiOF_5 are shifted from the 4*a* position to 24*e*, disordering on octahedron, thus allowing identification of the O and F atoms in the octahedron on a local scale. Interestingly, a disorder of $N1H_4$ groups or Rb cations has a displacive character in these compounds, while the dynamically disordered

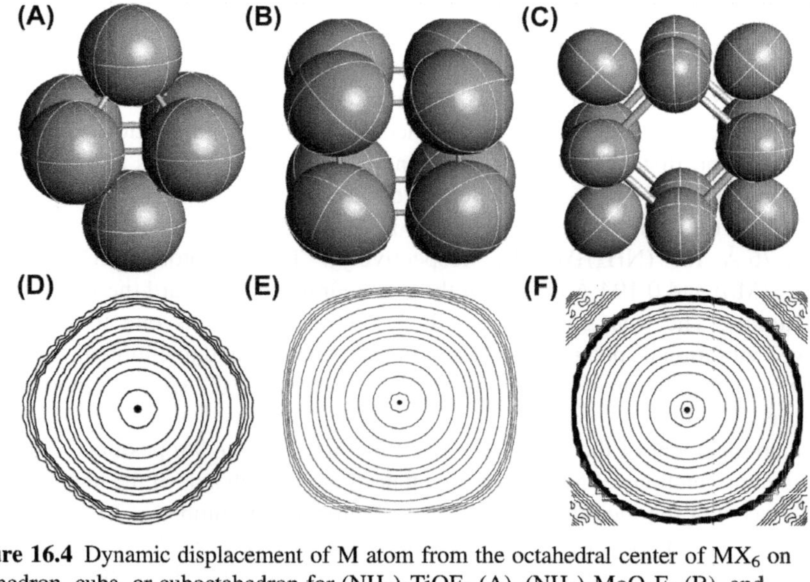

Figure 16.4 Dynamic displacement of M atom from the octahedral center of MX_6 on octahedron, cube, or cuboctahedron for $(NH_4)_3TiOF_5$ (A), $(NH_4)_3MoO_3F_3$ (B), and $(NH_4)_3WO_2F_5$ (C), respectively, and the corresponding electron density profiles (D–F).

MX_6 octahedra and $N2H_4$ tetrahedra contribute mainly to the entropy changes at PTs which are of the order–disorder type [87].

In seven-coordinated $(NH_4)_3NbOF_6$, the O atom occupies an axial position of pentagonal bipyramid and the Nb atom escapes the polyhedron center disordering on octahedron with 12 possible equivalent orientations that enables to determine the short Nb–O distance (1.751(1) Å) [85]. This oxide-fluoride is characterized by gigantic total entropy of four successive PTs $\Delta S \approx R\ln 136$ [27]. Disordering of a central atom on cuboctahedron in accordance with its displacement toward the octahedra edge was quite recently detected in tungsten double salt $(NH_4)_3WO_2F_5 = (NH_4)_2[WO_2F_4](NH_4F)$, which represents a unique example of PT from dynamic to static orientational disorder between two cubic modifications (Pm-$3m \rightarrow Pa3$) at 275K [88].

To discuss in detail entropy of PTs in ammonium fluoro- or oxyfluoroelpasolites (cryolites), it is necessary to determine their crystal structures of low-temperature phases. Until now, such structural determinations were unsuccessful because of ferroelastic twinning of crystals.

2.4 Critical and Noncritical Parameters of Phase Transitions

2.4.1 Models of Ordered and Disordered Structures

Deep understanding of nature and mechanism of PT is impossible without using group-theoretical methods [38,79] for detecting and clarification of irreducible representation [47,80] and OPs [36], the occurrence of which induce changes in the symmetry of parent phase. Apparently, due to complex mathematical theory it was not widely used by research. Fortunately, several programs and internet-based tools such as ISOTROPY [38], ISODISTORT [12], Bilbao Crystallographic Server [8] have appeared which greatly help with this purpose.

A distorted structure can be automatically decomposed by ISODISTORT into symmetry modes of parent structure with calculation of mode amplitudes, which can be useful to answer the question, in which atoms are significant/critical and responsible for the shifts or ordering [62]. Symmetry of distorted phases can be determined and classified before the experiment was carried out [3,5,6,34,35]. Especially this is topical for fluorocompounds because they usually have PT associated with ordering or shift of light atoms (fluorine, oxygen, etc.), and in this case there are small changes in the X-ray diffraction pattern. If so, then distorted phase can be fitted well by several structures with different space groups. For a second-order PT, a number of possible structures can be noticeably decreased and sometimes crystal structure of distorted phase can be determined as a sole variant. For example, low-temperature structure of $(NH_4)_2KWO_3F_3$ first was solved in $P2_1/n$ phase [54], however, second order of PT was not accounted. Further, critical OP was derived, which showed that PT Fm-$3m \leftrightarrow P2_1/n$ can be only of the first order. Correct reinvestigation showed that real crystal structure is associated with $Pmmn$ space group [55].

Final stage of PT investigation is the building a model of order–disorder and/or displacement of structural elements (ions, molecules, polyhedra). Special attention is paid to order–disorder processes in fluorides and oxide-fluorides because the model should meet several conditions: (1) all phases should consist of undistorted structural elements (rigid body), ie, disordering should not distort these elements; (2) a ratio between numbers of positions in disordered N_2 and ordered N_1 phases should be comparable with entropy changes $\Delta S = R\ln(N_2/N_1)$. However, sometimes it's difficult to fulfill these conditions and diffraction methods solely cannot solve this task. In this case the knowledge of critical OP is again beneficial. For example, the WO_3F_3-distorted octahedron in low-temperature phase of $K_3WO_3F_3$ retains disordered between several orientations forming polyhedron of the complex shape [56]. Using X-ray diffraction only it was impossible to say how many orientations N_2 remain in distorted phase, and the comparison of $\Delta S = R\ln(N_2/N_1)$ with experimental values was failed. In order to understand the mechanism of the PTs and the separation of critical atoms, the simulation of ordering process of the WO_3F_3 octahedron with symmetry C_{3v} was carried out. Since the W–O and W–F distances are inherently different, it is possible to imagine a vector directed along the threefold axis of the WO_3F_3 octahedron with coordinates (x, x, x) in the parent *Fm-3m* phase. Eight different orientations N_1 of this vector means eight different orientations of the WO_3F_3 octahedron in the cubic phase. Using the group-theoretical analysis and suggesting critical OP it was possible to simulate ordering of this vector and determine which orientations N_2 are retained below PT point. OPs leading to $\Delta S = R\ln(N_2/N_1)$ values not comparable with experimental entropies were omitted from the consideration. Finally, two PTs in $K_3WO_3F_3$ were successfully modeled, and experimental values ΔS at each PT agree well with theoretically calculated entropy. Thus, the method above helps to find proper critical parameters and build the correct model of order–disorder. It was also successfully used to resolve distorted phases of other related oxyfluorocompounds: $(NH_4)_2KWO_3F_3$ [55] and $(NH_4)_3WO_3F_3$ [57]. Building of order–disorder model using OP for all oxide-fluorides above gave in particular the understanding that the WO_3F_3 octahedron is similarly disordered in their parent phases; however, ordering sequences are different that can be associated with the NH_4/K ratio.

Modeling for $(NH_4)_3Nb(O_2)_2F_4$ in the above considered way revealed disordering of $Nb(O_2)_2F_4$ among 12 positions, which were associated with local C_{2v} symmetry of polyhedron [58]. So it was proved that F/O composition of polyhedron has a great influence on disordering process.

It should be noted that in some cases it is not necessary to solve crystal structure for estimation of critical OP values. For example, the study of $CsFe_2F_6$ [59] showed that magnitude of OP is connected with tilting cluster of five FeF_6 octahedra. At the same time such rotation leads to certain changes of cell parameters; therefore, OP can be easily linked with cell parameters and one value can be calculated from another.

Some examples of modeling ordered and disordered structures of oxide-fluorides are presented below, in which an entropic parameter, the temperature, and order of PTs were taking into account. As seen in Fig. 16.3, the total entropy change associated with two successive PTs *Fm-3m* ↔ *I4mm* ↔ *Cm* of ferroelectric and ferroelastic nature in cryolite $K_3WO_3F_3$ is significantly less than in ammonium cryolite $(NH_4)_3WO_3F_3$. A statistical model for the possible mechanism of such structural distortions was suggested using crystal structure data [56], which found critical OPs in the framework of unified approach [43].

It was found that for the formation of phase states the structure should consist, at least, of two subsystems, one of which is soft (ie, easily deformed) whereas the other is rigid, formed by K cations and WO_3F_3 anionic polyhedra, respectively. Upon cooling from *Fm-3m* phase, the first PT connected with the WO_3F_3 subsystem is accompanied by the appearance of small polarization because of anion charge ordering. Due to striction interaction the appearance of polarization should be accompanied by elongation of respective anion octahedra. Since neighboring WO_3F_3 and KO_3F_3 polyhedra have a common F/O vertex, the extension of the former will be accompanied by a contraction of the latter. This leads to the loss of the stability with respect to the ordering in the K cation subsystem, and the second PT occurs. It should be noted that the temperature T_1 of PT from cubic to tetragonal phase in two series of compounds $A_3WO_3F_3$ and $A_3MoO_3F_3$ (A = K, Rb, Cs) differs by ~60–80K; however, the temperature T_2 of PT from tetragonal to monoclinic phase differs only by 15–30K [69,70]. Thus, strong dependence of T_1 on the central atom M proves that first PT should be associated with the ordering of MO_3F_3 polyhedra.

Slight difference in T_2 values indicates that the second PT is due to the ordering of monoatomic cations in soft subsystems. All facts above are in accordance with suggested model and prove it. Moreover the proposed model allowed one to estimate the ratio between temperatures of two PTs in $K_3WO_3F_3$ $T_2/T_1 \sim 0.84$, which is in a good agreement with experimentally observed temperatures $T_2/T_1 = 414K/452K = 0.91$.

3. Nature of Phase Transitions

Depending on the chemical composition, fluoride and oxide-fluoride perovskite-like crystals undergo PTs of distinct ferroic nature [3,5,6,50], which were discussed in the series of reviews [2,71,73,75].

This presentation is aimed at the effects associated with the symmetry change of the crystal lattice, and therefore the attention will be paid only to ferroelastic and/or ferroelectric transformations connected with the spontaneous deformation and/or polarization. In accordance with the group-theoretical analysis, structural distortions at such transformations are connected mainly with the tilting of octahedral anionic groups or/and with the polar displacements of atoms [3,20].

A lot of fluorides and oxide-fluorides were found to undergo ferroelastic PTs which sometimes are accompanied by simultaneous appearance of polarization. Such phase transformations are known as multiferroic and are of considerable interest because of a possibility to control physical properties of material by two distinct or simultaneous external fields of different origin—electric field and pressure.

3.1 Phase Transitions in Fluorides

Fluorides $AM^{2+}F_3$ and $A_2A'M^{3+}F_6$ with the initial perovskite *Pm-3m* and elpasolite *Fm-3m* structure, respectively, are known to undergo PTs of only ferroelastic nature but at the same time can show a great variety of successions of transformations which can be controlled by the chemical and external pressure [3,5,6,20,50]. Structural deformations in these crystals are associated with one of the variants of octahedra tilting contrary to $AM^{4+}O_3$ and $A_2A'M^{6+}O_6$ oxides in which the additional shifts of A and O atoms give rise to macroscopic polarization [44,78].

Nonetheless, representatives of other fluoride families with octahedral anionic groups show ferroelectric properties at ambient conditions with rather high temperature of PTs into paraelectric phase or sometimes remain polarized up to melting point [71]. The point symmetry of the ferroelectric phases in different families is *mm*2, 4*mm*, or 4. The polarization origin is associated with the displacive or mixed, displacive and order–disorder, mechanism. The former takes place in the series of compounds with the general chemical formulations $Pb_5M^{3+}{}_3F_{19}$, $Na_2M^{2+}AlF_7$, and $SrM^{3+}F_5$. Fluorides $BaM^{2+}F_4$ (M = Mg, Mn, Fe, etc.) and $K_{3-x}Fe_5F_{15}$ were characterized as undergoing the latter mechanism [71]. Very rich as well as interesting succession of structural transformations was revealed upon heating in $Pb_5Al_3F_{19}$ crystal: ferroelectric (4*mm*) → antiferroelectric (4/*m*) → ferroelastic (2/*m*) → paraelastic (4/*m*) → prototype paraelectric (4/*mmm*). The ferroelectricity in compounds $Pb_5M_3F_{19}$ is connected with the displacement of Me^{3+} cation from the center of its octahedral site. But in $BaM^{2+}F_4$ fluorides, the appearance of spontaneous polarization is accompanied by simultaneous displacement of the Ba^{2+} cation and the tilting of octahedra. Complex mechanism in $K_{3-x}Fe_5F_{15}$ is associated with the shifts of Fe^{2+} and Fe^{3+} cations, which are disordered and ordered in para- and ferroelectric phase, respectively.

The most important point is that the polarization in some fluorides reaches the values comparable with that in oxide perovskites ($BaTiO_3$; $P_s \approx 30$ mC/cm^2), for example, at room temperature $P_s = 10$ mC/cm^2 in $BaM^{2+}F_4$ [71] and $P_s = 46$ mC/cm^2 in $Pb_5Cr_3F_{19}$ [1]. In the latter case, the estimation of P_s was performed without taking into account the opposite signs of polar displacements of two nonequivalent Cr^{3+} cations; thus, real spontaneous polarization will be of course considerably less.

3.2 Phase Transitions in Oxide-Fluorides

Contrary to fluorides and oxides, anionic polyhedra in oxyfluorometallates $A^+{}_2(A')^+MO_xF_{6-x}$ ($x = 1, 2, 3$) and $A^+{}_2MO_xF_{6-x}$ ($x = 1, 2$) are inhomogeneous in ligand composition and characterized by the low local symmetry (trigonal, tetragonal, orthorhombic). As a result, the MO_xF_{6-x} polyhedron in many cases has the inherent dipole moment associated with the displacements of M ion toward oxygens. On the other hand, in spite of these facts, oxide-fluorides form centrosymmetric structures including cubic phases, for example, in elpasolites–cryolites $A^+{}_2(A')^+MO_xF_{6-x}$ [20,50,71].

The origin of the high-lattice symmetry of oxide-fluorides can be associated generally with two reasons. First, all the vertices of the six-coordinated polyhedron are occupied with equal probability by fluorine and oxygen atoms that can be considered as their statistical disorder. In such a model, ligands can be also disordered by dynamic or static way between several equal crystallographic sites as it was discussed in the previous section of this chapter. Second, one more explanation of high symmetry was suggested in

Ref. [91], where initially ordered anionic polyhedra placed alongside each other were supposed to have relative orientations leading to canceling the dipole moments of separate octahedra on the local unit cell as well as macroscopic scales.

Both types of F/O ligands or anionic polyhedra disorder considered above are the reasons why oxide-fluorides keep rather often nonferroelectric structure even in distorted phases realized as the result of PTs.

However, ferroelectricity was believed to exist in cryolites and elpasolites $A^{+}{}_2(A')^{+}MO_xF_{6-x}$ ($x = 1, 2, 3$) with spherical monovalent cations and a central atom of a different valence. The existence of spontaneous polarization at room temperature was proved for only $A_2A'MO_3F_3$ compounds with the following cationic compositions A_2A': Rb_2Rb, K_2K, Cs_2Cs, Rb_2K, Cs_2Rb and central atoms M: W, Mo [71]. The optical observations showed that these crystals are simultaneously ferroelastics and both ferrophases transform upon heating into paraelectric–paraelastic phase through the succession of one or two PTs.

A cation and/or central atom substitution significantly effects on both the behavior and values of permittivity and tgδ in ammonium containing oxide-fluorides (Fig. 16.5). In accordance with small stepwise ε change at PT point, elpasolites–cryolites presented undergo ferroelastic transformations, except $(NH_4)_3MoO_3F_3$ and $(NH_4)_3TiOF_5$. For these two latter crystals, rather sharp and strong increase of ε at the first-order PTs from *Fm*-3*m* phase was observed. One could suppose that both transformations are of the ferroelectric nature and the ε peak in the values of 60–200 is associated with quasiceramic state of the samples prepared without heat treatment.

Indeed, the $Na_3MoO_3F_3$ ceramics sintered at high temperature showed very high permittivity ($\varepsilon \sim 10^6$) at *Fm*-3*m* → $P2_1$ PT [13]. On the other hand, ferroelectric transformation in the $K_3MoO_3F_3$ single crystal was not accompanied by peak anomaly of ε, only drastic stepwise increase on $\Delta\varepsilon \approx 200$ was observed [92]. In spite of such a

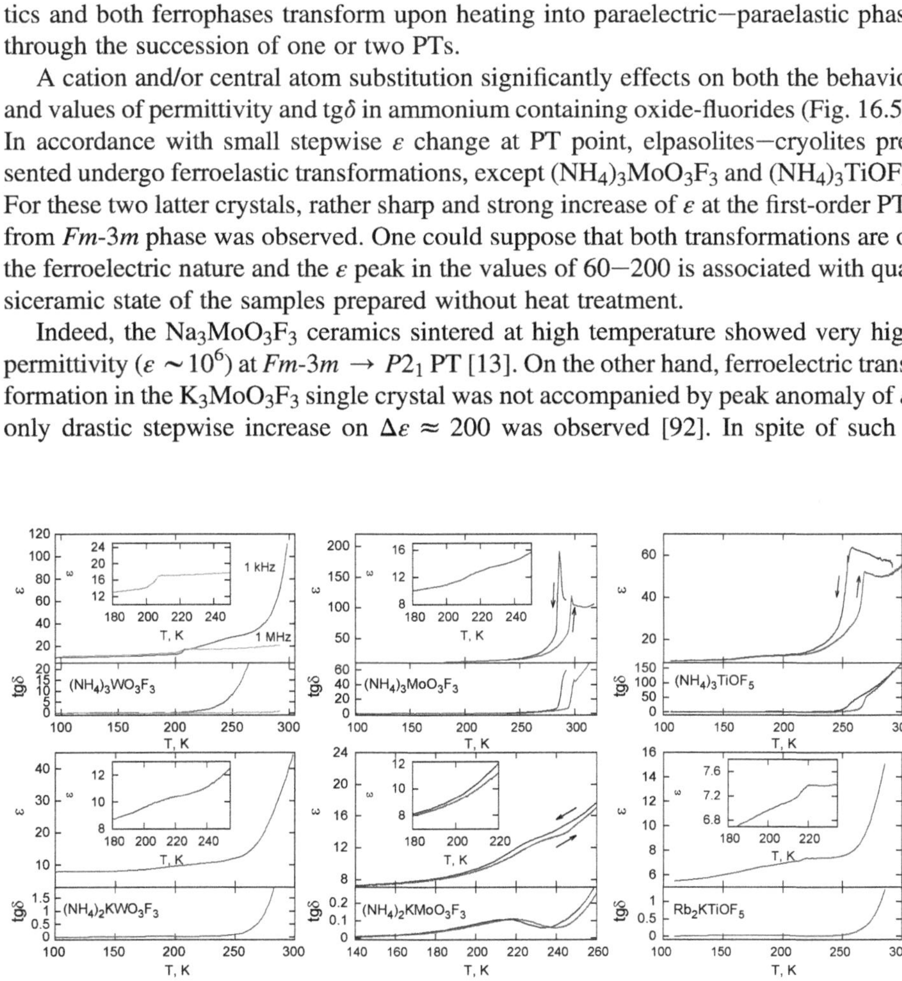

Figure 16.5 Temperature dependences of permittivity and dielectric losses in some oxide-fluorides.

Table 16.1 Spontaneous Polarization in Some Oxide-Fluorides at Room Temperature. $\Delta T = T_{PT} - RT$ Is a Difference Between Phase Transition Point (T_{PT}) and Room Temperature (RT)

Compound	P_s, μC/cm^2	$\Delta T = T_{PT} - RT$, K	Method	References
$Na_3MoO_3F_3$	0.02	530	Exp	[10]
	1		Calc	
$K_3MoO_3F_3$	30	250	Exp	[92]
$Rb_2KMoO_3F_3$	0.1	50	Calc	[70]
$(NH)_3MoO_3F_3$	3	5	Exp	[23]
$(NH_4)_2KVO_2F_4$	0.6	45	Exp	[65]

Exp, experimental measurements; *Calc*, estimation from ionic displacements.

behavior of permittivity, the ferroelectric nature of the $K_3MoO_3F_3$ crystal was confirmed by the dielectric hysteresis loop observation below PT from a cubic phase.

The values of spontaneous polarization determined in some oxide-fluorides are presented in Table 16.1.

One can see that the P_s values are widely ranged even for compounds with the same composition of anionic polyhedra. Comparing these data, first of all, it is necessary to take into account the difference $\Delta T = T_{PT} - RT$ between PT point and room temperature where P_s was determined. Very small spontaneous polarization in $Na_3MoO_3F_3$ was assumed to be associated with not sufficiently high density of the samples [10]. Patwe et al. [65] have supposed that the structure of $(NH_4)_2KVO_2F_4$ contains VO_2F_4 polyhedra of *cis*- as well as *trans*-configuration, and only the former having unbalanced dipole moment gives a contribution to the polar properties. Such an explanation is not conclusive because *trans*-configuration is less stable from the energetic point of view.

Ferroelectricity was found also in some other families of oxide-fluorides [71]. For example, the dielectric hysteresis loops were observed in single-layer compounds, $Bi_2TiO_4F_2$ and Bi_2MO_5F (M = Nb, Ta), showing ferroelectric–paraelectric transition in the narrow range of rather low temperatures: 280–300K.

Ferroelectric properties at room temperature were announced in NH_4TiOF_3 with the layered crystal lattice [51] but information about the existence of PT between ferro- and paraelectric phases in this compound was not presented.

4. Sensitivity to Pressure

PTs in ferroics are accompanied by anomalous behavior of many properties among which caloric effects (CEs) are of a special interest. Solids showing pronounced CEs connected with the reversible change of entropy ΔS_{CE} or temperature ΔT_{AD} under external field (magnetic H, electric E, mechanical stress σ, or hydrostatic pressure p)

variation at constant temperature and entropy, respectively, are considered as potential solid-state refrigerants in effective alternative cooling technologies [26]. The relation between extensive ΔS_{CE} or intensive ΔT_{AD} CE and the generalized ferroic OP Y (magnetization, polarization, elastic deformation) as well as conjugated external field X is described by following equations [26].

$$\Delta S_{CE} = \int \left(\frac{\partial Y}{\partial T}\right)_X dX; \quad \Delta T_{AD} = -\int \frac{T}{C_X} \left(\frac{\partial Y}{\partial T}\right)_X dX \qquad [16.2]$$

It is obvious that CEs reach the maximum values near PT point where the $(\partial Y/\partial T)_X$ derivative is the largest. Both CEs can be represented by another way. Because the total entropy S of a system is a thermodynamic state function, the extensive CE is the difference of two terms

$$\Delta S_{CE}(T;X) = S(T;X) - S(T;0), \qquad [16.3]$$

and the adiabatic change of external field leads to ΔT_{AD} as the total entropy must be conserved $S(T + \Delta T_{AD}; X) = S(T; 0)$. Thus, in order to show significant CE, materials should undergo PTs accompanied by large values of entropy change as well as susceptibility to external field dT/dX.

Currently magneto(MCE)-, electro(ECE) and baro(BCE)-caloric effects are the subjects of many investigations [26]; however, the advantage of the latter is a versatility associated with a possibility to realize it in materials of distinct physical nature. There are no examples of direct measurements of ECE and MCE in fluorides and oxide-fluorides. The reasons are those that these materials were prepared usually as ceramics with low dielectric strength and/or undergo antiferromagnetic transformations at rather low temperatures with small entropy change [50,68].

As evidenced by the foregoing, many fluorides and oxide-fluorides with octahedral or quasi-octahedral anionic groups are characterized by a high degree of disorder in the initial phase and, as a result, by very large values of the entropy of ferroelectric and/or ferroelastic PTs. And what about the effect of hydrostatic pressure on PTs in these compounds?

4.1 Bond Stress: Temperature–Pressure Phase Diagrams

The crystal phase stability is characterized by susceptibility of material to chemical (due to a variation in the composition) and external pressures both of which lead to the change in the interatomic distances giving rise to the change of relation between repulsive and attractive forces. The analysis of the former effect in fluorides and oxide-fluorides $A_2A'MO_xF_{6-x}$ [20,24] has shown that the temperature T_1 of transitions from *Fm*-3*m* phase, and baric coefficient dT_1/dp can be significantly influenced by varying anionic group and/or the set of cations (Table 16.2).

In individual cases, especially when 8*c* position is occupied by the large A = Cs atom, there is a possibility to stabilize the cubic symmetry *Fm*-3*m* even down to liquid helium temperatures. Such an effect can be explained considering bond

Table 16.2 Thermodynamic Parameters of the Transitions From the *Fm*-3*m* Phase of $A_2A'MO_xF_{6-x}$ Compounds. MO_xF_{6-x} Group: (1) GaF_6 and (2) WO_3F_3

A_2A'	MO_xF_{6-x}	a_0, Å	T_1, K	$\Delta S_1/R$	dT_1/dp, K/GPa	References
Cs_2Rb	1	9.129	–	–	–	[24]
	2	9.305	348			
Rb_2K	1	8.807	124	ln6	120	[30]
	2	8.917	290			[70]
Cs_2NH_4	1	9.167	161	ln8	53	[24]
	2	9.298	–	–	–	
$(NH_4)_3$	1	9.038	249	ln16	−12	[24]
	2	9.153	200	ln8	−2	
$(NH_4)_2K$	1	8.866	287	ln6	−19	[24]
	2	8.958	235	ln1.8	−11	

stress μ_B characterized by the relation between the ionic radii R of A′, M, and F/O atoms and the lattice parameter a_0 [20]

$$\mu_B = 1 - a_0 \Big/ \left[2\left(R_{A'} + 2R_{F/O} + R_M\right)\right] \qquad [16.4]$$

Because the sum of the ionic radii of atoms laying on the edge of the cubic cell of perovskite-like structure is larger than a_0, instantaneous position of F/O atoms is not on the edge. On the one hand, the change of bond stress at cation/anion substitution can be considered as a result of uniaxial stress change. On the other hand, the decrease in μ_B is equivalent to a decrease in the repulsive energy contribution to the potential energy of the crystal and leads, first, to a decrease in the vibration anisotropy of the F/O atoms and, second, to increase in the initial phase stability. The former statement can be proved by the electron density distribution in ligands of compounds with the WO_3F_3 anionic species (Fig. 16.6). The latter is evident from Fig. 16.7 showing almost

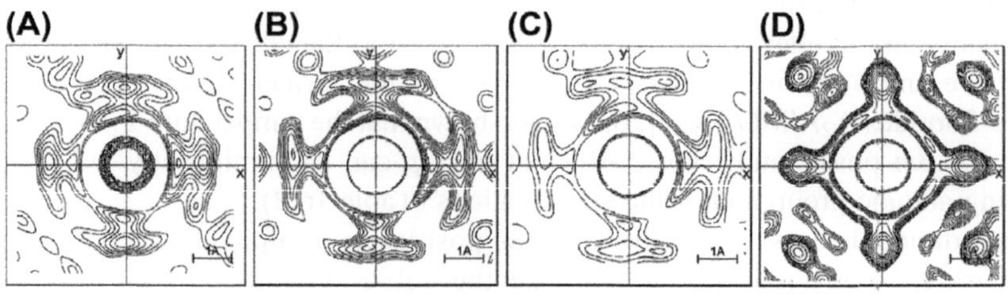

Figure 16.6 Electron density maps (0.4 e/Å^3 step) calculated for cubic phase of oxide-fluorides $K_3WO_3F_3$ (A), $(NH_4)_2KWO_3F_3$ (B), $(NH_4)_3WO_3F_3$ (C), $Cs_2NH_4WO_3F_3$ (D).

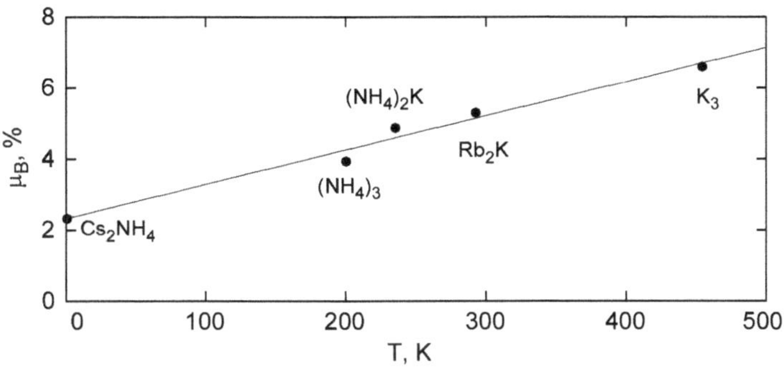

Figure 16.7 Dependence of the phase transition temperature on bond stress μ_B determined with Eq. [16.4].

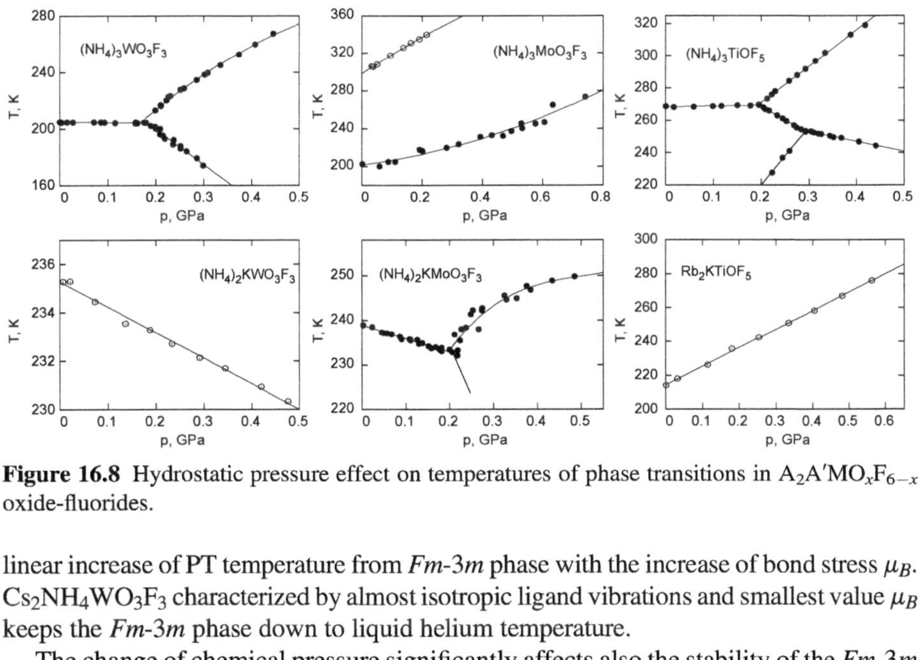

Figure 16.8 Hydrostatic pressure effect on temperatures of phase transitions in $A_2A'MO_xF_{6-x}$ oxide-fluorides.

linear increase of PT temperature from *Fm*-3*m* phase with the increase of bond stress μ_B. $Cs_2NH_4WO_3F_3$ characterized by almost isotropic ligand vibrations and smallest value μ_B keeps the *Fm*-3*m* phase down to liquid helium temperature.

The change of chemical pressure significantly affects also the stability of the *Fm*-3*m* phase of fluorides and oxide-fluorides against the external hydrostatic pressure: crystals are characterized by wide range of the baric coefficients dT_1/dp (Table 16.2).

Fig. 16.8 shows the *T*–*p* phase diagrams for the several pairs of oxide-fluorides designed by the replacement of cations in 4*a*, 4*b*, and 8*c* positions.

It is interesting that $(NH_4)_3WO_3F_3$ and $(NH_4)_3TiOF_5$ in spite of different anionic polyhedra show almost similar *T*(*p*) behavior except the second triple point found for oxyfluorotitanate at about 0.3 GPa. On the other hand, a substitution Mo → W in ammonium cryolites and Rb_2K → $(NH_4)_3$ in oxyfluorotitanates leads to gigantic increase of baric coefficient dT_1/dp: from −3K/GPa (W) to + 240K/GPa (Mo) and from 6K/GPa ($(NH_4)_3$) to 110K/GPa (Rb_2K) which can be due to different nature as well as mechanism of PTs in both groups of compounds as it was shown in Fig. 16.8.

It was previously shown that there is a significant difference between the largest values of the PT entropy at ambient pressure in fluorides ($\Delta S = R\ln16$) and oxide-fluorides ($\Delta S = R\ln8$). However, the succession of transformations in $(NH_4)_3TiOF_5$ cryolite including both triple points is characterized by much larger entropy ($R\ln16$–$R\ln24$) which indirectly confirms a possibility of the higher degree of disorder of the *Fm*-3*m* phase in oxide-fluorides [22].

4.2 Barocaloric Effect: Fluorides and Oxide-Fluorides as Promising Solid-State Refrigerants

Comparative analysis of the *T*–*p* phase diagrams and the results of the heat capacity measurements allow one to evaluate barocaloric efficiency of materials. Maximum value of extensive BCE is equal to PT entropy $\Delta S_{\mathrm{BCE}}^{\max} = \Delta S$ [24,26]. The highest value of intensive BCE determined as

$$\Delta T_{\mathrm{AD}}^{\max} = (\Delta S \cdot T)/C_L, \qquad [16.5]$$

where C_L is the lattice specific heat at temperature *T*, can be reached at the pressure,

$$p \geq (\Delta S \cdot T)/(C_{\mathrm{L}} \cdot dT/dp). \qquad [16.6]$$

Because of large entropy change at single PT ($R\ln16$), fluorides $(NH_4)_3MF_6$ (M = Ga, Fe, V) seem to be rather promising for implementing large BCE. However, small value of baric coefficient of $(NH_4)_3GaF_6$ (Table 16.2) is the reason of fairly high pressure (~1.6 GPa) corresponding to $\Delta T_{\mathrm{AD}}^{\max} \approx 19$K. On the other hand, only 0.26 GPa needs in order to realize $\Delta T_{\mathrm{AD}}^{\max} \approx 14$K in $Cs_2NH_4GaF_6$ showing lower ΔS and higher dT/dp compared to cryolite (Table 16.2).

Taking into account a high importance of the baric coefficient in caloric efficiency, $(NH_4)_3MoO_3F_3$, Rb_2KTiOF_5 and Rb_2KMF_6 (M = Ga, Fe) could be supposed to show significant BCE [24,26]. Fig. 16.9 shows the temperature and pressure

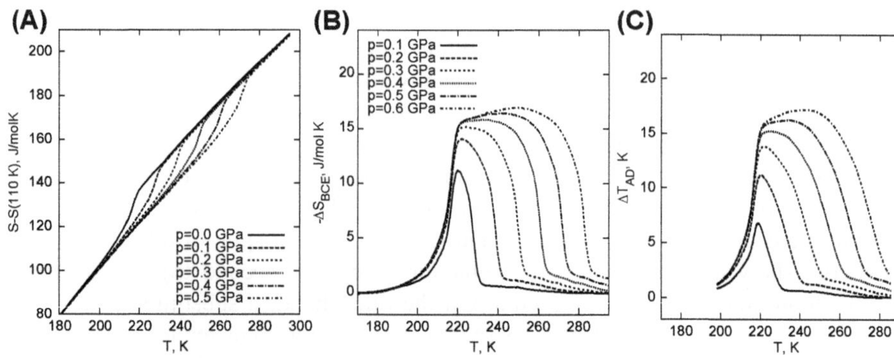

Figure 16.9 (A) Temperature dependence of the total entropy of Rb_2KTiOF_5 for different pressures. (B) Extensive (ΔS_{BCE}) and (C) intensive (ΔT_{AD}) BCE in Rb_2KTiOF_5.

Table 16.3 Caloric Characteristics of Some Ferroelastics, Ferromagnets, and Ferroelectrics on External Fields. RCP(*S*) and RCP(*T*) are Determined at 0.5 GPa [24,26]

Compound	T_{PT}, K	$-\Delta S_{CE}$, J/kg·K	ΔT_{AD}, K	Δp, GPa	ΔH, kOe	ΔE, kV/cm	RCP(*S*), J/kg	RCP(*T*), K^2
Rb_2KFeF_6	190	42	18	0.14			2760	1040
Rb_2KGaF_6	120	38	11	0.10				
Rb_2KTiOF_5	220	46	17	0.15			2180	760
$(NH_4)_3MoO_3F_3$	297	55	15	0.07			5200	1260
$Gd_5Si_2Ge_2$	270	30	21		100		610	630
FeRh	313	55	33		50		900	330
Gd	296	16	18		100		1600	1800
$PbZr_{0.95}Ti_{0.05}O_3$ (thin film 300 nm)	500	8	12			480	1080	1600

dependences of the total entropy $S(T, p)$ of Rb_2KTiOF_5 as well as extensive $\Delta S_{BCE}(T, p)$ and intensive $\Delta T_{AD}(T)$ CE determined using Eqs. [16.2] and [16.3].

It is seen that both values reach rather quickly the maximum with pressure increase and show the plateau-like behavior.

The data on BCE in some fluorides and oxide-fluorides are presented in Table 16.3 with MCE and ECE in some ferromagnets and ferroelectrics, which are considered as efficient working media for refrigeration cycles.

The values of the extensive effect in oxide-fluorides are close to the ΔS_{CE} value for very expensive material FeRh and exceed parameters of other presented compounds. The same situation also takes place for the intensive CE. However, it is necessary to take in mind that large values of ΔT_{AD} can be realized in fluorides at rather low pressure, and significant intensive MCE and ECE need high magnetic and electric fields.

The advantage of caloric efficiency of fluorides and oxide-fluorides is illustrated by the consideration of the integral characteristics known as relative cooling powers (RCP*s*) [24].

$$\mathrm{RCP}(S) = -\Delta S_{\mathrm{CE}}^{\max} \times \delta T_{\mathrm{FWHM}}, \qquad [16.7]$$

$$\mathrm{RCP}(T) = \Delta T_{\mathrm{AD}}^{\max} \times \delta T_{\mathrm{FWHM}}, \qquad [16.8]$$

where δT_{FWHM} is the full width at half maximum for $\Delta S_{CE}^{max}(T)$ and $\Delta T_{AD}^{max}(T)$ functions.

As it is seen from Table 16.3, FeRh is characterized by the smallest values of RCP(*S*) and RCP(*T*) in spite of the largest absolute values of $\Delta S_{CE}(T)$ and $\Delta T_{AD}(T)$. The RCP(*S*) parameter of fluorides and oxide-fluorides determined at $p = 0.5$ GPa exceeds that of ferromagnets measured even at $H = 100$ kOe. The anomalously large value of RCP(*S*) for $(NH_4)_3MoO_3F_3$ is due to the very large value of dT/dp (Table 16.2). The entropic and temperature parameters discussed allow ones to suppose that fluorides and oxide-fluorides can be considered as promising solid-state refrigerants.

5. Conclusions

The analysis of the comprehensive studies of fluorides and oxide-fluorides with the perovskite-related structure has allowed one to postulate the following:

1. The mechanism of structural distortions at PTs strongly depends on the composition of materials and degree of order–disorder of structural units in the initial phase. A wide variety of ligand disordering exists in accordance with the structural models considered.
2. A good correlation parameter of anharmonicity of critical ion vibrations with entropic parameter proves the existence of transformations with the "intermediate" between limited displacive and order–disorder types.
3. Chemical and hydrostatic pressures effect significantly on the order, type, and succession of structural transformations.

4. Very often ferroelastic and ferroelectric states are realized at the same temperature leading to formation of multiferroic phase.
5. Fluorides and oxide-fluorides undergoing order—disorder transformations accompanied by high susceptibility to hydrostatic pressure show high barocaloric parameters and can be considered as promising solid-state refrigerants.

Acknowledgments

The reported study was partially supported by RFBR, research project No. 15-02-02009 a.

References

[1] S.C. Abrahams, J. Albertsson, C. Svensson, J. Ravez, Structure of $Pb_5Cr_3F_{19}$ at 295 K, polarization reversal and the 595 K phase transition, Acta Cryst. B46 (1990) 497—502.
[2] S.C. Abrahams, J. Ravez, Dielectric and related properties of fluorine-octahedra ferroelectrics, Ferroelectrics 135 (1992) 21—37.
[3] K.S. Aleksandrov, A.T. Anistratov, B.V. Beznosikov, N.V. Fedoseeva, Phase Transitions of Halide ABX_3 Compounds, Nauka, Novosibirsk, 1981.
[4] K.S. Aleksandrov, S.V. Misyul, Phase transitions associated with rotational distortions of the structure in crystals related to perovskites, Kristallografiya 26 (1981) 1074—1085.
[5] K.S. Aleksandrov, B.V. Beznosikov, Perovskite-Like Crystals, Nauka, Novosibirsk, 1997.
[6] K.S. Aleksandrov, B.V. Beznosikov, Perovskites. Present and Future, SB RAS, Novosibirsk, 2004.
[7] E.V. Antipov, A.M. Abakumov, Synthesis, structure and superconducting/magnetic properties of Cu- and Mn-based oxyfluorides, Chapter 13, in: A. Tressaud (Ed.), Functionalized Inorganic Fluorides, Wiley-Blackwell, Chichester, 2010, pp. 383—422.
[8] M.I. Aroyo, J.M. Perez-Mato, D. Orobengoa, E. Tasci, G. de la Flor, A. Kirov, Crystallography online: Bilbao crystallographic server, Bulg. Chem. Commun. 43 (2011) 183—197.
[9] B.V. Beznosikov, S.V. Misyul, On crystal chemistry of halide compounds of the type $A_2B^+B^{3+}X_6$, Kristallografiya 23 (1978) 622—625.
[10] F.J. Brink, L. Norren, D.J. Goossens, R.L. Withers, Y. Liu, Ch.-N. Xu, A combined diffraction (XRD, electro and neutron) and electrical study of $Na_3MoO_3F_3$, J. Solid State Chem. 174 (2003) 450—458.
[11] A.D. Bruce, R.A. Cowley, Structural Phase Transitions, Taylor and Francis, Philadelphia, 1981.
[12] B.J. Campbell, H.T. Stokes, D.E. Tanner, D.M. Hatch, ISODISPLACE: an internet tool for exploring structural distortions, J. Appl. Cryst. 39 (2006) 607—614.
[13] J.-P. Chaminade, M. Cerverra-Marzal, J. Ravez, P. Hagenüller, Ferroelectric and ferroelastic behavior of the oxyfluoride $Na_3MoO_3F_3$, Mater. Res. Bull. 21 (1986) 1209—1214.
[14] B.D. Chapman, E.A. Stern, S.-W. Han, J.O. Cross, G.T. Seidler, V. Gavrilyatchenko, R.V. Vedrinskii, V.L. Kraizman, Diffuse x-ray scattering in perovskite ferroelectrics, Phys. Rev. B71 (2005) 020102.
[15] M. Couzi, S. Khairoun, A. Tressaud, Structural phase transitions in $Rb_2KM^{III}F_6$ Elpasolites. II. Raman scattering and group-theoretical studies of Rb_2KFeF_6 and Rb_2KYF_6, Phys. State Solidi A 98 (1986) 423—434.

[16] H. Faget, J. Grannec, A. Tressaud, V. Rodriguez, T. Roisnel, I.N. Flerov, M.V. Gorev, Neutron powder refinements of the three allotropic varieties of Rb_2KScF_6, Eur. J. Solid State Inorg. Chem. 33 (1996) 893–905.

[17] I. Flerov, M. Gorev, S. Melnikova, S. Misyul, V. Voronov, K. Aleksandrov, Phase-transitions in elpasolite Rb_2KScF_6, Fizika Tverdogo Tela 34 (1992) 2185–2195.

[18] I.N. Flerov, M.V. Gorev, Thermodynamic properties and phase transitions in halide perovskites-like ferroelastics. Izvestiya RAN, Seriya fizicheskaya 59 (9) (1995) 48–53.

[19] I. Flerov, M. Gorev, V. Voronov, Entropy of $Rb_2KB^{3+}F_6$ (B^{3+}: Se, In, Lu, Er, Ho) ferroelastics and the triple point on $T(R_B{}^{3+})$ diagram, Fizika Tverdogo Tela 38 (1996) 717–727.

[20] I. Flerov, M. Gorev, K. Aleksandrov, A. Tressaud, J. Grannec, M. Couzi, Phase transitions in elpasolites (ordered perovskites), Mater. Sci. Eng. R Rep. 24 (1998) 81–151.

[21] I.N. Flerov, M.V. Gorev, Entropy and the mechanism of phase transitions in elpasolites, Phys. Solid State 43 (2001) 127–136.

[22] I.N. Flerov, M.V. Gorev, V.D. Fokina, A.F. Bovina, N.M. Laptash, Calorimetric and X-ray investigations of perovskite-like oxyfluorides $(NH_4)_3WO_3F_3$ and $(NH_4)_3TiOF_5$, Phys. Solid State 46 (2004) 888–894.

[23] I.N. Flerov, V.D. Fokina, A.F. Bovina, E.V. Bogdanov, M.S. Molokeev, A.G. Kocharova, E.I. Pogorel'tsev, N.M. Laptash, Mechanism and nature of phase transitions in the $(NH_4)_3MoO_3F_3$ oxyfluoride, Phys. Solid State 50 (2008) 515–524.

[24] I.N. Flerov, M.V. Gorev, A. Tressaud, N.M. Laptash, Perovskite like fluorides and oxyfluorides: phase transitions and caloric effects, Crystallogr. Rep. 56 (2011a) 9–17.

[25] I.N. Flerov, I.N. Gorev, V.D. Fokina, A.F. Bovina, E.V. Bogdanov, E.I. Pogoreltsev, N.M. Laptash, Disorder and phase transitions in oxyfluoride $(NH_4)_3Ta(O_2)_2F_4$, J. Fluorine Chem. 132 (2011b) 713–718.

[26] I.N. Flerov, E.A. Mikhaleva, M.V. Gorev, A.V. Kartashev, Caloric and multicaloric effects in oxygen ferroics and multiferroics, Phys. Solid State 57 (2015) 429–441.

[27] V.D. Fokina, I.N. Flerov, M.V. Gorev, E.V. Bogdanov, A.F. Bovina, N.M. Laptash, Thermophysical studies of the phase transitions in $(NH_4)_3NbOF_6$ crystals, Phys. Solid State 49 (2007) 1548–1553.

[28] V.D. Fokina, M.V. Gorev, E.V. Bogdanov, E.I. Pogoreltsev, I.N. Flerov, N.M. Laptash, Thermal properties and phase transitions in $(NH_4)_3ZrF_6$, J. Fluor. Chem. 154 (2013) 1–6.

[29] R. Gautier, K.B. Chang, K.R. Poeppelmeier, On the origin of the differences in structure directing properties of polar metal oxyfluoride $[MO_xF_{6-x}]^{2-}$ ($x = 1$, 2) building units, Inorg. Chem. 54 (2015) 1712–1719.

[30] M.V. Gorev, I.N. Flerov, A. Tressaud, J. Grannec, Thermodynamic properties of the mixed elpasolites $Rb_2KGa_xSc_{1-x}F_6$ (x = 0.6–1.0), Phys. Solid State 39 (1997) 1647–1651.

[31] M.V. Gorev, I.N. Flerov, A. Tressaud, E. Durand, The p–T phase diagram of ammonium hexafluoroaluminate, J. Phys. Condens. Matter 14 (2002) 6447–6453.

[32] J. Granzin, H. Saalfeld, Crystal structure refinement of K_3HfF_7 and the disorder of the heptafluorohafnate complex, Zeitschrift fuer Kristallographie 183 (1988) 71–76.

[33] R. Haegele, W. Verscharen, D. Babel, Einkristall-Strukturdaten einiger fluoride und cyanide $A_2{}^IB^IM^{III}X_6$ der Elpasolithfamilie/single crystal structural data of some fluorides and cyanides $A_2{}^IB^IM^{III}X_6$ of the Elpasolite family, Z. Naturforsch. B 30 (1975) 462–464.

[34] D.M. Hatch, H.T. Stokes, Classification of octahedral tilting phases in the perovskitelike A_2BX_4 structure, Phys. Rev. B 35 (1987) 8509–8516.

[35] D.M. Hatch, H.T. Stokes, K.S. Aleksandrov, S.V. Misyul, Phase transitions in the perovskitelike A_2BX_4 structure, Phys. Rev. B 39 (1989) 9282–9288.

[36] D.M. Hatch, H.T. Stokes, The complete listing of order parameters for a crystalline phase transition: a solution to the generalized landau problem, Phys. Rev. B 65 (2002), 014113−1−4.
[37] K. Hirokawa, Y. Furukawa, Phase transitions and ionic motions in ammonium potassium hexafluoroaluminates with compositions on $(NH_4)_{3-x}K_xAlF_6$ ($x = 1, 2$) as studied by ^1H, ^{19}F, and ^{27}Al magnetic resonance, J. Phys. Chem. Solids 49 (1988) 1047−1056.
[38] C.J. Howard, H.T. Stokes, Structures and phase transitions in perovskites - a group-theoretical approach, Acta Cryst. A 61 (2005) 93−111.
[39] J.-L. Huang, X.-Y. Kuang, Y. Li, Investigation of optical spectra and local structure for $(CrF_6)^{3-}$ cluster in $Cs_2NaAlF_6:Cr^{3+}$ and $Cs_2NaGaF_6:Cr^{3+}$ systems, Chem. Phys. Lett. 458 (2008) 227−230.
[40] J.H. Hurst, J.C. Taylor, A neutron diffraction analysis of the disorder in ammonium heptafluorozirconate, Acta Cryst. B26 (1970) 2136−2137.
[41] J. Hutton, R.J. Nelmes, High-resolution studies of cubic perovskites by elastic neutron diffraction. II. $SrTiO_3$, $KMnF_3$, $RbCaF_3$ and $CsPbCl_3$, J. Phys. C Solid State Phys. 14 (1981) 1713−1736.
[42] J. Hutton, R.J. Nelmes, G.M. Meyer, V.R. Eiriksson, High-resolution studies of cubic perovskites by elastic neutron diffraction: $CsPbCl_3$, J. Phys. C Solid State Phys. 12 (1979) 5393−5410.
[43] M.P. Ivliev, S.V. Misyul, M.S. Molokeev, V.P. Sakhnenko, Pseudo-proper ferroelectric phase transitions in oxyfluoride $K_3WO_3F_3$, Phase Transit. 87 (2014) 592−602.
[44] N. Izyumskaya, Ya Alivov, H. Morkoç, Oxides, oxides, and more oxides: high-κ oxides, ferroelectrics, ferromagnetics, and multiferroics, Crit. Rev. Solid State Mater. Sci. 34 (2009) 89−179.
[45] K. Knox, D.W. Mitchell, The preparation and structure of K_2NaCrF_6, K_2NaFeF_6 and K_2NaGaF_6, J. Inorg. Nucl. Chem. 21 (1961) 253−258.
[46] K. Kobayashi, T. Matsuo, H. Suga, S. Khairoun, A. Tressaud, Phase-transition in ammonium Hexafluorovanadate(III), Solid State Commun. 53 (1985) 719−722.
[47] O.V. Kovalev, in: H.T. Stokes, D.M. Hatch (Eds.), Representations of the Crystallographic Space Groups: Irreducible Representations, Induced Representations and Corepresentations, Gordon and Breach, Amsterdam, 1993.
[48] T. Krahl, M. Ahrens, G. Scholz, D. Heidemann, E. Kemnitz, Structure of $(NH_4)_3GaF_6$ investigated by multinuclear magic-angle spinning NMR spectroscopy in comparison with rietveld refinement, Inorg. Chem. 47 (2008) 663−670.
[49] S.N. Krylova, A.N. Vtyurin, A. Bulou, A.S. Krylov, N.G. Zamkova, Lattice dynamics and Raman spectra in elpasolite Rb_2KScF_6—comparative analysis, Phys. Solid State 46 (2004) 1271−1279.
[50] M. Leblanc, V. Maisonneuve, A. Tressaud, Crystal chemistry and selected physical properties of inorganic fluorides and oxide-fluorides, Chem. Rev. 115 (2015) 1191−1254.
[51] Y. Liu, A. Kumar, Zh Fan, Y. Zhang, Q. Ke, K. Zeng, J. Wang, D.J. Singh, K.P. Ong, Ferroelectricity and dipole-dipole interactions in NH_4TiOF_3 mesocrystals, Appl. Phys. Lett. 102 (2013) 232903.
[52] E.G. Maksimov, Theoretical investigations of the ferroelectric transition, Phys. Usp. 52 (2009) 603−614.
[53] W. Massa, D. Babel, M. Epple, W. Rüdorff, Are fluoroelpasolites disordered − structural determination in single-crystals K_2NaCrF_6, Rb_2NaFeF_6 and Rb_2KFeF_6, Rev. Chim. Miner. 23 (1986) 508−519.
[54] M.S. Molokeev, A.D. Vasiliev, A.G. Kocharova, Crystal structures of room- and low-temperature phases in oxyfluoride $(NH_4)_2KWO_3F_3$, Powder Diffr. 22 (2007) 227−230.

[55] M.S. Molokeev, S.V. Misyul, Structural transformations and the critical and noncritical parameters during the phase transition in the $(NH_4)_2KWO_3F_3$ oxyfluoride, Phys. Solid State 53 (2011) 1672−1679.
[56] M.S. Molokeev, S.V. Misyul, V.D. Fokina, A.G. Kocharova, K.S. Aleksandrov, Structure transformations during phase transitions in the $K_3WO_3F_3$ oxyfluoride, Phys. Solid State 53 (2011) 834−839.
[57] M.S. Molokeev, S.V. Misyul, Processes of ordering of structural elements, critical and noncritical parameters of phase transitions in the $(NH_4)_3WO_3F_3$ crystal, Phys. Solid State 54 (2012a) 155−165.
[58] M.S. Molokeev, S.V. Misyul, Structural changes during phase transitions and the critical and noncritical order parameters in the $(NH_4)_3Nb(O_2)_2F_4$ crystal, Phys. Solid State 54 (2012b) 143−154.
[59] M.S. Molokeev, E.V. Bogdanov, S.V. Misyul, A. Tressaud, I.N. Flerov, Crystal structure and phase transition mechanisms in $CsFe_2F_6$, J. Solid State Chem. 200 (2013) 157−164.
[60] K. Moriya, T. Matsuo, H. Suga, S. Seki, On the phase transition of ammonium hexafluoro ferrate (III), Bull. Chem. Soc. Jpn. 50 (1977) 1920−1926.
[61] K. Moriya, T. Matsuo, H. Suga, S. Seki, On the phase transitions in ammonium hexafluoroaluminate (III), Bull. Chem. Soc. Jpn. 52 (1979) 3152−3162.
[62] M. Muller, R.E. Dinnebier, A.-C. Dippel, H.T. Stokes, B.J. Campbell, A symmetry-mode description of rigid-body rotations in crystalline solids: a case study of $Mg(H_2O)_6RbBr_3$, J. Appl. Cryst. 47 (2014) 532−538.
[63] T. Nakajima, B. Zemva, A. Tressaud, Advanced Inorganic Fluorides, Elsevier, Amsterdam, 2000.
[64] T. Nakajima, H. Groult, Fluorinated Materials for Energy Storage, Elsevier, Amsterdam, 2005.
[65] S.J. Patwe, S.N. Achary, K.G. Girija, C.G.S. Pillai, A.K. Tyagi, Ferroelectric and proton conducting behavior of a new elpasolite-related vanadium oxyfluoride $(NH_4,K)_3VO_2F_4$, J. Mater. Res. 25 (2010) 1251−1263.
[66] L. Pauling, The crystal structures of ammonium fluoferrate, fluo-aluminate and oxyfluomolybdate, J. Am. Chem. Soc. 46 (1924) 2738−2751.
[67] J. Pebler, K. Schmidt, R. Haegele, D. Babel, Untersuchung des Mossbauer-Effekts in Elpasolithen $A^I_2NaFeF_6$, Z. Naturforsch. B 31 (1976) 1039−1042.
[68] S.S. Pedro, J.C.G. Tedesco, F. Yokaichiya, P. Brandão, A.M. Gomes, S. Landsgesell, M.J.M. Pires, L.P. Sosman, A.M. Mansanares, M.S. Reis, H.N. Bordallo, $Cs_2NaAl_{1-x}Cr_xF_6$: a family of compounds presenting magnetocaloric effect, Phys. Rev. B 90 (2014) 064407.
[69] G. Peraudeau, J. Ravez, P. Hagenmuller, H. Arend, Study of phase transitions in $A_3MO_3F_3$ compounds (A = K, Rb, Cs; M = Mo, W), Solid State Commun. 27 (1978) 591−593.
[70] J. Ravez, G. Peraudeau, H. Arend, S.C. Abrahams, P. Hagenmuller, A new family of ferroelectric materials with composition $A_2BMO_3F_3$ (A, B = K, Rb, Cs, for $r_A > r_B$ and M = Mo, W), Ferroelectrics 26 (1980) 767−769.
[71] J. Ravez, The inorganic fluoride and oxyfluoride ferroelectrics, J. Phys. III EDP Sci. 7 (1997) 1129−1144.
[72] J. Ravez, Crystalline network influence on the variation of T_c with the F-O substitution rate in oxyfluorides derived from octahedral-type ferroelectric oxides, C.R. Acad. Sci. Paris T. 2 Ser. II (1999) 415−419.
[73] J. Ravez, Ferroelectricity in solid state chemistry, C.R. Acad. Sci. Paris Ser. IIC Chim. Chem. 3 (2000) 267−283.

[74] A.A. Shevyurev, L.D. Muradjan, V.E. Zavodnik, K.S. Aleksandrov, V.I. Simonov, Thermal vibrations of atoms in cubic phase of $KMnF_3$ at 198 and 293 K, Kristallografiya 25 (1980) 555–559.
[75] J.F. Scott, R. Blinc, Multiferroic magnetoelectric fluorides: why are there so many magnetic ferroelectrics? J. Phys. Condens. Matter 23 (2011) 113202.
[76] N. Sicron, B. Ravel, Y. Yacoby, E.A. Stern, F. Dogan, J.J. Rehr, Nature of the ferroelectric phase transition in $PbTiO_3$, Phys. Rev. B 50 (1994) 13168.
[77] U. Sliwczuk, R.U. Bartram, D.R. Gabbe, B.C. McCollum, Raman scattering of chromium-doped halide elpasolite crystals, J. Phys. Chem. Solids 52 (1991) 357–361.
[78] G.A. Smolenskii, V.A. Bokov, V.A. Isupov, N.N. Krainik, R.E. Pasynkov, M.S. Shur, Ferroelectrics and Antiferroelectrics, Nauka, Leningrad, 1971.
[79] H.T. Stokes, D.M. Hatch, J.D. Wells, Group-theoretical methods for obtaining distortions in crystals: applications to vibrational modes and phase transitions, Phys. Rev. B 43 (1991) 11010–11018.
[80] H.T. Stokes, B.J. Campbell, R. Cordes, Tabulation of irreducible representations of the crystallographic space groups and their superspace extensions, Acta Cryst. A 69 (2013) 388–395.
[81] J.S. Thrasher, S.H. Strauss, Inorganic fluorine chemistry: toward the 21st century, in: ACS Symposium Series 555, American Chemical Society, Washington, DC, 1994.
[82] A. Tressaud, S. Khairoun, I. Rabardel, T. Kobayashi, T. Matsuo, H. Suga, Phase transitions in ammonium hexafluorometallates(III), Phys. State Solidi A96 (1986) 407–414.
[83] A. Tressaud, Functionalized Inorganic Fluorides, Wiley-Blackwell, Chichester, 2010.
[84] A.A. Udovenko, N.M. Laptash, I.G. Maslennikova, Orientation disorder in ammonium elpasolites. Crystal structures of $(NH_4)_3AlF_6$, $(NH_4)_3TiOF_5$ and $(NH_4)_3FeF_6$, J. Fluorine Chem. 124 (2003) 5–15.
[85] A.A. Udovenko, N.M. Laptash, Orientational disorder in crystal structures of $(NH_4)_3ZrF_7$ and $(NH_4)_3NbOF_6$, J. Struct. Chem. 49 (2008a) 482–488.
[86] A.A. Udovenko, N.M. Laptash, Orientational disorder and crystals of $(NH_4)_3MoO_3F_3$ and $(NH_4)_3WO_3F_3$, Acta Cryst. B 64 (2008b) 305–311.
[87] A.A. Udovenko, N.M. Laptash, Dynamic orientational disorder in fluoroelpasolites, structural refinement of $(NH_4)_3AlF_6$, $(NH_4)_3TiOF_5$ and Rb_2KTiOF_5, Acta Cryst. B 67 (2011) 447–454.
[88] A.A. Udovenko, N.M. Laptash, Crystal structure and phase transition in $(NH_4)_3WO_2F_5$: from dynamic to static orientational disorder, Acta Cryst. B 71 (2015) 478–483, http://dx.doi.org/10.1107/S2052520615010835.
[89] V.G. Vaks, Introduction to the Microscopic Theory of Ferroelectrics, Nauka, Moscow, 1973.
[90] E.I. Voit, A.I. Voit, A.A. Mashkovsky, N.M. Laptash, V.Ya Kavun, Dynamic disorder in ammonium oxofluorotungstates $(NH_4)_2WO_2F_4$ and $(NH_4)_3WO_3F_3$, J. Struct. Chem. 47 (2006) 642–650.
[91] R.L. Withers, F.J. Brink, Y. Liu, L. Noren, Cluster chemistry in solid state: structural diffuse scattering, oxide/fluoride ordering and polar behaviour in transition metal oxyfluorides, Polyhedron 26 (2007) 290–299.
[92] Z.G. Ye, J. Ravez, J.-P. Rivera, J.-P. Chaminade, H. Schmid, Optical and dielectric studies on ferroelectric $K_3MoO_3F_3$ single crystals, Ferroelectrics 124 (1991) 281–286.

Group Theoretical Approach on Possible Magnetoelectric Fluorides

17

G. Nénert
PANalytical B.V., Almelo, Netherland

Chapter Outline

1. Introduction

In recent years, there has been a renewed interest in the coexistence and interplay of magnetism and electrical polarization [5,12,14]. This interest has been concentrated on multiferroics and magnetoelectric materials. In multiferroics, a spontaneous polarization coexists with a long-range magnetic order. In magnetoelectrics (we consider here only the linear effect), the polarization is induced by a magnetic field in a magnetically ordered phase [2]. In the Landau theory framework, multiferroics which are not magnetoelectric possess at least a coupling of the type P^2M^2 (P: polarization, M: total magnetization), while linear magnetoelectrics are characterized by terms like PM^2 or LMP (L: antiferromagnetic order parameter) [37]. Terms like P^2M^2 are of higher degree than PM^2 or LMP terms. Consequently, we expect a stronger interplay between the dielectric and magnetic properties in linear magnetoelectrics than in simple multiferroics (eg, $YMnO_3$ [30]). More complicated coupling terms can also characterize the magnetoelectric effect (eg, magnetic gradient) [19]. These kinds of terms are outside the purpose of the present contribution. In the search for materials exhibiting a strong coupling of magnetism and polarization, the most promising ones are multiferroics

Photonic and Electronic Properties of Fluoride Materials. http://dx.doi.org/10.1016/B978-0-12-801639-8.00017-9

exhibiting linear magnetoelectric properties. These materials are scarce. Thus, it is of interest to look for new magnetoelectric materials independently.

The idea of using symmetry analysis to predict magnetoelectric compounds is not new [17,18]. The first reported magnetoelectric compound, Cr_2O_3, was predicted to be magnetoelectric prior to any experimental evidence [1,11]. It is the same philosophy that we aim to take here.

In this chapter, we present a symmetry analysis of selected materials. All these materials should exhibit magnetoelectricity based on symmetry arguments. The first one is that there is a need to look for materials other than oxides if we are searching for new materials since magnetoelectric/multiferroic materials are scarce. The second reason is that polarization cannot exist in conducting materials. Thus, the high charge transfer in the fluorides makes them good candidates for experimental investigations. While oxides have been tremendously investigated, fluorides remain rather unfamiliar despite many efforts in the 1970s through 1990s. This is partly related to the traditional solid-state chemistry approach which is a difficult route for making fluorides. Recent development with simpler methods such as hydrothermal synthesis with CF_3COOH as the source for fluorine [22], hopefully will trigger and motivate more work on transition metal fluorides.

Several fluorides have been reported to crystallize in a polar structure. Consequently, in addition to magnetoelectric properties, several fluorides are potentially ferroelectric. When this is the case, we discuss this possibility in the light of known ferroelectrics related to the material under investigation. All the compounds discussed in this article have been the subject of detailed crystallographic and magnetic studies by means of neutron diffraction. We present below the results of our search for new magnetoelectric fluorides.

The vast majority of the reported magnetic structures can be described using a single propagation vector. Additionally, most of them have spin configurations which are described by one of the possible maximal magnetic symmetries compatible with its propagation vector. Consequently, by exploring the possible magnetic symmetries for a given space group, one can predict based on symmetry arguments, magnetoelectric and multiferroic properties of materials. In this chapter, we will consider the simplest cases where the magnetic propagation vectors are special points within the first Brillouin zone.

2. The $BaM^{III}F_5$ Family

The $BaMF_5$ family (M = Ti, V, Cr, Mn, Fe, Co, Al, Ga, In) exhibits a large range of crystal structures: monoclinic, orthorhombic, and tetragonal phases. Despite a rich chemistry and crystal chemistry, this family of materials remains only partly investigated and a lot of properties remain poorly understood. Nevertheless, there is enough evidence demonstrating the potential of this family of materials for exhibiting magnetoelectric and multiferroic properties. We present in this section several members of this family demonstrating promising properties.

2.1 The Ferroelectric $BaMF_5$ Phases

Several compositions among the $BaMF_5$ family belongs to the tetragonal system where M = Fe, V, Ti, Al, In ([13,16,27,34]). Furthermore, the physical properties of these phases can be tuned by replacing Ba by Sr such as in $Ba_{1-x}Sr_xFeF_5$ [34]. While there is a consensus that most of those phases crystallize in the tetragonal system, several compositions and in particular $SrAlF_5$ have been reported with several possible space groups leading to an unclear nature of their ground state [25]. While many phases belong to the centrosymmetric space groups *I 4/m* or *I 4₁/a*, the $Ba_{1-x}Sr_xFeF_5$ series belongs to the symmetry *I 4*. These phases exhibit phase transitions characterized by a divergence of their dielectric constants above room temperature with $T_C = 730K$ for $BaFeF_5$ and a linear dependence of this transition for $BaFe_{1-x}In_xF_5$, while this transition temperature varies in a nonlinear fashion for $Ba_{1-x}Sr_xFeF_5$. This is evidencing a ferroelectric state at room temperature for those phases. The crystal structure of $BaFeF_5$ is illustrated in Fig. 17.1.

This crystal structure exhibits two types of chains which are illustrated in Fig. 17.1B and C. The chain of type I (Fig. 17.1B) is a 1D chain decorated with additional FeF_6 octahedra running along the *c* direction. Each octahedral participating in the column of the chain is decorated by additional two FeF_6 octahedra present on each side of the column. Those type I chains form a pseudo-square lattice (Fig. 17.1A) which hosts in its center the type II chains as displayed in Fig. 17.1C. This second type of chains is also running along the *c*.

In addition to the ferroelectric properties being present at room temperature for $BaFeF_5$ and related phases, the magnetic properties of some of these materials have been investigated by magnetometry. $BaFeF_5$ undergoes a long-range magnetic order below $T_N = 35K$ [16]. Another related phase to $BaFeF_5$ exhibiting similar crystal structure is $BaVF_5$ which undergoes some long-range magnetic order below

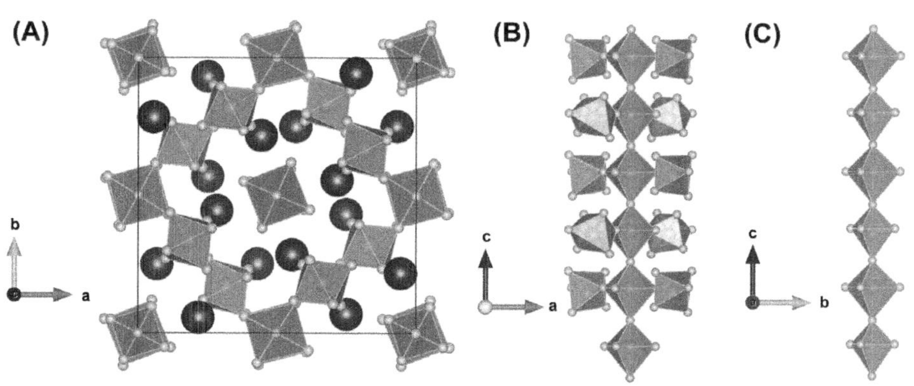

Figure 17.1 Crystal structure of $BaFeF_5$ (A) projected along the *c* direction and a representation of (B) type I $(FeF_6)_2$ decorated 1D chain and of (C) type II simple 1D FeF_6 octahedral chain propagating along the *c* direction.

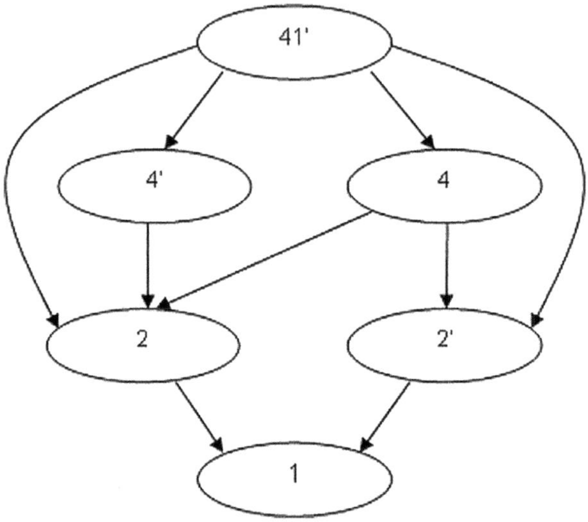

Figure 17.2 Possible magnetic point groups associated with the phases of the $BaFeF_5$ type.

$T_N = 20K$. The coexistence of some long-range magnetic order and ferroelectricity suggests that this type of materials is potentially exhibiting multiferroic properties.

Furthermore, an investigation of the possible magnetic point groups associated to the appearance of a long-range magnetic order in such type of materials is shown in Fig. 17.2. An inspection of the possible magnetic point groups shows that the possible ones are 4, 4′, 2, 2′, or 1. All these magnetic point groups allow for a linear magnetoelectric effect. Consequently, below their magnetic ordering temperatures, compositions such as $BaFeF_5$, $Ba_{1-x}Sr_xFeF_5$, $BaVF_5$, or $BaFe_{1-x}In_xF_5$ are expected to be not only multiferroic but also magnetoelectric. Other compositions within this family exhibiting the tetragonal symmetries are likely to present similar properties. We hope that these considerations will trigger the investigation of these phases.

2.2 *The Orthorhombic Phases:* P $2_12_12_1$ *Versus* C 222_1

As stated above, the $BaMF_5$ family exhibits several possible symmetries. In this section, we shall pay particular attention to the members crystallizing in the orthorhombic−piezoelectric phase $P\ 2_12_12_1$ where the representing composition is $BaGaF_5$. We show in Fig. 17.3 the crystal structure of this material.

The crystal structure of $BaGaF_5$ contains corner-sharing GaF_6 octahedra forming one-dimensional zigzag chains running along the crystallographic c axis as illustrated in Fig. 17.1B. These chains are surrounded by Ba atoms (in blue (black in print versions) in Fig. 17.1A) which form a distorted hexagonal network. The other chemical compositions reported in the space group $P\ 2_12_12_1$ are $BaAlF_5$ [26], $BaMnF_5$ [4], and $BaCrF_5$ [20]. As illustrated in Fig. 17.3, these materials are good representatives of low-dimensional magnetism due to the presence of this 1D chain running along the

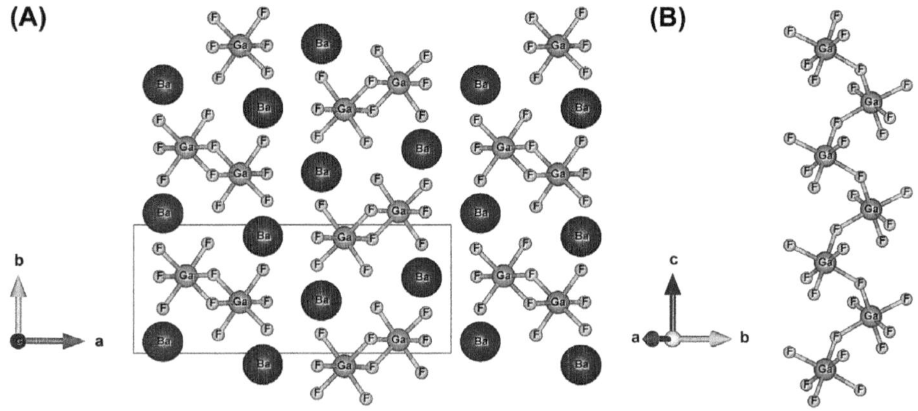

Figure 17.3 Crystal structure of $BaGaF_5$ crystallizing in the space group $P\ 2_12_12_1$. (A) Projected in the (ab) plane and (B) projected along the c axis showing the 1D chains running along that same direction.

c axis. As a result, the long-range magnetic order appears at low temperature such as in $BaCrF_5$ which exhibits a Néel temperature of 3.4K [20]. $BaCrF_5$ exhibits a Curie–Weiss temperature $|\theta|$ of 21K. In $BaMnF_5$, the corresponding temperature is 24K [24]. This is suggesting that $BaMnF_5$ is likely to exhibit a long-range magnetic order at a similar temperature than $BaCrF_5$. While the presence of 1D chain in the structure suggests pseudo-1D magnetism with some long-range order at low temperature, the frustration factor $f = \theta/T_N = 21/3.4 \sim 6.2$ in $BaCrF_5$ is significant. This suggests that magnetic frustration is present in this material. A closer look at the crystal structure shows that the angle F–M–F within the chain is close to 90 degrees (91.7 degrees in $BaCrF_5$ and even 89.5(3) degrees in $BaMnF_5$). This angle being close to 90 degrees suggests that the interaction within the chains is weak ferromagnetic in nature. Consequently, another nonnegligible interaction is the next nearest-neighbor interaction which is weak antiferromagnetic. This gives rise to a competition between the nearest- and next nearest-neighbor interactions resulting in a frustrated J_1–J_2 model as illustrated in Fig. 17.4.

The existence of an additional magnetic interaction besides J_1 is further confirmed by the determination of the magnetic exchange coupling within the chains in $BaCrF_5$ using the Fischer model. Holler et al. found experimentally that $J = -3.74(4)$K while

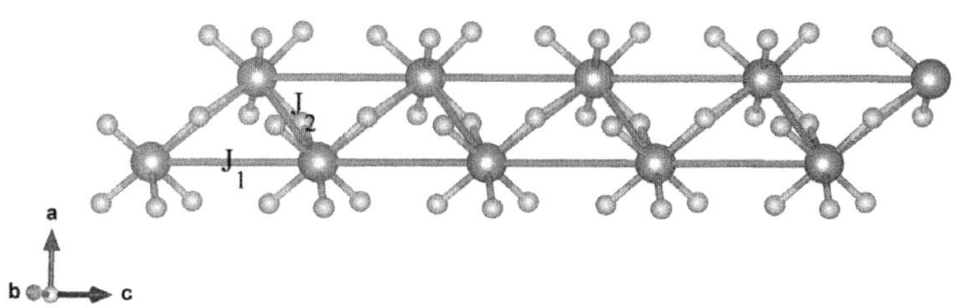

Figure 17.4 Frustrated antiferromagnetic J_1–J_2 chains in $BaCrF_5$ running along the *c* axis.

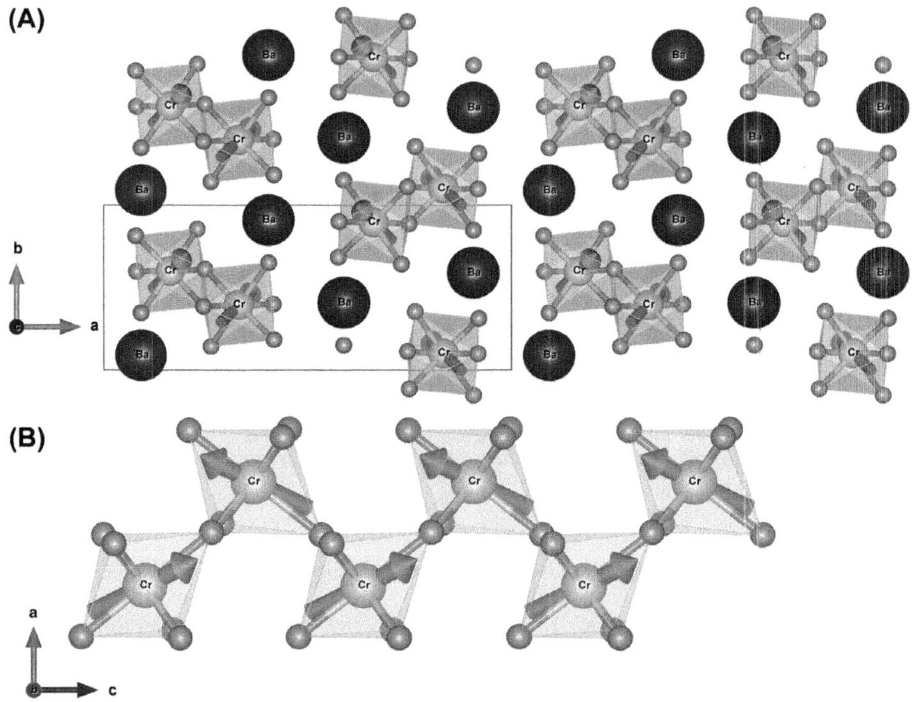

Figure 17.5 Representation of the magnetic structure observed in $BaCrF_5$ below $T_N = 3.4K$.

$|\theta| = 21K$ [20]. Using the mean-field theory relationship $\theta = (1/3)zS(S+1)(J/k_B)$, one finds $\theta = -9.35K$, while experimentally it was found to be 21K. This is suggesting that an additional interaction exists which is nonnegligible in strength giving rise to a long-range magnetic order below $T_N = 3.4K$.

Holler et al. have further investigated the magnetic structure of $BaCrF_5$ using powder neutron diffraction. We present in Fig. 17.5 a representation of its magnetic structure.

There is only one Cr^{3+} ion position in $BaCrF_5$ per unit cell occupying the Wyckoff position 4*a*. Consequently, there are four magnetic sites carrying out one spin S_j. We can define the following four magnetic vectors (one ferromagnetic and three antiferromagnetic ones):

$$M = S_1 + S_2 + S_3 + S_4$$

$$L_1 = S_1 - S_2 + S_3 - S_4$$

$$L_2 = S_1 - S_2 - S_3 + S_4$$

$$L_3 = S_1 + S_2 - S_3 - S_4$$

Table 17.1 Magnetic Components Classified by Irreducible Representation

Γ_1 (++++)	L_{1x}, L_{2y}, L_{3z}
Γ_2 (++−−)	M_x, L_{3y}, L_{2z}
Γ_3 (+−+−)	L_{3x}, M_y, L_{1z}
Γ_4 (+−−+)	L_{2x}, L_{1y}, M_z

Holler et al. have investigated the transformation properties of the different components of the magnetic vectors. We reproduce in Table 17.1 the results of their derivations [20].

In order to look for possible linear magnetoelectric coupling, one needs to search for the invariant of the type $L_iM_jP_k$ [37]. For this purpose, we have investigated the transformation properties of the polarization vector components P_i under the effect of the symmetry elements of the space group $P\ 2_12_12_1$. The results of the transformation properties are displayed in Table 17.2.

Experimentally, it is observed that $BaCrF_5$ exhibits a magnetic structure associated to the irreducible representation Γ_3. Taking into account this and the transformation properties of the electrical polarization, we find that terms like $(L_{3x}+L_{1z})M_xP_z$ and $(L_{3x}+L_{1z})M_zP_x$ are invariants and would allow for a linear magnetoelectric effect. Consequently, an induced electrical polarization may appear along P_x (P_z) if one applies a magnetic field along z (x).

To be noted, besides this allowed magnetoelectric effect in $BaCrF_5$ which is expected to be present also in $BaMnF_5$ and $SrNiF_5$ (same symmetry $P\ 2_12_12_1$ [4,15]), piezoelectric properties are allowed by symmetry. Thus, the magnetic ground state could be controlled not only by the application of an electric field but also by the application of an applied stress and vice versa. Additional composition which is expected to show such behavior (magnetoelectric coupled to piezoelectric effect) is $BaNiF_5$ which exhibits the symmetry $C\ 222_1$ [15]. Since the properties of materials are fully governed by the point group of their crystal structure, the point group 222 in $BaNiF_5$ is expected to show similar behavior to that of the other materials crystallizing in the $P\ 2_12_12_1$ symmetry.

Table 17.2 Transformation Properties of the Electrical Polarization Vector

	2_{1x}	2_{1y}	2_{1z}
P_x	+	−	−
P_y	−	+	−
P_z	−	−	+

Finally, the high temperature behavior of these phases has been not investigated besides the case of $BaAlF_5$. $BaAlF_5$ exhibits three different polymorphs as functions of temperature: α, β, and γ [26]. The α phase is isotypic to $BaGaF_5$ while the β phase crystallizes in the $P\,2_1/c$ space group and the γ phase in $P\,2_1$ symmetry. This γ phase can be stabilized at ambient conditions by quenching. The γ phase is pyroelectric and thus potentially ferroelectric. Consequently, it is not unreasonable to think that some of the phases among $BaMF_5$ (M = Cr, Mn) and $SrNiF_5$ could be potentially stabilized in this symmetry. The possible magnetic point groups associated to the symmetry $P\,2_1$ are 2, 2′, or 1. All these symmetries allow for a linear magnetoelectric effect in addition to likely ferroelectric state at room temperature with an induced additional electrical polarization contribution resulting from the appearance of a long-range magnetic order (in the case of magnetic point group 1).

3. Na_2SiF_6-Related Family: $AMM'F_6$ Materials (A = Na, Li; M, M′ = Mn, Cr, Ga, Fe)

Transition metal fluorides of general formula $A^+M^{2+}M^{3+}F_6$ ($A^+ = Li^+, Na^+, Ag^+$) exhibit several forms: trirutile, modified pyrochlore, tetragonal bronze, trigonal Na_2SiF_6, and much more [6–8,28,40]. The structures which are of interest here are the trigonal Na_2SiF_6 and the trirutile structure. The crystal structure of Na_2SiF_6 is presented in Fig. 17.6A. It consists of isolated SiF_6 octahedra which are running along the c axis of the *P321* symmetry.

Various compositions have been investigated due to their interesting and complex magnetic properties. We shall show in this contribution that the fluorides exhibiting an ordered Na_2SiF_6 crystal structure form a large playground to search for new magnetoelectric materials and interesting multiferroic properties.

In the ordered Na_2SiF_6 type, one sodium atom is replaced (namely Na1 position) by a transition metal ion M^{2+} and by Mn^{2+} in the case of $NaMnCrF_6$ [6]. This ordered phase is shown in Fig. 17.6B. Due to the cationic ordering, the octahedral CrF_6 are not

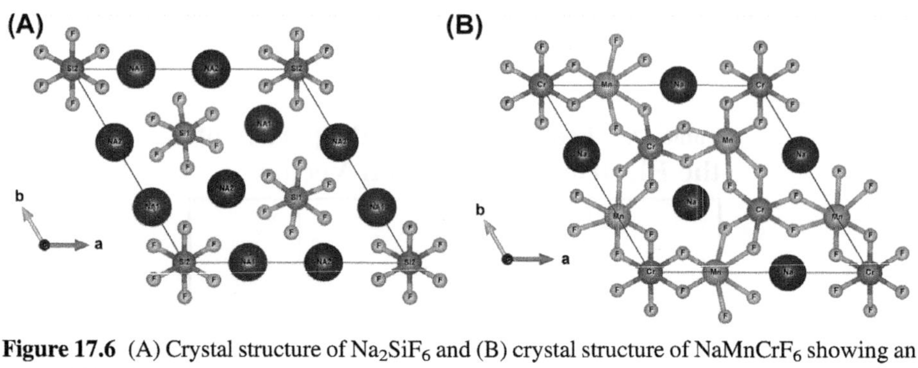

Figure 17.6 (A) Crystal structure of Na_2SiF_6 and (B) crystal structure of $NaMnCrF_6$ showing an ordered Na_2SiF_6 structure type.

isolated anymore but are connected between each other via heavily distorted MnF_6 octahedra. This distortion of the MnF_6 octahedra results in non-Mn−F equivalent bond distances but especially in an angle F−Mn−F of 157.9 degrees to be compared with the F−Cr−F angle which is 178.3 degrees. MnF_6 and CrF_6 octahedra share either an edge or a corner depending on the orientations of these octahedra. This gives rise to a 3D network which is heavily frustrated due to the presence of pseudohexagonal lattice in the system resulting from the cationic ordering (see Fig. 17.6B).

$NaCrMnF_6$ orders antiferromagnetically below $T_N = 21.0(5)$K [7]. As a result of this magnetic frustration, its ground state is characterized by a helimagnetic structure with a periodicity of about 278 Å (magnetic wave-vector $\mathbf{k} = (0\ 0\ 0.01821)$) [3]. It has been demonstrated by Mostovoy [29] that such ground state is expected to break the symmetry giving rise to an electrical polarization which is magnetically induced and strong magnetoelectric coupling such as in $TbMnO_3$ [23]. Bacmann et al. reported that the lower symmetry below the Néel temperature is P3 which would then suggest that an electrical polarization along the *c* direction appears [3]. Furthermore, whether the magnetic point group associated to this helimagnetic structure is 3 or 3′, both magnetic point groups allow for a linear magnetoelectric effect. Consequently, $NaMnCrF_6$ can be regarded as a very likely candidate for multiferroic and magnetoelectric properties.

Following those considerations (piezoelectric space group with frustrated magnetism), we have been looking at the reported compositions crystallizing in this ordered Na_2SiF_6 structure type. The reported compositions and their associated Néel temperature (if known) are tabulated in Table 17.3.

In order to gain some insight into the possible magnetic ground state of this family of materials, we have been exploring the possible magnetic structures. For this purpose, we explore the possible maximal magnetic space groups for the parent space group *P321* (#150), limiting ourselves to the special points within the first Brillouin zone. For this purpose, we use the program ISOTROPY [21]. We find six special points which are listed in Table 17.4.

Using the program k-the program k-Subgroupmag Subgroupmag from the Bilbao Crystallographic Server [32], one can explore the various possible magnetic symmetries for a given propagation vector. Doing such kind of analysis for the listed propagation vectors given in Table 17.4 allows us to investigate the possible magnetic point groups potentially exhibited by this family of materials in their magnetically ordered state. This allows us to discuss the possible linear magnetoelectric or/and multiferroic properties in the ordered Na_2SiF_6 structure type. We present in Figs. 17.7 and 17.8 the whole possible magnetic symmetries associated to the special points within the first Brillouin zone. To be noted: we do not show the magnetic symmetries associated to the wave-vector $\mathbf{k} = (0, 0, \frac{1}{2})$ as they are identical to the ones associated to the wave-vector $\mathbf{k} = (\frac{1}{2}, 0, \frac{1}{2})$ which are shown in Fig. 17.7C.

From Figs. 17.7 and 17.8, one can notice that the possible magnetic point groups of the ordered Na_2SiF_6 structure type are 32′, 32, 2′, 2, 3, and 1. All of these point groups allow for a linear magnetoelectric effect. Consequently, material compositions reported in Table 17.3 are good candidates for magnetoelectric properties. Furthermore, we notice that some of these point groups allow for a spontaneous electrical

Table 17.3 Reported Chemical Positions Exhibiting the Na_2SiF_6 Ordered Structure (Space Group *P321*) and Their Associated Magnetic Ordering Temperatures (If Known)

Composition	Magnetic Ordering Temperature	References
$LiMnGaF_6$	?	[38]
$LiMnCrF_6$	$T_N = 23(1)K$	[7]
α-$LiMnFeF_6$	$T_N = 158K$	[9]
β-$LiMnFeF_6$	$T_N = 113K$	[9]
$NaMnFeF_6$	$T_N = 44.6(5)K$	[36,10]
$NaMnCrF_6$	$T_N = 21.0(5)K$	[7]
$NaMn(Cr_{2/3}Ga_{1/3})F_6$	$T_N = 14(2)K$	[7]
$NaMn(Cr_{1/3}Ga_{2/3})F_6$	$T_N = 8(2)K$	[7]
$NaMnGaF_6$	0 (?)	[7]
$LiTiMnF_6$	?	[35]
$LiTiFeF_6$	?	[35]
$LiTiCoF_6$	?	[35]
$LiTiNiF_6$	?	[35]

Table 17.4 Special Propagation Vectors for the Space Group *P321* (#150)

Propagation Vector	Coordinates
GM	(0,0,0)
A	(0,0,1/2)
H	(1/3,1/3,1/2)
K	(1/3,1/3,0)
L	(1/2,0,1/2)
M	(1/2,0,0)

polarization (cases where the point group is 2′, 2, 3, or 1). Therefore, like in the case of $NaMnCrF_6$, a magnetically induced ferroelectric state could appear in the ground state of some compositions coupled to magnetoelectric properties. We hope that it will stimulate a renewed interest for investigating this family of materials.

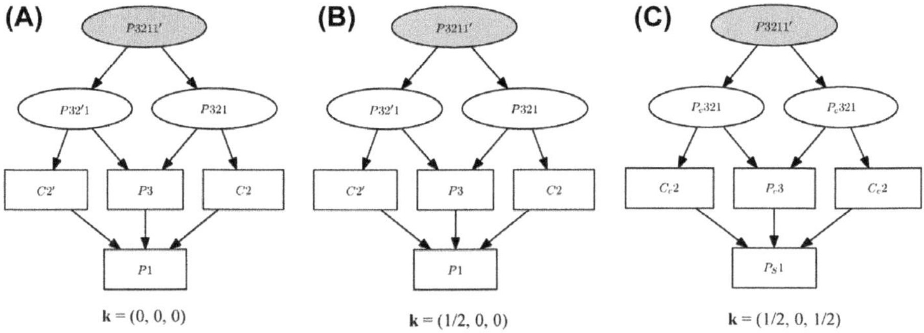

Figure 17.7 Possible magnetic symmetries associated to different special points of the first Brillouin zone for the space group *P3211′*. The different considered wave-vectors are (A) $\mathbf{k} = (0,0,0)$; (B) $\mathbf{k} = (1/2,0,0)$; and (C) $\mathbf{k} = (1/2,0,1/2)$.

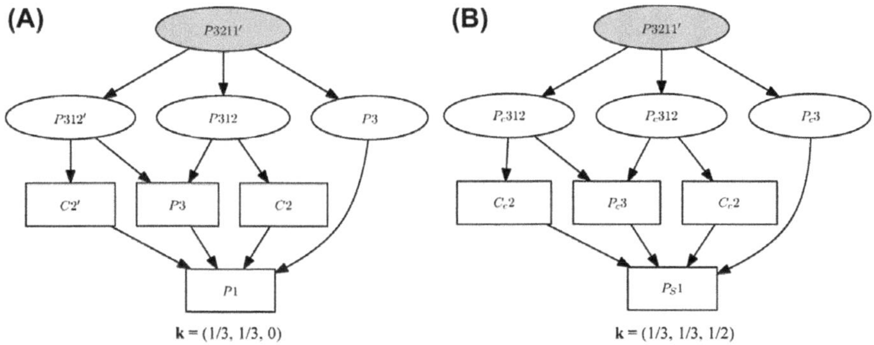

Figure 17.8 Possible magnetic symmetries associated to different special points of the first Brillouin zone for the space group *P3211′*. The different considered wave-vectors are (A) $\mathbf{k} = (1/3,1/3,0)$ and (B) $\mathbf{k} = (1/3,1/3,1/2)$.

4. Weberite

The weberite family $A_2B_2X_7$ shows a wide range of compositions and crystal structures. Weberite is a fluorite-related superstructure. The superstructure is given by stacking cubic close-packed cation layers. These layers alternate between the compositions A_3B and AB_3 and are parallel to the (011) planes [33,39]. The rich chemistry of this family gave rise to a lot of attention both for oxide and fluoride compositions ($X = O$ or F, respectively). Weberite is a natural mineral named after Theobald C.F. Weber (1823–86) having the ideal composition Na_2MgAlF_7. This mineral forms the aristotype of orthorhombic weberites with space group *Imma*. While a majority of weberite crystallizes in the orthorhombic space groups *Imma* or *Pnma*, a number of compositions exhibit the symmetry *C2/c* and $P3_121$ (# 152). In this chapter, dedicated to the prediction of magnetoelectric properties using group theoretical methods, we concentrate our attention to the trigonal weberites crystallizing in the space group $P3_121$.

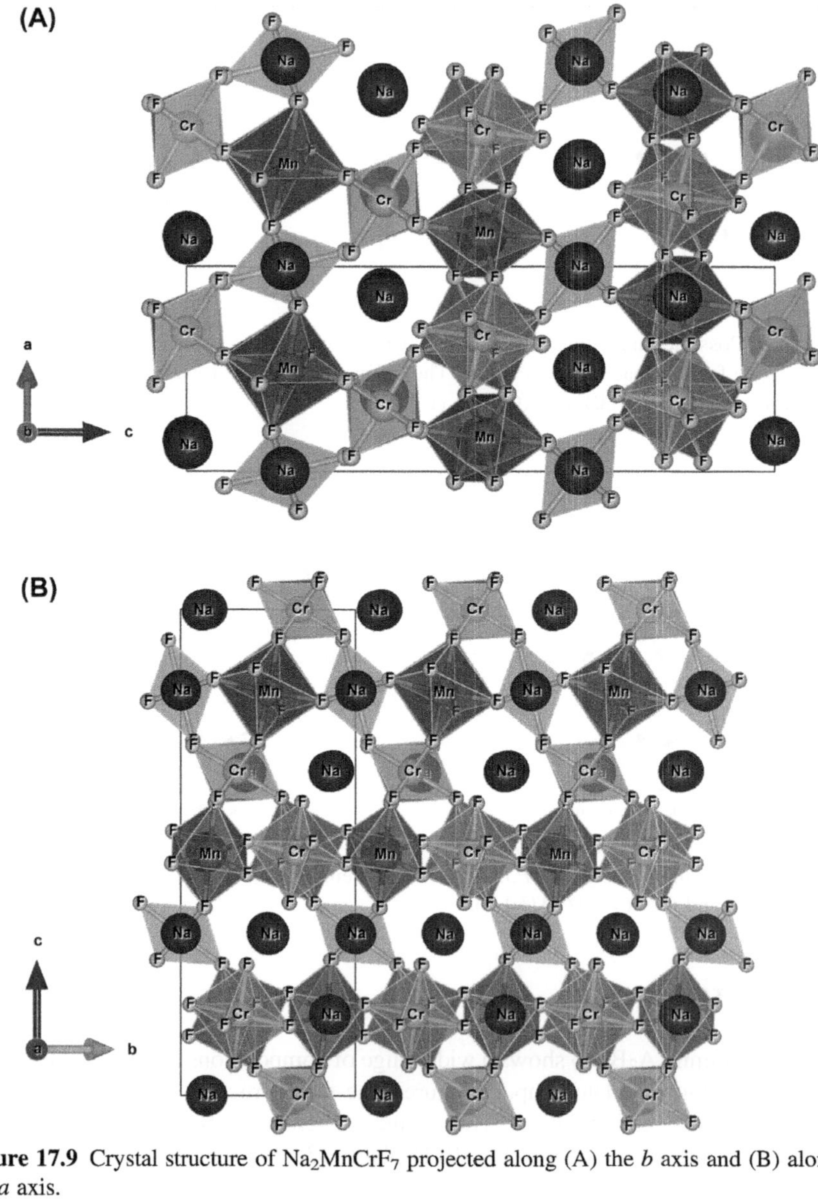

Figure 17.9 Crystal structure of Na_2MnCrF_7 projected along (A) the *b* axis and (B) along the *a* axis.

A representation of the crystal structure of such trigonal weberite is illustrated in Fig. 17.9 projected along *a* and *b* axis. This trigonal weberite exhibits several channels containing Na ions. The existence of these channels coupled to the presence of transition metal ions which could exhibit different oxidation state suggests that these trigonal weberites could be used as electrode material. One channel is propagating along the *a* direction while there is an additional one, 120 degrees away from the first one. This is suggesting the possibility for a high storage of Na^+ ions within this type of structure.

One key parameter to stabilize the trigonal weberite structure type is to use large transition metal M^{2+} ions such as Mn^{2+}. Known compositions exhibiting such symmetry are Na_2MnGaF_7, Na_2MnCrF_7, Na_2MnVF_7, and Na_2MnFeF_7. More precisely, the trigonal weberite structure is exhibited when $r(M^{3+})$ is around 0.6−0.65 Å, while $r(M^{2+}) \geq 0.8$ Å [39]. These are the cases for the chemical compositions mentioned just above. Consequently, one could imagine that composition such as $Na_2CdM^{3+}F_7$ with M^{3+} = Cr, V, Fe ($r(Cd^{2+}) = 0.95$ Å) would potentially exhibit the trigonal symmetry. From powder diffraction alone, it is difficult to disentangle between the orthorhombic and trigonal weberites as their patterns are very similar. And several compositions reported with an orthorhombic structure from powder data alone could actually well be trigonal. In particular, the size of the cation in the channel does not seem to be a key parameter for the stabilization of the trigonal structure. Consequently, additional compositions with Ag^+ ion instead of Na^+ would potentially present some additional trigonal weberites. The high spin value of Mn^{2+} is likely to give rise to some long-range magnetic order at low temperature. For instance, Na_2MnFeF_7 has been reported having a Néel temperature of 94K [31].

In order to investigate the potential magnetoelectric properties of this family of materials, we have taken a similar approach to the one shown in the previous section on the Na_2SiF_6-related family. The special propagation vectors within the first Brillouin zone of the space group *P3₁21* (#152) are identical to the ones associated to the space group *P321* (#150) (see Table 17.4).

Using the program *k-Subgroupmag*, we have been exploring the possible magnetic symmetries for the propagation vectors shown in Table 17.4. The symmetry-allowed

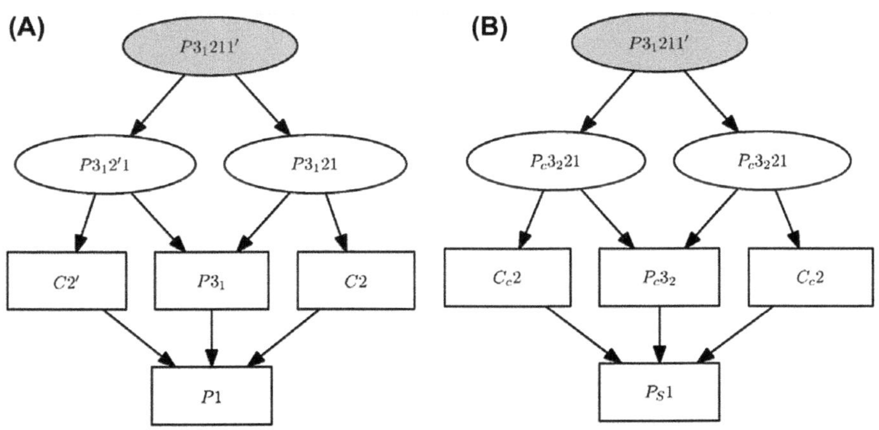

k =(1/2, 0, 0) **k** = (0, 0, 1/2)

Figure 17.10 Possible magnetic symmetries associated to different special points of the first Brillouin zone for the space group $P3_1211'$. The different considered wave-vectors are for (A) **k** = (1/2,0,0) and (B) **k** = (0,0,1/2).

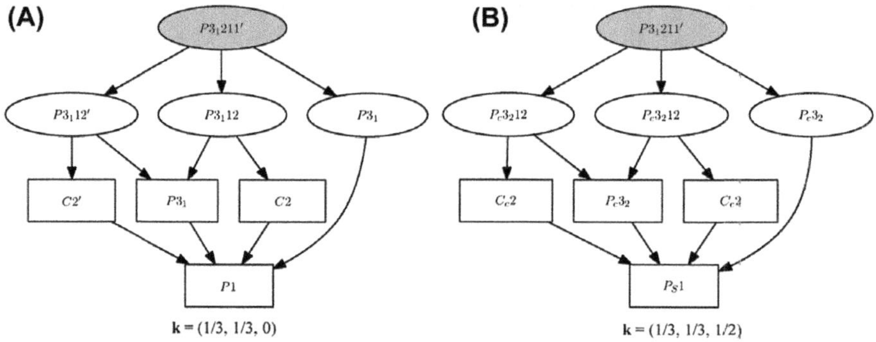

Figure 17.11 Possible magnetic symmetries associated to different special points of the first Brillouin zone for the space group $P3_1211'$. The different considered wave-vectors are for (A) $\mathbf{k} = (1/3,1/3,0)$; and (B) $\mathbf{k} = (1/3,1/3,1/2)$.

magnetic subgroups for the parent paramagnetic space group $P3_1211'$ are given in Figs. 17.10 and 17.11. The special points within the first Brillouin zone of $P3_121$, $\mathbf{k} = (0,0,0)$ and $\mathbf{k} = (1/2,0,1/2)$ are not discussed as they allow for magnetic space groups already generated by other wave-vectors.

From the inspections of Figs. 17.10 and 17.11, it appears that all the magnetic point groups susceptible to be exhibited by the trigonal fluorine weberites allow for a linear magnetoelectric effect. Furthermore, several of these point groups are compatible with a ferroelectric state, which would be induced by the appearance of a long-range magnetic order. This is the case of point groups 2, 2′, and 3. Consequently, compositions such as Na_2MnGaF_7, Na_2MnCrF_7, Na_2MnVF_7, and Na_2MnFeF_7 are potential candidates for linear magnetoelectric materials with a possibility for some additional multiferroicity of type II.

5. Conclusion

In that chapter, following up the very early work of Dzialoshinskii on predicting the magnetoelectric effect in Cr_2O_3, we intended to take a similar approach in order to point toward potential candidates for magnetoelectricity. The various families that we have been investigating exhibit a wide range of possible coupling between dielectric and magnetic properties. $BaFeF_5$ and related phases are linear magnetoelectric and multiferroic of type I where the ferroelectricity and the magnetic ordering temperature do not coincide. Other compositions, related to the BaGaF5 structural type are purely linear magnetoelectric compounds. Such materials are expected to present some magnetic frustration due to the presence of J_1–J_2 chains running along the c axis. Consequently, for some compositions, one cannot exclude the existence of more complex magnetic ground state.

Finally, families such as the ordered Na_2SiF_6 crystal structures or the trigonal weberites are crystal structures of high interest, as any composition exhibiting some long-range magnetic order would result in at least linear magnetoelectric if not magnetically induced ferroelectric state such as in $NaMnCrF_6$. With the development of new synthetic methods for transition metal fluorides, we expect a revival of the investigation of these fluorides and their associated properties.

References

[1] D.N. Astrov, Magnetoelectric effect in chromium oxide, Soviet Phys. JETP 13 (1961) 729–733.
[2] A. Authier (Ed.), International Tables for Crystallography Vol D Physical Properties of Crystals, Kluwer–Academic, Dordrecht, 2003.
[3] M. Bacmann, P. Wolfers, G. Courbion, Structure helimagnetique de $NaMnCrF_6$. Mat. Res. Bull. 15 (1980) 1479–1487.
[4] P. Bukovec, R. Hoppe, Zur Kenntnis von $BaMnF_5$: Eine Jahn-Teller-verzerrte Variante von $BaGaF_5$, Z. Anorg. Allg. Chem. 509 (1984) 138–144.
[5] S.-W. Cheong, M. Mostovoy, Multiferroics: a magnetic twist for ferroelectricity, Nat. Mater. 6 (2007) 13–20.
[6] G. Courbion, C. Jacobini, R. de Pape, La structure cristalline de $NaMnCrF_6$, Acta Cryst B33 (1977) 1405–1408.
[7] G. Courbion, G. Ferey, R. de Pape, Composés ferrimagnétiques fluorés de structure Na_2SiF_6, Mat. Res. Bull. 13 (1978) 967–973.
[8] G. Courbion, C. Jacoboni, R. de Pape, The dimorphism of $LiMnFeF_6$: a new kind of cationic order in the structural type Na_2SiF_6, J. Solid State Chem. 45 (1982) 127–134.
[9] G. Courbion, R. de Pape, J. Teillet, F. Varret, J. Pannetier, Magnetization, Mossbauer and neutron studies of α- and β-$LiMnFeF_6$: evidence of idle spin behavior in a ferrimagnet, J. Mag. Mag. Mater. 42 (1984) 217–232.
[10] G. Courbion, M. Leblanc, The magnetic structure of $NaMnFeF_6$, J. Mag. Mag. Mat 74 (1988) 158–164.
[11] I.E. Dzyaloshinskii, On the magneto-electrical effect in antiferromagnets, Sov. Phys. JETP 10 (1960) 628.
[12] W. Eerenstein, N.D. Mathur, J.F. Scott, Multiferroic and magnetoelectric materials, Nature 442 (2006) 759–765.
[13] S.M. Eicher, J.E. Greedan, Journal of Solid State Chemistry vol. 52 (1984) 12–21.
[14] M. Fiebig, Revival of the magnetoelectric effect, J. Phys. D Appl. Phys. 38 (2005) R123.
[15] T. Fleischer, R. Hoppe, Neues über Pentafluorometallate(III). $SrNiF_5$, $BaNiF_5$, $BaCoF_5$ und $CdCoF_5$, Z. Anorg. Allg. Chem. 490 (1982) 111–120.
[16] R. Georges, J. Ravez, R. Olazcuaga, P. Hagenmuller, Propriétés magnétiques de quelques phases de structure $BaFeF_5$, J. Solid State Chem. 9 (1974) 1–5.
[17] S. Goshen, D. Mukamel, H. Shaked, S. Shtrikman, Magnetic symmetry and ferroelectricity induced by antiferromagnetic transitions, J. Appl. Phys. 40 (1969) 1590.
[18] S. Goshen, D. Mukamel, H. Shaked, S. Shtrikman, Mechanism leading to ferroelectricity induced in centrosymmetric crystals by antiferromagnetic transitions, Phys. Rev. B 2 (1970) 4679.
[19] A.B. Harris, Landau analysis of the symmetry of the magnetic structure and magneto-electric interaction in multiferroics, Phys. Rev. B 76 (2007) 054447 (41 pp.).
[20] H. Holler, W. Kurtz, D. Babel, W. Knop, Die Kristall- und magnetische Struktur von $BaCrF_5$, Z. für Naturforschung B 37 (1982) 54–60.
[21] Isotropy Software Suite, iso.byu.edu.

[22] S.W. Kim, S.-H. Kim, P.S. Halasyamani, M.A. Green, K.P. Bhatti, C. Leighton, H. Dase, C.J. Fennie, $RbFe^{2+}Fe^{3+}F_6$: synthesis, structure, and characterization of a new charge-ordered magnetically frustrated pyrochlore-related mixed-metal fluoride, Chem. Sci. 3 (2012) 741–751.
[23] T. Kimura, T. Goto, H. Shintani, K. Ishizaka, T. Arima, Y. Tokura, Magnetic control of ferroelectric polarization, Nature 426 (2003) 55–58.
[24] P. Köhler, W. Massa, D. Reinen, B. Hofmann, R. Hoppe, Der Jahn-Teller-Effekt des Mn^{3+}-Ions in oktaedrischer Fluorkoordination. Ligandenfeldspektroskopische und magnetische Untersuchungen, Z. Anorg. Allg. Chem. 446 (1978) 131–158.
[25] F. Kubel, The crystal structures of $SrAlF_5$ and $Ba_{0.43(1)}Sr_{0.57(1)}AlF_5$, Z. Anorg. Allg. Chem. 624 (1998) 1481–1486.
[26] A. Le Bail, G. Ferey, A.-M. Mercier, A. de Kozak, M. Samouel, Structure determination of β- and γ-$BaAlF_5$ by X-ray and neutron powder diffraction: a model for the $\alpha \rightarrow \beta\ \gamma$ transitions, J. Solid State Chem. 89 (1990) 282–291.
[27] A. Le Bail, A.M. Mercier, Eur. J. Solid State Inorg. Chem. vol. 32 (1995) 15–24.
[28] M. Leblanc, V. Maisonneuve, A. Tressaud, Crystal chemistry and selected physical properties of inorganic fluorides and oxide-fluorides, Chem. Rev. 115 (2015) 1191–1254.
[29] M. Mostovoy, Ferroelectricity in spiral magnets, Phys. Rev. Lett. 96 (2006) 067601 (4 pp.).
[30] A.A. Nugroho, N. Bellido, U. Adem, G. Nénert, C. Simon, M.O. Tjia, M. Mostovoy, T.T.M. Palstra, Enhancing the magnetoelectric coupling in $YMnO_3$ by Ga doping, Phys. Rev. 75 (2007) 174435 (5 pp).
[31] J. Pebler, K. Schmidt, D. Babel, W. Verscharen, Mossbauer-spektroskopische Untersuchungen an Na_2MnFeF_7, Z. Naturforsch 32 b (1977) 369–372.
[32] J.M. Perez-Mato, S.V. Gallego, E.S. Tasci, L. Elcoro, G. de la Flor, M.I. Aroyo, Symmetry-based computational tools for magnetic crystallography, Annu. Rev. Mater. Res. 45 (2015) 217–248.
[33] B. Peschel, D. Babel, Die Kristallstrukturen der vanadium = weberite $Na_2M^{II}V^{III}F_7$ (M^{II} = Mn, Ni, Cu) und von $NaVF_4$, Z. anorg. Allg. Chem. 623 (1997) 1614–1630.
[34] J. Ravez, V. Andriamampianina, A. Simon, J. Grannec, S. Abrahams, Correlation between Curie temperature and cation radius in the ferroelectric $BaFeF_5$ family, Ferroelectrics 109 (1990) 33–38.
[35] T. Sekino, T. Endo, T. Sato, M. Shimada, High pressure synthesis of LiTiMF6 (M = Mn, Fe, Co, Ni) with trirutile, Na_2SiF_6 and $PbSb_2O_6$ structures, J. Solid State Chem. 88 (1990) 505–512.
[36] M. Tamine, Y. Calage, M. Leblanc, G. Ferey, F. Varret, A weakly coupled magnetic sublattive in $NaMnFeF_6$, observed by Mossbauer spectroscopy and magnetic measurements, Hyperfine Interact. 28 (1986) 529–532.
[37] J.-C. Tolédano, P. Tolédano, The Landau Theory of Phase Transitions, World Scientific, Singapore, 1987.
[38] W. Viebahn, Untersuchungen an quaternären Fluoriden $LiMe^{II}Me^{III}F_6$. Die Struktur von $LiMnGaF_6$, Z. Anorg. Allg. Chem. 413 (1975) 77–84.
[39] O. Yakubovich, V. Urusov, W. Massa, G. Frenzen, D. Babel, Structure of $Na_2Fe_2F_7$ and structural relations in the family of weberites $Na_2M^{II}M^{III}F_7$, Z. Anorg. Allg. Chem. 619 (1993) 1909–1919.
[40] R.F. Williamson, E.S. Arafat, K.N. Baker, C.H. Rhee, J.R. Sanders, T.B. Scheffler, H.S. Zeidan, W.O.J. Boo, Magnetic properties of the trirutile compound $LiMnVF_6$, Inorg. Chem. 24 (1985) 482–485.

Part Five

Superconductor properties

Modification of Magnetic and Electronic Properties, in Particular Superconductivity, by Low Temperature Insertion of Fluorine into Oxides

18

P. Slater, L. Driscoll
University of Birmingham, Birmingham, United Kingdom

Chapter Outline

1. Introduction

The electronic and magnetic properties of ceramic transition metal–containing oxides have attracted considerable interest for a range of technological applications. An important feature in the development of materials for use in such applications is the fine control of the properties. In this respect the control of the transition metal oxidation state is crucial, and traditionally this has been achieved through doping on the cation sites. Nowhere more is this true than in the field of high-temperature cuprate superconductors, the first example of which was alkaline earth–doped La_2CuO_4. While undoped La_2CuO_4 is a semiconductor, metallic character can be introduced through partial substitution of La^{3+} by alkaline earths (Ca^{2+}, Sr^{2+}, Ba^{2+}) to raise the average Cu oxidation state above 2+ [54]. Subsequently it was shown that this incorporation of alkaline earth cations could also lead to the observation of

Photonic and Electronic Properties of Fluoride Materials. http://dx.doi.org/10.1016/B978-0-12-801639-8.00018-0

superconductivity, with the superconducting transition temperature, T_c, dependent on the Cu oxidation state and hence on the level of doping [9,16,20]. Attempts to increase the T_c through replacement of La by Y led to much higher T_c values in the Y-Ba-Cu-O system. The superconducting phase in this case was shown to be $YBa_2Cu_3O_{7-x}$, which is a triple perovskite with both cation and oxygen vacancy ordering, [17,31]. This was the first superconductor to have a T_c above the temperature at which liquid N_2 boils (77 K). As for alkaline earth–doped La_2CuO_4, the T_c strongly depends on the Cu oxidation state and is a maximum (93 K) for x as close as possible to 0. For this $YBa_2Cu_3O_{7-x}$ system, it was therefore shown that the control of the oxygen deficiency, x, could be exploited to control the Cu oxidation state and hence the T_c, thus illustrating the potential manipulation of the anion sublattice for control over the superconducting properties.

Control of the oxygen content was also subsequently shown to induce superconductivity in La_2CuO_4, with high-pressure treatments leading to an increase in the Cu oxidation and superconductivity through the introduction of additional oxide ions, ie, La_2CuO_{4+x} [10,33]. The introduction of additional oxygen is possible in this system due to the nature of the structure adopted by La_2CuO_4. This material adopts the well-known K_2NiF_4 structure (also called the T structure by researchers in the superconductivity field), which can be classed as consisting of perovskite-type $[LaCuO_3]^-$ blocks separated by LaO^+ rock salt layers. This leads to a structure with significant interstitial space between the LaO^+ layers, into which additional anions may be inserted (Fig. 18.1). The ability to control the Cu oxidation state and superconducting

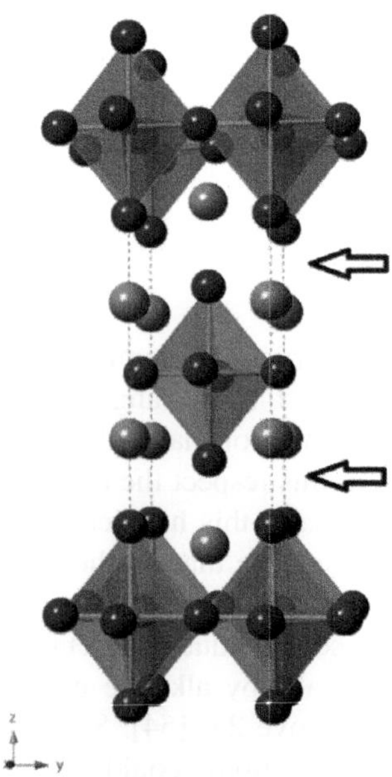

Figure 18.1 The structure of La_2CuO_4 (T structure) (Cu at the center of the octahedra). Regions where interstitial anions may be inserted are marked with arrows.

transition temperature via control of the oxygen content subsequently led to interest in low-temperature fluorination to achieve the same aim [2,19]. Moreover, following initial successful work on fluorination of La_2CuO_4 and $YBa_2Cu_3O_{7-x}$ using F_2, NF_3, this low-temperature fluorination process was shown to be applicable to other systems, leading to the synthesis of superconducting $Sr_2CuO_2F_{2+\delta}$ from fluorination of Sr_2CuO_3 [5]. A key aspect of this latter work was the conversion of a structure containing one-dimensional square planar Cu—O chains into one with two-dimensional Cu—O layers, of which the latter are a key requirement for superconducting cuprates. The synthesis and identification of superconductivity in this latter system therefore led to the great deal of interest and inspired a lot of work on the development of alternative solid-state fluorinating agents. In this chapter, we will discuss these low-temperature fluorination strategies, beginning with work on cuprate systems and then extending to more recent work on other transition metal systems. In these fluorination processes, charge compensation can be achieved in a number of ways, with a corresponding effect on the transition metal oxidation state, namely,

1. Replacement of one O^{2-} by one F^- leading to a reduction in the transition metal oxidation state.
2. Replacement of one O^{2-} by two F^- (the additional F^- entering vacant or interstitial sites) leading to no change in the transition metal oxidation state.
3. Insertion of F^- into vacant or interstitial sites leading to an increase in the transition metal oxidation state.
4. A combination of the above.

As will be detailed later, the actual charge compensation observed depends strongly on the fluorinating agents employed, as well as other factors, such as the fluorination temperature and the structure of the precursor oxide system. In terms of the latter, the fluorination process is favored when there are either vacant anion sites or available interstitial sites within the structure of the precursor oxide system.

2. Initial Studies of the Fluorination of Cuprates With F_2, NF_3 Gas, and Electrochemical Fluorination

The earliest reports of the use of F_2, NF_3 gas to introduce superconductivity in cuprates focused on studies of La_2CuO_4 and $YBa_2Cu_3O_{7-x}$. In the case of the latter system it was shown that low-temperature (typically <300°C) heat treatment with F_2, NF_3 gas on lightly oxygen-deficient systems $YBa_2Cu_3O_{7-x}$ ($x \leq 0.3$) led to increases in the superconducting transition temperature, T_c, up to values comparable ($\approx$90 K) to the fully oxygenated endmember ($x \approx 0$) [32,48,56,71,72]. This early work also showed that it was necessary to adopt this route involving low-temperature fluorination of a preprepared oxide system to prepare such F-containing materials, and that F could not simply be incorporated in $YBa_2Cu_3O_{7-x}$ by high-temperature synthesis using BaF_2 in the starting mixture [69]. This can be correlated with the high thermodynamic stability of BaF_2. Subsequent work examined the fluorination of the heavily oxygen-deficient system, $YBa_2Cu_3O_6$ (ie, $x = 1$), with this work showing the ability of this phase to accommodate large levels of F. In this work, XeF_2 (this decomposes on heating to give Xe and F_2 gas) was used as the fluorinating agent and the formation of $YBa_2Cu_3O_6F_2$ was reported with the linear Cu1 site units being converted to octahedral units (Fig. 18.2). Accompanying this change, $YBa_2Cu_3O_6F_2$ was shown to be superconducting with a T_c of 94 K [59]. In the case of La_2CuO_4, it was shown that similar low-temperature fluorination (200°C) using F_2 gas led to low levels of F insertion, $La_2CuO_4F_\delta$, along with the observation of superconductivity ($T_c \approx 40$ K) [19,73]. As noted in the introduction, this material adopts the K_2NiF_4-type structure (T structure), consisting of perovskite-type $[LaCuO_3]^-$ layers separated by LaO^+ rock salt layers. The fluorine inserts into interstitial positions within the rock salt layers (Fig. 18.1) [2]. Similar successful fluorination was achieved for alkaline earth–doped La_2CuO_4 [15]. In addition to successful fluorination of polycrystalline samples, as described earlier, the fluorination of thin-film La_2CuO_4 (grown on $SrTiO_3$ substrates) using F_2 gas has also been demonstrated leading to superconductivity with a T_c of ≈ 38 K [47].

Subsequent work investigated the fluorination of the related Nd_2CuO_4 system. This phase, although having a comparable general formula to La_2CuO_4, adopts a different structure, albeit still containing the requisite two-dimensional Cu–O layers necessary for superconductivity. In this Nd_2CuO_4 structure (called the T′ structure by researchers in the superconductivity field) there are $Nd_2O_2^{2+}$ fluorite-type layers separating CuO_2^{2-} layers (Fig. 18.3). The difference in structure on changing the rare earth from La to Nd can be correlated with lattice strain effects. In particular, for La_2CuO_4 there is a degree of lattice mismatch between the perovskite $[LaCuO_3]^-$ and LaO^+

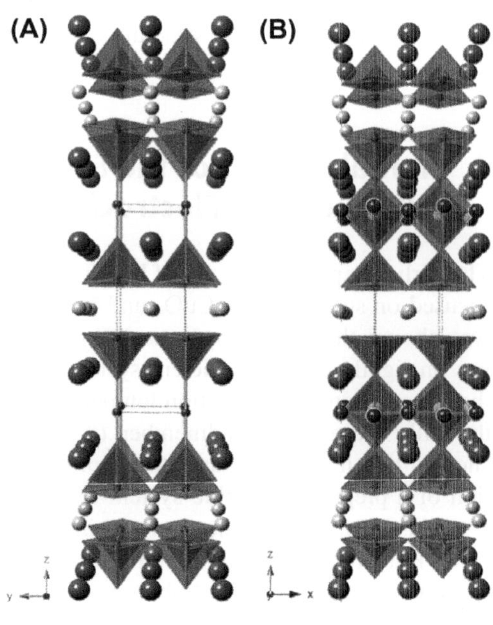

Figure 18.2 The structures of (A) $YBa_2Cu_3O_6$ and (B) $YBa_2Cu_3O_6F_2$. The linear Cu1 site units are converted to octahedral on fluorination.

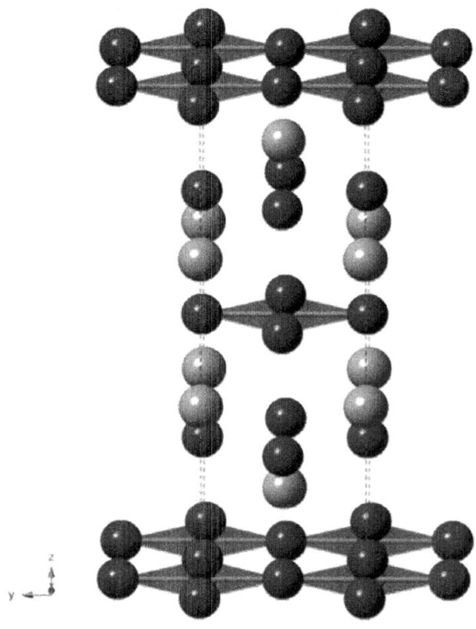

Figure 18.3 The structure of Nd_2CuO_4 (T′ structure) (Cu at the center of square planar units).

rock salt layers, such that the Cu—O bonds are under compression. In changing from La to a smaller lanthanide further strain would be introduced, such that for Nd the "La_2CuO_4" (T) structure is too unstable and so the alternative T′ structure is formed. Superconductivity can be introduced into Nd_2CuO_4 by electron doping, for example, partial Ce^{4+} substitution for Nd^{3+}. A similar result was observed for F doping to give $Nd_2CuO_{4-x}F_x$ synthesized via high-temperature reaction under an inert atmosphere [44]. Subsequent studies have investigated the low-temperature fluorination of Nd_2CuO_{4-x} with solid-state fluorinating reagents and will be discussed in the next section.

The fluorination of a range of Hg-based cuprates, $HgBa_2Ca_{n-1}Cu_nO_{2n+2+\delta}$, using F_2 gas has also been reported [55]. For $n = 1, 2, 3$, the F incorporation gave similar results to prior oxygenation work, with an increase in T_c to 96, 128, and 135 K, respectively.

Electrochemical fluorination of $YBa_2Cu_3O_{7-x}$ and La_2CuO_4 has also been achieved. Initial studies reported operation at elevated temperature (eg, 400°C for $YBa_2Cu_3O_{7-x}$ [74]), but more recently the electrochemical fluorination of La_2CuO_4 has been achieved at room temperature [75]. The latter work showed that the choice of electrolyte material was crucial, with the best results achieved for an *N,N*-dimethylformamide/$n$$Bu_4NBF_4$ electrolyte. The process leads to a thin surface layer of LaF_3 with superconducting ($T_c \approx 40$ K) fluorine intercalated $La_2CuO_4F_\delta$ ($\delta \approx 0.18$) below this surface layer.

The complete replacement of La by Sr or Ca is possible in La_2CuO_4 leading to the phases Sr_2CuO_3 and Ca_2CuO_3. These phases are oxygen-deficient variants of the K_2NiF_4 structure. Moreover the oxide ion vacancies are ordered, resulting in a change from layers of octahedral Cu to chains of square planar Cu (Fig. 18.4). In 1994, Al-Mamouri et al. investigated the fluorination of Sr_2CuO_3 with F_2 gas with a view in trying to incorporate F into the vacant sites between the square planar Cu—O chains [5]. This study showed that low-temperature ($\approx$210°C) reaction with F_2 gas was possible leading to the formation of a compound of composition $Sr_2CuO_2F_{2+\delta}$ ($\delta \approx 0.6$), in which all the channel sites were filled, transforming the Cu coordination from square planar to octahedral. This compound was isostructural with $La_2CuO_4F_\delta$ and had similar partial occupancy of the interstitial sites between the rock salt—type layers. The work indicated that in addition to fluorine insertion, partial fluorine substitution for oxygen had occurred on fluorination. The as-prepared fluorinated sample was nonsuperconducting, attributed to the high Cu oxidation state (ie, the sample was overdoped), but superconductivity could be induced through annealing under N_2 (<330°C) or H_2/N_2 (<220°C) to lower the F content and hence Cu oxidation state. The highest T_c was 46 K for $x \approx 0.3$, notably a higher T_c than observed for fluorinated or alkaline earth—doped La_2CuO_4. The low-temperature route is vitally important to synthesize this system: attempts to prepare this phase by high-temperature reaction between SrF_2, $SrCO_3$, and CuO were unsuccessful, leading to the formation of SrF_2 and $SrCuO_2$, which can be related to the high thermodynamic stability of SrF_2.

A key aspect of this low-temperature route is that it is topotactic, exploiting the structural similarity between Sr_2CuO_3 and $Sr_2CuO_2F_{2+\delta}$, and the fact that at low

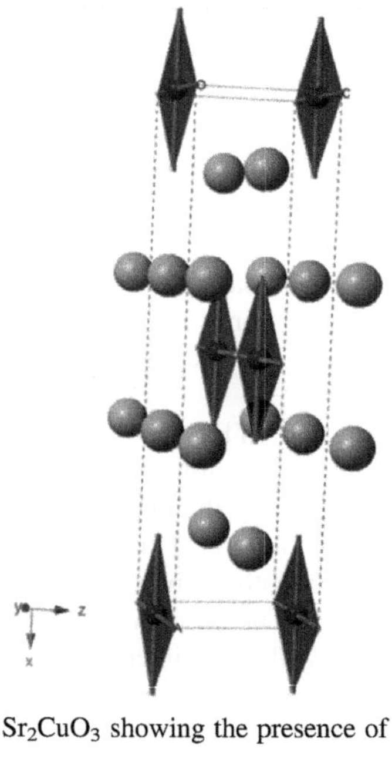

Figure 18.4 The structure of Sr_2CuO_3 showing the presence of square planar Cu—O chains.

temperatures the anions are mobile but the cations are not. Thus the substitution/insertion of fluorine is possible without breaking down the structure. The $Sr_2CuO_2F_{2+\delta}$ formed is, however, metastable, and heating to elevated temperatures (typically >400°C) leads to decomposition to SrF_2 and $SrCuO_2$.

One of the key issues with characterizing the structures of such oxide-fluoride compounds is the fact that O and F have near identical scattering lengths in the case of both X-rays and neutrons, and hence distinguishing between the O and F positions by either X-ray or neutron diffraction is difficult. Typically bond valence calculations and computer modeling studies are therefore employed to infer the F site. In this respect, such studies suggested that F occupied the apical and interstitial sites [5,41,43]. The location of F in the apical sites, thus leaving the two-dimensional Cu—O layers intact, is also consistent with the observation of superconductivity in this system.

The related Ca_2CuO_3, which is isostructural with Sr_2CuO_3, can also be prepared and similarly fluorinated at low temperature (240°C) with F_2 gas. However, the resulting $Ca_2CuO_2F_{2+\delta}$ ($\delta \approx 0.0$) was shown to adopt a different structure, namely the T′ structure adopted by Nd_2CuO_4. Thus in this material there are $[CuO_2]^{2-}$ layers which are separated by fluorite-type $[Ca_2F_2]^{2+}$ units (the F location was inferred through bond valence calculations) [6]. For this system, no superconductivity was observed due to an inability to incorporate extra F ($\delta \approx 0.0$) to raise the Cu oxidation state above 2.0+.

An interesting later discovery was made on investigating the reduction of $Sr_2CuO_2F_{2+\delta}$ under $10\%H_2/90\%N_2$ between 300 and 340°C for 5 min. In this work, the formation of $Sr_2CuO_2F_2$ (ie, $\delta = 0$) was achieved [46]. Somewhat unexpectedly this compound was shown to adopt the T′ structure as for $Ca_2CuO_2F_2$. This structural change will be discussed further in the next section.

The previously discussed studies invigorated research interest into the synthesis of cuprate superconductors containing fluorine, and it was shown that a number of additional systems could be synthesized by high-temperature, high-pressure routes (see for example Refs [3,58]). These systems, however, fall out of the scope of this chapter, which focuses on low-temperature routes.

As outlined earlier, gaseous fluorinating agents such as F_2 or NF_3 have been shown to be successful in fluorinating a range of cuprates; however, they do have a number of issues. These issues include the high toxicity of these gases, and their high reactivity, both of which cause issues in terms of their handling. The use of these gases therefore requires a specialist furnace setup (an example of which is shown in Fig. 18.5). Another key problem is the difficulty in controlling the level of fluorination. Thus typically these reactions are performed on small (<0.5 g) samples, and the samples are heated for short reaction times (typically less than 30 min). This is to ensure an even distribution of fluorine throughout the sample.

As a result of these difficulties, there has been considerable interest in low-temperature fluorination using solid-state fluorinating agents, which will be discussed in the next section.

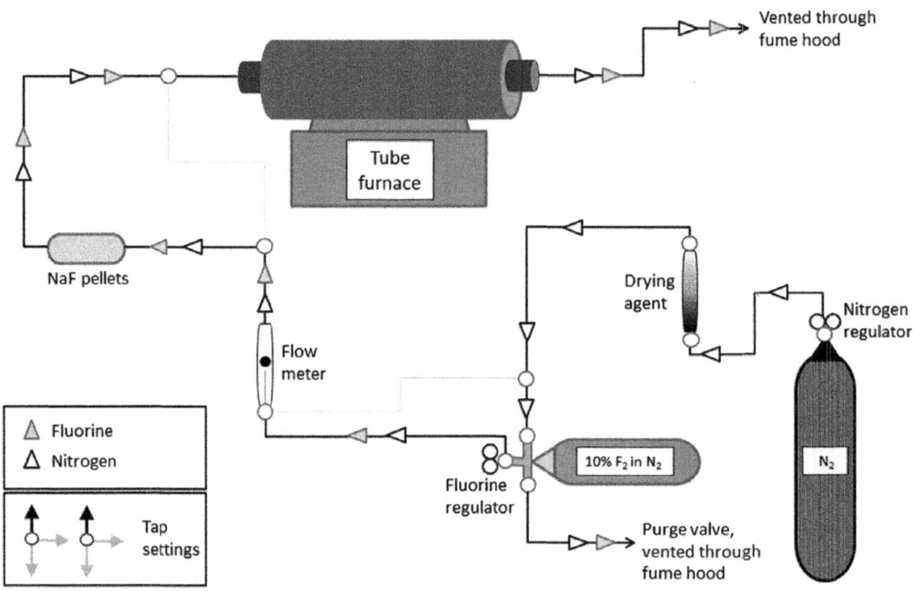

Figure 18.5 Example schematic diagram for a F_2 gas furnace (the N_2 line allows for purging out the furnace between experiments).

3. Fluorination of Cuprates With Solid-State Fluorinating Agents (CuF_2, ZnF_2, NiF_2, AgF_2, XeF_2, NH_4F, PVDF, PTFE)

A key benefit of solid-state fluorinating reagents is that they allow for potential control of the F content (by varying the amount of reagent added), while the ability to intimately mix with the sample allows the possibility for the fluorination of larger samples. One of the first solid-state reagents investigated for the low-temperature fluorination of cuprate materials was NH_4F. It was shown that low-temperature (225°C) reaction between Sr_2CuO_3 with NH_4F in air resulted in the formation of superconducting $Sr_2CuO_2F_{2+x}$ without the need for a postreductive anneal as for the F_2 gas–prepared samples (a large excess of NH_4F was employed, 4–6 moles of NH_4F per mole of Sr_2CuO_3) [60]. Interestingly despite the fact that NH_4F is nominally a nonoxidizing fluorinating agent and so reaction should simply occur via the substitution of one O^{2-} by two F^- (ie, $Sr_2CuO_3 + 2NH_4F \rightarrow Sr_2CuO_2F_2 + H_2O + 2NH_3$), the fact that a superconducting sample was observed suggests an average Cu oxidation state above 2+ and therefore some degree of oxidation, which was attributed to the presence of air. However, one issue noted with the use of NH_4F were the large SrF_2 impurities that resulted, most likely as a result of the formation of water in the reaction facilitating the decomposition of $Sr_2CuO_2F_{2+x}$. This study also investigated the synthesis of the mixed alkaline earth cation systems $Sr_{2-y}A_yCuO_2F_{2+x}$ (A = Ca, Ba). This work showed that the introduction of Ca led to an increase in the Sr/CaF_2 impurity levels and a decrease in T_c. In contrast the incorporation of Ba led to an increase in T_c. Specifically the T_c of 64 K for $Sr_{1.4}Ba_{0.6}CuO_2F_{2+x}$ represented the highest T_c observed for a material with the K_2NiF_4 structure.

With the issues regarding the observation of large Sr/Ca/BaF_2 impurities from the use of NH_4F, other solid-state fluorinating agents were also examined for the previously discussed systems. Low-temperature (225–250°C) reaction with transition metal difluorides (CuF_2, NiF_2, ZnF_2, AgF_2) was also shown to be able to produce superconducting $Sr_{2-y}A_yCuO_2F_{2+x}$, in these cases leading to much lower levels of alkaline earth fluoride impurities [61]. Such reagents operate via a metathesis style reaction with F from the difluoride entering the sample, and O replacing the F in the difluoride. As a result, the samples were shown to be a mixture of $Sr_{2-y}A_yCuO_2F_{2+x}$ and the transition metal oxide (CuO, NiO, ZnO, Ag_2O). As for samples prepared with F_2 gas, the as-synthesized samples were overdoped, indicating that reaction with these fluorinating agents is highly oxidizing. T_c values could be optimized through postsynthesis reduction, and this led to similar values to those observed for fluorination with NH_4F, confirming the beneficial effect of Ba doping on the superconducting properties of $Sr_{2-y}Ba_yCuO_2F_{2+x}$.

In order to try to eliminate this issue of the reagent contributing to the impurities, XeF_2 has also been investigated. In this case, while the reagent is a low-melting point solid, it decomposes on heating to give Xe and F_2 gases. The use of this reagent was shown to successfully produce $Sr_2CuO_2F_{2+x}$, although SrF_2 impurities were still observed [7]. In order to prevent loss of F_2 before reaction, the fluorination was performed in a closed Ni container.

The polymer-based reagents, poly(vinylidene fluoride), $-(CH_2CF_2)_n-$ (PVDF), poly(tetrafluoroethylene), $-(CF_2CF_2)_n-$ (PTFE), and perfluoropentadecane, have been shown to be successful fluorinating agents for these and other systems, with the former two systems attracting the most widespread use [63]. As for XeF_2, the use of these reagents eliminates issues with by-products from the fluorinating agents themselves, since these are burnt-off in the reaction, provided the reaction is performed under air or O_2. PVDF was shown to successfully produce $Ca_2CuO_2F_2$ by reaction with Ca_2CuO_3 (a stoichiometric mole ratio of 1:1 Ca_2CuO_3:PVDF (CH_2CF_2) monomer unit was employed) at 350°C. Although some CaF_2 impurity was observed, the level was much lower than that obtained through the reaction with NH_4F.

These reagents were also shown to be able to fluorinate the Sr-based system, Sr_2CuO_3, as well as the mixed Sr/Ca series, $Sr_{2-x}Ca_xCuO_3$ [38,65]. Interestingly rather than the formation of $Sr_2CuO_2F_{2+x}$ with the T structure, as observed for the other fluorinating agents (NH_4F, XeF_2, CuF_2, etc.), the reaction led to the formation of $Sr_2CuO_2F_2$ with the T′ structure, a phase that had only been made previously by reduction of F_2 gas–prepared $Sr_2CuO_2F_{2+x}$. These studies therefore indicate that all the $Sr_{2-x}Ca_xCuO_2F_2$ phases adopt the T′ structure, and it is only on the incorporation of additional anions to give $Sr_{2-x}Ca_xCuO_2F_{2+x}$ that the T structure is obtained. The formation of the T′ structure for $Sr_{2-x}Ca_xCuO_2F_2$ is contrary to modeling/lattice energy predictions which suggested that the Sr-rich series should adopt the T structure. This discrepancy may be related to the fact that these reactions are low-temperature syntheses leading to metastable compounds. Thus it has been proposed that the mechanism of the fluorination reaction may involve firstly the incorporation of F into the interstitial sites, rather than the channel vacancy sites as initially thought [38,65]. This result shows one of the potential difficulties with using computer modeling studies/lattice energy calculations on these oxide-fluoride phases. In particular, such studies rely on finding the most thermodynamically stable situation, whereas these oxide-fluoride phases are typically metastable and their synthesis involves low-temperature kinetically controlled processes. Thus the most thermodynamically stable structure is not necessarily the structure that is obtained.

The fluorination of Ruddlesden–Popper-type cuprates has also attracted attention. Ruddlesden–Popper phases have the general formula $A_{n+1}M_nO_{3n+1-x}$ and may be classed as intergrowths between perovskite and rock salt units. Thus La_2CuO_{4+x} can be classed as the $n = 1$ member of a series of Ruddlesden–Popper cuprates, with it containing a single perovskite layer. Phases can also be prepared containing a double perovskite layer, $n = 2$. The successful fluorination of such Ruddlesden–Popper cuprates, $La_{2-x}A_{1+x}Cu_2O_6$ (A = Ca, Sr) and $Nd_{2-x}Sr_{1-x}Cu_2O_{6-y}$, was demonstrated with a range of fluorinating reagents, F_2 gas, CuF_2, NH_4F [62]. This work highlighted differences between the fluorinating reagents, showing that for these systems different products were obtained due to the different oxidizing power of the reagents. Reaction with F_2 gas was shown to favor F insertion into vacant/interstitial sites, leading to a large increase in Cu oxidation state. In contrast reaction with NH_4F was shown in this case to be nonoxidative, with the Cu oxidation state maintained, indicating substitution of one O^{2-} by two F^-. Reaction with CuF_2 was shown to give products in between these two extremes, with a combination of substitution and insertion, and an increase in Cu oxidation state in between the two extremes, the degree of which was dependent on the temperature used (in this case the higher the fluorination temperature, the more prevalent the substitution reaction became).

These solid-state fluorination methods have also been applied to a range of other cuprate systems (for example, see the reviews Refs [34,51]). Superconductivity up to 55 K was achieved through fluorination of $La_{0.7}Sr_{1.3}CuO_{3.25}$ with CuF_2 [18], and similar results ($T_c = 44$ K) were observed for fluorination of $Nd_{0.7}Sr_{1.3}CuO_{3.25}$ [70]. High levels of F incorporation have also been reported from fluorination of La_2CuO_4 and Nd_2CuO_{4-x} using XeF_2 [2,30,36]. Fluorination of $HgBa_2CuO_{4+\delta}$ with XeF_2 could also be achieved with a maximum observed T_c of 97 K [1]. Similar successful fluorination of $HgBa_2Ca_2Cu_3O_{8+\delta}$ was also reported with a high T_c of 138 K, which increased to 166 K under high pressure [52,57].

4. Extension of Low-Temperature Fluorination Routes to Other Transition Metal–Containing Systems

Following the successful demonstration of these low-temperature routes to the fluorination of cuprate materials, they have since been extended to a range of other transition metal–containing systems. $Ba_{2-x}Sr_xPdO_3$ is isostructural with Sr_2CuO_3, and fluorination of these systems has also been investigated using PVDF (heat treatment at 300–350°C). The results showed the formation of $Ba_{2-x}Sr_xPdO_2F_2$ with the T′ structure and hence square planar Pd. This observation was correlated with the strong preference for Pd^{2+} (a d^8 ion) to adopt this geometry [8].

The fluorination of perovskite-related ferrites has also been investigated, driven by the ease and flexibility of fluorination with the polymers PVDF and PTFE. In these cases the fluorination was shown to be reductive even when performed under air. Thus fluorination of $SrFeO_{3-x}$ with PVDF or PTFE at 350–400°C has been used to produce $SrFeO_2F$ where the Fe oxidation state has been confirmed by Mössbauer spectroscopy to be Fe^{3+}. Moreover, this material was shown to display antiferromagnetic order at room temperature, with magnetization measurements indicated a high Neel temperature of ≈685 K [11,21]. Interestingly, $SrFeO_{3-x}F_x$ cannot be prepared single phase by fluorination of the brownmillerite-type phase $SrFeO_{2.5}$ with F_2 gas. In this case, a two-phase mixture of $SrFeO_2F$ and $SrFeO_{3-x}$ was obtained [67], indicating a disproportionation process occurring for "$SrFeO_{2.5}F_{0.5}$." In contrast, fluorination of the related brownmillerite phases, $Sr_2Co_2O_5$ and Sr_2CoFeO_5, using F_2 gas does not lead to disproportionation, with successful F insertion to give $Sr_2Co_2O_5F$ and Sr_2CoFeO_5F, respectively [67]. Similarly fluorination of $Sr_2Mn_2O_5$ has been reported using XeF_2 to give $Sr_2Mn_2O_{5-x}F_{1-x}$ [50].

Further work has similarly shown the successful synthesis of $Sr_{1-x}Ba_xFeO_2F$ from the low-temperature (350–400°C) reaction between PVDF and $Sr_{1-x}Ba_xFeO_{3-x}$ [21,39]. As for $SrFeO_2F$, Mössbauer spectroscopy indicated that all the Fe was Fe^{3+}, with magnetic ordering at room temperature and a similarly high T_N. The structural studies of $BaFeO_2F$ showed some interesting results. While X-ray and neutron diffraction studies indicated that the average structure of this phase was cubic, a large displacement of the Fe from the ideal perovskite B cation site was indicated, similar to that observed in the ferroelectric material $BaTiO_3$. The overall cubic symmetry of this phase could be explained by the random nature of these displacements [13,39]. Further work has shown that the F content can be varied in this system, by reducing the amount of polymer employed, thus illustrating the potential of these polymer reagents for the accurate control of the F content. Thus the synthesis of the system $BaFeO_{2.33}F_{0.33}$ has been reported by Clemens et al. [29]. The electrochemical fluorination of $BaFeO_{2.5}$ has also been demonstrated [28]. In this electrochemical synthesis experiment, a solid electrolyte, $La_{1-x}Ba_xF_{3-x}$ was utilized, and the cell was operated at 150°C to achieve sufficient ionic conduction in this electrolyte, and the results suggested the formation of a material with composition of approximately $BaFeO_{2.5}F_{0.5}$.

Subsequently two other polymorphs of $BaFeO_2F$ have been prepared by starting with different oxide precursor systems and exploiting the topotactic nature of the fluorination process. Thus hexagonal 6H-$BaFeO_3$ can be prepared from stoichiometric amounts of $BaCO_3$ and Fe_2O_3 by synthesis at temperatures below 950°C. Fluorination of this phase with PVDF at 350–400°C leads to the formation of 6H-$BaFeO_2F$ [23]. Another polymorph, 15R-$BaFeO_2F$ was prepared by similar fluorination of 15R-$BaFeO_{3-x}F_{0.2}$ with PVDF [24]. The parent phase, 15R-$BaFeO_{3-x}F_{0.2}$, was first reported by Sturza et al. [66] and can be prepared by reaction between stoichiometric amounts of $BaCO_3$, Fe_2O_3, and

BaF_2 at 950°C, an example of an F-containing perovskite system that is stable and hence can be prepared at elevated temperatures. For both the abovementioned systems, neutron diffraction and Mössbauer spectroscopy studies confirmed antiferromagnetic ordering at room temperature as for cubic $BaFeO_2F$. The structures of the three polymorphs are shown in Fig. 18.6. They differ in the linking of the Fe octahedra: for cubic $BaFeO_2F$ all the Fe octahedra are corner-linked, while for 6H-$BaFeO_2F$ and 15R-$BaFeO_2F$ there are different degrees of corner-linking and face-sharing of the Fe octahedra. Bond valence sum calculations suggested a degree of O/F ordering in these latter systems. These three examples highlight the potential to make use of the topotactic nature of these fluorination methods to produce new phases by varying the starting precursor oxide and hence "locking in" the cation order.

The fluorination of Ba doped $BiFeO_3$ with PVDF has also been examined [27]. While the parent oxide phases $Bi_{1-x}Ba_xFeO_{3-\delta}$ ($x = 0.2, 0.3$) were cubic, the fluorinated phases $Bi_{1-x}Ba_xFeO_{3-x}F_x$ displayed a large polar tetragonal distortion and were isostructural with $Pb(Ti,Zr)O_3$, suggesting potential applications as Pb-free ferroelectrics. As for $BaFeO_2F$, the Fe oxidation state was 3+ and these Bi-doped samples showed G-type antiferromagnetic ordering.

The fluorination of La-doped $SrMO_{3-\delta}$ (M = Fe, Co) has also been examined. Fluorination of these $Sr_{1-x}La_xMO_{3-\delta}$ phases with PVDF/PTFE at 400°C can be utilized to prepare the corresponding F-containing phases $Sr_{1-x}La_xMO_{3-x}F_x$ in which the transition metal cation's oxidation state has been reduced to 3+. As for $SrFeO_2F$, these systems display magnetic ordering at room temperature [12,22,25].

In addition to the previous work on perovskite systems, the fluorinations of a range of other $n = 1$ (K_2NiF_4-type) and $n = 2$ Ruddlesden—Popper systems have also

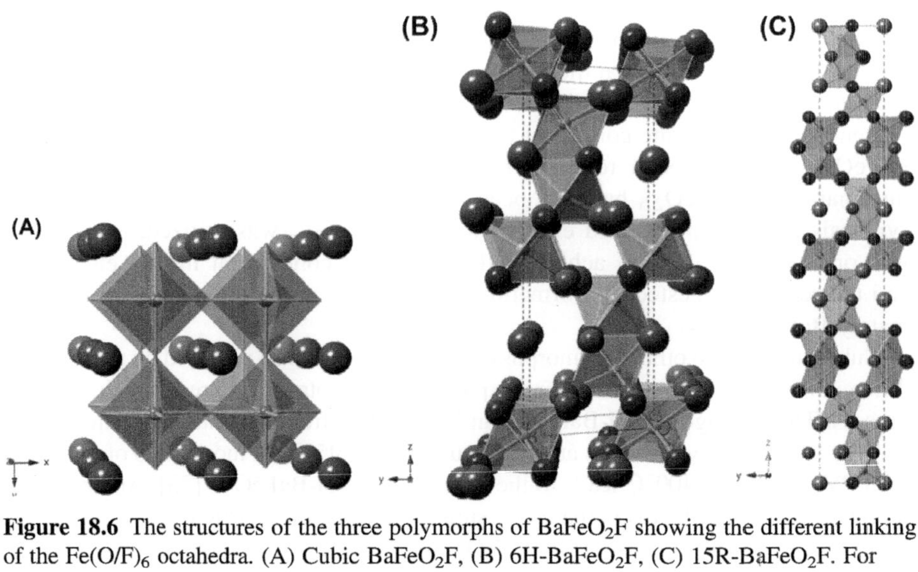

Figure 18.6 The structures of the three polymorphs of $BaFeO_2F$ showing the different linking of the $Fe(O/F)_6$ octahedra. (A) Cubic $BaFeO_2F$, (B) 6H-$BaFeO_2F$, (C) 15R-$BaFeO_2F$. For 6H-$BaFeO_2F$, and 15R-$BaFeO_2F$ there is evidence for some F ordering.

been investigated leading to the synthesis of a range of systems with high F occupancy of the interstitial sites [26]. The fluorination of the $n = 2$ Ruddlesden–Popper system, $Sr_3Fe_2O_{7-x}$, in particular, has been investigated by a number of researchers. Initial work on this system focused on the fluorination of a prereduced phase, $Sr_3Fe_2O_6$, with F_2 gas leading to fluorine insertion and hence an increase in the Fe oxidation state. The phase composition was determined to be $Sr_3Fe_2O_6F_{0.87}$ with the F inserting into the vacant anion sites [14]. Further work showed that fluorination of as-prepared $Sr_3Fe_2O_{7-\delta}$ (ie, without prereduction) was possible by reaction (350–400°C) with PTFE/PVDF. A 1:0.5 ratio of $Sr_3Fe_2O_{7-x}$:PTFE (CF_2CF_2) monomer unit led to the formation of $Sr_3Fe_2O_{5+x}F_{2-x}$ ($x \approx 0.44$) [68]. A similar synthesis could be achieved with PVDF [37]. In this later study, higher levels of PVDF were also investigated, and the results showed that it was possible to prepare two other high-F content phases, $Sr_3Fe_2O_4F_4$ and $Sr_3Fe_2O_3F_6$, showing the flexibility to control the F content through reaction with F-containing polymer reagents. In these latter two cases, F inserts into the interstitial sites within the rock salt–type SrO layers, which leads to a large expansion in the unit cell along the c direction. Mössbauer spectroscopy studies indicated the presence of only Fe^{3+} in these two high-F content phases, and all samples showed magnetic order at room temperature. The synthesis of the highest F content phase, however, led to high-SrF_2 impurities attributed to the thermal instability of this very high-F content phase.

Fluorinations of the related $La_2SrFe_2O_7$ and $LaSrFeO_4$ have also been reported using F_2 gas (240–290°C). As expected for reaction with F_2 gas, F insertion into the interstitial sites and oxidation of the Fe were observed leading to phases with approximate compositions of $La_2SrFe_2O_7F_2$ and $LaSrFeO_3F_3$ [51]. Similar fluorination of $La_2BaFe_2O_7$ (F_2 gas, 300°C) led to the formation of $La_2BaFe_2O_5F_4$ in which there was complete filling of the interstitial sites [35].

The fluorination of the Ruddlesden–Popper phase $Sr_3Ru_2O_7$ has also been reported. Reaction with F_2 gas was shown to give $Sr_3Ru_2O_7F_2$ corresponding to F insertion into interstitial sites (filling these sites) within the SrO rock salt–type layers with concomitant Ru oxidation [50]. Similar studies on the fluorination of $LaSrMnO_4$ were shown to produce $LaSrMnO_4F_{1.7}$, for which the interstitial sites are nearly all filled. Subsequent mixing of the latter with $LaSrMnO_4$ and heat treatment at 300°C led to the formation of $LaSrMnO_4F$ in which the interstitial sites were filled in alternate layers, suggesting that the fluorination process can occur in stages [4]. The staged nature of the fluorination process was also demonstrated in the fluorination of Sr_2TiO_4 to give $Sr_2TiO_3F_2$, which contains alternate layers of filled and empty interstitial sites (Fig. 18.7). This fluorination can be achieved either through the reaction with PVDF or ZnF_2 [63,64].

One of the key problems with these oxide-fluoride systems is their instability at high temperatures, which is the key reason why low-temperature routes are required for their synthesis. As noted earlier, this instability results from the high thermodynamic stability of alkaline earth fluorides (AF_2: A = Ca, Sr, Ba) and rare earth fluorides/oxide fluorides (ROF, RF_3: R = rare earth), and thus typically the oxide-fluoride phases will decompose at temperatures above 500°C to give such AF_2/ROF/RF_3 impurity phases. This low-temperature stability has therefore hindered progress in their electrical

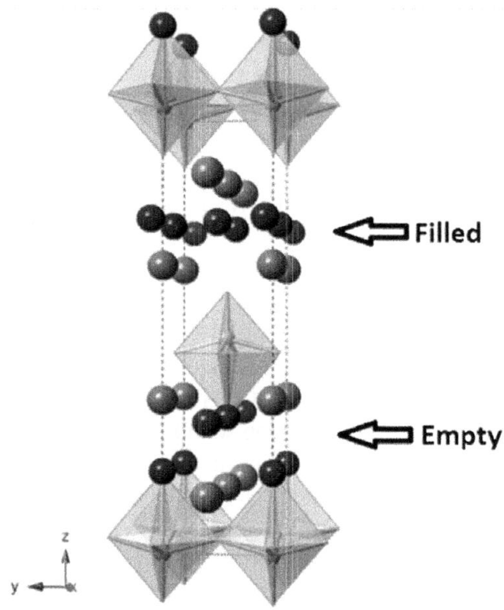

Figure 18.7 The structure of $Sr_2TiO_3F_2$ showing alternate layers of filled and empty interstitial sites. Bond valence sums indicate F occupancy of apical and interstitial sites (Ti are at the center of the octahedra).

characterization, since it precludes the preparation of dense sintered membranes. However, there have been advances in this respect through the preparation of thin films and the postfluorination of these with PVDF [76]. Two different approaches (spin coating and vapor transport) have been shown to be successful in fluorinating thin films of $SrFeO_{3-\delta}$ to produce $SrFeO_{3-x}F_x$. Unlike fluorination of bulk $SrFeO_{3-\delta}$ samples, where only the formation of the endmember $SrFeO_2F$ phase can be achieved, the use of thin films allowed fluorination to give a range of F contents by controlling the reaction temperature and amount of vacancies in the $SrFeO_{3-\delta}$ thin film. The thin-film $SrFeO_{3-\delta}$ was grown either by pulsed-laser deposition or molecular beam epitaxy on $SrTiO_3$ substrates and then fluorinated using either a spin coating (10 wt% solution of PVDF in dimethylformamide) or vapor transport method (the perovskite thin film is heated under N_2/Ar in a quartz tube with PVDF pellets upstream). Through this fluorination of thin films it has been proved that it is possible to measure changes in electronic conductivity on fluorination, showing a large reduction in the resistivity at room temperature on fluorination of the oxygen-deficient thin film [53]. Similarly such methodologies have also been successfully applied to the fluorination of $SrCoO_{3-\delta}$ thin films [45].

The reductive ability of fluorination with polymer reagents has been further extended by performing the fluorination under an inert atmosphere. In such cases, it has been shown that reductive fluorination can be achieved in systems for which fluorination in air is unsuccessful. Thus Kobayashi et al. have shown that it is possible to perform a strongly reductive low-temperature fluorination of $RbLaNb_2O_7$ systems using PTFE under an inert atmosphere [77]. This led to the substitution of one O^{2-} by one F^- to give $RbLaNb_2O_6F$. The resulting mixed valency for Nb led to a significant increase in the electronic conductivity. This reductive fluorination method has also been investigated for WO_3. The results indicated the formation of $WO_{3-x}F_x$ ($x \approx 0.45$) with evidence for superconductivity below 0.4 K [42].

All the studies discussed previously highlight the great flexibility and future potential of these low-temperature fluorination routes for the synthesis of new F-containing phases. Indeed, the extension of these routes to the fluorination of nitrides has also been illustrated, with the formation of $Ce_2MnN_3F_2$ from low-temperature heat treatment of Ce_2MnN_3 under F_2 [40]

5. Conclusions and Future Directions

While the development of low-temperature fluorination routes for oxide systems was inspired by their early use to synthesize superconducting cuprates, their applicability has been demonstrated for a wide range of other transition metal oxides. A range of reagents/processes have been employed from the use of gaseous reagents to solid-state reagents and electrochemical fluorination. Each of the methods have been shown to possess their own characteristics, and advantages and disadvantages, as highlighted in the previous sections and summarized in Table 18.1. The use of polymer reagents, such as PVDF and PTFE, has been attracting significant interest due to their ease of use, and the ability to control the F level (by the amount of polymer employed). Thus, whereas fluorination using F_2, NF_3 gas requires a quite specialist setup, the

Table 18.1 Summary of the Characteristics of the Different Low-Temperature Fluorination Methods

Fluorinating Method	Characteristics/Advantages	Disadvantages
F_2, NF_3 gas	Highly oxidative Generally favors fluorine insertion giving high oxidation states	Toxicity/handling difficulties of reagent gases: requirement for specialist equipment. Difficult to control level of fluorination, which also limits the quantity of sample that can be fluorinated.
NH_4F	Generally nonoxidative Favors substitution (two F^- replacing one O^{2-}) Some control of the F content through the amount of NH_4F added Reaction under inert atmospheres can give rise to reduction	Strong tendency to form AF_2 (A = alkaline earth) or ROF (R = rare earth) impurities.

Continued

Table 18.1 **Summary of the Characteristics of the Different Low-Temperature Fluorination Methods—cont'd**

Fluorinating Method	Characteristics/Advantages	Disadvantages
MF_2 (M = Cu, Zn, Ni, Ag)	In terms of oxidizing power, lies between that of NH_4F and F_2 gas. Thus these can give rise to both substitution and insertion processes Good control of the F content through the amount of MF_2 added Much lower AF_2 (A = alkaline earth) or ROF (R = rare earth) impurities than observed for NH_4F	MO impurities from the reagent added, although some research has shown that it is possible to perform the reaction in a sealed system with the precursor oxide and MF_2 kept separate (in such cases, however, it is difficult to control the F content).
XeF_2	Similar oxidizing power to F_2 gas Some control of F content through amount of XeF_2 added Typically lower AF_2 (A = alkaline earth) or ROF (R = rare earth) impurities than observed for NH_4F	If reaction not performed in a sealed tube, then can lose the reagent before it has a chance to react with the precursor oxide. High reagent cost.
Fluorinated polymers (PTFE, PVDF)	Favors substitution processes; either nonoxidative (two F^- replacing one O^{2-}) or reductive (one F^- replacing one O^{2-}) Good control of the F content through the amount of polymer added Typically gives significantly purer samples than other fluorination methods	If oxidation of the sample is required, then this fluorinating agent is typically not suitable.
Electrochemical fluorination	Highly oxidative Favors fluorine insertion giving high oxidation states	Tendency to give high F content impurity coating. Experimental setup more complicated than use of solid-state fluorinating reagents.

fluorination with PVDF/PTFE can be performed in any furnace (albeit it is advisable to perform with the furnace placed in/vented to a fume cupboard). The recent exploitation of these polymer-based fluorination routes to thin films offers great potential to investigate the effect of F incorporation on the electronic properties, as well as in terms of technological applications. In addition, an avenue of research that has not received attention is the use of combinations of these fluorination processes. Thus, for example, initial nonoxidative fluorination using PVDF/PTFE could be performed, followed by heat treatment under F_2 gas to provide some degree of postoxidation. Overall the ability to extend fluorination from reductive (polymer reagents) to oxidative (F_2) gives a wide avenue of research to ensure that these methods continue to be exploited in the armory of the synthetic solid-state chemist/material scientist.

References

[1] A.M. Abakumov, V.L. Aksenov, V.A. Alyoshin, E.V. Antipov, A.M. Balagurov, D.A. Mikhailova, S.N. Putilin, M.G. Rozova, Effect of fluorination on the structure and superconducting properties of the Hg-1201 phase, Phys. Rev. Lett. 80 (1998) 385–388.

[2] A.M. Abakumov, J. Hadermann, G. Van Tendeloo, R.V. Shpanchenko, P.N. Oleinikov, E.V. Antipov, Anion ordering in fluorinated La_2CuO_4, J. Solid State Chem. 142 (1999) 440–450.

[3] S. Adachi, T. Tatsuki, T. Tamura, K. Tanabe, Halooxocuprate superconductors and related compounds with the 02(n-1)n and 0222 structures, Chem. Mater. 10 (1998) 2860–2869.

[4] L.D. Aikens, R.K. Li, C. Greaves, The synthesis and structure of a new oxide fluoride, $LaSrMnO_4F$, with staged fluorine insertion, Chem. Commun. (2000) 2129–2130.

[5] M. Al-Mamouri, P.P. Edwards, C. Greaves, M. Slaski, Synthesis and superconducting properties of the strontium copper oxy-fluoride $Sr_2CuO_2F_{2+\delta}$, Nature 369 (1994) 382–384.

[6] M. Al Mamouri, P.P. Edwards, C. Greaves, P.R. Slater, M. Slaski, Synthesis and structure of the calcium copper oxyfluoride, $Ca_2CuO_2F_{2+\delta}$, J. Mater. Chem. 5 (1995) 913–916.

[7] E.I. Ardashnikova, S.V. Lubarsky, D.I. Denisenko, R.V. Shpanchenko, E.V. Antipov, G. Van Tendeloo, A new way of synthesis and characterisation of superconducting oxyfluoride $Sr_2Cu(O,F)_{4+\delta}$, Physica C 253 (1995) 259–265.

[8] T. Baikie, E.L. Dixon, J.F. Rooms, N.A. Young, M.G. Francesconi, $Ba_{2-x}Sr_xPdO_2F_2$ ($0 <= x <= 1.5$): the first palladium-oxide-fluorides, Chem. Commun. (2003) 1580–1581.

[9] J.G. Bednorz, K.A. Müller, Possible high T_c superconductivity in the Ba-La-Cu-O system, Z. Physik B Condens. Matter 64 (1986) 189–193.

[10] J. Beille, R. Cabanel, C. Chaillout, B. Chevalier, G. Demazeau, F. Deslandes, J. Etourneau, P. Lejay, C. Michel, J. Porvost, Superconductivity below 40 K in orthorhombic structure La_2CuO_4, Comptes Rendus de l Academie Des Sci. Ser. 304 (1987) 1097–1101.

[11] F.J. Berry, X. Ren, R. Heap, P. Slater, M.F. Thomas, Fluorination of perovskite-related $SrFeO_{3-d}$, Solid State Commun. 134 (2005) 621–624.

[12] F.J. Berry, X. Ren, R. Heap, P. Slater, M.F. Thomas, Fluorination of perovskite-related phases of composition $La_{1-x}Sr_xFe_{1-y}Co_yO_{3-d}$, J. Phys. Chem. Solids 69 (2008) 2032–2036.

[13] F.J. Berry, F.C. Coomer, C. Hancock, Ö. Helgason, E.A. Moore, P.R. Slater, A.J. Wright, M.F. Thomas, Structure and magnetic properties of the cubic oxide fluoride $BaFeO_2F$, J. Solid State Chem. 184 (2011) 1361–1366.

[14] G.S. Case, A.L. Hector, W. Levason, R.L. Needs, M.F. Thomas, M.T. Weller, Syntheses, powder neutron diffraction structures and Mossbauer studies of some complex iron oxyfluorides: $Sr_3Fe_2O_6F_{0.87}$, Sr_2FeO_3F and $Ba_2InFeO_5F_{0.68}$, J. Mater. Chem. 9 (1999) 2821–2827.

[15] M. Cassart, E. Grivei, J.P. Issi, E. Bensalem, B. Chevalier, C. Brisson, A. Tressaud, Improvement of the superconducting properties of the $La_{2-x}Ba_xCuO_4$ system ($0 \leq x \leq 0.15$) by low temperature fluorination, Physica C 213 (1993) 327–337.
[16] R.J. Cava, R.B. Vandover, B. Batlogg, E.A. Rietman, Bulk superconductivity at 36 K in $La_{1.8}Sr_{0.2}CuO_4$, Phys. Rev. Lett. 58 (1987) 408–410.
[17] R.J. Cava, B. Batlogg, C.H. Chen, E.A. Rietman, S.M. Zahurak, D. Werder, Oxygen stoichiometry, superconductivity and normal-state properties of $YBa_2Cu_3O_{7-\delta}$, Nature 329 (1987) 423–425.
[18] X.L. Chen, J.K. Liang, W.H. Tang, G. Wang, G.H. Rao, Superconductivity at 55 K in $La_{0.7}Sr_{1.3}Cu(O,F)_{4+\delta}$ with reduced CuO_2 sheets and apical anions, Phys. Rev. B 52 (1995) 16233–16236.
[19] B. Chevalier, A. Tressaud, B. Lepine, K. Amine, J.M. Dance, L. Lozano, E. Hickey, J. Etourneau, Stabilization of a new superconducting phase by low temperature fluorination of La_2CuO_4, Physica C 167 (1990) 97–101.
[20] C.W. Chu, P.H. Hor, R.L. Meng, L. Gao, Z.J. Huang, Y.Q. Wang, Evidence for superconductivity above 40 K in the La-Ba-Cu-O compound system, Phys. Rev. Lett. 58 (1987) 405–407.
[21] O. Clemens, R. Haberkorn, P.R. Slater, H.P. Beck, Synthesis and characterisation of the $Sr_xBa_{1-x}FeO_{3-y}$-system and the fluorinated phases $Sr_xBa_{1-x}FeO_2F$, Solid State Sci. 12 (2010) 1455–1463.
[22] O. Clemens, M. Kuhn, R. Haberkorn, Synthesis and characterization of the $La_{1-x}Sr_xFeO_{3-\delta}$ system and the fluorinated phases $La_{1-x}Sr_xFeO_{3-x}F_x$, J. Solid State Chem. 184 (2011) 2870–2876.
[23] O. Clemens, A.J. Wright, F.J. Berry, R.I. Smith, P.R. Slater, Synthesis, structural and magnetic characterisation of the fully fluorinated compound 6H–$BaFeO_2F$, J. Solid State Chem. 198 (2013) 262–269.
[24] O. Clemens, F.J. Berry, A.J. Wright, K.S. Knight, P.R. Slater, Synthesis, structural and magnetic characterisation of 15R–$BaFeO_2F$, J. Solid State Chem. 203 (2013) 218–226.
[25] O. Clemens, J.M. Perez-Mato, F.J. Berry, A.J. Wright, K.S. Knight, P.R. Slater, A neutron diffraction study on compounds of the system $La_{1-x}Sr_xFeO_{3-x}F_x$ ($x = 1, 0.8, 0.5, 0.2$) and their magnetic properties, J. Solid State Chem. 206 (2013) 158–169.
[26] O. Clemens, P.R. Slater, Topochemical modifications of mixed metal oxide compounds by low-temperature fluorination routes, Rev. Inorg. Chem. 33 (2013) 105–117.
[27] O. Clemens, R. Kruk, E. Patterson, C. Loho, C. Reitz, A.J. Wright, K.S. Knight, H. Hahn, P.R. Slater, Introducing a large polar tetragonal distortion into Ba-doped $BiFeO_3$ by low temperature fluorination, Inorg. Chem. 53 (2014) 12572–12583.
[28] O. Clemens, C. Rongeat, M.A. Reddy, A. Giehr, M. Fichtner, H. Hahn, Electrochemical fluorination of perovskite-type $BaFeO_{2.5}$, Dalton Trans. 43 (2014) 15771–15778.
[29] O. Clemens, Structural characterization of a new vacancy ordered perovskite modification found for Ba3Fe3O7F ($BaFeO_{2.333}F_{0.333}$): towards understanding of vacancy ordering for different perovskite-type ferrites, J. Solid State Chem. 225 (2015) 261–270.
[30] G. Corbel, J.P. Attfield, J. Hadermann, A.M. Abakumov, A.M. Alekseeva, M.G. Rozova, E.V. Antipov, Anion rearrangements in fluorinated $Nd_2CuO_{3.5}$, Chem. Mater. 15 (2003) 189–195.
[31] W.I.F. David, W.T.A. Harrison, J.M.F. Gunn, O. Moze, A.K. Soper, P. Day, J.D. Jorgensen, D.G. Hinks, M.A. Beno, L. Soderholm, D.W. Capone, I.K. Schuller, C.U. Sgre, K. Zhang, J.D. Grace, Structure and crystal chemistry of the high T_c superconductor $YBa_2Cu_3O_{7-x}$, Nature 327 (1987) 310–312.

[32] P.K. Davies, J.A. Stuart, D. White, C. Lee, P.M. Chaikin, M.J. Naughton, R.C. Yu, R.L. Ehrenkaufer, Fluorination of superconducting $Ba_2YCu_3O_{9-\delta}$, Solid State Commun. 64 (1987) 1441–1444.
[33] G. Demazeau, F. Tresse, T. Plante, B. Chevalier, J. Etourneau, C. Michel, M. Hervieu, B. Raveau, P. Lejay, A. Sulpice, R. Tournier, Evolution of the superconducting properties in the La_2CuO_4 lattice versus oxygen pressure, Physica C 153 (1988) 824–825.
[34] C. Greaves, M.G. Francesconi, Fluorine insertion in inorganic materials, Curr. Opin. Solid State Mater. Sci. 3 (1998) 132–136.
[35] N.M. Gurusinghe, J.C. Fones, J.F. Marco, F.J. Berry, C. Greaves, Fluorine insertion into the Ruddlesden Popper phase $La_2BaFe_2O_7$: the structure and magnetic properties of $La_2BaFe_2O_5F_4$, Dalton Trans. 43 (2014) 2038–2043.
[36] J. Hadermann, G. Van Tendeloo, A.M. Abakumov, M.G. Rozova, E.V. Antipov, HREM study of fluorinated Nd_2CuO_4, J. Solid State Chem. 157 (2001) 56–61.
[37] C.A. Hancock, T. Herranz, J.F. Marco, F.J. Berry, P.R. Slater, Low temperature fluorination of $Sr_3Fe_2O_{7-x}$ with polyvinylidine fluoride: an X-ray powder diffraction and Mössbauer spectroscopy study, J. Solid State Chem. 186 (2012) 195–203.
[38] R. Heap, Synthesis and Characterisation of K_2NiF_4-Related Oxides and Oxide Fluorides, University of Birmingham, 2006. Ph.D. thesis).
[39] R. Heap, P.R. Slater, F.J. Berry, O. Helgason, A.J. Wright, Synthesis and structural determination of the new oxide fluoride $BaFeO_2F$, Solid State Commun. 141 (2007) 467–470.
[40] D.A. Headspith, E. Sullivan, C. Greaves, M.G. Francesconi, Synthesis and characterisation of the quaternary nitride-fluoride $Ce_2MnN_3F_2$-delta, Dalton Trans. (2009) 9273–9279.
[41] J.P. Hill, N.L. Allan, W.C. Mackrodt, Atomistic simulation of $M_2CuO_2F_{2+\delta}$ (M=Ca, Sr), Chem. Commun. 24 (1996) 2703–2704.
[42] D. Hirai, E. Climent-Pascual, R.J. Cava, Superconductivity in $WO_{2.6}F_{0.4}$ synthesized by reaction of WO_3 with teflon, Phys. Rev. B 84 (2011) 174519.
[43] M.S. Islam, S. DArco, Defect chemistry of the oxyfluoride $Sr_2CuO_2F_{2+\delta}$: relationship to high T_c superconductivity, Chem. Commun. 19 (1996) 2291–2292.
[44] A.C.W.P. James, D.W. Murphy, S.M. Zahurak, Superconductivity at 27 K in fluorine-doped Nd_2CuO_4, Nature 338 (1989) 240–241.
[45] T. Katayama, A. Chikamatsu, Y. Hirose, Topotactic fluorination of strontium cobalt oxide epitaxial thin films, J. Sol Gel Sci. Technol. 73 (2015) 527–530.
[46] J.L. Kissick, C. Greaves, P.P. Edwards, V.M. Chekashenko, E.Z. Kurmaev, S. Bartkowski, M. Neumann, Synthesis, structure and XPS characterisation of the stoichiometric phase $Sr_2CuO_2F_2$, Phys. Rev. B 56 (1997) 2831–2835.
[47] G. Kong, M.O. Jones, J.S. Abell, P.P. Edwards, S.T. Lees, K.E. Gibbons, I. Gameson, M. Aindow, Microstructure of laser-ablated superconducting $La_2CuO_4F_x$ thin films on $SrTiO_3$, J. Mater. Res. 16 (2001) 3309–3316.
[48] J.R. LaGraff, E.C. Behrman, J.A.T. Taylor, F.J. Rotella, J.D. Jorgensen, L.Q. Wang, P.G. Mattocks, Magnetic and structural properties of $YBa_2Cu_3O_{7-x}F_y$ as prepared by a NF_3 gas anneal, Phys. Rev. B 39 (1989) 347–355.
[49] R.K. Li, C. Greaves, Double-layered ruthenate $Sr_3Ru_2O_7F_2$ formed by fluorine insertion into $Sr_3Ru_2O_7$, Phys. Rev. B 62 (2000) 3811–3815.
[50] M.V. Lobanov, A.M. Abakumov, A.V. Sidorova, M.G. Rozova, O.G. D'Yachenko, E.V. Antipov, J. Hadermann, G. Van Tendeloo, Synthesis and investigation of novel Mn-based oxyfluoride $Sr_2Mn_2O_{5-x}F_{1+x}$, Solid State Sci. 4 (2002) 19–22.
[51] E.E. McCabe, C. Greaves, Fluorine insertion reactions into pre-formed metal oxides, J. Fluorine Chem. 128 (2007) 448–458.

[52] M. Monteverde, C. Acha, M. Nunez-Regeiro, D.A. Pavlov, K.A. Lokshin, S.N. Putilin, E.V. Antipov, High pressure effects in fluorinated $HgBa_2Ca_2Cu_3O_{8+\delta}$, Europhys. Lett. 72 (2005) 458−464.

[53] E.J. Moon, Y.J. Xie, E.D. Laird, D.J. Keavney, C.Y. Li, S.J. May, Fluorination of epitaxial oxides: synthesis of perovskite oxyfluoride thin films, J. Amer. Chem. Soc. 136 (2014) 2224−2227.

[54] N. Nguyen, F. Studer, B. Raveau, Ternary mixed-valence copper oxides of the type K_2NiF_4 deficient in oxygen − progressive evolution of $La_{2-x}Sr_xCuO_{4-x/2+\delta}$ oxides from a semiconductor state to a semimetallic state, J. Phys. Chem. Solids 44 (1983) 389−400.

[55] G.B. Peacock, I. Gameson, M. Slaski, J.J. Capponi, P.P. Edwards, Fluorination of mercury-based high-temperature superconductors, Physica C 289 (1997) 153−160.

[56] C. Perrin, A. Dinia, O. Peña, M. Sergent, P. Burlet, J. Rossat-Mignod, Fluorination of $YBa_2Cu_3O_{6.7}$: powder neutron diffraction determination of fluorine sites and their influence on the superconducting properties, Solid State Commun. 76 (1990) 401−407.

[57] S.N. Putilin, E.V. Antipov, A.M. Abakumov, M.G. Rozova, K.A. Lokshin, D.A. Pavlov, A.M. Balagurov, D.V. Sheptyakon, M. Marezio, Effect of fluorination and high pressure on the structure and properties of the Hg-bearing superconducting Cu mixed oxides, Physica C 338 (2000) 52−59.

[58] P.M. Shirage, D.D. Shivagan, Y. Tanaka, Y. Kodama, H. Kito, A. Iyo, Simplest nontoxic double-layered cuprate $Ba_2CaCu_2O_4(O/F)_2$ superconductor with a transition temperature of 108 K, Appl. Phys. Lett. 92 (2008) 222501.

[59] R.V. Shpanchenko, M.G. Rozova, A.M. Abakumov, E.I. Ardashnikova, M.L. Kovba, S.N. Putilin, E.V. Antipov, O.I. Lebedev, G. Van Tendeloo, Inducing superconductivity and structural transformations by fluorination of reduced YBCO, Physica C 280 (1997) 272−280.

[60] P.R. Slater, P.P. Edwards, C. Greaves, I. Gameson, M.G. Francesconi, J.P. Hodges, M. Al-Mamouri, M. Slaski, Superconductivity up to 64 K in the copper oxyfluorides $Sr_{2-x}A_xCuO_2F_{2+\delta}$ (A=Ca, Ba) prepared using NH_4F as a fluorinating reagent, Physica C 241 (1995) 151−157.

[61] P.R. Slater, J.P. Hodges, M.G. Francesconi, P.P. Edwards, C. Greaves, I. Gameson, M. Slaski, An improved route to the synthesis of superconducting copper oxyfluorides $Sr_{2-x}A_xCuO_2F_{2+\delta}$ (A=Ca, Ba) using transition metal difluorides as fluorinating reagents, Physica C 253 (1995) 16−22.

[62] P.R. Slater, J.P. Hodges, M.G. Francesconi, C. Greaves, M. Slaski, Fluorination of the Ruddlesden-Popper type cuprates, $Ln_{2-x}A_{1+x}Cu_2O_{6-y}$ (Ln=La, Nd; A=Ca, Sr), J. Mater. Chem. 7 (1997) 2077−2083.

[63] P.R. Slater, PVDF as a reagent for the synthesis of K_2NiF_4-related inorganic oxide fluorides, J. Fluorine Chem. 117 (2002) 43−45.

[64] P.R. Slater, R.K.B. Gover, Synthesis and structure of the new oxide fluoride $Sr_2TiO_3F_2$ from the low temperature fluorination of Sr_2TiO_4: an example of a staged fluorine substitution/insertion reaction, J. Mater. Chem. 12 (2002) 291−294.

[65] P.R. Slater, R. Heap, Synthesis of $Ca_{2-x}Sr_xCuO_2F_2$ ($0 \leq X \leq 2$) with the T′-structure through Fluorination of $Ca_{2-x}Sr_xCuO_3$ with Poly(vinylidene Fluoride)/Poly(tetrafluoroethylene), 2009. University of Birmingham Chemistry Papers 1667.

[66] M. Sturza, H. Kabbour, S. Daviero-Minaud, D. Filimonov, K. Pokholok, N. Tiercelin, F. Porcher, L. Aldon, O. Mentre, Unprecedented robust antiferromagnetism in fluorinated hexagonal perovskites, J. Am. Chem. Soc. 133 (2011) 10901−10909.

[67] E. Sullivan, C. Greaves, Fluorine insertion reactions of the brownmillerite materials $Sr_2Fe_2O_5$, Sr_2CoFeO_5, and $Sr_2Co_2O_5$, Mater. Res. Bull. 47 (2012) 2541−2546.

[68] Y. Tsujimoto, K. Yamaura, N. Hayashi, K. Kodama, N. Igawa, Y. Matsushita, Y. Katsuya, Y. Shirako, M. Akaogi, E. Takayama-Muromachi, Topotactic synthesis and Crystal structure of a highly fluorinated Ruddlesden-Popper-type iron oxide, $Sr_3Fe_2O_{5+x}F_{2-x}$ ($x \sim 0.44$), Chem. Mater. 23 (2011) 3652–3658.

[69] A.K. Tyagi, S.J. Patwe, U.R.K. Rao, R.M. Iyer, Attempts to prepare fluorine containing $YBa_2Cu_3O_{6.5+\delta}$, Solid State Commun. 65 (1988) 1149–1151.

[70] J.L. Yang, J.K. Liang, G.H. Rao, Y.L. Qin, Y. Shi, W.H. Tang, Synthesis and superconductivity of $Nd_{0.7}Sr_{1.3}Cu(O,F)_{4-\delta}$ with $T_c = 44$ K, Physica C 270 (1996) 35–40.

[71] C. Perrin, O. Pena, M. Sergent, P.S. Christensen, G. Fonteneau, J. Lucas, Fluorination of high-T_c superconductors $YBa_2Cu_3O_x$- influence on the superconducting properties, Physica C 153 (1988) 934–935.

[72] A. Tressaud, B. Chevalier, B. Lepine, K. Amine, J.M. Dance, L. Lozano, J. Darriet, J. Etourneau, R. Tournier, A. Sulpice, P. Lejay, J.L. Diot, P. Maestro, Influence of low-temperature fluorination process on the characteristics of superconductors $YBa_2Cu_3O_{7-\delta}$ and $Bi_4(Ca,Sr)_6Cu_4O_{16+\delta}$, Solid State Ionics 32 (1989) 1109–1114.

[73] B.M. Tissue, K.M. Cirillo, J.C. Wright, M. Daeumling, D.C. Larbalestier, Conversion of $La_2CuO_{4-\delta}$ to a superconductor by treatment in fluorine gas, Solid State Commun 65 (1988) 51–54.

[74] J.L. MacManus, D.J. Fray, J.E. Evetts, Fluorination of $YBa_2Cu_3O_{7-x}$ by a solid-state electrochemical method, Physica C 184 (1991) 172–184.

[75] M.H. Delville, D. Barbut, A. Wattiaux, J.M. Bassat, M. Menetrier, C. Labrugere, J.C. Grenier, J. Etourneau, Electrochemical Fluorination of La_2CuO_4: A Mild "Chimie Douce" Route to Superconducting Oxyfluoride Materials, Inorg. Chem. 48 (2009) 7962–7969.

[76] T. Katayama, A. Chikamatsu, Y. Hirose, R. Takagi, H. Kamisaka, T. Fukumura, T. Hasegawa, Topotactic fluorination of strontium iron oxide thin films using polyvinylidene fluoride, J. Mater. Chem. C 2 (2014) 5350–5356.

[77] Y. Kobayashi, M. Tian, M. Eguchi, T.E. Mallouk, Ion-Exchangeable, Electronically Conducting Layered Perovskite Oxyfluorides, J. Am. Chem. Soc. 131 (2009) 9849–9855.

Superconductivity in Iron Oxypnictide Induced by F-Doping

19

Y. Kamihara
Keio University, Yokohama, Japan

H. Hosono
Tokyo Institute of Technology, Yokohama, Japan

Chapter Outline

1. Background and Recent Progress in Layered Oxypnictides

1.1 History of Superconductors

In 1911, zero resistivity in mercury was discovered by H. Kamerlingh Onnes [70]. Subsequently, perfect diamagnetism, which was measured by Meissner et al. [60], verified the existence of superconductivity (SC). Since these discoveries, tremendous efforts have been spent for searching high-superconducting temperature (T_c) superconductors. Several researchers have succeeded in finding new compounds with high-T_cs; ie, Nb_3Sn [59], MgB_2 [64], Cs_3C_{60} [25], Cu-based oxide superconductors [1,56,101]. In addition to various types of these superconductors, so-called Fe-based superconductors, which contain a typical magnetic element as a major constituent responsible to

Photonic and Electronic Properties of Fluoride Materials. http://dx.doi.org/10.1016/B978-0-12-801639-8.00019-2

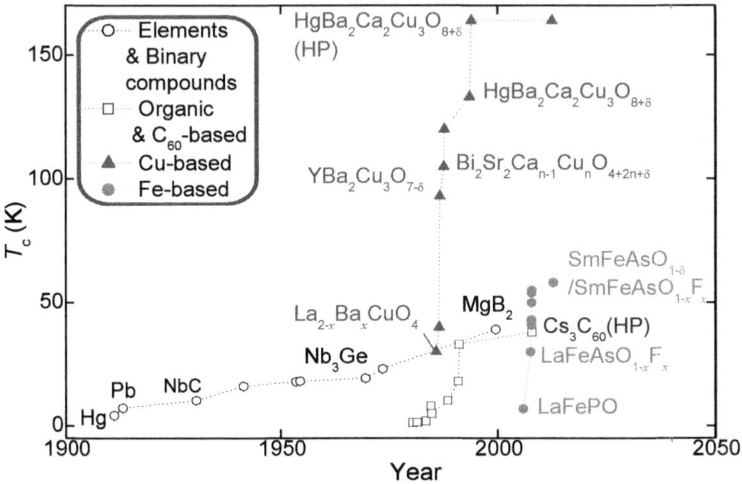

Figure 19.1 Highest superconducting transition temperatures (T_c) versus year for elements, binary compounds, organic materials, alkaline ion-doped C_{60}, cuprate (Cu-based layered oxide), Fe-based oxypnictide.

electrical conductivity with layered crystallographic structures, have been reported for 2006–15 [12,87].

A first representative Fe-based superconductor LaFePO, which shows T_c at temperatures (T) ~4 K, was discovered by our group in 2006 [35]. In 2005–08, we have examined a series of quaternary 3d transition-metal oxyphosphides, LaT_MPO (T_M: divalent transition-metal cations such as Mn^{2+}, Fe^{2+}, and Co^{2+}), with the expectation of finding a new correlated electron system [99,102,103]. These studies led to the next discovery of a high-T_c Fe-based superconductor, $LaFeAsO_{1-x}F_x$, with T_c ~26 K in 2008 [36]. Discovery of the Fe-based superconductors has invoked a recurrence of a new SC boom similar to that driven by discovery of layered Cu-based oxide superconductors [4].

Fig 19.1 shows progress of highest T_c for representative superconductors. Fe-based superconductors have the second highest transition temperature at ~58 K after Cu-based oxide superconductors [14]. Cu-based oxide superconductors and Fe-based superconductors possess potential for high-field magnet applications and superconducting power cable with inexpensive cooling cost (for a review, see Ref. [83]).

1.2 Crystallographic Characteristics

The mother compound LaFeAsO, a member of ReT_MPnO (Ln: rare earth elements, Pn: P, As), has a layered crystal structure belonging to the tetragonal *P4/nmm* space group. Although iron and iron-based compounds are representative ferromagnetic and/or ferri/antiferromagnetic (AF) materials, a considerable number of iron-based compounds have been reported to exhibit SC. Many researchers focus on this LaFeAsO material as a candidate for a new high-T_c superconductor. After our reports, series of iron-based superconducting materials have been reported. These materials are called "1111"-type compounds, which is named after the element ratio. Fig. 19.2 shows the crystallographic phases of iron-based superconducting materials. For 1111-type compounds, two layers with chemical formula units in the unit cell are represented by Re_2O_2 and Fe_2Pn_2. This crystal has a layered structure composed of an alternate stack of $(ReO)^+$ and $(FePn)^-$ layers (Fig. 19.2C). An Fe ion is coordinated with four *Pn* ions to form a distorted $FePn_4$ tetrahedron, which is linked with neighboring $FePn_4$ tetrahedra in an edge-sharing manner to comprise the Fe*Pn* layer. Each Fe forms a two dimensional square sublattice. From the fact that iron monophosphide Fe*Pn* exhibits metallic conduction, the Fe*Pn* layer in LaFe*Pn*O also works as a carrier conduction path. The carrier conduction layer is sandwiched by the carrier-blocking insulating *Ln*O layers. Therefore, it is natural to consider that the electron–electron interaction is strong, owing to the two dimensional carrier confinements in the Fe*Pn* layers.

Fe-based superconductors containing the common structural unit of Fe-square sublattice have also been discovered with variety of crystallographic structures. The Fe-based superconductors, which are illustrated in Fig. 19.2, contain the Fe*Pn* layer or Fe*Ch* (*Ch*: S, Se, Te) layers responsible to electrical conduction. The simplest one is Fe*Ch*, which was reported by Hsu et al. [28], composed of edge-sharing $FeCh_4$ tetrahedron and called "11"-type (Fig. 19.2A) and also has a layered crystal structure belonging to the tetragonal *P4/nmm* space group. Although the T_cs of 11-type compounds are lower than ~ 20 K [2], the T_c can be pushed up to temperatures $> \sim 30$ K for samples in which carrier-blocking layers are intercalated between carrier conducting Fe*Ch* layers [21,22,53,72]. It is noteworthy that extraordinary high-T_c superconductors have been reported for monolayered FeSe thin films epitaxially grown on $SrTiO_3$ substrates (eg, Ref. [20]). Rotter et al. reported SC of $Ba_{1-x}K_xFe_2As_2$, which is a member of the so-called "122"-type with $ThCr_2Si_2$-type structure belonging the tetragonal *I4/mmm* space group (Fig. 19.2D). 122-type superconductors consist of an alternate stack of Fe*Pn* layer and alkaline-earth square sublattices. The 122-type superconductors show smaller superconducting anisotropy and have more advantageous grain boundaries than those of other iron-based superconductors [44]. Among Fe-based superconductors, 122 superconductors have been demonstrated as the most applicable for magnet applications. (for a review, see Ref. [90]). Picher et al., Tapp et al., and Deng et al., reported superconducting LiFe*Pn* ($Pn = $ P, As) in 2008 [7,73,91]. LiFe*Pn* are categorized in "111"-type belonging to the tetragonal *P4/nmm* space group (Fig. 19.2B). Ogino et al. reported a superconducting layered oxypnictide with a perovskite oxide layer, $Sr_2ScFePO_3$ [68]. They also reported SC above 40 K in a layered oxypnictide with a thicker perovskite-related oxide layer, $Ca_4(Mg,Ti)_3Fe_2As_2O_{8-x}$. $Sr_2ScFePO_3$ and $Ca_4(Mg,Ti)_3Fe_2As_2O_{8-x}$ are categorized in the so-called "21113"-type (Fig. 19.2E) belonging *P4/nmm* space group and "43228"-type (Fig. 19.2F) belonging *I4/mmm* space group, respectively. While these Fe-based superconductors take tetragonal structures, Nohara et al. reported an Fe-based superconductor described as Ca-Fe-Pt-As belonging to lower symmetry P^{-1} space or *P4/n* space groups. Crystallographic structures of Ca-Fe-Pt-As are illustrated in Fig. 19.2G and H [33,52,66]. Among each Fe-based superconductor, only 1111-type shows bulk SC with $T_c > 50$ K (eg, Refs [14,46,77].

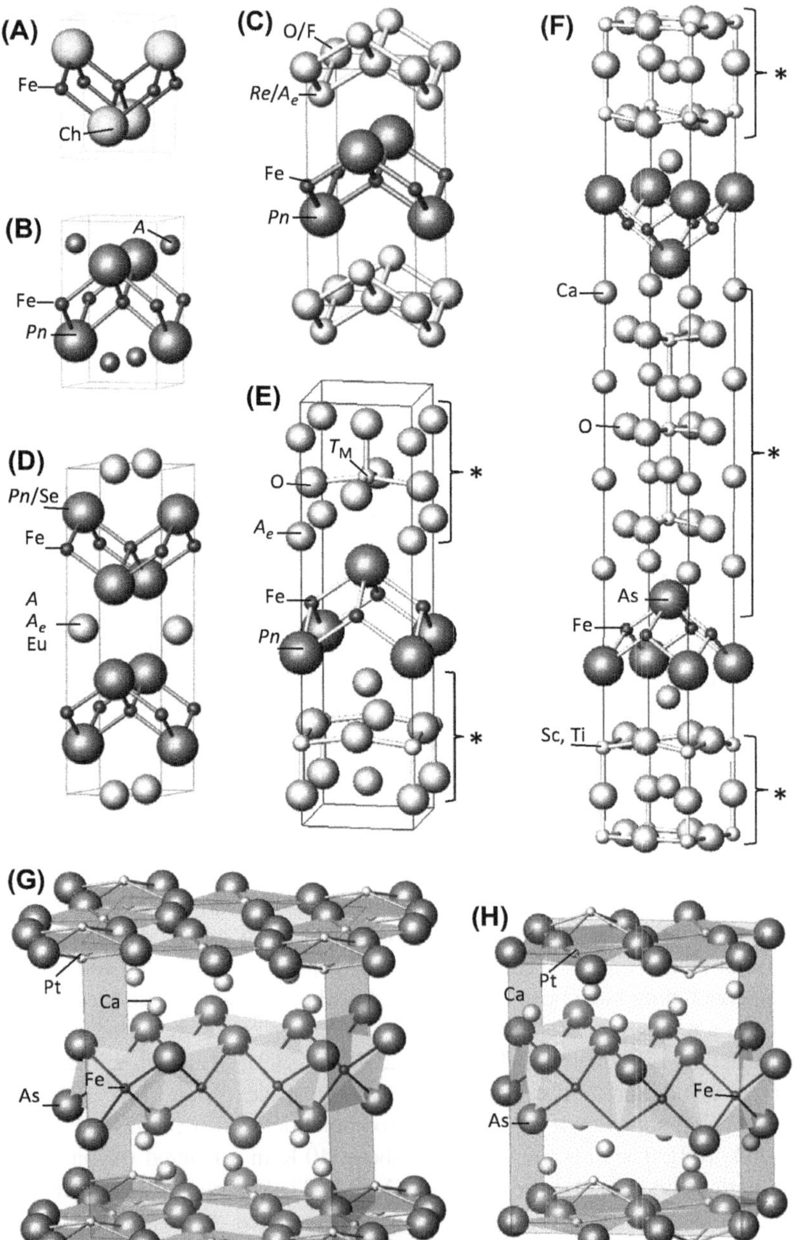

Figure 19.2 Crystallographic structures of Fe-based high-T_c superconductors. (A) Fe*Ch* (*Ch*: S, Se, Te), 11-type; (B) *A*Fe*Pn* (*A*: Li, Na, *Pn* = P, As), 111-type; (C) *Re*Fe*Pn*O (*Re*: rare earth ions)/*Ae*Fe*Pn*F (*Ae*: alkaline earth ion), 1111-type; (D) AFe_2Pn_2, $AeFe_2Pn_2$, $EuFe_2Pn_2$, AFe_2Se_2, 122-type; (E) $Ae_2T_MFePnO_3$ (T_M = Mg, Al, Ti, V), 21113-type; (F) $Ae_4T_{M3}Fe_2As_2O_8$, * denotes homologous structures. (G), (H) Ce-Fe-Pt-As.
Reproduced from Ref. [34].

1.3 Carrier Doping to Induce Superconductivity for 1111-Types

The composition, dopant species, and T_c values for representative 1111-type superconductors are summarized in Table 19.1. Although 1111-type contains Co-doped AeFe*Pn*F (Ae: alkaline earth) and La-doped CaFe*Pn*H, and EuFeAsF [58,63,92,109], we focus on $ReFePnO_{1-x}F_x$ in this chapter. In these Fe-based superconductors, each $ReFePnO_{1-x}F_x$ undergoes AF transition at low temperatures when they contain a magnetic sublattice of rare earth elements with unfilled 4f shells such as Sm [37,41]. Mother compounds of these superconductors crystallize in a tetragonal layered structure, which is composed of an alternate stack of carrier-conducting Fe*Pn* and carrier-blocking *Re*O layers. The former layer consists of edge-shared $FePn_4$ tetrahedra with an anti-PbO-type structure, which mainly contributes to the appearance of SC. A crystallographic phase transition from tetragonal to orthorhombic phase takes place at $T < 160$ K for each *Re*FeAsO. In addition to crystallographic phase transition, the AF phase of Fe is observed at slightly lower temperatures than the crystallographic phase transition temperatures ($T_{\text{ortho-tetra}}$). The orthorhombic Fe sublattice forms a magnetic lattice in *Re*FeAsO, while the Fe sublattice does not show clear AF ordering in *Re*FePO.

As shown in Table 19.1, possible characteristics of 1111-type are a variety of aliovalent and isovalent ion substitutions, which suppress AF ordering of iron. For instance, the replacement of O^{2-} with F^-, provides an extra positive charge in the La_2O_2 layer and a resulting negative charge is injected into the conducting Fe_2As_2 layer because of the presence of band offset between these two layers. This would lead to the modulation of the interactions through the change in carrier density, without seriously exerting crystallographic defects in the conduction layer. AF ordering of Fe ions in $LaFeAsO_{1-x}F_x$ is well suppressed by F-doping, and is converted into bulk superconducting phase. Hence, $LaFeAsO_{1-x}F_x$ with small F contents shows AF Fe and does not show bulk SC, T_cs of the $LaFeAsO_{1-x}F_x$ are sensitive to F contents. Although apparent T_c are often observed for polycrystalline $LaFeAsO_{1-x}F_x$ with

Table 19.1 Isovalent and/or Aliovalent Ion Substitution Method for LaFeAsO, and Maximum Superconducting Transition Temperatures (T_c)

Chemical Formula	*Tc* Onset (K)
$LaFeAsO_{1-x}F_x$	29 [36]
$LaFeAsO_{1-x}$	28 [77]
LaFeAsO: H	38.5 [61]
$LaFe_{1-x}Co_xAsO$	14 [81]
$LaFeAs_{1-x}P_xO$	12 [95]
LaFeAsO (HP)	21 [69]
$LaFeAsO_{0.89}F_{0.11}$ (HP)	43 [88]

Data are reproduced from Refs [36,61,69,77,81,88,95].

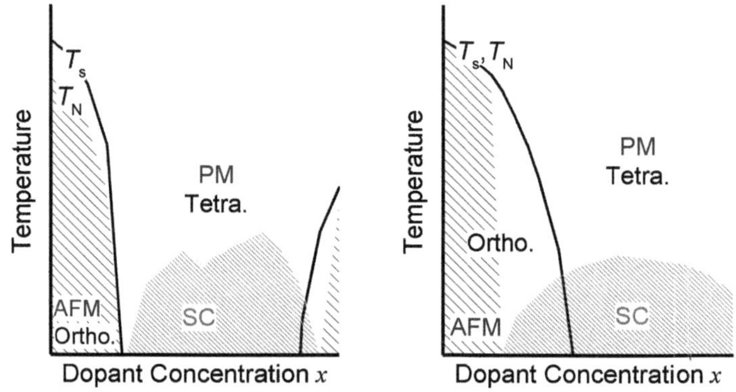

Figure 19.3 Schematic electronic and magnetic phase diagrams for Fe-based superconductors. Temperature versus dopant concentration. *Tetra*, tetragonal crystallographic phase; *Ortho*, orthorhombic crystallographic phase; *AFM*, antiferromagnetic; *PM*, paramagnetic; *SC*, superconducting phase; T_s, crystallographic transition temperatures; T_N, antiferromagnetic temperatures (Neel temperatures) of Fe.

$x < \sim 0.04$ [38,39], the observed T_cs are due to a trace superconducting phase. Both $LaFeAs_{1-x}P_xO$ and $LaFe_{1-x}Co_xAsO$ show bulk SC, although their maximum T_c are lower than those of $LaFeAsO_{1-x}F_x$ indicating that the randomness of the lattice, which is induced by substituted aliovalent and/or isovalent ions in the conduction layer, suppresses the maximum T_c for optimally doped samples. Although there are possible crystallographic phase segregation and magnetic phase separation due to magnetic ordering at nanoscale, the macroscopic magnetic and electronic phase diagrams of iron-based superconductors can be classified into two types, as demonstrated in Fig. 19.3. *Re*$FeAsO_{1-x}F_x$, *Re*$FeAsO_{1-x}H_x$, and $Ca_{1-x}La_xFeAsH$ [24,30,57,63] show the phase diagrams categorized in Fig. 19.3A, whereas phase diagrams of *Re*$FeAs_{1-x}P_xO$ and *Re*$Fe_{1-x}Co_xAsO$ are categorized in Fig. 19.3B [34].

2. Electronic Phase and Superconducting Mechanism in Layered Oxypnictides

2.1 Magnetic Structures of Mother Compounds

For undoped LaFeAsO which is a representative mother compound, AF ordering of Fe was verified using neutron diffraction [6] and ^{57}Fe Mössbauer spectroscopy (MS) [45]. These results led to a conclusion that Fe elements having a magnetic moment of 0.35 μ_B/Fe in undoped LaFeAsO exhibit AF ordering at temperatures below $\sim$140 K corresponding to a Néel temperature of the Fe magnetic sublattice (T_N(Fe)) of $\sim$140 K. In addition to the magnetic transition, a crystallographic phase transition from tetragonal to orthorhombic phase takes place at slightly higher temperature than T_N(Fe) [47,67]. In 2008, Ishibashi et al. first reported possible magnetic structures of LaFeAsO using ab initio electronic structure calculations based on the generalized gradient approximation in the density functional theory. Their calculation predicts that the ground state of LaFeAsO is AF with a stripe type magnetic moment alignment leading to orthorhombic symmetry of the crystal. In this particular magnetic state, the density of states (DOS) at the Fermi level is very small. On the other hand, LaFePO has turned out to be paramagnetic and a good metal [31]. They also report that a theoretical magnetic moment is $\sim 2\ \mu_B$/Fe for LaFeAsO. Regarding the magnetic moment of Fe, the theoretical value is five times larger than the experimental values. Although there is such discrepancy between experimental and theoretical approaches, it was surprising that both approaches give very similar results for the magnetic phase of LaFeAsO, simultaneously. Fig. 19.4 shows total DOS and projected DOS of Fe in LaFeAsO. Here, α and β indicate different Fe sites in a spin unrestricted unit cell.

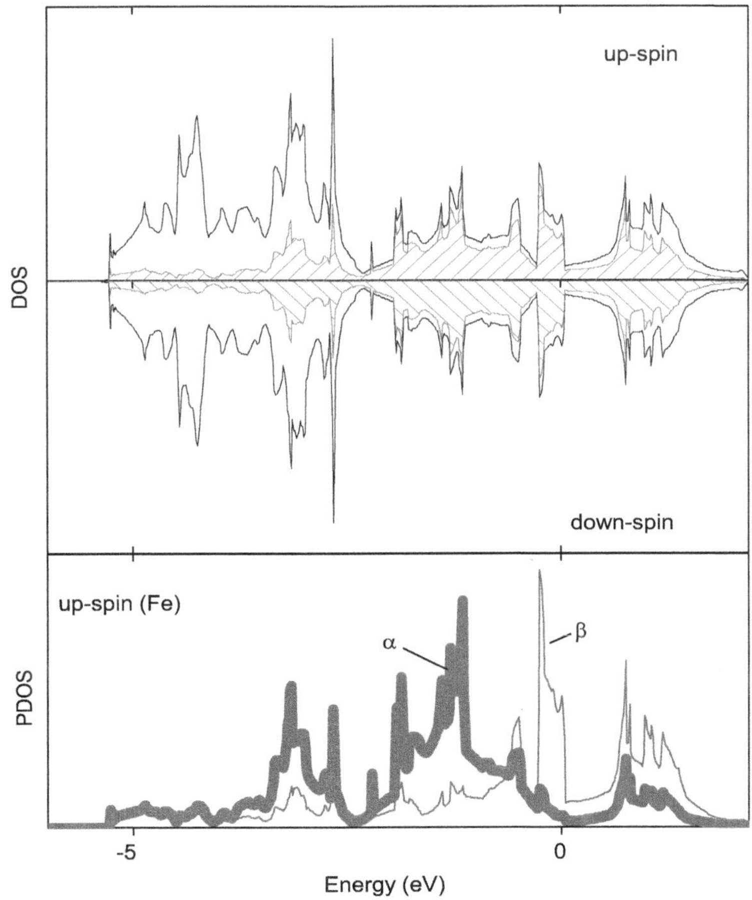

Figure 19.4 Up-spin and down-spin density of states (DOS) of LaFeAsO calculated with an effective Coulomb parameter $U-J=0$. Shadowed areas represent projected DOS (PDOS) of Fe. Bottom panels represent site-projected DOS of an Fe ion. α and β indicate different Fe sites in a spin unrestricted unit cell.
Reproduced from Ref. [40].

2.2 Atomic Scale Characterization of F-Dopant in $LaFeAsO_{1-x}F_x$

For $LaFeAsO_{1-x}F_x$, partial fluorine substitution is intended to produce one electron per substituted oxygen site. The fluorine had been only believed to be substituted for the oxygen sites and is invisible to conventional imaging techniques in 2008, while inspection of the dopant state was important because the existence and the state of the fluorine ions is a key for the emergence of SC in iron-based oxypnictide. Tohei et al. reported an electron micrography study of polycrystalline samples of fluorine-doped LaFeAsO using the scanning transmission electron microscopy (STEM) technique combined with angstrom resolved electron energy loss spectroscopy (EELS). As shown in Fig. 19.5, atomic-scale EELS measurements in conjunction

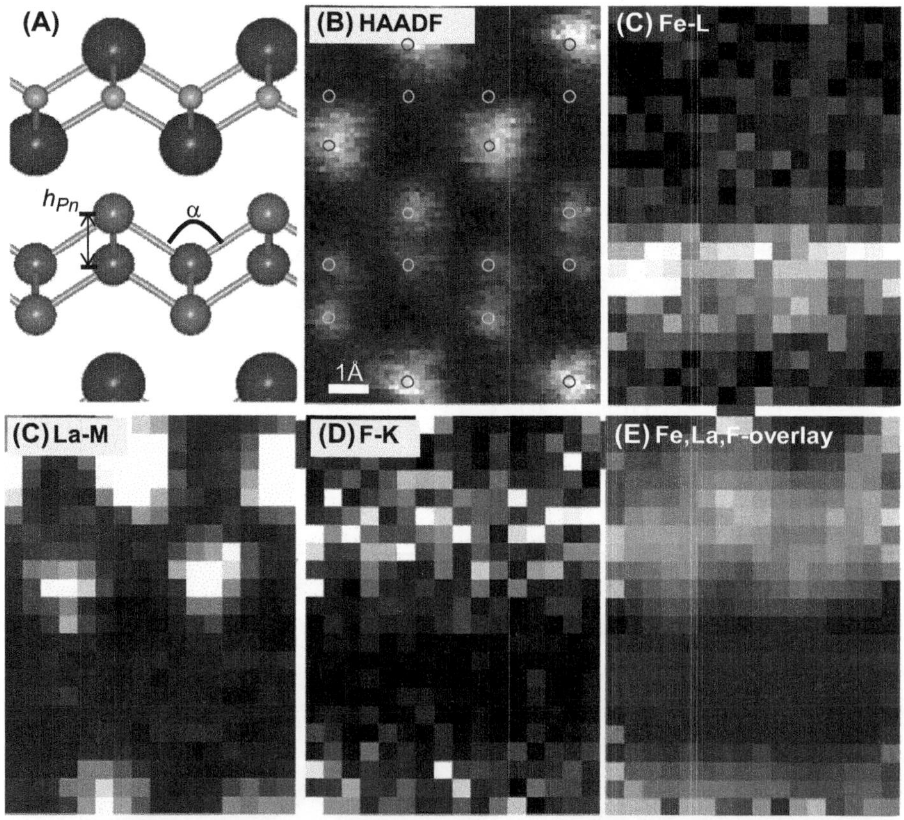

Figure 19.5 Atomic imaging of F-doped LaFeAsO obtained by scanning transmission electron microscopy and atomic electron energy loss spectroscopy (EELS) using atomic sized electron probe. (A) A crystallographic structure obtained by X-ray diffraction. h_{pn} and α denotes anion height from Fe-square lattice and horizontal As-Fe-As angle. (B) High-angle annular dark field (HAADF) image gives contrast proportional to atomic number. (C–D) EELS images of Fe L-edge, La M-edge, F K-edge. F K-edge line corresponds to the location of oxygen ions, demonstrating the random substitution of oxygen sites by F ions. (E) Fe, La, Fe-overlay. Reproduced from Ref. [94].

with Z-contrast observations by the high-angle annular dark field (HAADF) technique were performed. HAADF-STEM observations, in which heavier elements are imaged as brighter spots, demonstrate information about atomic arrangements through simple interpretation of the images obtained. The EELS measurements give element-specific quantitative data originating from the electronic or chemical state of the components. EELS mapping with a finely converged electron probe allows us to investigate the spatial distribution of the different elements at a scale of one atomic column. EELS measurements directly detect the doped fluorine ions in the nominal $LaFeAsO_{0.9}F_{0.1}$. The spatial distribution of fluorine ion dopant that is invisible by HAADF imaging was visualized at an atomistic scale, proving fluorine substitution into the oxygen sites [94].

T_c of ReFeAs$O_{1-x}F_x$ was raised substantially by partially replacing the O site with F ions and/or applying pressure. T_c of ReFeAs$O_{1-x}F_x$ correlates well with As-Fe-As angle along a-axis (α), which is defined in Fig. 19.5A. Such a relationship, which is the so-called Lee's plot [11,50,51], is partially supported theoretically using pnictogen height (h_{Pn}) from the Fe-layer [11,48], although the T_c of ReFeAs$O_{1-x}F_x$ in which Re site has randomness with two rare earth elements, exhibits different character [85]. 1111 compounds show the best agreement to Lee's empirical and Kuroki's theoretical relationship between T_c and local structures among the various iron-based superconductors. Maximum T_cs were observed for F-doped 1111 compounds with α for regular tetrahedron ($\sim 109.47°$). The local structures are verified using Rietveld analysis for X-ray and/or neutron diffraction patterns [3,32]. Atomic radial distribution function, which can be obtained by extended X-ray absorption fine-structure measurements, is also available to characterize lattice dynamics [105].

2.3 Transport Properties

In this section, the electrical properties of $SmFeAsO_{1-x}F_x$ are described as a representative example [41]. Fig. 19.6 shows electrical resistivity (ρ) as a function of temperature (T) for $SmFeAsO_{1-x}F_x$ with various x values, which reveal the following: (1) The $x = 0.037$, 0.040 samples are normal conductors at temperatures down to 1.8 K, whereas the $x = 0.045$, 0.046, 0.060, 0.069, and 0.083 samples are superconductors. For the $x = 0.045$ sample, ρ decreases below our detection limit, whereas ρ in the $x = 0.037$, 0.040 samples exhibit finite values at 1.8 K. Thus, it indicates that the $x = 0.037$, 0.040 samples are normal conductors, and the $x = 0.045$ sample is a superconductor at $T = 1.8$ K. (2) The $x = 0$ sample (undoped) exhibits an anomalous decrease or kink at 143 K (T_{anom}), denoted by an upward arrow that agrees with $T_{ortho\text{-}tetra}$. (3) With increasing x, T_{anom} decreases from 143 K for the undoped sample to 91 K for the $x = 0.040$ sample, and eventually the kink vanishes for the $x = 0.045$ sample. With a further increase in the F content, it reappears and shifts to a higher T, 161 K for $x = 0.046$ and 198K for $x = 0.083$. The sign of dT_{anom}/dx value changes at $x = \sim 0.045$. (4) For the $x = 0.040$ sample, $\rho - T$ curve shows a clear thermal hysteresis. (5) The $\rho - T$ curves show minima (T_{min}) at $x = 0.040$ and 0.045, designated by the open triangles, in addition to a kink at T_{anom}. Since $x = 0.040$ corresponds to a region where the magnetic phase changes to a superconducting phase by electron doping, the observed singularities imply that electrical property disorder exists in

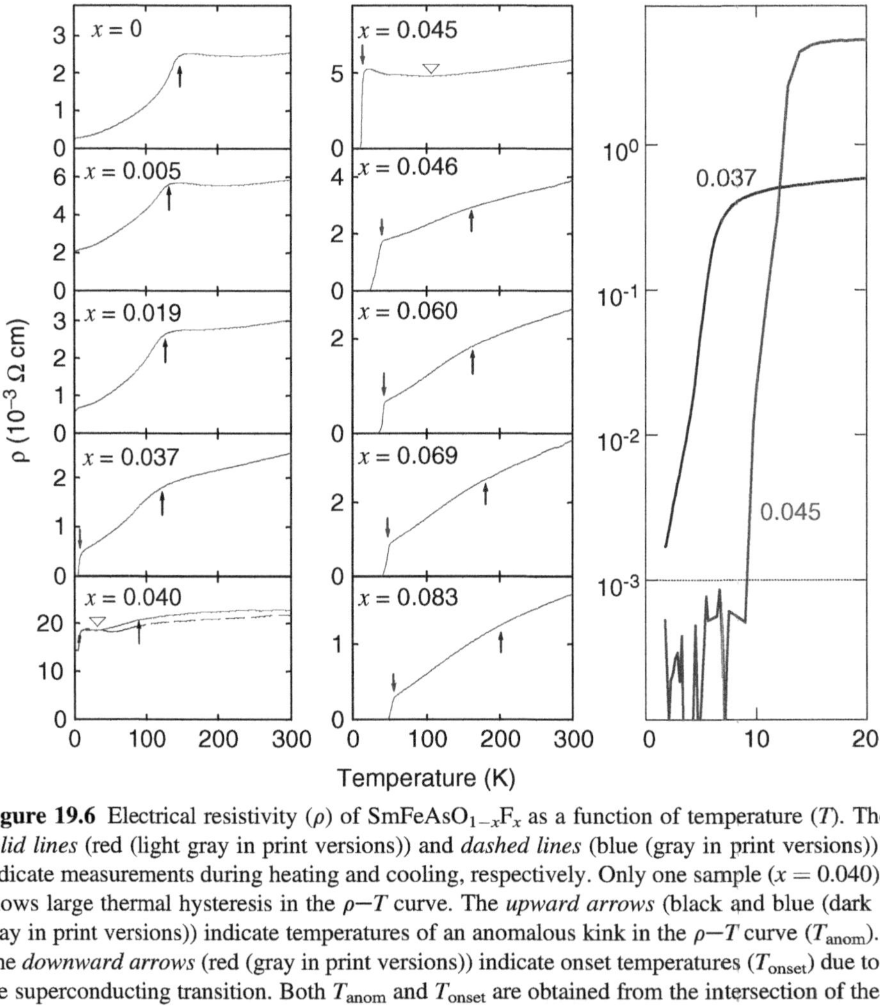

Figure 19.6 Electrical resistivity (ρ) of $SmFeAsO_{1-x}F_x$ as a function of temperature (T). The *solid lines* (red (light gray in print versions)) and *dashed lines* (blue (gray in print versions)) indicate measurements during heating and cooling, respectively. Only one sample ($x = 0.040$) shows large thermal hysteresis in the ρ–T curve. The *upward arrows* (black and blue (dark gray in print versions)) indicate temperatures of an anomalous kink in the ρ–T curve (T_{anom}). The *downward arrows* (red (gray in print versions)) indicate onset temperatures (T_{onset}) due to the superconducting transition. Both T_{anom} and T_{onset} are obtained from the intersection of the tangents to the ρ–T curve in a higher temperature state and halfway through the transition. *Broken lines* are a visual guide indicating T_{onset} for $x = 0.046$, 0.060, 0.069, and 0.083. The *downward triangles* indicate minima in the ρ–T curve. For ρ–T curves for $SmFeAsO_{1-x}F_x$, $x = 0.037$ and 0.045. The *dotted line* denotes the detection limit of our measurement. Reproduced from Ref. [41].

this concentration region. Therefore this observation implies that the electrical phase is inhomogeneous and segregated at $x \sim 0.040$.

2.4 ^{57}Fe Mössbauer Spectroscopy

Fig. 19.7 shows ^{57}Fe MS of undoped ($x = 0$) and F-doped ($x = 0.005$, 0.019, 0.037, 0.040, 0.045, 0.046, 0.060, and 0.069) samples of $SmFeAsO_{1-x}F_x$ at $T = 4.2$–300 K. In our expression, ^{57}Fe MS is illustrated by distribution maps with

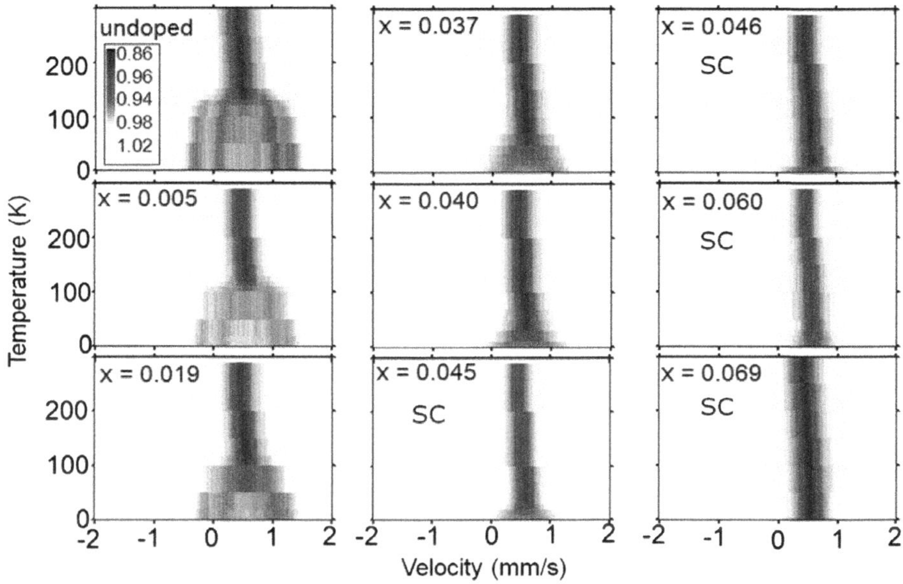

Figure 19.7 Temperature dependences of ^{57}Fe Mössbauer spectra for $SmFeAsO_{1-x}F_x$ ($x = 0$, 0.005, 0.019, 0.037, 0.040, 0.045, 0.048, 0.060, 0.069) illustrated by distribution maps with relative transmission-dependent colors. Red (dark gray in print versions) area denotes a paramagnetic component, while green (gray in print versions) area, which is broaden at low temperatures, shows spin a polarized component of Fe ions. Bulk superconducting samples are shown as SC.
Data reproduced from Ref. [42].

relative transmission-dependent colors. Provided that small quadruple splitting is negligible, both samples at room temperature (RT) are composed of a single absorption line, which is illustrated as a red-colored region in the figures, indicating paramagnetic Fe sublattice. MS of the undoped sample shows a paramagnetic sublattice at $T = 150-298$ K. However, with a decrease in T, the red-colored region vanishes at $T \leq 150$ K, and the green region becomes broader simultaneously. With further decrease in T, MS starts to undergo multiple splitting at $T \sim 140$ K, and a sextet split spectrum, which is illustrated by the disappearance of red-colored regions and broadening of green in colored regions, observed at 4.2 K. If the splitting is caused by an internal magnetic field (H_{int}), H_{int} starts to appear at $T \sim 140$ K. It increases with decreasing T and reaches 5.16(1) T at $T = 4.2$ K, using a conversion factor (CF = 15 T/μ_B) [37] corresponding to the magnetic moment of 0.34 μ_B/Fe at 4.2 K. An AF Fe sublattice (AF_{Fe}) is clearly observed in undoped SmFeAsO. This is a typical behavior of magnetic ordered Fe compounds [45]. The disappearance of red-colored regions and clear broadening of green-colored regions are also observed for normal conducting $x = 0.005$, 0.019, and 0.037 samples at lower temperature. Small H_{int} might be observed for normal conducting $x = 0.040$ sample, and superconducting $x = 0.045$, 0.046 samples, although clear sextet split spectra were hardly observed for $x = 0.040$, 0.045, and 0.046 samples. These results indicate that the bulk AF phase of the Fe sublattice does not coexist with the bulk superconducting phase. Provided that crystallographic phase segregation is excluded, the small H_{int} is mainly due to short range–ordered magnetic phase separation on a nanometer range [54,86].

2.5 Phase Diagrams

Fig. 19.8A shows lattice constants as a function of an F content (x_{VVR}) for $SmFeAsO_{1-x}F_x$, x_{VVR} are determined by the lattice constants using Vegard's

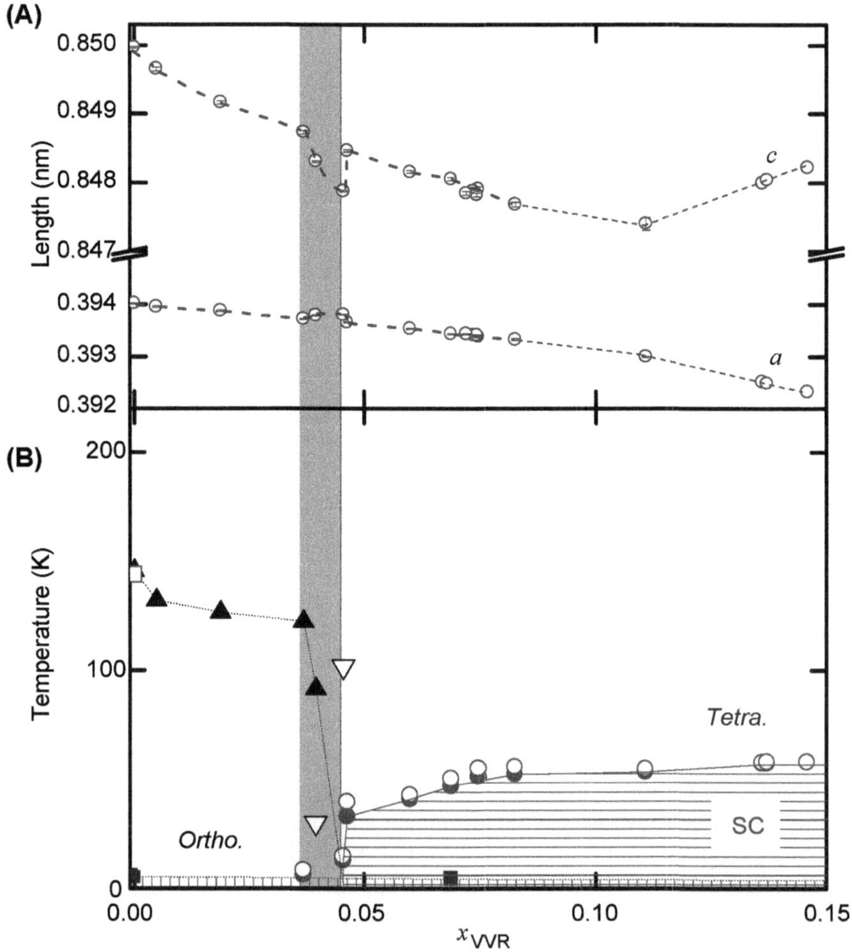

Figure 19.8 (A) Lattice constants (a, c) versus F content (x_{VVR}) at room temperature. *Solid and dotted lines* are visual guides. x_{VVR} are determined by the lattice constants using Vegard's volume rule (VVR) [5]. (a, c)-x_{VVR} curves change following to VVR at $x_{VVR} < 0.11$. (B) T_{anom} (*upward triangles*), T_{min} (*open downward triangle*), T_{onset} (*open circle*), T_c (*closed circle*), and T_N of Sm (*closed square*) versus x_{VVR}. The horizontal and the vertical striped-shadowed areas show the superconducting phase (SC) and the antiferromagnetic phase of Sm, respectively.
Data are reproduced from Ref. [41]. Plots at $x_{VVR} > 0.13$ are calculated from lattice constants, which are reported by Ref. [14]. The shadowed area, in which both T_{min} and trace T_c appears, indicate phase segregated state in polycrystalline samples.

volume rule (VVR) [5,8]. Many phase diagrams had been reported for polycrystalline $ReFeAsO_{1-x}F_x$ based on nominal F contents, although minor but finite second phases inevitably appear in the samples. The second phases are assigned to LnOF and binary arsenides. Indeed, the amounts of the second phases drastically increase with increasing nominal F contents $x_{nominal} > 0.16$ [14]. $(a, c)-x_{VVR}$ curves show clear discrepancy from linear behavior at $x_{VVR} > 0.13$. These experimental results suggest that a solubility limit of $ReFeAsO_{1-x}F_x$ exists under ambient pressure. It should be noted that nominal F contents are not identical to intrinsic F contents. VVR is an empirical, but systematic and verifiable way to determine the F content. x_{VVR} might be a better way to avoid confusing broad readers based on nominal values. Based on T_{anom} ($\sim T_{ortho\text{-}tetra}$), T_c, T_N(Fe), and T_N(Sm), we have constructed a phase diagram of $SmFeAsO_{1-x}F_x$ in terms of x_{VVR} and temperature. Fig. 19.8B, which has been proposed by us, is close to those by Hess et al. [26] and Fujioka et al. [14]. Both of the T_N(Fe) and crystallographic phase transition temperatures do not persist in the superconducting regime.

As shown in Fig. 19.9, such a relation between T_{anom} and SC phase is a common feature among $ReFeAsO_{1-x}F_x$ (Ln = La [54], Ce [108], Pr [78], Nd [75], Sm [41], Gd [79], $La_{0.5}Gd_{0.5}$ [85]). These results indicate that the relation among T_{anom}, the static magnetism, and T_c of $ReFeAsO_{1-x}F_x$ shows similar topology to that of copper-based high-T_c superconductors (for a review, see Ref. [49]). Minimum F contents, which are required for bulk SC, primarily depend on the rare earth ion's mass.

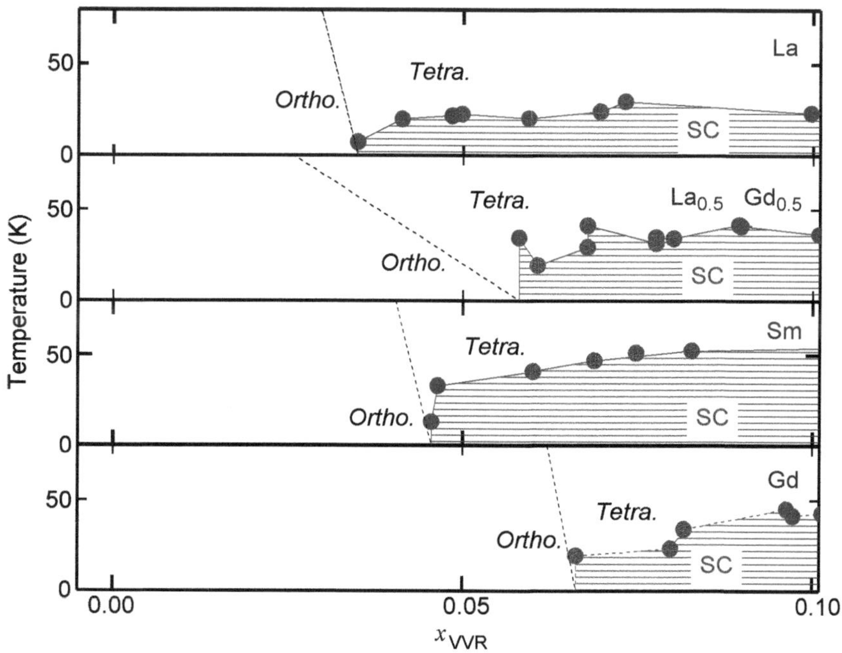

Figure 19.9 T_c versus F content (x_{VVR}) for $ReFeAsO_{1-x}F_x$ (Re = La, $La_{0.5}Gd_{0.5}$, Sm, and Gd). *Tetra*, tetragonal crystallographic phase; *Ortho*, orthorhombic crystallographic phase. Data are reproduced from Refs [36,41,79,85].

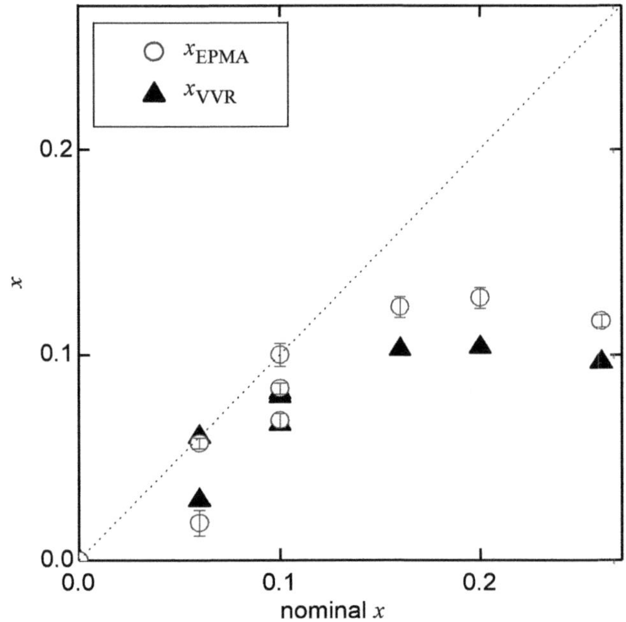

Figure 19.10 F contents (x) verified by electron probe microanalyzer (EPMA) (x_{EPMA}) and Vegard's volume rule (x_{VVR}) as function of nominal x for F-doped GdFeAsO. Data are reproduced from Ref. [79].

Fig. 19.10 shows that F contents depend on methods to quantify x for $GdFeAsO_{1-x}F_x$. An electron probe microscopy analyzer (EPMA), which allows microscopic X-ray fluorescence analysis, can quantify the F contents (x_{EPMA}) comparing to other elements, although calibrated curves are required for each element, each surface structure, and each chemical bonding feature. There is a discrepancy between x_{EPMA} and x_{VVR}, although the discrepancy is within 2 at.% for well-characterized polycrystalline samples.

3. Synthesis Procedures and Applications

1111 compounds demonstrate a highest T_c and very high superconducting upper critical field (H_{c2}) [29,62,82]. These attractive characteristics of 1111 compounds triggered research into new high-T_c superconductors as well as several attempts at applications. Fig. 19.11 shows yearly progress of transport critical current density (J_c) for superconducting wires and tapes made from iron-based superconductors. For Fe-based superconducting wires, such as 122 compounds, J_c attains values of $\sim$0.1 MA cm^{-2} under self-field, ie, without applied field [100] and $\sim$30 kA cm^{-2} under $\mu_0 H = 20$ T [19], while maximum J_c of 1111 compounds is lower than such a practical value. These iron-based superconducting wires were fabricated by the powder-in-tube (PIT)

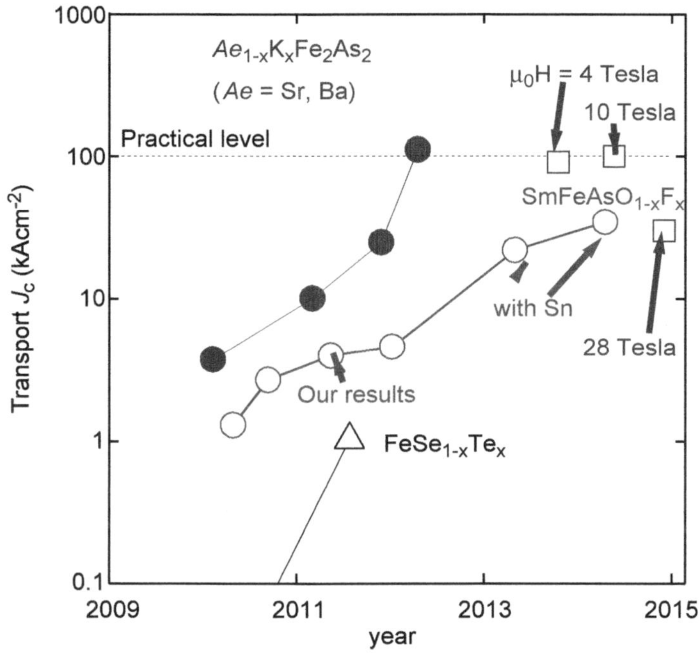

Figure 19.11 Transport superconducting critical current density J_c at 4.2 K for $Ae_{1-x}K_xFe_2Se_2$ (Ae = Sr, Ba), $SmFeAsO_{1-x}F_x$, and $FeSe_{1-x}Te_x$.
Data are reproduced from Refs [13,16–18,55,71,74,93,96,97,98,100,106,107].

technique in which the powders, which are synthesized for the superconducting core, are packed into a metal tube and the tubes are annealed after wire drawing. For practical applications, it is useful if a high J_c value is easy to scale up to the long length with productions. The PIT technique is the most promising method for the scale up with lower costs. The PIT technique is classified into two different processes: in situ and ex-situ. The former employs powder starting materials, whereas the latter employs a powder of synthesized superconducting materials.

Fig. 19.12 shows a flowchart for fabricating F-doped SmFeAsO superconducting bulk samples and wires. $SmFeAsO_{1-x}F_x$ (Sm1111) shows the highest T_c among 1111 compounds. For synthesizing Sm1111 exhibiting the highest T_c, conditions to be examined were control of F content in bulk polycrystalline samples. In this section, sample preparation procedures for polycrystalline Sm1111 and Sm1111 superconducting wires will be described. Polycrystalline Sm1111 were prepared by a two-step solid state reaction in a sealed silica tube using dehydrated Sm_2O_3 and a mixture of compound of SmAs, Fe_2As, and FeAs (SmAs-Fe_2As-FeAs powder) as starting materials [41]. Fujioka et al. reported polycrystalline Sm1111 with onset T_c at ~58.1 K using sintering at 1253 K with slow cooling at a rate of $-5\ K\,h^{-1}$ to 873 K [14]. Crystal structures, including lattice constants of the tetragonal main phase as well as impurity phases of the sintered powders, were examined by powder

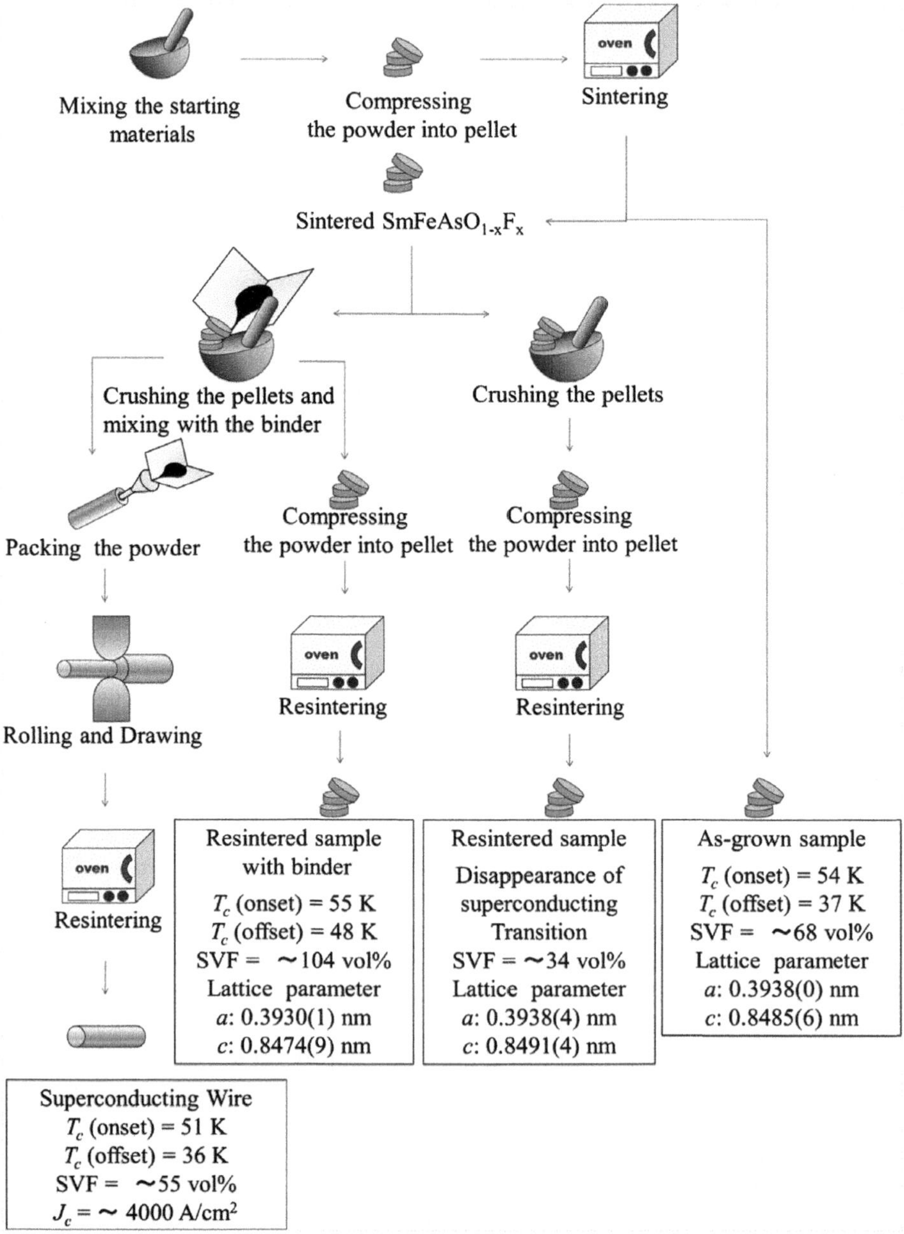

Figure 19.12 Flowchart for fabricating F-doped SmFeAsO superconducting bulk samples and wire. Superconducting transition temperatures (T_c), superconducting volume fractions (SVF), and lattice parameters are given for each sample.

X-ray diffraction (XRD) at RT using Cu Kα radiation from a rotating anode. Ex-situ PIT for Sm1111 was performed after polycrystalline Sm1111 preparation. In 2011, we demonstrated the first ex-situ fabrication of superconducting Sm-1111 wire by using a reactive binder [13]. The introduction of the reactive binder not only reduces the T_c lowering during resintering but also increases the superconducting volume fraction in polycrystalline Sm1111 samples. Although resintering reduces the F content in as grown Sm1111, adding the reactive binder compensates by sufficiently supplementing the F content [11].

In 2014, Zhang et al. reported that Sn-added Sm1111 wires show significantly enhanced J_c (c. [98,106]). Although it is a tough challenge for obtaining practical superconducting wires using Sm1111, several unknown approaches might enhance J_c, due to their variety of possible binders. Table 19.2 shows comparison between representative high-T_c superconductors that show practical transport J_cs at $T > 20$ K.

Table 19.2 Electronic Structures on Fermi Level, Maximum Onset T_c, Sensitivity to Impurity Scattering-Induced Pair Breaking (Impurity), Superconducting (SC) Gap Symmetry, Upper Critical Magnetic Flux Under Self-Field ($\mu_0 H_{c2}$), SC Anisotropy Ratio Defined by ($H_{c2\perp c}$)/($H_{c2\parallel c}$) (Anisotropy), and Superconducting Critical Angle (θ_{GB}) in Which Transport J_c in Bicrystal Grain Boundary Remains Half of Normal Intragrain J_c in Comparison with MgB_2 and Cuparates

	Fe-pnictides	MgB_2	Cuprates
Fermi level	Fe 3d 5-bands [27,90];	B 2p 2-bands [27,90]	Cu 3d single band [27,90]
Max onset T_c	58 K [12,14]	39 K [64]	134 K [80,89]
Impurity	Robust [27]	Sensitive [27]	Sensitive [27]
SC gap	Extended s-wave [27]	s-wave [27]	d-wave [27]
$\mu_0 H_{c2}$ (0)	70–200 T [23,62,104];	~40 T [10]	~120 T [65,76]; (YBCO)
Anisotropy	1–3 (122-type) [43,104]	2–6 [10]	5–7 (YBCO) [84]
	4–7 (1111-type) [15,62];		50–155 (BSCCO) [84]
θ_{GB}	8°–9° (Ba 122) [44]	–	~5° (YBCO) [9]

BSCCO, Bi-Sr-Ca-Cu-O superconductors including $Bi_2Sr_2CaCu_2O_y$ and $Bi_2Sr_2Ca_2Cu_3O_y$; YBCO, $YBa_2Cu_3O_{7-\delta}$
Data are reproduced from Refs [9,10,12,14,15,23,27,43,44,62,64,65,76,80,84,89,90,104].

4. Conclusions

1. Fe-based superconductor $ReFePnO_{1-x}F_x$ (1111) has the highest T_c at ~58 K next to Cu-based oxide superconductors. Crystal structures of 1111-type is made from alternate stack of insulating $(ReO)^+$ and carrier conducting $(FePn)^-$ layers. In mother compounds of 1111, Fe sublattices show itinerant antiferromagnetism. F-doping suppresses the long-range spin ordering of Fe ions in 1111-types by providing an extra negative charge into the conducting layers. Superconducting phases appear in $ReFePnO_{1-x}F_x$ ($x >$ ~0.04) with tetragonal crystallographic symmetry.
2. The superconducting phase diagram of 1111 has been revealed clearly using synchrotron XRDs, neutron diffractions, ^{57}Fe MS, and atomic-scale electron energy loss characterization. A discrepancy between intrinsic F contents and empirical F contents is within 2 at.%. Local structures, which are verified using Rietveld analysis, show empirical relation to T_c. The T_c of 1111-type compounds correlates with *Pn*-Fe-*Pn* bond angles (α), the so-called Lee's plot. The maximum T_c was observed for 1111 samples with α for regular tetrahedron. Lee's plot is justified by a theory based on spin fluctuation by Kuroki who proposed a relationship between T_c and anion height.
3. 1111 compounds exhibit very high superconducting upper critical fields (H_{c2}). This property is very attractive for power cable under high magnetic fields, although fabrication of practical superconducting wires is a tough challenge due to difficulty to control F contents during a processing for ex-situ powder in tube method.

F-doping method is an essential technology to obtain superconducting bulks and wires with $T_c > 50$ K in an ambient pressure. Although various synthesis methods had been tried for higher T_c, polycrystalline samples, the highest T_c is obtained for the 1111 system synthesized by the simple solid state reaction which obeys thermodynamics. Further understanding of intrinsic phase diagram and chemical composition will be the only way to achieve new milestones for practical material processing of Fe-based superconductors.

References

[1] J.G. Bednorz, K.A. Muller, Possible high T_c superconductivity in the Ba-La-Cu-O system, Z. Phys. B Condens. Matter 64 (1986) 189−193.

[2] E. Bellingeri, I. Pallecchi, R. Buzio, A. Gerbi, D. Marre, M.R. Cimberle, M. Tropeano, M. Putti, A. Palenzona, C. Ferdeghini, $T_c = 21$K in epitaxial $FeSe_{0.5}Te_{0.5}$ thin films with biaxial compressive strain, Appl. Phys. Lett. 96 (2010), 102512, 1−3.

[3] Bruker AXS, TOPAS, Version 3, 2005. Karlsruhe, Germany.

[4] P. Canfield, Still alluring and hard to predict at 100, Nat. Mater. 10 (2011) 259−261.

[5] A. Cox, M.J.L. Sangster, Vegard's rule and volumes of formation for impurity ions, J. Phys. C: Solid State Phys. 18 (1985) L1123−L1126.

[6] C. de la Cruz, Q. Huang, J.W. Lynn, J. Li, W. Ratcliff II, J.L. Zarestky, H.A. Mook, G.F. Chen, J.L. Luo, N.L. Wang, P. Dai, Magnetic order close to superconductivity in the iron-based layered $LaO_{1-x}F_xFeAs$ systems, Nature 453 (2008) 899−902.

[7] Z. Deng, X.C. Wang, Q.Q. Liu, S.J. Zhang, Y.X. Lv, J.L. Zhu, C.Q. Yu, R.C. Jin, A new "111" type iron pnictide superconductor LiFeP, Europhys. Lett. 87 (2009) 1−4, 37004.

[8] A.R. Denton, N.W. Ashcroft, Vegard's law, Phys. Rev. A 43 (1991) 3161–3164.
[9] D. Dimos, P. Chaudhari, J. Mannhart, Superconducting transport properties in $YBa_2Cu_3O_7$ bicrystals, Phys. Rev. B 41 (1990) 4038–4049.
[10] J.D. Fletcher, A. Carrington, O.J. Taylor, S.M. Kazakov, J. Karpinski, Temperature-dependent anisotropy of the penetration depth and coherence length of MgB_2, Phys. Rev. Lett. 95 (2005), 097005, 1–4.
[11] M. Fujioka, (In Japanese) Study on Iron-Based Mixed-anion Superconducting Wire by Reactive Solid State Binder Method, Keio University, 2012 (Ph.D. thesis).
[12] M. Fujioka, Iron-Based Superconductors, $SmFeAsO_{1-x}F_x$, Wiley Encyclopedia of Electronics Engineering, 2014, pp. 1–30.
[13] M. Fujioka, T. Kota, M. Matoba, T. Ozaki, Y. Takano, H. Kumakura, Y. Kamihara, Effective ex-situ fabrication of F-doped SmFeAsO wire for high transport critical current density, Appl. Phys. Express 4 (2011), 063102, 1–3.
[14] M. Fujioka, S.J. Denholme, T. Ozaki, H. Okazaki, K. Deguchi, S. Demura, H. Hara, T. Watanabe, H. Takeya, T. Yamaguchi, H. Kumakura, Y. Takano, Phase diagram and superconductivity at 58.1K in α-FeAs-free $SmFeAsO_{1-x}F_x$, Supercond. Sci. Technol. 26 (2013), 085023, 1–4.
[15] M. Fujioka, S.J. Denholme, M. Tanaka, H. Takeya, T. Yamaguchi, Y. Takano, The effect of exceptionally high fluorine doping on the anisotropy of single crystalline $SmFeAsO_{1-x}F_x$, Appl. Phys. Lett. 105 (2014), 102602, 1–4.
[16] Z. Gao, L. Wang, Y. Qi, D. Wang, X. Zhang, Y. Ma, H. Yang, H. Wen, Superconducting properties of granular $SmFeAsO_{1-x}F_x$ wires with $T_c = 52K$ prepared by the powder-in-tube method, Supercond. Sci. Technol. 21 (2008), 112001, 1–4.
[17] Z. Gao, L. Wang, C. Yao, Y. Qi, C. Wang, X. Zhang, D. Wang, C. Wang, Y. Ma, High transport critical current densities in textured Fe-sheathed $Sr_{1-x}K_xFe_2As_2 + Sn$ superconducting tapes, Appl. Phys. Lett. 99 (2011), 242506, 1–4.
[18] Z. Gao, K. Togano, A. Matsumoto, H. Kumakura, Achievement of practical level critical current densities in $Ba_{1-x}K_xFe_2As_2$/Ag tapes by conventional cold mechanical deformation, Sci. Rep. 4 (2014) 4065, 1–5.
[19] Z. Gao, K. Togano, A. Matsumoto, H. Kumakura, High transport J_c in magnetic fields up to 28 T of stainless steel/Ag double sheathed Ba122 tapes fabricated by scalable rolling process, Supercond. Sci. Technol. 28 (2015), 012001, 1–6.
[20] J.F. Ge, Z.L. Liu, C. Liu, C.L. Gao, D. Qian, Q.K. Xue, Y. Liu, J.F. Jia, Superconductivity above 100K in single-layer FeSe films on doped $SrTiO_3$, Nat. Mater. 14 (2014) 285–289.
[21] J. Guo, S. Jin, G. Wang, S. Wang, K. Zhu, T. Zhou, M. He, X. Chen, Superconductivity in the iron selenide $K_xFe_2Se_2$ ($0 \leq x \leq 1.0$), Phys. Rev. B 82 (2010), 180520, 1–4.
[22] J. Guo, H. Lei, F. Hayashi, H. Hosono, Superconductivity and phase instability of NH_3-free Na-intercalated $FeSe_{1-z}S_z$, Nat. Commun. 5 (2014) 4756, 1–6.
[23] A. Gurevich, To use or not to use cool superconductors? Nat. Mater. 10 (2011) 255–259.
[24] T. Hanna, Y. Muraba, S. Matsuishi, N. Igawa, K. Kodama, S. Shamoto, H. Hosono, Hydrogen in layered iron arsenides: indirect electron doping to induce superconductivity, Phys. Rev. B 84 (2011), 024521, 1–7.
[25] A.F. Hebard, M.J. Rosseinsky, R.C. Haddon, D.W. Murphy, S.H. Glarum, T.T.M. Palstra, A.P. Ramirez, A.R. Kortan, Superconductivity at 18K in potassium-doped C_{60}, Nature 350 (1991) 600–601.
[26] C. Hess, A. Kondrat, A. Narduzzo, J. Hamann-Borrero, R. Klingeler, J. Werner, B. Behr, G. Buchner, The intrinsic electronic phase diagram of iron-oxypnictide superconductors, Europhys. Lett. 87 (2009), 17005, 1–6.

[27] H. Hosono, K. Kuroki, Iron-Based Superconductors: current status of materials and pairing mechanism, Phys. C 514 (2015) 399–422.
[28] F.C. Hsu, J.Y. Luo, K.W. Yeh, T.K. Chen, T.W. Huang, P.M. Wu, Y.C. Lee, Y.L. Huang, Y.Y. Chu, D.C. Yan, M.K. Wu, Superconductivity in the PbO-type structure α-FeSe, Proc. Natl. Acad. Sci. U.S.A. 105 (2008) 14262–14264.
[29] F. Hunte, J. Jaroszynski, A. Gurevich, D.C. Larbalestier, R. Jin, A.S. Sefat, M.A. McGuire, B.C. Sales, D.K. Christen, D. Mandrus, Two-band superconductivity in $LaFeAsO_{0.89}F_{0.11}$ at very high magnetic fields, Nature 453 (2008) 903–905.
[30] S. Iimura, S. Matsuishi, H. Sato, T. Hanna, Y. Muraba, S.W. Kim, J.E. Kim, M. Takata, H. Hosono, Two-dome structure in electron-doped iron arsenide superconductors, Nat. Commun 3 (2012) 943, 1–6.
[31] S. Ishibashi, K. Terakura, H. Hosono, A possible ground state and its electronic structure of a mother material (LaOFeAs) of new superconductors, J. Phys. Soc. Jpn. 77 (2008), 053709, 1–4.
[32] F. Izumi, K. Momma, Three-dimensional visualization in powder diffraction, Solid State Phenomenum 130 (2007) 15–20.
[33] S. Kakiya, K. Kudo, Y. Nishikubo, K. Oku, E. Nishibori, H. Sawa, T. Yamamoto, T. Nozaka, M. Nohara, Superconductivity at 38K in iron-based compound with platinum-arsenide layers $Ca_{10}(Pt_4As_8)(Fe_{2-x}Pt_xAs_2)_5$, J. Phys. Soc. Jpn. 80 (2011), 093704, 1–4.
[34] Y. Kamihara, Current status of iron-based superconductors, Hyperfine Interact 203 (2012) 123–131.
[35] Y. Kamihara, H. Hiramatsu, M. Hirano, R. Kawamura, H. Yanagi, T. Kamiya, H. Hosono, Iron-based layered superconductor: LaOFeP, J. Am. Chem. Soc. 128 (2006) 10012–10013.
[36] Y. Kamihara, T. Watanabe, M. Hirano, H. Hosono, Iron-based layered superconductor $La[O_{1-x}F_x]FeAs$ (x = 0.05–0.12) with $T_c = 26K$, J. Am. Chem. Soc. 130 (2008) 3296–3297.
[37] Y. Kamihara, H. Hiramatsu, M. Hirano, Y. Kobayashi, S. Kitao, S. Higashitaniguchi, Y. Yoda, M. Seto, H. Hosono, Coexistence of superconductivity and antiferromagnetic ordering in the layered superconductor SmFePO, Phys. Rev. B 78 (2008), 184512, 1–7.
[38] Y. Kamihara, H. Hosono, (In Japanese) A Small history to discovery of iron based high T_c superconductors and summary, Rev. High Press. Sci. Technol. 19 (2009) 97–105.
[39] Y. Kamihara, H. Hosono, (In Japanese) Current status of iron based high T_c superconductors, Denshi-Zairyo (Electron. Parts Mater.) 49 (2010) 18–27.
[40] Y. Kamihara, T. Watanabe, T. Nomura, S.W. Kim, T. Kamiya, M. Hirano, H. Hosono, Electromagnetic properties of undoped LaFe*Pn*O (*Pn* = P, As), J. Phys. Conf. Ser. 150 (2009), 052090, 1–4.
[41] Y. Kamihara, T. Nomura, M. Hirano, J.E. Kim, K. Kato, M. Takata, Y. Kobayashi, S. Kitao, S. Higashitaniguchi, Y. Yoda, M. Seto, H. Hosono, Electronic and magnetic phase diagram of superconductors, $SmFeAsO_{1-x}F_x$, New J. Phys. 12 (2010), 033005, 1–14.
[42] Y. Kamihara, Y. Kobayashi, S. Kitao, Y. Yoda, M. Seto, H. Hosono, Magnetism of $SmFeAsO_{1-x}F_x$, American Physical Society March Meeting, Dallas, USA, 2011.
[43] M. Kano, Y. Kohama, D. Graf, F. Balakirev, A.S. Sefat, M.A. Mcguire, B.C. Sales, D. Mandrus, S.W. Tozer, Anisotropy of the upper critical field in a Co-doped $BaFe_2As_2$ single crystal, J. Phys. Soc. Jpn. 78 (2009), 084719, 1–5.
[44] T. Katase, Y. Ishimaru, A. Tsukamoto, H. Hiramatsu, T. Kamiya, K. Tanabe, H. Hosono, Advantageous grain boundaries in iron pnictide superconductors, Nat. Commun 2 (2011), 409, 1–6.

[45] S. Kitao, Y. Kobayashi, S. Higashitaniguchi, M. Saito, Y. Kamihara, M. Hirano, T. Mitsui, H. Hosono, M. Seto, Spin ordering in LaFeAsO and its suppression in superconductor $LaFeAsO_{0.89}F_{0.11}$ probed by Mössbauer spectroscopy, J. Phys. Soc. Jpn. 77 (2008), 103706, 1–3.
[46] H. Kito, H. Eisaki, A. Iyo, Superconductivity at 54K in F-Free $NdFeAsO_{1-y}$, J. Phys. Soc. Jpn. 77 (2008), 063707, 1–3.
[47] Y. Kohama, Y. Kamihara, M. Hirano, H. Kawaji, T. Atake, H. Hosono, Ferromagnetic spin fluctuation in $LaFeAsO_{1-x}F_x$, Phys. Rev. B 78 (2008), 020512, 1–4.
[48] K. Kuroki, H. Usui, S. Onari, R. Arita, H. Aoki, Pnictogen height as a possible switch between high-T_c nodeless and low-T_c nodal pairings in the iron-based superconductors, Phys. Rev. B 79 (2009), 224511, 1–16.
[49] P.A. Lee, N. Nagaosa, X.G. Wen, Doping a Mott insulator: Physics of high-temperature superconductivity, Rev. Mod. Phys. 78 (2006) 17–85.
[50] C.H. Lee, A. Iyo, H. Eisaki, H. Kito, M.T. Fernandez-Diaz, T. Ito, K. Kihou, H. Matsuhata, M. Braden, K. Yamada, Effect of structural parameters on superconductivity in fluorine-free $LnFeAsO_{1-y}$ (Ln = La, Nd), J. Phys. Soc. Jpn. 77 (2008), 083704, 1–4.
[51] C.H. Lee, K. Kihou, A. Iyo, H. Kito, P.M. Shirage, H. Eisaki, Relationship between crystal structure and superconductivity in iron-based superconductors, Solid State Commun. 152 (2012) 644–648.
[52] C. Löhnert, T. Stürzer, M. Tegel, R. Frankovsky, G. Friederichs, D. Johrendt, Superconductivity up to 35K in the iron platinum arsenides $(CaFe_{1-x}Pt_xAs)_{10}Pt_{4-y}As_8$ with layered structures, Angew. Chem. Int. Ed. 50 (2011) 9195–9199.
[53] X.F. Lu, N.Z. Wang, G.H. Zhang, X.G. Luo, Z.M. Ma, B. Lei, F.Q. Huang, X.H. Chen, Superconductivity in $LiFeO_2Fe_2Se_2$ with anti-PbO-type spacer layers, Phys. Rev. B 89 (2013), 020507, 1–5.
[54] H. Luetkens, H.H. Klauss, M. Kraken, F.J. Litterst, T. Dellmann, R. Klingeler, C. Hess, R. Khasanov, A. Amato, C. Baines, M. Kosmala, O.J. Schumann, M. Braden, J. Hamann-Borrero, N. Leps, A. Kondrat, G. Behr, J. Werner, B. Buchner, The electronic phase diagram of the $LaO_{1-x}F_xFeAs$ superconductor, Nat. Mater 8 (2009) 305–309.
[55] Y. Ma, L. Wang, Y. Qi, Z. Gao, D. Wang, X. Zhang, Development of powder-in-tube processed iron pnictide wires and tapes, IEEE Trans. Appl. Supercond. 21 (2011) 2878–2881.
[56] H. Maeda, Y. Tanaka, M. Fukutumi, T. Asano, A new high-T_c oxide superconductor without a rare earth element, Jpn. J. Appl. Phys. 27 (1988) L209–L210.
[57] S. Matsuishi, T. Hanna, Y. Muraba, S.W. Kim, J.E. Kim, M. Takata, S. Shamoto, R.I. Smith, H. Hosono, Structural analysis and superconductivity of $CeFeAsO_{1-x}H_x$, Phys. Rev. B 85 (2012), 014514, 1–6.
[58] S. Matsuishi, Y. Inoue, T. Nomura, H. Yanagi, M. Hirano, H. Hosono, Superconductivity induced by Co-doping in quaternary fluoroarsenide CaFeAsF, J. Am. Chem. Soc. 130 (2008) 14428–14429.
[59] B.T. Matthias, T.H. Geballe, S. Geller, E. Corenzwit, Superconductivity of Nb_3Sn, Phys. Rev. 95 (1954) 1435.
[60] W. Meissner, R. Ochsenfeld, Ein neuer Effekt bei Eintritt der Supraleitfahigkeit, Naturwissenschaften 21 (1933) 787–788.
[61] K. Miyazawa, S. Ishida, K. Kihou, P.M. Shirage, M. Nakajima, C.H. Lee, H. Kito, Y. Tomioka, T. Ito, H. Eisaki, H. Yamashita, H. Mukuda, K. Tokiwa, S. Uchida, A. Iyo, Possible hydrogen doping and enhancement of T_c (=35K) in a LaFeAsO-based superconductor, Appl. Phys. Lett. 96 (2010), 072514, 1–3.

[62] P.J.W. Moll, R. Puzniak, F. Balakirev, K. Rogacki, J. Karpinski, N.D. Zhigadlo, B. Batlogg, High magnetic-field scales and critical currents in SmFeAs(O, F) crystals, Nat. Mater. 9 (2010) 628–633.

[63] Y. Muraba, S. Matsuishi, H. Hosono, La-substituted CaFeAsH superconductor with $T_c = 47$K, J. Phys. Soc. Jpn. 83 (2014), 033705, 1–4.

[64] J. Nagamatsu, N. Nakagawa, T. Muranaka, Y. Zenitani, J. Akimitsu, Superconductivity at 39K in magnesium diboride, Nature 410 (2001) 63–64.

[65] K. Nakao, N. Miura, S. Uchida, H. Takagi, S. Tanaka, K. Kishio, J. Shimoyama, K. Kitazawa, K. Fueki, Study of the upper critical field H_{c2} for high-T_c superconductors in pulsed high magnetic fields, Jpn. J. Appl. Phys. 26 (1987) 1187–1188.

[66] N. Ni, J.M. Allred, B.C. Chan, R.J. Cava, High T_c electron doped $Ca_{10}(Pt_3As_8)(Fe_2As_2)_5$ and $Ca_{10}(Pt_4As_8)(Fe_2As_2)_5$ superconductors with skutterudite intermediary layers, Proc. Natl. Acad. Sci. U.S.A. 108 (2011) E1019–E1026.

[67] T. Nomura, S.W. Kim, Y. Kamihara, M. Hirano, P. Sushko, K. Kato, M. Takata, H. Shluger, A. Hosono, Crystallographic phase transition and high-T_c superconductivity in LaFeAsO: F, Supercond. Sci. Technol. 21 (2008), 125028, 1–9.

[68] H. Ogino, Y. Matsumura, Y. Katsura, K. Ushiyama, S. Horii, K. Kishio, J. Shimoyama, Superconductivity at 17K in $(Fe_2P_2)(Sr_4Sc_2O_6)$: a new superconducting layered pnictide oxide with a thick perovskite oxide layer, Supercond. Sci. Technol. 22 (2009), 075008, 1–4.

[69] H. Okada, K. Igawa, H. Takahashi, Y. Kamihara, M. Hirano, H. Hosnoso, K. Matsubayashi, Y. Uwatoko, Superconductivity under high pressure in LaFeAsO, J. Phys. Soc. Jpn. 77 (2008), 113712, 1–4.

[70] H.K. Onnes, Further experiments with liquid helium. G. On the electrical resistance of pure metals, etc. VI. On the sudden change in the rate at which the resistance of mercury disappears, Leiden Comm. 124c (1911).

[71] T. Ozaki, K. Deguchi, Y. Mizuguchi, Y. Kawasaki, T. Tanaka, T. Yamaguchi, H. Kumakura, Y. Takano, Fabrication of binary FeSe superconducting wires by diffusion process, J. Appl. Phys. 111 (2012), 112620, 1–3.

[72] U. Pachmayr, F. Nitsche, H. Luetkens, S. Kamusella, F. Bruckner, R. Sarkar, H. Klauss, D. Johrendt, Coexistence of 3d-ferromagnetism and superconductivity in $[(Li_{1-x}Fe_x)OH](Fe_{1-y}Li_y)Se$, Angew. Chem. Int. Ed. 54 (2015) 293–297.

[73] M.J. Pitcher, D.R. Parker, P. Adamson, S.J.C. Herkelrath, A.T. Boothroyd, R.M. Ibberson, M. Brunellid, S.J. Clarke, Structure and superconductivity of LiFeAs, Chem. Commun. (Cambridge) 45 (2008) 5918–5920.

[74] Y. Qi, L. Wang, D. Wang, Z. Zhang, Z. Gao, X. Zhang, Y. Ma, Transport critical currents in the iron pnictide superconducting wires prepared by the ex situ PIT method, Supercond. Sci. Technol. 23 (2010), 055009, 1–5.

[75] Y. Qiu, We Bao, Q. Huang, T. Yildirim, J.M. Simmons, M.A. Green, J.W. Lynn, Y.C. Gasparovic, J. Li, T. Wu, G. Wu, X.H. Chen, Crystal structure and antiferromagnetic order in $NdFeAsO_{1-x}F_x$ ($x = 0.0$ and 0.2) superconducting compounds from neutron diffraction measurements, Phys. Rev. Lett. 101 (2008), 257002, 1–4.

[76] B.J. Ramshaw, J. Day, B. Vignolle, D. LeBoeuf, P. Dosanjh, C. Proust, L. Taillefer, R. Liang, W.N. Hardy, D.A. Bonn, Vortex lattice melting and H_{c2} in underdoped $YBa_2Cu_3O_y$, Phys. Rev. B 86 (2012), 174501, 1–6.

[77] Z.A. Ren, G.C. Che, X.L. Dong, J. Yang, W. Lu, W. Yi, X.L. Shen, Z.C. Li, L.L. Sun, F. Zhou, Z.X. Zhao, Superconductivity and phase diagram in iron-based arsenic-oxides $ReFeAsO_{1-\delta}$ (Re = rare-earth metal) without fluorine doping, Europhys. Lett. 83 (2008), 17002, 1–4.

[78] C. Rotundu, D. Keane, B. Freelon, S. Wilson, A. Kim, P. Valdivia, R. Bourret-Courchesne, E. Birgeneau, Phase diagram of the $PrFeAsO_{1-x}F_x$ superconductor, Phys. Rev. B 80 (2009), 144517, 1−5.
[79] A. Sakurai, H. Shiozawa, M. Matoba, S. Matsuishi, H. Hosono, Y. Kamihara, (2016). (Unpublished).
[80] A. Schilling, M. Cantoni, J.D. Guo, H.R. Ott, Superconductivity above 130K in the Hg-Ba-Ca-Cu-O system, Nature 363 (1993) 56−58.
[81] A.S. Sefat, A. Huq, M.A. McGuire, R. Jin, B.C. Sales, D. Mandrus, L.M.D. Cranswick, P.W. Stephens, K.H. Stone, Superconductivity in $LaFe_{1-x}Co_xAsO$, Phys. Rev. B 78 (2008), 104505, 1−9.
[82] C. Senatore, R. Flukiger, M. Cantoni, G. Wu, R.H. Liu, X.H. Chen, Upper critical fields well above 100 T for the superconductor $SmFeAsO_{0.85}F_{0.15}$ with $T_c = 46K$, Phys. Rev. B 78 (2008), 054514, 1−7.
[83] J. Shimoyama, Potentials of iron-based superconductors for practical future materials, Supercond. Sci. Technol. 27 (2014), 044002, 1−7.
[84] J. Shimoyama, K. Kitazawa, K. Shimizu, S. Ueda, S. Horii, N. Chikumoto, K. Kishio, Generic guiding law between irreversibility field and anisotropy-chemical control of critical current of high temperature superconductors, J. Low Temp. Phys. 131 (2003) 1043−1052.
[85] H. Shiozawa, S. Inagi, M. Matoba, T. Atou, M. Itoh, Y. Kamihara, Relations between superconducting critical temperatures and crystallographic local structures in iron-based mixed anion compounds, JPS Conf. Proc. 1 (2014), 012026, 1−4.
[86] T. Shiroka, G. Lamura, S. Sanna, G. Prando, R. De Renzi, M. Tropeano, M.R. Cimberle, A. Martinelli, C. Bernini, A. Palenzona, R. Fittipaldi, A. Vecchione, P. Carretta, A.S. Siri, C. Ferdeghini, M. Putti, Long- to short-range magnetic order in fluorine-doped CeFeAsO, Phys. Rev. B 84 (2011), 195123, 1−8.
[87] G.R. Stewart, Superconductivity in iron compounds, Rev. Mod. Phys. 83 (2011) 1589−1652.
[88] H. Takahashi, K. Igawa, K. Arii, Y. Kamihara, M. Hirano, H. Hosono, Superconductivity at 43K in an iron-based layered compound $LaO_{1-x}F_xFeAs$, Nature 453 (2008) 376−378.
[89] N. Takeshita, A. Yamamoto, A. Iyo, H. Eisaki, Zero resistivity above 150K in $HgBa_2Ca_2Cu_3O_{8+\delta}$ at high pressure, J. Phys. Soc. Jpn. 82 (2013), 023711, 1−4.
[90] K. Tanabe, H. Hosono, Frontiers of research on iron-based superconductors toward their application, Jpn. J. Appl. Phys. 51 (2012), 010005, 1−17.
[91] J.H. Tapp, Z. Tang, B. Lv, K. Sasmal, B. Lorenz, P.C.W. Chu, A.M. Guloy, LiFeAs: an intrinsic FeAs-based superconductor with $T_c = 18K$, Phys. Rev. B 78 (2008), 060505, 1−4.
[92] M. Tegel, S. Johansson, V. Weis, I. Schellenberg, W. Hermes, R. Pottgen, D. Johrendt, Synthesis, crystal structure and spin-density-wave anomaly of the iron arsenide-fluoride SrFeAsF, Europhys. Lett. 84 (2008), 67007, 1−5.
[93] K. Togano, A. Matsumoto, H. Kumakura, Large transport critical current densities of Ag sheathed $(Ba,K)Fe_2As_2$+Ag superconducting wires fabricated by an ex-situ powder-in-tube process, Appl. Phys. Express 4 (2011), 043101, 1−3.
[94] T. Tohei, T. Mizoguchi, H. Hiramatsu, Y. Kamihara, H. Hosono, Y. Ikuhara, Direct imaging of doped fluorine in $LaFeAsO_{1-x}F_x$ superconductor by atomic scale spectroscopy, Appl. Phys. Lett. 95 (2009), 193107, 1−3.
[95] C. Wang, S. Jiang, Q. Tao, Z. Ren, Y. Li, L. Li, C. Feng, J. Dai, G. Cao, Z. Xu, Superconductivity in $LaFeAs_{1-x}P_xO$: effect of chemical pressures and bond covalency, Europhys. Lett. 86 (2009), 47002, 1−5.
[96] L. Wang, Y. Qi, D. Wang, Z. Gao, X. Zhang, Z. Zhang, C. Wang, Y. Ma, Low-temperature synthesis of $SmFeAsO_{0.7}F_{0.3-\delta}$ wires with a high transport critical current density, Supercond. Sci. Technol. 23 (2010), 075005, 1−4.

[97] C. Wang, C. Yao, X. Zhang, Z. Gao, D. Wang, C. Wang, H. Lin, Y. Ma, S. Awaji, K. Watanabe, Effect of starting materials on the superconducting properties of $SmFeAsO_{1-x}F_x$ tapes, Supercond. Sci. Technol. 25 (2012), 035013, 1–4.
[98] C. Wang, C. Yao, H. Lin, X. Zhang, Q. Zhang, D. Wang, Y. Ma, S. Awaji, K. Watanabe, Y. Tsuchiya, Y. Sun, T. Tamegai, Large transport J_c in Sn-added $SmFeAsO_{1-x}F_x$ tapes prepared by an ex situ PIT method, Supercond. Sci. Technol. 26 (2013), 075017, 1–6.
[99] T. Watanabe, H. Yanagi, T. Kamiya, Y. Kamihara, H. Hiramatsu, M. Hirano, H. Hosono, Nickel-based oxyphosphide superconductor with a layered crystal structure, LaNiOP, Inorg. Chem. 46 (2007) 7719–7721.
[100] J.D. Weiss, C. Tarantini, J. Jiang, F. Kametani, A.A. Polyanskii, E.E. Larbalestier, D.C. Hellstrom, High intergrain critical current density in fine-grain $Ba_{0.6}K_{0.4}Fe_2As_2$ wires and bulks, Nat. Mater. 11 (2012) 682–685.
[101] M.K. Wu, J.R. Ashburn, C.J. Torng, P.H. Hor, R.L. Meng, L. Gao, Z.J. Huang, Y.Q. Wang, C.W. Chu, Superconductivity at 93K in a new mixed-phase Y-Ba-Cu-O compound system at ambient pressure, Phys. Rev. Lett. 58 (1987) 908–910.
[102] H. Yanagi, R. Kawamura, T. Kamiya, Y. Kamihara, M. Hirano, T. Nakamura, H. Osawa, H. Hosono, Itinerant ferromagnetism in the layered crystals LaCoOX (X = P, As), Phys. Rev. B 77 (2008), 224431, 1–7.
[103] H. Yanagi, T. Watanabe, K. Kodama, S. Iikubo, S. Shamoto, T. Kamiya, M. Hirano, H. Hosono, Antiferromagnetic bipolar semiconductor LaMnPO with ZrCuSiAs-type structure, J. Appl. Phys. 105 (2009), 93916, 1–8.
[104] H.Q. Yuan, J. Singleton, F.F. Balakirev, S.A. Baily, G.F. Chen, J.L. Luo, N.L. Wang, Nearly isotropic superconductivity in $(Ba,K)Fe_2As_2$, Nature 457 (2009) 565–568.
[105] C.J. Zhang, H. Oyanagi, Z.H. Sun, Y. Kamihara, H. Hosono, Low-temperature lattice structure anomaly in the $LaFeAsO_{0.93}F_{0.07}$ superconductor by x-ray absorption spectroscopy: Evidence for a strong electron-phonon interaction, Phys. Rev. B 78 (2008), 214513, 1–5.
[106] Q. Zhang, C. Yao, H. Lin, X. Zhang, D. Wang, C. Dong, P. Yuan, S. Tang, Y. Ma, S. Awaji, K. Watanabe, Y. Tsuchiya, T. Tamegai, Enhancement of transport critical current density of $SmFeAsO_{1-x}F_x$ tapes fabricated by an ex-situ powder-in-tube method with a Sn-presintering process, Appl. Phys. Lett. 104 (2014), 172601, 1–5.
[107] X. Zhang, C. Yao, H. Lin, Y. Cai, Z. Chen, J. Li, C. Dong, Q. Zhang, D. Wang, Y. Ma, H. Oguro, S. Awaji, K. Watanabe, Realization of practical level current densities in $Sr_{0.6}K_{0.4}Fe_2As_2$ tape conductors for high-field applications, Appl. Phys. Lett. 104 (2014), 202601, 1–5.
[108] J. Zhao, Q. Huang, C. Cruz, S. Li, J. Lynn, Y. Chen, M. Green, G. Chen, G. Li, Z. Li, J. Luo, N. Wang, P. Dai, J. Zhao, Structural and magnetic phase diagram of $CeFeAsO_{1-x}F_x$ and its relation to high-temperature superconductivity, Nat. Mater. 7 (2008) 953–959.
[109] X. Zhu, F. Han, P. Cheng, G. Mu, B. Shen, H.H. Fang, L. Wen, Superconductivity in fluoride-arsenide $Sr_{1-x}La_xFeAsF$ compounds, Europhys. Lett. 85 (2009), 17011, 1–4.

Part Six

Ionic conductors

Ionic Conductivity of Nanocrystalline Metal Fluorides

20

M.A. Reddy, M. Fichtner
Helmholtz Institute Ulm for Electrochemical Energy Storage (HIU), Ulm, Germany

Chapter Outline

1. Introduction

Ionic materials which show the phenomenon of atomic transport are important in a number of technological applications where they are used as solid electrolytes or active materials in batteries and fuel cells, or in sensors.

A common application for fluoride ion conductors is as electrolyte in a chemical sensor for quantitative determination of fluoride ion concentrations, eg, in aqueous solutions, using LaF_3 as the electrolyte. There is a variety for applications as electrolyte also in other sensors and for determining species other than fluorides. A further field of interest has been the use of ion-conducting metal fluorides in electrochromic displays where, under the influence of an electric field, different colors of light are emitted. The colors are due to changes in the wavelength-dependent absorption of light when the fluoride material makes an anion exchange in the electrolyte. Lately, the use of fluorinated organic compounds and metal fluorides has gained considerable interest in the development of energy materials. In particular, fluorides have been introduced as electrode materials in Li-ion batteries due to the relatively small size of the anion and the high free energy of formation of the fluorides yielding high electrode potentials of a cathode. Also secondary batteries based on fluoride ion shuttle have been realized,

Photonic and Electronic Properties of Fluoride Materials. http://dx.doi.org/10.1016/B978-0-12-801639-8.00020-9

which showed reversibly working fluoride ion shuttle. The solid-state batteries were made by using pressed powder pellets of fine-grained material.

In the 1980s and before, the work on fluoride ion conductors had mainly focused on elucidating the mechanisms of the sometimes very high ionic conductivities of various classes of materials for fluoride ion transport. For such fundamental studies it may be necessary to work with well-defined samples, which are single crystals in the ideal case. The reason is that single crystals do not contain any grain boundaries which would interfere with the ionic transport and slow it down to a great extent, as will be explained later. In parallel, a second point of interest has always been the development of better materials, ie, ionic or so-called superionic conductors with optimized conductivities for fluoride ions.

While the granular materials usually have ionic conduction properties that differ strongly from those of single crystals, the properties of nanoscale ionic conductors can again differ greatly from those of their coarse-grained counterparts. The small particle diameter and the considerably higher fraction of atoms at the interface may let processes become paramount that are associated with the particular properties of the surfaces of ionic conductors [1].

The origin of the unusual properties at the nanoscale in general has been attributed to two major reasons. First, the dimension of the particles is similar to, or becomes smaller than, the critical length for certain phenomena such as the de Broglie wavelength for the electron, the mean free path of an exciton, the thickness of the space charge layer, and others. Second, surface effects can dominate the thermodynamics and energetics of the particles if these are very small, in the lower nanometer range. Properties affected are the crystal structure, surface morphology, and the chemical properties. The second factor can lead to nanocrystals adopting different morphologies to bulk crystals with different exposed lattice planes leading to an extraordinary surface chemistry and catalytic activity.

Nanomaterials with exceptionally high diffusion coefficients are often referred to as "nanoionics" and considerable effort has been spent to quantify the origin and the extent of the diffusion enhancement. This chapter deals with ionic conduction phenomena in metal fluorides. After a general introduction to ionic conduction and doping of solids, nanocrystalline effects will be discussed and different types of materials will be presented. As will be shown their properties depend on the level of doping and the influence of grain boundaries. The discussion will particularly refer to the influence of grain size.

2. Ionic Conduction in Solids

High ionic conductivities are usually related to liquid electrolytes. However, high ionic conductivities were found also in solid state in AgI and $RbAg_4I_5$ where the Ag^+ conductivity can reach 3×10^{-1} S/cm at near room temperature which are comparable to that of liquid electrolytes. Ionic conduction occurs in a solid due to the presence of point defects [2–4]. Two types of point defects can exist in a crystal (in thermodynamic equilibrium with the lattice), depending on the type of material and its structure: ion vacancies or interstitial ions. These native defects naturally present in the crystals are called intrinsic defects and their concentration depends on the temperature. They are usually organized in different types of disorder as the overall charge neutrality and the composition of a compound is maintained on the long range. There are four main types of disorder that can be found in solids:

- Schottky disorder: association of cation vacancies and anion vacancies (eg, in NaCl, LiF, or MgO)
- Frenkel disorder: association of cation vacancies and cation interstitials (eg, in AgI or FeO)
- Redox Frenkel disorder: the charge of a foreign interstitial cation is balanced by the reduction in the charge of one of the native species (common in electrode materials for batteries)
- Anti-Frenkel disorder: association of anion vacancies with anion interstitials (eg, CaF_2, ZrO_2, or UO_2).

2.1 Doping

Apart from fast ionic conductors, conductivity can be significantly improved for ordinary conductors by modifying the structure and/or composition of the materials by doping, ie, by the creation of new defects called extrinsic defects. The conduction mechanism is the same as for intrinsic defects and the conductivity can be improved by increasing the concentration of mobile charge carriers or decreasing the migration enthalpy.

Two types of doping methods are widely used for different types of materials: homogeneous doping and heterogeneous doping.

Homogeneous doping consists of adding an aliovalent cation/anion to an ionic conductor. The presence of these ions leads to the creation of point defects (vacancies or interstitials) to compensate the additional charges and maintain the electroneutrality. These compounds are also known as nonstoichiometric compounds [4,5].

Heterogeneous doping is mainly achieved by creating interfaces containing defects and/or providing diffusion paths [6]. Different types of interfaces are possible:

- between an ionic conductor and an insulator [7],
- between two different conductors [8,9],
- within a polycrystalline ionic conductor (grain boundaries) [10].

Extrinsic defects are generally present with a much higher concentration than intrinsic defects especially at low temperature and their concentration is temperature independent. For typical ionic conductors, the temperature dependence of the conductivity falls into three regions as displayed in Fig. 20.1. Each region can be described by the Arrhenius-type law given below [4].

$$\frac{\sigma = \frac{\sigma_0}{T}\exp(-E_a)}{kT}$$

where σ_0 is the preexponential factor and E_a the activation energy, k the Boltzmann constant, and T the absolute temperature

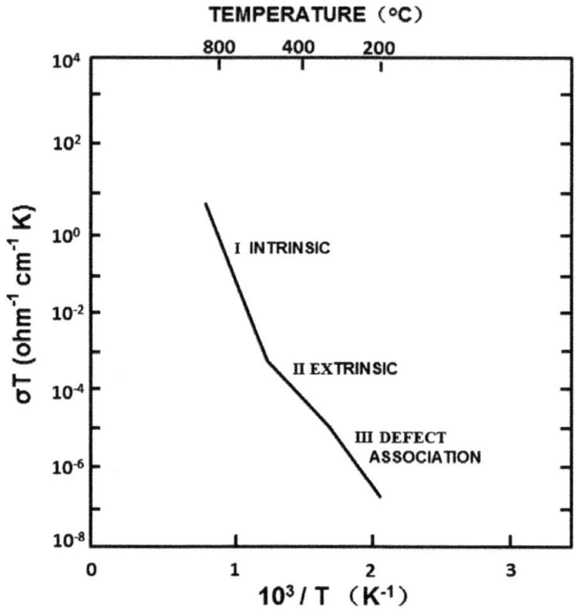

Figure 20.1 Conductivity of LiF as a function of the temperature. Reproduced with permission [4]. Copyright © 1981 Elsevier B.V.

At temperatures close to the melting point (region I), the conductivity is controlled by the intrinsic defects. The activation energy is quite high as it includes the enthalpy of migration and the enthalpy of defect formation. At moderate and low temperatures (region II), the conductivity is controlled by the extrinsic defects present at much higher concentration than the intrinsic defects. The activation energy is smaller as in region (I) as it mainly reflects the enthalpy of migration. In addition, for some compounds and at low temperatures, it is possible to have a third region (III) where the defect association influences the conductivity. The activation energy increases as the contribution of defect interactions should be added to the enthalpy of migration.

2.2 Nanocrystalline Effects in Ionic Conductors

The transport phenomenon is relatively well understood in the case of single crystals and polycrystalline solids with large particle size. However, in the case of nanocrystalline solids the fraction of atoms lying on the surface exponentially increases as the size decreases, in the nanometer regime [11]. In other words the volume of the disordered grain boundaries increases with decrease in particle size. The grain boundaries possess high defect densities and enhanced mobility. The grain boundary diffusion is often in orders of magnitude higher than that of corresponding bulk diffusion through the grains.

The high diffusion in the grain boundaries may not transform into higher ionic conductivity. In the case of polycrystalline materials the grain boundaries lying parallel to the current flux represent only a very small fraction. Derived from the ideal "brick model" in the case of polycrystalline sample with grain size of 1.0 μm and grain boundary core width of 1.0 nm, $\sigma_{GB} \approx 500\sigma_B$ is required to double overall conductance of the material considering all the grain boundaries are lying parallel to the current flux [12]. Whereas in the case of nanocrystalline solid, where the widths of the grain boundaries are comparable to the size of the grains, a significant fraction of atoms is expected to lie in the disordered region of grain boundaries and possesses large mobility. In such cases the enhanced diffusion of grain boundaries will transform into overall conductivity of the solid.

In addition to enhanced ionic conductivity of nanocrystalline solids due to large volume of defect grain boundaries, further enhancement in ionic conduction can be found if the grain size approaches the Debye length. In such cases, the space charge regions (formed near interfaces in ionic solids to compensate charged defects and impurities, which segregate to surfaces) overlap, and the defect density no longer reaches the bulk value, even at the center of the grain (Fig. 20.2) [6]. In this case, Maier described and calculated the supplementary enhancement, corresponding to the nanosize factor g,

$$g = 4\lambda/L(c_0 - c^*/c_0)^{1/2}$$

where c^* is the defect concentration in the center of the grain.

For a large defect concentration, $c_0 \gg c^*$, $g = 4\lambda/L$, and with a grain size of $L = 0.4\lambda$, the conductivity is enhanced by a supplementary order of magnitude [6,13]. Such enhancement in ionic conductivity was recently demonstrated for epitaxial heterolayers of solid fluoride ion conductors CaF_2/BaF_2 [9].

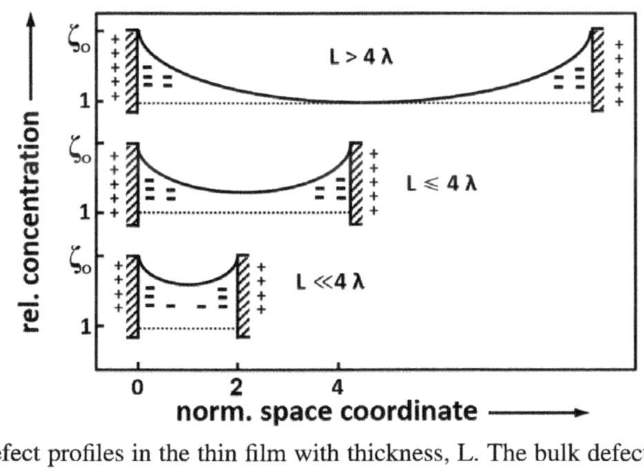

Figure 20.2 Defect profiles in the thin film with thickness, L. The bulk defect concentration is not reached when $L \ll 4\lambda$, where λ is the Debye length.

Reproduced with permission [6]. Copyright © 1995 Elsevier B.V.

3. Fluoride Ion Conductors

Fluorine is the element with the highest electronegativity in the periodic table. Fluoride ions are highly ionic, monovalent, and relatively small with high charge density. These basic properties enable fluoride ion to be highly mobile in some of the ionic solids. However, the conductivity depends strongly on the structure of the fluorides as the fluoride ions or vacancies should have a high diffusion rate in the crystals. There are two basic structure types which show high fluoride ion conductivity: tysonite and fluorite-type structures. Alkaline earth fluorides (MF_2) adopt fluorite-type structure and few rare earth fluorides (RF_3) crystallize in tysonite-type structure.

3.1 Fluorite-Type Fluoride Ion Conductors

The crystal structure of CaF_2 is shown in Fig. 20.3. It crystallizes in cubic structure with space group Fm-3m [14]. The structure is built up of cubic close packing of fluoride ions with cations occupying an eight-coordinated site and fluoride ions occupying four-coordinated site. Binary metal fluorides of type MF_2 (M = Ca, Sr, Ba, Pb, and Cd) and their alkaline, some rare earth–doped metal fluorides ($M_{1-x}A_xF_{2-x}$ and $M_{1-x}R_xF_{2+x}$), and ternary metal fluorides of AMF_4 type (A = K, M = Sb, Bi) crystallize in fluorite-type structure.

The conductivities of single crystals of alkaline earth fluorides are in the following order $CaF_2 < SrF_2 < BaF_2$ [15]. Fluoride ion conductivity of undoped fluorite-type compounds is relatively low at room temperature. However, conductivity can be substantially increased by doping. Two different approaches were adapted to improve the fluoride ion conductivity of fluorite-type structures, homogeneous and heterogeneous doping. Doping of alkaline earth fluorides (fluorite structure) by monovalent cations (Na, K) and trivalent cations (rare earth) has been widely studied to improve

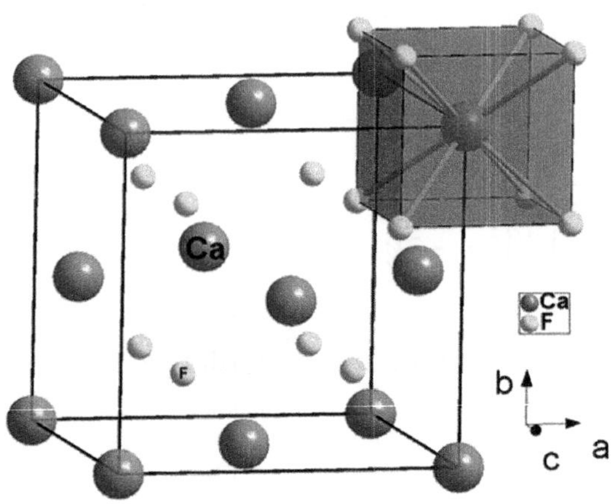

Figure 20.3 Crystal structure of CaF_2.

the ionic conductivity of fluorites [16,17]. Heterogeneous doping is mainly achieved by creating interfaces between two different conductors [9] within a polycrystalline ionic conductor (grain boundaries) [10,18].

3.1.1 Homogeneous Doping

Homogeneous doping is a classical approach adopted for the improvement of ionic conductivity of various ionic conductors. Doping of alkaline earth metals with alkali cations will increase the defect concentration, thereby enhancing the fluoride ion conductivity. However, the activation energy required for the intrinsic mobility of defects is still high for the lattice fluoride ions. On the other hand, when doped with trivalent cations, the excess fluoride ions will populate in different interstitial sites depending on the doping level. These interstitial sites have relatively low activation energy and contribute to high fluoride ion conductivity. For example, Ivanov-Shits et al. have studied single crystals of $Ba_{1-x}R_xF_{2+x}$ with different rare earth elements. For all rare earth elements, the conductivity increased with doping due to the decrease of the activation enthalpy. The high conductivity of 1.9×10^{-4} S/cm at 200°C was obtained for $Ba_{0.6}La_{0.4}F_{2.4}$ [19]. While there are many studies on fluoride ion conductivity in single crystals and polycrystalline materials with fluorite-type structure, here we focused only on nanostructured materials.

Sobolev et al. synthesized nonstoichiometric $Ca_{1-x}La_xF_{2+x}$ phase ($x \geq 0.1$) by mechanical milling of CaF_2 and LaF_3 single crystals. The crystallite sizes were in the range of 10–30 nm. They found that the ionic conductivity of nanocrystalline $Ca_{0.8}La_{0.2}F_{2.2}$ exceeds that of corresponding single crystals in the 200–250°C range [20].

Rongeat et al. studied the ionic conductivity of nanostructured $Ba_{1-x}La_xF_{2+x}$ ($0 \leq x \leq 0.55$) synthesized by mechanical milling using polycrystalline BaF_2 and LaF_3 as starting materials [21]. The crystallite sizes were in the range of 20–30 nm. Large improvement in the ionic conductivity was found compared to that in undoped BaF_2. The conductivity increases up to $x = 0.4$ then decreases with higher x values. Highest conductivity of 1.9×10^{-4} S/cm was obtained at 160°C for $Ba_{0.6}La_{0.4}F_{2.4}$ in the $Ba_{1-x}La_xF_{2+x}$ ($0 \leq x \leq 0.55$) series. A striking feature of nanostructured $Ba_{1-x}La_xF_{2+x}$ materials reported in this study is that they possess high ionic conductivity even compared to corresponding single crystals. Based on electrochemical impedance spectroscopy and AC conductivity analysis, it was shown that the ionic conduction in these materials is controlled by the motion of vacancies through the grain boundaries (Fig. 20.4). In addition, $Ba_{0.6}La_{0.4}F_{2.4}$ was demonstrated as a potential electrolyte for fluoride ion batteries. Düvel et al. synthesized solid solutions in the $Ba_{1-x}La_xF_{2+x}$ system from fluorite- to tysonite-type structure. They found that the fluorite-type structure can maintain up to $x = 0.775$ [22].

Another interesting fluoride ion conductor which adopts fluorite-type structure is β-PbF_2. The conductivity of β-PbF_2 is 5×10^{-7} S/cm at room temperature, which is high compared to isostructural alkaline earth MF_2. Doping with KF was found to

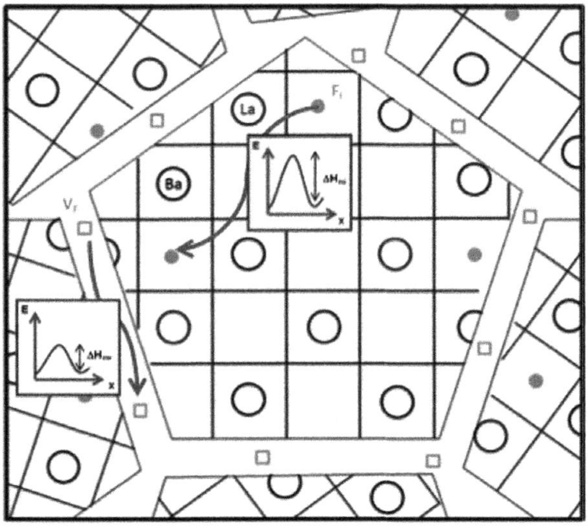

Figure 20.4 Depiction of ionic conductivity mechanism in nanostructured $Ba_{1-x}La_xF_{2+x}$ materials.
Reprinted with permission from Ref. [21]. Copyright (2013) American Chemical Society.

increase the ionic conductivity up to 1×10^{-3} S/cm [23]. Surprisingly, substitution of isovalent Sn^{2+} found to increase ionic conductivity (generally isovalent substitution will not change the defect concentration, hence isovalent substitution is not expected to increase the conductivity) [24]. PbF_2 forms cubic solid solutions $Pb_{1-x}Sn_xF_2$ with SnF_2 in the range $0 \leq x \leq 0.3$ and tetragonal solid solutions in the range $0.3 < x \leq 0.5$ [25]. Maximum conductivity was achieved with $x = 0.5$, ie, $PbSnF_4$ [26]. There are several phases of $PbSnF_4$ depending on the method of preparation. Mainly four types of $PbSnF_4$ exist: orthorhombic, o-$PbSnF_4$; tetragonal, α-$PbSnF_4$ (P4/nmm); tetragonal, β-$PbSnF_4$; and cubic, γ-$PbSnF_4$ (Fm-3m). However, all phases are related to fluorite-type structure. Ahmad et al. synthesized nanocrystalline $Pb_{1-x}Sn_xF_2$ ($0 < x \leq 0.6$) solid solutions by mechanical milling [27]. Surprisingly the cubic phase was maintained up to $x = 0.5$ and the tetragonal phase was obtained for $x = 0.6$. High ionic conductivity was found for $x = 0.5$, ie, $PbSnF_4$. An ionic conductivity of 2.28×10^{-3} S/cm was obtained at room temperature. Denes et al. synthesized $BaSnF_4$ by high-temperature solid-state reaction, which is isostructural to the layered tetragonal β-$PbSnF_4$ phase [28]. Tetragonal $BaSnF_4$ shows high ionic conductivity of 10^{-4} S/cm at room temperature [28]. Similar to $PbSnF_4$, cubic phase $BaSnF_4$ was synthesized by mechanical milling [29]. The conductivity of the cubic phase is 10^{-6} S/cm at room temperature, which is much lower than that of the tetragonal phase. However, the cubic $BaSnF_4$ transforms to tetragonal phase upon annealing at 460K for 8 h, which shows similar ionic conductivity to that of tetragonal $BaSnF_4$ synthesized by solid-state reaction. Despite high fluoride ionic conductivity of these compounds, they have narrow electrochemical stability window which limits their use as electrolytes for fluoride ion batteries. However, they are potential candidates for sensor applications.

3.1.2 Heterogeneous Doping

As mentioned earlier, ionic conductivity of solids can be enhanced by introducing interfaces or space charge regions by the dispersion of second phase which has the chemical affinity to the mobile ion. The dispersion of insulating second phase on the surface of ionic conducting solids absorbs the mobile cation and improves the defect concentration thereby increasing the ionic conductivity. However, it should be noted that the absorption of mobile ion will only increase the concentration of defects, and to get significantly high ionic conductivity the mobility of these defects should be high. For example, Liang studied the electrical properties of the two-phase system LiI–Al_2O_3 and found that the ionic conductivity increased significantly compared to the pure phases [30]. This work was a fundamental breakthrough in this line—it generated significant interest on composite (polyphase) electrolytes.

Similar studies were made on fluoride ion conductors. The conductivity of sintered polycrystalline CaF_2 and BaF_2 was increased by one to two orders of magnitude by the addition of Al_2O_3 [31]. Puin and Heitjans made nanocrystalline CaF_2 (9 nm) by the inert-gas condensation and studied frequency-dependent fluoride ion conductivity [32]. They found that conductivity was increased by more than four orders of magnitude higher than micron-sized polycrystalline sample. The large increase in ionic conductivity of nanocrystalline CaF_2 compared to polycrystalline solid was attributed to the formation of large fraction of interface regions in the nanocrystalline material [32,33]. The conductivity of CaF_2 could also be increased by the grain boundary activation using Lewis acids [34]. The adsorbed SbF_5 or BF_3 at the boundaries attract F ions and enrich defect concentration and significantly increase the F anion mobility similar to the LiI–Al_2O_3 system [30].

Recently, Sata et al. fabricated heterolayers composed of CaF_2 and BaF_2 by molecular-beam epitaxy method. These heterolayers showed significantly high ionic conductivity parallel to interfaces compared to pure films. Fig. 20.5 shows the Arrhenius plot for the ionic conductivity of CaF_2, BaF_2, and CaF_2/BaF_2 heterolayers with various interfacial densities. The ionic conductivity increases linearly with the interfacial density in the period (total thickness of $CaF_2 + BaF_2$ layer) range of 500–16 nm. The anomalous increase in ionic conductivity for periods less than 100 nm is related to the overlap of the space charge layers [9].

Later, Ruprecht et al. synthesized nanocrystalline CaF_2–BaF_2 composites by mechanical milling and studied their ionic conductivity [10,35]. They found unexpectedly high DC conductivity of about 0.1 mS/cm at 450K for CaF_2:BaF_2 nanocomposites. The origin for the observed high ionic conductivity was investigated by X-ray diffraction (XRD) and ^{19}F magic angle spinning (MAS) nuclear magnetic resonance (NMR) measurements at high spinning speeds. The XRD studies reveal the formation of mixed phases. Fig. 20.6 shows the static ^{19}F NMR (Fig. 20.6A) and ^{19}F MAS NMR spectra of CaF_2:BaF_2 nanocomposites. The static ^{19}F NMR shows the evolution of new narrow peak at 433K which was not seen in the spectra recorded at room temperature. The ^{19}MAS NMR shows a large distribution of F anions which reflects the highly defective structure of the CaF_2:BaF_2 nanocomposites. The high ionic conductivity was attributed to the formation of highly defective structure.

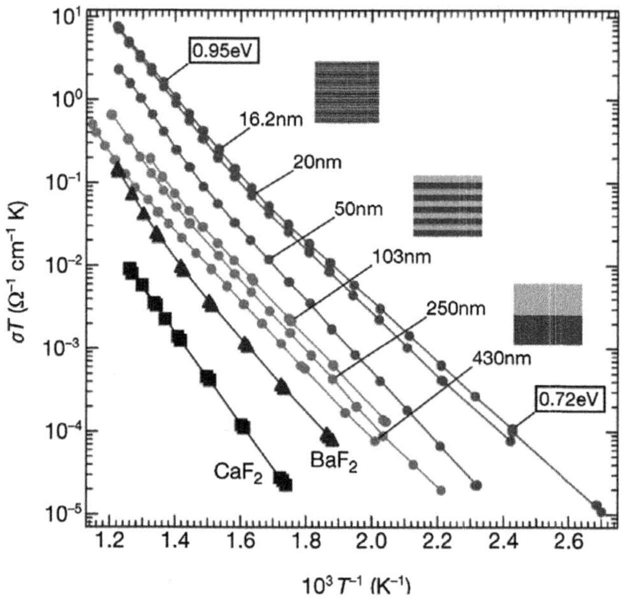

Figure 20.5 Parallel ionic conductivity of the CaF_2, BaF_2, and CaF_2–BaF_2 films. Data are shown for films with various periods and interfacial densities in the 430 ± 16 nm range. The overall thickness is approximately 500 nm in all cases. Different colors refer to different site regimes (*green* (light gray in print versions): semiinfinite space charge zones; *red* (gray in print versions): finite space charge zones).
Reprinted by permission from Macmillan Publishers Ltd: [NATURE] [9], copyright (2000).

3.2 Tysonite-Type Fluoride Ion Conductors

Rare earth fluorides (RF_3) with larger cations (La–Nd) adopt tysonite-type structure whereas the RF_3 with smaller cations (Sm–Lu, Y) adopt the orthorhombic β-YF_3 (Pnma)-type structure [36]. While there are only minor changes among the crystal structure of tysonite-type and β-YF_3, tysonite-type fluorides show high fluoride conductivity. In addition, there are two modifications within the tysonite-type structure-hexagonal (P63/mmc) [37] and trigonal phase (P-3c1) [38]. The trigonal modification is more common and discussed further. The following phases adopt the trigonal tysonite-type structure: RF_3 (R = La–Nd) and solid solutions on their basis $R_{1-x}M_xF_{3-x}$ (M = Ca, Sr, Ba and R = La–Nd) [39]. The crystal structure of LaF_3 is shown in Fig. 20.7. There are three crystallographically and dynamically distinct fluoride ions, commonly known as F1, F2, and F3. The ratio of these fluoride ions F1:F2:F3 is 12:4:2. The F2 and F3 ions have nearly similar chemical environment and are often treated structurally and dynamically equivalent (F2, 3 sublattices) [40]. However, high-resolution ^{19}F MAS NMR is able to distinguish F2 and F3 sites and the existence of three fluorine sites F1, F2, and F3 in the trigonal LaF_3 structure was confirmed [41].

The ionic conductivity in pure LaF_3 is anisotropic with a higher conductivity in parallel to the c-axis [42]. The anisotropy disappears at higher temperatures or after

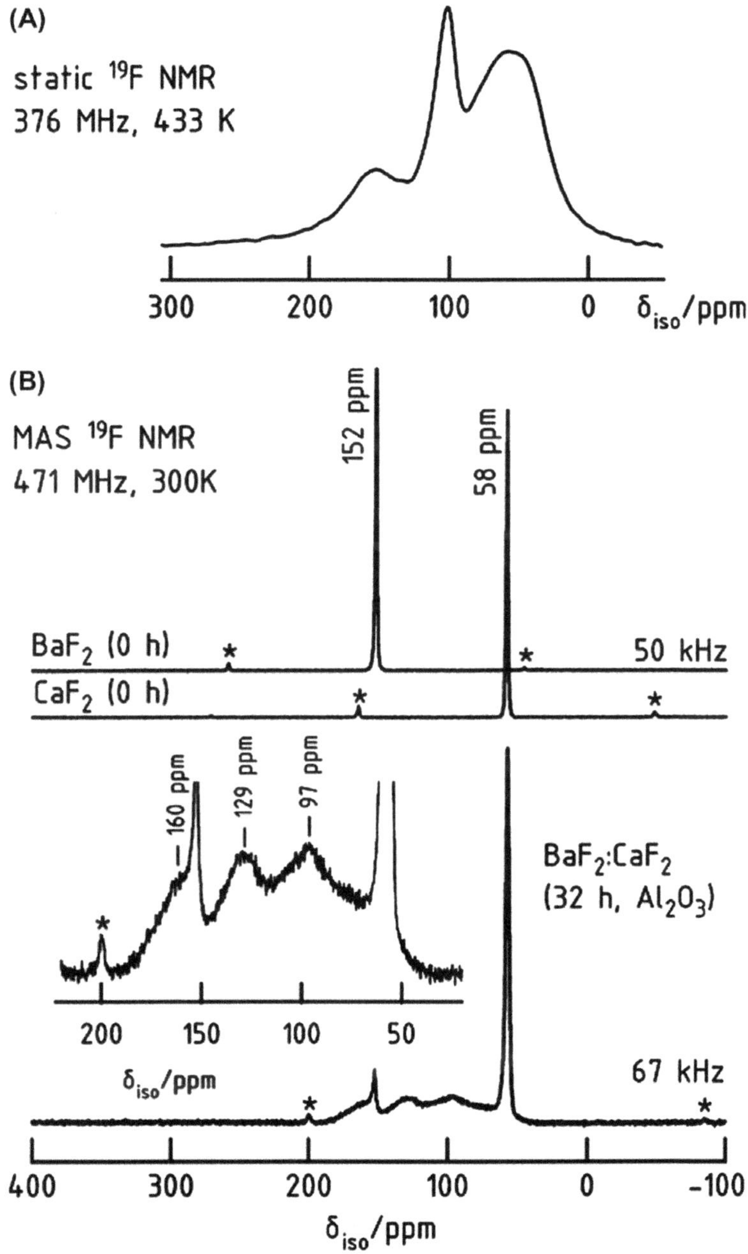

Figure 20.6 (A) Static ^{19}F nuclear magnetic resonance (NMR) spectrum of $(1-x)$ BaF_2: $xCaF_2$ with $x = 0.6$. (B) ^{19}F magnetic angle spinning NMR spectra of CaF_2 and BaF_2 recorded at 300 K using a spinning speed of 50 kHz as well as of the nanocomposite at a rotation frequency of 67 kHz. Inset shows the magnified part of the spectrum (*asterisks* denote spinning sidebands).

Reproduced with permission [18]. Copyright © 2011, Royal Society of Chemistry.

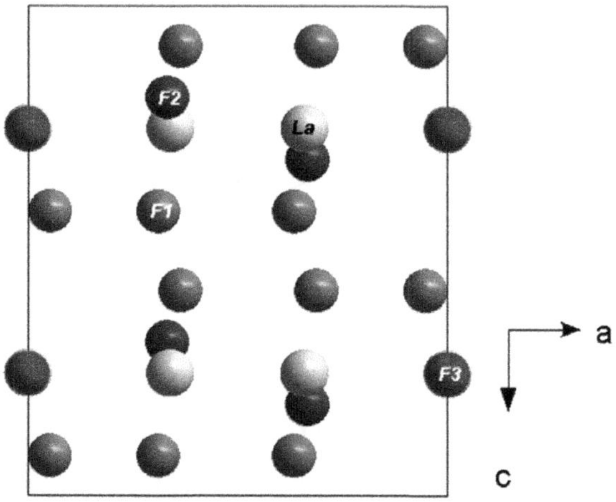

Figure 20.7 Crystal structure of LaF_3.

doping LaF_3 with aliovalent metal fluorides. The intrinsic defects in the tysonite-type structure are Schottky defects (cation vacancy associated with anion vacancy) and the anion conduction is a vacancy mechanism. The conduction mechanism in LaF_3 was well studied by various techniques [40,41,43,44]. Wang and Grey studied the conduction mechanism in polycrystalline LaF_3 and SrF_2-doped LaF_3 by variable temperature ^{19}F MAS NMR. For the first time the authors were able to distinguish the interaction between the F1 and F3 sites while earlier studies treated F2 and F3 sites as F2, 3 sublattices. According to Wang and Grey, at low temperatures the mobility is due to the exchange of fluoride ions among the F1 sites. Above 73°C, the exchange between F1 and F3 ions occurs, while the exchange among F1 sites is being faster (about one order of magnitude). The F2 ions remain rigid in the temperature range studied (until 266°C). Most likely the F2 sites participate in the exchange process above this temperature. In the case of SrF_2 doping ($La_{0.99}Sr_{0.01}F_{2.99}$), the exchange between F1 sites starts at even lower temperature (around 54°C) compared to undoped LaF_3. At 198°C the peak due to F3 ions completely disappears and the intensity of F2 peak started decreasing as a result of exchange between F2 and F1-3 sublattices. At 266°C, the peak due to F2 completely disappears. Thus it can be concluded that the anisotropy in the ionic conductivity observed for LaF_3 is related to the different exchange processes involved.

Most of the ionic conductivity studies published on tysonite-type compounds are based on single crystals of $R_{1-y}M_yF_{3-y}$ (R = La or Ce and M = Ba, Sr, or Ca), which are obtained by mixing RF_3 with MF_2 in different concentrations [45,46]. The solubility of aliovalent cation is low in tysonite-type compounds (up to 15 mol%) [47] compared to fluorite-type compounds (up to 55 mol%). Note that the solubility limit would exceed in the case of polycrystalline or nanocrystalline samples. The substitution of bivalent cations in place of trivalent cations leads to the formation of F vacancies. At low dopant concentrations, the conductivity increases due to the increase in defect concentration and activation energy decreases. The conductivity reaches a maximum at the percolation threshold (c. $y = 0.07$ for BaF_2 in LaF_3) [48]. Above this value, the conductivity decreases due to the formation of defect clusters which will impede the vacancy migration. Only few studies have been published on polycrystalline or nanocrystalline LaF_3 which are discussed later.

Sobolev et al. have prepared $La_{1-y}Ca_yF_{3-y}$ ($y = 0.15, 0.20$) by mechanical milling of CaF_2 and LaF_3 crystals [49]. The crystallite sizes were in the range of 10–30 nm. The ionic conductivity of 4.9×10^{-4} S/cm at 200°C was measured for $La_{0.8}Ca_{0.2}F_{2.8}$, but the activation energy was higher than the reported value for doped LaF_3 single crystals.

Reddy et al. studied the ionic conductivity of tysonite-type fluorides $La_{1-y}Ba_yF_{3-y}$ ($0 \leq y \geq 0.15$) synthesized by mechanical milling [50,51]. Fig. 20.8 shows the Arrhenius plot for the ionic conductivities of tysonite-type $La_{0.9}Ba_{0.1}F_{2.9}$ as synthesized by mechanical milling and sintered at 800°C for 2 and 20 h. The conductivities of these samples were compared with single crystal conductivities of same composition reported. Despite being nanocrystalline, the conductivities of as synthesized samples are two orders of magnitude less compared to single crystals. The conductivity increased after sintering at 800°C. This is in contrary to what was observed in nanocrystalline fluorite-type compounds [21]. In the case of fluorite-type $Ba_{0.6}La_{0.4}F_{2.4}$ the as synthesized sample shows high conductivity compared to single crystals and the conductivity decreases after sintering at 800°C (Fig. 20.8). While the observed high ionic conductivity was in the case of nanocrystalline fluorite-type compounds and could be explained by the formation of space charge regions, the low conductivity observed in tysonite-type compounds is not well understood at present.

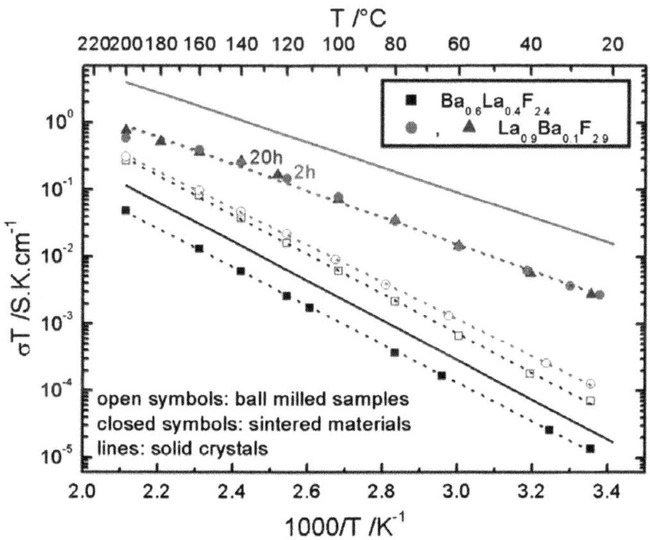

Figure 20.8 Arrhenius plot of the conductivities of tysonite-type $La_{0.9}Ba_{0.1}F_{2.9}$ (○ as-milled, ● sintered at 800°C for 2 h, and ▲ sintered for 20 h) and $Ba_{0.6}La_{0.4}F_{2.4}$ (□ as-milled and ■ sintered at 800°C for 2 h). Values reported in the literature for single crystals (*solid lines*) are given for comparison [48]. The *dotted lines* represent the linear fit obtained for calculating the activation energies.
Reprinted with permission from [51]. Copyright (2014) American Chemical Society.

4. Summary

Fluoride ion conductors regained interest since recent demonstration of rechargeable fluoride ion batteries, apart from their applications as ion selective electrodes and sensors. Similar to lithium-ion and silver-ion conductors, fluoride ion conductivity of metal fluorides can be significantly increased by homogeneous or heterogeneous doping. The fluoride ion conductivity of CaF_2 was increased in several orders of magnitude by reducing the crystallite size to few nanometers, by grain boundary modifications using Lewis acids, and modifying the interface by forming interlayers or milling with BaF_2. The increase in ionic conductivity of nanocrystalline CaF_2 was well understood by large fraction of defects at grain boundary or space charge effects. Despite large improvements in ionic conductivity of CaF_2, it is not significant enough to use as electrolyte in low-temperature fluoride ion batteries. On the other hand tysonite-type compounds show high intrinsic mobility, and consequently ionic conductivities more than 1×10^{-4} S/cm was achieved in single crystals at room temperature. However, corresponding nanocrystalline samples show significantly low ionic conductivity, which is in contrary to fluorite-type metal fluorides. The reason for such behavior is yet to be understood.

References

[1] A.V. Chadwick, S.L.P. Savin, Solid State Ionics 177 (2006) 3001–3008.
[2] L. Smart, E. Moore, in: Solid State Chemistry: An Introduction, CRC Press, Boca Raton, USA, 2005, pp. 201–280.
[3] R.A. Huggins, in: Advanced Batteries: Materials Science Aspects, Springer, New York, USA, 2009, pp. 339–373.
[4] C.G. Farrington, Sens. Actuators 1 (1981) 329–346.
[5] B.P. Sobolev, The Rare Earth Trifluorides, Parts 1&2; Inst. Estud. Catal., Barcelona, 2000 and 2001.
[6] J. Maier, Prog. Solid State Chem. 23 (1995) 171–263.
[7] S. Fujitsu, M. Miyayama, K. Koumoto, H. Yanagida, T. Kanazawa, J. Mater. Sci. 20 (1985) 2103–2109.
[8] U. Lauer, J. Maier, Solid State Ionics 51 (1992) 209–213.
[9] N. Sata, K. Eberman, K. Eberl, J. Maier, Nature 408 (2000) 946–949.
[10] B. Ruprecht, M. Wilkening, A. Feldhoff, S. Steuernagel, P. Heitjans, Phys. Chem. Chem. Phys. 11 (2009) 3071–3081.
[11] A.V. Chadwick, Diffus. Fundam. 2 (2005) 44.1–44.22.
[12] H.L. Tuller, Solid State Ionics 131 (2000) 143–157.
[13] P. Knauth, H. Tuller, J. Am. Ceram. Soc. 85 (2002) 1654–1680.
[14] A.K. Cheetham, B.E.F. Fender, M.J. Cooper, J. Phys. C Solid State Phys. 4 (1971) 3107.
[15] K.E.D. Wapenaar, J. Schoonman, J. Electrochem. Soc. 126 (1979) 667.
[16] N.I. Sorokin, B.P. Sobolev, Crystallogr. Rep. 52 (2007) 842–863.
[17] K.E.D. Wapenaar, J.L. Van Koesveld, J. Schoonman, Solid State Ionics 2 (1981) 145–154.
[18] B. Ruprecht, M. Wilkening, S. Steuernagel, P. Heitjans, J. Mater. Chem. 18 (2008) 5412–5416.

[19] A.K. Ivanov-Shits, N.I. Sorokin, P.P. Fedorov, B.P. Sobolev, Solid State Ionics 31 (1989) 269–280.
[20] B.P. Sobolev, I.A. Sviridov, V.I. Fadeeva, S.N. Sul'yanov, N.I. Sorokin, Z.I. Zhmurova, P. Herrero, A. Landa-Canovas, R.M. Rojas, Crystallogr. Rep. 50 (2005) 478–485.
[21] C. Rongeat, M. Anji Reddy, R. Witter, M. Fichtner, J. Phys. Chem. C 117 (2013) 4943–4950.
[22] A. Düvel, J. Bednarcik, V. Sepelák, P. Heitjans, J. Phys. Chem. C 118 (2014) 7117–7129.
[23] J.H. Kennedy, R.C. Miles, J. Electrochem. Soc. 123 (1976) 47–51.
[24] S. Yoshikado, Y. Ito, J.M. Reáu, Solid State Ionics 154–155 (2002) 503–509.
[25] G. Denes, Y.H. Yu, T. Tyliszczak, A.P. Hitchcock, J. Solid State Chem. 104 (1993) 239–252.
[26] M. Uno, M. Onitsuka, Y. Ito, S. Yoshikado, Solid State Ionics 176 (2005) 2493–2498.
[27] M.M. Ahmad, Y. Yamane, K. Yamada, S. Tanaka, J. Phys. D Appl. Phys. 40 (2007) 6020–6025.
[28] G. Denes, T. Birchall, M. Sayer, M.F. Bell, Solid State Ionics 13 (1984) 213–219.
[29] M.M. Ahmad, Y. Yamane, K. Yamada, J. Appl. Phys. 106 (2009) 074106.
[30] C.C. Liang, J. Electrochem. Soc. 120 (1973) 1298.
[31] S. Fujitsu, M. Miyayama, K. Koumoto, H. Yanagida, J. Mater. Sci. 20 (1985) 2103.
[32] W. Puin, P. Heitjans, Nanostructured Mater. 6 (1995) 885.
[33] W. Puin, S. Rodewald, R. Ramlau, P. Heitjans, J. Maier, Solid State Ionics 131 (2000) 159–164.
[34] Y. Saito, J. Maier, Solid State Ionics 86-88 (1996) 581–584.
[35] A. Düvel, B. Ruprecht, P. Heitjans, M. Wilkening, J. Phys. Chem. C 115 (2011) 23784.
[36] V. Trnovcová, L.S. Garashina, A. Skubla, P.P. Fedorov, R. Cicka, E.A. Krivandina, B.P. Sobolev, Solid State Ionics 157 (2003) 195.
[37] K. Schlyter, Ark. Kemi 5 (1952) 73.
[38] A. Rhandour, J.-M. Reáu, S. Matar, P. Hagenmuller, J. Solid State Chem. 64 (1986) 206.
[39] B.P. Sobolev, V.B. Alexandrov, P.P. Fedorov, K.B. Seiranian, N.L. Tkashenko, Kristallografiya 21 (1976) 96.
[40] A.F. Privalov, H.M. Vieth, I.V. Murin, J. Phys. Condens. Matter 6 (1994) 8237–8243.
[41] F. Wang, C.P. Grey, Chem. Mater. 9 (1997) 1068.
[42] J. Schoonman, G. Oversluizen, K.E.D. Wapenaar, Solid State Ionics 1 (1980) 211–221.
[43] N.I. Sorokin, B.P. Sobolev, Phys. Solid State 50 (2008) 416–421.
[44] M. Omari, J. El Senegas, J.-M. Reau, Solid State Ionics 107 (1998) 293–305. F. Wang, C.P. Grey, Chem. Mater. 9 (1997) 1068.
[45] I.V. Murin, O.V. Glumov, Y.V. Amelin, J. Appl. Chem. USSR 53 (1980) 1132–1135.
[46] N.I. Sorokin, E.A. Krivandina, Z.I. Zhmurova, B.P. Sobolev, M.V. Fominykh, V.V. Fistul, Phys. Solid State 41 (1999) 573–575.
[47] B.P. Sobolev, Crystallogr. Rep. 57 (2012) 434–454.
[48] A. Roos, F.C.M. van de Pol, R. Keim, J. Schoonman, Solid State Ionics 13 (1984) 191–203.
[49] B.P. Sobolev, I.A. Sviridov, V.I. Fadeeva, S.N. Sul'yanov, N.I. Sorokin, Z.I. Zhmurova, I.I. Khodos, A.S. Avilov, M.A. Zaporozhets, Crystallogr. Rep. 53 (2008) 868–880.
[50] M. Anji Reddy, M. Fichtner, J. Mater. Chem. 21 (2011) 17059–17062.
[51] C. Rongeat, M. Anji Reddy, R. Witter, M. Fichtner, ACS Appl. Mater. Interfaces 6 (2014) 2103–2110.

Nonstoichiometric Single Crystals $M_{1-x}R_xF_{2+x}$ and $R_{1-y}M_yF_{3-y}$ (M = Ca, Sr, Ba: R = Rare Earth Elements) as Fluorine-Ionic Conductive Solid Electrolytes

21

B.P. Sobolev, N.I. Sorokin, N.B. Bolotina
Shubnikov Institute of Crystallography of Russian Academy of Sciences, Moscow, Russia

Chapter Outline

Photonic and Electronic Properties of Fluoride Materials. http://dx.doi.org/10.1016/B978-0-12-801639-8.00021-0

1. Introduction

Neither high electric conductivity of PbF_2 crystals discovered by Faraday in 1834 nor certain theoretical advantages of F-ion batteries and accumulators over Li-ion devices have led so far to the development of industrial electrochemical devices based on fluorine-ionic conductive solid electrolytes (FCSEs). An extensive search for FCSEs is described in reviews [4,20,86]. Some prototypes of F-ion batteries were reported in Refs [3,36,37]. They all were heated to raise FCSE's conductivity (σ).

The objects of search for FCSEs with high σ are nonstoichiometric fluoride crystals with structural defects [51,56,75]. This work sums up our findings about highly conductive nonstoichiometric FCSEs with fluorite CaF_2- and tysonite LaF_3-type structures in the MF_2–RF_3 systems.

Fluorine-ionic conductivity of these types of crystals varies in homogeneity regions of nonstoichiometric phases of $M_{1-x}R_xF_{2+x}$ and $R_{1-y}M_yF_{3-y}$. A continuous transition of a dielectric crystal (components MF_2, RF_3) to superionic state ($M_{1-x}R_xF_{2+x}$, $R_{1-y}M_yF_{3-y}$) is caused by a remarkable structural imperfection when the chemical composition of the crystal changes. Searching strategy for fluoride superionics consists in studies of nonstoichiometry in the systems of inorganic fluorides [43,47].

Phase diagrams of the fluoride systems have been investigated in the Fluoride Materials Laboratory (**FMLab**) of the Institute of Crystallography, Moscow, since early 1970s due to the search for laser crystals. By 2000, phase diagrams of over 200 MF_m–RF_n systems ($m < n \leq 4$) had been obtained, 85% of them for the first time. The diversity of the fluoride materials was increased 40–50 times due to two-component crystals [44,47].

Phase diagrams of the MF_m–RF_n systems were reported in volume 1 of monograph [43]. Chemical and structural aspects of nonstoichiometry in fluorides as well as crystal growth and physical properties (including σ-values) were considered in volume 2 [44]. These data were used for the analysis of the MF_m–RF_n systems as sources of FCSEs.

Not all the 55 fluorides MF_m, RF_n are suitable as components of multicomponent materials; 27 of them are enough for lasers and FCSEs. They form 351 MF_m–RF_n systems.

The MF_2–RF_3 systems (M = Ca, Sr, Ba: R = rare earth elements (REEs)) contain nonstoichiometric phases of $M_{1-x}R_xF_{2+x}$ (CaF_2-type) and $R_{1-y}M_yF_{3-y}$ (LaF_3-type), which are stable against decomposition and have wide homogeneity regions. Only some of them, however, have sufficiently high σ-values to be used as FCSEs.

Defect crystal structures are directly related to σ. The structures of the first representatives of $M_{1-x}R_xF_{2+x}$ and $R_{1-y}M_yF_{3-y}$ were studied in the FMLab. Defect structures of over 50 fluorite-like crystals and 20 tysonite-like ones were determined. A detailed discussion of this issue goes beyond this chapter.

The first fluorite-like crystal of $Ca_{0.61}Ce_{0.39}F_{2.39}$ was studied by Alexandrov and Garashina [1] who discovered new sites for interstitial F^-_{int}. These fluorine ions are responsible for the formation of structure defect clusters that influence the conductivity.

Bevan et al. [5] described a fluorite-like cluster structure as "randomly dispersed, sometimes isolated, sometimes edge sharing in various ways, forming and reforming kaleidoscopically within fluorite matrix over the whole range of existence." These authors have clearly conveyed the complexity and variability of the crystals which are being studied until now.

Defect structures of tysonite-like $R_{1-y}M_yF_{3-y}$ crystals were studied later. The first crystal of $Gd_{0.8}Ca_{0.2}F_{2.8}$ was studied by Otroshchenko et al. [34].

Single-crystal form is the only technological form of FCSEs studied in this work. The form gives true characteristics of σ, being least affected by uncontrolled factors.

Optimization of the composition lies in research of σ-values by changing various M:R combinations with appropriate x,y-values to arrive at a maximal σ for the given M:R.

For $M_{1-x}R_xF_{2+x}$ phases the optimization was performed at 500K because of low σ-values at 293K. Crystals of $M_{0.5}La_{0.5}F_{2.5}$ (M = Sr, Ba) have $\sigma_{500\ K} = 6 \times 10^{-4}$ S/cm (activation enthalpy of ionic conductivity $\Delta H_\sigma = 0.6$–0.7 eV) that is record for fluorite-like FCSEs at this temperature. Optimization of the composition of $R_{1-y}M_yF_{3-y}$ phases at room temperature revealed single crystals having maximal $\sigma_{293K} = 5 \times 10^{-4}$ S/cm among the crystals studied [57,80].

The problem how to produce FCSEs for practical devices has not yet been solved. Taking running at $-30°$C into consideration, σ-values of FCSEs should be increased by about an order of magnitude from state-of-the-art.

This work is devoted to two anniversaries, which are epochal for these issues. They are 180 years from the discovery of the PbF_2 ionic conductivity by Faraday in 1834 and 100 years from the discovery of nonstoichiometry in CaF_2 and LaF_3-type structures by Vogt in 1914.

2. Nonstoichiometric Phases With Fluorite- and Tysonite-Type Structures

Nonstoichiometric phases in the CaF_2–YF_3 and CaF_2–CeF_3 systems were discovered by Vogt [89]. Two fluorite-like phases ($Ca_{1-x}Y_xF_{2+x}$, $Ca_{1-x}Ce_xF_{2+x}$) and an unknown one whose structural similarity to tysonite (Ce,La)F_3 was proved by optical methods were crystallized from melt. Vogt discovered (in current terminology) a nonstoichiometric tysonite-like $Y_{1-y}Ca_yF_{3-y}$ phase without using X-ray diffraction.

3. Fluorine-Ion Transport in Nonstoichiometric Fluorides With CaF_2- and LaF_3-Structures

Ion transport in the $Ca_{1-x}Y_xF_{2+x}$ ($0 < x < 0.01$) crystals was revealed by Ure [88]. He detected a rise in σ when the temperature and/or YF_3-content increased, but the functional $\sigma(x)$ dependence was not studied.

The research on σ-values in tysonites ($R_{1-y}M_yF_{3-y}$) started after their discovery in the MF_2—RF_3 systems [14,52].

A relationship between nonstoichiometry and high σ-values in fluorides was discovered "at the tip of the pen" [12]. While studying the MF_2—RF_3 systems, we found that ~77% of nonstoichiometric phases (33 CaF_2-type and 32 LaF_3-type) had maxima on the melting curves. Among ~5000 salt systems studied by then, only 33 phases had such maxima. All these phases were nonstoichiometric with high σ-values.

A relationship between high σ-values and nonstoichiometry of tysonite-like phases was found by Nagel and O'Keeffe [31] for single crystals of $Y_{0.73}Ca_{0.27}F_{2.73}$, $Er_{0.71}Ca_{0.29}F_{2.71}$, and a polycrystalline ingot of $La_{0.95}Sr_{0.05}F_{2.95}$. The compositions were selected according to Refs [14,52], with proper references.

A nonstoichiometric crystal (NSC) is not necessarily a superionic conductor; superionics have to be intentionally sought among nonstoichiometric phases.

4. The Search for Fluoride Systems Involving FCSEs

Phase analysis of the MF_m—RF_n ($m < n \leq 4$) systems [43] suggests that MF_2—RF_3 systems best meet the FCSE's formation conditions. Three families of such systems, namely, CaF_2—RF_3, SrF_2—RF_3, and BaF_2—RF_3, are selected. They include 45 systems formed by three MF_2 and 15 RF_3 (R = REEs without Pm and Sc).

Stoichiometry disturbance in these systems is significant. An average homogeneity region of fluorite- and tysonite-like phases covers ~60 mol% of the composition range in the systems of CaF_2—RF_3 [12,13,48], SrF_2—RF_3 [54,55] and BaF_2—RF_3 [58].

Results of the 41-year search for FCSEs in MF_2—RF_3 systems seem to be the modest. Maximal σ_{293K} increased from 2×10^{-5} [31] to 5×10^{-4} S/cm [56]. Apparently, either these systems do not contain crystals with higher σ_{293K} or the performed search is incomplete.

At the beginning of our research on FCSEs, the search was limited to nine $M_{1-x}R_xF_{2+x}$ phases of various (M,R) compositions (20% of all the phases). The following x-compositions were studied: 5 for $Ca_{1-x}R_xF_{2+x}$ (R = La, Gd, Er), 19 for $Sr_{1-x}R_xF_{2+x}$ (R = La, Ce, Gd, Yb), and 12 for $Ba_{1-x}R_xF_{2+x}$ (R = La, Yb). This was insufficient to estimate a potential of 45 fluorite-like phases as FCSEs. Previous search for tysonite-like FCSEs was limited to five $R_{1-y}M_yF_{3-y}$ phases of various (R,M) compositions (13%) and 23 y-compositions. This was also insufficient to estimate a potential of 39 tysonite-like phases as FCSEs.

Forty fluorite-like $M_{1-x}R_xF_{2+x}$ phases we have studied by now cover 89% of the phases different in (M,R) compositions. The σ-values are measured for 273 (M,R,x) compositions. These data are sufficient to characterize three specified families of the nonstoichiometric fluorites in terms of σ.

Studies of 29 tysonite-like $R_{1-y}M_yF_{3-y}$ phases cover 74% of the phases different in (R,M) compositions, except for (Eu,M) with M = Ca, Sr, Ba (partial reduction of europium) (R,Ba) with R = Sm, Tb, Dy (narrow homogeneity range) and (Yb,Ca), (Yb,Sr), and (Lu,Ca) (ordering). There are no tysonite-like phases in the BaF_2-RF_3 systems with R = Ho−Lu, Y. The σ-values are measured for 62 different (R,M,y) compositions. Our study allows assessing the $R_{1-y}M_yF_{3-y}$ tysonites as more promising FCSEs than the $M_{1-x}R_xF_{2+x}$ fluorites.

5. Defect Structures of Nonstoichiometric $M_{1-x}R_xF_{2+x}$ and $R_{1-y}M_yF_{3-y}$ Crystals

5.1 *Fluorite-Like* $M_{1-x}R_xF_{2+x}$ *Crystals*

Crystals of $M_{1-x}R_xF_{2+x}$ have substitutional lattice defects (R^{3+} cations occupy some M^{2+}-sites), anionic vacancies V_F^+ and interstitial ions F_{int}^-.

The lattice defects group into nanometer clusters. Two types of single defect clusters are formed: an octahedral $\{R_6F_{37}\}$ cluster [18,27], which is later modified to an octahedral-cubic $\{M_8R_6F_{69}\}$ cluster (**OCC**) [15], and a tetrahedral $\{R_4F_{26}\}$ cluster (**TC**) [30]. A single phase can contain OCCs and TCs simultaneously.

Temperature decrease leads to defect cluster ordering with the formation of ordered $M_mR_nF_{2m+3n}$ phases. This tendency grows with an increase in REE atomic number. Structural ordering of the $M_{1-x}R_xF_{2+x}$ crystals brings σ down [62,79].

Fig. 21.1 shows three isoconcentration series of the $M_{0.8}R_{0.2}F_{2.2}$ crystals, whose (M,R) compositions correlate with cluster type and conductivity as well. The cluster type is defined by interstitial F_{int}^- ions, which partially occupy either 48i (OCC) or 32f (TC) positions in Wyckoff's designation for the $Fm\bar{3}m$ group.

The dominating cluster type in the $Ca_{1-x}R_xF_{2+x}$ and $Sr_{1-x}R_xF_{2+x}$ crystals changes from TC for REEs in the head of the REE family (La−Nd) to OCC for REEs in its second half (Gd−Lu,Y). The only OCC type is observed in the $Ba_{1-x}R_xF_{2+x}$ crystals within the whole REE family except Lu [73].

The cluster cationic framework is built by R^{3+} ions. There is no direct relation between R^{3+} ions and cluster type. As appears from Fig. 21.1, the same R^{3+} cations can form different defect clusters in different MF_2 matrixes.

A "diffuse" phase transition is observed at heating of MF_2 fluorites [32]. It is due to displacements of some F^- from their basic positions (8c) to the interstitials (32f, 48i, 4b, etc.). The high-temperature form of MF_2 crystals has a high fluorine-ionic

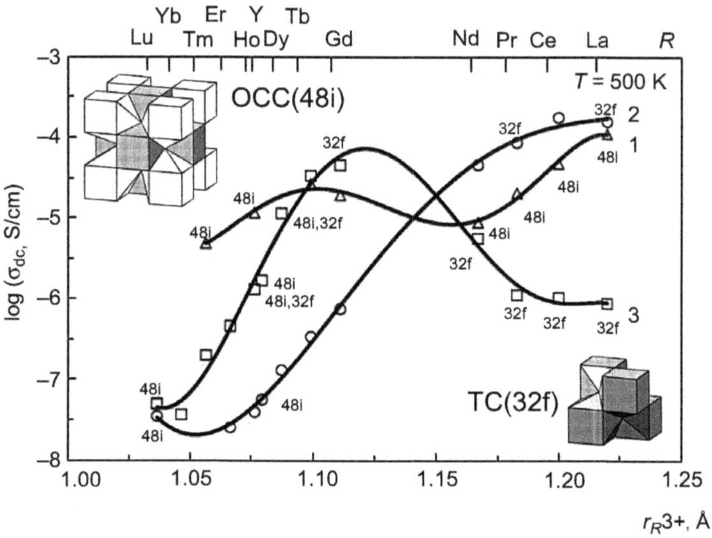

Figure 21.1 Relation between ionic conductivity σ and cluster structure of three isoconcentration series of the $Ca_{0.8}R_{0.2}F_{2.2}$ (1), $Sr_{0.8}R_{0.2}F_{2.2}$ (2), and $Ba_{0.8}R_{0.2}F_{2.2}$ (3) crystals. TC and OCC clusters are shown in the inset. *TC*, Tetrahedral cluster; *OCC*, octahedral-cubic cluster.

conductivity. Isomorphic replacements of M^{2+} by R^{3+} decrease formation temperatures of the conductive state in the $M_{1-x}R_xF_{2+x}$ crystals.

5.2 Tysonite-Like $R_{1-y}M_yF_{3-y}$ Crystals

Replacements of R^{3+} by M^{2+} lead to anionic V_F^+ vacancies, which take part in the ion transfer as charge carriers. Fluorine ions in two polymorphic forms of the tysonite structure are distributed over two positions (F1(4*f*), F2(2*b*), sp. gr. *P*6$_3$/*mmc*) for α-LaF_3 or three positions (F1(12*g*), F2(4*d*), F3(2*a*), sp. gr. *P* $\bar{3}$c1) for β-LaF_3. The relationship between σ-values and defect structures of $R_{1-y}M_yF_{3-y}$ is described in Ref. [4].

We performed structural studies of 20 tysonite-like $R_{1-y}M_yF_{3-y}$ crystals. New data were obtained about an incomplete defect ordering in $R_{0.715}Ca_{0.285}F_{2.715}$ (R = Y, Er) crystals [7].

The influence of anionic $V^+{}_F$ vacancies concentration on defect structures of $R_{1-y}M_yF_{3-y}$ was studied for the crystal family of $La_{1-y}Sr_yF_{3-y}$ ($y = 0$; 0.05; 0.07; 0.15) [8]. The structure versus conductivity changes are shown in Fig. 21.2.

Twin crystals of β-LaF_3 and β-$La_{0.95}Sr_{0.05}F_{2.95}$ (trigonal β-modification with sp. gr. *P* $\bar{3}$*c*1; $Z = 6$) have been studied. A morphotropic transition β↔α-$La_{1-y}Sr_yF_{3-y}$ at $y = 0.05{-}0.10$ was first observed by Bolotina et al. [8]. The symmetry of α-$La_{1-y}Sr_yF_{3-y}$ crystals changes to sp. gr. *P*6$_3$/*mmc* ($Z = 2$), while twinning disappears.

A sharp increase from intrinsic conductivity $\sigma_{293K} = 4 \times 10^{-9}$ S/cm in β-LaF_3 [80] to extrinsic conductivity $\sigma_{293K} = 3 \times 10^{-4}$ S/cm in β-$La_{0.95}Sr_{0.05}F_{2.95}$ [77] is due to

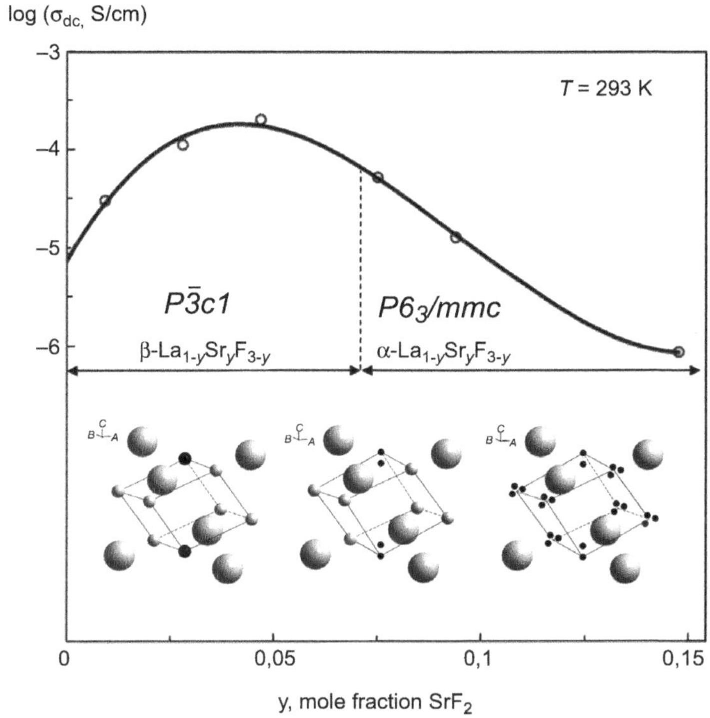

Figure 21.2 Top row: Conductivity changes in the tysonite-like $La_{1-y}Sr_yF_{3-y}$ crystals; bottom row: structure changes in the crystals of β-LaF_3 (left), β-$La_{0.95}Sr_{0.05}F_{2.95}$ (middle), and α-$La_{0.85}Sr_{0.15}F_{2.85}$ (right).

F1(12*g*)-vacancies accompanied with a discovered splitting of F3(2*a*)-positions [8]. The split fluorine sites (Fig. 21.2, middle structure fragment) confirm a disordering of the anionic motif when the stoichiometry is perturbed.

Further doping by SrF_2 up to α-$La_{0.85}Sr_{0.15}F_{2.85}$ increases disordering of the anionic motif (Fig. 21.2, right-hand structure fragment) and strongly decreases σ_{293K} to 1.4×10^{-6} S/cm [74] due to ion—ion interactions at a high SrF_2 concentration [10].

A "diffuse" phase transition in the tysonite-like crystals was discovered by Greis and Cader [17] for RF_3 with R = La—Gd. Unfortunately, this high-temperature transition has not been studied in detail. We assumed the $\alpha \leftrightarrow \beta$—$LaF_3$ transition to be similar to the "diffuse" phase transition in fluorite-like crystals because both of them are low-energy transformations and are due to a disordering of the anionic motif, as the fluorine-ionic conductivity increases.

So far, the room temperature conductivity in real crystals of LaF_3 is believed to be caused by "impurity-based" extrinsic V_F^+ vacancies only. Fluorine ions which form twin boundaries in the β-LaF_3 crystals are disordered as those in the conductive state of β-$La_{1-y}Sr_yF_{3-y}$ crystals, which can facilitate ion transport along the twin boundaries [8].

6. Chemical Engineering Aspects of Research on Fluorine-Ionic Conductivity in $M_{1-x}R_xF_{2+x}$ and $R_{1-y}M_yF_{3-y}$ Crystals

Because of the high-temperature chemical activity of fluorides, some requirements must be kept so that σ-data could be correct and comparable.

6.1 Single Crystals as the Most Preferable Technological Form of FCSEs

Technological forms of FCSEs can be single crystals, hot-pressing and cold-pressing ceramics, polycrystalline ingots, products of mechanochemical and solid-phase syntheses, films, etc. Even when the composition is fixed, both σ-values and $\sigma(T)$ behavior depend on the forms of FCSEs. Single-crystal form alone makes it possible to get true bulk σ-characteristics.

Single crystals (12 mm in diameter and up to 40 mm in length) were grown by Bridgman technique in a graphite crucible with 10–12 cells. Deoxidation was done by fluorination with pyrolized tetrafluorethylene, CF_4, or PbF_2. The middle parts of the crystal rods were cut, polished, and used for σ-measurements. Polished discs of the $Sr_{0.9}R_{0.1}F_{2.1}$ crystals are shown in Fig. 21.3 [46].

6.2 Uncontrolled Impurities and Conductivity

Oxygen impurity is extra hazardous for FCSEs with developed surfaces. A high-temperature reaction of the fluorides with water vapor is named pyrohydrolysis. Oxygen seems to be one of most uncontrolled impurities because of its various influences on the fluorides, a complicated analysis for oxygen and water abundance. A substitution of O^{2-} for $2F^-$ forms mobile V_F^+ vacancies. They can increase the conductivity but improved σ-values are nonreproducible. Crystal growth, annealing, and σ(T)-measurements should be carried out in fluorinated atmosphere or vacuum to remove oxygen. The mechanisms of pyrohydrolysis and methods of its suppression are discussed in Refs [45,49]. A high σ-dependence on cationic impurities requires a high chemical quality of the starting fluorides. Impurity-induced influence

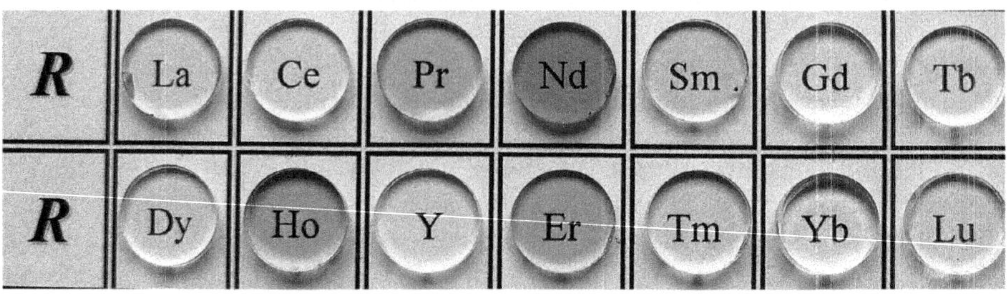

Figure 21.3 Single-crystalline disks of the concentration series $Sr_{0.9}R_{0.1}F_{2.1}$.

on the conductivity is shown by Bollmann [6] for "nominally pure" MF_2 and by Sorokin et al. [83] for "nominally pure" RF_3.

6.3 Unification of Thermal Conditions of Crystal Growth

Defect structure of the melt-grown $M_{1-x}R_xF_{2+x}$ and $R_{1-y}M_yF_{3-y}$ crystals can depend on cooling conditions. For $Ca_{1-x}Y_xF_{2+x}$ crystals, temperature-dependent structural changes are shown in situ by neutron diffraction analysis [19]. Monitoring of these changes is extremely difficult. Series of the $M_{1-x}R_xF_{2+x}$ and $R_{1-y}M_yF_{3-y}$ single crystals were grown concurrently using one multichannel crucible to provide comparative studies of $\sigma(T,x,y)$.

6.4 Compositional Irregularities of Two-Component Crystals

Cellular substructure in two-component crystals is due to incongruent melting of the $M_{1-x}R_xF_{2+x}$ and $R_{1-y}M_yF_{3-y}$ phases. Cell-boundary fluctuations of the dopant concentrations are small: ±0.005 mol fraction LuF_3 for $Ba_{0.97}Lu_{0.03}F_{2.03}$ and ±0.02 mol fraction HoF_3 for $Ca_{0.9}Ho_{0.1}F_{2.1}$ [87]. The upper fluctuation limit of the dopant has been determined using σ-measurements of $Ba_{0.5}Ce_{0.5}F_{2.5}$ (0.50 mol fraction of CeF_3) that differs at most from the congruent one (0.26 mol fraction CeF_3). The σ-value along [111] direction is ~30% higher than that along [110] and [100] directions [82]. This anisotropy of σ for $Ba_{0.5}Ce_{0.5}F_{2.5}$ crystal corresponds to ~0.04 mol fraction of CeF_3 deviation (from $\sigma(x)$-dependence for $Ba_{1-x}Ce_xF_{2+x}$). The cellularity-induced anisotropy of σ is smaller for most of cubic $M_{1-x}R_xF_{2+x}$ crystals. It is acceptable and does not prevent their usage as FCSEs.

6.5 Thermal Stability of $M_{1-x}R_xF_{2+x}$ and $R_{1-y}M_yF_{3-y}$ Phases

These nonstoichiometric phases can become ordered or decomposed upon cooling. Thermal stability of FCSEs was determined for practical purposes during the σ(T) studies. All the studied $M_{1-x}R_xF_{2+x}$ phases are stable up to 1100K except $Ba_{0.85}Lu_{0.15}F_{2.15}$ [67—69]. Most of the $R_{1-y}M_yF_{3-y}$ (M = Ca, Sr, Ba) phases are also stable up to 1100K except $R_{1-y}Ca_yF_{3-y}$ (R = Tb, Dy, Ho), which decompose from 723 to 823K [76].

Annealing impact (at 1173K during 14 days) on the defects and σ-values of the fluorite- and tysonite-like phases was proved for the $Sr_{1-x}Pr_xF_{2+x}$ and $Pr_{1-y}Sr_yF_{3-y}$. The defect structures, which were equilibrium at the temperature of annealing, were kept by quenching (~200 K/s). Their σ-values remained practically unchanged [84].

7. Techniques of σ_{dc} Measurements in $M_{1-x}R_xF_{2+x}$ and $R_{1-y}M_yF_{3-y}$ Crystals

Many new materials were discovered within the long period of fluorine-ionic conductivity studies. New methods and instruments appeared for σ-measurements that changed the knowledge about the conductivity phenomenon.

7.1 Impedance Spectroscopy

First σ-measurements were made at a single frequency (usually 10 kHz) that determined local movements of F^- ions. Translational movements of the ions are determined by measurements of direct current conductivity (σ_{dc}). Below the term σ_{dc} denotes impedance spectroscopy at wide-ranging frequencies.

The σ_{dc}-conductivity, which describes F-ion transport, was practically unknown before the 1980s. In this work, the impedance technique is systematically applied to the FCSEs. Systematic and reliable data on σ_{dc} in $M_{1-x}R_xF_{2+x}$ and $R_{1-y}M_yF_{3-y}$ were collected using the following techniques: 2- and 4-electrode methods of impedance spectroscopy over the frequency range of 10^{-1} to 10^7 Hz to select bulk resistance from total impedance; measurements in flow of inert gas (Ar, N_2) or vacuum (10^{-3}–10^{-2} Pa), and ultrasonic cleaning of crystalline surfaces [22,67,77,78]. Our data on σ_{dc} for the fluorite-like $Ba_{1-x}La_xF_{2+x}$ [24] and tysonite-like $La_{1-y}Ba_yF_{3-y}$ (Sorokin et al. [71,81]) single crystals were confirmed by Wapenaar et al. [90] and Roos et al. [39] who also used impedance technique.

7.2 Two Components of Ionic Conductivity

The temperature dependence $\sigma_{dc}(T)$ is determined by the Frenkel–Arrhenius equation:

$$\sigma_{dc}(T) = [A/T]\exp\{-\Delta H_\sigma/kT\},$$

where A is the preexponential conductivity factor and ΔH_σ is the activation enthalpy of the ion transport. The σ_{dc} value in the $M_{1-x}R_xF_{2+x}$ and $R_{1-y}M_yF_{3-y}$ phases is determined by concentration n_{mob} and mobility μ_{mob} of the charge carriers. In case of one-type conductive ions, the conductivity is expressed by the formula:

$$\sigma_{dc}(x,y,T) = qn_{mob}(x,y)[\mu_0(x,y)/T]\exp\{-\Delta H_h(x,y)/kT\},$$

where q is the F^- ion charge; μ_0 is the preexponential mobility factor (which depends on the frequency of hopping attempts of the charge carriers); ΔH_h is the activation enthalpy of hopping migration; x and y represent quantitative compositions of the crystals.

The n_{mob} in FCSEs with dopant-caused structural disorder is temperature-independent and given by stoichiometry perturbations. Maximum $n_{mob}(x,y)$ values are determined by homogeneity ranges of the nonstoichiometric phases, which rarely reach $x \sim 0.5$ (for $M_{1-x}R_xF_{2+x}$) and $y \sim 0.3$ (for $R_{1-y}M_yF_{3-y}$). It is incorrect to assume that n_{mob} is always equivalent to the content of substitutional cations $n_R(x)$ or $n_M(y)$. High-concentrated defects can interact, that changes an effective concentration n_{mob} and affects conductivity σ_{dc}. This interaction clearly shows itself in the formation of defect clusters in the $M_{1-x}R_xF_{2+x}$ crystals.

The mobility μ_{mob} in FCSEs depends not only on the nonstoichiometry degree (x,y). There is a more complicated dependence $\mu_{mob}(M,R,x,y,T)$. In case the (M,R,x,y) composition of $M_{1-x}R_xF_{2+x}$ or $R_{1-y}M_yF_{3-y}$ is fixed:

$$\mu_{mob}(T) = [\mu_0/T]\exp\{-\Delta H_h/kT\}.$$

Temperature rise leads to higher μ_{mob} and sharply increases σ_{dc}-values.

The relationship between n_{mob} and μ_{mob} is complicated. To separate their contributions to $\sigma_{dc} \sim n_{mob} \times \mu_{mob}$, either n_{mob} or μ_{mob} value is needed, which is independently determined in addition to σ_{dc} measured. The first, (n_{mob}), can be found from the structural models for the $M_{1-x}R_xF_{2+x}$ and $R_{1-y}M_yF_{3-y}$ crystals. The second one, (μ_{mob}), can be calculated using the Nernst–Einstein relation:

$$\mu_{mob} = \beta \nu_h d^2 q / kT,$$

where d is the hopping distance, β is the geometric factor (1/6 for cubic system). An average hopping frequency ν_h follows from the frequency dependence $\sigma(\nu)$ in the "hopping relaxation" model suggested by Almond et al. [2]:

$$\sigma(\nu) = \sigma_{dc}[1 + (\nu/\nu_h)^n],$$

where $0 \leq n \leq 1$. These generalities are differently implemented in nonstoichiometric fluorite and tysonite crystals. That is why possibilities to vary σ_{dc} through n_{mob} and μ_{mob} will be discussed separately for each of the two structure types.

7.3 Methods to Increase σ_{dc} in Nonstoichiometric Crystals

Temperature rise has an almost equal influence on the nonstoichiometric phases of both structure types (CaF_2 and LaF_3) as regards σ_{dc}. A temperature rise over the 210–1073K range increases σ-values of the $M_{1-x}R_xF_{2+x}$ phases up to 10^{11} times. The same happens for $R_{1-y}M_yF_{3-y}$ over the 130–1073K range when σ reaches ~1 S/cm.

Structural-chemical factor is defined by the structure type as well as by qualitative (M,R) and quantitative (x,y) compositions, which can change σ by many orders of magnitude. At 500K in $M_{1-x}R_xF_{2+x}$, the changes of σ can reach 10^8 times as compared to MF_2. At 293K in $R_{1-y}M_yF_{3-y}$, the σ-value increases up to 10^6 times as compared to RF_3. We widely used this approach to optimize compositions in terms of σ-values.

8. Experimental Studies of σ_{dc} in Fluorite-Like $M_{1-x}R_xF_{2+x}$ Single Crystals

8.1 Percolation Effect

The phenomenon of percolation was revealed while studying $\sigma_{dc}(x)$ in the range of 0–50 mol% RF_3 ($0 < x \leq 0.50$) for three families of fluorite-like phases: $Ca_{1-x}R_xF_{2+x}$, $Sr_{1-x}R_xF_{2+x}$, and $Ba_{1-x}R_xF_{2+x}$ [23–25,66–69] (Fig. 21.4).

The moment when the x-concentration in $M_{1-x}R_xF_{2+x}$ crystal reaches a percolation threshold $x = x_p$ is accompanied with a strong nonlinear growth of $\sigma_{dc}(x)$. The crystal

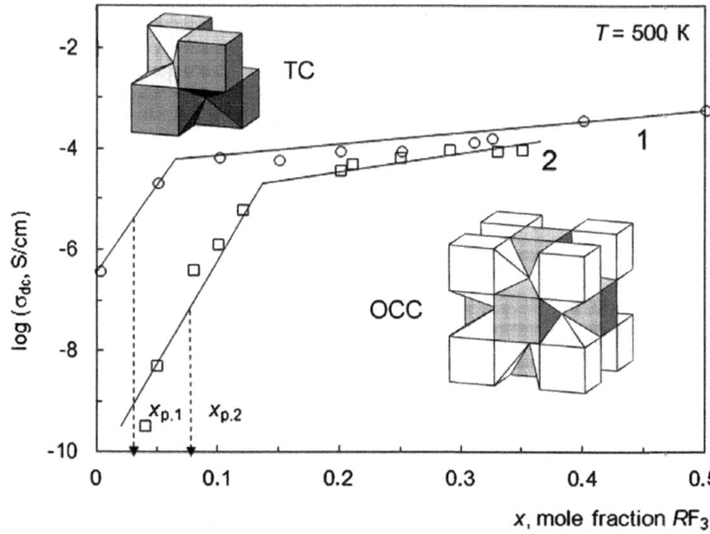

Figure 21.4 Dependences of σ_{dc} on the composition of the $Ba_{1-x}La_xF_{2+x}$ crystals with OCC clusters (1) and the $Ca_{1-x}Gd_xF_{2+x}$ crystals with TC clusters (2). *TC*, Tetrahedral cluster; *OCC*, octahedral-cubic cluster.

transfers into superionic state. Mobile charge carriers and low potential barriers of their transport ($\Delta H_\sigma < 0.5-0.6$ eV) are necessary for that. The phases with a significant percolation are $Ca_{1-x}R_xF_{2+x}$ (R = La–Tb) [25], $Sr_{1-x}R_xF_{2+x}$ (R = La–Nd) [23], and $Ba_{1-x}R_xF_{2+x}$ (R = La–Nd) [24].

8.2 "Defect Regions" Model

As shown by σ-measurements, the concentration of the composition-induced F^-_{int} defects is several times higher than that of the charge carriers. A model where a noticeable part of F^-_{int} defects is blocked is known as a (defect regions) DRs model [23–25].

Admittedly, fluorine-ion conductivity of nonstoichiometric superionics is due to defect clusters. Our data indicate the opposite. Anionic defects in the $M_{1-x}R_xF_{2+x}$ crystals (interstitial F^-_{int} ions and V_F^+ vacancies) are partially blocked in the clusters by high-charged R^{3+} cations.

Each DR is surrounded by a peripheral zone. It is a distorted fluorite-like matrix that accumulates mobile F^-_{int} ions (if they are) hereinafter called F^-_{mob}. The conductivity of $M_{1-x}R_xF_{2+x}$ is determined by the formation of conductive or nonconductive DRs (which contain or do not contain F^-_{mob}) and by a ratio between partial σ-values of DR and the matrix. Defect interactions either inside the cluster or with the crystalline matrix eliminate a direct relationship "the higher structural imperfection—the higher σ." Spatial overlapping of the conductive DRs at $x = x_p$ creates continuous paths for the fluorine ions which provide superionic conductivity.

8.3 Cluster Structure of $M_{1-x}R_xF_{2+x}$ and Defect Regions Model

To confirm an assumption that DR contains a cluster nucleus surrounded by a peripheral zone [23–25]; we should compare the cluster volume calculated from structural analysis and the DR volume calculated from independent electrophysical measurements [73]. The cluster volumes were calculated from the unit-cell values of $M_{1-x}R_xF_{2+x}$. Isolated OCC has volumes 700–800 Å^3 and TC, 190–210 Å^3 for various REEs.

The DR volumes were calculated from the percolation threshold x_p. On the assumption about one-type clusters in $M_{1-x}R_xF_{2+x}$, this threshold was determined in a two-level percolation model [63]. The σ_{dc} values are described there by the equations:

$$\sigma_{dc}(x) = \sigma_1[x - x_p]^{1.7} \quad \text{at } x > x_p$$

$$\sigma_{dc}(x) = \sigma_2[x_p - x]^{-0.72} \quad \text{at } x < x_p,$$

where σ_1 and σ_2 are the partial σ_{dc}-values of the peripheral DR and the matrix, respectively. The values of $x_{p,1} = 2–3$ mol% RF_3 for the $\{M_8R_6F_{69}\}$ clusters and $x_{p,2} = 7–8$ mol% RF_3 for the $\{R_4F_{26}\}$ clusters are found experimentally.

The DR volumes are divided into two groups: 3000–4000 Å^3 ($x_{p,1}$) and 500–700 Å^3 ($x_{p,2}$), which correlate with two clusters: OCC $\{M_8R_6F_{69}\}$ and TC $\{R_4F_{26}\}$, respectively. The volumes of both OCC and TC are substantially smaller, however, than the corresponding DR volumes. One intermediate layer of anionic polyhedra added between each cluster and the matrix [73] leads to practical coincidence between the new volumes and DR volume for both cluster types. It is, thereby, exactly the peripheral DR layer suggested by Ivanov-Shits et al. [23–25] long before these estimations.

8.4 Interstitial Mechanism of Ion Transport in $M_{1-x}R_xF_{2+x}$ Crystals

A block isomorphism occurs when OCC or TC substitute for $\{M_{14}F_{64}\}^{36-}$ or $\{M_4F_{23}\}^{15-}$, respectively, in the fluorite structure [61,64,73]. Either OCCs or TCs prevail in the $M_{1-x}R_xF_{2+x}$ crystals depending on their (M,R) compositions. The OCCs prevail for R of the yttrium subgroup of REEs in case M = Ca, Sr and for any R in case M = Ba. The $\{M_8R_6F_{69}\}$ OCC has an effective charge, relative to the fluorite lattice:

$$\{M_{14}F_{64}\}^{36-} \rightarrow \{M_8R_6F_{69}\}^{35-} + F^-_{mob}$$

The TCs prevail for R of the cerium subgroup of REEs and M = Ca, Sr. An effective charge of $\{R_4F_{26}\}$ TC is:

$$\{M_4F_{23}\}^{15-} \rightarrow \{R_4F_{26}\}^{14-} + F^-_{mob}$$

The F^-_{mob} ions, weakly connected with the clusters, occupy peripheral zones of DRs and ensure superionic conductivity. Phases of $Ca_{1-x}R_xF_{2+x}$ and $Sr_{1-x}R_xF_{2+x}$ with R of the cerium subgroup of REEs as well as $Ba_{1-x}R_xF_{2+x}$ with any REE contain mobile F^-_{mob} and show high values of $\sigma_{dc} \sim 10^{-5}$–10^{-4} S/cm at 500K. There is no percolation effect in the $Ca_{1-x}R_xF_{2+x}$ and $Sr_{1-x}R_xF_{2+x}$ phases with REEs of the yttrium subgroup, and they have not high σ-values (Fig. 21.1). The peripheral DR F^-_{mob} ions have a low mobility because of strong ion–ion interactions with DR cores (clusters) and/or steric hindrances when lattice parameters of the fluorite phase decrease for R near the end of the REE family.

Thus, the mechanism of ion transport in the $M_{1-x}R_xF_{2+x}$ crystals is as follows. Peripheral interstitial fluorine ions are charge carriers, whereas interstitial fluorine ions and fluorine vacancies in the nuclei of DRs do not participate in the ion transport. According to molecular dynamics, fluorine-ion transport is due to a noncollinear interstitial mechanism [16]. The F^-_{mob} ions hop between Wyckoff positions 8c and 4b of the $Fm\bar{3}m$ space group.

The hopping frequencies ν_h of the F^-_{mob} ions were found [21] using a real part of the complex conductivity $\sigma(\nu) = \mathrm{Re}[\sigma^*(\nu)]$ and an imaginary part of the complex electric modulus $\mathrm{Im}[M^*(\nu)]$ over the frequency range of 10^{-1} to -1.5×10^{10} Hz. The diagrams of the complex conductivity $Y^*(\nu)$ and complex modulus $M^*(\nu)$ reveal a spectrum of hopping frequencies for the ion charge carriers that reflects structural inhomogeneity of the $M_{1-x}R_xF_{2+x}$ crystals due to clusters.

8.5 Concentration and Mobility of F^-_{mob} Depending on $M_{1-x}R_xF_{2+x}$ Composition

The concentration of F^-_{mob} in $M_{1-x}R_xF_{2+x}$ with a percolation effect was calculated using structural data that evidence different-type clusters [73]:

$$n_{mob} = Zx/\left(pa^3\right),$$

where a is the unit-cell parameter. For OCC and TC, REE quantities are $p = 6$ and $p = 4$, respectively.

Temperature-independent F^-_{mob} concentration in the $M_{1-x}R_xF_{2+x}$ phases linearly grows, as RF_3 content increases (Fig. 21.5, curve 1). Such dependence does not explain the percolation effect in the $\sigma_{dc}(x)$ curves. Calculations show that $n_{mob} = 1 \times 10^{21}$–$3 \times 10^{21}$ cm^{-3} for the $M_{1-x}R_xF_{2+x}$ crystals with $x = 0.2$. This covers 20–30% of superstoichiometric F^-_{int} (as compared to MF_2) and 2–4% of all the anions. The percentage of n_{mob} with respect to all the anions in hypothetical $M_{1-x}R_xF_{2+x}$ phase without OCC and TC can be 5–14 times higher, that suggests a "crystal-chemical" activation for blocked F^-_{int} in order to increase the σ_{dc}-conductivity.

The carrier mobility μ_{mob} in the $M_{1-x}R_xF_{2+x}$ crystals was found from σ_{dc} and n_{mob}. The n_{mob}-values were calculated from structural models, taking different type clusters into account (see above). The composition dependences $\mu_{mob}(x)$ for some $M_{1-x}R_xF_{2+x}$ crystals are shown in Fig. 21.5 (curve 2). The maximal

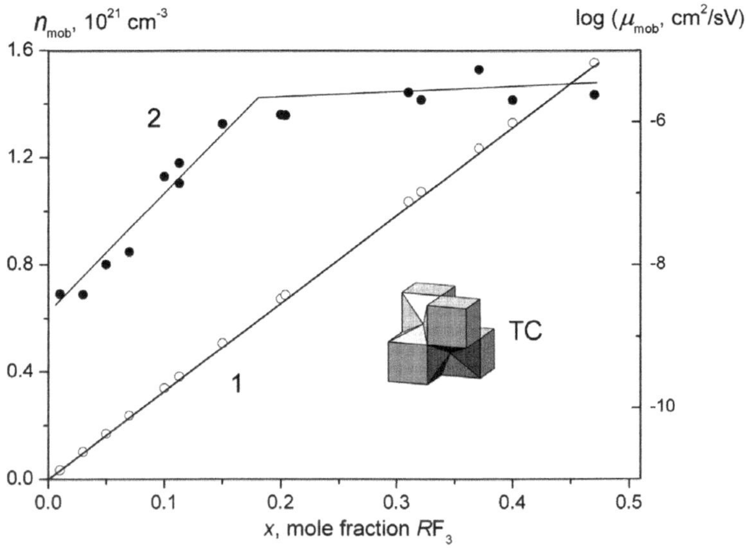

Figure 21.5 Dependences of concentration n_{mob} (1) and mobility μ_{mob} (2) on the composition of the $Sr_{1-x}La_xF_{2+x}$ crystals with TC clusters. *TC*, Tetrahedral cluster.

(at 500K) $\mu_{mob} = 1 \times 10^{-6}$–$2 \times 10^{-6}$ cm^2/Vs is reached in the $M_{1-x}La_xF_{2+x}$ crystals with $M =$ Sr, Ba and $x = 0.3$–0.5. As follows from Fig. 21.5, an essential contribution to the concentration-induced change of $\sigma_{dc}(x)$ in $M_{1-x}R_xF_{2+x}$ crystals is made by μ_{mob} of the charge carriers F^-_{mob}.

The hopping distance (d) of ionic carriers can be calculated from the Nernst–Einstein equation using known μ_{mob}- and ν_h-values determined from the $\sigma(\nu)$ curve:

$$d = [\mu_{mob} kT/\beta \nu_h q]^{1/2}.$$

The $d \sim 0.4$ nm, obtained for $Sr_{0.8}La_{0.2}F_{2.2}$ [21], agrees with the distance between interstitial sites in this crystal.

8.6 Relationship Between Activation Volume of Ion Migration and Chemical Composition of $M_{1-x}R_xF_{2+x}$ Crystals

The activation volume ΔV_m determines crystal lattice relaxation when the F^-_{mob} ions migrate. The less ΔV_m the easier ion transport is.

The activation volume $\Delta V_m = 1.2$ cm^3/mol for $Sr_{0.8}La_{0.2}F_{2.2}$ was found from a measured $\sigma(P)$ curve [60]. The conductivity of this sample decreases exponentially with the pressure increase over the range $P = 10^5$–6×10^8 Pa.

A linear correlation between ΔV_m and ΔH_σ in $M_{1-x}R_xF_{2+x}$ [65] allows one to calculate the ΔV_m-value using the ΔH_σ-value found from $\sigma_{dc}(T)$-dependence [23–25].

For isoconcentration series $Ca_{0.8}R_{0.2}F_{2.2}$, $Sr_{0.8}R_{0.2}F_{2.2}$, and $Ba_{0.8}R_{0.2}F_{2.2}$ (R = La–Lu, Y) the ΔV_m volumes are 2.1–3.2, 1.1–3.7, and 2.5–3.1 cm^3/mol, respectively.

Thus, superionic $M_{1-x}R_xF_{2+x}$ conductors have low values of $\Delta V_m \sim 1-2$ cm^3/mol. They are much lower than $\Delta V_m = 10-20$ cm^3/mol for dielectric ionic crystals [40].

8.7 "Diffuse" Phase Transition

A "diffuse" transition in fluorite MF_2 crystals [32] shows itself as a partial disorder of anionic lattice, similar to that in $M_{1-x}R_xF_{2+x}$ [13]. The $\sigma_{dc}(T)$ values reach 10^{-1} to 1 S/cm. There are few measured values of σ_{dc} for MF_2 and $M_{1-x}R_xF_{2+x}$ at $T > 1000$K. The $Ba_{0.75}Gd_{0.25}F_{2.25}$ crystal was taken to study the mechanism of superionic conductivity at the "diffuse" transition, since the BaF_2 matrix had a minimal transition temperature $T_{tr} \sim 1250$K.

According to Ref. [41], a "diffuse" transition in the MF_2 crystals is caused by critical values of $\mu_{mob} = 10^{-3}$to 10^{-2} cm^2/Vs at $T = T_{tr}$. Calculations [62] of μ_{mob} and ΔH_σ in $Ba_{0.75}Gd_{0.25}F_{2.25}$ at $T > T_{tr}$ yield $\mu_{mob} = 10^{-4}$ to 10^{-3} cm^2/Vs and $\Delta H_\sigma \sim 0.4$ eV, respectively, that confirms Schoonman's hypothesis. In $Ba_{0.75}Gd_{0.25}F_{2.25}$, the transition to the state with $\sigma_{dc} > 10^{-1}$ S/cm, comparable with σ_{dc} of ionic melts, occurs at $T_{tr} \sim 1000$K. An isomorphous substitution by 25% Gd^{3+} for Ba^{2+} stabilizes the high-temperature conductive state of $Ba_{0.75}Gd_{0.25}F_{2.25}$, bringing $T_{tr} \sim 250$K down (as compared to BaF_2).

The results of electrophysical [62], ultrasonic [9], and light scattering [33] studies suggest that introduction of R^{3+} cations into fluorite-like MF_2 matrix results in a T_{tr} decrease and stabilizes a high-conductive state with $\sigma_{dc} > 10^{-1}$ S/cm at lower temperatures.

8.8 Composition (M,R,x) Optimizing of $M_{1-x}R_xF_{2+x}$ Crystals With High σ_{dc} Values

Ionic radii of Ca^{2+}–Sr^{2+}–Ba^{2+} are "quantized" as 1.12–1.26–1.42 Å. Within 15 R^{3+} of the REE series from La^{3+} to Lu^{3+}, the ionic radii are also "quantized" but the intervals are smaller by an order of magnitude. Effective ionic radii are taken for coordination number 8 [42]. M^{2+} and R^{3+} cationic radii span the 0.977–1.42 Å interval, and isomorphic (M,R) substitutions are possible in this range.

Quantitative composition of fluorite-like phases varies over a wide x-range owing to specific (block) type of isomorphism. The blocks level the difference in M^{2+} and R^{3+} ionic radii, thus strongly increasing homogeneity ranges of $M_{1-x}R_xF_{2+x}$. Quantitative compositions of the crystals change continuously till the upper limits, which are 30–50 mol% of RF_3, depending on (M,R).

Fluorite-like $M_{1-x}R_xF_{2+x}$ FCSEs (M = Ca, Sr, Ba) have low $\sigma_{293K} < 10^{-7}$ S/cm at room temperature, and their compositions are, therefore, optimized at 500K. The obtained conductivity (σ_{500K}) and enthalpy (ΔH_σ) values are shown in Fig. 21.6. One can see that structure-chemical factor affects σ_{dc} at high temperatures. The σ_{500K} value increases by eight orders of magnitude, from 10^{-12} S/cm (typical dielectric)

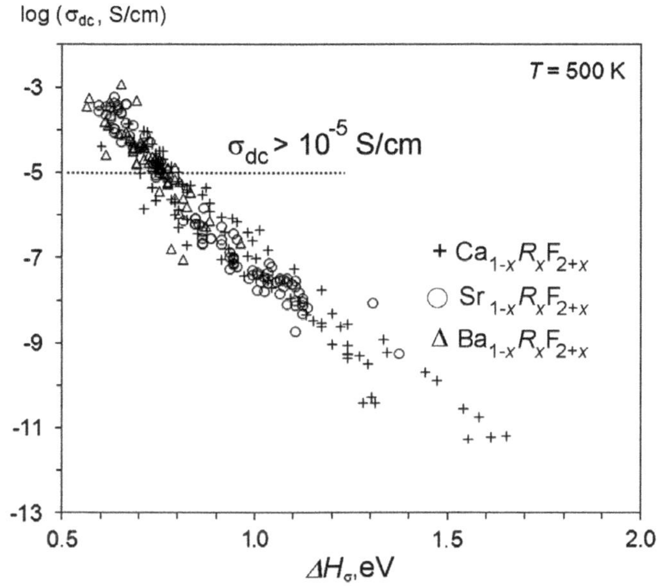

Figure 21.6 Dependence of conductivity σ_{500K} on enthalpy ΔH_σ for $Ca_{1-x}R_xF_{2+x}$, $Sr_{1-x}R_xF_{2+x}$, and $Ba_{1-x}R_xF_{2+x}$ crystals.

to 10^{-4} S/cm (superionic conductor) through composition variations, whereas ΔH_σ decreases from 1.7 to 0.5 eV.

The optimization at 500K yielded a considerable number of compositions with $\sigma_{dc} \geq 10^{-5}$ S/cm, namely (see Fig. 21.6) $Ca_{1-x}R_xF_{2+x}$ (R = La–Nd, Ho; $x \geq 0.35$), $Ca_{1-x}R_xF_{2+x}$ (R = Gd, Tb; $x \geq 0.2$), $Sr_{1-x}R_xF_{2+x}$ (R = La–Nd, $x \geq 0.15$), $Ba_{1-x}R_xF_{2+x}$ (R = La–Nd, $x \geq 0.05$), $Ba_{1-x}R_xF_{2+x}$ (R = Gd, Tb; $x \geq 0.1$) and $Ba_{1-x}Ho_xF_{2+x}$ ($x \geq 0.15$). Higher values of $\sigma_{500K} > 10^{-4}$ S/cm were reached for $Ca_{0.65}R_{0.35}F_{2.35}$ (R = Gd, Tb, Dy), $Sr_{1-x}R_xF_{2+x}$ (R = La–Nd, $x > 0.2$) and $Ba_{1-x}La_xF_{2+x}$ ($x > 0.2$).

Variations of σ_{dc} in isoconcentration series of the $M_{1-x}R_xF_{2+x}$ crystals were studied when R varied along the REE family. The σ_{dc} values for the series $Ca_{0.8}R_{0.2}F_{2.2}$, $Sr_{0.8}R_{0.2}F_{2.2}$, and $Ba_{0.8}R_{0.2}F_{2.2}$ (20 mol% RF_3) are shown in Fig. 21.1, as an example. A monotonous behavior of $\sigma_{dc}(R)$ curve is observed for $Sr_{0.8}R_{0.2}F_{2.2}$ and $Ba_{0.8}R_{0.2}F_{2.2}$, whereas this curve has a maximum between Gd and Tb for $Ca_{0.8}R_{0.2}F_{2.2}$. Among fluorite-like phases based on CaF_2, SrF_2, and BaF_2, the pairs Ca–Gd, Sr–La, and Ba–La determine phases with maximal σ_{dc} values, which are taken for the next optimization by fitting their quantitative (x) compositions.

Afterward, variations of σ_{dc} in the $M_{1-x}R_xF_{2+x}$ phases were studied when their (M,R) composition was fixed but the x-values varied within the homogeneity range. In the conductive phases of $Ca_{1-x}Gd_xF_{2+x}$, $Sr_{1-x}La_xF_{2+x}$, and $Ba_{1-x}La_xF_{2+x}$, maximal σ_{dc} at 500K are reached for the highest concentrations of RF_3: $Ca_{0.65}Gd_{0.35}F_{2.35}$ ($\sigma_{500K} = 1 \times 10^{-4}$ S/cm, $\Delta H_\sigma = 0.72$ eV), $Sr_{0.53}La_{0.47}F_{2.47}$ ($\sigma_{500K} = 6 \times 10^{-4}$ S/cm,

$\Delta H_\sigma = 0.68$ eV), and $Ba_{0.5}La_{0.5}F_{2.5}$ ($\sigma_{500K} = 6 \times 10^{-4}$ S/cm, $\Delta H_\sigma = 0.56$ eV). Maximal σ_{500K} for $M_{1-x}R_xF_{2+x}$ is 6×10^{-4} S/cm.

The optimization procedure was performed for 40 $M_{1-x}R_xF_{2+x}$ phases with different (M,R) compositions and 273 $M_{1-x}R_xF_{2+x}$ crystals with different (M,R,x) compositions. The σ_{dc}(T) dependences for different (M,R,x) compositions were obtained over the wide temperature range 210−1100K to select FCSE including 97 crystals with $\sigma_{500K} > 10^{-5}$ S/cm and 36 crystals with $\sigma_{500K} > 10^{-4}$ S/cm. At 500K, these $M_{1-x}R_xF_{2+x}$ FCSEs can be used in heated solid-state electrochemical units to solve specific problems of thermodynamics, solid-state ionics, environmental energetics, etc.

8.9 *Some Applications of Heated FCSEs With Fluorite-Like Structure*

Most common devices use single-crystal FCSEs and operate at 150−200°C. An amperometric sensor based on the $Ba_{0.69}La_{0.31}F_{2.31}$ single crystal was suggested by Jakes et al. [28] to analyze hydrogen fluoride (HF) in industrial atmospheres and in the air (heating to 230°C). "Fluoride ion batteries" [38] are based on $Ba_{0.6}La_{0.4}F_{2.4}$ (heating to 150°C), which is a product of mechanochemical synthesis. Single crystals of CaF_2 are used in high-temperature ($T > 1000$°C) galvanic cells to determine thermodynamic properties of fluorides, oxofluorides, oxides, sulfides, etc. Usage of the $M_{1-x}R_xF_{2+x}$ FCSEs should decrease the operation temperature down to 200°C [75].

9. Experimental Studies of σ_{dc} in Tysonite-Like $R_{1-y}M_yF_{3-y}$ Single Crystals

Tysonite-like phases $R_{1-y}Ca_yF_{3-y}$, which occur in each of the 15 CaF_2−RF_3 systems, can be ordered for R = Tb−Lu, Y [14,52]. The $R_{1-y}Sr_yF_{3-y}$ phases also occur in each of the 15 SrF_2−RF_3 systems [54]. The small number (9) of the $R_{1-y}Ba_yF_{3-y}$ phases in the BaF_2−RF_3 systems is due to instability of tysonite-like structure for R = Dy−Lu, Y and their morphotropic substitution by low-conductive compounds of the β-$BaTm_2F_8$-type [58].

The temperature $\sigma_{dc}(T)$-dependences (100−1100K) and concentration $\sigma(y)$-dependences (0−33 mol% MF_2) were studied for all $R_{1-y}Ca_yF_{3-y}$, $R_{1-y}Sr_yF_{3-y}$ and partially for $R_{1-y}Ba_yF_{3-y}$ phases [70,71,74,76,77,81]. As shown by Sorokin et al. [70,71,81], more promising σ_{dc}-values result from SrF_2 introduction into the RF_3 matrixes.

9.1 *Anisotropy of σ_{dc} in Tysonite-Like $R_{1-y}M_yF_{3-y}$ Crystals*

Anisotropy of σ_{dc} in crystals with tysonite-type structure is due to trigonal/hexagonal systems, in which RF_3 and $R_{1-y}M_yF_{3-y}$ are crystallized.

In LaF_3 at 293K, the ratio of σ-values along and perpendicular to the crystallographic c-axis is $\sigma_{||c}/\sigma_{\perp c} = 2$ [39]. The impact of orientation on σ_{dc} of tysonite-like crystals is studied for $R_{1-y}Sr_yF_{3-y}$ (R are REEs of cerium subgroup) with 2, 8, and 15 mol% SrF_2 [70,76]. In $R_{0.98}Sr_{0.02}F_{2.98}$ (R = Ce, Pr), $\sigma_{||c}/\sigma_{\perp c} = 3-4$ at 320–512K. $Ce_{0.92}Sr_{0.08}F_{2.92}$ and $Nd_{0.85}Sr_{0.15}F_{2.85}$ crystals do not show any σ_{dc} anisotropy within the experimental errors. Thus, the anisotropy of σ_{dc} in the RF_3 matrixes is suppressed to a quasi-isotropic state in nonstoichiometric $R_{1-y}M_yF_{3-y}$ crystals.

9.2 Vacancy Mechanism of Ion Transport in $R_{1-y}M_yF_{3-y}$ Crystals

The process of V_F^+ vacancy formation by R^{3+} substitution for M^{2+} was confirmed by measurements of $R_{1-y}M_yF_{3-y}$ densities. Kinks in curves $\sigma_{dc}(T)$ can be explained [29,70,74] by those V_F^+ vacancies in the crystallographic F1(12g), F2(4d), and F3(2a) sites of the $P\,\bar{3}c1$ symmetry group, which participate in the ion transport. Similar dynamic properties of the F^- ions in the F2(4d) and F3(2a) positions allow one to consider them as a common F2,3(4d + 2a) position. At temperatures <400–500K, the mobile F^- ions move through the F1(12g) positions. Elevated temperature is required for an exchange of V_F^+ vacancies between the F1(12g) and F2,3(4d + 2a) sites.

The $\sigma_{dc}(y)$ dependence has a maximum at 3–5 mol% MF_2 in $R_{1-y}M_yF_{3-y}$ (Fig. 21.7). A primary growth of σ_{dc} is explained by increasing n_{mob} when the M^{2+}-content (y) increases [10]. On the other hand, μ_{mob} decreases above 5 mol% MF_2 because of ion–ion interactions after a hopping enthalpy ΔH_h increase (Fig. 21.7), that leads to a drop of σ_{dc}.

To describe a nonlinear $\sigma_{dc}(y)$ dependence for $R_{1-y}M_yF_{3-y}$ (M = Ca, Sr, Ba) in more detail, the concentration n_{mob} and mobility μ_{mob} of the charge carriers must be studied as functions of the y-composition and defect structure.

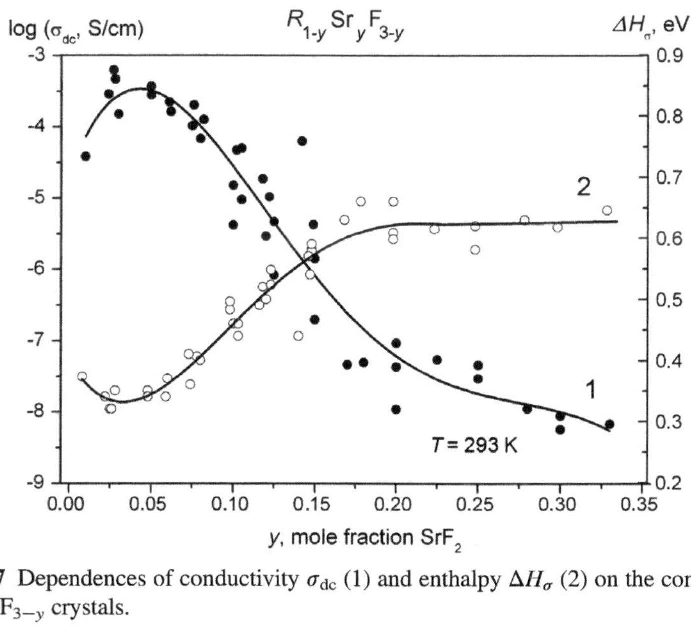

Figure 21.7 Dependences of conductivity σ_{dc} (1) and enthalpy ΔH_σ (2) on the composition of the $R_{1-y}Sr_yF_{3-y}$ crystals.

9.3 Concentration and Mobility of V_F^+ Vacancies in $R_{1-y}M_yF_{3-y}$ Phases

The concentration n_{mob} in $R_{1-y}M_yF_{3-y}$ is temperature independent for each fixed (R,M,y) composition. Assuming that all "extrinsic" V_F^+ participate in the ion transport, n_{mob} is calculated using unit-cell parameters [50]:

$$n_{mob} = 2Zy\Big/\sqrt{3}a^2c.$$

The n_{mob}-values for $R_{1-y}Sr_yF_{3-y}$ are presented in Fig. 21.8. For $La_{0.95}Sr_{0.05}F_{2.95}$, which is one of high-conductive tysonite-like FCSEs, $n_{mob} = 9.1 \times 10^{20}\ cm^{-3}$ (1.7% of the fluorine ions). The concentration behavior of ion carriers does not explain the maximum on the $\sigma_{dc}(y)$ curve.

The mobility μ_{mob} was calculated from σ_{dc} and n_{mob}:

$$\mu_{mob} = \sigma_{dc}/qn_{mob}.$$

The $\mu_{mob}(y)$ at 293K is given in Fig. 21.8 for the $R_{1-y}M_yF_{3-y}$ phases as an example. The plot of $\mu_{mob}(y)$ for $R_{1-y}M_yF_{3-y}$ has a maximum at 3–5 mol% MF_2. It is the mobility of F^-_{mob} that makes a significant contribution to $\sigma_{dc}(y)$ variations in these crystals.

It is found $\mu_{mob} = 1.9 \times 10^{-6}\ cm^2/sV$ at 293K for such a FCSE as $La_{0.95}Sr_{0.05}F_{2.95}$. This value falls (owing to ΔH_h) with the temperature decrease and reaches $\mu_{mob} = 2 \times 10^{-7}\ cm^2/sV$ at 255K. Fluorite-like $M_{0.8}R_{0.2}F_{2.2}$ (M = Sr, Ba; R = La–Nd) crystals have similar μ_{mob} at 500K.

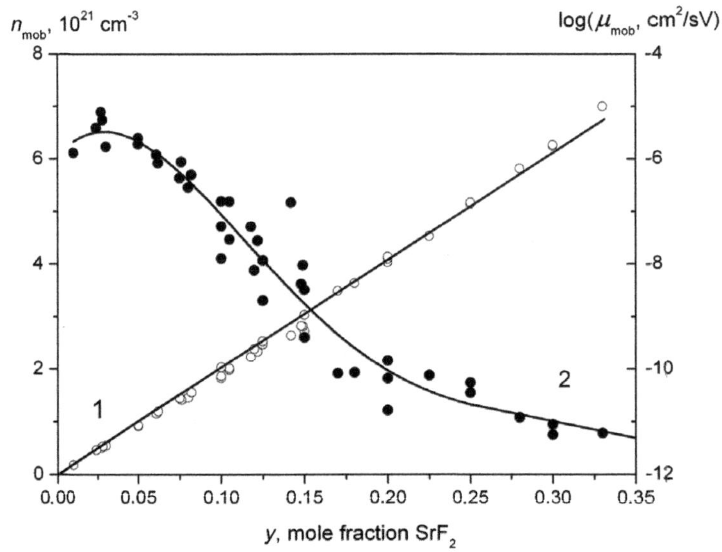

Figure 21.8 Dependences of concentration n_{mob} (1) and mobility μ_{mob} (2) on the composition of the $R_{1-y}Sr_yF_{3-y}$ crystals.

Average hopping frequency ν_h of the ion carriers was found by Sorokin and Sobolev [77] in 36 (from 62) $R_{1-y}M_yF_{3-y}$ single crystals of various (R,M,y) compositions from the $\sigma(\nu)$ dependences over the 110–410K range. The $\sigma(\nu)$ dependences for $R_{1-y}M_yF_{3-y}$ were explained in a hopping relaxation model. The hopping activation enthalpy ΔH_h was calculated from experimental $\nu_h(T)$-curve.

The conductivity activation enthalpy ΔH_σ is equal to hopping activation enthalpy ΔH_h for all the tysonite-like crystals studied [77]. According to Refs [10,26,29], the ion transport is caused by mobile V_F^+ vacancies in the crystallographic F1(12*g*) sites of sp. gr. $P\bar{3}c1$ (or F1(4*f*), sp. gr. $P6_3/mmc$). The hopping distances $d = 0.2–0.5$ nm (by the Nernst–Einstein equation) are consistent with an average distance between the F1(12*g*) positions in LaF_3 ($d = 0.27$ nm).

For fixed (R,M,y) composition, the $\sigma_{dc}(T)$ dependence is determined by the temperature behavior of the hopping frequency $\nu_h(T)$ or $\mu_{mob}(T) = \beta\nu_h(T)d^2q/kT$.

9.4 Composition (R,M,y) Optimizing of $R_{1-y}M_yF_{3-y}$ Crystals With High σ_{dc} Values

Isomorphous substitutions in the $R_{1-y}M_yF_{3-y}$ crystals open wide prospects to vary the (R,M,y) composition. The homogeneity ranges cover the MF_2 content from 0 to 33 mol% [43]. Crystals of 62 chemical compositions relating to 29 $M{:}R$ (from Ref. [39]) $R_{1-y}M_yF_{3-y}$ phases were studied to select optimal compositions with high σ_{dc} values [57,80].

Maximal σ_{dc} are observed for $R_{1-y}M_yF_{3-y}$ with 3–5 mol% MF_2, whose (M,R) optimization yields $R = $ Ce and $M = $ Sr.

The data on conductivity σ_{dc} (at 293K) and activation enthalpy ΔH_σ for 62 compositions of $R_{1-y}M_yF_{3-y}$ are summarized in Fig. 21.9. Substitutions R^{3+} for M^{2+} lead to $\sim 10^6$ times increase of σ_{293K} from 10^{-9} to 5×10^{-4} S/cm, accompanied by a drop of ΔH_σ from 0.7 to 0.3 eV.

There are 32 $R_{1-y}M_yF_{3-y}$ crystals whose fluorine-ionic conductivities are $\sigma_{dc} > 10^{-5}$ S/cm: $R_{0.95}Ba_{0.05}F_{2.95}$ ($R = $ La–Nd), $R_{1-y}Sr_yF_{3-y}$ ($R = $ La–Nd, $y = 0.01–0.12$), $R_{0.95}Ca_{0.05}F_{2.95}$ ($R = $ La–Nd), and $R_{1-y}Ca_yF_{3-y}$ ($R = $ Nd, Sm; $y = 0.01–0.10$), while 13 of them have conductivities $\sigma_{293K} > 10^{-4}$ S/cm.

Single crystals of $Ce_{0.97}Sr_{0.03}F_{2.97}$ have maximal $\sigma_{293K} = 5 \times 10^{-4}$ S/cm [80]. This is a record value for all nonstoichiometric fluorite- and tysonite-like phases studied at room temperature.

9.5 Room-Temperature Application Prospects for FCSEs Based on $R_{1-y}M_yF_{3-y}$ Crystals

Fluorine-selective electrode based on tysonite-like crystal $LaF_3{:}Eu^{2+}$ [11] has been the only device with FCSE for a long time. Then, models of chemical FCSE-based sensors appeared [53,71,72,81,85]. Models of F-ion batteries based on tysonite-like FCSEs are discussed by Potanin [35,36], Anji Reddy and Fichtner [3], and others. All the working models used polycrystalline FCSEs and compressed products of mechanochemical

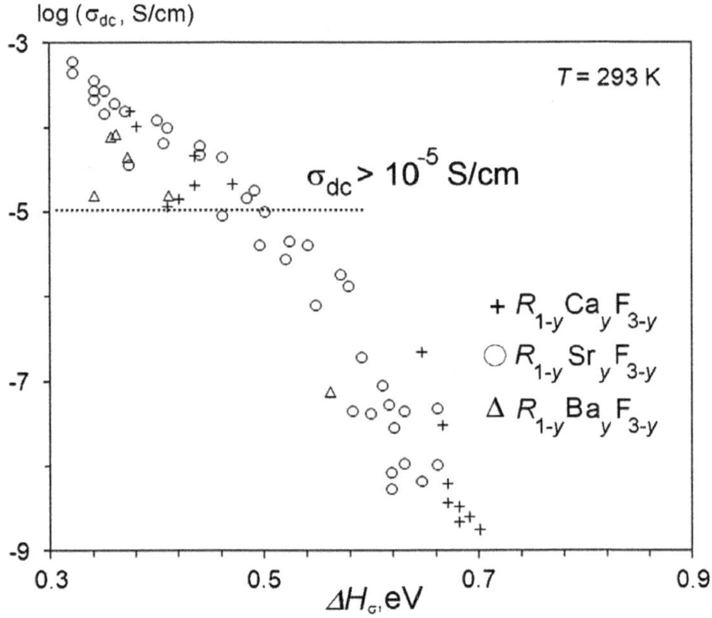

Figure 21.9 Dependence of conductivity σ_{293K} on enthalpy ΔH_σ for $R_{1-y}Ca_yF_{3-y}$, $R_{1-y}Sr_yF_{3-y}$, and $R_{1-y}Ba_yF_{3-y}$ crystals.

synthesis. This technique for FCSEs synthesis was suggested by Sobolev et al. [59]. Models of F-ion batteries as well as fluorine sensors operate in the 150–500°C temperature range to get higher σ_{dc}-values.

Now, the most promising application of tysonite-like FCSEs is fluorine-ion batteries. Galvanic reactions of metal oxidation by fluorine, which is the strongest oxidizer, belong to high-energetic chemical processes. The application of metals that provide high-charging R^{3+} cations (mostly REEs) increases specific energy of batteries. Potanin [35] estimated that a conversion to F-ion batteries can increase energy density by an order of magnitude, as compared to Li-ion batteries.

The required σ-values for room-temperature devices are not precisely determined since there are no tysonite-like FCSEs yet, whose σ_{293K}-values are sufficient for operating F-ion batteries. Approximate estimates were made from the analysis of high temperature–operating devices and prototypes based on FCSEs.

The requirements to σ_{293K} in tysonite-like FCSEs depend on the kind of devices. FCSEs with $\sigma_{293K} > 10^{-4}$ S/cm are required for fluorine sensors and F-ion batteries, whereas $\sigma_{293K} > 10^{-5}$ S/cm is sufficient for ion-selective electrodes.

In Fig. 21.9, a lower limit $\sigma_{293K} = 10^{-5}$ S/cm is indicated by a dotted horizontal line that is suitable for room temperature FCSE-based devices. The high σ_{293K} values of real $R_{1-y}M_yF_{3-y}$ crystals encourage progress in the field of technologically self-sufficient F-ion batteries operating at room temperature. These crystals can be considered as FCSE prototypes for such batteries.

10. Conclusions

The practical task has not yet been fulfilled, despite some available nonstoichiometric fluoride crystals of $R_{1-y}M_yF_{3-y}$ with record σ_{293K}-values. Room temperature is not the lower threshold for F-ion batteries and other functional FCSE-based devices.

An increase up to $\sigma_{293K} > 10^{-3}$ S/cm within an order of magnitude seems to be sufficient for F-ion batteries fit for work at $-30°C$. To fulfill this task, new two- and three-component nonstoichiometric fluoride crystals have to be grown. Their defect structures and compositions should be studied in connection with their ionic conductivity and based on the data on two-component compositions with high σ_{293K}, obtained and reported in this work.

References

[1] V.B. Alexandrov, L.S. Garashina, New data on the structure of solid solutions CaF_2–*TR*F_3, Dokl. Akad. Nauk SSSR 189 (1969) 307–310 (in Russian).

[2] D.P. Almond, G.K. Duncan, A.R. West, The determination of hopping rates and carrier concentrations in ionic conductors by a new analysis of ac-conductivity, Solid State Ionics 8 (1984) 159–164.

[3] M. Anji Reddy, M. Fichtner, Batteries based on fluoride shuttle, J. Mater. Chem. 21 (2011) 17059–17062.

[4] E.I. Ardashnikova, V.A. Prituzhalov, I.B. Kutsenok, Doping influence on defect structure and ionic conductivity of fluorine-containing phases, in: A.A. Tressaud (Ed.), Functionalized Inorganic Fluorides, Wiley and Sons Publ, Chippenham, Great Britain, 2010, pp. 423–468.

[5] D.J.M. Bevan, S.E. Ness, M.R. Taylor, On the crystal chemistry of Ca_2YbF_7 and other closely-related structures with cubooctahedral anion clusters, Eur. J. Solid State Chem. 25 (1988) 527–534.

[6] W. Bollmann, Incorporation of O^{2-} and OH^- ions in CaF_2 crystals by reaction with the surrounding atmosphere, Phys. Stat. Sol.(a) 57 (1980) 601–607.

[7] N.B. Bolotina, A.I. Kalyukanov, T.S. Chernaya, I.A. Verin, N.I. Sorokin, I.I. Buchinskaya, B.P. Sobolev, Ordering of tysonite structure in an as-grown $Er_{0.715}Ca_{0.285}F_{2.715}$ crystal and in a component of annealed two-phase crystal of the $Er_{0.67}Ca_{0.33}F_{2.67}$ composition, Crystallogr. Rep. 59 (2014) 504–512.

[8] N.B. Bolotina, T.S. Chernaya, A.I. Kalyukanov, I.A. Verin, N.I. Sorokin, L.E. Fykin, N.N. Isakova, B.P. Sobolev, Relationship between the defect structure and fluorine-ion conductivity of $La_{1-y}Sr_yF_{3-y}$ $(0 \leq y \leq 0.15)$ crystals according to X-ray and neutron diffraction data, Crystallogr. Rep. 60 (2015) 346–360.

[9] C.R.A. Catlow, J.D. Comins, F.A. Germano, R.T. Harley, W. Hayes, Effects of rare-earth dopants on the transition to the superionic state of fluorites, Phys. Lett. A 71 (1979) 97–98.

[10] M.A. Denecke, W. Gunsser, A.V. Privalov, I.V. Murin, NMR and XAS study on doped LaF_3, Solid State Ionics 52 (1992) 327–331.

[11] M. Frant, J. Ross, Electrode for sensing fluoride anion activity in solution, Science 154 (1966) 1553–1556.

[12] P.P. Fedorov, B.P. Sobolev, Conditions for the formation of maxima on the fusion curves of solid solutions in salt systems, Russ. J. Inorg. Chem. 24 (1979a) 574–575.

[13] P.P. Fedorov, B.P. Sobolev, Phase diagrams of the CaF_2–(Y,*Ln*)F_3 systems. Part 2, J. Less-Common Metals 63 (1979b) 31–44.
[14] L.S. Garashina, B.P. Sobolev, Hexagonal phases with variable compositions in the CaF_2–(Y,*Ln*)F_3 systems, Sov. Phys. Crystallogr. 16 (1971) 254–258.
[15] A.M. Golubev, V.I. Simonov, Superlattice structures based on fluorite, Sov. Phys. Crystallogr. 31 (1986) 281–284.
[16] I.Yu Gotlib, I.V. Murin, E.M. Piotrovskaya, Molecular dynamics simulation of the $Ba_{1-x}Gd_xF_{2+x}$ system in a wide temperature range, Solid State Ionics 159 (2003) 49–62.
[17] O. Greis, M.S.R. Cader, Polymorphism of high purity rare earth trifluorides, Thermochim. Acta 87 (1985) 145–150.
[18] O. Greis, J.M. Haschke, Rare earth fluorides, in: K.A. Gscheidner, L.R. Eyring (Eds.), Handbook on the Physics and Chemistry of Rare Earths, 5, 1982, pp. 387–460. Amsterdam, NY, Oxford.
[19] M. Hofmann, S. Hull, G.J. McIntyre, C.C. Wilson, A neutron diffraction study of the superionic transition in $(Ca_{1-x}Y_x)F_{2+x}$ with $x = 0.06$, J. Phys. Condens. Mater. 9 (1997) 845–857.
[20] A.K. Ivanov-Shits, I.V. Murin, Solid-State Ionics, vol. 2, Izd-vo S.-PbGU, St. Petersburg, 2010 (in Russian).
[21] A.K. Ivanov-Shits, N.I. Sorokin, A.I. Baranov, Frequency dependence of ionic conductivity in the $Sr_{0.8}La_{0.2}F_{2.2}$ crystal, Sov. Phys. Solid State 29 (1987) 3184–3186.
[22] A.K. Ivanov-Shits, N.I. Sorokin, P.P. Fedorov, B.P. Sobolev, Conductivity of $Sr_{1-x}La_xF_{2+x}$ $(0.03 \leq x \leq 0.40)$ solid solutions, Sov. Phys. Solid State 25 (1983) 1007–1010.
[23] A.K. Ivanov-Shits, N.I. Sorokin, P.P. Fedorov, B.P. Sobolev, Specific features of ion transport in nonstoichiometric $Sr_{1-x}R_xF_{2+x}$ phases ($R =$ La–Lu,Y) with the fluorite-type structure, Solid State Ionics 31 (1989a) 253–268.
[24] A.K. Ivanov-Shits, N.I. Sorokin, P.P. Fedorov, B.P. Sobolev, Specific features of ion transport in nonstoichiometric $Ba_{1-x}R_xF_{2+x}$ phases ($R =$ La–Lu) with the fluorite-type structure, Solid State Ionics 31 (1989b) 269–280.
[25] A.K. Ivanov-Shits, N.I. Sorokin, P.P. Fedorov, B.P. Sobolev, Specific features of ion transport in nonstoichiometric fluorite-type phases $Ca_{1-x}R_xF_{2+x}$ ($R =$ La–Lu, Y,Sc), Solid State Ionics 37 (1990) 125–137.
[26] M.G. Izosimova, A.I. Livshits, V.M. Buznik, P.P. Fedorov, E.A. Krivandina, B.P. Sobolev, Diffusion mechanism of fluorine ions in solid electrolytes having the tysonite structure, Sov. Phys. Solid State 28 (1986) 1482–1484.
[27] S.A. Kazanskii, Optically detected ESR of clusters of rare earth ions and yttrium in fluorite crystals, Sov. Phys. JETP 62 (1985) 727–733.
[28] D. Jakes, J. Kaplan, V. Trnovcova, B.P. Sobolev, Limiting-current type HF-gas sensor, in: III Asian Conf. On Solid State Ionics. Varanasi. Ext. Abstrs. 147, 1992.
[29] A.I. Livshits, V.M. Buznik, P.P. Fedorov, B.P. Sobolev, NMR investigation of anion mobility in defect phases with fluorite and tysonite structures in the CaF_2–LaF_3 system, Inorg. Mater. 18 (1982) 114–117.
[30] L.A. Muradian, B.A. Maximov, V.I. Simonov, Crystal structure of nonstoichiometric phases of the fluorite type, Russ. J. Coord. Chem. 12 (1986) 801–806.
[31] L.E. Nagel, M. O'Keeffe, Highly-conducting fluorides related to fluorite and tysonite, in: W. van Gool (Ed.), Fast Ion Transport in Solids, North-Holland Publ, Amsterdam, 1973, pp. 165–172.
[32] B.F. Naylor, Heat contents at high temperatures of magnesium and calcium fluorides, J. Amer. Chem. Soc. 67 (1945) 150–152.

[33] P.E. Ngoepe, J.D. Comins, Measurements of elastic constants in super-ionic $Ba_{1-x}La_xF_{2+x}$, J. Phys. C: Solid State Phys. 19 (1986) L267–L271.
[34] L.P. Otroshchenko, V.B. Alexandrov, B.A. Maximov, V.I. Simonov, B.P. Sobolev, Stabilization of highly symmetrical hexagonal form of tysonite in nonstoichiometric phase $Gd_{0.8}Ca_{0.2}F_{2.8}$, Crystallogr. Rep. 30 (1985) 658–662.
[35] A.A. Potanin, Solid-state chemical current source based on fluorine-ionic conductor with structure-type of lanthanum fluoride, J. Russ. Chem. Soc. (Mendeleev) 45 (2001) 58–63 (in Russian).
[36] A.A. Potanin, Solid-State Secondary Power Supply, Patent of the Russian Federation. RU2295178, 17.03.2007 (in Russian), 2005.
[37] A.A. Potanin, Solid-State Secondary Power Supply, US patent application. 20080102373/A1, 01.05.2008, 2008.
[38] C. Rongeat, M. Anji Reddy, R. Witter, M. Fichtner, Nanostructured fluorite-type fluorides as electrolytes for fluoride ion batteries, J. Phys. Chem. 117 (2013) 4943–4950.
[39] A. Roos, F.C.M. van de Pol, R. Keim, J. Schoonman, Ionic conductivity in tysonite-type solid solutions $La_{1-x}Ba_xF_{3-x}$, Solid State Ionics 13 (1984) 191–203.
[40] G.A. Samara, High-pressure studies of ionic conductivity in solids, Solid State Phys. 38 (1984) 1–80.
[41] J. Schoonman, Retarded ionic motion in fluorites, Solid State Ionics 1 (1980) 121–131.
[42] R.D. Shannon, Revised effective ionic radii and systematic studies of interatomic distances in halides and chalcogenides, Acta Crystallogr. A. 32 (1976) 751–767.
[43] B.P. Sobolev, The Rare Earth Trifluorides. Part 1. The High Temperature Chemistry of Rare Earth Trifluorides, Institute of Crystallography, Moscow, and Institut d'Estudis Catalans, Barcelona, 2000. www.books.google.ru/books/rare earth trifluorides.
[44] B.P. Sobolev, The Rare Earth Thrifluorides. Part 2. Introduction to Materials Science of Multicomponent Metal Fluoride Crystals, Institute of Crystallography, Moscow, and Institut d′Estudis Catalans, Barcelona, 2001. www.books.google.ru/books/rare earth trifluorides.
[45] B.P. Sobolev, Chemical aspects of crystal grows of multicomponent fluoride materials, Crystallogr. Rep. 47 (Suppl. 1) (2002) 63–75.
[46] B.P. Sobolev, D.N. Karimov, S.N. Sulyanov, Z.I. Zhmurova, Nanostructured crystal of fluorite phases $Sr_{1-x}R_xF_{2+x}$ (*R*—rare-earth elements) and their ordering. Part 1, Crystallogr. Rep. 54 (2009) 122–130.
[47] B.P. Sobolev, Nonstoichiometry in inorganic fluorides. Part 1, Crystallogr. Rep. 57 (2012) 434–454.
[48] B.P. Sobolev, P.P. Fedorov, Phase diagrams of the CaF_2–(Y,*Ln*)F_3 systems. Part 1, J. Less-Common Metals 60 (1978) 33–46.
[49] B.P. Sobolev, P.P. Fedorov, D.V. Steinberg, B.V. Sinitsyn, G.S. Shakhalamian, On the problem of polymorphism and fusion of lanthanide trifluorides: Part. 1, J. Solid State Chem. 17 (1976a) 191–199.
[50] B.P. Sobolev, V.B. Aleksandrov, P.P. Fedorov, K.B. Seiranyan, N.L. Tkachenko, Variable-composition phase with the LaF_3 structure in the MF_2–(Y,*Ln*)F_3 systems. Part 4, Sov. Phys. Crystallogr. 21 (1976b) 49–54.
[51] B.P. Sobolev, A.M. Golubev, P. Herrero, Fluorite $M_{1-x}R_xF_{2+x}$ phases ($M =$ Ca,Sr,Ba; $R =$ rare earth elements) as nanostructured materials, Crystallogr. Rep. 48 (2003) 141–161.
[52] B.P. Sobolev, E.G. Ippolitov, B.M. Zhigarnovskii, L.S. Garashina, Phase composition of the CaF_2–YF_3, SrF_2–YF_3 and BaF_2–YF_3, Inorg. Mater. 1 (1965) 334–340.
[53] B.P. Sobolev, E.A. Krivandina, I.V. Murin, O.V. Glumov, V.B. Bogolubov, B.S. Besprozvannii, T.A. Lugina, V.K. Avinnikov, Detector of Fluorine in Atmosphere, USSR patent No 311414, 31.11.1988, 1988.

[54] B.P. Sobolev, K.B. Seiranian, Phase diagrams of the SrF_2−(Y,*Ln*)F_3 systems. Part 2, J. Solid State Chem. 39 (1981) 17−24.
[55] B.P. Sobolev, K.B. Seiranian, Garashina, P.P. Fedorov, Phase diagrams of the SrF_2−(Y,*Ln*)F_3 systems. Part 1, J. Solid State Chem. 28 (1979) 51−58.
[56] B.P. Sobolev, N.I. Sorokin, Nonstoichiometry in inorganic fluorides: Part. 2, Crystallogr. Rep. 59 (2014) 807−830.
[57] B.P. Sobolev, N.I. Sorokin, E.A. Krivandina, Z.I. Zhmurova, 293K-conductivity optimization for single crystals of solid electrolytes with tysonite structure (LaF_3). Part 1, Crystallogr. Rep. 59 (2014) 550−562.
[58] B.P. Sobolev, N.L. Tkachenko, Phase diagrams of the BaF_2−(Y,*Ln*)F_3 systems, J. Less-Common Metals 85 (1982) 155−170.
[59] B.P. Sobolev, I.A. Sviridov, V.I. Fadeeva, S.N. Sul'yanov, N.I. Sorokin, Z.I. Zhmurova, P. Herrero, A. Landa-Canovas, R.M. Rojas, Mechanochemical synthesis of non-stoichiometric fluorite $Ca_{1-x}La_xF_{2+x}$ nanocrystals from CaF_2 and LaF_3 single crystals, Crystallogr. Rep. 50 (2005) 478−485.
[60] N.I. Sorokin, Conductivity at high hydrostatic pressures of a superionic SrF_2 crystal containing 20 mole % LaF_3, Sov. Phys. Solid State 32 (1990) 1832−1833.
[61] N.I. Sorokin, Schemes of isomorphic heterovalent cationic substitutions in solid solutions of fluorides with the fluorite structure, Sov. Phys. Crystallogr. 36 (1991) 832−834.
[62] N.I. Sorokin, Ionic conductivity of $Ba_{1-x}Gd_xF_{2+x}$ crystals at high temperatures, Sov. Phys. Solid State 34 (1992) 1008−1100.
[63] N.I. Sorokin, Determination of the percolation threshold of ionic migration in the non-stoichiometric phase $Sr_{1-x}La_xF_{2+x}$, Sov. Phys. Crystallogr. 38 (1993) 132−133.
[64] N.I. Sorokin, Crystallochemical model of ion-transport percolation in nonstoichiometric $M_{1-x}R_xF_{2+x}$ phases with the defect CaF_2-type structure, Crystallogr. Rep. 45 (2000a) 799−803.
[65] N.I. Sorokin, Activation volumes and activation enthalpies for various mechanisms of ion transfer in non-stoichiometric fluorides with the structure of fluorite and tysonite, Russ. J. Electrochem. 36 (2000b) 441−442.
[66] N.I. Sorokin, Superionic transport in fluoride solid solutions with a fluorite structure, Russ. J. Electrochem. 42 (2006) 744−759.
[67] N.I. Sorokin, M.W. Breiter, Anionic conductivity and thermal stability of single crystals of solid solutions based on SrF_2, Solid State Ionics 104 (1997a) 325−333.
[68] N.I. Sorokin, M.W. Breiter, Anionic conductivity and thermal stability of single crystals of solid solutions based on BaF_2, Solid State Ionics 99 (1997b) 241−250.
[69] N.I. Sorokin, M.W. Breiter, Anionic conductivity and thermal stability of single crystals of solid solutions based on CaF_2, Solid State Ionics 116 (1999) 157−165.
[70] N.I. Sorokin, M.V. Fominykh, E.A. Krivandina, Z.I. Zhmurova, B.P. Sobolev, Ion transport in $R_{1-x}Sr_xF_{3-x}$ ($R =$ La−Yb,Y) with a LaF_3 (tysonite) structure, Crystallogr. Rep. 41 (1996) 292−301.
[71] N.I. Sorokin, E.A. Krivandina, Z.I. Zhmurova, B.P. Sobolev, M.V. Fominykh, V.V. Fistul', Superionic conductivity of the heterovalent solid solutions $R_{1-x}M_xF_{3-x}$ ($R =$ REEs, $M =$ Ca,Ba) with tysonite-type structure, Phys. Solid State 41 (1999a) 573−575.
[72] N.I. Sorokin, E.A. Krivandina, Z.I. Zhmurova, E.F. Sudakova, YuV. Zemlerub, B.P. Sobolev, Optimization of transport properties of fluoride solid electrolytes with tysonite structure for HF detection device, in: Russ. Conf. "Sensor-2000." Sankt-Petersburg. 322, 2000.
[73] N.I. Sorokin, A.M. Golubev, B.P. Sobolev, Structural mechanisms of superionic conductivity in $M_{1-x}R_xF_{2+x}$ single crystals, Crystallogr. Rep. 59 (2014a) 238−247.

[74] N.I. Sorokin, B.P. Sobolev, The conductivity of single crystals $La_{1-x}Sr_xF_{3-x}$ solid solutions ($0 \leq x \leq 0.15$) with tysonite-structure, Crystallogr. Rep. 39 (1994) 810–814.
[75] N.I. Sorokin, B.P. Sobolev, Nonstoichiometric fluorides – solid electrolytes for electrochemical devices, Crystallogr. Rep. 52 (2007a) 842–863.
[76] N.I. Sorokin, B.P. Sobolev, Anionic high-temperature conduction in single crystals of nonstoichiometric phases $R_{1-y}M_yF_{3-y}$ (R = La–Lu, M = Ca,Sr,Ba) with the tysonite (LaF_3) structure, Russ. J. Electrochemistry 43 (2007b) 398–409.
[77] N.I. Sorokin, B.P. Sobolev, Frequency response of the low-temperature ionic conductivity of single crystals $R_{1-y}M_yF_{3-y}$ (R = La–Er, M = Ca,Sr,Ba,Cd), Phys. Solid State 50 (2008a) 416–421.
[78] N.I. Sorokin, B.P. Sobolev, Study of anionic conductors $R_{1-y}Sr_yF_{3-y}$ (R = Ce,Pr,Gd) and $M_{1-x}Y_xO_{2-x/2}$ (M = Zr,Hf) using four-electrode method, Russ. J. Electrochem. 44 (2008b) 1031–1035.
[79] N.I. Sorokin, B.P. Sobolev, Nanostructured crystals of fluorite phases $Sr_{1-x}R_xF_{2+x}$ (R are rare earth elements) and their ordering. Part 11, Crystallogr. Rep. 60 (2015) 959–962.
[80] N.I. Sorokin, B.P. Sobolev, E.A. Krivandina, Z.I. Zhmurova, Optimization of single crystals of solid electrolytes with tysonite (LaF_3) structure for conductivity at 293K. Part 2, Crystallogr. Rep. 60 (2015) 123–129.
[81] N.I. Sorokin, E.F. Sudakova, E.A. Krivandina, B.P. Sobolev, Thin films of the $La_{0.95}Sr_{0.05}F_{2.95}$ fluorine-ionic semiconductor: impedancemetry with platinum electrodes, Russ. J. Electrochem. 35 (1999b) 223–227.
[82] N.I. Sorokin, T.M. Turkina, B.P. Sobolev, Anion transport in $Ba_{0.5}Ce_{0.5}F_{2.5}$ single crystal with spatially oriented composition fluctuations, Sov. Phys. Crystallogr. 37 (1992) 848–850.
[83] N.I. Sorokin, Z.I. Zhmurova, E.A. Krivandina, B.P. Sobolev, Influence of purity of NdF_3 single crystals on their ionic conductivity, Crystallogr. Rep. 57 (2012) 461–462.
[84] N.I. Sorokin, Z.I. Zhmurova, E.A. Krivandina, B.P. Sobolev, Superionic conductivity of $Sr_{0.68}Pr_{0.32}F_{2.32}$ (type CaF_2) and $Pr_{0.85}Sr_{0.15}F_{2.85}$ (type LaF_3) crystals as-grown and after high-temperature annealing, Crystallogr. Rep. 59 (2014b) 88–92.
[85] E.F. Sudakova, E.Y. Alksnis, R.S. Perlovskii, E.A. Krivandina, B.P. Sobolev, Chemical sensor for F_2 and HF in gaseous atmospheres, in: Symp. On Solid State Ionics. November 30–December 4, 1992. Boston. Section U-6.5. 160, 1992.
[86] V. Trnovcova, N.I. Sorokin, P.P. Fedorov, L.S. Garashina, E.A. Krivandina, B.P. Sobolev, Defect structure and ionic transport in the MF_2–RF_3 (M = Ca, Sr; R = La–Lu, Y, Sc) solid solutions, Solid State Phenomena 90 (2003) 439–444.
[87] T.M. Turkina, P.P. Fedorov, B.P. Sobolev, The stability of plane crystallization front in growth of single crystals of solid solutions $M_{1-x}R_xF_{2+x}$ (M = Ca,Sr,Ba; R-rare earths) from the melt, Sov. Phys. Crystallogr. 31 (1986) 83–87.
[88] R.W. Ure, Ionic conductivity of calcium fluoride crystals, J. Chem. Phys. 26 (1957) 1363–1373.
[89] T. Vogt, Uber die Flusspat – Yttriumfluoritgruppe, Neues Jahrb. Mineral 2 (1914) 9–15.
[90] K.E.D. Wapenaar, J.L. van Koesveld, J. Schoonman, Conductivity enhancement in fluorite-structured $Ba_{1-x}La_xF_{2+x}$ solid solutions, Solid State Ionics 2 (1981) 145–154.

Index

'*Note:* Page numbers followed by "f" indicate figures, "t" indicate tables.'

O

Q

R

S

T

U

Edwards Brothers Malloy
Ann Arbor MI. USA
July 25, 2016